T0207080

New Mathematical Monographs

For information about Cambridge University Press mathematics publications visit http://publishing.cambridge.org/stm/mathematics/

Heights in Diophantine Geometry

The first half of the book is devoted to the general theory of heights and its applications, including a complete, detailed proof of the celebrated subspace theorem of W. M. Schmidt. The second part deals with abelian varieties, the Mordell–Weil theorem and Faltings's proof of the Mordell conjecture, ending wih a self-contained exposition of Nevanlinna theory and the related famous conjectures of Vojta. The book concludes with a comprehensive list of references. It is destined to be a definitive reference book on modern diophantine geometry, bringing a new standard of rigor and elegance to the field.

Professor ENRICO BOMBIERI is a professor of Mathematics at the Institute of Advanced Study, Princeton.

Dr WALTER GUBLER is a lecturer in Mathematics at the University of Dortmund.

HEIGHTS IN DIOPHANTINE GEOMETRY

Enrico Bombieri

Institute of Advanced Study, Princeton

Walter Gubler

University of Dortmund

CAMBRIDGE
UNIVERSITY PRESS

University Printing House, Cambridge CB2 8BS, United Kingdom

Cambridge University Press is part of the University of Cambridge.

It furthers the University's mission by disseminating knowledge in the pursuit of education, learning and research at the highest international levels of excellence.

www.cambridge.org
Information on this title: www.cambridge.org/9780521712293

First published 2006
Reprinted with corrections 2007

A catalogue record for this publication is available from the British Library

ISBN 978-0-521-84615-8 Hardback
ISBN 978-0-521-71229-3 Paperback

Contents

Preface

Diophantine geometry, the study of equations in integer and rational numbers, is one of the oldest subjects of mathematics and possibly the most popular part of number theory, for the professional mathematician and the amateur alike. Certainly, one of its main attractions is that, far from being a disconnected assembly of isolated results, it provides glimpses of a view which hints at a well-organized underlying structure.

Diophantine equations are of course determined by the underlying algebraic equations and therefore their associated algebraic geometry, obtained by dropping the condition that the solutions must be integers or rational numbers, plays a big role in their study. However, algebraic geometry is already not an easy subject. A pioneer and one of the founding fathers of algebraic geometry, the German mathematician Max Noether, after seeing the theory of algebraic curves with its elegance, simplicity, and also depth of results, and comparing it with the collection of the existing examples of algebraic surfaces at the time, for which nothing comparable could be found, used to say that algebraic curves were created by God and algebraic surfaces by the Devil. Only later, with the development of new tools, in particular the introduction of cohomological and topological methods, the theory of surfaces and higher-dimensional varieties over a field found a satisfactory status.

Of special importance for arithmetic was the development of algebraic geometry over fields of positive characteristic and p-adic fields, since the study of polynomial congruences leads very naturally to such problems. The next big step, the study of varieties over general rings (in contrast to fields), was done by Grothendieck in his monumental construction of the theory of schemes. This provided the basic setting for the study of diophantine equations from a geometric point of view. Bits and pieces of a theory were provided at an early stage (Weil's proof of the Mordell–Weil theorem is possibly the first example) and Weil's theory of heights, with its good arithmetic and geometric properties, was for a long time the main tool. However, the development of a consistent theory was hindered by two major obstacles.

An algebraic curve X over, for example, the ring $\mathbb{Z}$ of rational integers is, from the point of view of schemes, a two-dimensional object, an arithmetic surface, endowed with a morphism $f : X \to \operatorname{Spec}(\mathbb{Z})$. Ideally, we would like to find an analogue of the classical theory of algebraic surfaces which applies in this arithmetic setting.

This can be done only to some extent. First of all, global results require working with complete varieties, and a first problem was to compactify $\mathrm{Spec}(\mathbb{Z})$ and develop a good intersection theory for divisors. This step was brilliantly solved by Arakelov, using adeles and introducing metrics on the "fibre at infinity." Arakelov's work can be regarded as the start of a beautiful new theory, aptly named "arithmetic geometry." As an example, in arithmetic geometry the theory of heights is a special chapter of the much more precise arithmetic intersection theory.

Arakelov's theory did not solve all problems and major questions remain. In the "horizontal direction" given by the base $\mathrm{Spec}(\mathbb{Z})$, infinitesimal methods are no longer at our disposal and genuine new difficulties, with no counterpart in the classical theory, do appear. This is one of the major stumbling blocks for further progress. Thus at the present stage we may take a view half-way towards Max Noether's view: Arithmetic surfaces were also created by God, but their study encounters devilish difficulties.

Today, there are already good books devoted to the subject, and we can mention here Lang's [**169**], Serre's [**277**], the more expository but very comprehensive account of Lang's [**171**] and Hindry and Silverman's [**153**]. So, why a new book on diophantine geometry?

As is often the case, this book grew from introductory lectures at the graduate level, given over a decade ago at the Scuola Normale Superiore di Pisa and the Mathematisches Forschungsinstitut of the Eidgenössische Technische Hochschule in Zürich. An advanced knowledge of algebra or algebraic geometry was not a prerequisite of the courses. Thus the subject was developed mainly through classical lines, namely the theory of varieties over fields of characteristic 0 insofar as algebraic geometry was concerned, and the theory of heights for the number theoretic aspects.

Already with the initial rough notes, embracing the view that in order to learn tools it is best to use them in practice, it was decided to keep mathematical rigor as a strict requirement, supplying references whenever needed and making a clear distinction between a proof and a plausible argument. Examples, including unusual ones, and advanced sections in which deeper aspects of the theory were either developed or described, were included whenever possible. Rather than including this type of material as "exercises" at various levels of difficulty, often disguising good research papers as exercises, it was decided to include proofs and extended comments also for them. However, in the time needed to put the original material together, the subject matter continued to advance at a fast pace, whence the need for inclusion of additional interesting material, as well as substantial revisions of what had been done before.

In the final product, this book is basically divided into three parts. Chapters 1 to 7 develop the elementary theory of heights and its applications to the diophantine geometry of subvarieties of the split torus $\mathbb{G}_m^n$, including applications to diophantine approximation with proofs of Roth's theorem and Schmidt's subspace theorem and some unusual applications.

Chapters 8 to 11 deal with abelian varieties and the diophantine geometry of their subvarieties, ending with a detailed proof of Faltings's celebrated theorem establishing Mordell's conjecture for curves, following Vojta's proof as simplified in [**29**]. However, we felt that a proper treatment of Faltings's big theorem, namely his proof of Lang's conjectures about rational points on subvarieties of abelian varieties, was best done in the context of arithmetic geometry and with regrets we limited ourselves on this matter only to a few comments about the theorem itself and to some of its applications.

Chapters 12 to 14 are more speculative and at times straddle the borderline between diophantine geometry and arithmetic geometry. Chapter 12 deals with the so-called abc-conjecture over number fields, including a complete proof of Belyĭ's theorem and its application to Elkies's theorem, various examples, concluding with a finiteness result for the generalized Fermat equation, due to Darmon and Granville. Chapter 13, which is largely self contained, is an exposition of the classical Nevanlinna theory, with proofs of the first and second main theorems of Nevanlinna, and also Cartan's extension of them to the theory of meromorphic curves. Its purpose is to motivate the final Chapter 14 dealing with the well-known Vojta conjectures, which have spurred a great deal of work in the field.

Proofs are usually given in full detail, but of course it was not feasible to develop all algebra and algebraic geometry from scratch and they tend to be fairly condensed at times. To alleviate this, Appendix A summarizes all concepts of algebraic geometry needed in this book and Appendix B gathers the necessary facts about ramification in number theory and algebraic geometry. Both are provided with complete references to standard books and should help the reader in understanding which notions and notations we use. Finally Appendix C contains an account of Minkowski's geometry of numbers, with proofs, at least to the extent we need in this book.

Some sections in this book appear in small print. Their meaning is simply that they can be omitted in a first reading, either because they require more advanced knowledge of algebra and geometry, or because they deal with side topics not appearing elsewhere in the book. At the end of every chapter, the reader will find some bibliographical notes, containing both historical comments and references to additional literature. However, in no way do these references pretend to be complete and they only represent our personal choices for additional reading.

This book does not represent an introduction to diophantine geometry, nor a complete treatment of the theory of heights. Neither do we strive for maximum generality, and most of the book is concerned only with a number field as ground field, dealing only marginally with the function field case and even less with ground fields of positive characteristic. Also, we do not extend the theory to semiabelian varieties or non-split commutative linear groups, which are also quite important and lead to delicate questions.

The whole theory of effective diophantine approximation, and Baker's theory of logarithmic forms, are missing entirely from this book and relegated to a few comments at the end of Chapter 5. This is not due to a perception of lack of importance of the subject. Rather, an adequate treatment of the topic would have required a second large volume for this already large book.

The same can be said for arithmetic geometry, which no doubt deserves an advanced monograph by itself, also for the arithmetic theory of elliptic curves and abelian varieties, and for the arithmetic theory of modular functions and its applications to diophantine problems.

Our goal in writing this book was to provide, in addition to the existing literature, a wide selection of topics in the subject, containing foundational material with complete proofs, numerous examples, and additional material viewed as a bridge between the classical theory and arithmetic geometry proper. A fair portion of this book is meant to be accessible to a reader with only a basic course in algebra and algebraic geometry, but even the specialist in the field should be able to find interesting material in it. We made no serious attempt to reach completeness about the history of the subject, also referenced material (we never quote from secondary sources) is for this very reason mostly from literature in the English and French languages. Finally, although we attempted to put together a comprehensive bibliography, in no way do we pretend it to be complete. We apologize in advance for the inevitable omissions in our bibliography, regarding priorities and precursor works.

At the end of the book the reader will find an index of mathematical names in lexicographic order and an index of notations ordered by page number. The vanity index (index of authors mentioned in the text) has been omitted.

Terminology

We try to use standard terminology, but for convenience of the reader we gather here some of the most frequently used notation and conventions.

In set theory, $A \subset B$ means that A is a subset of B. In particular, A may be *equal* to B. If this case is excluded, then we write $A \subsetneq B$. The complement of A in B is denoted by $B \setminus A$ as we reserve $-$ for algebraic purposes. We denote the number of elements of A by $|A|$ (possibly ∞). The identity map is id.

A quasi-compact topological space is characterized by the Heine–Borel property for open coverings. In this book, a compact space is quasi-compact *and* Hausdorff.

We denote by $\mathbb{N}$ the set of natural numbers *with* 0 *included* and $\mathbb{Z}$ is the ring of rational integers. Then $\mathbb{Q}$, $\mathbb{R}$ and $\mathbb{C}$ are the fields of rational, real, and complex numbers. A positive number means $x > 0$, but we use $\mathbb{R}_+$ for the non-negative real numbers. The Kronecker symbol δ_{ij} is 0 for $i \neq j$ and 1 for $i = j$.

The real (resp. imaginary) part of a complex number z is denoted by $\Re z$ (resp. $\Im z$) and $\overline{z}$ is complex conjugation.

The floor function $\lfloor x \rfloor$, defined for $x \in \mathbb{R}$, is the largest rational integer $\leq x$. The ceiling function $\lceil x \rceil$ denotes the smallest rational integer $\geq x$.

The real functions on X are denoted by $\mathbb{R}^X$. For $f, g \in \mathbb{R}^X$, the Landau symbol $f = O(g)$ means $|f(x)| \leq Cg(x)$ for some unspecified positive constant C. If we want to emphasize the dependence of C on parameters $\varepsilon, L, \ldots$ we write $f = O_{\varepsilon,L,\ldots}(g)$. As a special case, $f = O(1)$ means that f is a bounded function on X. We also use, with the same meaning, the equivalent Vinogradov's symbol $f \ll g$ and $f \ll_{\varepsilon,L,\ldots} g$. The symbol $g \gg f$ is interpreted as $f \ll g$.

If X is a topological space and f, g are defined on a subset Y with an accumulation point x, then $f = O(g)$ for $y \to x$ means that $|f(y)| \leq Cg(y)$ holds for all $y \in Y$ contained in a neighbourhood of x. If this is true for all $C > 0$ (with neighbourhoods depending on C), then we use the Landau symbol $f = o(g)$ for $y \to x$. The asymptotic relation $f \sim g$ for $y \to x$ means that $f - g = o(|g|)$.

The Landau symbols and the Vinogradov symbol must be used with caution in presence of parameters, not just because the constant involved may depend on parameters, but especially because the neighbourhood in which the inequality holds will also depend on the parameters, an easily overlooked fact.

In number theory, we use $\mathrm{GCD}(a, b)$ for the greatest common multiple of a and b. As usual, $a|b$ means that a divides b. The number of primes up to x is $\pi(x)$.

The group of multiplicative units of a commutative ring with identity is denoted by $R^\times$. We use the symbol V^* to denote the dual of a vector space V. Rings and algebras are always assumed to be associative, fields are always commutative. If the rings have an identity, then we assume that ring homomorphisms send 1 to 1. The ideal generated by $g_1, \ldots, g_m$ is denoted by $[g_1, \ldots, g_m]$. The characteristic of a field K is $\mathrm{char}(K)$ and we write $\mathbb{F}_q$ for the finite field with q elements.

The ring of polynomials in the variable x with coefficients in K is denoted by $K[x]$. A monic polynomial has highest coefficient 1. The minimal polynomial of an algebraic number α over a field is assumed to be monic and its degree is the degree of α, denoted by $\deg(\alpha)$. If we consider the minimal polynomial over $\mathbb{Z}$ (or any factorial ring), then we replace monic by the assumption that the coefficients are coprime. We use $\mathbf{x}$ to denote a vector with entries x_i, thus $K[\mathbf{x}]$ is the ring of polynomials in the variables x_i. By $\overline{K}$ we denote a choice of an algebraic closure of the field K.

For the terminology used in algebraic geometry, the reader is referred to Appendix A.

The numbering in this book is by chapter (appendices in capitals), section, and statement, in progressive order. Equations are numbered separately by chapter (appendices in capitals) and statement in progressive order, with the label enclosed in parentheses. References to equations not occurring on the same page or the preceding page also give the page numbers; the first example is: (A.13) on page 558, occurring on page 15.

1 HEIGHTS

1.1. Introduction

This chapter contains preliminary material on absolute values and the elementary theory of heights on projective varieties. Most of this material is quite standard, although we have included some of the finer results on classical heights which are not usually treated in other texts.

In Section 1.2 we start with absolute values, and places are introduced as equivalence classes of absolute values. The definitions of residue degree and ramification index are given, as well as their basic properties and behaviour with respect to finite degree extensions. In Sections 1.3 and 1.4 we introduce normalized absolute values and the all-important product formula in number fields and function fields. Section 1.5 contains the definition of the absolute Weil height in projective spaces, the characterization of points with height 0, and a general form of Liouville's inequality in diophantine approximation. Section 1.6 studies the height of polynomials and Mahler's measure and proves Gauss's lemma and its counterpart at infinity, Gelfond's lemma. Section 1.7, which can be omitted in a first reading, elaborates further on various comparison results about heights and norms of polynomials, including an interesting result of Per Enflo on ℓ_1-norms.

The presentation of the material in this chapter is self contained with the exception of Section 1.2, where the basic facts about absolute values are quoted from standard reference books (N. Bourbaki **[47]**, Ch.VI, S. Lang **[173]**, Ch.XII, and N. Jacobson **[157]**, Ch.IX).

1.2. Absolute values

Definition 1.2.1. *An* ***absolute value*** *on a field K is a real valued function $|\ |$ on K such that:*

(a) $|x| \geq 0$ *and* $|x| = 0$ *if and only if* $x = 0$.

(b) $|xy| = |x|\ |y|$.

(c) $|x + y| \leq |x| + |y|$ *(triangle inequality).*

1.2.2. The **trivial absolute value** is equal to 1 except at 0. If an absolute value satisfies instead of the triangle inequality (c) the stronger condition

(c′) $|x + y| \leq \max\{|x|, |y|\}$,

then it is called **non-archimedean.** If (c′) fails to hold for some $x, y \in K$, then the absolute value is called **archimedean.** The distance of $x, y \in K$ is $|x - y|$. This metric induces a topology on K. In the non-archimedean case, we have an ultrametric distance and (c′) is called the **ultrametric triangle inequality** . If two absolute values define the same topology, they are called **equivalent** .

Proposition 1.2.3. *Two absolute values* $|\ |_1, |\ |_2$ *are equivalent if and only if there is a positive real number* s *such that*

$$|x|_1 = |x|_2^s$$

for $x \in K$.

Proof: See [**157**], Th.9.1 or [**173**], Prop.XII.1.1. □

1.2.4. A **place** v is an equivalence class of non-trivial absolute values. By $|\ |_v$ we denote an absolute value in the equivalence class determined by the place v. If the field L is an extension of K and v is a place of K, we write $w|v$ for a place w of L if and only if the restriction to K of any representative of w is a representative of v, and say that w extends v and, equivalently, that w lies over v. We also employ the notation $w|v$ (that is, w divides v), motivated by the fact that non-archimedean places in number fields correspond to prime ideals.

The **completion** of K with respect to the place v is an extension field K_v with a place w such that:

(a) $w|v$.

(b) The topology of K_v induced by w is complete.

(c) K is a dense subset of K_v in the above topology.

The completion exists and is unique up to isometric isomorphisms ([**157**], Th.9.7 or [**173**], Prop.XII.2.1). By abuse of notation, we shall denote the unique place w also by v.

Example 1.2.5. If the field is $\mathbb{Q}$, then there is only one archimedean place ∞ on $\mathbb{Q}$, given by the ordinary absolute value $|\ |$. We also write $|\ |_\infty$ for this absolute value (cf. [**157**], Th.9.4).

For a prime p we have the **p-adic absolute value** $|\ |_p$ determined as follows. Let $m/n \in \mathbb{Q}$ be a rational number and write it in the form

$$\frac{m}{n} = p^a \frac{m'}{n'},$$

where m', n' are integers coprime with p. Then we set

$$\left|\frac{m}{n}\right|_p = p^{-a}.$$

In fact, it suffices to define $| \ |_p$ by the conditions

$$|q|_p = \begin{cases} 1 & \text{for primes } q \neq p \\ \frac{1}{p} & \text{if } q = p. \end{cases}$$

The p-adic absolute values so defined give us a set of inequivalent representatives for all non-archimedean places on $\mathbb{Q}$ ([**157**], Th.9.5). The field $\mathbb{Q}_p$ of p-adic numbers is the completion of $\mathbb{Q}$ with respect to the place p. The compact subset $\mathbb{Z}_p$ of p-adic integers is the closure of $\mathbb{Z}$ in $\mathbb{Q}_p$ (for compactness, see [**47**], Ch.VI, §5, no.1, Prop.2). On the other hand, the completion of $\mathbb{Q}$ with respect to the archimedean place ∞ is $\mathbb{R}$. In full generality, we have the following well-known **Theorem of Ostrowski:**

Theorem 1.2.6. *The only complete archimedean fields are* $\mathbb{R}$ *and* $\mathbb{C}$.

Proof: [**157**], §9.5 or [**47**], Ch.VI, §6, no.4, Th.2. □

Proposition 1.2.7. *Let K be a field which is complete relative to an absolute value $| \ |_v$ and let L be a finite-dimensional extension field of K. Then there is a unique extension of $| \ |_v$ to an absolute value $| \ |_w$ of L. For any $x \in L$ the equation*

$$|x|_w = |N_{L/K}(x)|_v^{1/[L:K]}$$

holds, where $N_{L/K}$ is the norm from L to K. Moreover, the field L is complete with respect to $| \ |_w$.

Proof: [**157**], Th.s 9.8, 9.9, 9.12 or [**47**], Ch.VI, §8, no.7, Prop.10. □

Remark 1.2.8. Clearly, the preceding proposition implies that there is a unique extension to an absolute value on the algebraic closure of K. Note however that the last clause of this proposition need not hold for infinite-dimensional extensions; a well-known example is an algebraic closure of the p-adic field $\mathbb{Q}_p$ (cf. S. Bosch, U. Güntzer, and R. Remmert [**43**], 3.4.3).

1.2.9. Let K be a field with a non-archimedean place v and let L be a finite-dimensional field extension of K. Assume that w is a place of L with $w|v$. The ring

$$R_v := \{x \in K \mid |x|_v \leq 1\}$$

is called the **valuation ring** of v. The definition is obviously independent of the representative $| \ |_v$ of v. R_v is a local ring with unique maximal ideal $\mathfrak{m}_v := \{x \in K \mid |x|_v < 1\}$.

The **residue field** $k(v)$ is defined by $R_v/\mathfrak{m}_v$. The quotient map $R_v \to k(v), x \mapsto \overline{x}$ is called the **reduction**. Applying it to coefficients, it extends to polynomials and to power series.

The **residue degree** $f_{w/v}$ of L/K in w is the dimension of $k(w)$ over $k(v)$. Let $|\ |_w$ be an absolute value representing w and $|\ |_v$ the restriction of $|\ |_w$ to K. The **value group** $|K^\times|_v$ is a multiplicative subgroup of $|L^\times|_w$ and its index is called the **ramification index** $e_{w/v}$ of w in v.

The place v is called **discrete** if the value group $|K^\times|_v$ is cyclic. Then $\mathfrak{m}_v$ is a principal ideal and any principal generator is called a **local parameter**.

The following result is the very useful **Hensel's lemma**.

Lemma 1.2.10. *Let K be complete with respect to a non-archimedean place v. Let $f(t) \in K[t]$ be a monic polynomial with reduction $\overline{f}(t) = \overline{g}(t)\overline{h}(t)$ for some monic coprime polynomials $\overline{g}(t), \overline{h}(t) \in k(v)[t]$. Then there exist monic polynomials $G(t), H(t) \in R_v[t]$ with $F(t) = G(t)H(t)$ and $\overline{G}(t) = \overline{g}(t)$, $\overline{H}(t) = \overline{h}(t)$.*

Proof: For discrete valuations, we refer to [**157**], §9.11. The general case is proved in [**43**], 3.3.4. □

Proposition 1.2.11. *Let L/K and $w|v$ be as in 1.2.9.*

(a) *The residue degree and the ramification index do not change if we pass to completions.*

(b) *The product of the residue degree and the ramification index is at most $[L : K]$, with equality if v is discrete and K is complete relative to v.*

Proof: [**157**], §9.10 or [**173**], Prop.s XII.4.2, XII.6.1, and §XII.5. □

1.2.12. A **number field** is a finite-dimensional field extension of $\mathbb{Q}$. The ring of algebraic integers of K is denoted by O_K. Now let L be a locally compact field containing a number field K as a dense subset and assume that the topology is not discrete. Then it follows that L is complete because it is locally compact. The classification of non-discrete locally compact fields is well known and tells us that there is a place v of K such that L is the completion of K with respect to v. Moreover, if L is connected, then L is isomorphic to $\mathbb{R}$ or $\mathbb{C}$ or to a finite extension of $\mathbb{Q}_p$. The closure of O_K in L coincides with the valuation ring R_v of L.

On the other hand, every completion of a number field with respect to a non-archimedean place is a finite extension of $\mathbb{Q}_p$, hence locally compact. For details, we refer to [**47**], Ch.VI, §9, no.3, Th.1. The following result of Artin and Whaples is called the **approximation theorem**:

Theorem 1.2.13. *Let $|\ |_1, \ldots, |\ |_n$ be inequivalent non-trivial absolute values on a field K. Then for $x_1, \ldots, x_n \in K$ and $\varepsilon > 0$ there is $x \in K$ such that*

$$|x - x_k|_k < \varepsilon \quad (k = 1, \ldots, n).$$

Proof: [**157**], §9.2 or [**173**], Th.XII.1.2. □

1.3. Finite-dimensional extensions

Let K be a field with a fixed non-trivial absolute value $|\ |_v$.

Proposition 1.3.1. *Let L be a finite-dimensional field extension of K generated by a single element ξ. If $f(t)$ is the monic minimal polynomial of ξ over K and*

$$f(t) = f_1^{k_1}(t) \cdots f_r^{k_r}(t)$$

is its decomposition into different irreducible monic factors $f_j(t) \in K_v[t]$, then for each j there is an injective homomorphism

$$\iota\colon L \longrightarrow K_j := K_v[t]/(f_j(t))$$

of field extensions over K, given by $\xi \mapsto t$. There is a unique extension $|\ |_j$ of the absolute value of K_v to K_j. The absolute values $|\ |_j$ are pairwise inequivalent. Moreover, K_j is the completion of L with respect to $|\ |_j$ and the embedding ι. For any absolute value $|\ |_w$ extending $|\ |_v$ to L, there is a unique $j \in \{1, \dots, r\}$ such that the restriction of $|\ |_j$ to L is equal to $|\ |_w$.

Proof: Proposition 1.2.7 leads to the unique extension $|\ |_j$ of the absolute value of K_v. The map ι is a well-defined homomorphism of field extensions over K and the image of ι is dense in K_j. Hence K_j is the completion of L with respect to $|\ |_j$. If the restrictions of $|\ |_j$ and $|\ |_k$ are equivalent, then we have an isometric isomorphism of K_j onto K_k, leaving K_v fixed. Therefore, the images of ξ have to be roots of the same irreducible factor of $f(t)$ in $K_v[t]$, yielding $j = k$. Let $|\ |_w$ be an absolute value on L extending $|\ |_v$. The closure of K in L_w can be identified with K_v. Now ξ generates a finite-dimensional subfield of L_w over K_v which is complete by Proposition 1.2.7, therefore this subfield is L_w itself. Also ξ must be a root of some f_j, hence L_w is isomorphic to K_j over K_v. Moreover, we can assume that L is fixed under this isomorphism. Then it is clear from Proposition 1.2.7 that $|\ |_w$ is equal to the restriction of $|\ |_j$ to L. □

Corollary 1.3.2. *If L is a finite-dimensional separable field extension of K, then*

$$\sum_{w|v} [L_w : K_v] = [L : K],$$

where the sum ranges over all places w of L with $w|v$.

Proof: By the primitive element theorem (see N. Jacobson [**156**], §4.14), there is an element ξ of L which generates L over K. Proposition 1.3.1 implies the formula. □

Remark 1.3.3. If the extension is not separable, we still have $\sum_{w|v}[L_w : K_v] \leq [L : K]$. If L is generated by a single element over K, this is clear from Proposition 1.3.1. For the general case, we use induction on the degree.

Definition 1.3.4. *The number $[L_w : K_v]$ is called the **local degree** of L/K in w.*

Corollary 1.3.5. *Let L/K be a finite-dimensional Galois extension with Galois group $G = \mathrm{Gal}(L/K)$ and let $|\ |_{w_0}$, $|\ |_w$ be absolute values on L extending $|\ |_v$. Then there is an element $\sigma \in G$ with*

$$|x|_w = |\sigma(x)|_{w_0} \quad \text{for} \quad x \in L.$$

The completions L_w and L_{w_0} are isomorphic over K_v. However, they need not to be isomorphic over L.

Proof: As in the proof of Corollary 1.3.2, there is an element ξ of L with $L = K(\xi)$. If $f(t)$ is the minimal polynomial of ξ over K, then L_w is obtained by adjoining a root of $f_j(t)$ to K_v in a fixed splitting field of f over K_v, where $f_j(t)$ is an irreducible factor of $f(t)$ in $K_v[t]$. Since L is a Galois extension, all roots of f are contained in L_w, therefore $L_w = L_{w_0}$ as a field. Then the absolute values $|\ |_{w_0}$ and $|\ |_w$ correspond to embeddings ι_0 and ι of L into L_{w_0} over K. There is a unique $\rho \in \mathrm{Gal}(L_{w_0}/K_v)$ with $\iota = \rho \circ \iota_0$, given by $\iota_0(\xi) \mapsto \iota(\xi)$. If $|\ |$ is the unique absolute value of L_{w_0} extending the one of K_v and if σ is the unique element of G with $\rho \circ \iota_0 = \iota_0 \circ \sigma$, then

$$|x|_w = |\iota(x)| = |\rho \circ \iota_0(x)| = |\iota_0 \circ \sigma(x)| = |\sigma(x)|_{w_0} \quad \text{for } x \in L. \qquad \square$$

1.3.6. Let K be a field with a fixed non-trivial absolute value $|\ |_v$. We consider a finite-dimensional separable extension field L of K and a place w of L with $w|v$. For any $x \in L$, we define

$$\|x\|_w := |N_{L_w/K_v}(x)|_v$$

and

$$|x|_w := |N_{L_w/K_v}(x)|_v^{1/[L:K]}.$$

We know from Proposition 1.2.7 that the restriction of $|N_{L_w/K_v}|_v^{1/[L_w:K_v]}$ to L is a representative of w extending $|\ |_v$. The obvious inequality $[L_w : K_v] \leq [L : K]$ implies that $|\ |_w$ is an absolute value representing w. If v is not archimedean or $[L_w : K_v] = 1$, we have that $\|\ \|_w$ is also an absolute value representing w. On the other hand, if v is archimedean and $[L_w : K_v] = 2$, we have $L_w = \mathbb{C}$ and $K_v = \mathbb{R}$. Assume that the restriction of $|\ |_v$ to $\mathbb{Q}$ is the ordinary absolute value; then $\|\ \|_w$ is not an absolute value because the triangle inequality is not satisfied.

Lemma 1.3.7. *Let $x \in K \setminus \{0\}$ and $y \in L \setminus \{0\}$. With the notation above*

$$\sum_{w|v} \log |x|_w = \log |x|_v,$$

$$\sum_{w|v} \log \|y\|_w = \log |N_{L/K}(y)|_v.$$

Proof: Corollary 1.3.2 implies the first statement. There is an element ξ of L with $L = K(\xi)$. With the notation of Proposition 1.3.1, we have $k_1 = \cdots = k_r = 1$ and an isomorphism

$$L \otimes_K K_v \xrightarrow{\sim} \prod_{j=1}^{r} K_v[t]/(f_j(t))$$

of K_v-algebras, given by $\xi \longrightarrow (t)_{j=1,\dots,r}$ (this is a form of the Chinese remainder theorem). By Proposition 1.3.1 we get

$$N_{L/K}(y) = \prod_{w|v} N_{L_w/K_v}(y),$$

proving the second claim. □

1.3.8. If K is a number field, the archimedean absolute values of K are determined by the embeddings $\sigma : K \longrightarrow \mathbb{C}$ of K into the complex numbers. There are exactly $[K : \mathbb{Q}]$ such embeddings. An embedding σ is said to be real if $\sigma(K)$ is in the real subfield $\mathbb{R}$ of $\mathbb{C}$, and complex otherwise. If σ is a complex embedding, composition with complex conjugation yields a conjugate embedding $\overline{\sigma}$, and it is clear that σ and $\overline{\sigma}$ determine the same archimedean absolute value. Conversely, if σ and σ' are two embeddings of K in $\mathbb{C}$ determining the same absolute value, we have $\sigma' = \sigma$ or $\sigma' = \overline{\sigma}$. All this is immediate from Proposition 1.3.1, because K has a primitive element over $\mathbb{Q}$.

The completion of K at an archimedean place is isometric to either $\mathbb{R}$ or $\mathbb{C}$. Accordingly, the set of archimedean places is subdivided into **real places** and **complex places.**

Example 1.3.9. Let p be an odd prime and $K = \mathbb{Q}(\zeta)$ with ζ a primitive pth root of unity. Our goal in this example is to determine all extensions of an absolute value of $\mathbb{Q}$ to K.

The minimal polynomial of ζ over $\mathbb{Q}$ is given by

$$f(t) = t^{p-1} + t^{p-2} + \cdots + 1,$$

which is proved by applying Eisenstein's criterion to $f(t+1)$.

To begin with, we determine the extensions of the ordinary absolute value $|\ |_\infty$ of $\mathbb{Q}$ to K. The irreducible factors of $f(t)$ in $\mathbb{R}[t]$ have degree 2. By Proposition 1.3.1, there are exactly $\frac{p-1}{2}$ extensions of $|\ |_\infty$; all archimedean absolute values of K are associated to the $(p-1)/2$ pairs of complex conjugate embeddings of K and the local degree is equal to 2.

Next, we consider the extensions of the non-archimedean absolute value associated to a prime number q. Suppose first that $q \neq p$. We need to decompose $f(t)$ into irreducible factors over $\mathbb{Q}_q$. There is a smallest number $r \geq 1$ with $p | q^r - 1$, determined by the property that $\mathbb{F}_{q^r}$ is the smallest field of characteristic q containing a non-trivial pth root of unity (note also that, by Fermat's little theorem, $r | p-1$). In that case, the field $\mathbb{F}_{q^r}$

contains all pth roots of unity. Hence $f(t)$ is a product of $\frac{p-1}{r}$ distinct irreducible factors in $\mathbb{F}_q[t]$, each degree r. The same is true in $\mathbb{Q}_q[t]$ by Hensel's lemma (see Lemma 1.2.10). We conclude again by Proposition 1.3.1 that there are exactly $\frac{p-1}{r}$ extensions of $| \ |_q$ to K and the local degrees are equal to r. It is obvious that ζ remains a unit in any completion of K with respect to such an absolute value and its representative in the residue field is also a non-trivial primitive pth root of unity. By Proposition 1.2.11, the residue degree is equal to r and the ramification index is 1.

It remains to consider the place p. As before, Eisenstein's criterion shows that the polynomial $f(t)$ is irreducible in $\mathbb{Q}_p[t]$. Then there is only one extension $| \ |_v$ of $| \ |_p$ to K and the local degree is $p-1$. The minimal polynomial of $\zeta - 1$ over $\mathbb{Q}_p$ is $f(t+1)$ and so $N_{K_v/\mathbb{Q}_p}(\zeta - 1) = p$. Proposition 1.2.7 implies

$$|\zeta - 1|_v = p^{-1/(p-1)}.$$

By Proposition 1.2.11, the ramification index is equal to $p-1$ and the residue degree is equal to 1.

1.3.10. In the final part of this section, we handle finite-dimensional field extensions without separability assumptions. It turns out that it suffices to adjust the exponents in the normalization 1.3.6. Since we focus almost exclusively on number fields, the reader may skip the rest of this section in a first reading.

Let K be a field with absolute values $| \ |_v$ and let L/K be a finite-dimensional field extension. Our goal is to generalize Proposition 1.3.1 describing the extensions of $| \ |_v$ to the field L.

Since $L \otimes_K K_v$ is a finite-dimensional K_v-algebra, the structure theorem of commutative artinian rings ([157], Th.7.13) gives uniquely determined ideals R_j, which are local K_v-algebras with maximal ideals $\mathfrak{m}_j$ and such that

$$L \otimes_K K_v = \prod_{j=1}^{r} R_j. \tag{1.1}$$

We have natural embeddings of L and K_v into the residue field $K_j = R_j/\mathfrak{m}_j$ of R_j. By Proposition 1.2.7, there is a unique extension $| \ |_j$ of $| \ |_v$ to K_j. Clearly, L is dense in K_j, whence K_j is the completion of L with respect to this absolute value.

Proposition 1.3.11. *The restrictions of $| \ |_j$, $j = 1, \ldots, r$, to L are all extensions of $| \ |_v$ to absolute values on L and they are pairwise inequivalent.*

Proof: Suppose the restrictions to L of $| \ |_j$ and $| \ |_k$ are equivalent. Then there is an isomorphism $\varphi : K_j \xrightarrow{\sim} K_k$ which is the identity on L and K_v. Let φ_j be the canonical isomorphism of $L \otimes_K K_v$ onto K_j. Then $\varphi = \varphi_k \circ \varphi_j^{-1}$ (check on L and K_v), which is possible only if $j = k$. This proves the last clause of our claim.

Let $| \ |_w$ be an extension of $| \ |_v$ to L. The closure of K in L_w will be identified with K_v. Since L is finite dimensional over K, it follows that LK_v is a closed subfield of L_w, from which we conclude that $LK_v = L_w$. By the universal property of the tensor product, there is a homomorphism of $L \otimes_K K_v$ onto L_w. The kernel of this homomorphism is a maximal ideal, hence equal to $\mathfrak{m}_j \times \prod_{k \neq j} R_k$ for some j. Therefore, K_j is isomorphic to

L_w as a K_v-algebra, and by Proposition 1.2.7 we infer that it is in fact an isometry. This shows that the restriction of $| \ |_j$ to L is $| \ |_w$, as we wanted. $\square$

1.3.12. Now we are ready to define the correct normalizations of the absolute values as above. We have seen that the places w of L with $w|v$ are in one-to-one correspondence with the local K_v-algebras in (1.1), thus we may write

$$L \otimes_K K_v = \prod_{w|v} T_w \tag{1.2}$$

and identify the residue field of T_w with the completion L_w. For $y \in L$, we set

$$\|y\|_w := |N_{L_w/K_v}(y)|_v^{[T_w:L_w]}$$

and

$$|y|_w := \|y\|_w^{1/[L:K]}.$$

With these modifications the analogue of Lemma 1.3.7 still holds, namely:

Lemma 1.3.13. *If* $x \in K \setminus \{0\}$ *and* $y \in L \setminus \{0\}$, *then*

$$\sum_{w|v} \log |x|_w = \log |x|_v \,, \tag{1.3}$$

$$\sum_{w|v} \log \|y\|_w = \log |N_{L/K}(y)|_v \,. \tag{1.4}$$

Proof: Formula (1.4) is a trivial consequence of Proposition 1.2.7 and (1.2). If we set $y = x$ in (1.4), then (1.3) follows immediately from (1.2). $\square$

1.4. The product formula

The product formula over $\mathbb{Q}$ may be stated and proved as a consequence of the factorization of a non-zero rational number into a product of primes and a unit. In spite of its simplicity and essentially trivial nature, it plays a fundamental role and its importance cannot be overstated. The fact that it involves all places, including the places at ∞, means that, from the geometrical point of view, we are dealing with a complete variety. In the case considered here, the general fibre of the variety is a point and everything is quite simple. However, the best interpretation of the product formula and its generalizations is found in the framework of Arakelov theory.

1.4.1. Let K be a field and M_K be a set of non-trivial inequivalent absolute values on K such that the set

$$\{| \ |_v \in M_K \mid |x|_v \neq 1\}$$

is finite for any $x \in K \setminus \{0\}$. We identify the elements of M_K with the corresponding places and say that M_K satisfies the **product formula** if

$$\prod_{v \in M_K} |x|_v = 1$$

for any $x \in K \setminus \{0\}$.

We shall also refer to

$$\sum_{v \in M_K} \log |x|_v = 0$$

as the product formula for $x \neq 0$.

If L/K is a finite-dimensional extension and M_K is a set of places with associated normalized absolute values satisfying the product formula, we obtain a set of places M_L consisting of representatives $| \ |_w$ of $w|v$ for $v \in M_K$, normalized as in 1.3.6 and 1.3.12.

Proposition 1.4.2. *The set of places M_L so normalized again satisfies the product formula.*

Proof: Let $x \in L^\times$. We need to check that $|x|_w \neq 1$ only for finitely many $w \in M_L$. Since x is algebraic over K, we have

$$x^n + a_{n-1}x^{n-1} + \cdots + a_0 = 0 \tag{1.5}$$

for suitable $a_i \in K$. By assumption, we have $|a_i|_v \in \{0, 1\}$ up to finitely many $v \in M_K$. Since there are only finitely many $w \in M_L$ lying over a given v (Corollary 1.3.2), we have $|a_i|_w \leq 1$, up to finitely many $w \in M_L$. Clearly, there are only finitely many archimedean places in M_K and hence in M_L. Thus it is enough to consider non-archimedean $w \in M_L$ and then the ultrametric inequality applied to (1.5) shows that $|x|_w \leq 1$ whenever all coefficients $|a_i|_w \leq 1$. The same argument applied to $1/x$ completes the proof that $|x|_w = 1$ up to finitely many $w \in M_L$.

Once this is done, it is immediate from Lemmas 1.3.7 and 1.3.13 that the normalized set of absolute values on L satisfies the product formula. □

1.4.3. By Example 1.2.5, we get

$$M_{\mathbb{Q}} := \{| \ |_p \mid p \text{ prime number or } p = \infty\},$$

normalized as follows. If $p = \infty$, then $| \ |_p$ is the ordinary absolute value on $\mathbb{Q}$, and, if p is prime, then the absolute value is the p-adic absolute value on $\mathbb{Q}$, with $|p|_p = 1/p$.

Let K be a number field and let M_K be the associated set of places and normalized absolute values, obtained from the above construction applied to the extension $K/\mathbb{Q}$.

Proposition 1.4.4. *If K is a number field, M_K satisfies the product formula.*

Proof: By the above discussion, we can assume that K is equal to $\mathbb{Q}$ and it is obviously enough to show the product formula for a prime number x

$$\prod_{p \in M_{\mathbb{Q}}} |x|_p = |x|_x \, |x|_\infty = \frac{1}{x} x = 1.$$

□

In this book, whenever we talk about M_K of a number field it will always be the set so constructed from $M_{\mathbb{Q}}$. This is important for our normalizations, which we repeat: If $p = \infty$, then $| \ |_p$ is the ordinary absolute value on $\mathbb{Q}$, and, if p is prime, then the absolute value is the p-adic absolute value on $\mathbb{Q}$, with $|p|_p = 1/p$. In either case, we have

$$|x|_v := |N_{K_v/\mathbb{Q}_p}(x)|_p^{1/[K:\mathbb{Q}]} \tag{1.6}$$

for $x \in K$ and $v|p$.

As an application of our previous considerations in this chapter, we prove the following refinement of Theorem 1.2.13 for a number field K, called the **strong approximation theorem**:

Theorem 1.4.5. *Let $(| \ |_v)_{v \in S}$ be representatives for a finite set S of non-archimedean places of the number field K, let $x_v \in K_v$ for every $v \in S$, and let $\varepsilon > 0$. Then there is $x \in K$ with $|x - x_v|_v < \varepsilon$ for all $v \in S$ and $|x|_v \leq 1$ for all non-archimedean $v \notin S$.*

Proof: There is no loss of generality in assuming that $x_v \in K$, because, by definition of completion, K is dense in K_v. By Proposition 1.2.3, we may also assume that the absolute values extend the p-adic absolute values $| \ |_p$ from Example 1.2.5. By Corollary 1.3.2, there are only finitely many places lying over a natural prime number, hence we may enlarge S to the set of all places lying over a finite set S_0 of prime numbers, taking $x_v = 0$ at every new place v introduced by doing so. For any $x \in K^\times$, Proposition 1.4.2 shows that $|x|_v \neq 1$ only for finitely many places v. Now take x to be the approximation to x_v $(v \in S)$ obtained from Theorem 1.2.13 with $\varepsilon = 1$. Then there is a finite set S_1 of prime numbers, disjoint from S_0, such that $|x|_v = 1$ for all places v of K, which do not lie over $S_0 \cup S_1$. By the Chinese remainder theorem, there is $m \in \mathbb{Z}$ with $|m - 1|_p < \delta$ for $p \in S_0$ and $|m|_p < \delta$ for $p \in S_1$. If we choose $\delta > 0$ sufficiently small, then the approximation mx satisfies the conclusion of Theorem 1.4.5. □

1.4.6. Function fields (see A.4.11) are also important examples in diophantine geometry, where the product formula holds and we devote the rest of this section to its discussion. In order to understand the background from algebraic geometry, the reader may consult the material on divisors in Sections A.8 and A.9. Since the focus of this book is mostly on number fields, we may skip the proofs in a first reading.

Let X be a projective irreducible variety over a field K and let us fix an ample line bundle L (see A.6.10). We denote by $\deg(Z)$ the degree of a cycle Z with respect to L. Since the function field does not change by passing to the normalization (see A.12.6), we may and shall assume that X is regular in codimension 1 (see A.8.10). For any prime divisor Z of X, the local ring $\mathcal{O}_{X,Z}$ is a discrete valuation ring and the valuation of $f \in K(X)^\times$ is the order of f at Z. The latter is denoted by $\mathrm{ord}_Z(f)$. Since the degree of a principal divisor is 0, we have

$$\sum_Z \mathrm{ord}_Z(f) \deg(Z) = 0. \tag{1.7}$$

We normalize the absolute value corresponding to ord_Z by

$$|f|_Z := c^{\mathrm{ord}_Z(f)\deg(Z)},$$

where c is some fixed number, $0 < c < 1$.

Proposition 1.4.7. *The absolute values $|\ |_Z$, Z prime divisor of X, are not trivial, inequivalent, and satisfy the product formula.*

Proof: Since $\mathcal{O}_{X,Z}$ is a discrete valuation ring, the absolute value $|\ |_Z$ is not trivial. Let Y, Z be different prime divisors. We may think of X as embedded in projective space. There is a hyperplane not containing Y, Z. Let H be the corresponding very ample divisor on X and let $n \in \mathbb{N}$ be so large that $nH + Y$ is very ample (see A.6.10). By the same argument as above, there is an effective divisor H' not containing Y and Z such that

$$H' = nH + Y + \mathrm{div}(f)$$

for some $f \in K(X)^\times$. We have $|f|_Z = 1$ and $|f|_Y = c^{-\deg(Y)}$. We conclude that $|\ |_Y$ and $|\ |_Z$ are inequivalent. The product formula is a consequence of (1.7). □

Example 1.4.8. Let C be an irreducible projective curve over K. The curve is regular if and only if it is normal (see A.8.10). So let C be regular (note however that this does not mean that C is smooth, see A.13.3). Then C is a regular model for the function field $K(C)$. The prime divisors of C are in one-to-one correspondence with orbits of points under $\mathrm{Gal}(\overline{K}/K)$ (see A.2.7). The order at a prime divisor is the order at any point in the associated orbit.

This example fits with the preceding considerations if $L = O([P_0])$ for some point $P_0 \in C(K)$ (and L is of course automatically ample). The product formula follows from the fact that any rational function $f \in K(C)^\times$ has the same number of zeros and poles.

1.4.9. On the function field $K(X)$, we shall always use the set of absolute values considered above. We denote it by M_X to emphasize the role of the model X. Obviously, the choice of the constant c is irrelevant. Usually, we shall choose $c = 1/e$.

Lemma 1.4.10. *The following statements hold:*

(a) *Any finitely generated field over K is a function field of an irreducible projective normal variety over K.*

(b) *Two irreducible varieties are birationally equivalent over K if and only if they have isomorphic function fields over K.*

(c) *If L is a finite-dimensional extension of the function field $K(X)$ of an irreducible projective variety X over K, then there are an irreducible projective normal variety Y over K, and a finite surjective morphism $\varphi : Y \longrightarrow X$, such that $L \cong K(Y)$ and the inclusion $K(X) \subset L$ corresponds to $\varphi^\sharp : K(X) \longrightarrow K(Y)$.*

(d) *In (c), there is a distinguished choice for Y called the* **normalization of X in L**, *uniquely characterized up to isomorphisms by the following property: Given a dominant morphism $\varphi' : Y' \to X$ of an irreducible normal K-variety Y' to X and a homomorphism $\rho : K(Y) \hookrightarrow K(Y')$ over $K(X)$, then there is a unique dominant morphism $\psi : Y' \to Y$ with $\psi^\sharp = \rho$.*

Proof: Let L be a finitely generated field over K with generators $x_1, \dots, x_n$. Let $\mathfrak{I}$ be the kernel of the homomorphism

$$K[t_1, \dots, t_n] \longrightarrow L$$

given by $t_i \mapsto x_i$. Denote the corresponding closed subvariety of $\mathbb{A}^n$ by X. Since $\mathfrak{I}$ is prime, X is an irreducible variety. Obviously, $K(X)$ is isomorphic to L. The closure $\overline{X}$ of X in $\mathbb{P}^n_K$ is a projective variety. The normalization of $\overline{X}$ is a projective normal variety (see A.12.7) with function field L (see A.12.6, A.12.7). This proves (a).

For (b), see A.11.4.

To prove (c) and (d), a generalization of the construction in A.12.6 leads to the normalization of X in L. If X were affine, then we would take the integral closure of $K[X]$ in L. For any variety X, we glue the normalizations of the affine open charts to get the normalization of X (see A. Grothendieck [**135**], 6.3.9). The morphism from the normalization to X is finite ([**135**], 6.3.10) and hence projectivity of X implies projectivity of the normalization (see A.12.7). □

Remark 1.4.11. Note that a curve regular in codimension 1 is regular and hence determined up to isomorphism by its function field. For a higher-dimensional function field $K(X)$, there is no canonical choice for the model X. Even if X is smooth, we may blow up a point to get another smooth model X' for $K(X)$ and it is clear that $M_X \subsetneq M_{X'}$. Hence we always fix a model when dealing with higher-dimensional function fields.

If $L/K(X)$ is a finite extension, then Lemma 1.4.10 (d) shows that the normalization $\varphi : Y \to X$ of X in L is a canonical model for the function field L. Let $\varphi' : Y' \to X$ be any finite surjective morphism of an irreducible projective variety Y' onto X with $K(Y') = L$ and with $K(X) \hookrightarrow L$ equal to $(\varphi')^\sharp$. We claim that $M_Y = M_{Y'}$.

We first show that we may replace Y' by its normalization Y'' (in L) without changing the set of places. Indeed, the normalization morphism $\pi : Y'' \to Y'$ is finite and birational. Since Y' and Y'' are projective, the valuative criterion of properness (see A.11.10) shows that π induces an isomorphism outside of subsets of codimension ≥ 2 in Y'' and Y', hence $M_{Y'} = M_{Y'}$. So we may assume Y' normal and then Lemma 1.4.10 (d) yields a unique dominant morphism $\psi : Y' \to Y$ factoring through φ'. The morphism ψ is proper (see A.6.15) and has finite fibres, hence ψ is finite (see A.12.4). Since $K(Y') = K(Y) = L$, the morphism ψ is also birational (Lemma 1.4.10 (b)) and the same argument as above shows that ψ is an isomorphism in codimension 1, hence $M_Y = M_{Y'}$.

We see that the set of places of L is well determined by X and in the following examples we will show that M_Y is the set of places of L extending the places of M_X.

Example 1.4.12. Let us consider a finite-dimensional field extension of the function field $K(C)$ of an irreducible projective regular curve C over the ground field K. By Lemma 1.4.10, there is an irreducible projective regular curve C' over K and a morphism $\varphi : C' \to C$ such that the extension corresponds to the extension $K(C')/K(C)$ induced by φ. We know from the above that the order at a closed irreducible subset Z of dimension 0 induces an absolute value $|\ |_v \in M_C$. Since C is regular, Cartier divisors can be identified with Weil divisors (cf. A.8.21) and $\varphi^*(Z)$ is well defined. We have

$$\varphi^*(Z) = \sum_{Z'} m_{Z'} Z',$$

where Z' ranges over all irreducible closed subsets of C' lying over Z and where $m_{Z'}$ denotes the multiplicity in Z'. Note that for $f \in K(C)^\times$ we have $\mathrm{ord}_{Z'}(f) = m_{Z'} \mathrm{ord}_Z(f)$. Thus Z' induces a place v' on C' with $v'|v$. Its ramification index and residue degree are

$$e_{v'/v} = m_{Z'}, \qquad f_{v'/v} = [K(Z') : K(Z)].$$

The projection formula for proper intersection products (see W. Fulton [**125**], Prop.2.5(c)) gives

$$Z.\varphi_*(C') = \varphi_*(\varphi^*(Z)),$$

hence

$$[K(C') : K(C)] = \sum_{Z'} m_{Z'}[K(Z') : K(Z)] = \sum_{v'} e_{v'/v} f_{v'/v}.$$

By Remark 1.3.3 and Proposition 1.2.11, we see that all places v' dividing v are induced by the "fibre points" Z' and the local degree satisfies

$$[K(C')_{v'} : K(C)_v] = m_{Z'}[K(Z') : K(Z)].$$

Example 1.4.13. In order to extend the above example to higher dimensions, we use the language of schemes. Let us consider a finite-dimensional extension of a function field $K(X)$. By Lemma 1.4.10, we may identify this extension with an extension $K(X')/K(X)$ induced by a finite surjective morphism $\varphi : X' \to X$ of irreducible projective varieties over K and regular in codimension 1.

Let Z be a prime divisor on X with corresponding place v. To study the places v' of X' with $v'|v$, we may assume that X, and hence X', are affine. The fibre over the generic point ζ of Z is the affine scheme

$$\varphi^{-1}(\zeta) = \mathrm{Spec}\left(K[X'] \otimes_{K[X]} K(\zeta)\right),$$

where $K(\zeta)$ is the residue field of $\mathcal{O}_{X,\zeta}$. By the structure theorem of finite-dimensional algebras ([**157**], Th.7.13), we have

$$K[X'] \otimes_{K[X]} K(\zeta) \cong \prod_{\xi \in \varphi^{-1}(\zeta)} \mathcal{O}_{\varphi^{-1}(\zeta),\xi}.$$

Note that $K[X'] \otimes_{K[X]} \mathcal{O}_{X,\zeta}$ is a finitely generated module over the discrete valuation ring $\mathcal{O}_{X,\zeta}$. Since it is also torsion free (as a subring of $K(X')$), it is free of rank $[K(X') : K(X)]$. We conclude that

$$[K(X') : K(X)] = \sum_{\xi \in \varphi^{-1}(\zeta)} [\mathcal{O}_{\varphi^{-1}(\zeta),\xi} : K(\zeta)].$$

Clearly, the order at any point $\xi \in \varphi^{-1}(\zeta)$ yields a place v' of $K(X')$ with $v'|v$. Indeed, denoting by t_ζ a local parameter in $\mathcal{O}_{X,\zeta}$, we have

$$\mathrm{ord}_\xi(f) = \mathrm{ord}_\xi(t_\zeta)\mathrm{ord}_\zeta(f),$$

for any $f \in K(X)^\times$. Moreover, we verify that

$$e_{v'/v} = \mathrm{ord}_\xi(t_\zeta), \qquad f_{v'/v} = [K(\xi) : K(\zeta)].$$

Now, using $\mathcal{O}_{\varphi^{-1}(\zeta),\xi} = \mathcal{O}_{X',\xi}/\langle t_\zeta \rangle$, we see that

$$[\mathcal{O}_{\varphi^{-1}(\zeta),\xi} : K(\zeta)] = e_{v'/v} f_{v'/v}.$$

Therefore, as in the preceding example, we conclude that all places of $K(X')$ dividing v are induced by points of $\varphi^{-1}(\zeta)$.

Moreover, let L be an ample line bundle on X. Then the absolute values are normalized by

$$|f|_v := c^{\operatorname{ord}_v(f) \deg_L(v)} \qquad (v \in M_{K(X)}, f \in K(X))$$

to satisfy the product formula for some constant c. If we use the normalizations from 1.3.6 on $K(X')$ and the equation

$$\deg_{\varphi^* L}(w) = [K(w) : K(v)] \deg_L(v)$$

obtained from the projection formula (A.13) on page 558, then the above implies

$$|g|_w = c_1^{\operatorname{ord}_w(g) \deg_{\varphi^* L}(w)} \qquad (w \in M_{K(X')}, g \in K(X'))$$

for $c_1 := c^{1/[K(X'):K(X)]}$. Note that $\varphi^* L$ is ample (cf. A.12.7) and hence the normalizations on $K(X')$ fit with 1.4.6 for the new constant c_1.

1.5. Heights in projective and affine space

1.5.1. We denote by $\overline{\mathbb{Q}}$ a choice of an algebraic closure of $\mathbb{Q}$. Let us consider the projective space $\mathbb{P}^n_{\overline{\mathbb{Q}}}$ with standard global homogeneous coordinates $\mathbf{x} = (x_0 : x_1 : \cdots : x_n)$. Let $P \in \mathbb{P}^n_{\overline{\mathbb{Q}}}$. We now define a function, called **height,** on algebraic points of $\mathbb{P}^n_{\overline{\mathbb{Q}}}$, which may be considered as a measure of the "algebraic complication" needed to describe P. This is a fundamental notion at the basis of diophantine geometry.

Let P be a point of $\mathbb{P}^n_{\overline{\mathbb{Q}}}$ represented by a homogeneous non-zero vector $\mathbf{x}$ with coordinates in a number field K. Then we set

$$h(\mathbf{x}) := \sum_{v \in M_K} \max_j \log |x_j|_v.$$

Lemma 1.5.2. *$h(\mathbf{x})$ is independent of the choice of K.*

Proof: Let L be another number field containing the coordinates $x_0, \ldots, x_n$ of $\mathbf{x}$. We can assume that $K \subset L$. Then

$$\sum_{w \in M_L} \max_j \log |x_j|_w = \sum_{v \in M_K} \sum_{w|v} \max_j \log |x_j|_w.$$

Our claim now follows from the first formula of Lemma 1.3.7. □

Lemma 1.5.3. *$h(\mathbf{x})$ is independent of the choice of coordinates.*

Proof: Let $\mathbf{y}$ be another coordinate vector. By the preceding lemma, we may assume that $x_0, \dots, x_n, y_0, \dots, y_n \in K$. There is $\lambda \in K$, $\lambda \neq 0$, with $\mathbf{y} = \lambda \mathbf{x}$, hence

$$h(\mathbf{y}) = \sum_{v \in M_K} \max_j \log |y_j|_v = \sum_{v \in M_K} \log |\lambda|_v + \sum_{v \in M_K} \max_j \log |x_j|_v.$$

Thus we get $h(\mathbf{y}) = h(\mathbf{x})$ by the product formula. □

These two lemmas show that the height so defined depends only on the point P and not on the choice of coordinates of a homogeneous vector representing P.

Definition 1.5.4. *We call $h(\mathbf{x})$* ***the absolute logarithmic height*** *(briefly,* ***height****) of P and we denote it by $h(P)$. We also use the* ***multiplicative height*** *$H(P) = e^{h(P)}$.*

Example 1.5.5. If the coordinates $x_0, \dots, x_n$ of $P \in \mathbb{P}^n_{\overline{\mathbb{Q}}}$ can be chosen in $\mathbb{Q}$, we can assume that they are integers and that $x_0, \dots, x_n$ have no common factors. If we take such a representative for the coordinates of P, then the non-archimedean places give no contribution to the height, and we obtain

$$h(P) = \max_j \log |x_j|_\infty.$$

1.5.6. A similar notion holds for affine space. Let $\mathbb{A}^n_{\overline{\mathbb{Q}}}$ be the affine space of dimension n over $\overline{\mathbb{Q}}$, together with the usual embedding in $\mathbb{P}^n_{\overline{\mathbb{Q}}}$ given by

$$P = (x_1, \dots, x_n) \mapsto (1 : x_1 : \dots : x_n);$$

then we define $h(P)$ as the height of the image of P.

1.5.7. It should always be clear from the context whether we are dealing with points in affine or projective space, and there should be no problem in using the same notation h for heights in affine or projective space.

In performing local calculations, it proves to be convenient to introduce the function

$$\log^+ t = \max(0, \log t)$$

on the positive real numbers, extended by $\log^+ 0 = 0$. Then it is immediate that the height on affine space is given by

$$h(x_1, \dots, x_n) = \sum_{v \in M_K} \max_j \log^+ |x_j|_v.$$

As a special case, the **height of an algebraic number** α is

$$h(\alpha) = \sum_{v \in M_K} \log^+ |\alpha|_v$$

with K any number field with $\alpha \in K$.

1.5.8. Since any point in projective space admits a homogeneous representative with one coordinate equal to 1, it is clear that the height so defined is never negative. The next result, **Kronecker's theorem**, characterizes the case of equality.

Theorem 1.5.9. *The height of* $\zeta \in \overline{\mathbb{Q}}^{\times}$ *is* 0 *if and only* ζ *is a root of unity.*

Proof: Let K be a number field and let $\zeta \in K^{\times}$. If ζ is a root of unity, then its absolute values are all equal to 1, and hence its height is 0.

Conversely, assume $h(\zeta) = 0$. Then $|\zeta|_v \leq 1$ for every $v \in M_K$; in particular, ζ is an algebraic integer (for a formal argument, see the successive Remark 1.5.11). Let d be the degree of ζ and let $\boldsymbol{\zeta} = (\zeta_1, \ldots, \zeta_d)$ be a full set of conjugates of ζ. Now consider, for every positive integer m, the elementary symmetric functions $s_i(\boldsymbol{\zeta}^m)$, $i = 0, \ldots, d$, of $\zeta_1^m, \ldots, \zeta_d^m$. Since ζ is an algebraic integer, we have $s_i(\boldsymbol{\zeta}^m) \in \mathbb{Z}$ for every m.

Since $|\zeta_j|_v = 1$ for every j and v, and since $s_i(\boldsymbol{\zeta}^m)$ is the sum of $\binom{d}{i}$ terms each of which is a product of factors not exceeding 1 in absolute value, it is now clear that

$$\sum_{i=0}^{d} |s_i(\boldsymbol{\zeta}^m)| \leq \sum_{i=0}^{d} \binom{d}{i} = 2^d.$$

There are only finitely many possibilities for the vector of such symmetric functions, and by Dirichlet's pigeon-hole principle there are two integers m, n with $m > n$ and with the same vector of symmetric functions.

Obviously, this is the same as saying that $\boldsymbol{\zeta}^m = \pi(\boldsymbol{\zeta}^n)$ for some permutation π of $\{1, \ldots, d\}$ and by iterating this relation we find that $\zeta_i^{m^k} = \zeta_{\pi^k(i)}^{n^k}$. If we take k such that π^k is the identity, we conclude that $\zeta^{m^k - n^k} = 1$ with $m^k - n^k > 0$. □

1.5.10. We recall here some basic facts about S-integers and S-units in a number field K. Let $S \subset M_K$ be a finite set of places, which includes the set S_∞ of all archimedean places of K. An element $x \in K$ is an ***S*-integer** if $|x|_v \leq 1$ for $v \notin S$. The S-integers of K form a subring $O_{S,K}$ of K. The units in $O_{S,K}$ are called the ***S*-units** of K and form a group $U_{S,K}$. An element $x \in O_{S,K}$ is an S-unit if and only if $|x|_v = 1$ for all $v \notin S$.

Remark 1.5.11. An easy application of the non-archimedean triangle inequality and Gauss's lemma (see Lemma 1.6.3) shows that an S_∞-integer is the same as an algebraic integer in K. If $S = S_\infty$, we simply talk about the integers and the units of the number field K. The units of K are the algebraic integers $x \in K$ with norm $N_{K/\mathbb{Q}}(x) = \pm 1$, as we see from writing the norm as a product of conjugates of x.

1.5.12. We consider the following homomorphism

$$\phi : U_{S,K} \to \mathbb{R}^{|S|}, \qquad x \mapsto (\log |x|_v)_{v \in S}$$

of groups. By taking the logarithm of the product formula, we see that the image of ϕ is contained in the hyperplane $\sum_{v\in S} y_v = 0$, $\mathbf{y} \in \mathbb{R}^{|S|}$. By Kronecker's theorem, the kernel of ϕ is the group μ_K of roots of unity in K. This is part of **Dirichlet's unit theorem**:

Theorem 1.5.13. *Let S be as in 1.5.10. The image of ϕ is a lattice of maximal rank $|S|-1$ in the hyperplane $\sum_{v\in S} y_v = 0$. Hence $U_{S,K} \cong \mu_K \times \mathbb{Z}^{|S|-1}$.*

We will not prove this result here, and refer instead to W. Narkiewicz [**215**], Th.3.6, or S. Lang [**172**], p.104.

1.5.14. The **Segre embedding**

$$\mathbb{P}^n_{\overline{\mathbb{Q}}} \times \mathbb{P}^m_{\overline{\mathbb{Q}}} \longrightarrow \mathbb{P}^{(n+1)(m+1)-1}_{\overline{\mathbb{Q}}}$$

is given by

$$(\mathbf{x},\mathbf{y}) \longmapsto \mathbf{x}\otimes\mathbf{y} := (x_i y_j)\,,$$

where the pairs ij are, for example, ordered lexicographically (see A.6.4). An easy calculation shows that

$$h(\mathbf{x}\otimes\mathbf{y}) = h(\mathbf{x}) + h(\mathbf{y}),$$

using $\max_{ij} |x_i y_j|_v = \max_i |x_i|_v \cdot \max_j |y_j|_v$ for every v.

This notion extends in an obvious fashion to finite products of projective spaces with an arbitrary number of factors.

Proposition 1.5.15. *If $P_1,\ldots,P_r$ are points of $\mathbb{A}^n_{\overline{\mathbb{Q}}}$, then*

$$h(P_1+\cdots+P_r) \le h(P_1)+\cdots+h(P_r)+\log r.$$

Proof: Let $\mathbf{x}^{(k)}$ be coordinate vectors of P_k, $k=1,\ldots,r$, which we assume to be in a suitable number field K. Then

$$h(P_1+\cdots+P_r) = \sum_{v\in M_K} \max_j \log^+ |x_j^{(1)}+\cdots+x_j^{(r)}|_v.$$

If v is not archimedean, then

$$|x_j^{(1)}+\cdots+x_j^{(r)}|_v \le \max_k |x_j^{(k)}|_v.$$

If instead v is archimedean, we use the triangle inequality for the ordinary absolute value getting

$$|x_j^{(1)}+\cdots+x_j^{(r)}|_v \le |r|_v \max_k |x_j^{(k)}|_v.$$

Then Lemma 1.3.7 implies

$$\sum_{v|\infty} \log |r|_v = \log r$$

leading to

$$h(P_1 + \cdots + P_r) \leq \log r + \sum_{v \in M_K} \max_{j,k} \log^+ |x_j^{(k)}|_v.$$

The obvious fact

$$\max_{j,k} \log^+ |x_j^{(k)}|_v \leq \sum_k \max_j \log^+ |x_j^{(k)}|_v$$

concludes the proof. □

1.5.16. The following considerations show that the inequality in Proposition 1.5.15 cannot be improved upon in general.

Let $\alpha_1, \ldots, \alpha_r$ be algebraic numbers. By the preceding Proposition 1.5.15, we have

$$h(\alpha_1 + \cdots + \alpha_r) \leq h(\alpha_1) + \cdots + h(\alpha_r) + \log r.$$

Now suppose that equality occurs for some $r \geq 2$. Looking at the proof above, we must have

$$\log^+ |\alpha_1 + \cdots + \alpha_r|_v = \log^+ (|r|_v \max_k |\alpha_k|_v) = \log |r|_v + \log^+ \max_k |\alpha_k|_v$$

for any archimedean prime v. This is equivalent to the two conditions

$$\max_k |\alpha_k|_v \geq 1$$

and

$$|r|_v \max_k |\alpha_k|_v = |\alpha_1 + \cdots + \alpha_r|_v.$$

Hence $\alpha_1 = \cdots = \alpha_r$ and we conclude directly

$$h(\alpha_1 + \cdots + \alpha_r) = h(r\alpha_1) \leq \log r + \sum_{v \in M_K} \log^+ |\alpha_1|_v = \log r + h(\alpha_1).$$

On the other hand, the equality assumption implies $h(r\alpha_1) = \log r + rh(\alpha_1)$, hence $h(\alpha_1) = 0$ because $r \geq 2$. Thus by Kronecker's theorem in 1.5.9, α_1 is a root of unity and we get $h(r\alpha_1) = h(r) = \log r$.

Another example yielding almost equality in Proposition 1.5.15 is obtained taking $\alpha_i = l/(l + a_i)$ with $1 \leq a_i \leq N$ and distinct a_i and with $l = N!t + 1$, t any positive integer. Then the numbers $l, l + a_1, \ldots, l + a_r$ are coprime in pairs and an easy calculation shows that $h(\alpha_i) = \log(l + a_i)$ and $h(\sum \alpha_i) = \sum \log(l + a_i) + \log(\sum \alpha_i)$. Hence $h(\alpha_1 + \cdots + \alpha_r) > h(\alpha_1) + \cdots + h(\alpha_r) + \log r - \varepsilon$ for sufficiently large t.

The following result, quite useful in practice, expresses the fact that the height is invariant by Galois conjugation.

Proposition 1.5.17. *Let P be a point of affine or projective space with coordinates (x_j) in $\overline{\mathbb{Q}}$. If $\sigma \in \mathrm{Gal}(\overline{\mathbb{Q}}/\mathbb{Q})$ and if the point $\sigma(P)$ is given by the coordinates $(\sigma(x_j))$, then $h(P) = h(\sigma(P))$.*

Proof: We choose a finite-dimensional Galois extension K of $\mathbb{Q}$ containing all coordinates. Let $|\ |_p$ be an element of $M_{\mathbb{Q}}$ and $|\ |$ be an extension to an absolute value of K. The composition of $|\ |$ and σ is again an extension. Thus we have an action of $\mathrm{Gal}(K/\mathbb{Q})$ on the absolute values of K extending $|\ |_p$. Therefore, σ permutes the extensions and we have

$$\sum_{v|p} \max_j \log |x_j|_v = \sum_{v|p} \max_j \log |\sigma(x_j)|_v. \qquad \square$$

Lemma 1.5.18. *If $\alpha \in K \setminus \{0\}$, and $\lambda \in \mathbb{Q}$, then $h(\alpha^\lambda) = |\lambda| \cdot h(\alpha)$. In particular, $h(1/\alpha) = h(\alpha)$.*

Proof: If $\lambda \geq 0$, the result is clear by definition of height. Thus we need only consider $\lambda = -1$. For any absolute value $|\ |_v$ of K, we have

$$\log |\alpha|_v = \log^+ |\alpha|_v - \log^+ |1/\alpha|_v.$$

If we sum over v, the left-hand side is 0 by the product formula and the right-hand side equals $h(\alpha) - h(1/\alpha)$. $\square$

1.5.19. Let $S \subset M_K$ be a finite set of places. For $\alpha \in K \setminus \{0\}$, we have

$$\sum_{v \in S} \log |\alpha|_v \leq h(\alpha).$$

If we use $1/\alpha$ instead of α, then the preceding lemma shows that

$$\sum_{v \in S} \log |\alpha|_v \geq -h(\alpha).$$

This proves the so-called **fundamental inequality**

$$-h(\alpha) \leq \sum_{v \in S} \log |\alpha|_v \leq h(\alpha). \tag{1.8}$$

1.5.20. Now let L be a finite-dimensional field extension of K and consider a finite set S of places $w \in M_L$. A classical problem of diophantine approximation is that of approximating an element $\alpha \in L$ by elements $\beta \in K$, at all places $w \in S$. Classically, this is done with absolute values normalized relative to the field K rather than L, i.e. with $\|\ \|_w$ as in 1.3.6. In order to emphasize that this normalization depends on $v \in M_K$, we shall use the notation $\|\ \|_{w,K}$, hence for $x \in L$ we have

$$\|x\|_{w,K} = |N_{L_w/K_v}(x)|_v.$$

With this normalization relative to K, the fundamental inequality applied to $\alpha - \beta$ gives

$$-h(\alpha - \beta) \leq \sum_{w \in S} \log \|\alpha - \beta\|_{w,K}^{1/[L:K]} \leq h(\alpha - \beta)$$

under the assumption $\alpha \neq \beta$. Now applying Proposition 1.5.15 we find **Liouville's inequality**:

Theorem 1.5.21. *If $\alpha \in L$ and $\beta \in K$ with $\alpha \neq \beta$, then*

$$(2H(\alpha)H(\beta))^{-[L:K]} \leq \prod_{w \in S} \|\alpha - \beta\|_{w,K} \leq (2H(\alpha)H(\beta))^{[L:K]}.$$

The left-hand side inequality is a general formulation of the familiar Liouville inequality in diophantine approximation.

1.5.22. Heights can be introduced in any field with a product formula. We indicate the necessary changes. Let F be a field with a set M_F of non-trivial inequivalent absolute values, satisfying the product formula. This field F will play the role of $\mathbb{Q}$ in our previous considerations.

Let K/F be a finite-dimensional field extension and consider all places w with $w|v$ for some $v \in M_F$, together with corresponding absolute values $|\ |_w$ normalized as in 1.3.12. Then the set M_K of such absolute values satisfies the product formula, because of (1.4) on page 9. As before, this yields a non-negative height on $\mathbb{P}^n_{\overline{F}}$, independent of the choice of coordinates. On the other hand, Kronecker's theorem does not hold in general, as the following example shows.

Example 1.5.23. Let K be a field and let $F = K(X)$ be the function field of an irreducible projective variety X over K, which is regular in codimension 1 (see 1.4.9). Let $P \in \mathbb{P}^n(F)$, hence $P = (f_0 : \cdots : f_n)$ for certain rational functions f_i on X. Then

$$h(P) = -\sum_Z \deg(Z) \min_j \operatorname{ord}_Z(f_j),$$

where Z ranges over all prime divisors and the degree is with respect to a fixed ample class. In particular, the height of a rational function $f \in K(X)^\times$ is

$$h(f) = h((1 : f)) = -\sum_Z \deg(Z) \min(0, \operatorname{ord}_Z(f)).$$

Thus $h(f) = 0$ if and only if f has no poles. By $h(f) = h(f^{-1})$, this is equivalent to $\operatorname{div}(f) = 0$.

If X is normal, a function without poles is regular (R. Hartshorne **[148]**, Proposition I.6.3A), hence constant on the irreducible components of $X_{\overline{K}}$. We conclude that in this case $h(f) = 0$ if and only if f is locally constant on X (use A.6.15).

1.6. Heights of polynomials

Definition 1.6.1. *The **height of a polynomial***

$$f(t_1, \ldots, t_n) = \sum_{j_1, \ldots, j_n} a_{j_1 \ldots j_n} t_1^{j_1} \cdots t_n^{j_n} = \sum_{\mathbf{j}} a_{\mathbf{j}} \mathbf{t}^{\mathbf{j}}$$

with coefficients in a number field K is the quantity

$$h(f) = \sum_{v \in M_K} \log |f|_v,$$

where

$$|f|_v := \max_j |a_j|_v \tag{1.9}$$

is the **Gauss norm** *for any place* v.

Proposition 1.6.2. *Let* $f(t_1, \ldots, t_n)$ *and* $g(s_1, \ldots, s_m)$ *be polynomials in different sets of variables. Then*

$$h(fg) = h(f) + h(g).$$

Proof: Note that the height of a polynomial is equal to the height of the vector of coefficients in appropriate projective space. Then the claim follows from 1.5.14. □

We will need estimates for $h(fg)$ in terms of $h(f)$ and $h(g)$, without assuming different sets of variables for f and g. For finite places we have **Gauss's lemma**.

Lemma 1.6.3. *If* v *is not archimedean, then* $|fg|_v = |f|_v |g|_v$.

Proof: The inequality $|fg|_v \leq |f|_v |g|_v$ is immediate because v is not archimedean. Let us assume first that $f(t)$ and $g(t)$ are polynomials in one variable t. We denote by c_j the coefficient

$$\sum_{j=k+l} a_k b_l$$

of $f(t)g(t)$. Without loss of generality, we can assume that $|f|_v = 1$, $|g|_v = 1$. Suppose $|fg|_v < 1$. Let j be the smallest index with $|a_j|_v = 1$. Since $|c_j|_v < 1$ and $|a_k|_v < 1$ for $k < j$, we get $|b_0|_v < 1$. Now we apply the above formula for the coefficient c_{j+l} and conclude $|b_l|_v < 1$ by induction. This contradiction proves the lemma in the one-variable case. For several variables, let d be an integer larger than the degree of fg. The Kronecker substitution

$$x_j = t^{d^{j-1}} \qquad (j = 1, \ldots, n)$$

reduces the problem to the one-variable case. □

1.6.4. Gauss's lemma applies to every non-archimedean absolute value of a field. The archimedean case is more complicated and will be handled below.

If $f(t_1, \ldots, t_n)$ is a polynomial with complex coefficients, we define $|f|_\infty$ as in (1.9), namely the maximum of the euclidean absolute value $|\ |$ of the coefficients of f.

Another very useful quantity in studying polynomials is the **Mahler measure**

$$M(f) := \exp\left(\int_{\mathbb{T}^n} \log|f(e^{i\theta_1}, \ldots, e^{i\theta_n})| \, d\mu_1 \cdots d\mu_n\right),$$

where we have abbreviated $\mathbb{T}$ for the unit circle $\{e^{i\theta} \mid 0 \leq \theta < 2\pi\}$ equipped with the standard measure $d\mu = (1/2\pi)d\theta$. Its main advantage is the multiplicativity property

$$M(fg) = M(f)M(g).$$

Let $f(t) = a_d t^d + \cdots + a_0$ be a polynomial with complex coefficients and factorization

$$f(t) = a_d\,(t - \alpha_1) \cdots (t - \alpha_d).$$

Now we note that the mean value of $\log|t - \alpha|$ on the unit circle is $\log^+|\alpha|$. In fact, for $|\alpha| > 1$ the function $\log|t - \alpha|$ is harmonic in the unit disk, therefore its mean value on the unit circle is its value at the centre, namely $\log|\alpha| = \log^+|\alpha|$. If instead $|\alpha| < 1$, the function $\log|1 - \alpha\bar{t}|$ is harmonic in the unit disk and coincides with $\log|t - \alpha|$ on the unit circle, while its value at the centre is 0, that is $\log^+|\alpha|$. Finally the case $|\alpha| = 1$ is deduced by continuity.

We have shown that $M(t - \alpha) = \log^+|\alpha|$. If we combine this with the multiplicativity property of the Mahler measure, we obtain **Jensen's formula**:

Proposition 1.6.5.

$$\log M(f) = \log|a_d| + \sum_{j=1}^{d} \log^+|\alpha_j|.$$

The following result shows the connexion between the Mahler measure and the height and gives a bound for the absolute norm of an algebraic number.

Proposition 1.6.6. *Let $\alpha \in \overline{\mathbb{Q}}$ and let f be the minimal polynomial of α over $\mathbb{Z}$. Then*

$$\log M(f) = \deg(\alpha)h(\alpha).$$

In particular

$$\log|N_{\mathbb{Q}(\alpha)/\mathbb{Q}}(\alpha)| \leq \deg(\alpha)h(\alpha).$$

Proof: Let $d = \deg(\alpha)$ and write

$$f(t) = a_d t^d + \cdots + a_0.$$

We choose a number field K which contains α and is a Galois extension over $\mathbb{Q}$, with Galois group G. Then the list $(\sigma\alpha)_{\sigma\in G}$ contains every conjugate of α exactly $[K:\mathbb{Q}]/d$ times. Gauss's lemma gives

$$|a_d|_v \prod_{\sigma\in G} \max\left(1, |\sigma\alpha|_v\right)^{d/[K:\mathbb{Q}]} = 1 \tag{1.10}$$

for any non-archimedean $v \in M_K$.

We have

$$
\begin{aligned}
[K:\mathbb{Q}]\,h(\alpha) &= \sum_{v\in M_K}\sum_{\sigma\in G}\log^+|\sigma\alpha|_v && \text{(by Proposition 1.5.17)}\\
&= \sum_{v|\infty}\sum_{\sigma\in G}\log^+|\sigma\alpha|_v - \frac{[K:\mathbb{Q}]}{d}\sum_{v\nmid\infty}\log|a_d|_v && \text{(by (1.10))}\\
&= \frac{[K:\mathbb{Q}]}{d}\sum_{v|\infty}\left(\log|a_d|_v + \sum_{j=1}^{d}\log^+|\alpha_j|_v\right),
\end{aligned}
$$

where in the last step we have used the product formula and collected the elements $\sigma\alpha$ into the conjugates α_j, $j=1,\ldots,d$, of α. By Jensen's formula, this proves the first claim.

By the second formula of Lemma 1.3.7, we also have

$$
\log|N_{\mathbb{Q}(\alpha)/\mathbb{Q}}(\alpha)| \le \sum_{v|\infty}\sum_{j=1}^{d}\log^+|\alpha_j|_v
$$

and the second claim follows from the preceding computation. □

The following lemma is useful in estimates. Let $f(t) = a_d t^d + \cdots + a_d$ be a polynomial of degree d with complex coefficients, and for $1 \le p < \infty$ denote by $\ell_p(f)$ the norm

$$
\ell_p(f) := \left(\sum_{j=0}^{d}|a_j|^p\right)^{1/p}.
$$

For $p=\infty$, we set $\ell_\infty(f) = \max|a_j| = |f|_\infty$.

Lemma 1.6.7. *If $f(t)$ is as above, then $M(f) \le \ell_1(f)$. Moreover*

$$
\binom{d}{\lfloor d/2\rfloor}^{-1}\ell_\infty(f) \le M(f) \le \ell_2(f) \le (d+1)^{1/2}\ell_\infty(f).
$$

Proof: The first inequality is obvious from the definition of $M(f)$ and the pointwise bound $|f(e^{i\theta})| \le \ell_1(f)$ on $\mathbb{T}$.

Next, by convexity, we get

$$
M(f) \le \left(\int_{\mathbb{T}}|f(e^{i\theta})|^2\,\mathrm{d}\mu\right)^{1/2}.
$$

By Parseval's formula, the right-hand side equals

$$
\ell_2(f) = \left(\sum_{j=0}^{d}|a_j|^2\right)^{1/2} \le (d+1)^{1/2}\ell_\infty(f).
$$

Finally, we remark that

$$\left|\frac{a_{d-r}}{a_d}\right| = \left|\sum_{j_1<\cdots<j_r} \alpha_{j_1}\cdots\alpha_{j_r}\right|,$$

hence

$$|a_{d-r}| \le \binom{d}{r} |a_d| \prod_{j=1}^{d} \max(1, |\alpha_j|).$$

By Jensen's formula, we conclude that

$$|a_{d-r}| \le \binom{d}{r} M(f). \qquad \square$$

The following consequence, **Northcott's theorem**, is very important.

Theorem 1.6.8. *There are only finitely many algebraic numbers of bounded degree and bounded height.*

Proof: Let α be algebraic of degree d and height $h(\alpha) \le \log H$. Let $f(t) = a_d t^d + \cdots + a_0$ be the minimal polynomial of α over $\mathbb{Z}$. By Proposition 1.6.6, we have $M(f) \le H^d$. Also, Lemma 1.6.7 shows that $\max |a_i| \le 2^d M(f)$. Therefore, the coefficients of f are bounded by $(2H)^d$. Since there are $d+1$ integer coefficients for each f, they give rise to not more than $(2\lfloor(2H)^d\rfloor+1)^{d+1}$ distinct polynomials f. Since each f has d roots, the number of algebraic integers of degree d and height at most H is at most $d(2\lfloor(2H)^d\rfloor + 1)^{d+1} \le (5H)^{d^2+d}$. $\square$

For later use, we prove here a result of K. Mahler [**188**], which gives a bound for the discriminant in terms of the Mahler measure.

Proposition 1.6.9. *Let $f(x) = a_d x^d + \cdots + a_0$ be a polynomial with real or complex coefficients, with roots $\alpha_1, \ldots, \alpha_d$. Let*

$$D = a_d^{2d-2} \prod_{i>j} (\alpha_i - \alpha_j)^2$$

be its discriminant. Then

$$|D| \le d^d\, M(f)^{2d-2}.$$

In particular, if $f(x)$ is the minimal polynomial over $\mathbb{Z}$ of an algebraic number ξ of degree d, it holds

$$\frac{1}{d} \log |D| \le \log d + (2d-2) h(\xi).$$

Proof: We write D as the product of a_d^{2d-2} and the square of a Vandermonde determinant (see B.1.10 and Remark B.1.5) and estimate the determinant using

Hadamard's inequality,* obtaining

$$|D| = |a_d|^{2d-2} \left| \det \begin{pmatrix} 1 & \alpha_1 & \dots & \alpha_1^{d-1} \\ 1 & \alpha_2 & \dots & \alpha_2^{d-1} \\ \vdots & \vdots & \dots & \vdots \\ 1 & \alpha_d & \dots & \alpha_d^{d-1} \end{pmatrix} \right|^2 \leq |a_d|^{2d-2} \prod_{i=1}^{d} \Big(\sum_{j=0}^{d-1} |\alpha_i^j|^2 \Big).$$

The right-hand side of this inequality does not exceed

$$|a_d|^{2d-2} d^d \prod_{1=1}^{d} \max(1, |\alpha_i|)^{2d-2}$$

and the first statement follows from Jensen's formula, Proposition 1.6.5.

The second statement is also clear from Proposition 1.6.6. □

Lemma 1.6.10. *Let $f(t_1, \dots, t_n)$ be a polynomial with complex coefficients and partial degrees $d_1, \dots, d_n$. Then*

$$\prod_{j=1}^{n} (d_j + 1)^{-1/2} M(f) \leq \ell_\infty(f) \leq \prod_{j=1}^{n} \binom{d_j}{\lfloor d_j/2 \rfloor} M(f).$$

Proof: The same proof as in Lemma 1.6.7 holds for the inequality on the left. We prove the other assertion by induction on n. We can write uniquely

$$f(t_1, \dots, t_n) = \sum_{j=0}^{d_n} f_j(t_1, \dots, t_{n-1})\, t_n^j$$

for certain polynomials $f_j(t_1, \dots, t_{n-1})$. By definition, it holds

$$\log M(f) = \int_{\mathbb{T}^{n-1}} \log M\big(f(e^{i\theta_1}, \dots, e^{i\theta_{n-1}}, t)\big)\, d\mu_1 \cdots d\mu_{n-1}$$

and this is not smaller than

$$\int_{\mathbb{T}^{n-1}} \log \max_j |f_j(e^{i\theta_1}, \dots, e^{i\theta_{n-1}})|\, d\mu_1 \cdots d\mu_{n-1} - \log \binom{d_n}{\lfloor d_n/2 \rfloor}$$

by Lemma 1.6.7, and which in turn is not smaller than

$$\max_j \int_{\mathbb{T}^{n-1}} \log |f_j(e^{i\theta_1}, \dots, e^{i\theta_{n-1}})|\, d\mu_1 \cdots d\mu_{n-1} - \log \binom{d_n}{\lfloor d_n/2 \rfloor}.$$

*The inequality states that a determinant of a real matrix is majorized by the product of the euclidean lengths of its rows. Geometrically, it says that the volume of a parallelopiped generated by real vectors $\mathbf{v}_1, \dots, \mathbf{v}_n$ of given length is maximal when the vectors $\mathbf{v}_i$ are pairwise orthogonal, which is quite easy to prove. The result also holds for complex matrices. Hadamard's proof of 1893 can be found, among an interesting analysis of extremal cases with entries ± 1 (the so-called Hadamard's matrices), in **[143]**. The result was known much earlier to Lord Kelvin and was proved by T. Muir in 1885, see **[209]**, p.32.

We conclude that

$$\binom{d_n}{\lfloor d_n/2 \rfloor} M(f) \geq \max_j M(f_j)$$

and the induction hypothesis implies the claim. □

As remarked above, Lemma 1.6.7 leads to **Gelfond's lemma**:

Lemma 1.6.11. *Let $f_1, \ldots, f_m$ be complex polynomials in n variables and set $f := f_1 \cdots f_m$. Then*

$$2^{-d} \prod_{j=1}^{m} \ell_\infty(f_j) \leq \ell_\infty(f) \leq 2^d \prod_{j=1}^{m} \ell_\infty(f_j),$$

where d is the sum of the partial degrees of f.

Proof: Let $\left(d_1^{(j)}, \ldots, d_n^{(j)}\right)$ be the partial degrees of f_j. By carrying out the multiplication, we see that

$$\ell_\infty(f) \leq C \prod_{j=1}^{m} \ell_\infty(f_j)$$

with

$$C = \prod_{j=1}^{m-1} \prod_{k=1}^{n} \left(1 + d_k^{(j)}\right) \leq 2^d.$$

In the other direction, Lemma 1.6.10 implies

$$\prod_{j=1}^{m} \ell_\infty(f_j) \leq \left(\prod_{j=1}^{m} \prod_{k=1}^{n} \binom{d_k^{(j)}}{\lfloor d_k^{(j)}/2 \rfloor} \right) \left(\prod_{k=1}^{n} \left(1 + \sum_{j=1}^{m} d_k^{(j)} \right)^{1/2} \right) \ell_\infty(f).$$

The next lemma completes the proof.

Lemma 1.6.12. *Let $a \leq A, b \leq B$ and d be natural numbers. Then $\binom{A}{a}\binom{B}{b} \leq \binom{A+B}{a+b}$ and $\binom{d}{\lfloor d/2 \rfloor}(d+1)^{1/2} \leq 2^d$.*

Proof: The first statement is a trivial consequence of the identity

$$(1+t)^A (1+t)^B = (1+t)^{A+B}.$$

For the second claim (which also follows from a straightforward application of Stirling's formula), we proceed by induction. The inequality is obviously satisfied for $d = 0$ and $d = 1$. Set $C_d := \binom{d}{\lfloor d/2 \rfloor}(d+1)^{1/2}$ and let $m \in \mathbb{N}$; then

$$C_{2m+1}/C_{2m} = 2\left(1 - \frac{1}{2m+2}\right)^{1/2} < 2$$

and

$$C_{2m+2}/C_{2m} = 4\left(1 - \frac{1}{(2m+2)^2}\right)^{1/2} < 4.$$

The induction hypothesis implies the second statement. □

Gelfond's and Gauss's lemma together with Lemma 1.3.7 give us

Theorem 1.6.13. *Let $f_1, \ldots, f_m$ be polynomials in n variables with coefficients in $\overline{\mathbb{Q}}$ and let d be the sum of the partial degrees of $f := f_1 \cdots f_m$. Then*

$$-d\log 2 + \sum_{j=1}^{m} h(f_j) \leq h(f) \leq d\log 2 + \sum_{j=1}^{m} h(f_j).$$

Remark 1.6.14. For the upper bound, only the sum d' of the partial degrees of $f_1 \cdots f_{m-1}$ does matter. In fact, the proof of Gelfond's lemma shows

$$|f|_v \leq \left(\prod_{j=1}^{m-1} \prod_{k=1}^{n} |1 + d_k^{(j)}|_v \right) \prod_{j=1}^{m} |f_j|_v \leq |2|_v^{d'} \prod_{j=1}^{m} |f_j|_v$$

for any archimedean place v of a number field containing all the coefficients. Then

$$h(f) \leq \sum_{j=1}^{m} h(f_j) + \sum_{j=1}^{m-1} \sum_{k=1}^{n} \log(1 + d_k^{(j)}) \leq \sum_{j=1}^{m} h(f_j) + d' \log 2,$$

which is often important for applications.

1.6.15. We conclude this section by mentioning an interesting question raised by D.H. Lehmer ([**180**], p.476), known today as the **Lehmer conjecture** (in Lehmer's paper, this was addressed as a problem rather than a conjecture). If $\alpha \neq 0$ is algebraic with minimal polynomial f, the Mahler measure of α is $M(\alpha) := M(f)$. By Proposition 1.6.6, we have $M(\alpha) = H(\alpha)^{\deg(\alpha)}$. Now the question raised by Lehmer is whether there is an absolute constant c such that $M(\alpha) \geq c > 1$ for $\alpha \in \overline{\mathbb{Q}}^{\times}$ not a root of unity. Alternatively, $h(\alpha) \geq c/d$ for some absolute constant c.

The last inequality in the proof of Lemma 1.6.7 with $r = 0$ or d shows that $h(\alpha) \geq (\log 2)/d$ unless α is a unit. The same argument also shows that

$$h(\alpha) \geq -\log 2 + \frac{1}{d} \log \ell_\infty(f)$$

for all α. In particular, $h(\alpha) \geq (\log 2)/d$ if $\ell_\infty(f) \geq 2^{d+1}$. Since there are only finitely many polynomials f of degree d with integer coefficients and with $\ell_\infty(f) < 2^{d+1}$, by Kronecker's theorem 1.5.9 we deduce that there is $c(d) > 0$ such that any α of degree d, not a root of unity, satisfies $h(\alpha) \geq c(d)$. Thus in studying Lehmer's problem we may assume that d is arbitrarily large.

The algebraic number α with minimal polynomial $x^{10}+x^9-x^7-x^6-x^5-x^4-x^3+x+1$ has $M(\alpha) = 1.17628081825991\ldots$ and is conjectured to yield the infimum of the Mahler measure of an algebraic number.[†]

[†] This polynomial already appears in Lehmer's paper *loc. cit.*, with a slightly different numerical value which we have corrected here.

If f is not reciprocal (a reciprocal polynomial $f(z)$ satisfies $f(z) = \pm z^{\deg(f)} f(1/z)$), a nice theorem by C.J. Smyth [287] states that the minimum of $M(\alpha)$ occurs for the cubic number with minimal equation $x^3 - x - 1$. This non-reciprocal number α is about $\alpha = 1.32471795724474\ldots$. In the general case, for large d we have E. Dobrowolski's theorem [90]

$$M(\alpha) \geq 1 + c\left(\frac{\log\log d}{\log d}\right)^3,$$

following from Theorem 4.4.1.

1.7. Lower bounds for norms of products of polynomials

We elaborate here further on the question of lower bounds for norms of products of polynomials. The interesting question is to obtain lower bounds which are proportional to the product of the norms, as in Gelfond's lemma of the preceding section. It turns out that for certain natural norms the constants involved in such lower bounds depend only on the degrees of the polynomials, not on the number of variables. This section will not be needed at other places of the book.

1.7.1. Let us denote by $\ell_p(f)$ the ℓ_p-norm of the coefficients of a complex polynomial f. We shall prove that for $p = 1$ and $p = 2$ the ℓ_p-norm has the properties mentioned above. This result extends to all p, $1 \leq p < \infty$, but we will not prove this extension here. The more difficult case $p = 1$ is due to Enflo, who used it in his work on invariant subspaces of bounded operators in Banach spaces.

Theorem 1.7.2. *Let $d, e \in \mathbb{N}$. Then there is a constant $C(d,e) > 0$ such that*

$$\ell_1(fg) \geq C(d,e)\ell_1(f)\ell_1(g)$$

for complex polynomials f, g of degree d, e in several variables.

Proof: (H.L. Montgomery) For $k \in \mathbb{N}$, we define

$$C(d,e,k) := \inf \frac{\ell_1(P^k Q)}{\ell_1(P)^k \ell_1(Q)},$$

where the infimum ranges over all homogeneous polynomials P, Q of degree d, e.

We shall use in the sequel Euler's formula

$$\sum_j t_j \frac{\partial f}{\partial t_j} = df$$

for a homogeneous polynomial $f \in \mathbb{C}[t_1, \ldots, t_n]$ of degree d, and the formula

$$\sum_j \ell_1\left(\frac{\partial f}{\partial t_j}\right) = \sum_j \ell_1\left(t_j \frac{\partial f}{\partial t_j}\right) = d\ell_1(f).$$

Both are proved directly by looking at each monomial in f.

Lemma 1.7.3. *The following two estimates hold:*

$$C(d,0,k+1) \geq C(d-1,dk,1)\,C(d,0,k) \qquad \textit{if } d \geq 1,$$
$$C(d,e,k) \geq \frac{e}{2kd+e}\,C(d,e-1,k+1) \quad \textit{if } e \geq 1.$$

Proof: Let f be homogeneous of degree d. We compute

$$\begin{aligned}
\ell_1\left(\frac{\partial}{\partial t_j} f^{k+1}\right) &= (k+1)\,\ell_1\left(f^k \frac{\partial f}{\partial t_j}\right) \\
&\geq (k+1)\,C(d-1, dk, 1)\,\ell_1\left(\frac{\partial f}{\partial t_j}\right)\ell_1\left(f^k\right) \\
&\geq (k+1)C(d-1, dk, 1)\,C(d, 0, k)\,\ell_1\left(\frac{\partial f}{\partial t_j}\right)\ell_1(f)^k\,.
\end{aligned}$$

Summing over j, we find

$$(k+1)\,d\,\ell_1\left(f^{k+1}\right) \geq (k+1)\,C(d-1, dk, 1)\,C(d, 0, k)\,d\,\ell_1(f)^{k+1}$$

proving the first statement.

In a similar fashion, for f and g homogeneous of degrees d and e, we also have

$$\begin{aligned}
C(d, e-1, k+1)\,\ell_1(f)^{k+1}\,\ell_1\left(\frac{\partial g}{\partial t_j}\right) &\leq \ell_1\left(f^{k+1}\frac{\partial g}{\partial t_j}\right) \\
&= \ell_1\left(f\frac{\partial}{\partial t_j}(f^k g) - kf^k g\frac{\partial f}{\partial t_j}\right) \\
&\leq \ell_1(f)\,\ell_1\left(\frac{\partial}{\partial t_j}(f^k g)\right) + k\,\ell_1\left(f^k g\right)\ell_1\left(\frac{\partial f}{\partial t_j}\right).
\end{aligned}$$

Summing over j, we obtain

$$\begin{aligned}
C(d, e-1, k+1)\,e\,\ell_1(f)^{k+1}\,\ell_1(g) &\leq (dk+e)\,\ell_1(f)\,\ell_1\left(f^k g\right) + kd\,\ell_1\left(f^k g\right)\ell_1(f) \\
&= (2kd+e)\,\ell_1(f)\,\ell_1\left(f^k g\right).
\end{aligned}$$

After cancelling a factor $\ell_1(f)$, we get the claim. □

The proof of Enflo's theorem is now easy. There is no loss of generality in assuming that f and g are homogeneous polynomials. We order the triples (d, e, k) lexicographically. Proceeding by induction, we prove that $C(d, e, k) > 0$. If $d = 0$ or $k = 0$, then $C(d, e, k) = 1$. So let us assume that $d > 0, k > 0$. By Lemma 1.7.3, it is enough to show the claim for a smaller triple, and we are done. This gives Theorem 1.7.2, with $C(d, e) := C(d, e, 1) > 0$. □

1.7.4. The double induction in the proof of the theorem is very expensive for the final estimates. Let us compute some of the constants so obtained. We define recursively $\Gamma(d, e, k)$ as follows

$$\Gamma(d, e, k) = \begin{cases} 1 & \text{if } d = 0 \text{ or } k = 0 \\ \Gamma(d-1, d(k-1), 1)\,\Gamma(d, 0, k-1) & \text{if } e = 0 \text{ and } dk \neq 0 \\ \frac{e}{2kd+e}\,\Gamma(d, e-1, k+1) & \text{if } dek \neq 0 \end{cases}$$

and hence $\Gamma(d,e,k) \le C(d,e,k)$. For example

$$\begin{aligned}
\Gamma(d,0,1) &= 1 & C(d,0,1) &= 1 \\
\Gamma(1,1,1) &= 1/3 & C(1,1,1) &= \frac{1}{2} \\
\Gamma(2,2,1) &= 1/34020 \\
\Gamma(3,3,1) &= 1/(3.840584... \times 10^{95}) \\
\Gamma(4,4,1) &= 1/(2.089942... \times 10^{13529}) \\
\Gamma(5,5,1) &= 1/(6.562189... \times 10^{19906418})
\end{aligned}$$

and the computer took too much time for $\Gamma(6,6,1)$.

1.7.5. The solution for the case $p = 2$ uses the concept of **hypercube representation** of a polynomial. The usual way of writing a homogeneous polynomial of degree d in n variables is to represent it in the form

$$f(t_1,\dots,t_n) = \sum_{i_1+\dots+i_n=d} \cdots \sum a_{i_1\dots i_n} t_1^{i_1} \cdots t_n^{i_n}.$$

The sum here runs over the lattice points in the hyperplane $i_1 + \dots + i_n = d$ of the n-dimensional cube $0 \le i_\nu \le d$, $\nu = 1,\dots,n$. Note that the number of lattice points in this cube is $(d+1)^n$, growing exponentially in n for fixed d.

There is another way of writing the same polynomial, namely

$$f(t_1,\dots,t_n) = \frac{1}{d!} \sum_{i_1=1}^{n} \cdots \sum_{i_d=1}^{n} \frac{\partial^d f}{\partial t_{i_1} \cdots \partial t_{i_d}} t_{i_1} \cdots t_{i_d};$$

we define this as the hypercube representation of f, since now the sum is indexed by the lattice points of the d-dimensional cube $1 \le i_\delta \le n$, $\delta = 1,\dots,d$. The number of lattice points in this cube is n^d, which grows polynomially in n for fixed d.

The hypercube representation of a polynomial is very convenient if we want to study polynomials of low degree in a large number of variables.

1.7.6. Define for $p \ge 1$

$$[f]_p := \frac{1}{d!} \left(\sum_{i_1=1}^{n} \cdots \sum_{i_d=1}^{n} \left| \frac{\partial^d f}{\partial t_{i_1} \cdots \partial t_{i_d}} \right|^p \right)^{1/p}.$$

If we compare this norm with the ℓ_p-norm of the coefficients, simple combinatorics lead to

$$\left(\frac{1}{d!}\right)^{1-\frac{1}{p}} \ell_p(f) \le [f]_p \le \ell_p(f). \tag{1.11}$$

1.7.7. Let $d, e \in \mathbb{N}$. A **shuffle of type** (d,e) is a pair (K,L), where K and L are disjoint subsets of $\{1,\dots,d+e\}$ of cardinality d and e. The set of shuffles of type (d,e) will be denoted by $\mathrm{sh}(d,e)$. Its cardinality is equal to $\binom{d+e}{d}$. For $\mathbf{x} = (x_1,\dots,x_{d+e}) \in [0,1]^{d+e}$, we define $\mathbf{x}_K := (x_{k_1},\dots,x_{k_d})$, where $\{k_1,\dots,k_d\} = K$ and $k_1 < \dots < k_d$.

Let $k_p(d,e)$ be the largest constant such that

$$[fg]_p \geq k_p(d,e)\,[f]_p[g]_p$$

holds for all homogeneous polynomials f, g of degree d, e. Moreover, we define $c_p(d,e)$ as the largest constant for which

$$\left\| \sum_{(K,L)\in \mathrm{sh}(d,e)} F(\mathbf{x}_K)G(\mathbf{x}_L) \right\|_p \geq c_p(d,e)\,\|F\|_p\,\|G\|_p$$

holds for all symmetrical functions $F \in L^p([0,1]^d)$, $G \in L^p([0,1]^e)$, with $\|\ \|_p$ denoting the L^p-norm.

Lemma 1.7.8. *The constants* $c_p(d,e)$ *and* $k_p(d,e)$ *are related by*

$$c_p(d,e) = \binom{d+e}{d} k_p(d,e).$$

Proof: Let $f(t_1,\ldots,t_n)$ be a homogeneous polynomial of degree d and let F be the symmetrical step function on $[0,1)^d$ given by

$$F(x_1,\ldots,x_d) = n^{d/p}\frac{1}{d!}\frac{\partial^d f}{\partial t_{i_1}\cdots\partial t_{i_d}}$$

for $\frac{i_1-1}{n} \leq x_1 < \frac{i_1}{n},\ldots,\frac{i_d-1}{n} \leq x_d < \frac{i_d}{n}$. Also, let $g(t_1,\ldots,t_n)$ be a homogeneous polynomial of degree e and define G in the same way as F.

Then we verify that

$$[f]_p = \|F\|_p, \qquad [g]_p = \|G\|_p, \qquad [fg]_p = \frac{d!e!}{(d+e)!}\left\| \sum_{(K,L)\in \mathrm{sh}(d,e)} F(\mathbf{x}_K)G(\mathbf{x}_L) \right\|_p.$$

The rest of the proof is an approximation argument. Consider the discretization i/n, $i = 1,\ldots,n$ of $[0,1]$; given continuous F, G on $[0,1]^d$ and $[0,1]^e$, we approximate F, G by step functions as above and construct corresponding polynomials f, g. As $n \to \infty$, these functions are dense in $L^p([0,1]^d)$ and $L^p([0,1]^e)$. □

Proposition 1.7.9. *The constant* $c_2(d,e)$ *is*

$$c_2(d,e) = \binom{d+e}{d}^{1/2}.$$

Proof: Let F, G be symmetrical L^2-functions as in 1.7.7. Then

$$\left\| \sum_{(K,L)\in \mathrm{sh}(d,e)} F(\mathbf{x}_K)G(\mathbf{x}_L) \right\|_2^2 = \sum_{\substack{(K,L)\in \mathrm{sh}(d,e)\\ (K',L')\in \mathrm{sh}(d,e)}} \int_{[0,1]^{d+e}} F(\mathbf{x}_K)G(\mathbf{x}_L)\,\overline{F(\mathbf{x}_{K'})G(\mathbf{x}_{L'})}\,\mathrm{d}\mathbf{x}$$

and this is equal to

$$\binom{d+e}{d}\|F\|_2^2\,\|G\|_2^2 + \sum_{(K,L)\neq(K',L')} \int_{[0,1]^{d+e}} F(\mathbf{x}_K)\overline{G(\mathbf{x}_{L'})}\;\overline{F(\mathbf{x}_{K'})}G(\mathbf{x}_L)\,\mathrm{d}\mathbf{x}.$$

The integral is not negative, as we verify as follows. We have

$$\int_{[0,1]^{d+e}} F(\mathbf{x}_K)\overline{G(\mathbf{x}_{L'})}\,\overline{F(\mathbf{x}_{K'})}\,G(\mathbf{x}_L)\,d\mathbf{x} =$$

$$\int_{[0,1]^{d+e}} F(\mathbf{x}_{K\cap K'},\mathbf{x}_{K\cap L'})\overline{G(\mathbf{x}_{K\cap L'},\mathbf{x}_{L\cap L'})}\times$$

$$\overline{F(\mathbf{x}_{K\cap K'},\mathbf{x}_{L\cap K'})}G(\mathbf{x}_{L\cap K'},\mathbf{x}_{L\cap L'})\,d\mathbf{x}_{K\cap K'}\,d\mathbf{x}_{K\cap L'}\,d\mathbf{x}_{L\cap K'}\,d\mathbf{x}_{L\cap L'}.$$

Now we integrate first with respect to $d\mathbf{x}_{K\cap L'}\,d\mathbf{x}_{L\cap K'}$. By Fubini's theorem, we obtain

$$\left|\int F(\mathbf{x}_{K\cap K'},\mathbf{z})\overline{G(\mathbf{z},\mathbf{x}_{L\cap L'})}\,d\mathbf{z}\right|^2.$$

This proves non-negativity and $c_2(d,e) \geq \binom{d+e}{e}^{1/2}$.

The choices

$$F(x_1,\ldots,x_d) = \cos(2\pi x_1)\cdots\cos(2\pi x_d), \quad G(x_1,\ldots,x_e) = \sin(2\pi x_1)\cdots\sin(2\pi x_e),$$

also show, by orthogonality, that the constant $\binom{d+e}{e}^{1/2}$ is sharp. □

Corollary 1.7.10. *Let f, g be complex polynomials of degree d, e. Then:*

(a) $k_2(d,e) = \binom{d+e}{e}^{-1/2}$.

(b) $\ell_2(fg) \geq \frac{1}{\sqrt{(d+e)!}}\,\ell_2(f)\,\ell_2(g)$.

Proof: We may suppose that f, g are homogeneous. The first claim follows from Lemma 1.7.8 and Proposition 1.7.9. The second follows from (a) and (1.11) on page 31. □

1.8. Bibliographical notes

The material in the first five sections of this chapter is quite standard and was mainly taken from S. Lang [**169**] and J.-P. Serre [**277**]. However, the reader must be warned that our normalization for absolute values does not always agree with the normalization used by other authors. The rationale for our normalization is that the degree $[K:\mathbb{Q}]$ does not appear in the first formula in Lemma 1.3.7, and therefore it is absent in the definition of the absolute logarithmic height. This leads to formulas invariant by field extensions.

The proof of Gelfond's lemma in Section 1.6 follows K. Mahler [**187**], where the important Mahler's height is introduced. The inequality $M(f) \leq \|f\|_{L^2(\mathbb{T})}$ appearing in the proof of Lemma 1.6.7 can be found in E. Landau [**164**], Satz 443, with a somewhat different proof.

Section 1.7 is mostly from B. Beauzamy, E. Bombieri, P. Enflo, and H.L. Montgomery [**18**].

2 WEIL HEIGHTS

2.1. Introduction

In this chapter we study heights from a geometric point of view.

We begin with the important Section 2.2 introducing local Weil heights associated to Cartier divisors on a projective variety X, and studying their properties. These considerations are given here only for projective varieties, where the treatment is simpler.

Section 2.3 studies global Weil heights and their equivalence classes up to bounded functions.

In Section 2.4, we study the height on a projective variety induced by the height in the ambient projective space and in particular we prove the important Northcott's theorem on the finiteness of the number of points of bounded degree and bounded height in a fixed projective space.

These three sections are very important for the handling of heights in diophantine geometry and are required from Chapter 9 onwards.

In Section 2.5, which contains new material, the notion of presentation of a projective variety is introduced and explicit comparison theorems for the heights of a variety X in two different projective embeddings are given, in terms of presentations of these embeddings. This section may be skipped in a first reading. It will be used only partially in Section 11.7 and implicitly in questions dealing with effectivity.

Sections 2.6 and 2.7 extend the results obtained on local and global Weil heights to the associated heights of locally bounded metrized line bundles on a complete variety. They will be also used in the second half of the book.

Section 2.8 studies heights on Grassmann varieties and their properties. We need it only for Section 2.9, where we state the important Siegel's lemma in a strong form, as a consequence of Minkowski's geometry of numbers. For a quick tour, the reader may take from the last two sections only the elementary version of Siegel's lemma over $\mathbb{Z}$ given in 2.9.1 and its Corollary 2.9.2 over number fields, where the constants are not made explicit, but which is quite often enough for applications.

2.2. Local heights

The reader should be familiar with the concept of Cartier divisors and its connexion to meromorphic sections of line bundles, as in A.8.

In this section we introduce local heights associated to Cartier divisors on a projective variety X. However, in order to define them properly we need additional data beyond the divisor D itself, namely a realization $O(D) = O(D_+) \otimes O(-D_-)$ with base-point-free line bundles $O(D_\pm)$ coming with given sets of generating global sections. The set of Cartier divisors equipped with these additional data forms a monoid, and the local heights so defined behave functorially with respect to this monoid. This removes the need of working modulo bounded functions when studying Weil heights, a point of crucial importance for applications because it allows precise estimates.

2.2.1. Let K be a field and let us fix an absolute value $|\ |$ on $\overline{K}$. Let X be a projective variety over K, which for simplicity we assume here to be irreducible.

Let D be a Cartier divisor on X with associated line bundle $O(D)$ and meromorphic section s_D. For construction of $O(D)$ and s_D, see A.8.18. Note that the associated Cartier divisor $D(s_D)$ of s_D is equal to D.

There are base-point-free line bundles L, M on X such that $O(D) \cong L \otimes M^{-1}$ (cf. A.6.10 (a)). Now choose generating global sections $s_0, \ldots, s_n$ of L and $t_0, \ldots, t_m$ of M, and call the data

$$\mathcal{D} = (s_D;\ L, \mathbf{s};\ M, \mathbf{t})$$

a **presentation** of the Cartier divisor D.

2.2.2. For $P \notin \operatorname{supp}(D)$, we define

$$\lambda_{\mathcal{D}}(P) := \max_k \min_l \log \left| \frac{s_k}{t_l s_D}(P) \right| .$$

We use the notation $t_l s_D$ for $t_l \otimes s_D$ and s_k/s' for $s_k \otimes (s')^{-1}$. Hence $s_k/(t_l s_D)$ is a rational function on X.

We call $\lambda_{\mathcal{D}}(P)$ the **local height** of P relative to the presentation $\mathcal{D}$ and, by abuse of language, relative to D. In fact, it depends on the choice of s_D as well as on L, M and their generating sections. The local height is a real-valued function defined outside of the support of the divisor D.

Example 2.2.3. Let f be a non-zero rational function on X with Cartier divisor $D := D(f)$. Then $O(D) = O_X$ and f is a meromorphic section of

$O(D)$. Thus there is a local height λ_f relative to D, given by the presentation $(f;\, O_X, 1;\, O_X, 1)$. For $P \notin \operatorname{supp}(D)$, we have

$$\lambda_f(P) = -\log |f(P)|.$$

If g is another non-zero rational function on X, then $\lambda_{fg} = \lambda_f + \lambda_g$ and $\lambda_{f^{-1}} = -\lambda_f$.

2.2.4. Let D_1 and D_2 be Cartier divisors with presentations

$$\mathcal{D}_i = (s_{D_i};\, L_i, \mathbf{s}_i;\, M_i, \mathbf{t}_i)$$

and local heights $\lambda_{\mathcal{D}_i}$. Then $\mathbf{s}_1\mathbf{s}_2 = (s_{1k}s_{2k'})$, $\mathbf{t}_1\mathbf{t}_2 = (t_{1l}t_{2l'})$ are generating global sections of $L_1 \otimes L_2$, $M_1 \otimes M_2$, and we define $\lambda_{\mathcal{D}_1+\mathcal{D}_2}$ as the local height relative to the presentation

$$\mathcal{D}_1 + \mathcal{D}_2 = (s_{D_1}s_{D_2};\, L_1 \otimes L_2, \mathbf{s}_1\mathbf{s}_2;\, M_1 \otimes M_2, \mathbf{t}_1\mathbf{t}_2)$$

of the divisor $D_1 + D_2$. It is obvious that with this presentation we have

$$\lambda_{\mathcal{D}_1+\mathcal{D}_2}(P) = \lambda_{\mathcal{D}_1}(P) + \lambda_{\mathcal{D}_2}(P)$$

for $P \in X$, $P \notin \operatorname{supp}(D_1) \cup \operatorname{supp}(D_2)$.

2.2.5. If $\lambda_{\mathcal{D}}$ is a local height with presentation $(s_D;\, L, \mathbf{s};\, M, \mathbf{t})$, then $\lambda_{-\mathcal{D}}$ is defined by the presentation $(s_D^{-1};\, M, \mathbf{t};\, L, \mathbf{s})$ and we have

$$\lambda_{-\mathcal{D}}(P) = -\lambda_{\mathcal{D}}(P)$$

for $P \in X \setminus \operatorname{supp}(D)$. With these operations, the space of local heights is an abelian group.

2.2.6. Another important operation on presentations is the pull-back. If

$$\mathcal{D} = (s_D;\, L, \mathbf{s};\, M, \mathbf{t})$$

is a presentation of D on X and $\pi : Y \longrightarrow X$ is a dominant morphism of irreducible projective varieties over K, then

$$\pi^*\mathcal{D} = (\pi^* s_D;\, \pi^* L, \pi^* s;\, \pi^* M, \pi^* \mathbf{t})$$

is a presentation of $\pi^* D$. We have $\lambda_{\pi^*\mathcal{D}}(P) = \lambda_{\mathcal{D}}(\pi(P))$ for every $P \in Y$ such that $\pi(P) \notin \operatorname{supp}(D)$. More generally, this works for a morphism $\pi : Y \to X$ of irreducible projective varieties such that $\pi(Y)$ is not contained in $\operatorname{supp}(D)$.

We consider an affine variety U over K.

Lemma 2.2.7. *Let $h_j \in K[U]$, $j = 1, \ldots, N$, be without common zero in U. Then the ideal generated by the functions h_j is equal to $K[U]$.*

Proof: Choose a closed embedding $U \longrightarrow \mathbb{A}^n_K$ and let $I(U)$ be the ideal of U in $K[t_1, \ldots, t_n]$. The K-algebra $K[U]$ can be identified with $K[t_1, \ldots, t_n]/I(U)$. Let $\mathfrak{J}$ be the inverse image of the ideal generated by $h_1, \ldots, h_m$ under the projection

$$K[t_1, \ldots, t_n] \longrightarrow K[t_1, \ldots, t_n]/I(U).$$

We claim that $\mathfrak{J}$ is equal to $K[t_1, \dots, t_n]$. In fact, the polynomials in $\mathfrak{J}$ have no common zero, and our claim follows from Hilbert's Nullstellensatz in A.2.2. □

Definition 2.2.8. *The set $E \subset U(\overline{K})$ is **bounded** in U if for any $f \in K[U]$ the function $|f|$ is bounded on E.*

Lemma 2.2.9. *Let $\{f_1, \dots, f_N\}$ be generators of $K[U]$ as a K-algebra. If*

$$\sup_{P \in E} \max_{j=1,\dots,N} |f_j(P)| < \infty$$

holds, then E is bounded.

Proof: Let $f \in K[U]$. Then we can write $f = p(f_1, \dots, f_N)$ with p a polynomial with coefficients in K. Let C be the number of monomials in p and let d be the degree of p. We define

$$\delta := \begin{cases} 1 & \text{if the absolute value is archimedean} \\ 0 & \text{otherwise.} \end{cases}$$

Then, with $|p|$ the Gauss norm of p from 1.6.3, we find

$$\sup_{P \in E} |f(P)| \leq C^\delta |p| \max\left(1, \sup_{P \in E} \max_{j=1,\dots,N} |f_j(P)|\right)^d < \infty, \tag{2.1}$$

concluding the proof. □

Lemma 2.2.10. *If $\{U_l\}$ is a finite affine open covering of the affine K-variety U and if E is bounded in U, then there are bounded subsets E_l of U_l such that $E = \bigcup_l E_l$.*

Proof: It is enough to prove the claim for a refinement of $\{U_l\}$. Hence we can assume that there are regular functions h_l on U such that $U_l = \{x \in U \mid h_l \neq 0\}$, see A.2.10. By Lemma 2.2.7 there are regular functions g_l on U such that $\sum_l g_l h_l = 1$. If C is the cardinality of the covering and δ is as before, then

$$\inf_{P \in E} \max_l |h_l(P)| \geq C^{-\delta} \left(\sup_{P \in E} \max_l |g_l(P)|\right)^{-1} > 0. \tag{2.2}$$

We define

$$E_l := \{P \in E \mid |h_l(P)| = \max_k |h_k(P)|\}.$$

Obviously, $E_l \subset U_l(\overline{K})$ and $E = \bigcup_l E_l$. Let $f_1, \dots, f_N$ be a set of generators of $K[U]$. Then $f_1, \dots, f_N, 1/h_l$ are generators of $K[U_l]$. By Lemma 2.2.9, it is enough to show that $|1/h_l|$ is bounded on E_l. In fact, the bound

$$\sup_{P \in E_l} |1/h_l(P)| \leq C^\delta \sup_{P \in E} \max_k |g_k(P)| < \infty. \tag{2.3}$$

follows from (2.2). □

Theorem 2.2.11. *Let X be a projective variety over K and let $\mathcal{D}$, $\mathcal{D}'$ be two presentations of the Cartier divisor D. Then*

$$|\lambda_{\mathcal{D}} - \lambda_{\mathcal{D}'}| \le \gamma$$

for some constant $\gamma < \infty$.

Proof: By 2.2.4, we see that $\lambda_{\mathcal{D}} - \lambda_{\mathcal{D}'}$ is a local height relative to the presentation $\mathcal{D} - \mathcal{D}'$ of the zero divisor. Therefore, the left-hand side of the inequality extends to a well-defined real function on X. Moreover, it is enough to show the claim for $D = 0$ and $\mathcal{D}' = (1;\, L, 1;\, M, 1)$. Then $\mathcal{D}$ has the form $(1;\, L, \mathbf{s};\, L, \mathbf{t})$. We need to find γ as above such that

$$-\gamma \le \max_k \min_l \log \left| \frac{s_k}{t_l}(P) \right| \le \gamma.$$

To this end, it suffices to obtain only the right-hand of this inequality, because we can interchange the role of $\mathbf{s}$ and $\mathbf{t}$.

Now choose a closed embedding of X into $\mathbb{P}^N_K$ with standard coordinates $(x_0 : \cdots : x_N)$, let U_i be the affine open subset $\{x \in X \mid x_i \neq 0\}$ of X, and let U_{il} be the affine open subset $\{x \in U_i \mid t_l(x) \neq 0\}$. The restrictions of $g_{kl} := s_k/t_l$ to U_{il} are regular functions. The functions $f_{ij} := x_j/x_i$, $j = 0, \ldots, N$, generate $K[U_i]$ as a K-algebra (see A.2.10). Then define sets E_i by

$$E_i := \{P \in X(\overline{K}) \mid |x_i(P)| = \max_j |x_j(P)|\}.$$

It is clear that, if $P \in E_i$, we have

$$\max_j |f_{ij}(P)| = 1, \tag{2.4}$$

hence E_i is bounded in U_i (Lemma 2.2.9). Thus we can apply Lemma 2.2.10 to U_i, E_i and the covering $\{U_{il}\}$, obtaining bounded subsets E_{il} of U_{il} such that $E_i = \bigcup_l E_{il}$ and

$$\sup_{P \in E_{il}} \max_k |g_{kl}(P)| < \infty.$$

Since the sets E_{il} cover $X(\overline{K})$, we get the claim. □

2.2.12. Since Hilbert's Nullstellensatz is effective, the constant γ in Theorem 2.2.11 is effectively computable in terms of presentations of $\mathcal{D}$ and $\mathcal{D}'$. An effective version of the Nullstellensatz can be found in D. Masser and G. Wüstholz [**195**], Th.IV.

Remark 2.2.13. For the purpose of giving a precise meaning to the words "effectively computable," we need a closer look at the bounds in the results above.

In Lemma 2.2.9, there are finitely many elements $p_a \in K$ and $d \in \mathbb{N}$ such that

$$\sup_{P \in E} |f(P)| \le \max_a |p_a| \max\left(1, \sup_{P \in E} \max_j |f_j(P)|\right)^d.$$

The elements p_a may be chosen to be the coefficients of the polynomial p in (2.1) on page 37 if the absolute value is not archimedean, while in the archimedean case it suffices to add to this list C times the coefficients of the list, where C is the number of coefficients. Note also that the list of elements p_a so obtained and the degree d depend only on the geometric data $(U, f, f_1, \ldots, f_N)$, but not on E, nor on the absolute value.

In the situation of Lemma 2.2.10, the bound of $f \in K[U_l]$ is again of the same type, namely

$$\sup_{P \in E_l} |f(P)| \leq \max_m |p_m| \max\left(1, \sup_{P \in E} \max_j |f_j(P)|\right)^d,$$

where again $f_1, \ldots, f_N$ are generators of $K[U]$, the finitely many elements $p_m \in K$, and d, depend only on geometric data (f, the covering, generators) but not on E, the absolute value $|\ |$, or the decomposition $\{E_l\}$. This is clear by applying the above result to $f \in K[U_l]$ with generators $f_1, \ldots, f_N, 1/h$ and then again to every $g_l \in K[U]$ in (2.3) on page 37.

If we apply these remarks to the proof of Theorem 2.2.11 and use (2.4), then we may choose

$$\gamma = \max_m \log^+ |p_m|$$

for a certain finite set of elements $p_m \in K$, independent of the absolute value $|\ |$ and determined exclusively in terms of geometric data.

2.3. Global heights

In this section, starting from the local heights previously defined, we consider the case in which K is a number field and define global heights.

2.3.1. Let X be an irreducible projective variety defined over K.

We consider a Cartier divisor D on X with presentation

$$\mathcal{D} = (s_D;\ L, \mathbf{s};\ M, \mathbf{t}).$$

Let F be a number field with $K \subset F \subset \overline{K}$ and let $P \in X(F) \setminus \operatorname{supp}(D)$. For $v \in M_F$, we define the **local height**

$$\lambda_{\mathcal{D}}(P, v) := \max_k \min_l \log \left| \frac{s_k}{t_l s_D}(P) \right|_v$$

using our normalizations from 1.3.6. Let $p \in M_{\mathbb{Q}}$ be the restriction of v to $\mathbb{Q}$ and let $|\ |_u$ be an absolute value on $\overline{K}$ such that the restriction to K is equivalent to $|\ |_v$ and such that the restriction to $\mathbb{Q}$ is equal to $|\ |_p$. The existence of $|\ |_u$ follows from Proposition 1.3.1. Then

$$\lambda_{\mathcal{D}}(P, v) = \frac{[F_v : \mathbb{Q}_p]}{[F : \mathbb{Q}]} \lambda_{\mathcal{D}}(P, u),$$

where $\lambda_{\mathcal{D}}(P, u)$ is the local height relative to the absolute value $|\ |_u$ from 2.2.2. This allows us to apply the results from Section 2.2 to $\lambda_{\mathcal{D}}(P, v)$.

Example 2.3.2. The hyperplane $\{x_0 = 0\}$ in $\mathbb{P}^n_K$ has the presentation

$$\mathcal{D} = (x_0;\, O_{\mathbb{P}^n}(1), x_0, \ldots, x_n;\, O_{\mathbb{P}^n}, 1).$$

For $P \in \mathbb{P}^n(F)$ with $x_0(P) \neq 0$ and $v \in M_F$ the corresponding local height is

$$\lambda_{\mathcal{D}}(P, v) = \max_k \log \left|\frac{x_k}{x_0}(P)\right|_v$$

and the product formula becomes

$$h(P) = \sum_{v \in M_F} \lambda_{\mathcal{D}}(P, v).$$

This explains the name local height. This notion will be extended later to arbitrary divisors.

2.3.3. We go back to the general case in 2.3.1. Let $\lambda_{\mathcal{D}}$ be a local height relative to the presentation $\mathcal{D} = (s_D;\, L, \mathbf{s};\, M, \mathbf{t})$ of a Cartier divisor D on X. For $P \in X$ there are s_j and t_l such that $s_j(P) \neq 0$, $t_l(P) \neq 0$. Therefore, we can find a non-zero meromorphic section s of $O(D)$ such that P is not contained in the support of the Cartier divisor $D(s)$. Then $\mathcal{D}(s) = (s;\, L, \mathbf{s};\, M, \mathbf{t})$ is a presentation of $D(s)$ and we have

$$\lambda_{\mathcal{D}(s)} = \lambda_{\mathcal{D}} + \lambda_f,$$

where f is the rational function s/s_D. If F is a finite extension $K \subset F \subset \overline{K}$ such that $P \in X(F)$, the local height $\lambda_{\mathcal{D}(s)}(P, v)$ is finite for any $v \in M_L$, because P is not in the support of $D(s)$. Then we define the **global height** of P relative to $\lambda := \lambda_{\mathcal{D}}$ by

$$h_\lambda(P) := \sum_{v \in M_F} \lambda_{\mathcal{D}(s)}(P, v).$$

The next result justifies the definition and the name global height.

Proposition 2.3.4. *The global height h_λ is independent of the choices of F and of the section s.*

Proof: By Lemma 1.3.7, the global height is independent of F. Its independence from the choice of s can be verified as follows. Let t be another non-zero meromorphic section of $O(D)$ with $P \notin \operatorname{supp}(D(t))$. Then 2.2.4 and 2.2.5 show that

$$\lambda_{\mathcal{D}(s)}(P, v) - \lambda_{\mathcal{D}(t)}(P, v) = \lambda_{s/t}(P, v)$$

for any $v \in M_F$. On the other hand, the product formula shows that the global height of P relative to $\lambda_{s/t}$ is 0, proving the claim. □

Remark 2.3.5. As an immediate consequence the global height relative to the natural local height of a non-zero rational function is identically 0. It is also clear that the map $\lambda \mapsto h_\lambda$ is a group homomorphism.

Theorem 2.3.6. *Let λ, λ' be local heights relative to Cartier divisors D, D' with $D - D'$ a principal divisor. Then $h_\lambda - h_{\lambda'}$ is a bounded function.*

Proof: By Remark 2.3.5, we can assume $D = D' = 0$ and $\lambda' = 0$, hence we need only to show that h_λ is a bounded function for any local height relative to the zero divisor. Theorem 2.2.11 and Remark 2.2.13 give us a family $\{\gamma_v\}_{v \in M_K}$ of non-negative real numbers, almost all 0, such that

$$|\lambda(P,u)|_u \leq \gamma_v$$

for any $P \in X$ and any place u on $\overline{K}$ with $u|v$. As before, let F be a finite extension $K \subset F \subset \overline{K}$ such that $P \in X(F)$. By 2.3.1, we obtain

$$|\lambda(P,w)| \leq \frac{[F_w:\mathbb{Q}_p]}{[F:\mathbb{Q}]}\gamma_v$$

for any $w \in M_F$, which divides $v \in M_K$ and $p \in M_{\mathbb{Q}}$. By Corollary 1.3.2, we have

$$\sum_{w|v} [F_w : K_v] = [F : K]$$

and we get

$$|h_\lambda(P)| \leq \sum_{w \in M_F} |\lambda(P,w)| \leq \sum_{v \in M_K} \frac{[K_v:\mathbb{Q}_p]}{[K:\mathbb{Q}]}\gamma_v < \infty. \qquad \square$$

2.3.7. There is an isomorphism of the group of Cartier divisors modulo principal Cartier divisors onto $\mathrm{Pic}(X)$, given by $\mathrm{cl}(D) \mapsto \mathrm{cl}(O(D))$. Let us denote the real functions on X by $\mathbb{R}^X$ and the subspace of bounded functions by $O(1)$. Let $\mathbf{c} \in \mathrm{Pic}(X)$ and choose a Cartier divisor D with $\mathbf{c} = \mathrm{cl}(O(D))$ and a local height λ relative to D. By Theorem 2.3.6, the image $\mathbf{h_c}$ of h_λ under the projection

$$\mathbb{R}^X \longrightarrow \mathbb{R}^X/O(1)$$

is independent of the choice of D and λ. A representative of $\mathbf{h_c}$ is called a **height function** associated to $\mathbf{c}$.

In other words, an isomorphism class of line bundles determines a real-valued height function up to bounded functions. We note however that considering only equivalence classes of heights modulo bounded functions, as propounded by Weil, although it has attractive functorial properties, it also has the great disadvantage of throwing away the finer properties of heights needed to prove the deeper theorems of diophantine geometry. A better point of view is offered in the next sections.

Theorem 2.3.8. *The map*

$$\mathbf{h} : \mathrm{Pic}(X) \longrightarrow \mathbb{R}^X/O(1),$$

given by $\mathbf{c} \mapsto \mathbf{h_c}$, *is a homomorphism. If* $\varphi : Y \to X$ *is a morphism of irreducible projective varieties over* K, *then*

$$\mathbf{h}_{\varphi^*\mathbf{c}} = \mathbf{h_c} \circ \varphi$$

for any $\mathbf{c} \in \mathrm{Pic}(X)$.

Proof: The first claim follows from Remark 2.3.5 and Theorem 2.3.6. The second one is an immediate consequence of 2.2.6. □

It is quite trivial, but important, to remark that a base-point-free line bundle has always a non-negative height function. A more general result is the following

Proposition 2.3.9. *Let D be an effective Cartier divisor on X. Then there is a local height λ relative to D such that, for any $P \notin \mathrm{supp}(D)$ and for any place u of $\overline{K}$, it holds $\lambda(P, u) \geq 0$.*

Proof: There are base-point-free line bundles L, M on X such that $O(D) \cong L \otimes M^{-1}$. Choose generating global sections $t_0, \ldots, t_l$ of M. We can complete $s_D t_0, \ldots, s_D t_l$ to a family $s_0, \ldots, s_k$ of generating global sections of L. The local height given by the presentation

$$\mathcal{D} = (s_D;\ L, \mathbf{s};\ M, \mathbf{t})$$

is non-negative outside of the support of D. □

2.3.10. The results of Sections 2.2 and 2.3 extend immediately to varieties which are not necessarily irreducible. Here we must be careful to require that all meromorphic sections considered are invertible, i.e. not identically 0 on any irreducible component of X. For the functorial property of 2.2.6, we must assume that no irreducible component of Y is mapped into the support of D, in order to guarantee a well-defined pull-back of the Cartier divisor.

2.3.11. We may introduce global heights for any field with product formula as long as we work with properly normalized absolute values (see 1.3.6 for a perfect field and 1.3.12 in general). Then all results of this section continue to hold.

2.3.12. We may also replace the ground field K by $\overline{K}$. Then all geometric data as varieties, morphisms, line bundles, and sections are defined over a sufficiently large number field K and there is no problem about considerations with global heights relative to the ground field K. Since the global height does not depend on the ground field, it also makes sense to consider it as a global height over the algebraically closed field $\overline{K}$.

2.4. Weil heights

In this section we consider global heights given by a morphism of a projective variety to a projective space. In fact, we will see that any global height is the difference of two such Weil heights. We will formulate Theorem 2.3.8 and Northcott's theorem in terms of Weil heights. The results are based on the previous sections.

Let X be a projective variety X over $\overline{\mathbb{Q}}$.

Definition 2.4.1. *Let $\varphi : X \to \mathbb{P}^n_{\overline{\mathbb{Q}}}$ be a morphism over $\overline{\mathbb{Q}}$. The* **Weil height** *of $P \in X(\overline{\mathbb{Q}})$ relative to φ is defined by $h_\varphi(P) := h \circ \varphi(P)$, with h the usual height on $\mathbb{P}^n_{\overline{\mathbb{Q}}}$.*

2.4.2. If $\psi : X \to \mathbb{P}^m_{\overline{\mathbb{Q}}}$ is another morphism over $\overline{\mathbb{Q}}$, the **join** $\varphi \# \psi$ is the morphism

$$X \to \mathbb{P}^{(n+1)(m+1)-1}_{\overline{\mathbb{Q}}}, \qquad x \mapsto (\varphi_j(x)\psi_k(x)),$$

with the lexicographic ordering on pairs (i,j).
It may be viewed as the composition of the graph morphism $G(\psi) : X \to X \times \mathbb{P}^m_{\overline{\mathbb{Q}}}$, the product map $\varphi \times \mathrm{id} : X \times \mathbb{P}^m_{\overline{\mathbb{Q}}} \to \mathbb{P}^n_{\overline{\mathbb{Q}}} \times \mathbb{P}^m_{\overline{\mathbb{Q}}}$, and the Segre embedding $\mathbb{P}^n_{\overline{\mathbb{Q}}} \times \mathbb{P}^m_{\overline{\mathbb{Q}}} \to \mathbb{P}^{(n+1)(m+1)-1}_{\overline{\mathbb{Q}}}$ (cf. A.6.4).

Remark 2.4.3. If φ is a closed embedding, then $\varphi \# \psi$ is a closed embedding. In order to prove this claim, note that $G(\psi)$ is always a closed embedding (see A. Grothendieck [**134**], Cor.5.4.3). If φ is a closed embedding, then $\varphi \times \mathrm{id}$ is a closed embedding ([**134**], Prop.4.3.1). The Segre embedding is also a closed embedding (cf. A.6.4). Since the composition of closed embeddings remains a closed embedding ([**134**], Prop.4.2.5), we conclude that $\varphi \# \psi$ is a closed embedding.

The following proposition formalizes a remark already made in 1.5.14 about the height in Segre embeddings.

Proposition 2.4.4. *If $\varphi : X \to \mathbb{P}^n_{\overline{\mathbb{Q}}}$ and $\psi : X \to \mathbb{P}^m_{\overline{\mathbb{Q}}}$ are morphisms over $\overline{\mathbb{Q}}$, then*

$$h_{\varphi \# \psi} = h_\varphi + h_\psi.$$

2.4.5. We claim that every Weil height may be viewed as a global height in the sense of Section 2.3. There is a linear form $\ell = \ell_0 x_0 + \cdots + \ell_n x_n$, which does not vanish identically on any irreducible component of X. Then it follows from Example 2.3.2 and 2.2.6 that h_φ is the global height relative to the presentation $\varphi^*(\ell; O_{\mathbb{P}^n_{\overline{\mathbb{Q}}}}(1), x_0, \ldots, x_n; O_{\mathbb{P}^n_{\overline{\mathbb{Q}}}}, 1)$.

2.4.6. Conversely, we can write every global height as a difference of two Weil heights. Let h_λ be the global height relative to the presentation

$$\mathcal{D} = (s; L, s_0, \ldots, s_n; M, t_0, \ldots, t_m).$$

We consider the morphisms

$$\varphi : X \to \mathbb{P}^n_{\overline{\mathbb{Q}}}, \qquad x \mapsto (s_0(x) : \cdots : s_n(x))$$

and

$$\psi : X \to \mathbb{P}^m_{\overline{\mathbb{Q}}}, \qquad x \mapsto (t_0(x) : \cdots : t_m(x))$$

as in A.6.8. Then it follows from the independence of h_λ from s and 2.4.5 that

$$h_\lambda = h_\varphi - h_\psi.$$

2.4.7. Note that in 2.4.6 we may even assume that φ and ψ are closed embeddings into projective spaces. This follows from Remark 2.3.5 and Proposition 2.4.4, choosing any closed embedding θ of X into some projective space over $\overline{\mathbb{Q}}$ and replacing φ, ψ by $\varphi\#\theta$, $\psi\#\theta$.

Theorem 2.4.8. *If $\varphi : X \to \mathbb{P}^n_{\overline{\mathbb{Q}}}$ and $\psi : X \to \mathbb{P}^m_{\overline{\mathbb{Q}}}$ are morphisms over $\overline{\mathbb{Q}}$ with $\varphi^* O_{\mathbb{P}^n}(1) \cong \psi^* O_{\mathbb{P}^m}(1)$, then $h_\varphi - h_\psi$ is a bounded function.*

Proof: Using 2.4.5, this is a reformulation of Theorem 2.3.6. □

Our next result is the general version of **Northcott's theorem**, which is both simple and fundamental.

Theorem 2.4.9. *Let X be a projective variety defined over the number field K and let $h_{\mathbf{c}}$ be a height function associated to an ample class $\mathbf{c} \in \mathrm{Pic}(X)$. Then the set*

$$\{P \in X(\overline{K}) \mid h_{\mathbf{c}}(P) \leq C, \, [K(P) : K] \leq d\}$$

is finite for any constants $C, d \in \mathbb{R}$.

Proof: There is $m \in \mathbb{N}$ such that $m\mathbf{c}$ is very ample. By Theorem 2.3.8, $mh_{\mathbf{c}}$ is a height function associated to $m\mathbf{c}$. Therefore, we can assume without loss of generality that $\mathbf{c}$ is very ample. By Theorem 2.4.8, it is enough to prove the statement for $X = \mathbb{P}^n_{\overline{\mathbb{Q}}}$ and $\mathbf{c} = \mathrm{cl}(O_{\mathbb{P}^n}(1))$, i.e. for the standard height on $\mathbb{P}^n_{\overline{\mathbb{Q}}}$.

Let $U := \{x_j \neq 0\}$ be a standard affine subset of $\mathbb{P}^n_{\overline{\mathbb{Q}}}$. We have to show that there are only finitely many points P in $U(\overline{\mathbb{Q}})$ with $h(P) \leq C$ and $[K(P) : K] \leq d$. The height of $P \in U$ is an upper bound for the heights of the coordinates. Therefore, the case $n = 1$ implies the general statement. This is Theorem 1.6.8, ending the proof. □

Remark 2.4.10. Clearly, we may also introduce Weil heights for any field with product formula and all results above remain true with the exception of Northcott's theorem. We may use Example 1.5.23 as a counterexample if the field is infinite.

Example 2.4.11. The following example shows that Weil heights in the geometric case may be interpreted in terms of intersection theory, as a degree function. This is conceptually very important, because it allows us to use the intuition and methods of algebraic geometry in dealing with heights.

The corresponding result in the arithmetic case lies much deeper and requires intersection theory in the setting of arithmetic algebraic geometry (see Example 2.7.20).

Let X be an irreducible regular projective variety over an arbitrary field K, and let deg be the degree of cycles corresponding to a fixed embedding of X into a projective space $\mathbb{P}^n_K$. By Proposition 1.4.7, we have a canonical set of absolute values on $K(X)$ satisfying the product formula. A point $P \in \mathbb{P}^n_K(K(X))$ is given by coordinates $f_0, \dots, f_n \in K(X)$. Let φ be the rational map

$$X \dashrightarrow \mathbb{P}^n_K, \qquad x \mapsto \varphi(x) = (f_0(x) : \cdots : f_n(x)).$$

Let $x_0, \ldots, x_n$ be the coordinates of $\mathbb{P}^n_K$, viewed as global sections of $O_{\mathbb{P}^n}(1)$. Choose $j \in \{0, \ldots, n\}$ such that $x_j|_{\varphi(X)} \neq 0$. Then the vector $(f_0, \ldots, f_n)$ is proportional to

$$(\varphi^* x_0/\varphi^* x_j, \ldots, \varphi^* x_n/\varphi^* x_j) \in K(X)^{n+1}$$

and we may assume that they are equal. By Example 1.5.23, we have

$$h(P) = -\sum_Z \min_{i=0,\ldots,n} \operatorname{ord}_Z(f_i) \deg Z$$

$$= \sum_Z \left(\operatorname{ord}_Z(\varphi^* x_j) - \min_{i=0,\ldots,n} \operatorname{ord}_Z(\varphi^* x_i) \right) \deg Z,$$

where the sums range over all prime divisors Z of X. By the valuative criterion of properness (cf. A.11.10), the domain U of φ has a complement of codimension at least 2. The local ring associated to a prime divisor was introduced in A.8.7. By choosing a trivialization of $(\varphi|_U)^* O_{\mathbb{P}^n}(1)$ at a generic point of Z, we may view $\varphi^*(x_i)$ as regular functions in Z. Therefore, we have

$$\min_{i=0,\ldots,n} \operatorname{ord}_Z(\varphi^* x_i) = 0$$

and thus

$$h(P) = \sum_Z \operatorname{ord}_Z(\varphi^* x_j) \deg Z.$$

Since $X \backslash U$ is of codimension at least 2, the restriction map induces an isomorphism

$$\operatorname{Pic}(X) \xrightarrow{\sim} \operatorname{Pic}(U)$$

(because on a regular variety Cartier divisors and Weil divisors can be identified, cf. A.8.21). So it makes sense to view $(\varphi|_U)^* O_{\mathbb{P}^n}(1)$ as an element of $\operatorname{Pic}(X)$, which we simply denote by $\varphi^* O_{\mathbb{P}^n}(1)$. It follows that

$$h(P) = \deg \varphi^* O_{\mathbb{P}^n}(1),$$

where the right-hand side denotes the degree of any divisor of a non-zero meromorphic section of $\varphi^* O_{\mathbb{P}^n_K}(1)$ (see A.9.26). If Y is a projective variety over $K(X)$ and $\iota : Y \to \mathbb{P}^n_{K(X)}$ is a closed embedding over $K(X)$ into projective space, then $P \in Y(K(X))$ induces a rational map

$$\varphi : X \dashrightarrow Y$$

as above, and we have

$$h_\iota(P) = \deg \varphi^* O_Y(1),$$

where $O_Y(1)$ is the pull-back of $O_{\mathbb{P}^n}(1)$ to Y.

2.5. Explicit bounds for Weil heights

This is a somewhat technical section, the reading of which can be omitted at first. Its ultimate purpose is to give a meaning to the phrase "effectively computable," which otherwise would be only a hollow claim, devoid of true mathematical significance.

The main tool in this section is the concept of presentation of a closed embedding of a projective variety in projective space. The basic idea can be described as follows. Let $X \to \mathbb{P}^n_{\overline{\mathbb{Q}}}$ be a projective algebraic variety over the algebraically closed field $\overline{\mathbb{Q}}$, embedded in projective space $\mathbb{P}^n_{\overline{\mathbb{Q}}}$. It is well known that every rational function on X is then induced by restriction of a rational function in the ambient space $\mathbb{P}^n_{\overline{\mathbb{Q}}}$. On the other hand, we often need to compare situations relative to different embeddings. The point of view taken in this section is therefore the following. Since we are dealing with the function field $\overline{\mathbb{Q}}(X)$ of X, we are allowed to choose a hypersurface in $\mathbb{P}^{r+1}_{\overline{\mathbb{Q}}}$ as a birational model. The homogeneous coordinate ring S is the quotient of a polynomial ring by a principal ideal. This allows us to introduce a height in S. Thus fixing this choice gives us a reference description of elements in S.

This being done, we consider an arbitrary closed embedding $X \to \mathbb{P}^n_{\overline{\mathbb{Q}}}$. Then a presentation of the embedding $X \to \mathbb{P}^n_{\overline{\mathbb{Q}}}$, relative to the reference ring S, consists in expressing the rational functions $(x_i/x_j)|_X$ as elements of S. Now the problem of comparing heights relative to different embeddings can be solved by comparing corresponding presentations. This leads to very explicit comparison estimates for heights. The details are as follows.

2.5.1. Let X be an irreducible projective variety over $\overline{\mathbb{Q}}$ of dimension r. There is a $\overline{\mathbb{Q}}$-morphism

$$\pi : X \longrightarrow \mathbb{P}^{r+1}_{\overline{\mathbb{Q}}}$$

such that X is mapped birationally onto a hypersurface (cf. A.11.5 and A.11.6). We denote by $z_0, \ldots, z_{r+1}$ the standard coordinates of $\mathbb{P}^{r+1}_{\overline{\mathbb{Q}}}$. Then we may assume that the hypersurface is given by an irreducible homogeneous polynomial f of degree d of the form

$$f(z_0, \ldots, z_{r+1}) = f_0 + f_1 z_{r+1} + \cdots + f_{d-1} z_{r+1}^{d-1} + z_{r+1}^d,$$

where $f_i \in \overline{\mathbb{Q}}[z_0, \ldots, z_r]$ is homogeneous of degree $d-i$, $f(0, \ldots, 0, 1) \neq 0$ and d is the degree of X with respect to $\pi^* O_{\mathbb{P}^{r+1}}(1)$ (cf. A.11.7). This situation is fixed for the whole section.

2.5.2. Let S be the homogeneous coordinate ring of $\pi(X)$. We have

$$S = \overline{\mathbb{Q}}[z_0, \ldots, z_{r+1}]/\mathfrak{J},$$

where $\mathfrak{J}$ is the homogeneous ideal generated by f. Let $\overline{z}_i$ be the image of z_i in S $(0 \leq i \leq r+1)$ and note that $\overline{z}_{r+1}$ is integral over $\overline{\mathbb{Q}}[\overline{z}_0, \ldots, \overline{z}_r]$. The variables $\overline{z}_0, \ldots, \overline{z}_r$ are algebraically independent, because the transcendence degree of $\overline{\mathbb{Q}}(\pi(X)) = \overline{\mathbb{Q}}(X)$ is r (cf. A.4.11). By abuse of notation, we denote them again by $z_0, \ldots, z_r$. The minimal polynomial of $\overline{z}_{r+1}$ over the polynomial ring $\overline{\mathbb{Q}}[z_0, \ldots, z_r]$ is equal to $f(z_0, \ldots, z_r, \cdot)$, since

$$0 = f_0 + f_1 \overline{z}_{r+1} + \cdots + f_{d-1} \overline{z}_{r+1}^{d-1} + \overline{z}_{r+1}^d. \tag{2.5}$$

The elements $1, \overline{z}_{r+1}, \ldots, \overline{z}_{r+1}^{d-1}$ form a basis of S over $\overline{\mathbb{Q}}[z_0, \ldots, z_r]$ and so we have an isomorphism of $\overline{\mathbb{Q}}$-vector spaces

$$S \xrightarrow{\sim} \{p \in \overline{\mathbb{Q}}[z_0, \ldots, z_{r+1}] \mid \deg_{z_{r+1}}(p) < d\}.$$

By means of this map, we define the height of an element of S as the height of the corresponding polynomial.

2.5.3. For $l \in \mathbb{N}$, there are uniquely determined $q_{lj} \in \overline{\mathbb{Q}}[z_0, \ldots, z_r]$ $(j = 0, \ldots, d-1)$ such that

$$\overline{z}_{r+1}^l = \sum_{j=0}^{d-1} q_{lj} \overline{z}_{r+1}^j. \tag{2.6}$$

The polynomials q_{lj} are homogeneous of degree $l-j$ (elements of negative degree are 0), and $q_{lj} = \delta_{lj}$ for $0 \leq l \leq d-1$, where δ_{lj} is Kronecker's symbol. We may now assume that $l \geq d$. Equation (2.5) shows

$$\overline{z}_{r+1}^{l} = -\sum_{k=0}^{d-1} f_k \overline{z}_{r+1}^{k+l-d} = -\sum_{j=0}^{d-1}\sum_{k=0}^{d-1} f_k\, q_{k+l-d,j}\, \overline{z}_{r+1}^{j},$$

leading to the recursive formula

$$q_{lj} = -\sum_{k=0}^{d-1} f_k\, q_{k+l-d,j}, \tag{2.7}$$

where $j = 0, \ldots, d-1$. Let F be a number field containing the coefficients of $f_0, \ldots, f_{d-1}$ and for $v \in M_F$ define δ_v to be 1 if v is archimedean and 0 otherwise. The recursion (2.7) yields a bound

$$|q_{lj}|_v \leq \left| \binom{d+r+1}{r+1} \right|_v^{\delta_v} |f|_v \max_{l'=l-d,\ldots,l-1} |q_{l'j}|_v$$

for the Gauss norms. Here we have used that f_k has $\binom{d-k+r}{r}$ summands and

$$\sum_{k=0}^{d} \binom{d-k+r}{r} = \binom{d+r+1}{r+1}. \tag{2.8}$$

By induction we obtain

$$|q_{lj}|_v \leq \left| \binom{d+r+1}{r+1} \right|_v^{(l-d)\delta_v} |f|_v^{l-d+1} \tag{2.9}$$

and thus Lemma 1.3.7 leads to

$$h(q_{lj}) \leq (l-d+1)h(f) + (l-d)\log \binom{d+r+1}{r+1}.$$

Let $\varphi : X \to \mathbb{P}^n_{\overline{\mathbb{Q}}}$ be a closed embedding over $\overline{\mathbb{Q}}$ and let $x_0, \ldots, x_n$ be the standard coordinates of $\mathbb{P}^n_{\overline{\mathbb{Q}}}$. Let $\mathbf{p}$ be a vector with entries $p_i \in S$, $i = 0, \ldots, n$, homogeneous of degree $d(\mathbf{p})$.

Definition 2.5.4. *The vector* $\mathbf{p}$ *is said to be a* ***presentation*** *of* φ *if the following conditions are satisfied:*

(a) *If* $l \in \{0, \ldots, n\}$ *and* $x_l|_X \neq 0$, *then we have* $p_l \neq 0$.

(b) *If* l *as in (a) and* $i \in \{0, \ldots, n\}$, *then*

$$\frac{p_i}{p_l} = \left.\frac{x_i}{x_l}\right|_X$$

in $\overline{\mathbb{Q}}(X)$.

2.5.5. The number $d(\mathbf{p})$ is called the **degree** of $\mathbf{p}$. Consider the vector whose entries are given by all the coefficients of $p_0, \ldots, p_n$. The height of the corresponding point in appropriate projective space is called the **height** of the presentation, denoted by $h(\mathbf{p})$. The existence of a presentation of φ is an obvious consequence of $\overline{\mathbb{Q}}(X) = \overline{\mathbb{Q}}(\pi(X))$.

Lemma 2.5.6. *Let $\varphi_j : X \to \mathbb{P}^{n_j}_{\overline{\mathbb{Q}}}$, $j = 1, \ldots, k$, be closed embeddings over $\overline{\mathbb{Q}}$ with presentations $\mathbf{p}^{(j)}$ and let $n := (n_1 + 1) \cdots (n_k + 1) - 1$. Then the join $\varphi_1 \# \cdots \# \varphi_k$ gives a closed embedding*

$$\varphi : X \longrightarrow \mathbb{P}^n_{\overline{\mathbb{Q}}}, \qquad P \longmapsto (\varphi_{1i_1}(P) \cdots \varphi_{ki_k}(P))_{i_j \in \{0,\ldots,n_j\}}.$$

It has a presentation $\mathbf{p}$ defined by

$$p_{\mathbf{i}} := p^{(1)}_{i_1} \cdots p^{(k)}_{i_k} \qquad (i_j \in \{0, \ldots, n_j\})$$

of degree $d(\mathbf{p}) = d(\mathbf{p}^{(1)}) + \cdots + d(\mathbf{p}^{(k)})$ and height

$$h(\mathbf{p}) \le \sum_{j=1}^{k} h(\mathbf{p}^{(j)}) + r \sum_{j=1}^{k-1} \log\left(6 + 6\, d(\mathbf{p}^{(j)})/r\right) + C \cdot (k-1)$$

with

$$C = (d-1)h(f) + d(d + r + 1).$$

Proof: By Remark 2.4.3, φ is a closed embedding. Also, $\mathbf{p}$ is a presentation of φ of degree

$$d(\mathbf{p}) = d(\mathbf{p}^{(1)}) + \cdots + d(\mathbf{p}^{(k)}).$$

To prove the estimate for the height, by induction we may assume $k = 2$. We have the decomposition

$$p^{(j)}_{i_j} = \sum_{m=0}^{d-1} p^{(j)}_{i_j m}\, \overline{z}^{m}_{r+1} \qquad (j = 1, 2),$$

whence

$$p_{\mathbf{i}} = \left(\sum_{m_1=0}^{d-1} p^{(1)}_{i_1 m_1}\, \overline{z}^{m_1}_{r+1} \right) \left(\sum_{m_2=0}^{d-1} p^{(2)}_{i_2 m_2}\, \overline{z}^{m_2}_{r+1} \right).$$

Equation (2.6) on page 46 leads to the decomposition

$$p_{\mathbf{i}} = \sum_{m=0}^{d-1} p_{\mathbf{i}m}\, \overline{z}^{m}_{r+1}$$

with

$$p_{\mathbf{i}m} := \sum_{m_1 + m_2 = m} p^{(1)}_{i_1 m_1} p^{(2)}_{i_2 m_2} + \sum_{l=d}^{2d-2} \sum_{\substack{m_1 + m_2 = l \\ m_1, m_2 \le d-1}} p^{(1)}_{i_1 m_1}\, p^{(2)}_{i_2 m_2}\, q_{lm}.$$

Let F be a number field extension of $\mathbb{Q}$ containing the coefficients of $f_0, \ldots, f_{d-1}$ and of all $p^{(j)}_{i_j m_j}$, $j = 1, 2$, and for $v \in M_F$ define δ_v as in 2.5.3. Then we verify

$$|p_{\mathbf{i}m}|_v \le |B|_v^{\delta_v}\, |p^{(1)}_{i_1}|_v\, |p^{(2)}_{i_2}|_v \max_{l=d,\ldots,2d-2} (1, |q_{lm}|_v), \tag{2.10}$$

where B is an upper bound for

$$\sum_{m_1=0}^{m} \binom{r + d(\mathbf{p}^{(1)}) - m_1}{r} + \sum_{l=d}^{2d-2} \sum_{m_1 = l-d+1}^{d-1} \binom{r + d(\mathbf{p}^{(1)}) - m_1}{r} \binom{r + l - m}{r}.$$

Thereby we have used the fact that number of monomials of degree D in $r+1$ variables is equal to $\binom{r+D}{D}$. We use the estimates

$$\binom{r + d(\mathbf{p}^{(1)}) - m_1}{r} < \frac{1}{r!}\left(r + d(\mathbf{p}^{(1)})\right)^r < \left(3 + 3\,d(\mathbf{p}^{(1)})/r\right)^r$$

and

$$\binom{r + l - m}{r} \leq 2^{r+l-m}$$

to conclude that

$$B := d\,2^{r+2d}\left(3 + 3\,d(\mathbf{p}^{(1)})/r\right)^r$$

is such an upper bound. From (2.9) on page 47 and (2.10), we get

$$h(\mathbf{p}) \leq h(\mathbf{p}^{(1)}) + h(\mathbf{p}^{(2)}) + \log B + (d-1)h(f) + (d-2)\log\binom{r+d+1}{d+1}.$$

With the above value for B, we have

$$h(\mathbf{p}) \leq h(\mathbf{p}^{(1)}) + h(\mathbf{p}^{(2)}) + r\log(6 + 6\,d(\mathbf{p}^{(1)})/r)) + C$$

with

$$C := (d-1)h(f) + d(d+r+1).$$

□

Remark 2.5.7. If we work over a fixed number field K and with an irreducible reduced projective variety, then the constructions in 2.5.1 to 2.5.3 can be done over K and every K-morphism to projective space has a presentation defined over K. Moreover, Lemma 2.5.6 remains valid.

2.5.8. We use the following notation. For a multi-index $\boldsymbol{\alpha} = (\alpha_0, \ldots, \alpha_N) \in \mathbb{N}^{N+1}$, we set

$$|\boldsymbol{\alpha}| := \alpha_0 + \cdots + \alpha_N$$

and for $\mathbf{x} = (x_0, \ldots, x_N) \in \overline{\mathbb{Q}}^{N+1}$ we define

$$\mathbf{x}^{\boldsymbol{\alpha}} := x_0^{\alpha_0} \cdots x_N^{\alpha_N}.$$

If Y is a closed subvariety of $\mathbb{P}^N_{\overline{\mathbb{Q}}}$, we denote by $\mathcal{J}_Y$ the ideal sheaf of Y.

Proposition 2.5.9. *Let $\varphi : X \to \mathbb{P}^n_{\overline{\mathbb{Q}}}$, $\psi : X \to \mathbb{P}^m_{\overline{\mathbb{Q}}}$ be closed embeddings over $\overline{\mathbb{Q}}$, with corresponding presentations $\mathbf{p}$, $\mathbf{q}$. We assume*

$$\varphi^* O_{\mathbb{P}^n}(1) \cong \psi^* O_{\mathbb{P}^m}(1).$$

There is a positive integer k_ψ such that if $k \geq k_\psi$, then

$$H^1\left(\mathbb{P}^m_{\overline{\mathbb{Q}}}, \mathcal{J}_{\psi X} \otimes O_{\mathbb{P}^m}(k)\right) = 0.$$

If $k \geq k_\psi$ and $\chi(k) := \dim\left(H^0\left(X, \psi^* \mathcal{O}_{\mathbb{P}^m}(k)\right)\right)$ and $P \in X$, then

$$\begin{aligned} h_\varphi(P) - h_\psi(P) \leq (n+1)\chi(k)\Big(& h(\mathbf{p}) + h(\mathbf{q}) \\ & + r\log(6 + 6\,d(\mathbf{p})/r) + r\log(6 + 6\,d(\mathbf{q})/r) \\ & + \frac{1}{k}\log((n+1)\chi(k)) + C\Big), \end{aligned}$$

where

$$C := (d-1)h(f) + d(d+r+1).$$

Proof: In order to understand the proof, the reader should be familiar with some basic facts from cohomology of sheaves, as given in A.10.

The existence of k_ψ is a well-known result (see A.10.27). There is a short exact sequence

$$0 \longrightarrow \mathcal{J}_{\psi X} \longrightarrow \mathcal{O}_{\mathbb{P}^m} \longrightarrow \psi_* \mathcal{O}_X \longrightarrow 0$$

of coherent sheaves on $\mathbb{P}^m_{\overline{\mathbb{Q}}}$. Tensoring with $\mathcal{O}_{\mathbb{P}^m}(k)$ yields a short exact sequence

$$0 \longrightarrow \mathcal{J}_{\psi X} \otimes \mathcal{O}_{\mathbb{P}^m}(k) \longrightarrow \mathcal{O}_{\mathbb{P}^m}(k) \longrightarrow (\psi_* \mathcal{O}_X) \otimes \mathcal{O}_{\mathbb{P}^m}(k) \longrightarrow 0.$$

The projection formula gives

$$(\psi_* \mathcal{O}_X) \otimes \mathcal{O}_{\mathbb{P}^m_{\overline{\mathbb{Q}}}}(k) \cong \psi_* \psi^* \mathcal{O}_{\mathbb{P}^m_{\overline{\mathbb{Q}}}}(k),$$

as we have verified in Example A.10.20. Using A.10.25, the first terms of the long exact cohomology sequence are

$$\begin{aligned} 0 \longrightarrow H^0\left(\mathbb{P}^m_{\overline{\mathbb{Q}}}, \mathcal{J}_{\psi X} \otimes \mathcal{O}_{\mathbb{P}^m}(k)\right) \longrightarrow H^0(\mathbb{P}^m_{\overline{\mathbb{Q}}}, \mathcal{O}_{\mathbb{P}^m}(k)) \longrightarrow H^0(X, \psi^* \mathcal{O}_{\mathbb{P}^m}(k)) \longrightarrow \\ \longrightarrow H^1(\mathbb{P}^m_{\overline{\mathbb{Q}}}, \mathcal{J}_{\psi X} \otimes \mathcal{O}_{\mathbb{P}^m}(k)) \longrightarrow \cdots . \end{aligned}$$

The last cohomology group is 0 by the choice of k, and we infer that the map

$$H^0\left(\mathbb{P}^m_{\overline{\mathbb{Q}}}, \mathcal{O}_{\mathbb{P}^m}(k)\right) \longrightarrow H^0\left(X, \psi^* \mathcal{O}_{\mathbb{P}^m}(k)\right)$$

is surjective. By assumption, the invertible sheaves $\varphi^* \mathcal{O}_{\mathbb{P}^n}(1)$ and $\psi^* \mathcal{O}_{\mathbb{P}^m}(1)$ are isomorphic and we may identify them. Let $\mathbf{x} = (x_0 : \cdots : x_n)$ and $\mathbf{y} = (y_0 : \cdots : y_m)$ be the standard coordinates of $\mathbb{P}^n_{\overline{\mathbb{Q}}}$ and $\mathbb{P}^m_{\overline{\mathbb{Q}}}$, and choose $B \subset \{\boldsymbol{\beta} \in \mathbb{N}^{m+1} \mid |\boldsymbol{\beta}| = k\}$ such that

$$\left(\mathbf{y}^{\boldsymbol{\beta}}|_X\right)_{\boldsymbol{\beta} \in B}$$

is a basis of $H^0(X, \psi^* \mathcal{O}_{\mathbb{P}^m}(k))$. There are uniquely determined $a_{i\boldsymbol{\beta}} \in \overline{\mathbb{Q}}$ such that

$$x_i^k|_X = \sum_{\boldsymbol{\beta} \in B} a_{i\boldsymbol{\beta}}\, \mathbf{y}^{\boldsymbol{\beta}}|_X. \tag{2.11}$$

Let $P \in X(\overline{\mathbb{Q}})$. Choose a number field F containing $x_i(P), y_j(P)$ and $a_{i\beta}$ for all i, j, β. By Proposition 2.4.4, we obtain, for the k-fold join $\psi^{(k)} := \psi\#\cdots\#\psi$, the equation

$$\begin{aligned} k(h_\varphi(P) - h_\psi(P)) &= \sum_{v \in M_F} \log \max_i |x_i^k(P)|_v - h_{\psi^{(k)}}(P) \\ &= \sum_{v \in M_F} \log \max_i |x_i^k(P)|_v - \sum_{v \in M_F} \log \max_{|\beta|=k} |\mathbf{y}^\beta(P)|_v . \end{aligned}$$

By (2.11), the triangle inequality and Lemma 1.3.7, we deduce

$$h_\varphi(P) - h_\psi(P) \leq \frac{1}{k} h(a) + \frac{1}{k} \log \chi(k), \tag{2.12}$$

where a is the matrix $(a_{i\beta})$ and $h(a)$ is the height of the matrix viewed as a point in appropriate projective space.

We take the ratio of equations (2.11) with indices i and l and deduce, using the definition of presentation, that

$$\sum_{\beta \in B} a_{l\beta}\, p_i^k\, \mathbf{q}^\beta = \sum_{\beta \in B} a_{i\beta}\, p_l^k\, \mathbf{q}^\beta \quad (i, l \in \{0, \ldots, n\}). \tag{2.13}$$

Conversely, assume that $(a_{i\beta})$ is a non-trivial solution of this equation. Then we have

$$(x_i|_X)^k \sum_{\beta \in B} a_{l\beta}\, (\mathbf{y}|_X)^\beta = (x_l|_X)^k \sum_{\beta \in B} a_{i\beta}\, (\mathbf{y}|_X)^\beta .$$

Let i be such that $x_i|_X$ is not identically 0. Then the last displayed equation shows that the rational function on X defined by

$$g := \frac{\sum_{\beta \in B} a_{i\beta}\, \mathbf{y}^\beta|_X}{x_i^k|_X}$$

does not depend on the index i. We claim that g is constant. To prove this, it suffices to show that g is a regular function (use that X is projective and A.6.15). Indeed, since $x_0, \ldots, x_n$ generate $O_{\mathbb{P}^n}(1)$, we see that for any point $P \in X(\overline{\mathbb{Q}})$, there is an index i such that $x_i(P) \neq 0$, hence g is regular at P, as asserted.

This proves that the space of solutions of (2.13) is spanned by the matrix $a = (a_{i\beta})$ given by (2.11).

Our next task is to estimate $h(\mathbf{a})$. Since a scalar factor does not change the height, we may estimate the height of any non-trivial solution of (2.13). By Lemma 2.5.6, we have a natural presentation of $\varphi^{(k)}\#\psi^{(k)}$ in terms of $\mathbf{p}$ and $\mathbf{q}$. The elements $p_i^k \mathbf{q}^\beta$ of S are entries of that presentation. The decomposition

$$p_i^k \mathbf{q}^\beta = \sum_{j=0}^{d-1} c_{\beta ij}\, \overline{z}_{r+1}^j$$

with uniquely determined $c_{\beta ij} \in \overline{\mathbb{Q}}[z_0, \ldots, z_r]$ leads to the system of equations

$$\sum_{\beta \in B} c_{\beta ij}\, a_{l\beta} - \sum_{\beta \in B} c_{\beta lj}\, a_{i\beta} = 0, \tag{2.14}$$

with $i, l \in \{0, \ldots, n\}$ and we have $j \in \{0, \ldots, d-1\}$.

Let $c_{\beta ij} = \sum c_{\beta ij\alpha}\, z_0^{\alpha_0} \cdots z_r^{\alpha_r}$, so that the coefficients $c_{\beta ij\alpha}$ of the polynomials $c_{\beta ij}$ form a matrix $\mathbf{c}$ with

$$h(\mathbf{c}) \leq k\,(h(\mathbf{p}) + h(\mathbf{q}) + r\log(6 + 6d(\mathbf{p})/r) + r\log(6 + 6d(\mathbf{q})/r) + C) \tag{2.15}$$

again by Lemma 2.5.6. Moreover, (2.14) is equivalent to the linear system of equations

$$\sum_{\beta\in B} (c_{\beta ij\alpha}\, a_{l\beta} - c_{\beta lj\alpha}\, a_{i\beta}) = 0$$

indexed by $i, l, j, \boldsymbol{\alpha}$ and unknowns $a_{i\beta}$. Let A denote the matrix associated to this linear system; its entries are either 0 or $\pm c_{\beta ij\alpha}$. The number of unknowns is $(n+1)|B|$ and, as remarked before, the space of solutions has dimension 1. Therefore, the rank R of the matrix A is

$$R = (n+1)\chi(k) - 1.$$

Let A' be a $R \times (R+1)$ submatrix of A of full rank R. Since A and A' have the same kernel, we look for a non-zero solution $\mathbf{a}$ of $A' \cdot \mathbf{a} = 0$. We consider $\mathbf{a}$ as a vector $(a_0, \ldots, a_R)$ and A' as a matrix of the form $(A'_{\mu\nu})_{\mu\in\{1,\ldots,R\},\,\nu\in\{0,\ldots,R\}}$. Obviously, the vector $\mathbf{a}$ with ρth entry

$$a_\rho := (-1)^{\rho+1}\det(A'_{\mu\nu})_{\mu\in\{1,\ldots,R\},\,\nu\in\{0,\ldots,R\}\setminus\{\rho\}}$$

is a non-zero solution of $A' \cdot \mathbf{a} = 0$. The estimate

$$\max_{\rho=0,\ldots,R} |a_\rho|_v \leq |R!|_\delta^{\delta_v} \max_{\beta,i,j,\alpha} |c_{\beta ij\alpha}|_v^R$$

and (2.15) lead to

$$h(\mathbf{a}) \leq Rk\,(h(\mathbf{p}) + h(\mathbf{q}) + r\log(6 + 6d(\mathbf{p})/r) + r\log(6 + 6d(\mathbf{q})/r) + C) + \log(R!).$$

By (2.12) and the definition of R, we now get

$$h_\varphi(P) - h_\psi(P) \leq (n+1)\chi(k)\Big(h(\mathbf{p}) + h(\mathbf{q})$$
$$+ r\log(6 + 6d(\mathbf{p})/r) + r\log(6 + 6d(\mathbf{q})/r) + C + \frac{1}{k}\log((n+1)\chi(k))\Big)$$

proving the proposition. □

The upper bound in Proposition 2.5.9 is quite explicit except for $\chi(k)$. The next lemmas will handle this problem.

Lemma 2.5.10. *Let $\psi : X \to \mathbb{P}^m_{\overline{\mathbb{Q}}}$ be a closed embedding over $\overline{\mathbb{Q}}$ with presentation $\mathbf{q}$ and let $k \geq k_\psi$, $\chi(k)$ be as in Proposition 2.5.9. Then*

$$\chi(k) \leq \binom{kd(\mathbf{q}) + r + 1}{r+1} - \binom{kd(\mathbf{q}) - d + r + 1}{r+1}.$$

Proof: Let $y_0, \ldots, y_m$ be the standard coordinates of $\mathbb{P}^m_{\overline{\mathbb{Q}}}$. We have seen in the proof of Proposition 2.5.9 that the linear map

$$H^0(\mathbb{P}^m_{\overline{\mathbb{Q}}}, \mathcal{O}_{\mathbb{P}^m}(k)) \longrightarrow H^0(X, \psi^*\mathcal{O}_{\mathbb{P}^m}(k))$$

is surjective. Choose $B \subset \{\boldsymbol{\beta} \in \mathbb{N}^{m+1} \mid |\boldsymbol{\beta}| = k\}$ such that

$$\left(\mathbf{y}^{\boldsymbol{\beta}}|_X\right)_{\boldsymbol{\beta}\in B}$$

is a basis of $H^0(X, \psi^* O_{\mathbb{P}^m_{\overline{\mathbb{Q}}}}(k))$. The above monomials are linearly independent if and only if the polynomials $(\mathbf{q}^\beta)_{\beta \in B}$ are linearly independent, by definition of a presentation. Therefore

$$\chi(k) \leq \dim S_{kd(\mathbf{q})} = \binom{kd(\mathbf{q}) + r + 1}{r + 1} - \binom{kd(\mathbf{q}) - d + r + 1}{r + 1},$$

because by 2.5.2 the space $S_{kd(\mathbf{q})}$ is isomorphic to the vector space of homogeneous polynomials $p(z_0, \ldots, z_{r+1})$ of degree $kd(\mathbf{q})$ satisfying $\deg_{z_{r+1}}(p) < d$ and because of (2.9) on page 47. □

The next result is a slight generalization of a result of Mumford. Note that by base change A.10.28, we may just as well work over the field of complex numbers. For details and a proof of this lemma, we refer to A. Bertram, L. Ein, and R. Lazarsfeld [22].

Lemma 2.5.11. *Let Y be an irreducible smooth closed subvariety of $\mathbb{P}^m_{\overline{\mathbb{Q}}}$ defined over $\overline{\mathbb{Q}}$ and let $c := \min(1 + \dim(Y), \operatorname{codim}(Y, \mathbb{P}^m_{\overline{\mathbb{Q}}}))$. Then*

$$H^i(\mathbb{P}^m_{\overline{\mathbb{Q}}}, \mathcal{J}_Y \otimes O_{\mathbb{P}^m_{\overline{\mathbb{Q}}}}(k)) = 0$$

for $i \geq 1$ and $k \geq c(\deg(Y) - 1) - \dim(Y)$.

Lemma 2.5.12. *If $\psi : X \to \mathbb{P}^m_{\overline{\mathbb{Q}}}$ is a closed embedding over $\overline{\mathbb{Q}}$ with presentation $\mathbf{q}$, then*

$$\deg(\psi X) \leq d(\mathbf{q})^r d.$$

Proof: The Hilbert polynomial of ψX has degree r and its leading coefficient is $\deg(\psi X)/r!$, see A.10.33. For large k, the Hilbert polynomial at k equals the left-hand side of the inequality in Lemma 2.5.10. On the other hand, the right-hand side of that inequality is also a polynomial of degree r in k, with leading coefficient

$$\frac{d(\mathbf{q})^r}{(r+1)!}((r+1) + \cdots + 1) - \frac{d(\mathbf{q})^r}{(r+1)!}((r+1-d) + \cdots + (1-d)) = \frac{d(\mathbf{q})^r d}{r!}.$$

The conclusion follows. □

2.5.13. Now we summarize the results of this section. Let X be a smooth irreducible projective variety over $\overline{\mathbb{Q}}$. Let $r = \dim(X)$ and let $\pi : X \to \mathbb{P}^{r+1}_{\overline{\mathbb{Q}}}$ be a morphism over $\overline{\mathbb{Q}}$, mapping X birationally onto a hypersurface given by a homogeneous polynomial f of degree d, as in 2.5.1.

Assume that $\varphi : X \to \mathbb{P}^n_{\overline{\mathbb{Q}}}$ and $\psi : X \to \mathbb{P}^m_{\overline{\mathbb{Q}}}$ are closed embeddings over $\overline{\mathbb{Q}}$ with $\varphi^* O_{\mathbb{P}^n}(1) \cong \psi^* O_{\mathbb{P}^m}(1)$ and corresponding presentations $\mathbf{p}$ and $\mathbf{q}$. We also assume $d(\mathbf{q}) \geq 1$.

Theorem 2.5.14. *For each $P \in X$, it holds*

$$h_\varphi(P) - h_\psi(P) \leq C_1 (n+1) d(\mathbf{q})^{r^2+r} \Big(h(\mathbf{p}) + h(\mathbf{q}) + r \log(6 + 6 d(\mathbf{p})/r) + r \log(6 + 6 d(\mathbf{q})/r) + \log(n+1) + C_2 \Big)$$

with

$$C_1 = d \frac{(d+1)^r (r+1)^r}{r!}, \qquad C_2 = (d-1)h(f) + d(d+r+1) + r + 1.$$

Proof: Let $k := d\,(r+1)\,d(\mathbf{q})^r$; then $k \geq k_\psi$ by Lemmas 2.5.11 and 2.5.12. We have

$$\chi(k) \leq \binom{kd(\mathbf{q})+r+1}{r+1} - \binom{kd(\mathbf{q})+r+1-d}{r+1}$$
$$\leq d\binom{kd(\mathbf{q})+r}{r} \leq \frac{d}{r!}\,(kd(\mathbf{q})+r)^r,$$

where the first step comes from Lemma 2.5.10 and the second step uses (2.8) on page 47. From the definition of k, this implies

$$\chi(k) \leq d\,\frac{(d+1)^r(r+1)^r}{r!}\,d(\mathbf{q})^{r^2+r}.$$

An easy majorization shows that $k^{-1}\log\chi(k) \leq r+1$, and the result follows easily from Proposition 2.5.9. □

2.6. Bounded subsets

In order to show that a Weil height is determined by the isomorphism class of a line bundle, we have introduced in Section 2.2 bounded sets in affine varieties. In this section, we extend this concept first to arbitrary varieties and thereafter to several absolute values. Implicitly, this was already used in the proof of Theorem 2.2.11, which now becomes more transparent. This section is used in 2.7 to study locally bounded metrics on line bundles. Moreover, it is basic for proving the Chevalley–Weil theorem in Section 10.3.

2.6.1. Let K be a field and let us fix an embedding $K \subset \overline{K}$ into an algebraic closure. For the moment, we fix an absolute value $|\ |$ on $\overline{K}$.

We have defined in 2.2.8 bounded subsets of $U(\overline{K})$ for an affine variety U. We extend this notion to an arbitrary variety X over K.

Definition 2.6.2. *A subset $E \subset X(\overline{K})$ is called* **bounded** *in X, if there is a finite covering $\{U_i\}_{i\in I}$ of X by affine open subsets, and sets E_i with $E_i \subset U_i(\overline{K})$, such that E_i is bounded in U_i and $E = \bigcup_{i\in I} E_i$.*

Remark 2.6.3. If E is bounded in X, Lemma 2.2.10 shows that for any finite covering $\{U_i\}_{i\in I}$ of X by affine open subsets there is a subdivision

$$E = \bigcup_{i\in I} E_i, \quad E_i \subset U_i(\overline{K}),$$

such that each E_i is bounded in U_i.

It is easy to prove that the image of a bounded set under a morphism is again bounded. Moreover, if Y is a closed subvariety of X and $E \subset Y(\overline{K})$, then E is bounded in Y if E is bounded in X. The details of the proofs will be left to the reader.

Example 2.6.4. In this example, we assume that K is locally compact with respect to $| \ |$ (for example, the completion of a number field with respect to a place, cf. 1.2.12). We consider on $X(K)$ the topology induced locally by open balls with respect to closed embeddings into affine spaces and the maximum norm. Then the topology is locally compact and independent of the embeddings. It depends only on the place v represented by $| \ |$ and is called the v-topology on $X(K)$. A subset E of $X(K)$ is bounded in X if and only if E is relatively compact in $X(K)$.

In order to prove this statement, suppose first that E is bounded in X. By definition, E may be covered by finitely many closed balls in affine spaces. Since these balls are compact, we conclude that E is relatively compact.

On the other hand, let E be relatively compact in X. Passing to the closure, we may assume E compact. Then E may be covered by finitely many open balls in affine spaces. By Lemmas 2.2.9 and 2.2.10, E is bounded in X.

Example 2.6.5. The set $\mathbb{P}^n(\overline{K})$ is bounded in projective space $\mathbb{P}^n_K$. We can use the affine covering $X_i := \{\mathbf{x} \in \mathbb{P}^n_K \mid x_i \neq 0\}$, $i \in \{0, \ldots, n\}$, and the decomposition $E_i := \{\mathbf{x} \in \mathbb{P}^n_K \mid |x_i| = \max_{j=0,\ldots,n} |x_j|\}$ of E. By Remark 2.6.3, the set of $\overline{K}$-rational points is bounded in any projective variety. Implicitly, we have already used these facts in the proof of Theorem 2.2.11.

Proposition 2.6.6. *If X is a complete variety over K, then $X(\overline{K})$ is bounded in X. More generally, the inverse image of a bounded subset under a proper morphism remains bounded.*

Proof: By Chow's lemma in A.9.37, there is a projective variety Y over K and a surjective birational morphism $Y \longrightarrow X$. Using Remark 2.6.3 and Example 2.6.5, $X(\overline{K})$ has to be bounded in X.

More generally, let $\varphi : X' \to X$ be a proper morphism of arbitrary varieties over K and let $E \subset X(\overline{K})$ be bounded in X. By Chow's lemma again, the proof is reduced to the case of a projective morphism and hence to $X' = X \times \mathbb{P}^n_K$ with φ the first projection. By Remark 2.6.3, we may further assume X affine and then the same arguments as in Example 2.6.5 prove that $\varphi^{-1}(E)$ is bounded in $X \times \mathbb{P}^n_K$. □

Remark 2.6.7. It is trivial that any subset of a bounded subset is bounded again. However, we may not pass from X to an open subset. For example, the set $E = \{\mathbf{x} \in \mathbb{P}^n(\overline{K}) \mid x_0 \neq 0\}$ is bounded in $\mathbb{P}^n_K$ by Example 2.6.5 but it is certainly not bounded in the affine space $\{\mathbf{x} \in \mathbb{P}^n_K \mid x_0 \neq 0\}$. Thus the notion of bounded subset is not a local one and some care is needed with using it.

Definition 2.6.8. *A real function f on a K-variety X is called **locally bounded** if $f(E)$ is bounded for every bounded E in X.*

Remark 2.6.9. The locally bounded functions on X form an $\mathbb{R}$-algebra. Proposition 2.6.6 shows that a real function on a complete K-variety is locally bounded if and only if it is bounded.

2.6.10. To apply this later on to the theory of heights, we need a generalization to several absolute values. Let M_K be a set of places on K. For every $v \in M_K$, an absolute value $|\ |_v$ is fixed in the equivalence class of v. We assume that

$$\{v \in M_K \mid |\alpha|_v \neq 1\}$$

is finite for every $\alpha \in K \setminus \{0\}$. Let M be a set of places on $\overline{K}$. We assume that every $u \in M$ restricts to a $v \in M_K$ and we denote by $|\ |_u$ the unique extension of $|\ |_v$ to an absolute value representing u.

Definition 2.6.11. *Let U be an affine K-variety and let $(E^u)_{u \in M}$ be a family of subsets of $U(\overline{K})$. The family is said to be* ***M-bounded in*** *U if for any $f \in K[U]$ the quantity*

$$C_v(f) := \sup_{u \in M, u|v} \sup_{P \in E^u} |f(P)|_u$$

is finite for every $v \in M_K$ and $C_v(f) > 1$ for only finitely many v.

If M has only one element, then $E \subset U(\overline{K})$ is M-bounded in U if and only if E is bounded in U in the sense of Definition 2.2.8.

Remark 2.6.12. We note that Lemmas 2.2.9 and 2.2.10 extend to the situation with several absolute values instead of one. Indeed, if we replace in their formulations E by a family $(E^u)_{u \in M}$ of subsets and "bounded" by "M-bounded," then these statements continue to hold.

The straightforward generalization of the proofs is left to the reader. Implicitly, this was already used in the proof of Theorem 2.2.11.

Definition 2.6.13. *Still under the hypothesis of 2.6.10, let X be a K-variety and let $(E^u)_{u \in M}$ be a family of subsets of $X(\overline{K})$. The family is called* ***M-bounded in*** *X if there is a finite covering $\{U_i\}_{i \in I}$ of X by affine open subsets and, for any $u \in M$, a decomposition*

$$E^u = \bigcup_{i \in I} E_i^u, \quad E_i^u \subset U_i(\overline{K}),$$

such that $(E_i^u)_{u \in M}$ is M-bounded in U_i in the sense of Definition 2.6.11 for every $i \in I$.

Remark 2.6.14. If we make the same changes in Remark 2.6.3 as proposed in Remark 2.6.12, then Remark 2.6.3 still holds. If M has only one element, then $E \subset X(\overline{K})$ is M-bounded if and only if E is bounded in the sense of Definition 2.6.2.

Definition 2.6.15. *A subset $E \subset X(\overline{K})$ is called* ***M-bounded in*** *X if the constant family $(E)_{u \in M}$ is M-bounded.*

Example 2.6.16. The set $\mathbb{P}^n(\overline{K})$ is M-bounded in $\mathbb{P}^n_K$. The same covering and the same decomposition as in Example 2.6.5 work. Note that the old E_i now depend on $u \in M$. Even by starting with a single set, we have to work with families.

Proposition 2.6.17. *A complete K-variety is M-bounded. More generally, the inverse image of an M-bounded family of subsets under a proper morphism is M-bounded.*

Proof: We leave it to the reader to make the obvious adjustments in the proof of Proposition 2.6.6. □

Definition 2.6.18. *A real function f on $X \times M$ is called* ***locally M-bounded*** *if, for any M-bounded family $(E^u)_{u \in M}$ in X, there is for every $v \in M_K$ a non-negative real number γ_v, with $\gamma_v \neq 0$ only for finitely many $v \in M_K$, such that for all $u \in M$ with $v|u$ we have*

$$|f(E^u, u)| \leq \gamma_v.$$

Example 2.6.19. If a is a regular function on an affine variety, then the function $(x, u) \mapsto |a(x)|_u$ is not necessarily locally bounded because infinitely many bounds may be different from 0. But, if a is nowhere vanishing, then the function $(x, u) \mapsto \log |a(x)|_u$ is locally bounded. This is the typical situation where we apply this notion.

2.7. Metrized line bundles and local heights

This section contains additional material for local and global heights. It will be used only in Sections 9.5 and 9.6 and may be skipped in a first reading. In Sections 2.2–2.5, we have studied Weil heights coming from morphisms to projective space. However, as remarked by A. Néron and as we will see in 9.5, the canonical height functions on abelian varieties are not of this shape. To deal with this situation, a local height is associated to every locally bounded metric of a line bundle on an arbitrary variety. We will extend the results from Sections 2.2 and 2.3 to this framework.

Let K be a field with a fixed embedding into its algebraic closure $\overline{K}$. For the moment, we fix an an absolute value $|\ |$ on $\overline{K}$. From Definition 2.7.12 on, we deal with several absolute values. In order to define global heights, we will assume in 2.7.16–2.7.19 that the product formula holds.

Definition 2.7.1. *Let L be a line bundle on the K-variety X. A* **metric** *on L is a norm $\| \ \|$ on every fibre L_x, $x \in X$,* i.e. *a a real function not identically zero such that*

$$\|\lambda v\| = |\lambda| \cdot \|v\| \quad (\lambda \in \overline{K}, v \in L_x).$$

The pair $(L, \| \ \|)$ is called a **metrized line bundle** *usually denoted by $\overline{L}$. The metric is said to be* **locally bounded** *if $\log \|s\|$ is locally bounded on U for every open subset U of X and every nowhere vanishing section $s \in L(U)$.*

Example 2.7.2. Let f be a regular function on X. Then almost by definition $\log |f|$ is locally bounded on $\{x \in X \mid f(x) \neq 0\}$. The **trivial metric** on O_X is characterized by $\|1\| = |1|$. We conclude that the trival metric is locally bounded.

2.7.3. Let $\overline{L} = (L, \| \ \|), \overline{M} = (M, \| \ \|)$ be metrized line bundles on the K-variety X. Then the tensor product $\overline{L} \otimes \overline{M}$ is the metrized line bundle $(L \otimes M, \| \ \|)$, where the metric on $L \otimes M$ is given by

$$\|v \otimes w\| := \|v\| \cdot \|w\| \quad (v \in L_x, w \in M_x)$$

for any $x \in X$. Two metrized line bundles on X are called **isometric** if there is an isomorphism which is fibrewise an isometry. The isometry classes of line bundles on X form a group $\widehat{\mathrm{Pic}}(X)$. The identity is O_X with the trivial metric and the inverse of $\overline{L}$ is $(L^{-1}, \| \ \|^{-1})$. The locally bounded metrized line bundles obviously form a subgroup.

For a morphism $\varphi : X' \to X$ of varieties over K, we define the **pull-back** $\varphi^*(\overline{L}) = (\varphi^* L, \| \ \|)$ as the metrized line bundle on X' with metric on $\varphi^*(L)$ characterized by $\|\varphi^*(s)\| = \|s\| \circ \varphi$ for any open subset U of X and $s \in L(U)$. The pull-back induces a group homomorphism from $\widehat{\mathrm{Pic}}(X)$ to $\widehat{\mathrm{Pic}}(X')$. By Remark 2.6.3, the pull-back of a locally bounded metrized line bundle remains locally bounded.

Example 2.7.4. On $O_{\mathbb{P}^n_K}(1)$, we have the **standard metric** characterized by

$$\|\ell(\mathbf{x})\| = \frac{|\ell(\mathbf{x})|}{\max_{j=0,\ldots,n} |x_j|}$$

for any linear form ℓ in the coordinates $x_0, \ldots, x_n$. If $| \ |$ is archimedean, we often use the **Fubini–Study metric**

$$\|\ell(\mathbf{x})\|_2 = \frac{|\ell(\mathbf{x})|}{\left(\sum_{j=1}^n |x_j|^2\right)^{1/2}}.$$

We claim that both metrics are locally bounded. Let s be a nowhere vanishing section of $O_{\mathbb{P}^n_K}(1)$ over an open subset U of $\mathbb{P}^n_K$ and let E be bounded in U. By Remark 2.6.3, we may assume that U is contained in a standard open subset $\{\mathbf{x} \in \mathbb{P}^n_K \mid x_j \neq 0\}$. Then $s = f x_j$ for an invertible regular function f on U.

By definition, $\log|f|$ is bounded on E. For $\log\|x_j\|$ (resp. $\log\|x_j\|_2$), we have the upper bound 0. A lower bound is easily obtained by using $|x_i/x_j|$ bounded on E for every $i \in \{0,\ldots,n\}$. This proves the claim.

Proposition 2.7.5. *Every line bundle on an arbitrary variety over K admits a locally bounded metric.*

Proof: For simplicity, we first prove the claim for a projective variety X. We may assume that our line bundle L is generated by global sections $s_0,\ldots,s_m$ because every element in $\mathrm{Pic}(X)$ is the difference of two very ample ones (see A.6.10). Then the morphism $\varphi : X \to \mathbb{P}_K^m$, given by $\varphi(x) = (s_0(x) : \cdots : s_m(x))$, satisfies $\varphi^* O_{\mathbb{P}^m}(1) = L$. The pull-back of the standard metric (cf. Example 2.7.4) is locally bounded. For further reference, we note that

$$\|s_i(x)\| = \min_j \left|\frac{s_i}{s_j}(x)\right|. \tag{2.16}$$

Now let L be any line bundle on an arbitrary variety X over K. We cover X by finitely many affine trivializations $\{U_i\}_{i=1,\ldots,m}$ of L. Let s_i be a nowhere vanishing section of L over U_i. A first try to define the metric of L on U_j would be formula (2.16) with j restricted to the ones with $x \in U_j$. Clearly, $\log\|s_i\|$ would be bounded from above by 0 on U_i. But poles of s_j along $U_i \setminus U_j$ would make it impossible to give a lower bound on every bounded subset of U_i. The following smoothing process avoids this problem.

Let $x_{i1},\ldots,x_{in_i}$ be coordinates on U_i. Note first that $x_{i1},\ldots,x_{in_i},x_{j1},\ldots,x_{jn_j}$ are coordinates on $U_i \cap U_j$. This is clear by realizing $U_i \cap U_j$ as the closed affine subvariety of $U_i \times U_j$ given by intersection with the diagonal (which is closed by definition of a variety).

For notational simplicity, we add $x_{i0} = 1$ to the coordinates on every U_i. We consider the transition function $g_{ji} = \frac{s_j}{s_i}$ of L on $U_i \cap U_j$. Because g_{ji} may be written as a polynomial in the coordinates $\mathbf{x}_i, \mathbf{x}_j$, we have a constant C_1 and $r \in \mathbb{N}$ such that

$$|g_{ji}(x)| \leq C_1 \max_{k,l}(|x_{ik}|,|x_{jl}|)^r \leq C_1 \max_k |x_{ik}|^r \cdot \max_l |x_{jl}|^r \tag{2.17}$$

for every $x \in U_i \cap U_j$. Note that we may choose C_1 and r independently of i,j. Now we set

$$\|s_i(x)\| = \min_j \max_l |g_{ij}(x) x_{jl}^r| \tag{2.18}$$

for every $x \in U_i$. Again j ranges over all elements of $\{0,\ldots,m\}$ with $x \in U_j$. If we use the cocycle rule $g_{hj} = g_{hi} g_{ij}$, it is clear that

$$\|s_h(x)\| = |g_{hi}(x)| \cdot \|s_i(x)\|$$

for every $x \in U_h \cap U_i$. Hence (2.18) characterizes a well-defined metric of L on X.

To prove that this metric is locally bounded, let E be a bounded subset of an open subset U of X and let s be a nowhere vanishing section of L over U. By Remark 2.6.3, we may assume that $E \subset U_i$ for some $i \in \{1, \ldots, m\}$. By definition, it is clear that $\log |s/s_i|$ is bounded on E. Since

$$\log \|s\| = \log \left| \frac{s}{s_i} \right| + \log \|s_i\|$$

on E, it follows that we may assume $U = U_i, s = s_i$. There is a constant C_2 such that

$$\max_{k=0,\ldots,n_i} |x_{ik}| \leq C_2 \tag{2.19}$$

for every $x \in E$. This leads to the upper bound

$$\log \|s_i(x)\| \leq \log \max_l |g_{ii}(x) x_{il}^r| \leq r \log C_2.$$

Using $x_{i0} = x_{j0} = 1$, $g_{ij} g_{ji} = 1$, formulas (2.17) and (2.19), we get

$$\max_l |g_{ij}(x) x_{jl}^r| \geq \frac{1}{C_1} \left(\max_k |x_{ik}| \right)^{-r} \geq \frac{1}{C_1 C_2^r}$$

leading to the lower bound

$$\log \|s_i(x)\| \geq -\log C_1 - r \log C_2$$

on E. This proves the claim. □

Remark 2.7.6. It is sometimes useful to impose additional requirements on the metrics. In the archimedean case, it is often convenient to require that the functions $\|s\|$ be C^∞ for every open subset U and every nowhere vanishing $s \in L(U)$. In this case, we should work with the metric $\| \; \|_2$ on $O_{\mathbb{P}^n_K}(1)$, because the standard metric is not differentiable.

For a non-archimedean absolute value $| \; |$ on $\overline{K}$ with place u, it is natural to assume that $\|s(x)\| \in |K(x)|$ for every $x \in X$. Moreover, we may assume that the functions $\|s\|$ are continuous with respect to the u-topology on $U(\overline{K})$ defined in Example 2.6.4.

All the results above remain valid for this kind of metrics. In particular, such metrics always exist on any line bundle. In the non-archimedean case, the metric constructed in the proof of Proposition 2.7.5 has these additional properties. In the archimedean case, the existence of a C^∞-metric follows from a partition of unity argument.

In the non-archimedean case, continuity is not so important because the u-topology is totally disconnected. If K is not locally compact, then continuity does not necessarily imply local boundedness. At any rate, the only relevant property for our purposes is that the metrics we use are locally bounded.

2.7.7. In the situation of Example 2.7.4, we consider the presentation

$$\mathcal{D} = (\ell(\mathbf{x}); O_{\mathbb{P}^n_K}(1), x_0, \dots, x_n; O_{\mathbb{P}^n_K}(1))$$

of the hyperplane $D = \operatorname{div}(\ell(\mathbf{x}))$. Then the local height $\lambda_{\mathcal{D}}$ is given by

$$\lambda_{\mathcal{D}}(P) = \max_k \log \left| \frac{x_k}{\ell(\mathbf{x})} \right|$$

for any $P = \mathbf{x} \notin \operatorname{supp}(D)$ (as in Example 2.3.2). We conclude that

$$\lambda_{\mathcal{D}}(P) = -\log \|\ell(\mathbf{x})\|$$

depends only on D and the standard metric $\| \ \|$ on $O(D) = O_{\mathbf{P}^n}(1)$. This is used in what follows to generalize the concept of local heights, replacing the presentation by a suitable locally bounded metric on $O(D)$.

2.7.8. A **Néron divisor** $\widehat{D}$ on the K-variety X is a Cartier divisor D on X with a locally bounded metric on the line bundle $O(D)$. The corresponding locally bounded metrized line bundle is denoted by $O(\widehat{D})$. Note that D induces a canonical meromorphic section s_D of $O(D)$ and we have $D = D(s_D)$ (cf. A.8.18).

The Néron divisors on X form a group with composition law

$$\widehat{D} + \widehat{E} = \left(D + E, O(\widehat{D}) \otimes O(\widehat{E})\right).$$

Let $\varphi : X' \to X$ be a morphism of K-varieties such that no irreducible component of X' is mapped into $\operatorname{supp}(D)$. Then the pull-back $\varphi^*(D)$ is a well-defined Cartier divisor on X' (see A.8.26) and we define the **pull-back** $\varphi^*(\widehat{D})$ as the Néron divisor

$$\varphi^*(\widehat{D}) = \left(\varphi^*(D), \varphi^* O(\widehat{D})\right).$$

Definition 2.7.9. *The* ***local height*** *associated to the Néron divisor* $\widehat{D} = (D, \| \ \|)$ *on the variety* X *is given by*

$$\lambda_{\widehat{D}}(P) := -\log \|s_D(P)\|, \quad P \in X \setminus \operatorname{supp}(D).$$

Proposition 2.7.10. *Let* $\widehat{D} = (D, \| \ \|)$ *be a Néron divisor on the* K*-variety* X.

(a) *If* $\widehat{E}$ *is a Néron divisor on* X, *then*

$$\lambda_{\widehat{D}+\widehat{E}}(P) = \lambda_{\widehat{D}}(P) + \lambda_{\widehat{E}}(P), \quad P \notin \operatorname{supp}(D) \cup \operatorname{supp}(E).$$

(b) *If* $\varphi : X' \to X$ *is a* K*-morphism such that no irreducible component of the* K*-variety* X' *is mapped into* $\operatorname{supp}(D)$, *then*

$$\lambda_{\widehat{D}} \circ \varphi(P') = \lambda_{\varphi^*(\widehat{D})}(P'), \quad P' \in X' \setminus \varphi^{-1}\operatorname{supp}(D).$$

(c) *If f is a rational function on X, not identically zero on any irreducible component, and $\widehat{D}(f)$ denotes the Néron divisor with trivial metric on $O(D(f)) = O_X$, then*

$$\lambda_{\widehat{D}(f)}(P) = -\log|f(P)|, \quad P \notin \mathrm{supp}(D).$$

(d) *If $\| \ \|$ is another locally bounded metric on $O(D)$, then*

$$\lambda_{(D,\| \ \|')}(P) - \lambda_{(D,\| \ \|)}(P) = \log \rho,$$

where ρ is the norm of $1 \in \Gamma(X, O_X)$ with respect to the locally bounded metric $\| \ \|/\| \ \|'$.

Proof: (a)–(c) are immediate from the definitions and (d) follows from (a). □

Proposition 2.7.11. *Let $\mathcal{D} = (s_D;\, L, \mathbf{s};\, M, \mathbf{t})$ be a presentation of the Cartier divisor $D = D(s_D)$ (cf. 2.2.1). Then there is a unique locally bounded metric on $O(D) = L \otimes M^{-1}$ given on a local section s by*

$$\|s(x)\| = \min_k \max_l \left| \frac{st_l}{s_k}(x) \right|.$$

The local height $\lambda_{\mathcal{D}}$ (cf. 2.2.2) is equal to the local height $\lambda_{(D,\| \ \|)}$ with respect to the Néron divisor $(D, \| \ \|)$.

Proof: By linearity, we may assume that $O(D) = L$ and $t_0, \ldots, t_m = 1$. Then the metric is the pull-back of the standard metric on $O_{\mathbb{P}^n_K}(1)$ and we apply (2.16) on page 59. □

Again, we generalize our considerations to several absolute values. Let us use the same assumptions and notation as in 2.6.10.

Definition 2.7.12. *An **M-metric** on a line bundle L is a family $(\| \ \|_u)_{u \in M}$ such that $\| \ \|_u$ is a metric with respect to the absolute value $| \ |_u$ satisfying the following compatibility condition:*

For every $x \in X$ and $u_1, u_2 \in M$ with the same restriction to $K(x)$, the equation $\| \ \|_{u_1} = \| \ \|_{u_2}$ holds on $L_x(K(x))$.

*An M-metric is called **locally bounded** if for every nowhere vanishing section s of L on an open subset U of X, the function $(x, u) \mapsto \log \|s(x)\|_u$ on $U \times M$ is locally M-bounded.*

2.7.13. Now a Néron divisor $\widehat{D}$ is a Cartier divisor D with a locally bounded M-metric on $O(D)$. All results from 2.7.2 to 2.7.11 hold in this context, replacing metrics by M-metrics. We leave the details to the reader. Proposition 2.7.10 (d) and Proposition 2.6.17 give our main result:

Theorem 2.7.14. *Let D be a Cartier divisor on a complete variety X over K and let $(\| \ \|_u)_{u \in M}$, $(\| \ \|'_u)_{u \in M}$ be locally M-bounded metrics on $O(D)$ giving* *rise to the local heights*

$$\lambda_{\widehat{D}}(P,u) := -\log \|s_D(P)\|_u, \quad \lambda_{\widehat{D}'}(P,u) := -\log \|s_D(P)\|'_u.$$

For $v \in M_K$, there are constants $\gamma_v \in \mathbb{R}$, with $\gamma_v \neq 0$ only for finitely many v, such that

$$|\lambda_{\widehat{D}'}(P,u) - \lambda_{\widehat{D}}(P,u)| \leq \gamma_v$$

for all $P \in X \setminus \mathrm{supp}(D)$.

2.7.15. We apply these results to a non-zero rational function f on an irreducible regular projective variety X over K. Then we have

$$\mathrm{div}(f) = \sum_{j=1}^{n} m_j Y_j$$

for prime divisors Y_j. By Proposition 2.3.9, there is a non-negative local height λ_j relative to Y_j. For any place u on $\overline{K}$, $\exp(-\lambda_j(P,u))$ measures the u-distance from $P \in X(\overline{K})$ to Y_j. If we choose the local height λ_j relative to a presentation as in Proposition 2.3.9, then this distance is even continuous with respect to the u-topology on $X(\overline{K})$ defined as in Example 2.6.4. By the generalization of Proposition 2.7.10 and Theorem 2.7.14, there is a family $(\gamma_v)_{v \in M_K}$ of non-negative real numbers, with $\gamma_v = 0$ for all but finitely many v, such that

$$\sum_{j=1}^{n} m_j \lambda_j(P,u) - \gamma_v \leq -\log|f(P)|_u \leq \sum_{j=1}^{n} m_j \lambda_j(P,u) + \gamma_v$$

for any $u \in M$, $u|v \in M_K$, and any $P \in X(\overline{K}) \setminus \mathrm{supp}(\mathrm{div}(f))$. This is **Weil's theorem of decomposition.**

2.7.16. In order to define global heights, we assume that the absolute values of M_K satisfy the product formula (cf. 1.4). Let F/K be a subextension of $\overline{K}/K$ and denote by M_F the set of places w of the field F with $w|v$ for some $v \in M_K$. We normalize the absolute value $| \ |_w$ as in 1.3.6 and 1.3.12. Then for $u \in M$ we have

$$| \ |_w = | \ |_u^{[T_w:K_v]/[F:K]},$$

where T_w is the completion F_w if F/K is separable.

2.7.17. Let $\overline{L}$ be a locally bounded M-metrized line bundle on the K-variety X. Our goal is to define the **associated global height function** $h_{\overline{L}}$.

Let $P \in X(\overline{K})$. We choose a finite subextension F/K of $\overline{K}/K$ with $P \in X(F)$. For every $w \in M_F$, let us choose any $u \in M$ with $u|w$. Then we consider the $| \ |_w$-norm

$$\| \ \|_w := \| \ \|_u^{[T_w:K_v]/[F:K]}$$

on the fibre $L_P(F)$. By the compatibility condition in the definition of an M-metric, this norm is independent of the choice of u.

There is an invertible meromorphic section s of L with $P \notin \operatorname{supp}(D(s))$ (because there is an open dense trivialization of L in a neighborhood of P). Then the M-metric on L yields a Néron divisor $\widehat{D}(s)$ and we set

$$\lambda_{\widehat{D}(s)}(P,w) := -\log \|s(P)\|_w, \quad h_{\overline{L}}(P) := \sum_{w \in M_F} \lambda_{\widehat{D}(s)}(P,w).$$

By (1.4) on page 9 (resp. Lemma 1.3.7), the definition of $h_{\overline{L}}(P)$ is independent of the choice of F. From the product formula and Proposition 2.7.10 (c), it is clear that $h_{\overline{L}}(P)$ does not depend on the choice of s.

Proposition 2.7.18. *The global height functions above have the following properties:*

(a) $h_{\overline{L}}$ *depends only on the isometry class of* $\overline{L}$.

(b) *If* $\overline{L_1}, \overline{L_2}$ *are locally bounded* M*-metrized line bundles on* X*, then*

$$h_{\overline{L_1} \otimes \overline{L_2}} = h_{\overline{L_1}} + h_{\overline{L_2}}.$$

(c) *Let* $\varphi : X' \to X$ *be a morphism of* K*-varieties. Then* $h_{\varphi^* \overline{L}} = h_{\overline{L}} \circ \varphi$.

(d) *Let* $\mathcal{D}$ *be a presentation of a Cartier divisor* D *and let* $\widehat{\mathcal{D}}$ *be the associated Néron divisor from Proposition 2.7.11. Then the global height with respect to* $\mathcal{D}$ *in 2.3.3 is equal to* $h_{O(\widehat{\mathcal{D}})}$.

(e) *If* X *is complete (or more generally* M*-bounded), then* $h_{\overline{L}}$ *does not depend on the choice of the locally bounded* M*-metric up to bounded functions.*

Proof: Property (a) is obvious and (b), (c) follow from Proposition 2.7.10. Claim (d) is an immediate consequence of Proposition 2.7.11 and (e) follows from Theorem 2.7.14 or from Proposition 2.7.10 (d). □

Remark 2.7.19. By the generalization of Proposition 2.7.5 to several absolute values, every line bundle L admits a locally bounded M-metric. On complete varieties, it follows from Proposition 2.7.18 that the corresponding global height depends only on the isomorphism class of L up to bounded functions. Moreover, Theorem 2.3.8 holds for arbitrary complete varieties X, Y over K.

In practice, metrics often arise from integral models, as it will transpire in the following example which assumes that the reader is familiar with the basics of scheme theory. As the example plays no further role in this book, the reader may skip it without problems.

Example 2.7.20. Let R be a Dedekind domain with quotient field K. For every maximal ideal $\wp_v$ of R, we fix a discrete valuation $|\ |_v$ on K with $\wp_v = \{x \in R \mid |x|_v < 1\}$. Again, M is the set of absolute values on $\overline{K}$ extending those of M_K.

We consider a line bundle $\mathcal{L}$ on a flat proper reduced scheme $\mathcal{X}$ over R. The generic fibre $X = \mathcal{X}_K$ is a complete variety over K with line bundle $L := \mathcal{L}_K$. The goal is to describe the natural M-metric $\| \ \|_{\mathcal{L}}$ on L induced from $\mathcal{L}$.

Let $x \in X$. Then there is a finite subextension F/K of $\overline{K}/K$ with $x \in X(F)$. The integral closure R_F of R in F is again a Dedekind domain ([**157**], Th.10.7). By the valuative criterion of properness ([**148**], Th.II.4.7), there is a unique morphism $\overline{x} : \operatorname{Spec}(R_F) \to \mathcal{X}$ mapping the generic point $\{0\}$ to x. Let $u \in M$ restricting to F to the valuation w on F with corresponding prime ideal $\wp_w$. There is a local nowhere vanishing section s of $\mathcal{L}$ in $\overline{x}(\wp_w)$. Then s is also defined and non-zero in x (because $\overline{x}(\wp_w)$ is in the closure of x in $\mathcal{X}$) and we set

$$\|s(x)\|_{\mathcal{L},u} := 1. \tag{2.20}$$

If we replace s by another local nowhere vanishing section s' in $\overline{x}(\wp_w)$, then s'/s is a unit in the localization of R_F in $\wp_w$ (consider the stalk of the sheaf of sections of $\mathcal{L}$ in $\overline{x}(\wp_w)$). We conclude that (2.20) determines a well-defined M-metric on L.

To prove that this M-metric is locally bounded, let $(E^u)_{u \in M}$ be an M-bounded family in an open subset U of X and let $s \in L(U)$ be nowhere vanishing. We may cover $\mathcal{X}$ by finitely many affine trivializations $\mathcal{U}_i$ of $\mathcal{L}$ with nowhere vanishing $s_i \in \mathcal{L}(U_i)$. For $u \in M$, we define

$$E_i^u := \{x \in E^u \mid \overline{x}(\wp_w) \in \mathcal{U}_i\},$$

using the notation from above. Clearly, E_i^u is contained in the generic fibre U_i of $\mathcal{U}_i$ and we have $E^u = \bigcup_i E_i^u$. For $x \in E_i^u$, we have $\|s_i(x)\|_{\mathcal{L},u} = 1$ and hence

$$\log \|s(x)\|_{\mathcal{L},u} = \log \left| \frac{s}{s_i}(x) \right|_u .$$

So it is enough to prove that $(E_i^u)_{u \in M}$ is M-bounded in $U \cap U_i$. The coordinates on $U \cap U_i$ are given by the coordinates of U and U_i (cf. proof of Proposition 2.7.5). Clearly, the coordinates of U_i are M-bounded on $(E_i^u)_{u \in M}$. Let $x_1, \ldots, x_n$ be a set of coordinates on $\mathcal{U}_i$ over R. Then they are also coordinates of U_i and $|x_j|_u \leq 1$ on E_i^u for $j = 1, \ldots, n$ (using $\overline{x}(\wp_w) \in \mathcal{U}_i$). By the generalization of Lemma 2.2.10, $(E_i^u)_{u \in M}$ has to be M-bounded in $U \cap U_i$ proving that our M-metric is locally bounded.

Let s be an invertible meromorphic section of L. The M-metric $\| \ \|_{\mathcal{L}}$ induces a Néron divisor $\widehat{D}$ on $D = \operatorname{div}(s)$ and a local height

$$\lambda_{\widehat{D}}(x, u) := -\log \|s(x)\|_{\mathcal{L},u}$$

for $x \in X \setminus \operatorname{supp}(D)$ and $u \in M$.

Now let $x \in X(K) \setminus \operatorname{supp}(D)$ and let $u \in M$ with $u|v \in M_K$. We will give a geometric interpretation of $\lambda_{\widehat{D}}(x, u)$ in terms of intersection multiplicities. By renormalization, we may assume that $\log |K^\times| = \mathbb{Z}$.

There is a similar intersection theory of Cartier divisors on $\mathcal{X}$ with cycles, as described in Section A.9 (cf. [**125**], 20.2). By flatness, we easily deduce that X is dense in $\mathcal{X}$ and that s extends uniquely to an invertible meromorphic section $s_{\mathcal{L}}$ of $\mathcal{L}$. On the other hand, the closure $\overline{\{x\}}$ of x in $\mathcal{X}$ is a one-dimensional prime cycle on $\mathcal{X}$. Since $x \notin \operatorname{supp}(D)$, the proper intersection product

$$D(s_{\mathcal{L}}).\overline{\{x\}} = \operatorname{div}\left(s_{\mathcal{L}}|_{\overline{\{x\}}}\right)$$

is a well-defined zero-dimensional cycle $\mathcal{Z}$ on $\mathcal{X}$. For $v \in M_K$, let $\mathcal{Z}_v$ be the part of $\mathcal{Z}$ contained in the fibre $\mathcal{X}_{\wp_v}$, which is a variety over the residue field and let $\deg_v(\mathcal{Z})$ be the degree of $\mathcal{Z}_v$. We claim that

$$\lambda_{\widehat{D}}(x,u) = \deg_v\left(D(s_{\mathcal{L}}).\overline{\{x\}}\right).$$

Indeed, by projection formula, we have

$$\deg_v\left(D(s_{\mathcal{L}}).\overline{\{x\}}\right) = \deg_v\left(\overline{x}^* D(s_{\mathcal{L}}).\mathrm{Spec}(R_F)\right).$$

Using Proposition 2.7.10, we get

$$\lambda_{\widehat{D}}(x,u) = \lambda_{\overline{x}^*\widehat{D}}(\{0\},u),$$

thus we may assume $\mathcal{X} = \mathrm{Spec}(R_F), \mathcal{L} = O_{\mathcal{X}}$. Then $s \in K^\times$ and by our normalization of $|\ |_v$, we get

$$\deg_v(\operatorname{div}(s_{\mathcal{L}})) = -\log|s|_v = \lambda_{\widehat{D}}(x,u),$$

as claimed.

2.8. Heights on Grassmannians

Let M denote the set of absolute values on $\overline{\mathbb{Q}}$ extending $M_{\mathbb{Q}}$.

There are other choices for the height on $\mathbb{P}^n_{\overline{\mathbb{Q}}}$ (see Example 2.7.4 and Remark 2.7.6). In Arakelov theory, the following height is more natural.

Definition 2.8.1. *For* $\mathbf{x} \in \overline{\mathbb{Q}}^{n+1}$ *and* $u \in M$, *we set*

$$H_u(\mathbf{x}) := \begin{cases} \max_j |x_j|_u & \text{if } u \text{ is non-archimedean,} \\ \left(\sum_{j=0}^n |x_j|_u^2\right)^{1/2} & \text{if } u \text{ is archimedean.} \end{cases}$$

Let $F \subset \overline{\mathbb{Q}}$ *be a number field. Then, for* $\mathbf{x} \in F^{n+1}$ *and* $w \in M_F$, *we define*

$$H_w(\mathbf{x}) := H_u(\mathbf{x})^{[F_w:\mathbb{Q}_p]/[F:\mathbb{Q}]},$$

where $p \in M_{\mathbb{Q}}$ *and* $u \in M$ *are such that* $w|p$ *and* $u|w$.

2.8.2. Now we define the **Arakelov height** for $P \in \mathbb{P}^n_{\overline{\mathbb{Q}}}(F)$ with representative $\mathbf{x} \in F^{n+1}$ by

$$h_{\mathrm{Ar}}(P) := \sum_{w \in M_F} \log H_w(\mathbf{x}).$$

As always, the **multiplicative Arakelov height** is defined by $H_{\mathrm{Ar}}(P) := \exp(h_{\mathrm{Ar}}(P))$.

2.8.3. Again by the product formula, $h_{\mathrm{Ar}}(P)$ is independent of the choice of the representative $\mathbf{x}$. It follows from Corollary 1.3.2 that the definition $h_{\mathrm{Ar}}(P)$ does not depend on the extension F, i.e. h_{Ar} is a well-defined function on $\mathbb{P}^n_{\overline{\mathbb{Q}}}$. It is easy to see that h_{Ar} differs from the old height h by a bounded function on $\mathbb{P}^n_{\overline{\mathbb{Q}}}$. In the sense of 2.3.7, they are equivalent.

The preceding remark also follows from Proposition 2.7.18, because h_{Ar} is the global height associated to the locally bounded metrized line bundle on $O_{\mathbb{P}^n}(1)$ using the Fubini–Study metrics at the archimedean places and the standard metrics at the non-archimedean places (cf. 2.7.4).

2.8.4. Let W be an m-dimensional subspace of $\overline{\mathbb{Q}}^n$. The mth exterior power $\wedge^m W$ is a one-dimensional subspace of $\wedge^m \overline{\mathbb{Q}}^n$. Therefore, we may view W as a point P_W of the projective space $\mathbb{P}(\wedge^m \overline{\mathbb{Q}}^n)$. The latter may be identified with projective space of dimension $\binom{n}{m}$ by using standard coordinates.

Definition 2.8.5. *$h_{\mathrm{Ar}}(W) := h_{\mathrm{Ar}}(P_W)$ is called the* ***Arakelov height of W***.

Definition 2.8.6. *Let A be an $n \times m$ matrix of rank m with entries in $\overline{\mathbb{Q}}$. Then the* ***Arakelov height $h_{\mathrm{Ar}}(A)$ of A*** *is defined as the Arakelov height of the subspace of $\overline{\mathbb{Q}}^n$ spanned by the columns of A.*

Remark 2.8.7. We can be quite explicit in defining $h_{\mathrm{Ar}}(A)$. Let $I \subset \{1, \ldots, n\}$ with $|I| = m$. We denote by A_I the $m \times m$ submatrix of A formed with the ith rows, $i \in I$, of A. Then the point in $\mathbb{P}(\wedge^m \overline{\mathbb{Q}}^n)$ corresponding to A is given by the coordinates $\det(A_I)$, where I ranges over all subsets of $\{1, \ldots, n\}$ of cardinality m. Let $F \subset \overline{\mathbb{Q}}$ be a number field containing all entries of A and set

$$H_u(A) := \begin{cases} \max_I |\det(A_I)|_u & \text{if } u \text{ is non-archimedean,} \\ \left(\sum_I |\det(A_I)|^2 \right)^{1/2} & \text{if } u \text{ is archimedean.} \end{cases}$$

For $w \in M_F$, we now set

$$H_w(A) = H_u(A)^{[F_w:\mathbb{Q}_p]/[F:\mathbb{Q}]},$$

where $p \in M_{\mathbb{Q}}$ and $u \in M$ are such that $w|p$ and $u|w$. Then we have

$$h_{\mathrm{Ar}}(A) = \sum_{w \in M_F} \log H_w(A).$$

It is also clear that $h_{\mathrm{Ar}}(AG) = h_{\mathrm{Ar}}(A)$ for any invertible $m \times m$ matrix G with entries in $\overline{\mathbb{Q}}$.

Proposition 2.8.8. *Let $u \in M$ be an archimedean place. Then*

$$H_u(A) = |\det(A^* A)|_u^{1/2},$$

where $A^* = \overline{A}^t$ is the adjoint.

Proof: This follows from the well-known Binet formula

$$\det(A^*A) = \sum_I |\det(A_I)|^2.$$

For completeness, we give the proof for the case of complex matrices. Let $u : \mathbb{C}^m \longrightarrow \mathbb{C}^n$ be the corresponding linear map and let u^* be the adjoint map. We have

$$\wedge^m(u^*) \circ \wedge^m(u) = \wedge^m(u^* \circ u).$$

In the canonical basis, the matrix of $\wedge^m(u)$ (resp. $\wedge^m(u^*)$) has only one column (resp. one row) and its entries are $\det(A_I)$ (resp. $\overline{\det(A_I)}$), where I ranges over all subsets of $\{1, \dots, n\}$ of cardinality m. The matrix of $\wedge^m(u^* \circ u)$ is a 1×1 matrix with entry $\det(A^* A)$, proving Binet's formula. □

Remark 2.8.9. Let A be $n \times m$ submatrix of rank m with entries in $\overline{\mathbb{Q}}$ and let B, C be complementary submatrices of type $n \times m_1$ and $n \times m_2$, respectively. For any $w \in M_F$, we have

$$H_u(A) \leq H_u(B)\, H_u(C)$$

and thus

$$h_{\mathrm{Ar}}(A) \leq h_{\mathrm{Ar}}(B) + h_{\mathrm{Ar}}(C).$$

For a non-archimedean u, this follows easily from Laplace's expansion. If instead u is archimedean, it is a consequence of Proposition 2.8.8 and Fischer's inequality

$$\det \begin{pmatrix} B^*B & B^*C \\ C^*B & C^*C \end{pmatrix} \leq \det(B^*B)\det(C^*C)$$

from linear algebra (see e.g. L. Mirsky [**206**], Th.13.5.5), which extends Hadamard's inequality.

Proposition 2.8.10. *Let W be an m-dimensional subspace of $\overline{\mathbb{Q}}^n$ and let $W^\perp$ be its annihilator in the dual $(\overline{\mathbb{Q}}^n)^* = \overline{\mathbb{Q}}^n$. Then $h_{\mathrm{Ar}}(W^\perp) = h_{\mathrm{Ar}}(W)$.*

Proof: Let V be the vector space $\overline{\mathbb{Q}}^n$. Any element x of $\wedge^m(V)$ defines a linear map $\psi(x) : y \mapsto x \wedge y$ from $\wedge^{n-m}V$ to $\wedge^n(V)$, or in other words an element $\varphi(x)$ of $\wedge^n(V) \otimes \wedge^{n-m}(V^*)$. The map

$$\varphi : \wedge^m(V) \longrightarrow \wedge^n(V) \otimes \wedge^{n-m}(V^*)$$

is an isomorphism, which maps each element of the canonical basis of $\wedge^m(V)$ to $\pm$ an element of the canonical basis of $\wedge^n(V) \otimes \wedge^{n-m}(V^*)$. The line $\wedge^m(W)$ is mapped by φ to the line $\wedge^n(V) \otimes \wedge^{n-m}(W^\perp)$, since, for any non-zero x in $\wedge^m(W)$, the kernel of $\psi(x)$ is the subspace of $\wedge^{n-m}(V)$ generated by the elements of the form $w \wedge z$, with $w \in W$ and $z \in \wedge^{n-m-1}(V)$. We conclude that the coordinates of $\wedge^m(W)$ in $\mathbb{P}(\wedge^m(V))$ are, up to a sign, equal to the coordinates of $\wedge^{n-m}(W^\perp)$ in $\mathbb{P}(\wedge^{n-m}(V^*))$. This proves the claim. □

Definition 2.8.11. *Let A be an $m \times n$ matrix of rank m with entries in $\overline{\mathbb{Q}}$. Then the* ***Arakelov height of A*** *is defined by*

$$h_{\mathrm{Ar}}(A) := h_{\mathrm{Ar}}(A^t),$$

where A^t is the transpose and $h_{\mathrm{Ar}}(A^t)$ is the Arakelov height as in Definition 2.8.6.

Corollary 2.8.12. *With A as in 2.8.11, the Arakelov height of the space of solutions of $A \cdot \mathbf{x} = 0$ equals $h_{\mathrm{Ar}}(A)$.*

Proof: Note that the rows of A form a basis of

$$\{A \cdot \mathbf{x} = 0\}^{\perp}$$

and use Proposition 2.8.10. □

Another interesting and important property, which we give without proof, is a theorem of W.M. Schmidt [**271**], Ch.I, Lemma 8A, and independently of T. Struppeck and J.D. Vaaler [**294**]. We do not need it in the sequel.

Theorem 2.8.13. *Let V, W be subspaces of $\overline{\mathbb{Q}}^n$. Then*

$$h_{\mathrm{Ar}}(V+W) + h_{\mathrm{Ar}}(V \cap W) \leq h_{\mathrm{Ar}}(V) + h_{\mathrm{Ar}}(W).$$

2.8.14. The following local considerations hold more generally over the completion $\overline{\mathbb{Q}}_u$ with respect to $u \in M$. Note that this field is algebraically closed ([**43**], Prop.3.4.1/3). For any $\varphi : \mathrm{Hom}((\overline{\mathbb{Q}}_u)^n, (\overline{\mathbb{Q}}_u)^m)$ and $u \in M$, we define a dual local height by

$$H_u^*(\varphi) := \sup\{H_u(\varphi(\mathbf{x})) \mid \mathbf{x} \in (\mathbb{Q}_u)^n,\ H_u(\mathbf{x}) \leq 1\}.$$

If we identify φ with a $M \times N$ matrix $A = (a_{kl})$, we find that, if u is not archimedean, we have

$$H_u^*(A) = \max |a_{kl}|_u.$$

If instead u is archimedean, let A^* denote the adjoint of A and let

$$0 \leq \lambda_1 \leq \lambda_2 \leq \cdots \leq \lambda_N$$

denote the eigenvalues of the positive semi-definite matrix A^*A. Using standard facts about operator norms, we get

$$H_u^*(A) = |\lambda_N|_u^{1/2}.$$

Definition 2.8.15. *Let $\psi \in PGL(n+1, \overline{\mathbb{Q}}_u)$ and let A be a representative of ψ in $GL(n+1, \overline{\mathbb{Q}}_u)$. The* ***distorsion factor*** *of ψ is*

$$\eta_u(\psi) = H_u^*(A) H_u^*(A^{-1}).$$

2.8.16. Let $u \in M$. Now define for $\mathbf{x}$, $\mathbf{y} \in (\overline{\mathbb{Q}}_u)^{n+1} \backslash \{0\}$ the **projective distance** of $\mathbf{x}$, $\mathbf{y}$ with respect to u to be

$$\delta_u(\mathbf{x}, \mathbf{y}) = \frac{H_u(\mathbf{x} \wedge \mathbf{y})}{H_u(\mathbf{x}) H_u(\mathbf{y})}.$$

This gives a map

$$\delta_u : \mathbb{P}^n(\overline{\mathbb{Q}}_u) \times \mathbb{P}^n(\overline{\mathbb{Q}}_u) \longrightarrow [0,1].$$

Definition 2.8.17. *If $F \subset \overline{\mathbb{Q}}$ is a number field and $\mathbf{x}, \mathbf{y} \in F^{n+1}$, then for $w \in M_F$ we define*

$$\delta_w(\mathbf{x}, \mathbf{y}) = \delta_u(\mathbf{x}, \mathbf{y})^{[F_w:\mathbb{Q}_p]/[F:\mathbb{Q}]},$$

where $p \in M_{\mathbb{Q}}$ and $u \in M_{\overline{\mathbb{Q}}}$ are such that $w|p$ and $u|v$.

Proposition 2.8.18. *For $u \in M$, the projective distance δ_u is a metric on $\mathbb{P}^n(\overline{\mathbb{Q}}_u)$.*

Proof: It is clear that $\delta_u(P,Q) = \delta_u(Q,P)$ and $\delta_u(P,Q) = 0$ if and only if $P = Q$. It remains to verify the triangle inequality.

Consider first the case of a non-archimedean u. By homogeneity and continuity, it is enough to show that

$$H_u(\mathbf{x} \wedge \mathbf{z}) \leq \max(H_u(\mathbf{x} \wedge \mathbf{y}), H_u(\mathbf{y} \wedge \mathbf{z})) \tag{2.21}$$

for $\mathbf{x}, \mathbf{y}, \mathbf{z} \in F^{n+1}$ with $H_u(\mathbf{x}) = H_u(\mathbf{y}) = 1$ and $H_u(\mathbf{z}) \leq 1$, where $F \subset \overline{\mathbb{Q}}_u$ is a number field. We argue by contradiction, hence we assume

$$H_u(\mathbf{x} \wedge \mathbf{z}) > \max(H_u(\mathbf{x} \wedge \mathbf{y}), H_u(\mathbf{y} \wedge \mathbf{z})); \tag{2.22}$$

in particular, we have

$$H_u(\mathbf{y} \wedge \mathbf{z}) < H_u(\mathbf{x} \wedge \mathbf{z}) \leq H_u(\mathbf{z}). \tag{2.23}$$

Let $\nu \in F$ be such that $H_u(\mathbf{z}) = |\nu|_u$. Then (2.23) shows that the reductions of $\mathbf{y}$ and $\nu^{-1}\mathbf{z}$ are linearly dependent over the residue field of F. This clearly implies that there are $\mu \in F$ and $\mathbf{z}' \in L^{n+1}$ such that

$$\mathbf{z} = \mu\mathbf{y} + \mathbf{z}',$$

with $|\mu|_u = H_u(\mathbf{z})$ and $H_u(\mathbf{z}') < H_u(\mathbf{z})$.

We have $\mathbf{x} \wedge \mathbf{z}' = \mathbf{x} \wedge \mathbf{z} - \mu\mathbf{x} \wedge \mathbf{y}$, and $H_u(\mu\mathbf{x} \wedge \mathbf{y}) < H_u(\mathbf{x} \wedge \mathbf{z})$ by equation (2.22). Therefore, the ultrametric triangle inequality for H_u shows that

$$H_u(\mathbf{x} \wedge \mathbf{z}') = H_u(\mathbf{x} \wedge \mathbf{z}) > 0. \tag{2.24}$$

Moreover, it is obvious that $\mathbf{y} \wedge \mathbf{z} = \mathbf{y} \wedge \mathbf{z}'$. This and (2.24) show that, if (2.22) holds for $\mathbf{x}, \mathbf{y}, \mathbf{z}$, it also holds for $\mathbf{x}, \mathbf{y}, \mathbf{z}'$, for some $\mathbf{z}'$ with $H_u(\mathbf{z}') < H_u(\mathbf{z})$. By repeating this process, and noting that the absolute value $|\ |_u$ is discrete on F, we can also make $H_u(\mathbf{z}')$ arbitrarily small. This contradicts (2.24) and proves (2.21).

If instead u is archimedean, we may work in $\mathbb{C}$, with the usual euclidean absolute value $|\ |$. An easy calculation shows that for $\mathbf{x}, \mathbf{y} \in \mathbb{C}^{n+1} \setminus \{\mathbf{0}\}$ we have

$$\delta_u(\mathbf{x}, \mathbf{y}) = \sqrt{1 - \left(\frac{|\langle \mathbf{x}, \mathbf{y} \rangle|}{\|\mathbf{x}\|\,\|\mathbf{y}\|}\right)^2},$$

where $\langle \ , \ \rangle$ is the standard scalar product in $\mathbb{C}^{n+1}$ with norm $\| \ \|$. We need to show that

$$(1-|\langle \mathbf{x},\mathbf{z}\rangle|^2)^{1/2} \le (1-|\langle \mathbf{x},\mathbf{y}\rangle|^2)^{1/2} + (1-|\langle \mathbf{y},\mathbf{z}\rangle|^2)^{1/2} \tag{2.25}$$

whenever $\|\mathbf{x}\| = \|\mathbf{y}\| = \|\mathbf{z}\| = 1$.

Replacing $\mathbf{x}$ and $\mathbf{y}$ by $e^{i\theta}\mathbf{x}$ and $e^{i\phi}\mathbf{y}$ for suitable θ and ϕ, we may assume that the scalar products $\langle \mathbf{x},\mathbf{y}\rangle$ and $\langle \mathbf{y},\mathbf{z}\rangle$ are real numbers. We identify $\mathbb{C}^{n+1}$ with $\mathbb{R}^{2n+2}$ equipped with the euclidean scalar product $\langle \ , \ \rangle_{\mathbb{R}} = \Re\langle \ , \ \rangle$; then it is enough to prove (2.25) in real euclidean space $\mathbb{R}^{2n+2}$, with the standard euclidean scalar product. Let $\mathbf{u},\mathbf{v},\mathbf{w}$ be the three vectors in $\mathbb{R}^{2n+2}$ corresponding to $\mathbf{x},\mathbf{y},\mathbf{z}$, and let $\alpha = \angle(\mathbf{u},\mathbf{v})$, $\beta = \angle(\mathbf{v},\mathbf{w})$, $\gamma = \angle(\mathbf{u},\mathbf{w})$ be the angles formed by these vectors. Since the scalar product of two vectors of length 1 is the cosine of their angle, the inequality to be proven becomes

$$|\sin\gamma| \le |\sin\alpha| + |\sin\beta|.$$

We may also assume that $0 \le \alpha,\beta \le \pi/2$, because for our purposes we are free to replace $\mathbf{v}$ and $\mathbf{w}$ by $-\mathbf{v}$ and $-\mathbf{w}$. The spherical distance now gives $\gamma \le \alpha+\beta$. If $\alpha+\beta > \pi/2$, we have $\sin\alpha + \sin\beta > 1$, and the inequality to be shown is trivial. If instead $\alpha+\beta \le \pi/2$, then

$$|\sin\gamma| \le \sin(\alpha+\beta) = \cos\beta\sin\alpha + \cos\alpha\sin\beta \le \sin\alpha + \sin\beta. \qquad \square$$

We note without proof the following useful inequality.

Proposition 2.8.19. *Suppose $u \in M$, $P, Q \in \mathbb{P}^n(\overline{\mathbb{Q}}_u)$ and $\psi \in PGL(n+1,\overline{\mathbb{Q}}_u)$. Then*

$$\eta_u(\psi)^{-1}\delta_u(P,Q) \le \delta_u(\psi(P),\psi(Q)) \le \eta_u(\psi)\delta_u(P,Q).$$

For the proof, we refer to K.K. Choi and J.D. Vaaler [66].

2.8.20. Let $\mathbf{O}$ be the point $(1:0)$ on $\mathbb{P}^1_{\overline{\mathbb{Q}}}$. The Arakelov height of $P \in \mathbb{P}^1_{\overline{\mathbb{Q}}}$ is now given by the elegant formula

$$h_{\mathrm{Ar}}(P) = \sum_{w\in M_F} \log\frac{1}{\delta_w(P,\mathbf{O})}.$$

We give an application of the projective metrics so introduced. We have the **projective Liouville inequality**:

Theorem 2.8.21. *Let $F \subset \overline{\mathbb{Q}}$ be a number field and let α, β be different elements of F. Let $\boldsymbol{\alpha} = (1:\alpha)$ and $\boldsymbol{\beta} = (1:\beta)$ be the corresponding points on $\mathbb{P}^1_{\overline{\mathbb{Q}}}$. Then*

$$\prod_{w\in M_F} \delta_w(\boldsymbol{\alpha},\boldsymbol{\beta}) = (H_{\mathrm{Ar}}(\boldsymbol{\alpha})H_{\mathrm{Ar}}(\boldsymbol{\beta}))^{-1}.$$

In particular, for every $w \in M_F$ it holds

$$\delta_w(\boldsymbol{\alpha}, \boldsymbol{\beta}) \geq (H_{\mathrm{Ar}}(\boldsymbol{\alpha}) H_{\mathrm{Ar}}(\boldsymbol{\beta}))^{-1}.$$

Proof: Note that $H_w(\boldsymbol{\alpha} \wedge \boldsymbol{\beta}) = |\alpha - \beta|_w$ and then the claim follows from the product formula. □

2.9. Siegel's lemma

In 1929 C.L. Siegel, in the course of his work on diophantine approximation and transcendency, formally stated what is known today as **Siegel's lemma**

Lemma 2.9.1. *Let a_{ij}, $i = 1, \ldots, M$, $j = 1, \ldots, N$ be rational integers, not all 0, bounded by B and suppose that $N > M$. Then the homogeneous linear system*

$$\begin{array}{llllll} a_{11}x_1 & + a_{12}x_2 & + & \cdots & + \ a_{1N}x_N & = 0 \\ a_{21}x_1 & + a_{22}x_2 & + & \cdots & + \ a_{2N}x_N & = 0 \\ \cdot & \cdot & \cdot & \cdots & \cdot \quad \cdot & \cdot \\ a_{M1}x_1 & + a_{M2}x_2 & + & \cdots & + \ a_{MN}x_N & = 0 \end{array}$$

has a solution $x_1, \ldots, x_N$ in rational integers, not all 0, bounded by

$$\max_i |x_i| \leq \lfloor (NB)^{\frac{M}{N-M}} \rfloor.$$

Siegel's lemma and its numerous variants have become a fundamental tool in diophantine approximation and transcendency.

Proof: The $M \times N$ matrix (a_{ij}) is denoted by A. Clearly, we may assume that no row is identically 0. For a positive integer k, let us consider the set

$$T := \{\mathbf{x} \in \mathbb{Z}^N \mid 0 \leq x_i \leq k, \ i = 1, \ldots, N\}.$$

We denote by S_m^+ the sum of the positive entries in the mth row of A, and similarly by S_m^- the sum of the negative entries. Then for $\mathbf{x} \in T$ and $\mathbf{y} := A\mathbf{x}$ we have

$$kS_m^- \leq y_m \leq kS_m^+.$$

Let

$$T' := \{\mathbf{y} \in \mathbb{Z}^M \mid kS_m^- \leq y_m \leq kS_m^+, \ m = 1, \ldots, M\}.$$

Writing $B_m := \max_n |a_{mn}|$, we have $S_m^+ - S_m^- \leq NB_m$ and we conclude that the set T' has at most $\prod_m (NkB_m + 1)$ elements. Now we choose k so that T has more elements than T', namely

$$\prod_m (NkB_m + 1) < (k+1)^N. \tag{2.26}$$

If we choose k to be the integer part of $\prod_m (NB_m)^{\frac{1}{N-M}}$ and use $NkB_m + 1 < NB_m(k+1)$, then we easily verify that inequality (2.26) is fulfilled.

By Dirichlet's pigeon-hole principle, there are two different points $\mathbf{x}', \mathbf{x}'' \in T$ with $A\mathbf{x}' = A\mathbf{x}''$. The point $\mathbf{x} := \mathbf{x}' - \mathbf{x}''$ is a solution of $A\mathbf{x} = 0$ in integers, with $\max_n |x_n| \leq k$. □

Corollary 2.9.2. *Let K be a number field of degree d contained in $\mathbb{C}$ with $|\ |$ the usual absolute value on $\mathbb{C}$. Let $M, N \in \mathbb{N}$, $0 < M < N$. Then there are positive constants C_1, C_2 such that for any non-zero $M \times N$ matrix A with entries $a_{mn} \in O_K$, there is $\mathbf{x} \in O_K^N \setminus \{\mathbf{0}\}$ with $A \cdot \mathbf{x} = \mathbf{0}$ and*

$$H(\mathbf{x}) \leq C_1 (C_2 NB)^{\frac{M}{N-M}},$$

where $B := \sup_{\sigma,m,n} |\sigma(a_{mn})|$ with σ ranging over the embeddings of K into $\mathbb{C}$.

Proof: Let $\omega_1, \ldots, \omega_d$ be a $\mathbb{Z}$-basis of the ring of algebraic integers O_K. The entries of A may be written in the form

$$a_{mn} = \sum_{j=1}^{d} a_{mn}^{(j)} \omega_j, \quad a_{mn}^{(j)} \in \mathbb{Z}. \tag{2.27}$$

Using $x_n = \sum_{k=1}^{d} x_n^{(k)} \omega_k$, we get

$$(A \cdot \mathbf{x})_m = \sum_{n=1}^{N} \sum_{j,k=1}^{d} a_{mn}^{(j)} \omega_j \omega_k x_n^{(k)} = \sum_{l=1}^{d} \sum_{n=1}^{N} \sum_{j,k=1}^{d} a_{mn}^{(j)} b_{jk}^{(l)} x_n^{(k)} \omega_l,$$

where $\omega_j \omega_k = \sum_{l=1}^{d} b_{jk}^{(l)} \omega_l$. Let A' be the $(Md) \times (Nd)$ matrix

$$A' := \left(\sum_{j=1}^{d} a_{mn}^{(j)} b_{jk}^{(l)} \right)$$

with rows indexed by (m,l), columns indexed by (n,k), and let $\mathbf{y} \in \mathbb{Z}^{Nd}$ be the vector $(x_n^{(k)})$. Then the Siegel lemma above gives a non-zero integer solution $\mathbf{y}$ of $A' \cdot \mathbf{y} = \mathbf{0}$ with

$$H(\mathbf{y}) \leq \left(Nd^2 \max_{m,n,j} |a_{mn}^{(j)}| \max_{j,k,l} |b_{jk}^{(l)}| \right)^{\frac{M}{N-M}}.$$

Let σ be ranging over the $d = [K : \mathbb{Q}]$ different embeddings of K into $\mathbb{C}$. By conjugating (2.27) with all σs and using that $(\sigma(\omega_j))$ is an invertible $d \times d$-matrix (because the square of the determinant is the discriminant, see B.1.14 and Remark B.1.15), we get

$$\max_j |a_{mn}^{(j)}| \leq C_2' \max_\sigma |\sigma(a_{mn})|$$

for a suitable positive constant C_2'. Then $x_n = \sum_{k=1}^d x_n^{(k)} \omega_k$ yields $H(\mathbf{x}) \leq C_1 H(\mathbf{y})$ and using $C_2 := C_2' d^2 \max_{j,k,l} |b_{jk}^{(l)}|$, we get the claim. $\square$

2.9.3. The goal of this section is to give an improved version of Siegel's lemma for a given number field K of degree d. Often, the simple version above is sufficient for applications and the reader may skip the remaining part of this section in a first reading.

After stating the result in Theorem 2.9.4, we give a series of immediate corollaries, which are quite useful in practice; even in the case $K = \mathbb{Q}$, we get an improvement of the elementary form of Siegel's lemma proved above. The proof of the generalized Siegel's lemma will be done in several steps. First, we recall for completeness some basic results in the geometry of numbers, referring to Appendix C for the proofs. Then we use Minkowski's second main theorem to construct a "small" basis in the range of a $N \times M$ matrix A of rank M. In order to find a "small" basis of solutions of our original equation $A\mathbf{x} = 0$, we apply this result to the matrix A' whose columns are formed by any basis of solutions. As A and A' have the same Arakelov height, we will get the generalized Siegel's lemma. Finally, we give a relative version where the entries are in a finite extension F of K but the solutions are required to be in K.

We now state Siegel's lemma over a number field K of degree d and discriminant $D_{K/\mathbb{Q}}$, in the form given by E. Bombieri and J.D. Vaaler [35].

Theorem 2.9.4. *Let A be an $M \times N$ matrix of rank M with entries in K. Then the K-vector space of solutions of $A\mathbf{x} = 0$ has a basis $\mathbf{x}_1, \ldots, \mathbf{x}_{N-M}$, contained in O_K^N, such that*

$$\prod_{l=1}^{N-M} H(\mathbf{x}_l) \leq |D_{K/\mathbb{Q}}|^{\frac{N-M}{2d}} H_{\mathrm{Ar}}(A).$$

Remark 2.9.5. Here $H(\mathbf{x})$ is the multiplicative homogeneous height, so that we consider $\mathbf{x}$ as a point in $\mathbb{P}^{N-1}(K)$. Hence there is no deep information contained in the statement that we can choose our solutions in O_K^N, because we can replace any solution by a scalar multiple without changing the height.

Remark 2.9.6. The Arakelov height of a $M \times N$ matrix of rank M has been introduced in Definition 2.8.11. We recall that, if W is the subspace of K^N spanned by the rows of A, then $H_{\mathrm{Ar}}(A)$ is the multiplicative Arakelov height of the line $\wedge^M W$ in the projective space $\mathbb{P}(\wedge^M K^N)$. The difference with the usual height consists only in using the L^2-local height, instead of the L^∞-local height, at the archimedean places.

Sometimes it is not practical to assume always that A has maximal rank M. This can be obviated as follows. For any $M \times N$ matrix of rank R with entries in K, let W be the subspace of K^N spanned by the rows of A and define

$$H_{\mathrm{Ar}}^{\mathrm{row}}(A) := H_{\mathrm{Ar}}(W) = H_{\mathrm{Ar}}(\wedge^R W),$$

where the line $\wedge^R W$ is viewed as a point of the projective space $\mathbb{P}(\wedge^R K^N)$.

Corollary 2.9.7. *Let A be an $M \times N$ matrix over K of rank R. Then there is a basis $\mathbf{x}_1, \ldots, \mathbf{x}_{N-R}$ of the kernel of A, contained in O_K^N, such that*

$$\prod_{l=1}^{N-R} H(\mathbf{x}_l) \leq |D_{K/\mathbb{Q}}|^{\frac{N-R}{2d}} H_{\mathrm{Ar}}^{\mathrm{row}}(A).$$

Proof: There is an $R \times N$ submatrix A' of rank R. Then A and A' have the same kernel and the same Arakelov height. The result follows by applying Theorem 2.9.4 to A'. □

2.9.8. If A_m is the mth row of A, then Remark 2.8.9 shows that

$$H_{\mathrm{Ar}}^{\mathrm{row}}(A) \leq \prod_m H_{\mathrm{Ar}}(A_m),$$

where m ranges over R linearly independent rows of A. We denote by $H(A)$ the ***multiplicative height of the matrix*** A as a point of $\mathbb{P}_K^{NM-1}$. Then the obvious inequality

$$H_{\mathrm{Ar}}(A_m) \leq \sqrt{N} H(A)$$

proves

Corollary 2.9.9. *With the same assumptions as in Corollary 2.9.7, it holds that*

$$\prod_{l=1}^{N-R} H(\mathbf{x}_l) \leq |D_{K/\mathbb{Q}}|^{\frac{N-R}{2d}} (\sqrt{N} H(A))^R.$$

In particular, there is a non-zero solution $\mathbf{x} \in O_K^N$ of $A\mathbf{x} = 0$ with

$$H(\mathbf{x}) \leq |D_{K/\mathbb{Q}}|^{\frac{1}{2d}} \left(\sqrt{N} H(A)\right)^{\frac{R}{N-R}}.$$

If we compare with the original Siegel's lemma in the case $K = \mathbb{Q}$, we have improved the result by replacing the factor N by $\sqrt{N}$ in the final estimate.

2.9.10. The proof of Theorem 2.9.4 uses geometry of numbers over the adeles, and we shall refer to Appendix C for details. Here we will limit ourselves to basic definitions and results.

Let K_v be the completion of the number field K with respect to the place $v \in M_K$ and let $v|p \in M_{\mathbb{Q}}$. Then K_v is a locally compact group (see 1.2.12) with Haar measure uniquely determined up to a scalar. We normalize this Haar measure as follows:

(a) if v is non-archimedean, β_v denotes the Haar measure on K_v normalized so that

$$\beta_v(R_v) = |D_{K_v/\mathbb{Q}_p}|_p^{1/2},$$

where R_v is the valuation ring of K_v and $D_{K_v/\mathbb{Q}_p}$ is the discriminant;

(b) if $K_v = \mathbb{R}$, then β_v is the ordinary Lebesgue measure;

(c) if $K_v = \mathbb{C}$, then β_v is twice the ordinary Lebesgue measure.

2.9.11. Let N be a positive integer. For every archimedean $v \in M_K$, let S_v be a non-empty convex, symmetric, open subset of K_v^N. By symmetric, we mean $S_v = -S_v$. For each non-archimedean $v \in M_K$, let S_v be a K_v-lattice in K_v^N, namely a non-empty compact and open R_v-submodule of K_v^N. We assume that $S_v = R_v^N$ for all but finitely many v. Then the set

$$\Lambda := \{\mathbf{x} \in K^N \mid \mathbf{x} \in S_v \text{ for every non-archimedean } v\}$$

is a K-lattice in K^N, that is a finitely generated O_K-module, which generates K^N as a vector space (cf. C.2.6). Moreover, the image Λ_∞ of Λ under the canonical embedding of K^N into $E_\infty := \prod_{v|\infty} K_v^N$ is an $\mathbb{R}$-lattice in E_∞ (cf. C.2.7). This is the familiar notion of a lattice, meaning that Λ_∞ is a discrete subgroup of the $\mathbb{R}$-vector space E_∞ and that E_∞/Λ_∞ is compact.

Definition 2.9.12. *The* ***n*th successive minimum** *of the non-empty convex, symmetric, open subset* $S_\infty := \prod_{v|\infty} S_v$ *of* E_∞ *with respect to the lattice* Λ_∞ *is*

$$\lambda_n := \inf\{t > 0 \mid tS_\infty \text{ contains } n \ K\text{-linearly independent vectors of } \Lambda_\infty\}.$$

The adelic version of Minkowski's second theorem is (cf. Theorem C.2.11).

Theorem 2.9.13. *The successive minima satisfy*

$$(\lambda_1\lambda_2\cdots\lambda_N)^d \prod_{v\in M_K} \beta_v(S_v) \le 2^{dN}.$$

2.9.14. For the proof of Siegel's lemma 2.9.4 we choose the sets S_v as follows. First, let Q_v^N be the unit cube in K_v^N of volume 1 with respect to the Haar measure β_v. Explicitly, this is given by

$$Q_v^N := \begin{cases} \max \|x_n\|_v < \frac{1}{2} & \text{if } v \text{ is real} \\ \max \|x_n\|_v < \frac{1}{2\pi} & \text{if } v \text{ is complex} \\ \max \|x_n\|_v \le 1 & \text{if } v \text{ is non-archimedean} \end{cases}$$

using the normalization from 1.3.6 with respect to $K/\mathbb{Q}$. Let A be an $N \times M$ matrix of rank M with entries in K. Then we set

$$S_v := \{\mathbf{y} \in K_v^M \mid A\mathbf{y} \in Q_v^N\}.$$

If v is archimedean, then S_v is a non-empty, convex, symmetric, bounded open subset of K_v^M; with respect to the injective map $\mathbf{x} \mapsto A\mathbf{x}$, the image of S_v is a linear slice of the cube Q_v^N. If instead v is non-archimedean, then it is easy to show, as we will verify below, that S_v is a K_v-lattice in K_v^M.

In order to apply Minkowski's second theorem, we need to estimate volumes from below.

Proposition 2.9.15. *With the notation of 2.9.14 and for an archimedean $v \in M_K$, it holds*

$$\beta_v(S_v) \geq \|\det(A^*A)\|_v^{-\frac{1}{2}},$$

where $A^* = \overline{A}^t$ is the transpose conjugate of A.

Proof: If v is real, this is Theorem C.3.8 for $n_1 = \cdots = n_N = 1$. If v is complex, we write $A = U + iV$ and $\mathbf{y} = \mathbf{u} + i\mathbf{v}$ for real $U, V, \mathbf{u}, \mathbf{v}$. We identify K_v^M with $\mathbb{R}^{2M}$ by means of $\mathbf{y} = \mathbf{u} + i\mathbf{v}$ and we proceed similarly with $K_v^N = \mathbb{R}^{2N}$. Then the map $\mathbf{y} \mapsto A\mathbf{y}$ is given by the real $2N \times 2M$ matrix

$$A' = \begin{pmatrix} U & -V \\ V & U \end{pmatrix}$$

and also

$$Q_v^N = \left\{(\mathbf{u}, \mathbf{v}) \in \mathbb{R}^{2N} \mid u_j^2 + v_j^2 < \frac{1}{2\pi}\right\}.$$

Now we apply Theorem C.3.8 with $n_1 = \cdots = n_N = 2$ to get

$$\beta_v(S_v) \geq \det(A'^t A')^{-\frac{1}{4}}.$$

Since $A \mapsto A'$ is a ring homomorphism from the complex $N \times M$ matrices to the real $2N \times 2M$ matrices, we conclude that

$$\det(A'^t A') = \det((A^*A)') = \det(A^*A)^2,$$

thereby proving the claim. □

The corresponding result for non-archimedean $v \in M_K$ is more precise, uses the discriminant $D_{K_v/\mathbb{Q}_p}$ (see B.1.14), and is easier to prove. We have:

Proposition 2.9.16. *Let v be a non-archimedean place of K lying over the prime p. Then, with the notation of 2.9.14, it holds*

$$\beta_v(S_v) = |D_{K_v/\mathbb{Q}_p}|_p^{\frac{M}{2}} \left(\max_I \|\det(A_I)\|_v\right)^{-1},$$

where I ranges over all subsets of $\{1, \ldots, N\}$ of cardinality M and A_I is the $M \times M$ matrix formed by the ith rows of A with $i \in I$.

Proof: Choose a subset J of cardinality M such that $\|\det(A_J)\|_v$ is maximal. Without loss of generality, we may assume that $J = \{1, \ldots, M\}$. Then $W = AA_J^{-1}$ has the form

$$W = \begin{pmatrix} I_M \\ W' \end{pmatrix},$$

where I_M is the $M \times M$ unit matrix and W' is an $(N-M) \times M$ matrix. We have

$$\|\det(W_I)\|_v \leq 1$$

for any subset I of cardinality M, because $\|\det(A_J)\|_v$ was chosen to be maximal. Therefore, applying this with $I = \{1, \ldots, l-1, l+1, \ldots, M, M+j\}$, we see that

$$\|w_{M+j,l}\|_v = \|\det(W_I)\|_v \leq 1.$$

This means that all entries of W are in the valuation ring R_v and proves

$$A_J S_v = \{\mathbf{y} \in K_v^M \mid W\mathbf{y} \in Q_v^N\} = R_v^M.$$

Under the linear transformation $\mathbf{y}' = A_J^{-1}\mathbf{y}$ on K_v^M, the volume transforms by a factor $\|\det(A_j)\|_v^{-1}$ (cf. C.1.3). Hence

$$\beta_v(S_v) = \|\det(A_j)\|_v^{-1}\beta_v(R_v^M) = \|\det(A_j)\|_v^{-1}|D_{K_v/\mathbb{Q}_p}|_p^{\frac{M}{2}},$$

completing the proof. □

Remark 2.9.17. The proof of the above proposition shows immediately that S_v is a K_v-lattice in K_v^M for every non-archimedean v and $S_v = R_v^M$ for all but finitely many v.

We are now ready to prove:

Proposition 2.9.18. *Let A be an $N \times M$ matrix of rank M with entries in K. Then the image of A has a basis $\mathbf{x}_1, \ldots, \mathbf{x}_M$ with*

$$\prod_{m=1}^{M} H(\mathbf{x}_m) \leq \left(\frac{2}{\pi}\right)^{\frac{Ms}{d}} |D_{K/\mathbb{Q}}|^{\frac{M}{2d}} H_{\mathrm{Ar}}(A),$$

where s is the number of complex places of K.

Proof: We apply Minkowski's second theorem from 2.9.13 to the sets S_v defined in 2.9.14. Then Propositions 2.9.15, 2.9.16, 2.8.8, show that

$$\prod_v \beta_v(S_v) \geq \prod_{v \nmid \infty} |D_{K_v/\mathbb{Q}}|_p^{\frac{M}{2}} H_{\mathrm{Ar}}(A)^{-d}.$$

As shown in (C.4) on page 606, we have

$$\prod_{v \nmid \infty} |D_{K_v/\mathbb{Q}_p}|_p = |D_{K/\mathbb{Q}}|^{-1},$$

whence

$$\prod_v \beta_v(S_v) \geq |D_{K/\mathbb{Q}}|^{-\frac{M}{2}} H_{\mathrm{Ar}}(A)^{-d}.$$

Now Minkowski's second theorem yields

$$\lambda_1 \cdots \lambda_M \leq 2^M |D_{K/\mathbb{Q}}|^{\frac{M}{2d}} H_{\mathrm{Ar}}(A). \tag{2.28}$$

It remains to connect the successive minima with the basis we want to find. With our specific sets S_v, we form the K-lattice Λ as in 2.9.11 and identify Λ with its image Λ_∞ in E_∞. Let $\mathbf{y} \in K^M$ be a lattice point in λS for some $\lambda > 0$ and let $\mathbf{x} := A\mathbf{y}$. Then the definition of $S = \prod_{v|\infty} S_v$ gives $\max_n \|x_n\|_v < \lambda/2$ if v is real, $\max_n \|x_n\|_v < \lambda^2/(2\pi)$ if v is complex, and $\max_n \|x_n\|_v \leq 1$ if v is not archimedean. Thus we have

$$H(A\mathbf{y}) < \frac{\lambda}{2}\left(\frac{2}{\pi}\right)^{\frac{s}{d}}. \tag{2.29}$$

There are linearly independent lattice points $\mathbf{y}_1, \ldots, \mathbf{y}_M \in K^M$ such that $\mathbf{y}_m \in \lambda_m \overline{S}$, for $m = 1, \ldots, M$. Then (2.28) and (2.29) prove what we want, with $\mathbf{x}_m = A\mathbf{y}_m$. □

Proof of Theorem 2.9.4: Let A' be an $N \times (N-M)$ matrix whose columns form a basis of the kernel of A. Clearly, A' has rank $N-M$ and the image of A' is the same as the kernel of A. By Corollary 2.8.12, A and A' have the same Arakelov height. Now Proposition 2.9.18 applied to A' gives a basis $\mathbf{x}_1, \ldots, \mathbf{x}_{N-M}$ of the kernel of A such that

$$\prod_{l=1}^{N-M} H(\mathbf{x}_l) \leq \left(\frac{2}{\pi}\right)^{\frac{Ms}{d}} |D_{K/\mathbb{Q}}|^{\frac{M}{2d}} H_{\mathrm{Ar}}(A).$$

This completes the proof of Theorem 2.9.4. □

Finally, we give a relative version of Siegel's lemma.

Theorem 2.9.19. *Let K be a number field of degree d and discriminant $D_{K/\mathbb{Q}}$ and let F be a finite-dimensional field extension of K of degree $r := [F : K]$. Let A be an $M \times N$ matrix with entries in F and assume $rM < N$. Then there exist $N - rM$ K-linearly independent vectors $\mathbf{x}_l \in O_K^N$ such that*

$$A\mathbf{x}_l = 0, \qquad l = 1, 2, \ldots, N - rM$$

and

$$\prod_{l=1}^{N-rM} H(\mathbf{x}_l) \leq |D_{K/\mathbb{Q}}|^{\frac{N-rM}{2d}} \prod_{i=1}^{M} H_{\mathrm{Ar}}(A_i)^r,$$

where A_i is the ith row of A.

Proof: Let $\omega_1, \ldots, \omega_r$ be a basis of F/K. For the entries of $A = (a_{mn})$, we have

$$a_{mn} = \sum_{j=1}^{r} a_{mn}^{(j)} \omega_j$$

for uniquely determined $a_{mn}^{(j)} \in K$. Let $A^{(j)}$ be the $M \times N$ matrix with entries $a_{mn}^{(j)}$. Then for $\mathbf{x} \in K^N$, the equation $A\mathbf{x} = 0$ is equivalent to the system of equations $A^{(j)}\mathbf{x} = 0$, for $j = 1, \ldots, r$. Let A' denote the associated $rM \times N$ matrix. Then we are looking for solutions of $A'\mathbf{x} = 0$, $\mathbf{x} \in K^N$. Let $\sigma_1, \ldots, \sigma_r$

be the distinct embeddings of F into $\overline{K}$ over K and let $\Omega \times \Omega$ be the $r \times r$ matrix with entries the $M \times M$ matrices $\Omega_{ij} = \sigma_i(\omega_j) I_M$, where I_M is the $M \times M$ unit matrix. Thus Ω is an $rM \times rM$ matrix built up by r^2 blocks of $M \times M$ matrices. By Remark B.1.15, we have $D_{F/K} = \det(\sigma_i(\omega_j))^2$, whence Ω is invertible and its inverse is again formed by r^2 blocks of multiples of I_M. From our definitions, we also see that

$$A'' := \begin{pmatrix} \sigma_1 A \\ \vdots \\ \sigma_r A \end{pmatrix} = \Omega A'. \tag{2.30}$$

We apply Corollary 2.9.7 to A'. If R denotes the rank of A' (and hence also of A''), we get a basis $\mathbf{x}_1, \ldots, \mathbf{x}_{N-R}$ of the kernel of A' over K, contained in O_K, such that

$$\prod_{l=1}^{N-R} H(\mathbf{x}_l) \leq |D_{K/\mathbb{Q}}|^{\frac{N-R}{2d}} H_{\mathrm{Ar}}^{\mathrm{row}}(A').$$

By (2.30) and the fact that Ω is a non-singular matrix, it is also clear that A'' and A' have the same kernels and

$$H_{\mathrm{Ar}}^{\mathrm{row}}(A') = H_{\mathrm{Ar}}^{\mathrm{row}}(A'').$$

By 2.9.8, we conclude that

$$\prod_{l=1}^{N-R} H(\mathbf{x}_l) \leq |D_{K/\mathbb{Q}}|^{\frac{N-R}{2d}} \prod_i H_{\mathrm{Ar}}(A_i''),$$

where i ranges over R linearly independent rows of A''. We rearrange our basis $\mathbf{x}_l$ by increasing height. Then

$$\prod_{l=1}^{N-rM} H(\mathbf{x}_l) \leq \left(\prod_{l=1}^{N-R} H(\mathbf{x}_l) \right)^{\frac{N-rM}{N-R}} \leq |D_{K/\mathbb{Q}}|^{\frac{N-rM}{2d}} \prod_{i=1}^{rM} H_{\mathrm{Ar}}(A_i'').$$

The theorem follows easily using $H_{\mathrm{Ar}}(\sigma_j A_i) = H_{\mathrm{Ar}}(A_i)$. □

2.10. Bibliographical notes

An account of the geometric theory of heights can be found in Lang's monograph [**169**]. A. Weil [**324**] was the first in his thesis in 1927 to study heights in a geometric setting and their functorial properties, and Siegel used general heights associated to ample divisors on curves in his work on integral points on curves. Weil's treatment again is based on Hilbert's Nullstellensatz (an effective version of the Nullstellensatz is in [**195**], Th.IV).

Our exposition differs from Lang's in several respects. We use systematically the notion of presentation of a Cartier divisor, so to emphasize how local heights are determined not just by the divisor itself but in fact from other data. Global heights are treated first only in the case of projective varieties, using the join operation on projective embeddings and the fact that very ample line bundles generate the Picard group to obtain the group structure on global heights. This explicit approach to heights is less elegant than a more abstract one, but is sufficiently constructive to be usable for the explicit estimates in Section 2.5, which appear to be new.

The general statement of Northcott's theorem is in a basic foundational paper by A. Weil [**328**], following earlier work of D.G. Northcott [**227**], [**228**]. The general theory of local heights is due to A. Néron [**218**]. We present it in Section 2.7, but instead of using his quasi-functions, we use the modern language of metrized line bundles promoted by Arakelov theory.

A systematic theory of heights for subspaces of linear spaces was developed for the first time by W.M. Schmidt in [**263**]. Our exposition is mainly based on [**35**]. Propositions 2.8.8 and 2.8.10 were communicated to us by Oesterlé. The distorsion factor 2.8.15 is introduced in E. Bombieri, A.J. Van der Poorten, and J.D. Vaaler [**37**]; the projective distance is of course much older.

Already in 1909 A. Thue [**298**] used the pigeon-hole principle to find small integer solutions to linear systems of equations. C.L. Siegel [**283**] gave the bound $\lceil (NB)^{M/(N-M)} \rceil$, but it turns out that the cleaner bound $\lfloor (NB)^{M/(N-M)} \rfloor$ requires only a minor modification in his proof, see A. Baker's monograph [**14**]. The improvement $\lfloor (\sqrt{N}\, B)^{M/(N-M)} \rfloor$ obtained here seems to be of no consequence for the applications we will make in this book.

Minkowski's second theorem in the adelic setting is due to R.B. McFeat [**198**]. If we ask for solutions in $\overline{\mathbb{Q}}$, then we can get a bound independent of the discriminant $D_{K/\mathbb{Q}}$ in Theorem 2.9.4, see D. Roy and J.L. Thunder [**247**], [**248**] (note that this cannot be done for solutions in the field K). This is quite important in some cases, see the bibliographical notes to Chapter 7.

3 LINEAR TORI

3.1. Introduction

This short chapter contains simple but basic material about the algebraic group $\mathbb{G}_m^n$ over a field K of characteristic 0. Rather than developing a theory in a more general context (as it will be done in Section 8.2) and deriving our results as special cases, our aim in this section has been to give a down-to-earth elementary treatment of $\mathbb{G}_m^n$, its subgroups and maximal subgroups of subvarieties of $\mathbb{G}_m^n$. In order to read this chapter, the reader should be familiar with basic concepts about varieties as provided by the first sections of Appendix A. We will also apply Siegel's lemma in 2.9.4 over $\mathbb{Q}$ to get effective bounds for certain normalization matrices. The theory of linear tori will be used in Chapter 4.

3.2. Subgroups and lattices

3.2.1. As an affine variety, we identify $G := \mathbb{G}_m^n$ with the Zariski open subset

$$x_1 x_2 \cdots x_n \neq 0$$

of affine space $\mathbb{A}_K^n$, with the obvious multiplication

$$(x_1, x_2, \ldots, x_n) \cdot (y_1, y_2, \ldots, y_n) = (x_1 y_1, x_2 y_2, \ldots, x_n y_n).$$

The element $\mathbf{1}_n = (1, 1, \ldots, 1)$ is the identity of the group structure.

3.2.2. An **algebraic subgroup** of G is a Zariski closed subgroup, and a **linear torus** H is an algebraic subgroup which is geometrically irreducible. Note that we require that an algebraic subgroup is always defined over the fixed ground field K of characteristic 0. We will later show that any algebraic subgroup is defined by polynomials with coefficients in $\mathbb{Q}$ (Corollary 3.2.15).

Let H_1, H_2 be algebraic subgroups of $\mathbb{G}_m^{n_1}$ and $\mathbb{G}_m^{n_2}$. Then $\varphi : H_1 \to H_2$ is called a **homomorphism** (of algebraic subgroups) if φ is a morphism of algebraic varieties which is also a group homomorphism. In Corollary 3.2.8, we prove that a linear torus in G of dimension r is isomorphic to $\mathbb{G}_m^r$.

A **torus coset** is simply a coset gH of a linear torus H of positive dimension.

A **torsion coset** is a coset εH, where H is a linear torus and ε a torsion point in G, i.e. a point of finite order in G. The linear torus H may be trivial, hence a torsion point is the simplest example of a torsion coset. For a torus coset (resp. torsion coset), we do not assume that g (resp. ε) is a K-rational point, hence a torus coset or a torsion coset need not be defined over K.

3.2.3. For $\mathbf{i} \in \mathbb{Z}^n$, we abbreviate $\mathbf{x}^{\mathbf{i}} = x_1^{i_1} \cdots x_n^{i_n}$. Let $\mathbf{e}_1 = (1, 0, \ldots, 0)^t$, $\ldots, \mathbf{e}_n = (0, 0, \ldots, 1)^t$ be column vectors (^{t} is the transpose), which we identify with the usual basis of $\mathbb{Z}^n$. Let A be an $n \times n$ matrix with columns $A\mathbf{e}_i = (a_{1i}, \ldots, a_{ni})^t \in \mathbb{Z}^n$ for $i = 1, \ldots, n$ and let $\varphi_A : G \longrightarrow G$ be the map defined by

$$\varphi_A(\mathbf{x}) := (\mathbf{x}^{A\mathbf{e}_1}, \ldots, \mathbf{x}^{A\mathbf{e}_n}) = (x_1^{a_{11}} \cdots x_n^{a_{n1}}, \ldots, x_1^{a_{1n}} \cdots x_n^{a_{nn}}).$$

Then we have $\varphi_{AB} = \varphi_B \circ \varphi_A$, making it clear that, if $\det(A) = \pm 1$, then φ_A is an isomorphism with inverse $\varphi_{A^{-1}}$. The group of such matrices is denoted by $GL(n, \mathbb{Z})$ and $SL(n, \mathbb{Z})$ is the subgroup with $\det(A) = 1$.

Definition 3.2.4. *An isomorphism φ_A is classically called a* ***monoidal transformation****.*

3.2.5. If $\mathbf{a} = a_1\mathbf{e}_1 + \cdots + a_n\mathbf{e}_n \in \mathbb{Z}^n$, we have

$$\mathbf{x}^{A\mathbf{a}} = ((\mathbf{x}^{A\mathbf{e}_1})^{a_1} \cdots (\mathbf{x}^{A\mathbf{e}_n})^{a_n}) = (\varphi_A(\mathbf{x}))^{\mathbf{a}}.$$

If we apply this with A^{-1}, $A\mathbf{a}$, in place of A, $\mathbf{a}$, we find

$$\mathbf{x}^{\mathbf{a}} = \mathbf{x}^{A^{-1}A\mathbf{a}} = (\varphi_{A^{-1}}(\mathbf{x}))^{A\mathbf{a}}. \tag{3.1}$$

3.2.6. Let Λ be a subgroup of $\mathbb{Z}^n$. We say that Λ is a **lattice** if it is a subgroup of rank n. If Λ is a subgroup, it spans a linear space $V_\Lambda := \Lambda \otimes_{\mathbb{Z}} \mathbb{R} \subset \mathbb{R}^n$. Then $\widetilde{\Lambda} = V_\Lambda \cap \mathbb{Z}^n$ is a subgroup which contains Λ as a subgroup of finite index $\rho(\Lambda) := [\widetilde{\Lambda} : \Lambda]$. The subgroup is called **primitive** if $\rho(\Lambda) = 1$.

It is easy to see that the subgroup Λ determines an algebraic subgroup

$$H_\Lambda := \{\mathbf{x} \in G \mid \mathbf{x}^{\boldsymbol{\lambda}} = 1 \;\forall \lambda \in \Lambda\}$$

of G. The following result describes the structure of H_Λ as a direct product of algebraic subgroups F and M_Λ, which means that multiplication gives an isomorphism $F \times M_\Lambda \to H_\Lambda$ of algebraic subgroups.

Proposition 3.2.7. *Let Λ be a subgroup of $\mathbb{Z}^n$ of rank $n - r$. Then H_Λ is an algebraic subgroup of G of dimension r, which is the direct product of F and M_Λ, where F is a finite algebraic subgroup of order $\rho(\Lambda)$ and $M_\Lambda \subset H_\Lambda$ is a linear torus equal to the connected component of the identity of H_Λ.*

Proof: By the theorem of elementary divisors (N. Bourbaki [**49**], Ch.VII, §4, no.3, Th.1), there is a basis $\mathbf{b}_1, \ldots, \mathbf{b}_n$ of $\mathbb{Z}^n$ and elements $\lambda_1, \ldots, \lambda_{n-r} \in \mathbb{Z} \setminus \{0\}$ such that $\lambda_1\mathbf{b}_1, \ldots, \lambda_{n-r}\mathbf{b}_{n-r}$ is a basis of Λ. Using a monoidal transformation to change coordinates, we may assume that $\mathbf{b}_1, \ldots, \mathbf{b}_n$ is the standard basis. Then $H_\Lambda = F \times \mathbb{G}_m^r$ with

$$F = \{\mathbf{x} \in \mathbb{G}_m^{n-r} \mid x_1^{\lambda_1} = 1, \ldots, x_{n-r}^{\lambda_{n-r}} = 1\}$$

and all assertions are easy to prove. □

The following two corollaries are immediate from Proposition 3.2.7 and its proof.

Corollary 3.2.8. *For a subgroup Λ of $\mathbb{Z}^n$ of rank $n-r$, the following properties are equivalent:*

(a) *H_Λ is a linear torus;*

(b) *H_Λ is isomorphic to $\mathbb{G}_m^r$;*

(c) *H_Λ is irreducible;*

(d) *Λ is primitive.*

Corollary 3.2.9. *If $\Lambda \subset \Lambda'$ are subgroups of $\mathbb{Z}^n$ of the same rank $n-r$ and if $H_\Lambda = FM_\Lambda$ is a direct product decomposition as in Proposition 3.2.7, then $M_\Lambda = M_{\Lambda'}$ and there is a direct product $H_{\Lambda'} = F'M_\Lambda$ in the sense of Proposition 3.2.7 such that $F' \subset F$.*

For the purpose of obtaining effective estimates, we need to bound the entries of the matrix A used in the proof of the preceding proposition. For the rest of this chapter, $\| \ \|$ will denote the ℓ^1-norm of a vector $\mathbf{x}$. For a matrix A, we denote by $\|A\|$ the maximum of the ℓ^1-norms of its columns.

Proposition 3.2.10. *Let Λ be a subgroup of $\mathbb{Z}^n$ of rank $n-r$ and suppose that Λ has $n-r$ independent vectors of norm at most d. Then there are a finite subgroup Φ of $\mathbb{G}_m^{n-r}$ and a matrix $A \in SL(n, \mathbb{Z})$ with $\|A\| \leq n^3 d^{n-r}$ and $\|A^{-1}\| \leq n^{2n-1} d^{(n-1)^2}$, such that $\varphi_A(\Phi \times \mathbb{G}_m^r) = H_\Lambda$.*

We use a lemma of Mahler, which gives control of a basis of a lattice from the knowledge of a maximal set of independent vectors.

Lemma 3.2.11. *Let M be a subgroup of $\mathbb{Z}^n$ of rank m and let $\boldsymbol{\lambda}_1, \ldots, \boldsymbol{\lambda}_m \in M$ be independent vectors with norm at most d, generating a subgroup $\Lambda \subset M$. For $i = 1, \ldots, m$, let V_i be the real span of $\boldsymbol{\lambda}_1, \ldots, \boldsymbol{\lambda}_i$ and define $M_i = M \cap V_i$, hence $M = M_m$. Then:*

(a) $[M : \Lambda] \leq d^m$;

(b) *there are $\mathbf{v}_1, \ldots, \mathbf{v}_m \in M$ such that for every i the vectors $\mathbf{v}_1, \ldots, \mathbf{v}_i$ form a basis of M_i and $\|\mathbf{v}_i\| \leq md$.*

Proof: Since M_i is a discrete subgroup of $\mathbb{R}^n$, it has a positive ℓ^1-distance from V_{i-1} and there is $\mathbf{v}_i = t_1\boldsymbol{\lambda}_1 + \ldots + t_i\boldsymbol{\lambda}_i \in M_i$ of minimal distance to V_{i-1} with

$t_k \in \mathbb{R}$. We may replace $\mathbf{v}_i$ by any vector in $\mathbf{v}_i + M_{i-1}$, hence we may assume $|t_k| \leq \frac{1}{2}$ for $k = 0, \ldots, i-1$. We note that $\mathbf{v}_i$ and $t_i \boldsymbol{\lambda}_i$ have the same distance to V_{i-1}, hence we must have $0 < |t_i| \leq 1$. Therefore, we have

$$\|\mathbf{v}_i\| \leq \|\boldsymbol{\lambda}_1\| + \ldots + \|\boldsymbol{\lambda}_i\| \leq md.$$

The vectors $\mathbf{v}_i$, $i = 1, \ldots, m$, form the required basis of M.

For the assertion about the index, we may assume that $M = \widetilde{\Lambda}$. Then $[M : \Lambda] = \rho(\Lambda)$. The vectors $\boldsymbol{\lambda}_j$, $j = 1, \ldots, m$ have norm bounded by d and generate Λ as a lattice in V_m. *A fortiori*, these vectors have euclidean length bounded by d and span a parallelopiped in V_m of volume at most d^m. Thus $\rho(\Lambda) \leq d^m$. □

Remark 3.2.12. The proof yields the better bound $\|\mathbf{v}_i\| \leq \max(1, i/2)d$.

Proof of Proposition 3.2.10: Let $H = H_\Lambda$ and let $\boldsymbol{\lambda}_i \in \Lambda$, $i = 1, \ldots, n-r$, be independent vectors of norm at most d. Define

$$M := \{\mathbf{x} \in \mathbb{Z}^n \mid \boldsymbol{\lambda}_i^t \cdot \mathbf{x} = 0,\ i = 1, \ldots, n-r\},$$

where $\boldsymbol{\lambda}^t$ is the transpose of $\boldsymbol{\lambda}$. Then M is a primitive subgroup of $\mathbb{Z}^n$ of rank r. By Siegel's lemma (use Theorem 2.9.4 and 2.9.8), there are independent vectors $\mathbf{x}_1, \ldots, \mathbf{x}_r$ in M with height

$$H(\mathbf{x}_1) \cdots H(\mathbf{x}_r) \leq d^{n-r}.$$

Since $H(\mathbf{x}_i) \geq 1$ ($\mathbf{x}_i \neq 0$ has integer coordinates), we get *a fortiori* $\|\mathbf{x}_i\| \leq nd^{n-r}$ for $i = 1, \ldots, r$.

For some choice of unit vectors, $\mathbf{x}_1, \ldots, \mathbf{x}_r$ and $\mathbf{e}_{h_1}, \ldots, \mathbf{e}_{h_{n-r}}$ are independent in $\mathbb{Z}^n$. By Lemma 3.2.11, we obtain a basis $\mathbf{v}_1, \ldots, \mathbf{v}_n$ of $\mathbb{Z}^n$ such that $\|\mathbf{v}_i\| \leq n^2 d^{n-r}$ and moreover $\mathbf{v}_1, \ldots, \mathbf{v}_r$ is a basis of M, because M is primitive.

The matrix $A = (\mathbf{v}_n, \ldots, \mathbf{v}_1)^t$ has determinant ± 1, satisfies $\|A\| \leq n^3 d^{n-r}$, and we have

$$A\Lambda = (\Lambda_0, \underbrace{\mathbf{0}, \mathbf{0}, \ldots, \mathbf{0}}_{r \text{ times}})^t$$

with Λ_0 a lattice in $\mathbb{Z}^{n-r}$. Hence H_{Λ_0} is a finite subgroup of $\mathbb{G}_m^{n-r}$ and $\varphi_A(H_{\Lambda_0} \times \mathbb{G}_m^r) = H_\Lambda$ by (3.1) on page 83. Replacing $\mathbf{v}_1$ by $-\mathbf{v}_1$, we may assume $\det(A) = 1$.

Finally, A^{-1} is the matrix whose entries are the cofactors of A. By Hadamard's inequality (see footnote in the course of the proof of Proposition 1.6.9), these entries are majorized by the product of the norms of $n-1$ row vectors of A, hence by $(n^2 d^{n-r})^{n-1}$. This gives $\|A^{-1}\| \leq n^{2n-1} d^{(n-1)^2}$. □

3.2.13. Let X be a Zariski closed subvariety of G. We say that an algebraic subgroup H of G is **maximal** in X if $H \subset X$ and H is not contained in a larger algebraic subgroup in X.

Proposition 3.2.14. *Let X be a Zariski closed subvariety of G, defined by polynomial equations $f_i(\mathbf{x}) := \sum a_{i,\boldsymbol{\lambda}}\mathbf{x}^{\boldsymbol{\lambda}} = 0$ $(i = 1,\dots,m)$ and let $\mathcal{L}_i$ be the set of exponents appearing in the monomials in f_i. Let H be a maximal algebraic subgroup of G contained in X. Then $H = H_\Lambda$, where Λ is generated by vectors of type $\boldsymbol{\lambda}_i' - \boldsymbol{\lambda}_i$ with $\boldsymbol{\lambda}_i', \boldsymbol{\lambda}_i \in \mathcal{L}_i$, for $i = 1,\dots,m$.*

Proof: We may suppose that X is not G, otherwise there is nothing to prove. Let $\overline{K}$ be an algebraic closure of K. The restriction of the monomial $\mathbf{x}^{\boldsymbol{\lambda}}$ to H is a character $\chi_{\boldsymbol{\lambda}}$ of H with values in $(\overline{K})^\times$. For any such character χ, define

$$\mathcal{L}_{i,\chi} = \{\boldsymbol{\lambda} \mid \boldsymbol{\lambda} \in \mathcal{L}_i,\ \chi_{\boldsymbol{\lambda}} = \chi\}.$$

Since $H \subset X$, we have linear relations

$$\sum_{\chi}\Bigg(\sum_{\boldsymbol{\lambda}\in\mathcal{L}_{i,\chi}} a_{i,\boldsymbol{\lambda}}\Bigg)\chi = 0.$$

By Artin's theorem on linear independence of characters ([**173**], Ch.VIII,Th.4.1), or directly using the Vandermonde determinant, this must be a trivial relation and hence

$$\sum_{\boldsymbol{\lambda}\in\mathcal{L}_{i,\chi}} a_{i,\boldsymbol{\lambda}} = 0 \tag{3.2}$$

for every i, χ. By definition of $\mathcal{L}_{i,\chi}$, the group H is contained in the subgroup given by the system of equations

$$\mathbf{x}^{\boldsymbol{\lambda}} = \mathbf{x}^{\boldsymbol{\lambda}'}, \qquad \boldsymbol{\lambda},\, \boldsymbol{\lambda}' \in \mathcal{L}_{i,\chi}$$

for varying i, χ. Conversely, by (3.2) this subgroup is contained in X and hence coincides with H because H is maximal in X. □

The following consequence shows that algebraic subgroups of G always determine a subgroup of $\mathbb{Z}^n$.

Corollary 3.2.15. *Every algebraic subgroup H of G is of type H_Λ for some subgroup Λ of $\mathbb{Z}^n$.*

Proof: It suffices to apply the above Proposition 3.2.14 choosing $X = H$. □

Corollary 3.2.16. *Let H be an algebraic subgroup of G. Then H is a smooth variety over K. If H is irreducible, then H is a linear torus.*

Proof: By Corollary 3.2.15, there is a subgroup Λ of $\mathbb{Z}^n$ with $H = H_\Lambda$. By Proposition 3.2.7, we see that $H_{\overline{K}}$ is smooth and hence H is smooth by A.7.14. Moreover, Corollary 3.2.8 proves the last claim. □

Another consequence is the converse of 3.2.3.

Proposition 3.2.17. *If $\varphi : \mathbb{G}_m^n \to \mathbb{G}_m^m$ is a homomorphism of linear tori with coordinates $\mathbf{x}$ and $\mathbf{y}$, then there are $\boldsymbol{\mu}_1,\dots,\boldsymbol{\mu}_m \in \mathbb{Z}^n$ with*

$$\varphi(\mathbf{x}) = (\mathbf{x}^{\boldsymbol{\mu}_1},\dots,\mathbf{x}^{\boldsymbol{\mu}_m}).$$

In particular, any automorphism of $\mathbb{G}_m^n$ *is a monoidal transformation.*

Proof: It suffices to prove the claim for $m = 1$, that is for a character $\varphi : \mathbb{G}_m^n \to \mathbb{G}_m$. Consider its graph $X \subset \mathbb{G}_m^n \times \mathbb{G}_m \cong \mathbb{G}_m^{n+1}$, namely the locus of points $(\mathbf{x}, \varphi(\mathbf{x}))$. We denote by $(\mathbf{x}, y)$ the standard coordinates on $\mathbb{G}_m^{n+1} = \mathbb{G}_m^n \times \mathbb{G}_m$. Then X is a linear torus of codimension 1 and *a fortiori* coincides with its maximal algebraic subgroup. By Corollaries 3.2.15 and 3.2.8, X is defined by a non-trivial single monomial equation $(\mathbf{x}, y)^{\boldsymbol{\lambda}} = 1$, with $\boldsymbol{\lambda} \in \mathbb{Z}^{n+1}$. This equation is identically satisfied by setting $y = \varphi(\mathbf{x})$, therefore

$$\varphi(\mathbf{x}) = \mathbf{x}^{\boldsymbol{\mu}}$$

for some $\boldsymbol{\mu} \in \mathbb{Q}^n$. Since φ is a morphism, we must have $\boldsymbol{\mu} \in \mathbb{Z}^n$, and the result follows. □

Proposition 3.2.18. *Let* $\varphi : H_1 \to H_2$ *be a homomorphism of algebraic subgroups* H_1, H_2 *of* $\mathbb{G}_m^n$ *and* $\mathbb{G}_m^m$. *Then* $\varphi(H_1)$ *is an algebraic subgroup of* $\mathbb{G}_m^m$.

Proof: It is enough to prove the claim for K algebraically closed, because Corollary 3.2.15 shows that every algebraic subgroup is defined by polynomials with coefficients in $\mathbb{Q}$. We have to prove that $\varphi(H_1)$ is Zariski closed in H_2. By Proposition 3.2.7, we may assume that H_1 is a linear torus and so we may suppose that $H_1 = \mathbb{G}_m^n$ (Corollary 3.2.8). Replacing H_2 by the Zariski closure of $\varphi(H_1)$, which is a linear torus of dimension r and hence isomorphic to $\mathbb{G}_m^r$, it remains to show that a dominant homomorphism $\varphi : \mathbb{G}_m^n \to \mathbb{G}_m^m$ of linear tori is surjective.

By Proposition 3.2.17, there are $\boldsymbol{\mu}_1, \ldots, \boldsymbol{\mu}_m \in \mathbb{Z}^n$ with $\varphi(\mathbf{x}) = (\mathbf{x}^{\boldsymbol{\mu}_1}, \ldots, \mathbf{x}^{\boldsymbol{\mu}_m})$. Let B the matrix with columns $\boldsymbol{\mu}_1, \ldots, \boldsymbol{\mu}_m$, then generalizing 3.2.3 we set $\varphi(\mathbf{x}) = \varphi_B(\mathbf{x})$. Since $\varphi(\mathbb{G}_m^n)$ is dense, it is immediate that B has rank m. By linear algebra, there are $R \in GL(m, \mathbb{Q})$ and $S \in GL(n, \mathbb{Q})$ with $B = RES$, where $E = (I_m, \mathbf{0}, \ldots, \mathbf{0})$ with I_m the unit matrix of rank m. If we had $R \in GL(m, \mathbb{Z})$ and $S \in GL(n, \mathbb{Z})$, then this would immediately lead to surjectivity of $\varphi_B = \varphi_R \circ \varphi_E \circ \varphi_S$. In general, the argument gives a $k \in \mathbb{Z} \setminus \{0\}$ such that for all $\mathbf{y} \in \mathbb{G}_m^m$ there is an $\mathbf{x} \in \mathbf{G}_m^n$ with $\varphi_B(\mathbf{x}) = \mathbf{y}^k$, which is enough to prove surjectivity. □

Theorem 3.2.19. *The following statements hold:*

(a) *The map* $\Lambda \mapsto H_\Lambda$ *is a bijection between subgroups of* $\mathbb{Z}^n$ *and algebraic subgroups of* G.

(b) *Let* Λ, M *be subgroups of* $\mathbb{Z}^n$. *Then* $H_\Lambda H_M = H_{\Lambda \cap M}$ *and* $H_\Lambda \cap H_M = H_{\Lambda + M}$.

Proof of (a): Corollary 3.2.15 shows that this map is surjective. In order to prove that it is also injective, we argue as follows. Suppose Λ, M are two subgroups of $\mathbb{Z}^n$ and that $H_\Lambda = H_M$. We have $\mathbf{x}^{\boldsymbol{\lambda}} = 1$ for $\boldsymbol{\lambda} \in \Lambda$ and $\mathbf{x}^{\boldsymbol{\mu}} = 1$ for $\boldsymbol{\mu} \in M$, therefore $H_\Lambda \subset H_{\Lambda+M}$. Since the reverse inclusion is obvious, we deduce that $H_\Lambda = H_{\Lambda+M}$. Now Proposition 3.2.7 shows that $\mathrm{rank}(\Lambda) = \mathrm{rank}(\Lambda + M)$ and $\rho(\Lambda) = \rho(\Lambda+M)$, which clearly implies that $\Lambda = \Lambda+M$. Thus M is a subgroup of Λ, and equality follows by symmetry.

Proof of (b): Using multiplication as a homomorphism and Proposition 3.2.18, we conclude that $H_\Lambda H_M$ is an algebraic subgroup of G. Thus $H_\Lambda H_M$ is the smallest algebraic subgroup containing both H_Λ and H_M. The correspondence between subgroups of $\mathbb{Z}^n$ and algebraic subgroups of G reverses inclusion relations. Since $\Lambda \cap M$ is the largest subgroup contained in both Λ and M, we have the first assertion of (b). The proof of the second assertion is the same, because the intersection of two algebraic subgroups of G is again an algebraic subgroup of G. □

3.3. Subvarieties and maximal subgroups

In this section we study algebraic subgroups and their cosets contained in a closed subvariety X of $G := \mathbb{G}_m^n$ defined over a number field K.

3.3.1. We say that X is defined by polynomials of degree at most d, if X is the set of zeros of a finite collection of polynomials $f_i(\mathbf{x})$ of degree at most d, with coefficients in $\overline{K}$. The **essential degree** $\delta(X)$ of X is the minimum integer $d \geq 1$, such that X is defined by polynomials of degree at most d.

Note that, if X is defined over $\overline{K}$ by polynomials of degree at most d, then it is also defined by polynomials of degree at most d with coefficients in K. In fact, let $K' \supset K$ be a field of definition for the polynomials f_i; by considering the traces $\mathrm{Tr}(\omega f_i(\mathbf{x}))$ with $\omega \in K'$, we obtain a set of equations defined over K (see A.4.13).

Proposition 3.3.2. *Let φ_A be a monoidal transformation or more generally any homomorphism $\varphi_A : \mathbb{G}_m^m \to \mathbb{G}_m^n$ induced by a matrix A as in 3.2.3. Then*

$$\delta\left(\bigcup_{i=1}^{k} X_i\right) \leq k \max_{i=1,\ldots,k} \delta(X_i)$$

$$\delta\left(\bigcap_{i=1}^{k} X_i\right) \leq \max_{i=1,\ldots,k} \delta(X_i)$$

$$\delta(\varphi_A^{-1}(X)) \leq \|A\|\, \delta(X).$$

Moreover, if X is irreducible of degree d (i.e. its closure in $\mathbb{P}_K^n$ has degree d), then $\delta(X) \leq d$.

Proof: If X_i is defined by polynomials $f_{ij}(\mathbf{x}), j \in J_i$, then $\bigcup X_i$ is defined by polynomials

$$f_{1j_1}(\mathbf{x}) \cdots f_{kj_k}(\mathbf{x}) \quad (j_i \in J_i).$$

This proves the first inequality. The second inequality is obvious and the third follows easily from the definition of $\|A\|$ given before Proposition 3.2.10.

For the last statement, it suffices to consider all linear cones of dimension $n-1$ over X, and note that each cone, of codimension 1 in $\mathbb{A}^n$ and of degree at most d, is defined by a single polynomial equation of the same degree. $\square$

3.3.3. If X is defined by polynomials of degree at most $d \geq 1$ and has pure dimension $r = \dim(X)$, by Bézout's theorem (see **[125]**, Ex.8.4.6) every irreducible component of X has degree at most d^{n-r} and in particular is defined by polynomials of degree at most d^{n-r}.

Proposition 3.3.4. *The number of algebraic subgroups H of G with $\delta(H) \leq d$ does not exceed $(4ed)^{n^2}$. If H is such a subgroup and H' is an algebraic subgroup of H of finite index, then $\delta(H') \leq nd$ and H/H' is a finite group of order at most d^n.*

Proof: Let H be an algebraic subgroup of G of dimension r defined by polynomials of degree at most d. Then Proposition 3.2.14 shows that H is defined by monomial equations $\mathbf{x}^{\boldsymbol{\lambda}} = 1$ with $\boldsymbol{\lambda} = (\lambda_1, \ldots, \lambda_n)$, a vector which is the difference of two vectors, with non-negative components and norm at most d; in particular, the norm of $\boldsymbol{\lambda}$ is bounded by $\|\boldsymbol{\lambda}\| \leq d$. By Corollary 3.2.15, $H = H_\Lambda$ with Λ a subgroup of $\mathbb{Z}^n$ of rank $n-r$, generated by elements $\boldsymbol{\lambda}$ as above. Let also $H' = H_M$, so that $\Lambda \subset M$ has finite index in M.

By Lemma 3.2.11 and Corollary 3.2.9, we have $|H/H'| = [M : \Lambda] \leq d^n$ and M has a basis of vectors of norm at most nd, hence H' is defined by polynomials of degree at most nd. Moreover, Λ has a basis $\mathbf{v}_i$ $(i = 1, \ldots, n-r)$ such that $\|\mathbf{v}_i\| \leq nd$. The number of such vectors does not exceed

$$2^n \binom{nd+n}{n} \leq 2^n \frac{(2nd)^n}{n!} < (4ed)^n,$$

because $n! > (n/e)^n$. It follows from this that the number of subgroups Λ does not exceed $(4ed)^{n^2}$. $\square$

Proposition 3.3.5. *Let X be a closed subvariety of G. Then every algebraic subgroup $H \subset X$ is contained in a maximal algebraic subgroup $H' \subset X$.*

Proof: Without loss of generality, we may assume that every algebraic subgroup H_1 with $H \subset H_1 \subset X$ has the same dimension as H, say r, and that $r < n$. In order to prove the proposition, we need to show that there is no infinite chain $H_1 \subsetneq H_2 \subsetneq H_3 \subsetneq \cdots$ of algebraic subgroups contained in the subvariety X. By Theorem 3.2.19, we have $H_i = H_{\Lambda_i}$ for certain subgroups Λ_i of $\mathbb{Z}^n$ of rank $n-r$, satisfying $\Lambda_1 \supsetneq \Lambda_2 \supsetneq \Lambda_3 \supsetneq \cdots$. The intersection $M = \bigcap \Lambda_i$ is a subgroup of rank strictly less than $n-r$ (note that $[\Lambda_i : \Lambda_{i+1}] \geq 2$ for every i), therefore the corresponding subgroup H_M has dimension at least $r+1$.

Let H' be the Zariski closure of $\bigcup H_i$. Then H' is an algebraic subgroup of G and it is the smallest algebraic subgroup containing all subgroups H_i. By Theorem 3.2.19, we conclude that $H' = H_M$, because M is the largest subgroup contained in every Λ_i. Since each $H_i \subset X$, the Zariski closure of $\bigcup H_i$ is also contained in X and we conclude that $H_1 \subset H_M \subset X$ with $\dim(H_M) > \dim(H_1)$. This is a contradiction. □

The same argument proves:

Proposition 3.3.6. *The torsion points of an algebraic subgroup H of G are Zariski dense in H.*

Proof: By Corollary 3.2.15, there is a subgroup Λ of $\mathbb{Z}^n$ with $H = H_\Lambda$. Let $H_i = \{\mathbf{x} \in H \mid x_1^{i!} = 1, \ldots, x_n^{i!} = 1\}$. Then $\bigcup H_i$ is the set of torsion points of H. By Theorem 3.2.19, we have $H_i = H_{\Lambda_i}$ with $\Lambda_i = i! \cdot \mathbb{Z}^n + \Lambda$, and $M = \bigcap \Lambda_i$ is Λ using the theorem of elementary divisors ([**49**], Ch.VII, §4, no.3, Th.1). As noted in the proof of the preceding proposition, the Zariski closure of $\bigcup H_i$ is H_M. □

Definition 3.3.7. *Let $X \subset G$ be a closed subvariety of G defined by polynomials of degree at most d. We define*

$$X^* = X - \bigcup\{all\ torsion\ cosets\ \subset X\},$$
$$X^\circ = X - \bigcup\{all\ torus\ cosets\ \subset X\}.$$

Theorem 3.3.8. *Let $X \subset G$ be a closed subvariety of G defined by polynomials of degree at most $d \geq 1$. Then:*

(a) *Every torsion coset in X is contained in a maximal torsion coset $\varepsilon H \subset X$.*

(b) *If εH is a maximal torsion coset in X, then H is defined by polynomials of degree at most nd, and hence the number of such linear tori does not exceed $(4end)^{n^2}$.*

(c) *For any linear torus H the number of maximal torsion cosets $\varepsilon H \subset X$ is finite and bounded in terms of d and n alone.*

In particular, X^* is a Zariski open subset of X.

Proof of (a) *and* (b)*:* Statement (a) is immediate by considering dimensions.

To prove (b), we may assume that $\varepsilon = 1$ (note that replacing X by $\varepsilon^{-1}X$ requires a base change from K to $K(\varepsilon)$, but using the remark in 3.3.1 we get a set of defining equations over K. Let $\widetilde{H}$ be the maximal subgroup of $\mathbb{G}_m^n$ with $H \subset \widetilde{H} \subset X$. Then $\dim(\widetilde{H}) = \dim(H)$ (otherwise H would not be maximal) and Proposition 3.2.14 shows that $\widetilde{H}$ is defined by polynomials of degree at most d.

It follows that H is the connected component of the identity of $\widetilde{H}$ and (b) follows from Proposition 3.3.4. □

The proof of (c) is harder and will be postponed to Chapter 4.

Theorem 3.3.9. *Let X be a closed subvariety of G defined over K by polynomials of degree at most $d \geq 1$. Then:*

(a) *Every torus coset contained in X is contained in a maximal torus coset $gH \subset X$.*

(b) *If gH is a maximal torus coset in X, then H is defined by polynomials of degree at most nd, and hence the number of such linear tori does not exceed $(4end)^{n^2}$.*

(c) *For any non-trivial linear torus H of dimension r, the union*

$$X(H) := \bigcup_{gH \subset X} gH$$

of all torus cosets $gH \subset X$ has the following structure. There is $A \in SL(n, \mathbb{Z})$ such that

$$X(H) = \varphi_A(\widetilde{X(H)} \times \mathbb{G}_m^r),$$

where $\widetilde{X(H)}$ is a closed subvariety of $\mathbb{G}_m^{n-r}$ defined over K by polynomials of degree bounded by $n^3 \delta(H)^{n-r} d$.

In particular, X° is a Zariski open subset of X.

Proof: The proof of (a) and (b) is the same as in the previous theorem. For the proof of (c), let Λ define the linear torus H, so that Λ is a primitive subgroup of $\mathbb{Z}^n$ of rank $n-r$ (Corollary 3.2.8). By Proposition 3.2.14, Λ is generated by vectors of norm bounded by $\delta(H)$.

By Proposition 3.2.10, we may assume that $H = \mathbf{1}_{n-r} \times \mathbb{G}_m^r$ and replace X by $\widetilde{X} := \varphi_A^{-1}(X)$. By Proposition 3.3.2, $\widetilde{X}$ is defined over K by polynomials of degree at most $n^3 \delta(H)^{n-r} d$.

Let $\widetilde{f}_i(x_1, \ldots, x_n) = 0$ be a set of defining equations for $\widetilde{X}$. To say that $gH \subset \widetilde{X}$ means that

$$\widetilde{f}_i(g_1, \ldots, g_{n-r}, x_{n-r+1}, \ldots, x_n) = 0$$

is identically satisfied in $x_{n-r+1}, \ldots, x_n$. This yields a finite set of polynomial equations in $g_1, \ldots, g_{n-r}$, again of degree at most $n^3 \delta(H)^{n-r} d$, defining a closed subvariety $\widetilde{X(H)} \subset \mathbb{G}_m^{n-r}$ with the desired properties. This proves (c). □

Remark 3.3.10. The proof gives a matrix $A \in SL(n, \mathbb{Z})$ with $\|A\| \leq n^3 \delta(H)^{n-r}$ and $\|A^{-1}\| \leq n^{2n-1} \delta(H)^{(n-1)^2}$.

3.4. Bibliographical notes

The material in this chapter follows closely the treatment given in Schmidt's paper [**273**], with several additions. For the general theory of linear algebraic groups, we refer to A. Borel [**41**] and M. Demazure and P. Gabriel [**85**].

4 SMALL POINTS

4.1. Introduction

In this chapter, we study the distribution of algebraic points of small height on subvarieties of $\mathbb{G}_m^n$.

The general case is handled by Zhang's theorem and its uniform version, discussed in Section 4.2, which presupposes knowledge of the results in the previous Chapter 3. In Section 4.3, we give an alternative non-constructive proof using an equidistribution theorem.

The special case in which the subvariety coincides with the ambient variety $\mathbb{G}_m$ is a fascinating question of D.H. Lehmer, and we shall prove in Section 4.4 Dobrowolski's well-known theorem, which provides a good lower bound for the height of algebraic numbers which are not roots of unity.

In this circle of ideas, we shall also give a proof of a result of F. Amoroso and R. Dvornicich providing an absolute positive lower bound for the height of algebraic numbers, not roots of unity, in abelian extensions of $\mathbb{Q}$.

In Section 4.5, we give examples of infinite algebraic extensions of number fields which have only finitely many numbers of bounded height as in Northcott's theorem. In a similar way, we discuss in Section 4.6 infinite algebraic extensions of $\mathbb{Q}$ such that the height has a positive lower bound outside of the roots of unity.

4.2. Zhang's theorem

In 1992 S. Zhang obtained a surprising result on the height of algebraic points on curves in $\mathbb{G}_m^n$. He showed that for any geometrically irreducible curve $C \subset \mathbb{G}_m^n$, not a torsion coset, there is a positive lower bound for the height of non-torsion algebraic points in C. This was new even in the simplest case of the equation $x + y = 1$. Later, he extended this to general subvarieties of $\mathbb{G}_m^n$.

As we shall see in Section 5.2, there are interesting applications to the problem of obtaining good upper bounds for the number of solutions of the unit equation.

We identify $\mathbb{G}_m$ with the affine line punctured at the origin, together with the usual multiplication. Definitions and notation will be as in Chapter 3, with the additional assumption that K is a number field. Note also that the statement of Theorem 3.3.8, (c) has not yet been proved at this stage.

The **standard height** of $\mathbf{x} = (x_1, \dots, x_n) \in \mathbb{G}_m^n$ is

$$\widehat{h}(\mathbf{x}) := \sum_{i=1}^{n} h(x_i)$$

with h the absolute Weil height of algebraic numbers. The height $\widehat{h}$ has the following properties:

(a) Homogeneity and symmetry: $\widehat{h}(P^m) = |m| \cdot \widehat{h}(P)$ for $m \in \mathbb{Z}$.

(b) Non-degeneracy: $\widehat{h}(P) = 0$ if and only if P is a torsion point.

(c) Triangle inequality: $\widehat{h}(PQ^{-1}) \leq \widehat{h}(P) + \widehat{h}(Q)$.

(d) Finiteness: There are only finitely many points $P \in \mathbb{G}_m^n$ such that $[\mathbb{Q}(P) : \mathbb{Q}]$ and $\widehat{h}(P)$ are both bounded.

These properties are clear from the corresponding properties of the Weil height, with (b) and (d) following from the theorems of Kronecker and Northcott (see 1.5.9, 2.4.9). Thus $\widehat{h}(PQ^{-1})$ is a translation invariant semidistance $d(P, Q)$ on $\mathbb{G}_m^n$ and actually a translation invariant distance on $\mathbb{G}_m^n/\mathrm{tors}$.

Proposition 4.2.1. *Let $\varphi_A : \mathbb{G}_m^n \xrightarrow{\sim} \mathbb{G}_m^n$ be a monoidal transformation determined by $A \in GL(n, \mathbb{Z})$. Then the height $\widehat{h} \circ \varphi_A$ is equivalent to the height $\widehat{h}$, in the sense that there are two positive constants c_1, c_2 such that*

$$c_1 \widehat{h}(\mathbf{x}) \leq \widehat{h}(\varphi_A(\mathbf{x})) \leq c_2 \widehat{h}(\mathbf{x})$$

for every $\mathbf{x} \in \mathbb{G}_m^n$.

Proof: We have, with the notation of 3.2.3 and 3.2.10

$$\widehat{h}(\varphi_A(\mathbf{x})) = \sum_{i=1}^{n} h(\mathbf{x}^{A\mathbf{e}_i}) \leq \sum_{i=1}^{n} \sum_{j=1}^{n} |a_{ji}|\, h(x_j) \leq n\, \|A\| \widehat{h}(\mathbf{x}).$$

This proves the second half of the inequality, with $c_2 = n\|A\|$.

Now we apply this inequality with $\varphi_{A^{-1}}(\mathbf{x})$ in place of $\mathbf{x}$ and by replacing A by A^{-1}, we obtain the first half of the inequality, with $c_1 = 1/(n\|A^{-1}\|)$. □

A torsion coset in X is said to be **maximal** in X if it is not contained in a larger torsion coset in X. A similar definition applies to torus cosets in X.

We have **Zhang's theorem**:

Theorem 4.2.2. *Let X be a closed subvariety of $\mathbb{G}_m^n$ defined over a number field K and let X^* be the complement in X of the union of all torsion cosets $\varepsilon H \subset X$. Let $f_i \in K[\mathbf{x}]$ be a set of polynomials of degree at most d defining X. Then:*

(a) *The number of maximal torsion cosets in X is finite and bounded in terms of d and n alone. Moreover, every maximal torsion coset has the form ζH, where the orders of the torsion points ζ and the essential degrees of the linear tori H are also bounded in terms of d and n.*

(b) *The height of points $P \in X^*$ has a positive lower bound, depending on n, d, $[K : \mathbb{Q}]$, and $\max h(f_i)$.*

The results (a) *and* (b) *are effective.*

There is a slightly different uniform version for torus cosets.

Theorem 4.2.3. *Let $\widehat{h}$ be the standard height on $\mathbb{G}_m^n$, with the associated semi-distance $d(P,Q) = \widehat{h}(PQ^{-1})$. Let X be a closed subvariety of $\mathbb{G}_m^n$ defined over a number field K by polynomials of degree at most d, and let X° be the complement in X of all its torus cosets $gH \subset X$, $\dim(H) \geq 1$. Then:*

(a) *X° is Zariski open in X, and $X \setminus X^\circ$ is defined by polynomials of degree bounded in terms of n and d alone.*

(b) *There are a positive constant $\gamma(d,n)$ and a positive integer $N(d,n)$, depending only on d and n, with the following property: Let $Q \in \mathbb{G}_m^n$. Then*

$$\{ P \in X^\circ \mid d(P,Q) \leq \gamma(d,n)\}$$

is a finite set of cardinality at most $N(d,n)$. Moreover, for every point P in this set, the estimate $[K(P,Q) : K(Q)] \leq N(d,n)$ holds.

Remark 4.2.4. The closed set $X \setminus X^\circ$ can be effectively determined.

To see this, note that $X \setminus X^\circ$ is the union of the subvarieties $X(H)$ (see Theorem 3.3.9 (c)), where H runs over all non-trivial tori that can appear in maximal torus cosets $gH \subset X$. By Theorem 3.3.9 (c) and Remark 3.3.10, we can find $A \in SL(n,\mathbb{Z})$, with entries bounded in terms of d and n, such that $X(H) = \varphi_A(Y \times \mathbb{G}_m^r)$, where $r = \dim(H) \geq 1$ and Y is defined by polynomials of degree bounded in terms of d and n. The proof there also shows that their height is bounded in terms of the height of the polynomial equations used to define X. By Northcott's theorem (see Theorem 2.4.9), Y can be effectively determined.

Remark 4.2.5. The constants $\gamma(d,n)$ and $N(d,n)$ are effective and hence the finite set of points in (b) can be effectively determined for every $Q \in \mathbb{G}_m^n$.

Remark 4.2.6. There is no uniform version of Theorem 4.2.2. For example, if a, b are not roots of unity or 0, the equation $1 + ax + by = 0$ in $\mathbb{G}_m^2$ has a non-torsion solution $\xi = (a^{-1}\rho, b^{-1}\rho^2)$ with ρ a primitive cubic root of unity. We have $\widehat{h}(\xi) = h(a) + h(b) > 0$, and we can make it arbitrarily small by choosing a and b, for example $a = b = 2^{1/m}$ with $m \to \infty$.

Remark 4.2.7. D. Zagier [**336**] has obtained the optimal lower bound

$$\widehat{h}(\boldsymbol{\xi}) \geq \frac{1}{2} \log \left(\frac{1 + \sqrt{5}}{2} \right)$$

for non-torsion solutions $\boldsymbol{\xi}$ of $1 + x + y = 0$. Equality is attained for x or y, a primitive 10th root of unity, and this minimum is isolated.

We need the following result:

Lemma 4.2.8. *Let $f(x_1, \ldots, x_n)$ be a polynomial with integer coefficients, of degree d and height $H(f)$ and let $p > e\binom{d+n}{n}H(f)$ be a prime number. Let $\boldsymbol{\xi} = (\xi_1, \ldots, \xi_n)$ be an algebraic point with $f(\xi_1, \ldots, \xi_n) = 0$ and $f(\xi_1^p, \ldots, \xi_n^p) \neq 0$. Then $\widehat{h}(\boldsymbol{\xi}) \geq 1/(pd)$.*

Proof: We may assume that the coefficients of f have no common divisor greater than 1. Let K be a number field containing all coordinates ξ_i.

By Fermat's little theorem, we have

$$f^p(x_1, \ldots, x_n) = f(x_1^p, \ldots, x_n^p) + p\, g(x_1, \ldots, x_n),$$

where $g(x_1, \ldots, x_n) \in \mathbb{Z}[x_1, \ldots, x_n]$ has degree at most pd. Since by hypothesis we have $f(\xi_1, \ldots, \xi_n) = 0$, we get

$$f(\xi_1^p, \ldots, \xi_n^p) = -p\, g(\xi_1, \ldots, \xi_n). \tag{4.1}$$

For any $\zeta \in K \setminus \{0\}$ the product formula yields

$$\sum_{v \in M_K} \log |\zeta|_v = 0\,.$$

We apply this with $\zeta := f(\xi_1^p, \ldots, \xi_n^p)$ and estimate terms as follows.

If $v | p$, we have by (4.1) and the fact that g has integer coefficients

$$\log |\zeta|_v = \log |p\, g(\xi_1, \ldots, \xi_n)|_v \leq \log |p|_v + pd \sum_{i=1}^{n} \log^+ |\xi_i|_v + \log |f|_v,$$

because the Gauss norm equals 1 (the coefficients of f have no non-trivial common divisor and v is not archimedean).

At the other places v, we have, with $\varepsilon_v = [K_v : \mathbb{R}]/[K : \mathbb{Q}]$ if v is archimedean and $\varepsilon_v = 0$ if v is not archimedean

$$\log |\zeta|_v = \log |f(\xi_1^p, \ldots, \xi_n^p)|_v \leq pd \sum_{i=1}^{n} \log^+ |\xi_i|_v + \log |f|_v + \varepsilon_v \log \binom{d+n}{n},$$

because the number of monomials in f does not exceed $\binom{d+n}{n}$.

Summing over all $v \in M_K$ and using $\sum_{v|p} \log |p|_v = -\log p$ from Lemma 1.3.7, we infer

$$0 = \sum_{v \subset M_K} \log |\zeta|_v \leq -\log p + pd\,\widehat{h}(\boldsymbol{\xi}) + h(f) + \log \binom{d+n}{n}. \qquad \square$$

The following consequence of Lemma 4.2.8 is due to W.M. Schmidt.

Corollary 4.2.9. *Let $f(\mathbf{x})$ be as before and let m be a positive integer all of whose prime factors are greater than $e\binom{d+n}{n}H(f)$. Suppose $f(\boldsymbol{\xi}) = 0$. Then either $f(\boldsymbol{\xi}^m) = 0$ or $\widehat{h}(\boldsymbol{\xi}) \geq 1/(md)$.*

Proof: The easy proof is by induction on the number of prime factors of m, writing $m = pm'$. If $f(\boldsymbol{\xi}^{m'}) \neq 0$, we apply the corollary inductively with m' in place of m. If instead $f(\boldsymbol{\xi}^{m'}) = 0$, we apply Lemma 4.2.8 to the point $\boldsymbol{\xi}^{m'}$ in place of $\boldsymbol{\xi}$ and note that $\widehat{h}(\boldsymbol{\xi}^{m'}) = m'\widehat{h}(\boldsymbol{\xi})$. $\square$

4.2.10. *Proof of Theorem 4.2.2* (b) *and the following partial result of 4.2.2* (a)*:*

4.2.2 (a$'$) *The union of torsion cosets of X is contained in finitely many maximal torsion cosets of the form ζM, where the orders of the torsion points ζ and the essential degrees of the linear tori M are bounded in terms of $d, n, [K:\mathbb{Q}]$, and* $\max h(f_i)$.

We say that a quantity is **controlled** if it is a function of $d, n, [K:\mathbb{Q}]$, and $\max h(f_i)$, which, in principle, can be made explicit.

We prove (a$'$) and (b) simultaneously by induction on the rank n of the ambient linear torus. The claim is trivial for $n = 0$. Now let $n \geq 1$. If $X = \mathbb{G}_m^n$, we have nothing to prove. So we may assume that there is a non-zero f_k in our list of defining polynomials. Let us consider

$$f(\mathbf{x}) := \prod_{\sigma} f_k(\sigma\mathbf{x}) = \sum_{\mathbf{m}} a_{\mathbf{m}} \mathbf{x}^{\mathbf{m}},$$

where σ ranges over all embeddings of K in $\overline{\mathbb{Q}}$. It is a polynomial of degree at most $D = [K:\mathbb{Q}]d$ with coefficients in $\mathbb{Q}$. Let q be the product of all primes $p \leq e\binom{D+n}{n}H(f)$. Let $\boldsymbol{\xi} \in X$. We apply Corollary 4.2.9 with $m = qj+1$, for $j = 0, \ldots, \binom{D+n}{n} - 1$. If $f(\boldsymbol{\xi}^{qj+1}) \neq 0$ for some j, we obtain a controlled positive lower bound for $\hat{h}(\boldsymbol{\xi})$. If instead $f(\boldsymbol{\xi}^{qj+1}) = 0$ for every j, we get

$$\sum_{\mathbf{m}} a_{\mathbf{m}} (\boldsymbol{\xi}^{\mathbf{m}})^{qj+1} = 0.$$

We view this as a homogeneous linear system with coefficients $(\boldsymbol{\xi}^{\mathbf{m}})^{qj}$ and unknowns $a_{\mathbf{m}}\boldsymbol{\xi}^{\mathbf{m}}$, so that its determinant must be 0. This is a Vandermonde determinant, and looking at its factorization we see that

$$\boldsymbol{\xi}^{q\mathbf{m}} = \boldsymbol{\xi}^{q\mathbf{m}'}$$

for some $\mathbf{m} \neq \mathbf{m}'$. Hence $\boldsymbol{\xi}$ belongs to a subgroup H of $\mathbb{G}_m^n$ defined by polynomials of degree at most qD.

Let $H = FM$ be a direct product decomposition as in Proposition 3.2.7 with M a linear torus of rank r and F a finite abelian group, and let $\boldsymbol{\xi} \in \gamma M$, $\gamma \in F$. By Proposition 3.3.4, the essential degree of M and the order of F is controlled. Let $\boldsymbol{\zeta} \in F$ with $\boldsymbol{\xi} \in \boldsymbol{\zeta} M$. Since the order of $\boldsymbol{\zeta}$ is controlled, we conclude that

$$X' := \boldsymbol{\zeta}^{-1} X \cap M$$

is defined over the extension $K' := K(\boldsymbol{\zeta})/K$, also of controlled degree. Now we bring M in normal form using Proposition 3.2.10. There is a monoidal transformation φ_A with controlled $\|A\|$ and $\|A^{-1}\|$ such that $\varphi_{A^{-1}}(M) = \mathbf{1}_{n-r} \times \mathbb{G}_m^r$. Then $Y := \varphi_{A^{-1}}(X') \cap (\mathbf{1}_{n-r} \times \mathbb{G}_m^r)$ is given by the polynomials $f_i(\boldsymbol{\zeta} A(\mathbf{1}_{n-r} \times \mathbf{y}))$ as a subvariety of $\mathbb{G}_m^r$. They have controlled degree and controlled height. Since $r < n$, induction proves that $Y \setminus Y^* = \bigcup \boldsymbol{\zeta}_l' M_l'$ for torsion points $\boldsymbol{\zeta}_l'$, of controlled number and order and for linear tori M_l' of controlled essential degree. Moreover, the height has a controlled positive lower bound on Y^*. Using Proposition 4.2.1 and $\hat{h}(\boldsymbol{\xi}) = \hat{h}(\boldsymbol{\zeta}^{-1}\boldsymbol{\xi})$, we see that either $\hat{h}(\boldsymbol{\xi})$ has a controlled positive lower bound or $\boldsymbol{\xi}$ is contained in some $\boldsymbol{\zeta}\varphi_A(\mathbf{1}_{n-r} \times \boldsymbol{\zeta}_l' M_l') \subset X$. Clearly, the latter is a torsion coset of the form $\boldsymbol{\zeta}_\mu M_\mu$, where the torsion point $\boldsymbol{\zeta}_\mu$ has controlled order and the linear torus M_μ has controlled essential degree. The number of such torsion cosets is also controlled. It remains to show that

$$X \setminus X^* = \bigcup_\mu \boldsymbol{\zeta}_\mu M_\mu.$$

Let $\boldsymbol{\xi} \in X \setminus X^*$. Then $\boldsymbol{\xi}$ is contained in a torsion coset. By Proposition 3.3.6, there is a sequence $(\boldsymbol{\xi}_n)$ of torsion points in X converging to $\boldsymbol{\xi}$. Because of the positive lower bound outside the $\boldsymbol{\zeta}_\mu M_\mu$, we conclude that all $\boldsymbol{\xi}_n$ and hence $\boldsymbol{\xi}$ are contained in $\bigcup \boldsymbol{\zeta}_\mu M_\mu$. This proves the claim. □

4.2.11. *Proof of Theorem 4.2.3:* Note that Theorem 4.2.3 (a) is part of Theorem 3.3.9, hence it remains to show (b). Now it will be convenient to say that a quantity is **controlled** if it admits a bound depending only on d and n and which, in principle, can be made explicit. However, no effort will be made here to give explicit calculations for such bounds. To prove (b), it is enough to show the existence of controlled constants $\gamma > 0$ and $N > 0$ such that the set

$$E_\gamma := \{P \in X^\circ \mid \hat{h}(P) \leq \gamma\}$$

has at most N points. Then the first claim follows immediately by applying translation with Q^{-1} and the last statement is also clear because conjugation over $K(Q)$ does not change X and heights.

We proceed by induction on n. By convention, $\mathbb{G}_m^0$ is the trivial torus $\{1\}$, thus the claim is obvious for $n = 0$. For the induction step, we may assume that X is a proper subvariety of $\mathbb{G}_m^n$. Let $\mathcal{M}$ be again the set of monomials of degree bounded by d and let $r := |\mathcal{M}| = \binom{d+n}{n}$. For $s \in \{1, \ldots, r\}$, consider the $s \times r$ matrix

$$A_s(\mathbf{x}) := \begin{pmatrix} \mathbf{x}_1^{\mathbf{m}} \\ \vdots \\ \mathbf{x}_s^{\mathbf{m}} \end{pmatrix}_{\mathbf{m} \in \mathcal{M}},$$

where $\mathbf{x}_1, \ldots, \mathbf{x}_s$ are points of $\mathbb{G}_m^n$. Let Y_s be the closed subvariety of $(\mathbb{G}_m^n)^s$, given by the vanishing of all $s \times s$ minors of the matrix $A_s(\mathbf{x})$.

Note that Y_s is universal in the sense that it is defined over $\mathbb{Q}$ and depends only on d, n, and s. If we apply Theorem 4.2.2 (a$'$) and (b), then the constants depend only on d and n. Hence the number of maximal torsion cosets in Y_s is controlled and they have the form εH, where the order of the torsion point ε and the essential degree of H are also controlled. Moreover, the height has a controlled positive lower bound on Y_s^*.

Let γ be a sufficiently small positive constant, which is controlled and which we will define precisely later in the course of the proof. For the moment, we just assume that $s\gamma$ is smaller than the above lower bound on Y_s^* for every $s = 0, \ldots, r$. Clearly, we may assume that E_γ is not empty. Now let us choose

$$s := 1 + \max_{\boldsymbol{\xi}_1, \ldots, \boldsymbol{\xi}_r \in E_\gamma} \operatorname{rank} A_r(\boldsymbol{\xi}_1, \ldots, \boldsymbol{\xi}_r).$$

Note that the rank of A_r is at most $r - 1$ on X^r since any equation of X yields a linear relation among the columns of the matrix, hence $2 \leq s \leq r$.

We fix $\boldsymbol{\xi}_1^*, \ldots, \boldsymbol{\xi}_{s-1}^* \in E_\gamma$ such that $A_{s-1}(\boldsymbol{\xi}_1^*, \ldots, \boldsymbol{\xi}_{s-1}^*)$ has maximal rank $s-1$. Our rank condition implies that

$$S^* := \boldsymbol{\xi}_1^* \times \cdots \times \boldsymbol{\xi}_{s-1}^* \times \mathbb{G}_m^n$$

is not contained in Y_s, otherwise the Laplace expansion of the $s \times s$ minors of the matrix $A_s(\boldsymbol{\xi}_1^*, \ldots, \boldsymbol{\xi}_{s-1}^*, \mathbf{x}_s)$ with respect to the last row would show that all $(s-1) \times (s-1)$ minors of $A_{s-1}(\boldsymbol{\xi}_1^*, \ldots, \boldsymbol{\xi}_{s-1}^*)$ vanish. Hence the projection p_s to the last factor does not map $S^* \cap Y_s$ onto $\mathbb{G}_m^n$. For any $\boldsymbol{\xi} \in E_\gamma$, the point $(\boldsymbol{\xi}_1^*, \ldots, \boldsymbol{\xi}_{s-1}^*, \boldsymbol{\xi})$ is contained in a maximal torsion coset εH. Now we fix the torsion coset εH and also some $\boldsymbol{\xi}_s^* \in E_\gamma$ with $(\boldsymbol{\xi}_1^*, \ldots, \boldsymbol{\xi}_s^*) \in \varepsilon H$.

We have seen that $p_s(S^* \cap \varepsilon H)$ is properly contained in $\mathbb{G}_m^n$. Let $(\mathbf{x}/\varepsilon)^{\boldsymbol{\lambda}}, \boldsymbol{\lambda} \in \Lambda$, be a set of defining equations of εH of controlled degree. Then $S^* \cap \varepsilon H$ is defined by setting $\mathbf{x}_i = \boldsymbol{\xi}_i^*, i = 1, \ldots, s-1$ in these equations. We conclude that

$$p_s(S^* \cap \varepsilon H) = \{\mathbf{x} \in \mathbb{G}_m^n \mid \mathbf{x}^{\boldsymbol{\lambda}_s} = \varepsilon^{\boldsymbol{\lambda}} \prod_{i=1}^{s-1} (\boldsymbol{\xi}_i^*)^{-\boldsymbol{\lambda}_i},$$
$$\boldsymbol{\lambda} = (\boldsymbol{\lambda}_1, \ldots, \boldsymbol{\lambda}_s) \in \Lambda\} = \boldsymbol{\xi}_s^* H',$$

where H' is the algebraic subgroup of $\mathbb{G}_m^n$ given by the equations $\mathbf{x}^{\boldsymbol{\lambda}_s} = 1, \boldsymbol{\lambda} \in \Lambda$. Let M be the connected component of the identity of H'. By Proposition 3.2.7 and Proposition 3.3.4, the essential degree of M is controlled and there are torsion points $\boldsymbol{\rho}_j$ of $\mathbb{G}_m^n$, whose number and orders are also controlled, such that

$$H' = \bigcup_j \boldsymbol{\rho}_j M.$$

We conclude that

$$p_s(S^* \cap \varepsilon H) = \bigcup_j \boldsymbol{\eta}_j M$$

with $\boldsymbol{\eta}_j = \boldsymbol{\xi}_s^* \boldsymbol{\rho}_j$. Now let $X' := \boldsymbol{\eta}_j^{-1} X \cap M$. Since M is a linear torus of dimension $n' < n$, the idea is to apply our induction hypothesis to X' in M. The only obstacle is that M has not the normal form $\mathbb{G}_m^{n'}$. As noted in 4.2.10, this can be obviated by using Proposition 3.2.10 to bring M in normal form, and Proposition 4.2.1 to control the change in height induced by the monoidal transformation. Thus the induction step gives controlled positive constants γ' and N' such that the number of points $\boldsymbol{\xi}' \in (X')^\circ$ with $\hat{h}(\boldsymbol{\xi}') \leq \gamma'$ is bounded by N'.

For $\boldsymbol{\xi} \in E_\gamma$ with $(\boldsymbol{\xi}_1^*, \ldots, \boldsymbol{\xi}_{s-1}^*, \boldsymbol{\xi})$ contained in our fixed maximal torsion coset εH of Y_s, we have $\boldsymbol{\xi} \in p_s(S^* \cap \varepsilon H)$, thus $\boldsymbol{\xi}$ is contained in a torus coset $\eta_j M$. Then $\boldsymbol{\xi}' := \boldsymbol{\eta}_j^{-1} \boldsymbol{\xi} \in X'$ and

$$\widehat{h}(\boldsymbol{\xi}') \leq \widehat{h}(\boldsymbol{\eta}_j) + \widehat{h}(\boldsymbol{\xi}) = \widehat{h}(\boldsymbol{\xi}_s^*) + \widehat{h}(\boldsymbol{\xi}) \leq 2\gamma.$$

Finally, we may assume $\gamma \leq \gamma'/2$. By the inductive step, we conclude that the number of $\boldsymbol{\xi} \in E_\gamma \cap \boldsymbol{\eta}_j M$ with $(\boldsymbol{\xi}_1^*, \ldots, \boldsymbol{\xi}_{s-1}^*, \boldsymbol{\xi}) \in \varepsilon H$ is bounded by N'. Since the number of torus cosets $\boldsymbol{\eta}_j M$ is bounded by a controlled N'', there are at most $N'N''$ points $\boldsymbol{\xi} \in E_\gamma$ with $(\boldsymbol{\xi}_1^*, \ldots, \boldsymbol{\xi}_{s-1}^*, \boldsymbol{\xi}) \in \varepsilon H$.

If $\boldsymbol{\xi}$ ranges over E_γ, then we have seen that the points

$$(\boldsymbol{\xi}_1^*, \ldots, \boldsymbol{\xi}_{s-1}^*, \boldsymbol{\xi})$$

are contained in various maximal torus cosets εH of Y_s. For every such εH, we fix $\boldsymbol{\xi}_s^* \in E_\gamma$ with $(\boldsymbol{\xi}_1^*, \ldots, \boldsymbol{\xi}_s^*) \in \varepsilon H$ and apply our preceding procedure. If N''' is the controlled number of such maximal torsion cosets, then E_γ contains at most $N = N'N''N'''$ points. Since N is controlled, this proves the induction step. □

4.2.12. *Completion of the proof of Theorem 4.2.2* (a): By Theorem 3.3.8 (b), we already know that every maximal torsion coset in X has the form εH with $\delta(H) \leq nd$.

In proving Zhang's theorem 4.2.2 (a), it suffices to deal with the **maximal torsion points** of X, namely maximal torsion cosets of dimension 0. To see this, we use Theorem 3.3.9 as follows. We have to find all maximal torsion cosets $\varepsilon H \subset X$. Assume that $\dim(H) \geq 1$. Then εH is also a torus coset. By Theorem 3.3.8 (b), we can fix the linear torus H. Now, using Theorem 3.3.9 (c), the maximal torus cosets εH in X are in one-to-one correspondence with the maximal torsion points of $\widetilde{X(H)}$. Since $\widetilde{X(H)}$ is defined by polynomials of degree bounded in terms of $\delta(X)$ and n, our claim follows.

Thus, by our preceding considerations, it suffices to deal only with maximal torsion points. A maximal torsion point ε cannot belong to a torus coset $gH \subset X$, otherwise we could take $g = \varepsilon$ and ε would not be maximal. Hence every maximal torsion point belongs to X°. Since ε is a torsion point it has semidistance 0 from the identity 1, because $d(\varepsilon, 1) = \widehat{h}(\varepsilon) = 0$. By Theorem 4.2.3 (b), the number of such points ε and their order are at most $N(d,n)$, as asserted, proving what we want for maximal torsion points. □

4.2.13. *Completion of the proof of Theorem 3.3.8* (c)*:* This is immediate from Theorem 4.2.2 (a) in the case of number fields. In general, we argue as follows. Let $\varepsilon_i H_i$ be a list of all maximal torsion cosets in X. They are defined over $\overline{\mathbb{Q}}$ (Corollary 3.2.15). Now consider the Zariski closure Y of the union of all cosets $\bigcup \varepsilon_i H_i$. Then $Y \subset X$, thus all $\varepsilon_i H_i$ are maximal torsion cosets of Y. Since Y is defined over $\overline{\mathbb{Q}}$, there are only finitely many $\varepsilon_i H_i$. □

4.3. The equidistribution theorem

Another approach is due to L. Szpiro, E. Ullmo, and S. Zhang [**296**], and Yu. F. Bilu [**24**]. The idea is that points of small height under the action of Galois conjugation tend to be equidistributed with respect to a suitable measure.

For a in the multiplicative group $\mathbb{C}^\times$ of $\mathbb{C}$, let δ_a be the usual Dirac measure at a and for $\xi \in \overline{\mathbb{Q}}$ let

$$\delta_\xi = \frac{1}{[\mathbb{Q}(\xi) : \mathbb{Q}]} \sum_{\sigma : \mathbb{Q}(\xi) \to \mathbb{C}} \delta_{\sigma\xi}$$

be the probability measure supported at all complex conjugates of ξ, with equal mass at each point.

In order to understand the following considerations, we recall some basic facts from functional analysis and measure theory.

Let X be a locally compact Hausdorff space and let $C_c(X)$ be the space of complex-valued continuous functions on X with compact support, endowed with the supremum norm. Then the Riesz representation theorem says that for every continuous linear functional Λ on this normed space there exists a unique complex regular Borel measure μ such that

$$\Lambda(f) = \int_X f \, \mathrm{d}\mu, \qquad f \in C_c(X).$$

Moreover, the operator norm of Λ equals $|\mu|(X)$, where $|\mu|$ is the total variation of the measure μ. For details, we refer to W. Rudin [**252**], Th.6.19.

It is also clear that every complex regular Borel measure μ on X yields a continuous linear functional on $C_c(X)$. Therefore, the space of complex regular Borel measures on X is the dual of $C_c(X)$ in the sense of functional analysis, and we denote it by $C_c(X)^*$. The weak-* topology on $C_c(X)^*$ is the coarsest topology on $C_c(X)$ such that, for every $f \in C_c(X)$, the linear functional $\mu \mapsto \int_X f \, \mathrm{d}\mu$ is continuous. The Banach–Alaoglu theorem (see W. Rudin [**251**], 3.15) says that the unit ball

$$\{\mu \in C_c(X)^* \mid |\mu|(X) \leq 1\}$$

is weak-* compact.

In what follows, we apply these concepts to $X = \mathbb{C}^\times$.

We have the following result of Bilu:

Theorem 4.3.1. *Let $(\xi_i)_{i\in\mathbb{N}}$ be an infinite sequence of distinct non-zero algebraic numbers such that $h(\xi_i) \to 0$ as $i \to \infty$. Then the sequence $(\boldsymbol{\delta}_{\xi_i})_{i\in\mathbb{N}}$ converges in the weak-* topology to the uniform probability measure $\mu_\mathbb{T} := \mathrm{d}\theta/(2\pi)$ on the unit circle $\mathbb{T} := \{e^{i\theta} \mid 0 \leq \theta < 2\pi\}$ in $\mathbb{C}$.*

Proof: By the weak-* compactness of the unit ball in $C_c(\mathbb{C}^\times)^*$, it is enough to show that any convergent subsequence of the sequence $(\boldsymbol{\delta}_{\xi_i})_{i\in\mathbb{N}}$ has limit $\mu_\mathbb{T}$. Thus we may assume that the measures $\boldsymbol{\delta}_{\xi_i}$ converge in the weak-* topology to a Borel measure μ, and we have to show that $\mu = \mu_\mathbb{T}$.

Let μ be a weak-* limit of the measures $\boldsymbol{\delta}_{\xi_i}$. Let a_{0i} and d_i be the leading coefficient and degree of a minimal equation for ξ_i. Since the ξ_i are distinct, Northcott's theorem in 1.6.8 shows that $d_i \to \infty$.

As in 1.5.7, $\log^+$ is the maximum of 0 and $\log$. By Propositions 1.6.5 and 1.6.6 and the hypothesis $h(\xi_j) \to 0$, we have

$$h(\xi_i) = \frac{1}{d_i} \log |a_{0i}| + \frac{1}{d_i} \sum_\sigma \log^+ |\sigma\xi_i| \to 0 \tag{4.2}$$

as $i \to \infty$; this implies $\log |a_{0i}| = o(d_i)$ and $\sum_\sigma \log^+ |\sigma\xi_i| = o(d_i)$, where σ ranges over all embeddings of $\mathbb{Q}(\xi_i)$ into $\mathbb{C}$.

By weak-* convergence we deduce

$$\frac{1}{d_i} \sum_\sigma f(\sigma\xi_i) \log^+ |\sigma\xi_i| \to \int_\mathbb{C} f(z) \log^+ |z| \, \mathrm{d}\mu(z)$$

for any continuous function $f(z)$ with compact support in $\mathbb{C}^\times$. Thus (4.2) shows that

$$\int_\mathbb{C} f(z) \log^+ |z| \mathrm{d}\mu(z) = 0$$

and μ must be supported in the unit disk $|z| \leq 1$. Since $h(1/\xi_i) = h(\xi_i) \to 0$, working with the sequence $(1/\xi_i)_{i\in\mathbb{N}}$ we deduce in a similar fashion that μ is supported in $|z| \geq 1$. Thus any limit measure μ has support in the unit circle $\mathbb{T}$. Moreover, by (4.2) and weak-$*$ convergence again, the mass of μ does not go to ∞ and μ is a probability measure.

Let D_i be the discriminant of a minimal equation for ξ_i. By Proposition 1.6.9, we have

$$\frac{1}{d_i} \log |D_i| \leq \log d_i + (2d_i - 2)h(\xi_i).$$

Therefore, we get

$$0 \leq \log |D_i| = (2d_i - 2) \log |a_{0i}| + \sum_{\sigma \neq \sigma'} \log |\sigma\xi_i - \sigma'\xi_i| = o(d_i^2). \tag{4.3}$$

By (4.2), (4.3), we easily deduce that μ is a continuous measure, in other words the measure of a point is 0. If not, there are a point $a \in \mathbb{T}$ and a constant $c > 0$ such that, for any $\varepsilon > 0$, there are at least cd_i conjugates $\sigma\xi_i$ with

$$|\sigma\xi_i - a| < \varepsilon/2, \tag{4.4}$$

as soon as i is large enough along the sequence for which $\mu_i \to \mu$. The contribution to (4.3) of σ, σ' verifying (4.4) is then $\leq -c^2 \log(1/\varepsilon)d_i^2$. The remaining terms contribute not more than $O(d_i^2) + 2d_i \sum \log^+ |\sigma\xi_i| = O(d_i^2)$. This contradicts (4.3).

We claim that the energy integral

$$I(\mu) := \int_{\mathbb{T}^2} \log \frac{1}{|x-y|} \, d\mu(x) \, d\mu(y) \leq 0 \tag{4.5}$$

is not positive. To see this, let $\phi(x)$ be a positive continuous non-decreasing function on $[0,1]$ such that $\phi(x) = 0$ for $x \leq 1/2$ and $\phi(x) = 1$ for $x \geq 1$, and set $\phi_\varepsilon(x) = \phi(x/\varepsilon)$, where $0 < \varepsilon < 1$.

We have already observed that $\log |a_{0i}| = o(d_i)$, hence (4.3) shows that

$$\lim_{i \to \infty} \frac{1}{d_i^2} \sum_{\sigma \neq \sigma'} \log \frac{1}{|\sigma\xi_i - \sigma'\xi_i|} = 0.$$

A fortiori, we get

$$\limsup_{i \to \infty} \frac{1}{d_i^2} \sum_{\sigma \neq \sigma'} \phi_\varepsilon(|\sigma\xi_i - \sigma'\xi_i|) \log \frac{1}{|\sigma\xi_i - \sigma'\xi_i|} \leq 0,$$

because $\log(1/t) > 0$ if $\phi_\varepsilon(t) < 1$, while $\phi_\varepsilon(t) \leq 1$ always. By weak-* convergence, we infer that

$$\int_{\mathbb{T}^2} \phi_\varepsilon(|x-y|) \log \frac{1}{|x-y|} \, d\mu(x) \, d\mu(y) \leq 0$$

for $0 < \varepsilon \leq 1$. Since μ is a continuous measure, the diagonal has measure 0 and monotone convergence shows that (4.5) holds.

By appealing to well-known results of potential theory (see e.g. E. Hille [**152**], vol.II, §16), it is known that the energy integral is minimized by a unique regular probability measure $\mu_{\mathbb{T}}$ on $\mathbb{T}$. By symmetry, $\mu_{\mathbb{T}}$ must be the Haar measure on $\mathbb{T}$. It is well known that $I(\mu_{\mathbb{T}}) = 0$, indeed $\exp(-I(\mu_{\mathbb{T}}))$ is the logarithmic capacity, or transfinite diameter, of $\mathbb{T}$, which is equal to 1. This and (4.5) prove that $\mu = \mu_{\mathbb{T}}$. □

Another way of concluding the proof, which in its discrete version is Bilu's argument, is by Fourier analysis. Let $\mu * \nu$ be the convolution of two regular Borel measures on $\mathbb{T}$, that is the unique regular Borel measure determined by

$$\int_{\mathbb{T}} f(x)\, \mathrm{d}(\mu * \nu)(x) = \int_{\mathbb{T}^2} f(xy)\, \mathrm{d}\mu(x)\, \mathrm{d}\nu(y).$$

We consider the Fourier coefficients

$$\widehat{\mu}(n) = \int_{\mathbb{T}} x^{-n} \mathrm{d}\mu(x).$$

Clearly, we have $(\widehat{\mu * \nu})(n) = \widehat{\mu}(n)\widehat{\nu}(n)$ for $n \in \mathbb{Z}$. We apply this with our limit measure μ and with ν equal to the composition of μ with the inversion $x \mapsto x^{-1}$ on $\mathbb{T}$. Then the energy integral can be written as

$$I(\mu) = -\int_{\mathbb{T}} \log|1 - x| \mathrm{d}(\mu * \nu)(x).$$

Since μ is a real measure, we see that

$$\widehat{\nu}(n) = \overline{\widehat{\mu}(n)}.$$

The nth Fourier coefficient of $-\log|1 - e^{i\theta}|$ is 0 if $n = 0$, and $1/(2|n|)$ if $n \neq 0$ (expand $-\log(1 - z)$ in Taylor series), so the energy integral is

$$I(\mu) = \sum_{n \neq 0} \frac{|\widehat{\mu}(n)|^2}{2|n|} \geq 0.$$

Equality holds only if $\widehat{\mu}(n) = 0$ for $n \neq 0$, hence (4.5) proves that μ is the uniform measure on $\mathbb{T}$ (if a regular Borel measure has all its Fourier coefficients 0, it must be the zero measure). □

Remark 4.3.2. We have stated Bilu's theorem with the condition that the algebraic numbers ξ_i are all distinct. The proof of the theorem shows that the only thing that matters here is that $d_i \to \infty$. By the assumption $h(\xi_i) \to 0$ and Kronecker's theorem in 1.5.9, we may relax the hypothesis of the theorem to $h(\xi_i) \to 0$ and the condition that no root of unity in the sequence $(\xi_i)_{i \in \mathbb{N}}$ is repeated infinitely often.

4.3.3. Now we give a second proof of Zhang's theorem 4.2.2 (b). In the form given here, this proof is non-constructive (it depends on compactness arguments), hence it gives only the existence of a positive lower bound, but not an explicit dependence on n, d, $[K : \mathbb{Q}]$, and $\max h(f_i)$.

We proceed by induction on n, the claim being trivial for $n = 0$.

Suppose we have an infinite sequence of distinct points $\boldsymbol{\xi}_i \in X^*$ with $\widehat{h}(\boldsymbol{\xi}_i) \to 0$. We begin by mimicking the construction at the beginning of Section 4.3 by defining the probability measure

$$\boldsymbol{\delta}_{\boldsymbol{\xi}} = \frac{1}{[\mathbb{Q}(\boldsymbol{\xi}) : \mathbb{Q}]} \sum_{\sigma:\mathbb{Q}(\boldsymbol{\xi})\to\mathbb{C}} \delta_{\sigma\boldsymbol{\xi}}$$

associated to the Galois orbit of $\boldsymbol{\xi}$, and considering a weak-* limit measure μ of the sequence $(\boldsymbol{\delta}_{\boldsymbol{\xi}_i})_{i\in\mathbb{N}}$, as in the proof of Theorem 4.3.1. The same argument given there then shows that μ is supported in $\mathbb{T}^n$.

For any non-trivial character $\chi(\mathbf{x}) = x_1^{m_1} \cdots x_n^{m_n}$ of $(\mathbb{C}^\times)^n$, consider the associated sequence $(\chi(\boldsymbol{\xi}_i))_{i\in\mathbb{N}}$. Clearly, $h(\chi(\boldsymbol{\xi}_i)) \leq (\max|m_j|)\,\widehat{h}(\boldsymbol{\xi}_i) \to 0$.

Case I: For every non-trivial character χ, the sequence $(\chi(\boldsymbol{\xi}_i))_{i\mathbb{N}}$ ultimately consists of distinct elements.

In this case, we claim that

$$\int_{\mathbb{T}^n} \chi(\mathbf{x})\,\mathrm{d}\mu = 0. \tag{4.6}$$

We prove this as follows. By Theorem 4.3.1, the measure μ_χ determined by the sequence $(\chi(\boldsymbol{\xi}_i))_{i\in\mathbb{N}}$ is the uniform measure on $\mathbb{T} = \chi(\mathbb{T}^n)$. Let us fix $c > 0$ and let $f \in C_c(\mathbb{C}^\times)$ be the identity $f(x) = x$ in the neighborhood $\big|\log|x|\big| < c$ of $\mathbb{T}$. Then the function

$$f_\chi(\mathbf{x}) = f(x_1^{m_1}) \cdots f(x_n^{m_n})$$

has compact support in $(\mathbb{C}^\times)^n$ and coincides with $\chi(\mathbf{x})$ in a neighborhood of $\mathbb{T}^n$. By weak-* convergence, we have

$$\int_{\mathbb{T}^n} \chi(\mathbf{x})\,\mathrm{d}\mu = \int_{(\mathbb{C}^\times)^n} f_\chi(\mathbf{x})\,\mathrm{d}\mu = \lim_{i\to\infty} \frac{1}{[\mathbb{Q}(\boldsymbol{\xi}_i) : \mathbb{Q}]} \sum_{\sigma:\mathbb{Q}(\boldsymbol{\xi}_i)\to\mathbb{C}} f_\chi(\sigma\boldsymbol{\xi}_i). \tag{4.7}$$

We would like to replace $f_\chi(\mathbf{x})$ by $f(\chi(\mathbf{x}))$ in the last sum, but this step requires justification, because $f(\chi(\mathbf{x}))$ is not compactly supported in $(\mathbb{C}^\times)^n$. Let $m = \max|m_j|$ and $M = \max|f|$. By definition of f, we have $f_\chi(\mathbf{x}) = f(\chi(\mathbf{x}))$ if $\big|\log|x_j|\big| \leq c/m$ for every j, and $|f_\chi(\mathbf{x})| \leq M^n$ in any case.

We have

$$\frac{1}{[\mathbb{Q}(\boldsymbol{\xi}_i) : \mathbb{Q}]} \sum_{\sigma:\mathbb{Q}(\boldsymbol{\xi}_i)\to\mathbb{C}} f_\chi(\sigma\boldsymbol{\xi}_i) = \frac{1}{[\mathbb{Q}(\chi(\boldsymbol{\xi}_i)) : \mathbb{Q}]} \sum_{\tau:\mathbb{Q}(\chi(\boldsymbol{\xi}_i))\to\mathbb{C}} f(\tau\chi(\boldsymbol{\xi}_i))$$
$$+ \frac{1}{[\mathbb{Q}(\boldsymbol{\xi}_i) : \mathbb{Q}]} \sum_{\sigma:\mathbb{Q}(\boldsymbol{\xi}_i)\to\mathbb{C}} \left(f_\chi(\sigma\boldsymbol{\xi}_i) - f(\chi(\sigma\boldsymbol{\xi}_i))\right).$$

A typical summand in the last sum is 0 unless $\big|\log|\sigma\xi_{ij}|\big| > c/m$ for some j, and in any case does not exceed $M + M^n$. Thus the last sum does not exceed

$$\sum_{j=1}^{n} \frac{1}{[\mathbb{Q}(\boldsymbol{\xi}_i):\mathbb{Q}]} \sum_{\substack{\sigma:\mathbb{Q}(\boldsymbol{\xi}_i)\to\mathbb{C} \\ |\log|\sigma\xi_{ij}||>c/m}} (M+M^n) \le \sum_{j=1}^{n} \frac{1}{[\mathbb{Q}(\xi_{ij}):\mathbb{Q}]} \sum_{\substack{\sigma:\mathbb{Q}(\xi_{ij})\to\mathbb{C} \\ |\log|\sigma\xi_{ij}||>c/m}} (M+M^n).$$

Since $h(\boldsymbol{\xi}_i) \to 0$, by (4.2) on page 102 and Remark 4.3.2, this tends to 0 as $i \to \infty$, and proves

$$\begin{aligned} \lim_{i\to\infty} \frac{1}{[\mathbb{Q}(\boldsymbol{\xi}_i):\mathbb{Q}]} \sum_{\sigma:\mathbb{Q}(\boldsymbol{\xi}_i)\to\mathbb{C}} f_\chi(\sigma\boldsymbol{\xi}_i) &= \lim_{i\to\infty} \frac{1}{[\mathbb{Q}(\chi(\boldsymbol{\xi}_i)):\mathbb{Q}]} \sum_{\tau:\mathbb{Q}(\chi(\boldsymbol{\xi}_i))\to\mathbb{C}} f(\tau\chi(\boldsymbol{\xi}_i)) \\ &= \int_{\mathbb{T}} x \, \mathrm{d}\mu_\chi(x) \qquad \text{(by weak-* convergence)} \\ &= 0. \qquad \text{(by Theorem 4.3.1)} \end{aligned}$$

In view of (4.7), this proves (4.6).

As in the proof of Bilu's theorem, it is clear that μ is a probability measure. Now the characters $\chi(\mathbf{x})$ restricted to $\mathbb{T}^n$ form an orthonormal basis of $L^2(\mathbb{T}^n)$, whence (4.6) shows that the restriction of μ to $\mathbb{T}^n$ is the uniform measure on $\mathbb{T}^n$, because they have the same Fourier coefficients.

In particular, $\mathbb{T}^n$ is contained in the union of the conjugates of X over $\mathbb{Q}$. Since torsion points are Zariski dense in $\mathbb{G}_m^n$ (Proposition 3.3.6), this contradicts the assumption that X is a proper algebraic subvariety of $\mathbb{G}_m^n$.

Case II: There is a non-trivial character χ such that the sequence $(\chi(\boldsymbol{\xi}_i))_{i\in\mathbb{N}}$ has an element ε_0 occurring infinitely many times.

Since $h(\chi(\boldsymbol{\xi}_i)) \to 0$, we have $h(\varepsilon_0) = 0$ and ε_0 is a root of unity by Kronecker's theorem in 1.5.9. Let $\boldsymbol{\varepsilon}$ be a torsion point such that $\chi(\boldsymbol{\varepsilon}) = \varepsilon_0$ and replace X by $\boldsymbol{\varepsilon}^{-1}X$ and $\{\boldsymbol{\xi}_i\}$ by $\{\boldsymbol{\varepsilon}^{-1}\boldsymbol{\xi}_i\}$. Now $(\boldsymbol{\varepsilon}^{-1}X)^* = \boldsymbol{\varepsilon}^{-1}(X^*)$ and multiplication by a torsion point does not change the height; therefore, there is no loss of generality in assuming that $\varepsilon_0 = 1$. Further, going to an infinite subsequence of the sequence $(\boldsymbol{\xi}_i)_{i\in\mathbb{N}}$ if needed, we may also assume that there is a torsion point $\boldsymbol{\varepsilon}'$ such that $\{\boldsymbol{\varepsilon}'\boldsymbol{\xi}_i\}$ is contained in the connected component of the identity of the kernel of χ, say H. Now H is a proper subtorus of $\mathbb{G}_m^n$ and we may replace X, $\mathbb{G}_m^n$ by $\boldsymbol{\varepsilon}'X \cap H$ and H, and then use induction. □

Remark 4.3.4. As it stands, this proof does not lead to an effective form of Theorem 4.2.2. However, it is not difficult to show that there is a lower bound depending only on d, n, $[K:\mathbb{Q}]$, and $\max h(f_i)$ for a set of defining equations of X.

We verify this as follows:

Let us fix d, n, $[K:\mathbb{Q}]$, and H. By Northcott's theorem (see Theorem 2.4.9), there are only finitely many polynomials in n variables, of degree at most d, with coefficients in K and height at most H.

Therefore, the set of varieties X in $\mathbb{P}^n_K$, defined by polynomials of degree at most d, with coefficients in K and height at most H, is a finite set. Now the required lower bound depending only on the fixed data is the minimum of the lower bound obtained for a given variety X, when X varies over this finite set.

The following immediate consequence of Theorem 4.3.1 is worth noting. We recall that an algebraic number ξ is called **totally real** if $\mathbb{Q}(\xi)$ has only real embeddings into $\mathbb{C}$.

Corollary 4.3.5. *Let $\kappa > 0$. If $\xi \neq 0$ is algebraic, not a root of unity, and has at least $\kappa \deg(\xi)$ real conjugates, then $h(\xi) \geq c(\kappa) > 0$ with $c(\kappa)$ independent of ξ.*

In particular, totally real algebraic numbers other than 0 and ± 1 have height bounded below by an absolute positive constant.

Remark 4.3.6. A. Schinzel [**258**] obtained the sharp lower bound

$$h(\xi) \geq \frac{1}{2}\log\left(\frac{1+\sqrt{5}}{2}\right) = 0.2406059\ldots$$

for ξ ranging over all totally real algebraic numbers $\neq 0, \pm 1$, with minimum attained for $\xi = \pm(1 \pm \sqrt{5})/2$.

Further examples of totally real numbers η with small height can be obtained by noticing that, if ξ is totally real, then $\eta - \eta^{-1} = \xi$ yields a totally real η of degree not exceeding $2\deg(\xi)$. C.J. Smyth [**288**] used this process, starting with $\xi_0 = 1$, to construct a sequence of totally real numbers $\xi_1 = (1+\sqrt{5})/2, \ldots$ of small height, with $h(\xi_n)$ accumulating at $\lambda = 0.2732831\ldots$.

Moreover, he proved that the heights of totally real numbers are dense in the interval (λ, ∞). It is conceivable that λ is the smallest limit point of $h(\xi)$ for totally real ξ.

The minimum above is isolated, and subsequently Smyth [**288**] determined the first four smallest values of $h(\xi)$ for totally real ξ. They are attained at the points ξ_1, ξ_2, ξ_3, and $2\cos(2\pi/7)$.

4.4. Dobrowolski's theorem

In this section, we prove the following **theorem of Dobrowolski** alluded to in 1.6.15. We have:

Theorem 4.4.1. *Let α be an algebraic number of degree d, not a root of unity or 0. Then*

$$h(\alpha) \geq \frac{c}{d}\left(\frac{\log\log(3d)}{\log(3d)}\right)^3$$

for an absolute constant $c > 0$.

We begin with a simple lemma. We assume throughout that $\alpha \neq 0$ has degree d. Let $\alpha_1, \ldots, \alpha_d$ denote a full set of conjugates of α. We have:

Lemma 4.4.2. *We may assume that α is a unit and $\mathbb{Q}(\alpha) = \mathbb{Q}(\alpha^p)$ for every prime p not dividing d. In particular, α^p has degree d and $\alpha_1^p, \ldots, \alpha_d^p$ is a full set of conjugates of α^p. Moreover, the algebraic integers α_i^p, for varying i and p, $p = 1$ or a prime not dividing d, are all distinct.*

Proof: If α is not a unit, we have the easy estimate $h(\alpha) \geq (\log 2)/d$ (see 1.6.15), which is stronger than the lower bound stated in Dobrowolski's theorem.

Let $\xi = \alpha^p$ and $K = \mathbb{Q}(\xi)$. The polynomial $x^p - \xi$ is irreducible over a finitely generated field K, unless $\xi = \eta^p$ for some $\eta \in K$. This is an old result by Abel and is a special case of the well-known theorem* of Vahlen–Capelli on the reducibility of the polynomial $x^m - a$ (see L. Rédei [**239**], Th.427, Th.428 or [**49**], Ch.V, §11, ex.5). A simple direct proof is as follows. The roots of $x^p - \xi = 0$ are $\lambda\xi^{1/p}$ with λ a pth root of unity. If $x^p - \xi$ were reducible over K with a monic irreducible factor $g(x)$ of degree $s < p$, then looking at the last coefficient of $g(x)$ we would have $\lambda'\alpha^s \in K$ for some pth root of unity λ'. There is an integer m with $ms \equiv 1 \mod p$, hence $\eta := (\lambda')^m\alpha \in K$ and $\xi = \eta^p$, as asserted.

Now suppose that p is not a divisor of d. Let $s := [\mathbb{Q}(\alpha) : K]$. We cannot have $s = p$, because p does not divide d. Hence the polynomial $x^p - \xi$ is reducible over K, and $\alpha^p = \xi = \eta^p$ with $\eta \in K$. In particular, $h(\alpha) = h(\eta)$. If we had $s > 1$, then $\deg(\eta) < \deg(\alpha)$, and the theorem for α would follow from the theorem in lower degree. Thus we may assume $s = 1$, which is the first statement of the lemma.

Finally, assume that $\alpha_i^p = \alpha_j^q$ for two distinct primes p and q not dividing d. By the first part of the lemma, we have that $\alpha_1^q, \ldots, \alpha_d^q$ is a permutation of $\alpha_1^p, \ldots, \alpha_d^p$, thus $\alpha_i^q = \alpha_{\sigma(i)}^p$ for some permutation σ of $\{1, \ldots, d\}$. This yields, if m is the order of σ

$$\begin{aligned}
\alpha_i^{q^m} &= (\alpha_i^q)^{q^{m-1}} = (\alpha_{\sigma(i)}^p)^{q^{m-1}} \\
&= (\alpha_{\sigma(i)}^q)^{pq^{m-2}} = (\alpha_{\sigma^2(i)}^p)^{pq^{m-2}} \\
&= \cdots = (\alpha_{\sigma^{m-1}(i)}^q)^{p^{m-1}} = (\alpha_{\sigma^m(i)}^p)^{p^{m-1}} = \alpha_i^{p^m},
\end{aligned}$$

* The theorem states that $x^m - a$ is reducible over a field K of characteristic 0 if and only if $a = b^p$ for some $b \in K$ and a prime p with $p|m$, or $a = -4b^4$ and $4|m$.

hence $\alpha^{q^m - p^m} = 1$ and α is a root of unity, which was excluded from the beginning. □

Our next step is the construction of a polynomial $F(x) \in \mathbb{Z}[x]$ of degree at most D, vanishing at α to order at least m. Here D and m are large parameters (in the end going to ∞) and we want some control on the height of $F(x)$.

Lemma 4.4.3. *Let α be an algebraic number of degree $d \geq 2$. Let us fix ε with $0 < \varepsilon < 1$ and suppose that $dm \leq (1-\varepsilon)D$. Then there is a polynomial $F(x) \in \mathbb{Z}[x]$ of degree at most D, not identically 0, vanishing at α to order at least m and such that*

$$h(F) \leq \frac{dm^2}{D - dm}\left(\log \frac{D}{m} + 1\right) + \frac{Ddm}{D - dm} h(\alpha) + o(D)$$

as $D \to \infty$.

Proof: This follows from Siegel's lemma, Corollary 2.9.7, noting that in our case the matrix $\mathcal{A}$ is

$$\mathcal{A} = \left(\binom{j}{h}\alpha^{j-h}\right)$$

with m rows indexed by h, $0 \leq h < m$, and $D+1$ columns indexed by j, $j = 0, 1, \ldots, D$. Using the easy bound

$$\log \binom{a}{b} \leq b\left(\log \frac{a}{b} + 1\right)$$

we get $h(\mathcal{A}_h) \leq m(\log(D/m) + 1) + Dh(\alpha)$ for the hth row of $\mathcal{A}$. □

4.4.4. In what follows, α will satisfy the conditions of Lemma 4.4.2. Since α is a unit, it is an algebraic integer (see Remark 1.5.11). Let

$$f(x) = \prod_{i=1}^{d}(x - \alpha_i) \in \mathbb{Z}[x]$$

be its minimal polynomial. For every prime p not dividing d let

$$f_p(x) = \prod_{i=1}^{d}(x - \alpha_i^p)$$

and let e_p be the multiplicity of f_p as a factor of $F(x)$. Note that by Lemma 4.4.2 the polynomials f, f_p are irreducible in $\mathbb{Z}[x]$ and are all distinct. It follows, denoting by $\prod'$ a product over primes not dividing d, that

$$F(x) = f(x)^m {\prod_p}' f_p(x)^{e_p} G(x)$$

for some $G(x) \in \mathbb{Z}[x]$ with $G(\alpha^p) \neq 0$ for every prime p not dividing d. Moreover, by Lemma 4.4.2, we have $f(\alpha^p) \neq 0$ and $f_q(\alpha^p) \neq 0$ if $p \neq q$.

Proof of Dobrowolski's theorem: By the remarks in 1.6.15, we may assume that d is large.

We differentiate e_p times the polynomial $F(x)$, divide by $e_p!$, and specialize x to α^p, obtaining an algebraic integer

$$\eta := f(\alpha^p)^m f_p'(\alpha^p)^{e_p} \prod_{q \neq p}{}' f_q(\alpha^p)^{e_q} G(\alpha^p) = \frac{1}{e_p!}\left(\frac{\mathrm{d}}{\mathrm{d}x}\right)^{e_p} F(\alpha^p).$$

By 4.4.4, this is a non-zero algebraic integer of degree at most d.

On the other hand

$$f(x^p) \equiv f(x)^p \mod p\,\mathbb{Z}[x],$$

whence, specializing x to α, the algebraic integer $f(\alpha^p)$ is divisible by p. Thus the norm $N_{\mathbb{Q}(\alpha)/\mathbb{Q}}(\eta)$ is a non-zero rational integer divisible by p^{dm}.

An upper bound for the norm comes from the expression of η in terms of $F(x)$ just given and Lemma 4.4.3. We claim that

$$\left| N_{\mathbb{Q}(\alpha)/\mathbb{Q}}\left(\frac{1}{e_p!}\left(\frac{\mathrm{d}}{\mathrm{d}x}\right)^{e_p} F(\alpha^p)\right)\right| \leq \left((D+1)\binom{D}{e_p} H(F)\right)^d H(\alpha)^{dpD}, \quad (4.8)$$

where $H(\alpha) = \exp(h(\alpha))$ is the multiplicative height.

To see this, note first that

$$H\left(\frac{1}{e_p!}\left(\frac{\mathrm{d}}{\mathrm{d}x}\right)^{e_p} F\right) \leq \binom{D}{e_p} H(F). \quad (4.9)$$

Next, for any polynomial $G \in \overline{\mathbb{Q}}[x]$ of degree D and any $\beta \in \overline{\mathbb{Q}}$, we have

$$H(G(\beta)) \leq (D+1) H(G) H(\beta)^D. \quad (4.10)$$

Next, it is clear that

$$H(\alpha^p) = H(\alpha)^p. \quad (4.11)$$

Then the stated bound (4.8) for the norm follows from the last statement of Proposition 1.6.6 and (4.9), (4.10), and (4.11).

We compare with the lower bound p^{dm}, take logarithms, divide by d, and use Lemma 4.4.3 to estimate $h(F)$, obtaining

$$\begin{aligned} m \log p \leq e_p \left(\log \frac{D}{e_p} + 1\right) + \frac{dm^2}{D - dm}\left(\log \frac{D}{m} + 1\right) \\ + \left(pD + \frac{Ddm}{D - dm}\right) h(\alpha) + o(D). \end{aligned}$$

We define $\gamma \le 1-\varepsilon$ and γ_p by $m = \gamma D/d$ and $e_p = \gamma_p D/d$. Then the inequality above simplifies to

$$\begin{aligned}\frac{\gamma \log p}{d} \le \frac{\gamma_p}{d}\left(\log \frac{d}{\gamma_p} + 1\right) + \frac{\gamma^2}{(1-\gamma)d}\left(\log \frac{d}{\gamma} + 1\right)\\ + \left(p + \frac{\gamma}{1-\gamma}\right) h(\alpha) + o(1),\end{aligned} \tag{4.12}$$

with the $o(1)$ term going to 0 as $D \to \infty$.

It remains to optimize inequality (4.12) by choosing a range for the prime p, the parameter $\gamma < 1$, and an optimal γ_p. In what follows, we shall assume that d is large and deal with estimates asymptotic with respect to d.

It is clearly convenient to make sure that γ_p is as small as possible, and to this end we note that

$$\sum_p{}' \gamma_p \le 1,$$

because each $f_p^{e_p}$ divides F and f_p has degree d. In particular, if our set of primes consists of the primes in an interval $[Y_0, Y]$ not dividing d, and $N > 0$ is the number of such primes, there exists $p \in [Y_0, Y]$ such that p does not divide d and $\gamma_p \le 1/N$. Since $x(\log(d/x)+1)$ increases with x for $0 < x < d$, we see that (4.12) can be replaced by

$$\frac{\gamma \log p}{d} \le \frac{1}{Nd}(\log(Nd)+1) + \frac{\gamma^2}{(1-\gamma)d}\left(\log \frac{d}{\gamma} + 1\right) + \left(p + \frac{\gamma}{1-\gamma}\right) h(\alpha) + o(1).$$

Now we let $D \to \infty$, getting rid of the $o(1)$ term, and deduce *a fortiori*

$$\frac{\gamma \log Y_0}{d} \le \frac{1}{Nd}(\log(Nd)+1) + \frac{\gamma^2}{(1-\gamma)d}\left(\log \frac{d}{\gamma} + 1\right) + \left(Y + \frac{\gamma}{1-\gamma}\right) h(\alpha), \tag{4.13}$$

with N the number of primes, not dividing d, in $[Y_0, Y]$.

In order to optimize (4.13), we consider the parameters Y_0, Y, γ as functions of $d \to \infty$.

We begin by estimating N. The number of primes dividing d is less than $\lfloor \log d / \log 2 \rfloor$, therefore the prime number theorem (H. Koch [**162**], Theorem 1.7.3) shows that

$$N = \pi(Y) - \pi(Y_0) + O(\log d) \sim \frac{Y}{\log Y},$$

provided $Y_0 = o(Y)$ and $\log d = o(Y/\log Y)$ as $Y \to \infty$, which we shall assume. We do not want to choose Y too large (in fact, $Y = o(d)$ will suffice), nor Y_0 too small, and it is quite reasonable to choose $Y_0 = \lfloor Y/\log Y \rfloor$, ensuring that $\log Y_0 \sim \log Y$, and

$$\log Y = o(\log d), \tag{4.14}$$

ensuring that $\log(Yd) \sim \log d$. Moreover, we do not want γ to be too small, and in fact we shall need

$$\log(1/\gamma) = o(\log d) \tag{4.15}$$

as $d \to \infty$. Then (4.13) becomes, as $Y \to \infty$

$$\frac{(1-o(1))\gamma \log Y}{d} \leq (1+o(1))\frac{\log Y}{Yd}\log d + (1+o(1))\frac{\gamma^2}{d}\log d + (1+o(1))Yh(\alpha).$$

We rewrite this as

$$h(\alpha) \geq (1-o(1))\left(\frac{\gamma \log Y}{Yd} - \frac{(\log Y)(\log d)}{Y^2 d} - \frac{\gamma^2 \log d}{Yd}\right)$$

and proceed to optimize the two remaining parameters γ and Y, as functions of d. Optimization with respect to γ gives

$$\gamma = \frac{\log Y}{2\log d},$$

which is compatible with condition (4.15), yielding

$$h(\alpha) \geq (1-o(1))\left(\frac{(\log Y)^2}{4Yd\log d} - \frac{(\log Y)(\log d)}{Y^2 d}\right).$$

Finally, optimization with respect to Y occurs with

$$Y \sim \frac{4(\log d)^2}{\log\log d},$$

which is compatible with condition (4.14), and conclude with

$$h(\alpha) \geq (1-o(1))\frac{1}{8d}\left(\frac{\log\log d}{\log d}\right)^3$$

as $d \to \infty$. This proves Dobrowolski's theorem. □

Remark 4.4.5. The constant $1/8$ given here is far from being optimal for the method given here. A more careful evaluation in Lemma 4.4.3 yields easily the constant 1, as in E. Dobrowolski's original paper [**90**].

The remaining part of this section deals with further results about Lehmer's conjecture and may be skipped in a first reading. We will use it only in Section 4.6. A natural extension of Lehmer's conjecture to the higher dimensional case is:

Conjecture 4.4.6. *There is $c(n) > 0$ with the following property. let $\alpha_1, \ldots, \alpha_n$ be multiplicatively independent non-zero algebraic numbers. Then*

$$h(\alpha_1)\cdots h(\alpha_n) \geq \frac{c(n)}{[\mathbb{Q}(\alpha_1, \ldots, \alpha_n) : \mathbb{Q}]}.$$

For $n = 1$, this reduces to Lehmer's conjecture.

A significant extension of Dobrowolski's theorem has been obtained by F. Amoroso and S. David [**9**] in this context. They prove:

Theorem 4.4.7. *There is a positive constant* $c'(n)$ *with the following property. Let* $\boldsymbol{\alpha}$ *be as in 4.4.6 and let* $D = [\mathbb{Q}(\alpha_1, \dots, \alpha_n) : \mathbb{Q}]$. *Then*

$$h(\alpha_1) \cdots h(\alpha_n) \geq \frac{c'(n)}{D} (\log(3D))^{-n\kappa(n)},$$

where $\kappa(n) = (n+1)(n+1)!^n - 1$.

In fact, they prove this result in the more precise form in which the degree D is replaced by the smallest degree $\omega_{\mathbb{Q}}(\boldsymbol{\alpha})$ of a hypersurface defined over $\mathbb{Q}$ containing the point $\boldsymbol{\alpha}$.

As a corollary, we obtain the validity of Lehmer's conjecture for any α not a root of unity such that $\mathbb{Q}(\alpha)$ is a Galois extension of $\mathbb{Q}$.

We will not prove this result here and instead refer the interested reader to the original paper of Amoroso and David.

4.4.8. Recall that a field extension $K/\mathbb{Q}$ is called **abelian** if it is a Galois extension with an abelian Galois group. We conclude this section with a nice result of F. Amoroso and R. Dvornicich **[11]**, which provides a uniform positive lower bound for the height of algebraic numbers, not a root of unity or 0, in abelian extensions of $\mathbb{Q}$.

Theorem 4.4.9. *Let* $K/\mathbb{Q}$ *be an abelian extension and let* $\alpha \in K$, α *not a root of unity or* 0. *Then*

$$h(\alpha) \geq \frac{\log(5/2)}{10}.$$

Remark 4.4.10. Amoroso and Dvornicich obtain the more precise lower bound $\log(5)/12$ and give an example with height $\log(7)/12$.

4.4.11. For $m \geq 3$ let ζ_m be a primitive mth root of unity and denote by $C_m = \mathbb{Q}(\zeta_m)$ the mth cyclotomic field of degree $\varphi(m)$, and by O_m the ring of integers of C_m. By the Kronecker–Weber theorem (see L.C. Washington **[322]**, Th.14.1), any finite abelian extension K of $\mathbb{Q}$ is contained in a cyclotomic extension of $\mathbb{Q}$. Thus, in proving 4.4.9, there is no loss of generality in assuming that $K = C_m$ for some m.

Lemma 4.4.12. *Let* K *be a number field and let* w *be a non-archimedean place of* K. *Then, for any* $\alpha \in K \setminus \{0\}$, *there exists an algebraic integer* $\beta \in K \setminus \{0\}$, *such that* $\alpha\beta$ *is an algebraic integer and*

$$|\beta|_w = 1/\max(1, |\alpha|_w).$$

Proof: Let S_0 be the set of non-archimedean places v of K for which $|\alpha|_v > 1$ and set $\xi_v = 1/\alpha$. If, moreover, $w \notin S_0$, set $\xi_w = 1$. Define $S = S_0 \cup \{w\}$. By the strong approximation theorem (see Theorem 1.4.5), for any $\varepsilon > 0$, there is $\beta \in K \setminus \{0\}$ such that

$$|\beta - \xi_v|_v < \varepsilon$$

for every $v \in S$, and also $|\beta|_v \leq 1$ for a non-archimedean v not in S.

Since we are dealing with ultrametric absolute values, by definition of S_0 and ξ_v we see that, if ε is sufficiently small, we have $|\beta|_v = |1/\alpha|_v \leq 1$ for $v \in S_0$, and $|\beta|_w = 1$ if $w \notin S_0$. Hence $|\beta|_w = \min(1, |1/\alpha|_w) = 1/\max(1, |\alpha|_w)$ and also $|\beta|_v \leq 1$ and $|\alpha\beta|_v \leq 1$ for $v \in S$. If instead $v \notin S$ is a non-archimedean place, we have $|\alpha|_v \leq 1$ by definition of S_0, and again $|\beta|_v \leq 1$ and $|\alpha\beta|_v \leq 1$ for a non-archimedean v not in S. Hence β and $\alpha\beta$ are both algebraic integers, as noted in Definition 1.5.10. $\square$

Lemma 4.4.13. *Let p be a rational prime. Then there exists a non-trivial $\sigma = \sigma_p \in \mathrm{Gal}(C_m/\mathbb{Q})$ with the following property:*

(a) *If $\mathrm{GCD}(p,m) = 1$, then p divides $\gamma^p - \sigma\gamma$ for any $\gamma \in O_m$.*

(b) *If $\mathrm{GCD}(p,m) = p$, then p divides $\gamma^p - \sigma\gamma^p$ for any $\gamma \in O_m$. Moreover, if $\sigma\gamma^p = \gamma^p$, there exists an mth root of unity ζ such that $\zeta\gamma$ is contained in the proper cyclotomic subfield $C_{m/p}$ of C_m.*

Proof: We recall that the ring of integers of C_m is $\mathbb{Z}[\zeta_m]$ (see [**322**], Ch.1, Prop.1.2). Thus we can write $\gamma = f(\zeta_m)$ for some $f \in \mathbb{Z}[x]$. Suppose first that p does not divide m. Let $\sigma \in \mathrm{Gal}(C_m/\mathbb{Q})$ be defined by $\sigma\zeta_m = \zeta_m^p$. Then

$$\gamma^p \equiv f(\zeta_m^p) \equiv f(\sigma\zeta_m) \equiv \sigma\gamma \mod p,$$

proving (a).

If instead p divides m, we argue as follows. The Galois group $\mathrm{Gal}(C_m/C_{m/p})$ is cyclic of order p or $p-1$ according as p^2 divides m or not. Let σ be a generator; then $\sigma\zeta_m = \lambda_p\zeta_m$ for some primitive p-root of unity λ_p. A similar calculation as before yields

$$\gamma^p \equiv f(\zeta_m^p) \equiv f(\sigma\zeta_m^p) \equiv \sigma\gamma^p \mod p,$$

which is the first part of statement (b). Finally, if $\sigma\gamma^p = \gamma^p$, we have $\sigma\gamma = \lambda_p^a\gamma$ for some integer a. It follows that $\sigma(\gamma/\zeta_m^a) = \gamma/\zeta_m^a$ and γ/ζ_m^a belongs to the fixed field $C_{m/p}$. □

Proof of Theorem 4.4.9: Let $\alpha \in C_m \setminus \{0\}$, α not 0 or a root of unity. Let p be a prime. Since $h(\alpha) = h(\zeta\alpha)$ for any root of unity ζ, replacing α by $\zeta\alpha$ we may also assume that $\zeta\alpha$ is never contained in a proper cyclotomic subfield of C_m.

Case I: p does not divide m. Let σ be the element of $\mathrm{Gal}(C_m/\mathbb{Q})$ as in Lemma 4.4.13 and let v be any place dividing p. By Lemma 4.4.12, there is an algebraic integer $\beta \in C_m$ such that $\alpha\beta$ is an algebraic integer and $|\beta|_v = 1/\max(1, |\alpha|_v)$. Then by Lemma 4.4.13 we have

$$|(\alpha\beta)^p - \sigma(\alpha\beta)|_v \le |p|_v, \qquad |\beta^p - \sigma\beta|_v \le |p|_v. \tag{4.16}$$

Let us write $\eta := \alpha^p - \sigma\alpha$. By the ultrametric inequality, the bounds (4.16) and the identity

$$\alpha^p - \sigma\alpha = \beta^{-p}((\alpha\beta)^p - \sigma(\alpha\beta) + (\sigma\beta - \beta^p)\sigma\alpha),$$

we have

$$\begin{aligned}
|\eta|_v &= |\beta|_v^{-p}\, |(\alpha\beta)^p - \sigma(\alpha\beta) + (\sigma\beta - \beta^p)\sigma\alpha|_v \\
&\le |\beta|_v^{-p} \max(|(\alpha\beta)^p - \sigma(\alpha\beta)|_v, |\beta^p - \sigma\beta|_v\, |\sigma\alpha|_v) \\
&\le |p|_v\, |\beta|_v^{-p} \max(1, |\sigma\alpha|_v) \\
&= |p|_v \max(1, |\alpha|_v)^p \max(1, |\sigma\alpha|_v).
\end{aligned}$$

Moreover, $\eta \ne 0$ because α is not a root of unity. Otherwise, by induction on i, we would have $\alpha^{p^i} - \sigma^i\alpha = 0$ and, taking i to be the order of σ in $\mathrm{Gal}(C_m/\mathbb{Q})$, α would be a root of unity, which was excluded by hypothesis.

Now we apply the product formula to η. Let as usual $\varepsilon_v = 0$ if v is not archimedean and $\varepsilon_v = [(C_m)_v : \mathbb{Q}_v]/[C_m : \mathbb{Q}]$ if $v|\infty$. If $v|p$, we have shown that

$$\log|\eta|_v \leq \log|p|_v + p\log^+|\alpha|_v + \log^+|\sigma\alpha|_v.$$

If instead v does not divide p, we have trivially

$$\log|\eta|_v \leq p\log^+|\alpha|_v + \log^+|\sigma\alpha|_v + \varepsilon_v \log 2.$$

Hence summing over all places and using the product formula, we get

$$\begin{aligned}
0 &= \sum_v \log|\eta|_v \\
&\leq \sum_{v|p} \log|p|_v + \sum_v (p\log^+|\alpha|_v + \log^+|\sigma\alpha|_v + \varepsilon_v \log 2) \\
&= -\log p + ph(\alpha) + h(\sigma\alpha) + \log 2 = -\log(p/2) + (p+1)h(\alpha).
\end{aligned}$$

Therefore, in Case I, we have

$$h(\alpha) \geq \frac{\log(p/2)}{p+1}.$$

Case II: p divides m. We proceed in a similar way as in Case I, working now with $\eta := \alpha^p - \sigma\alpha^p$. The application of Lemma 4.4.13, proceding as before but using this time the identity

$$\alpha^p - \sigma\alpha^p = \beta^{-p}(\alpha\beta)^p - \sigma(\alpha\beta)^p + (\sigma\beta^p - \beta^p)\sigma\alpha^p),$$

shows that, if $v|p$, then

$$|\eta|_v \leq |p|_v \max(1, |\alpha|_v)^p \max(1, |\sigma\alpha|_v)^p.$$

If instead v does not divide p, we have the trivial estimate

$$\log|\eta|_v \leq p\log^+|\alpha|_v + p\log^+|\sigma\alpha|_v + \varepsilon_v \log 2.$$

Note that $\eta \neq 0$, otherwise Lemma 4.4.13 shows that there would be a root of unity $\zeta \in C_m$ such that $\zeta\alpha \in C_{m/p}$, which was excluded at the start. Hence the application of the product formula yields, much in the same way as in the preceding case, the inequality

$$h(\alpha) \geq \frac{\log(p/2)}{2p}.$$

Theorem 4.4.9 follows by considering the prime $p = 5$. □

We denote by a bar complex conjugation on C_m, namely the automorphism determined by $\zeta_m \mapsto \zeta_m^{-1}$.

Corollary 4.4.14. *Let γ be an algebraic integer in C_m. If $\overline{\gamma}/\gamma$ is not a root of unity, it holds that*

$$\frac{1}{\varphi(m)} \log|N_{C_m/\mathbb{Q}}(\gamma)| \geq \frac{\log(5/2)}{10},$$

where φ is the Euler φ-function.

Proof: By Theorem 4.4.9, Lemma 1.3.7, and the product formula, we have

$$\begin{aligned}\frac{\log(5/2)}{10} &\leq \sum_v \log^+ |\overline{\gamma}/\gamma|_v = \sum_{v \nmid \infty} \log^+ |\overline{\gamma}/\gamma|_v \\ &\leq - \sum_{v \nmid \infty} \log |\gamma|_v = \sum_{v|\infty} \log |\gamma|_v \\ &= \frac{1}{\varphi(m)} \log \left| N_{C_m/\mathbb{Q}}(\gamma) \right|.\end{aligned}$$

This proves what we want. □

The next result, apart from the numerical constant, is a well-known result of C.J. Smyth **[287]**.

Theorem 4.4.15. *Let $\alpha \neq 0$ be an algebraic integer of degree d and assume that α^{-1} is not a conjugate of α. Then*

$$h(\alpha) \geq \frac{c}{d}$$

for an absolute constant c.

Remark 4.4.16. The result of Smyth shows that the optimal constant c is $c = \log(\theta_0)$, where $\theta_0 > 1$ is the smallest Pisot–Vijaraghavan number, namely the real root of the cubic equation $x^3 - x - 1 = 0$. The method of Smyth is based on techniques of complex function theory and is quite different from the algebraic method followed here.

Proof: We have already remarked in 1.6.15 that such an estimate holds, with the constant $c = \log 2$, if α is not an algebraic integer. Thus we may assume that α is an algebraic integer of degree $d \geq 2$.

Let $f(x) \in \mathbb{Z}[x]$ be the minimal polynomial of α and let p be a prime number. We set $\gamma := f(\zeta_p)$ and apply Corollary 4.4.14 to the algebraic integer γ. To this end, we need to verify that $\overline{\gamma}/\gamma$ is not a root of unity, at least for p large enough.

Suppose the contrary. Then $\overline{\gamma}/\gamma$ must be a root of unity in C_p, hence $f(\zeta_p) = \pm\zeta_p^j f(\zeta_p^{-1})$ for some integer j, which we may assume to be in the range $-1 \leq j \leq p-2$. It follows that

$$\zeta_p^d f(\zeta_p) = \pm\zeta_p^j f^*(\zeta_p),$$

where $f^*(x) = x^d f(x^{-1})$ is the reciprocal polynomial of f. Consider the polynomial

$$g(x) = x^{\max(d-j,0)} f(x) \mp x^{\max(j-d,0)} f^*(x).$$

Then $g(x)$ is not identically 0 unless $j = d$ and $f(x) = \pm f^*(x)$, which is excluded because α^{-1} is not a conjugate of α.

Clearly, $g(x)$ has degree at most $d + |d-j| \leq \max(2d+1, p-2)$. On the other hand, $g(x)$ has degree at least $p-1$ because ζ_p is a root of $g(x)$. This is a contradiction if $p \geq 2d+3$.

By Corollary 4.4.14, we deduce

$$\frac{1}{p-1} \log \left| N_{C_p/\mathbb{Q}}(f(\zeta_p)) \right| \geq \frac{\log(5/2)}{10}.$$

Finally, by 1.6.15 and Proposition 1.6.6, we have

$$\begin{aligned}\lim_{p\to\infty} \frac{1}{p-1} \log\left|N_{C_p/\mathbb{Q}}(f(\zeta_p))\right| &= \lim_{p\to\infty} \frac{1}{p-1} \sum_{h=1}^{p-1} \log\left|f(e^{2\pi i h/p})\right| \\ &= \int_0^1 \log\left|f(e^{2\pi i\theta})\right| \, d\theta \\ &= \log M(f) = dh(\alpha). \end{aligned}$$

□

4.5. Remarks on the Northcott property

In this section, we consider only sets of algebraic numbers contained in a fixed algebraic closure $\overline{\mathbb{Q}}$.

4.5.1. We say that a set $\mathcal{A}$ of algebraic numbers has the **Northcott property** (N) if for every positive real number T the set

$$\mathcal{A}(T) = \{\alpha \in \mathcal{A} \mid h(\alpha) \leq T\}$$

is finite. The Northcott theorem states that the set of all algebraic numbers of degree at most d has property (N) (see Theorem 1.6.8).

We may ask if property (N) holds for other interesting sets. For example, does it hold for the field $\mathbb{Q}^{(d)}$, the composite field of all number fields of degree at most d over $\mathbb{Q}$? Although this question remains open in general, we shall show that this is the case if $d = 2$. More generally, we show that property (N) holds for the maximal abelian subfield of $\mathbb{Q}^{(d)}$.

4.5.2. Let K be a number field and denote by $K^{(d)}$ the compositum of all extension fields F/K of degree at most d over K. Then $K^{(d)}$ is normal over K. We also denote by $K^{(d)}_{ab}$ the compositum of all finite abelian extensions L/K with $K \subset L \subset K^{(d)}$. Since the compositum of two finite abelian extensions is again a finite abelian extension, $K^{(d)}_{ab}$ is the union of all finite abelian extensions over K. In particular, $K^{(d)}_{ab}/K$ is a Galois extension, and it is the maximal abelian subfield of $K^{(d)}_{ab}$. If $d \geq 2$, the fields $K^{(d)}$ and $K^{(d)}_{ab}$ have infinite degree over K.

We recall that a Galois extension F/K is called of **exponent dividing** $n \in \mathbb{N}$ if the order of every element of $\mathrm{Gal}(F/K)$ divides n.

If L is a finite extension of K with $[L : K] \leq d$ and Galois closure F, then $[F : K] \leq d!$ and hence F has exponent dividing $d!$. Since the compositum of Galois extensions of exponent dividing n is obviously a Galois extension of exponent dividing n, we conclude that $K^{(d)}$ and hence also $K^{(d)}_{ab}$ are Galois extensions of exponent dividing $d!$.

In the following result, we will prove that the local degrees of $K^{(d)}/K$ are bounded. This is a motivation to consider the Northcott property (N) for the field $K^{(d)}$, which is an open problem. However, we will prove in Theorem 4.5.4 that $K^{(d)}_{ab}$ has property (N).

Proposition 4.5.3. *Let v be any place of M_K and let w be an extension of v to $K^{(d)}$ and let K_v and $K_w^{(d)}$ be the corresponding completions. Then the local degree $[K_w^{(d)} : K_v]$ is bounded in terms of d and $[K : \mathbb{Q}]$ alone, independently of v, w.*

Proof: It is enough to consider non-archimedean places. Let us fix an algebraic closure Ω_v of K_v and let p be the residue characteristic of v. By results of M. Krasner **[163]**, the number of subextensions of Ω_v/K_v is precisely known. We only use here that the number of extensions of degree at most d is finite and bounded only in terms of d and $[K_v : \mathbb{Q}_p]$. Therefore, the degree of their compositum is bounded only in terms of d and $[K_v : \mathbb{Q}_p] \leq [K : \mathbb{Q}]$. Since $K_w^{(d)}$ may be embedded in such a compositum, the result follows. □

Theorem 4.5.4. *Property* (N) *holds for the field* $K_{ab}^{(d)}$ *, for any* $d \geq 2$.

Corollary 4.5.5. *The field* $K^{(2)}$ *has property* (N).

Proof: Obvious, because $K^{(2)} = K_{ab}^{(2)}$. □

Corollary 4.5.6. *For any* $m \geq 2$*, the field* $\mathbb{Q}(\sqrt[m]{1}, \sqrt[m]{2}, \sqrt[m]{3}, \dots)$ *has property* (N).

Proof: Let $K = \mathbb{Q}(\sqrt[m]{1})$. Then each field $K(\sqrt[m]{a})$ is of degree at most m and abelian over K. Therefore, their compositum $F = \mathbb{Q}(\sqrt[m]{1}, \sqrt[m]{2}, \sqrt[m]{3}, \dots)$ is abelian over K and a subfield of $K_{ab}^{(m)}$. By Theorem 4.5.4, $K_{ab}^{(m)}$ has the Northcott property and the same holds for its subfield F. □

Proof of Theorem 4.5.4: In what follows, we abbreviate $D = d!$. We may enlarge the number field K, hence we may suppose that K contains the field $\mathbb{Q}(\sqrt[D]{1})$ generated by roots of unity of order D. Let us fix a positive real number T and let $\alpha \in K_{ab}^{(d)}$ satisfy $h(\alpha) \leq T$. As a subfield of an abelian field, $L = K(\alpha)$ is automatically a finite abelian extension of K. By 4.5.2, L/K has exponent dividing D.

Let p be a prime, unramified in K and let v be a place of K above p. For the following considerations, the reader is assumed to be familiar with the notation and results from ramification theory developed in B.2.18 and B.2.19. Let $e = e_{w/v}$ be the ramification index of a place w of L lying over v. Since $\mathrm{Gal}(L/K)$ operates transitively on these places, e does not depend on the choice of w. If $p > d$, then our remarks on the exponent show that p does not divide the order of $\mathrm{Gal}(L/K)$. Hence w will be tamely ramified over v and the inertia group of w over v is cyclic of order e (see B.2.18 (d), (e)) proving that e divides D.

Now let $\theta = p^{1/e}$ for some choice of the root, and consider the field $L(\theta)$ with a place u lying over w. As a compositum of two abelian extensions of exponent dividing D, we note that $L(\theta)/K$ is also abelian of exponent dividing D. Using the theory of Eisenstein polynomials (see J.-P. Serre **[276]**, Ch.I, §6) and v unramified over p, we deduce that the ramification index of $u|_{K(\theta)}$ over v is e and the residue degree is 1. By Abhyankar's lemma (**[215]**, Cor.4, p.236), this and $u|_{K(\theta)}$ tamely ramified over v imply that u is unramified over w. By Proposition 1.2.11, we conclude $e_{u/v} = e$.

Let $I \subset \mathrm{Gal}(L(\theta)/K)$ be the inertia group of u over v, a group of order e. Since $L(\theta)/K$ is abelian, all the inertia groups above v are equal to I. Define U as the fixed field of I. Then U is normal over K and U/K is unramified over v (see B.2.18 (d)). By

Galois theory, $[L(\theta):U] = |I| = e$. Since $u|_U$ is unramified over p, we see again by the theory of Eisenstein polynomials that $u|_{U(\theta)}$ has ramification index $e = [U(\theta):U]$ over $u|_U$, proving in particular that $U(\theta) = L(\theta)$. It follows that $\alpha \in U(\theta)$ and we may write

$$\alpha = \beta_0 + \beta_1\theta + \cdots + \beta_{e-1}\theta^{e-1}, \qquad \beta_i \in U.$$

The conjugates of θ over U are $\zeta^r\theta$, where ζ is a primitive eth root of unity and $r = 0, 1, \ldots, e-1$. Therefore, the trace $\mathrm{Tr}_{U(\theta)/U}(\theta^j)$ vanishes if j is not a multiple of e and equals e if $j = 0$. Hence

$$\beta_j = \frac{1}{e}\,\mathrm{Tr}_{U(\theta)/U}(\alpha\theta^{-j}) = \frac{1}{ep^{j/e}}\sum_{r=0}^{e-1}\alpha_r\zeta^{-rj},$$

where α_r are certain conjugates of α. Note that Proposition 1.5.17 yields $h(\alpha_r\zeta^{-rj}) = h(\alpha) \leq T$ for $0 \leq r \leq e-1$. By a standard inequality about the height of a sum (see Proposition 1.5.15), we find

$$h(\beta_j p^{j/e}) \leq \log e + \sum_r h(\alpha_r) + \log e \leq 2\log D + DT. \tag{4.17}$$

As before, let u be any place of $U(\theta) = L(\theta)$ above v and use the same letter to denote the associated discrete valuation normalized by $u(L(\theta)^\times) = \mathbb{Z}$. Since $\beta_j \in U$, we have that $u(\beta_j)$ is divisible by e. Suppose now $1 \leq j \leq e-1$. Then $u(p^{j/e}) = j$ is not divisible by e, whence $u(\beta_j p^{j/e}) \neq 0$. This shows that $|u(\beta_j p^{j/e})| \geq u(p^{1/e}) = 1$.

Let us abbreviate $\gamma = \beta_j p^{j/e}$ and suppose that $\gamma \neq 0$. Letting δ_u be the local degree $\delta_u := [U(\theta)_u : \mathbb{Q}_p]$, the choice of our normalizations in 1.3.6 leads to

$$\big|\log|\gamma|_u\big| \geq -\frac{1}{e}\log|p|_u = \frac{\delta_u \log p}{e[U(\theta):\mathbb{Q}]}.$$

Thus we have

$$2h(\gamma) = h(\gamma) + h(\gamma^{-1}) \geq \sum_{u|v}\big|\log|\gamma|_u\big| \geq \frac{1}{e[U(\theta):\mathbb{Q}]}\left(\sum_{u|v}\delta_u\right)\log p.$$

By Corollary 1.3.2, we have $\sum \delta_u = [U(\theta):K]$. We conclude that, if $\gamma \neq 0$, then

$$2h(\gamma) \geq \frac{1}{e[K:\mathbb{Q}]}\log p.$$

Comparing with (4.17), we derive that either $\beta_j = 0$ or

$$\log p \leq 2e[K:\mathbb{Q}](2\log D + DT).$$

Let S be the set of rational primes containing all prime divisors of the discriminant $D_{K/\mathbb{Q}}$ and all primes $p \leq \exp(2e[K:\mathbb{Q}](2\log D + DT))$. If $v \in M_K$ is lying over a prime $p \notin S$, then B.2.13 shows that v is unramified over p. Hence our considerations above yield that we must have $\beta_j = 0$ for $1 \leq j \leq e-1$. This means that the algebraic number α lies in U, which is an abelian extension of K of exponent dividing D and unramified over v. Hence $K(\alpha)$ is unramified above any $p \notin S$. This implies that $K(\alpha)$ is of bounded degree over K, as we will show in Example 10.5.11. Here, we give a direct argument based on Hermite's discriminant theorem: Recall that a cyclic extension is a Galois extension with cyclic Galois group. Writing $\mathrm{Gal}(K(\alpha)/K)$ as a direct product of cyclic groups of order dividing D, we see that $K(\alpha)$ is the compositum of cyclic extensions of K of degree at most D, each unramified ouside S. On the other hand, the power to which a prime divides the discriminant of a number field of bounded degree is itself bounded (use Theorem B.2.12 and Corollary 1.3.2). Hence the discriminants of these cyclic extensions of K are bounded. We conclude by Hermite's discriminant theorem in B.2.14 that there are only finitely many such cyclic fields. Hence there are only finitely many distinct fields $K(\alpha)$ and, since α has bounded height, Northcott's theorem, as in 1.6.8, shows that only finitely many possibilities for α can occur. □

4.6. Remarks on the Bogomolov property

Again, we work always inside a fixed algebraic closure $\overline{\mathbb{Q}}$ of $\mathbb{Q}$.

4.6.1. We say that a set $\mathcal{A}$ of algebraic numbers has the **Bogomolov property** (B) if there exists a positive real number T_0 such that $\mathcal{A}(T_0)$ consists of all roots of unity in $\mathcal{A}$. We have already seen in Theorem 4.4.9 that the infinite cyclotomic extension of $\mathbb{Q}$ generated by all roots of unity has property (B). Another example of a field with property (B) is the field of totally real numbers, see Corollary 4.3.5.

We will give an extension of Bilu's theorem in 4.3.1 and its Corollary 4.3.5 to a p-adic setting and deduce from this some new cases of infinite algebraic field extensions with property (B).

4.6.2. For simplicity, we shall consider here only normal extensions L of $\mathbb{Q}$. Given such an extension, we denote by $S(L)$ the set of rational primes p such that L may be embedded in some finite extension L_p of $\mathbb{Q}_p$. We may also assume that the closure of L in L_p is again L_p, in which case, since L is normal, the residual degree f_p and ramification index e_p of the extension $L_p/\mathbb{Q}_p$ do not depend on the given embedding (using Corollary 1.3.5).

Theorem 4.6.3. *If $S(L)$ is not empty, then the field L has property* (B). *More precisely*

$$\liminf_{\alpha \in L} h(\alpha) \geq \frac{1}{2} \sum_{p \in S(L)} \frac{\log p}{e_p(p^{f_p}+1)}. \tag{4.18}$$

Remark 4.6.4. The lim inf is with respect to the directed system of finite subsets of L. If the sum on the right-hand side of (4.18) diverges, then L has property (N). Thus the question arises whether there are infinite extensions L where this occurs. We have been unable to find such examples, and we consider it unlikely that this can occur for an infinite extension.

By Proposition 4.5.3, $S(\mathbb{Q}^{(d)})$ is the set of all prime numbers for every $d \in \mathbb{N}$.

Example 4.6.5. Let us say that a non-zero algebraic number α is **totally p-adic** if the rational prime p splits completely in the field $\mathbb{Q}(\alpha)$, meaning that all local degrees of places over p are 1. Then the field L of all totally p-adic algebraic numbers is normal and $p \in S(L)$. Hence L has the Bogomolov property. This may be considered as the p-adic analog of results of Schinzel and Smyth for totally real algebraic numbers alluded to in Remark 4.3.6.

Example 4.6.6. Let $p_1, \ldots, p_m$ be distinct rational primes and let L be the field of all totally p-adic algebraic numbers for $p = p_1, \ldots, p_m$. Then it is clear that $p_i \in S(L)$ for

$i = 1, \ldots, m$. We can show that in this field L, we have

$$\liminf_{\alpha \in L} h(\alpha) \leq \sum_{i=1}^{m} \frac{\log p_i}{p_i - 1}.$$

This shows that the lower bound given by (4.18) is of the correct order of magnitude, insofar as the contribution of primes with $f_p = e_p = 1$ is concerned. We will not prove this result here, and refer instead to E. Bombieri and U. Zannier **[40]**.

4.6.7. *Proof of Theorem 4.6.3:* We shall prove a general lower bound for the height of an algebraic number, of which Theorem 4.6.3 will be an easy corollary.

Let K be a Galois extension of $\mathbb{Q}$, let $\alpha \in K \setminus \{0\}$, and denote by $\alpha_1, \alpha_2, \ldots, \alpha_d$ a full set of conjugates over $\mathbb{Q}$, satisfying a minimal equation

$$a_d x^d + a_{d-1} x^{d-1} + \cdots + a_0 = 0$$

over $\mathbb{Z}$, of discriminant D.

Fix a rational prime p and denote by v an extension to K, of residue degree f_p and ramification index e_p, of the usual valuation $-\log |\ |_p$ in $\mathbb{Q}_p$. By reordering the conjugates, we may assume

$$v(\alpha_1) \geq \cdots \geq v(\alpha_r) \geq 0 > v(\alpha_{r+1}) \geq \cdots \geq v(\alpha_d).$$

By Gauss's lemma in 1.6.3 applied to

$$a_d \prod_{i=1}^{d} (x - \alpha_i) = a_d x^d + a_{d-1} x^{d-1} + \cdots + a_0,$$

we find

$$v(a_d) + \sum_{i=1}^{d} \min(0, v(\alpha_d)) = \min_i v(a_i),$$

from which it follows that

$$v(a_d) = -\sum_{i=r+1}^{d} v(\alpha_i), \tag{4.19}$$

because the coefficients a_i are integers without common factors.

In order to evaluate $v(D)$ from below, we consider first the contribution to the product coming from terms with $v(\alpha_j) < 0$. We have

$$v\left(\prod_{j=r+1}^{d} \prod_{i=1}^{j-1} (\alpha_i - \alpha_j) \right) \geq \sum_{j=r+1}^{d} (j-1)\, v(\alpha_j),$$

yielding the lower bound

$$v(D) \geq (2d-2)\, v(a_d) + 2 \sum_{i<j\leq r} v(\alpha_i - \alpha_j) + 2 \sum_{j=r+1}^{d} (j-1)\, v(\alpha_j).$$

We substitute (4.19) in the right-hand side of this inequality and obtain from the formula for the discriminant in Proposition 1.6.9 the inequality

$$v(D) \geq 2 \sum_{i<j\leq r} v(\alpha_i - \alpha_j) - 2 \sum_{j=r+1}^{d} (d-j)\, v(\alpha_j). \tag{4.20}$$

Consider now the reductions of α_i, $i \leq r$, modulo the maximal ideal of the valuation ring of v. They are elements of the finite field $\mathbb{F}_q$ with $q = p^{f_p}$. For $x \in \mathbb{F}_q$, let N_x be the number of conjugates α_i with reduction x. Suppose $i < j \leq r$. If α_i and α_j have the same reduction, we have $v(\alpha_i - \alpha_j) > 0$, hence $v(\alpha_i - \alpha_j) \geq 1/e_p$, and otherwise we have $v(\alpha_i - \alpha_j) \geq 0$; note that the number of pairs (i, j) with $i < j$ and such that α_i and α_j have the same reduction x is $N_x(N_x - 1)/2$. If instead $j > r$, we have $v(\alpha_j) < 0$, hence $v(\alpha_j) \leq -1/e_p$.

In view of these remarks, we deduce from (4.20) that

$$v(D) \geq \frac{1}{e_p} \sum_{x\in\mathbb{F}_q} N_x(N_x - 1) + \frac{1}{e_p}(d-r)(d-r-1). \tag{4.21}$$

A more elegant formulation of (4.21) is obtained by defining the reduction of an element with negative valuation to be ∞. With this convention, we have

$$N_\infty = d - r, \qquad \sum_{x\in\mathbb{F}_q\cup\{\infty\}} N_x = d.$$

Therefore, introducing the **normalized variance**

$$V_p(\alpha; K) := \frac{1}{d^2} \sum_{x\in\mathbb{F}_q\cup\{\infty\}} \left(N_x - \frac{d}{q+1}\right)^2,$$

we rewrite (4.21) as

$$v(D) \geq \frac{d^2}{e_p}\left(V_p(\alpha; K) + \frac{1}{q+1}\right) - \frac{d}{e_p}. \tag{4.22}$$

This estimate is useful only in the range $q < d$, but, since D is a non-zero rational integer, we have $v(D) \geq 0$ in any case. Thus from (4.22) it follows that

$$\log|D| \geq d^2 \sum_{q<d} \frac{1}{e_p}\left(V_p(\alpha; K) + \frac{1}{q+1} - \frac{1}{d}\right) \log p, \tag{4.23}$$

where the sum ranges over all primes p with $q = p^f < d$. On the other hand, by Proposition 1.6.9, we have

$$\log|D| \leq d \log d + (2d-2) d\, h(\alpha). \tag{4.24}$$

Combining (4.23) and (4.24), we finally obtain:

Theorem 4.6.8. *Let K be a Galois extension of $\mathbb{Q}$. For a non-archimedean place v of K lying over the rational prime p let f_p and e_p be the residue degree and ramification index of v over p and write $q := p^{f_p}$. Let $\alpha \in K \setminus \{0\}$ be of degree d. Then*

$$h(\alpha) \geq -\frac{\log d}{2d-2} + \frac{d}{2d-2} \sum_{q<d} \frac{1}{e_p}\left(V_p(\alpha; K) + \frac{1}{q+1} - \frac{1}{d}\right) \log p.$$

4.6.9. *Completion of the proof of Theorem 4.6.3:* For $\alpha \in L$, we apply Theorem 4.6.8 with $K = L$. Since $V_p(\alpha; K) \geq 0$ in any case and we may restrict the sum to finitely many primes $p \in S(L)$, the proof is completed by noting that, by Northcott's theorem from 1.6.8, in any infinite sequence of distinct algebraic numbers of bounded height, the degrees must go to ∞. Thus $d \to \infty$ if we want to estimate $\liminf h(\alpha)$ in L. □

Remark 4.6.10. Theorem 4.6.8 implies an equidistribution theorem for elements of an infinite sequence (α_n) of algebraic numbers with height tending to 0. In particular, for any sequence (α_n) along which $h(\alpha_n) \to 0$, we have that, if p is unramified in the Galois closure of α_n, then $q := p^{f_p} \to \infty$ and

$$\frac{1}{\deg^2(\alpha_n)} \sum_{x \in \mathbb{F}_q \cup \{\infty\}} \left(N_x - \frac{\deg(\alpha_n)}{q+1} \right)^2 \log p \to 0. \tag{4.25}$$

This may be regarded as an analog of Bilu's theorem from 4.3.1. See also R. Rumely [**254**] for related results in a p-adic and adelic setting.

4.7. Bibliographical notes

Theorem 4.2.2 (a) in the special case of a linear equation $a_1x_1 + a_2x_2 + \cdots + a_nx_n = 1$ is quite old, the prototype going back to a theorem of H.B. Mann [**191**]. If $n = 2$, Lang [**169**], p.201, attributes it to Y. Ihara, J.-P. Serre, and J. Tate, and gives Tate's proof. Explicit results are in J.H. Conway and A.J. Jones [**71**] and R. Dvornicich and U. Zannier [**94**]. The methods in these papers are based on studying the action of the absolute Galois group of $\mathbb{Q}$ on torsion points of high order. The heuristic argument behind such methods is that, if the order of torsion points in X is unbounded, then X contains the closure, in the usual complex topology, of some non-trivial analytic subgroup of $\mathbb{T}^n$. The Zariski closure of this analytic subgroup then provides a non-trivial linear torus contained in X.

H.P. Schlickewei [**262**] proved that the number of non-degenerate solutions of $a_1x_1 + \cdots + a_nx_n = 1$ in roots of unity is at most $2^{4(n+1)!}$ for arbitrary complex numbers $a_1, \ldots, a_n$.

Shou-Wu Zhang's results can be found in a series of papers [**339**], [**340**], [**341**]. Theorem 4.2.3 is due to E. Bombieri and U. Zannier [**38**], building on earlier ideas of D. Zagier [**336**] and W.M. Schmidt [**272**]. Lemma 4.2.8 is inspired by [**90**]. The inductive proof in [**38**] yields extraordinarily small values for $\gamma(d, n)$ expressed by towers of exponentials of length n, but W.M. Schmidt [**273**] later obtained explicit values for $\gamma(d, n)$ and $N(d, n)$ requiring only double exponentials. Much better lower bounds, with a dependence on d in $\gamma(n, d)$ of type $d^{-c(n)}$, have been obtained by S. David and P. Philippon [**83**], with deeper methods of arithmetic geometry beyond the scope of this book.

Further results going beyond Zagier's lower bound in Remark 4.2.7 can be found in C. Doche [**91**]. The proof of Theorem 4.3.1 is a modification of an argument of Bilu [**24**] and the alternative argument is a suggestion of J. Bourgain.

Dobrowolski's theorem has been slightly improved by R. Louboutin [**183**], who obtains a constant $9/4$ instead of the constant $1/8$ given here, by a different method. The argument given here can also be refined to give the same constant $9/4$, by using the full force of Siegel's lemma in 2.9.4 (including the use of successive minima).

The higher dimensional version of Dobrowolski's theorem is due to Amoroso and David [**9**] (see also [**10**] for a correction and further results).

The presentation of the Amoroso–Dvornicich theorem and its application to Smyth's theorem follows closely [**11**]. A relative version of this result has been obtained by Amoroso and Zannier in [**12**].

The remarks about the Northcott property and the Bogomolov property can be found in a paper of Bombieri and Zannier [**40**].

5 THE UNIT EQUATION

5.1. Introduction

Let K be a number field. A classical and important problem is that of determining the units u of K such that $1 - u$ is also a unit. More generally, let Γ be a finitely generated subgroup of $K^\times \times K^\times$. The **unit equation in Γ** is the equation

$$x + y = 1$$

to be solved with $(x, y) \in \Gamma$. A basic result, going back to Siegel, Mahler, and Lang, asserts that this equation has only finitely many solutions. In Section 5.2, we shall give a complete proof of this result based on the uniform Zhang theorem in 4.2.3 and obtain a uniform bound for the number of solutions. This is applied in Section 5.3 to give an upper bound for the number of integer solutions of the Thue–Mahler equation and of a hyperelliptic equation.

The important problem of finding explicit upper bounds for the height of solutions of a unit equation requires different methods. In Section 5.4, we give just some results. We refer to A. Baker's monograph **[14]** and J.-P. Serre **[277]**, §8.3, for an approach using Baker's theory of linear forms in logarithms.

We may also consider a linear torus G over a field of characteristic 0, a finitely generated subgroup Γ of G and study the set $C \cap \Gamma$, where C is a geometrically irreducible algebraic curve in G. Lang proved that, if $C \cap \Gamma$ is an infinite set, then C is a translation of a subtorus of G. Lang conjectured and Liardet proved that the same conclusion holds if we replace Γ by its division group, that is the group Γ' consisting of all points $y \in G$ such that $y^n \in \Gamma$ for some n (we use multiplicative notation in G). We will give a sketch of an effective version of this theorem in the special case when the field is a number field, see Theorem 5.4.5.

Similar statements can be made for G a commutative algebraic group with no $\mathbb{G}_a$ components (that is, a semiabelian variety) and replacing C by a subvariety X of G, but they are far more difficult to prove; indeed, even the simplest case of a curve in an abelian variety turns out to be equivalent to Mordell's conjecture. The latter will be proved in Chapter 11 and for the semiabelian case we refer to the bibliographical notes in Section 11.11.

5.2. The number of solutions of the unit equation

We have the following nice result of F. Beukers and H.P. Schlickewei [**23**]:

Theorem 5.2.1. *There are absolute computable constants C_1, C_2 with the following property. Let Γ be a subgroup of $\overline{\mathbb{Q}}^\times \times \overline{\mathbb{Q}}^\times$ with* $\operatorname{rank}_{\mathbb{Q}}(\Gamma) = r < \infty$, *where* $\operatorname{rank}_{\mathbb{Q}}(\Gamma)$ *is the maximum number of multiplicatively independent elements in Γ. Then the equation*

$$x + y = 1, \qquad (x, y) \in \Gamma$$

has at most $C_1 \cdot C_2^r$ solutions.

This result improves bounds C^{r^2} and $(Cr)^r$ previously obtained by Schlickewei and Schmidt. Beukers and Schlickewei give the values $C_1 = C_2 = 256$.

5.2.2. It is an interesting problem to determine the maximum number of solutions of the equation $x + y = 1$ with (x, y) in a group Γ of rank r. In this vein, we may remark the following. Suppose Γ is a subgroup of $K^\times \times K^\times$, where $K^\times$ is the multiplicative group of a number field K. If we take cosets in Γ/tors of the subgroup of fourth powers, we are led to finding K-rational points on curves $ax^4 + by^4 = 1$, which have genus 3.

It has been conjectured by L. Caporaso, J. Harris, and B. Mazur [**56**] that the number of K-rational points on a curve of genus $g \geq 2$ is bounded solely in terms of K and g. Since we have 4^r cosets, this argument suggests that perhaps $C_2 \leq 4$.

5.2.3. In many applications the group Γ is the group $(U_{S,K})^2$, where S is a finite set of places, containing all infinite places, of a number field K and $U_{S,K}$ is the group of units of the ring of S-integers of K. By Dirichlet's unit theorem from 1.5.13, $U_{S,K}$ is finitely generated of rank $|S| - 1$ and it is possible to determine effectively a set of generators of $U_{S,K}$.

We will give below some examples of Γ with a large number of solutions of the unit equation.

Example 5.2.4. The following simple argument yields an example of a subgroup of $\mathbb{Q}^\times \times \mathbb{Q}^\times$ with a large number of solutions.

Let $N \geq 2$ be a positive integer. Let M be the number of positive integers up to x whose prime factors do not exceed $x^{1/N}$. It is clear that

$$M \geq \frac{\pi(x^{1/N})^N}{N!}.$$

Since $\pi(y) > y/\log y$ for $y \geq 17$ (see B. Rosser and L. Schoenfeld [**245**], Th.1, Cor.1, p.69) and $N! \leq \frac{1}{2}N^N$, we see that $M > 2x/(\log x)^N$ if $x \geq 17^N$. Consider the M^2 sums $n' + n$, where $n', n \leq x$ are positive integers whose prime factors do not exceed $x^{1/N}$. Since $n' + n \leq 2x$, one sum must occur at least $M^2/(2x) > 2x/(\log x)^{2N}$

times. In other words, there is an integer b such that the equation $n' + n = b$ has at least $2x/(\log x)^{2N}$ solutions.

It follows that, if Γ is the subgroup of $\mathbb{Q}^\times \times \mathbb{Q}^\times$ generated on each factor by all primes up to $x^{1/N}$ and by b, then the unit equation in Γ has at least $2x/(\log x)^{2N}$ solutions, provided $x \geq 17^N$.

This group Γ has rank r equal to either $2\pi(x^{1/N})$ or $2\pi(x^{1/N}) + 2$, hence we have $r \sim 2Nx^{1/N}/\log x$ as $x^{1/N}$ tends to ∞. If we make the asymptotically optimal choice

$$N = \left\lfloor \frac{\log x}{2 \log\log x} - \frac{\log x}{2(\log\log x)^2} \right\rfloor,$$

we verify that the number of solutions is at least

$$\frac{x}{(\log x)^{2N}} = e^{(c+o(1))\frac{\sqrt{r}}{\sqrt{\log r}}}$$

with $c = \sqrt{2}/e$.

Example 5.2.5. Consider the equation $au + bv = 1$ for non-zero algebraic numbers a, b to be solved with $(u, v) \in \Gamma$. This may be reduced to the unit equation by enlarging the range of solutions to the group generated by Γ and (a, b). This procedure will be used later.

Here, we are interested directly in the equation $ax^m + by^m = 1$ for varying m, corresponding to a group $\Gamma = (x, y)^{\mathbb{Z}}$ of rank 1. We want to find a, b, x, y such that it has the maximum number of solutions for $m \in \mathbb{Z}$.

We may assume that $m = 0$ is a solution. Suppose that $m = 1$ is also a solution, so the equation becomes $(y-1)x^m + (1-x)y^m - (y-x) = 0$. Here we must exclude $x = 1$, $y = 1$, and $x = y$ which correspond to degenerate cases.

If we fix two other solutions, say m_1 and m_2, we can eliminate y and obtain an equation for x. In general, this leads to pairs (x, y) such that we have four solutions, namely $m = 0,\ 1,\ m_1,\ m_2$.

Note however that there are special cases. If $m_1 = 2$, the equation degenerates into $(x-1)(y-1)(x-y) = 0$, so $m_1 = 2$ must be excluded. Also, if $m_1 = 3$, the values $m_2 = 4, 5, 6, 7, 9$ must be excluded, because they lead to degenerate cases or a group Γ of rank 0.

However, taking $m_1 = 4$ and $m_2 = 6$ gives the equation $x^6 + x^5 + 2x^4 + 3x^3 + 2x^2 + x + 1 = 0$. For any root ξ of this equation, we see that taking $\eta = -1/(1 + \xi + \xi^3)$, which is another root of the same equation, we have

$$\frac{\eta - 1}{\eta - \xi}\xi^m + \frac{1 - \xi}{\eta - \xi}\eta^m = 1$$

for the six values $m = 0,\ 1,\ 4,\ 6,\ 13,\ 52$.

Other examples are obtained by letting the Galois group of the equation (which is of order 6, generated by $\xi \to 1/\xi$ and $\xi \to \eta$) act on $\Gamma = (\xi, \eta)^{\mathbb{Z}}$, and also going to a division group. It is conceivable that 6 is the maximum number of solutions and that any group of rank 1 with six solutions is obtained in this way.

The above problem is closely connected to finding zeros of a linear recurrence: Let $u_{m+1} = Au_m + Bu_{m-1} + Cu_{m-2}$ be a linear recurrence of the third order, which we assume non-degenerate in the sense that the roots β_i $(i = 1, 2, 3)$ of the associated characteristic equation $x^3 - Ax^2 - Bx - C$ are distinct and non-zero. We may consider the recurrence also in the negative direction by solving for u_{m-2}. Then the general solution of the recurrence is given by

$$u_m = C_1\beta_1^m + C_2\beta_2^m + C_3\beta_3^m \quad (m \in \mathbb{Z}).$$

Let $a = -C_1/C_3$, $b = -C_2/C_3$, $x = \beta_1/\beta_3$ and $y = \beta_2/\beta_3$. Then solving the equation $ax^m + by^m = 1$ in the group $\Gamma = (x, y)^{\mathbb{Z}}$ of rank 1 is equivalent to finding the zeros of the recurrence $\{u_m \mid m \in \mathbb{Z}\}$.

Example 5.2.6. For a prime p, consider the cyclotomic field $C_p = \mathbb{Q}(\sqrt[p]{1})$ and the corresponding unit equation. Here we choose $\Gamma = U \times U$, where U is the group of units in the ring of algebraic integers of C_p.

If $u + v = 1$ and u, v are not real, then $\overline{u} + \overline{v} = 1$ is another solution of the unit equation. By Kronecker's theorem in 1.5.9, $\varepsilon := \overline{u}/u$ and $\varepsilon' := \overline{v}/v$ are roots of unity in C_p. Solving the system $u + v = 1$, $\varepsilon u + \varepsilon' v = 1$, we get $u = (\varepsilon' - 1)/(\varepsilon' - \varepsilon)$, $v = (1 - \varepsilon)/(\varepsilon' - \varepsilon)$. Conversely, given distinct roots of unity ε, ε' in C_p, not equal to 1, we obtain a solution u, v of the unit equation. Thus the number of non-real solutions of the unit equation in C_p is $(p-1)(p-2)$.

Example 5.2.7. The number of solutions of the unit equation in the maximal real subfield K_p of C_p is much larger. A computer search using cyclotomic units produced three solutions for K_5, 42 solutions for K_7, 570 solutions for K_{11}, 1830 solutions for K_{13}, 11 700 solutions for K_{17}, and 28 398 solutions for K_{19}.

Example 5.2.8. The following example gives an equation $u + v = 1$ with at least 2532 solutions u, $v \in U$, for a certain group U of rank 5. Let $K = \mathbb{Q}(\alpha)$ with α the real root $\alpha > 1$ of the Lehmer equation

$$x^{10} + x^9 - x^7 - x^6 - x^5 - x^4 - x^3 + x + 1 = 0.$$

This equation has another real root $1/\alpha$ and eight non-real roots all of absolute value 1; we shall refer to the map $\alpha \mapsto 1/\alpha$ as real conjugation in $\mathbb{Q}(\alpha)$. The Mahler measure of α is $M(\alpha) = \alpha = 1.17628081825991\ldots$, and it is widely conjectured to be the infimum of the Mahler measure of an algebraic number, not a root of unity – the so-called Lehmer conjecture (see 1.6.15).

The group U of units of K has rank 5: $U = \{\pm 1\} \times < \alpha, 1 - \alpha, 1 + \alpha, 1 + \alpha + \alpha^2, 1 + \alpha - \alpha^3 >$. Now an extensive computer search for solutions of the corresponding unit equation produced a remarkable total of 2532 solutions.

The following is a plot of the 2532 points $(\log |u|, \log |u'|)$, where u is a real unit and u' is the real conjugate of u.

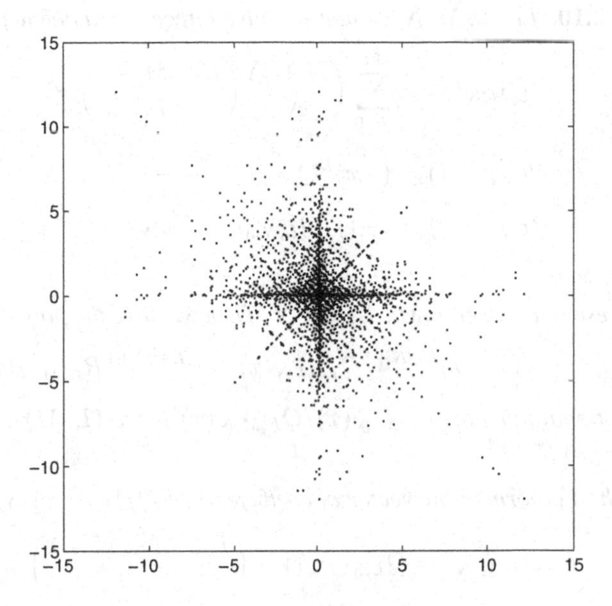

The proof of Theorem 5.2.1 is obtained by means of a Padé approximation method, which originates in the work of Thue, Siegel, and Baker.

Lemma 5.2.9. *Let $f(x) \in K[[x]]$ be a formal power series with coefficients in a field K. Let L, M be positive integers. Then there are polynomials $P(x) \in K[x]$, $Q(x) \in K[x]$ of degrees at most L and M and with Q not identically 0, such that*

$$P(x) - Q(x)f(x) = x^{L+M+1}R(x) \tag{5.1}$$

for some formal power series $R(x) \in K[[x]]$.

*The quotient $P(x)/Q(x)$ is uniquely determined and is called the (L, M)-***Padé approximant** *of $f(x)$.*

Proof: Equation (5.1) is equivalent to solving a system of $L+M+1$ homogeneous linear equations in $L+M+2$ unknowns, namely the coefficients of the polynomials P and Q. This proves the existence of a non-trivial solution $P(x)$, $Q(x)$, and non-triviality implies that Q is not identically 0. In order to show uniqueness, if $\widetilde{P}(x)$, $\widetilde{Q}(x)$ is another solution, then $\widetilde{Q}(x)P(x) - Q(x)\widetilde{P}(x)$ is a polynomial of degree at most $L+M$ divisible by x^{L+M+1}, hence is identically 0. □

We are interested in the special case in which $f(x) = (1-x)^n$, $n \in \mathbb{N}$. Clearly, we may assume $n \geq L+1$, otherwise the (L, M)-Padé approximant is $(1-x)^n$.

Theorem 5.2.10. *Let L, M, N be non-negative integers and define polynomials*

$$Q_{L,M,N}(x) = \sum_{j=0}^{M} \binom{N+j}{N} \binom{L+M-j}{L} x^j,$$

$$P_{L,M,N}(x) = (-x)^L Q_{N,L,M}(1 - \frac{1}{x}),$$

$$R_{L,M,N}(x) = (-1)^L Q_{L,N,M}(1-x)$$

in $\mathbb{Z}[x]$, of respective degree M, L, and N. Then we have the polynomial identity

$$P_{L,M,N}(x) - (1-x)^{L+N+1} Q_{L,M,N}(x) = x^{L+M+1} R_{L,M,N}(x), \tag{5.2}$$

hence the rational function $P_{L,M,N}(x)/Q_{L,M,N}(x)$ is the (L, M)-Padé approximant of $(1-x)^{L+N+1}$.

Moreover, the ℓ_1-norm of the vector of coefficients of $Q_{L,M,N}(x)$ is

$$\ell_1(Q_{L,M,N}) = Q_{L,M,N}(1) = \binom{L+M+N+1}{M}.$$

Finally, we have the identities, with $'$ denoting the derivative

$$(P_{L,M,N}(x))' = -(L+M+N+1) P_{L-1,M,N}(x),$$

$$((1-x)^{L+N+1} Q_{L,M,N}(x))' = -(L+M+N+1)(1-x)^{L+N} Q_{L-1,M,N}(x)$$

$$(x^{L+M+1} R_{L,M,N}(x))' = -(L+M+N+1) x^{L+M} R_{L-1,M,N}(x).$$

Proof: We have the following classical transformation of a hypergeometric integral, due to Kummer (A. Erdélyi, W. Magnus, F. Oberhettinger, and F. Tricomi [**100**], I, (29), p.106)

$$\begin{aligned}
&\int_0^1 t^M (t-1)^N (t-x)^L \mathrm{d}t \\
&= x^{L+M+1} \left(\int_0^1 + \int_1^{1/x} \right) u^M (xu-1)^N (u-1)^L \mathrm{d}u \\
&= x^{L+M+1} \int_0^1 u^M (u-1)^L (xu-1)^N \mathrm{d}u \\
&\quad + (-1)^N (1-x)^{L+N+1} \int_0^1 v^N (1-v)^L (1-(1-x)v)^M \mathrm{d}v,
\end{aligned}$$

where we have performed the changes of variables $t = xu$ in the first equation, and $xu = 1 - (1-x)v$ in the second equation.

This hypergeometric identity determines explicitly the (L, M)-Padé approximant of $(1-x)^{L+N+1}$. We define polynomials P, Q, R of precise degrees L, M, N

by means of

$$P(x) = \int_0^1 t^M (1-t)^N (t-x)^L \mathrm{d}t$$
$$Q(x) = \int_0^1 v^N (1-v)^L (1-(1-x)v)^M \mathrm{d}v$$
$$R(x) = (-1)^L \int_0^1 u^M (1-u)^L (1-xu)^N \mathrm{d}u.$$

Then our identity is

$$P(x) - (1-x)^{L+N+1} Q(x) = x^{L+M+1} R(x), \tag{5.3}$$

showing, by checking degrees, that $P(x)/Q(x)$ is the (L, M)-Padé approximant of $(1-x)^{L+N+1}$.

By a familiar evaluation of Euler's beta integral, we have

$$\begin{aligned} Q(x) &= \int_0^1 v^N (1-v)^L ((1-v)+xv)^M \mathrm{d}v \\ &= \sum_{j=0}^{M} \binom{M}{j} \left(\int_0^1 v^{N+j} (1-v)^{L+M-j} \mathrm{d}v \right) x^j \\ &= \sum_{j=0}^{M} \binom{M}{j} \frac{(N+j)!(L+M-j)!}{(L+M+N+1)!} x^j \\ &= D^{-1} \sum_{j=0}^{M} \binom{N+j}{N} \binom{L+M-j}{L} x^j, \end{aligned} \tag{5.4}$$

where we have abbreviated

$$D = \frac{(L+M+N+1)!}{L!M!N!}. \tag{5.5}$$

We define $(P_{L,M,N}, Q_{L,M,N}, R_{L,M,N}) := (DP, DQ, DR)$. Note that by (5.4) and (5.5) the polynomial $Q_{L,M,N}$ has positive integral coefficients. In particular, the ℓ_1-norm of $Q_{L,M,N}$ is $DQ(1)$, hence

$$\ell_1(Q_{L,M,N}) = D \int_0^1 v^N (1-v)^L \mathrm{d}v = \binom{L+M+N+1}{M}.$$

The uniqueness of Padé approximants now can be used to obtain relations between two Padé approximants associated to a triple (L, M, N) and to a permutation.

If in (5.3) we make the change of variable $x \mapsto 1 - x$, we verify that

$$(P_{L,M,N}(1-x), Q_{L,M,N}(1-x), R_{L,M,N}(1-x)) = \\ (-1)^L (P_{L,N,M}(x), R_{L,N,M}(x), Q_{L,N,M}(x)),$$

while making the change of variable $x \mapsto 1/x$, we find

$$(x^N R_{L,M,N}(1/x), x^M Q_{L,M,N}(1/x), x^L P_{L,M,N}(1/x)) = \\ ((-1)^{N+L} P_{N,M,L}(x), Q_{N,M,L}(x), (-1)^{N+L} R_{N,M,L}(x)).$$

By composing these changes of variable and permuting (L, M, N), we infer

$$P_{L,M,N}(x) = (-x)^L Q_{N,L,M}\left(1 - \frac{1}{x}\right), \quad R_{L,M,N}(x) = (-1)^L Q_{L,N,M}(1-x).$$

We easily see that differentiating (5.2) on page 130 yields an $(L-1, M, N)$-approximant of $(1-x)^{L+M}$. Thus the final identities follow by uniqueness of Padé approximants. This completes the proof. □

We need another important property of a triple (P, Q, R) as in Theorem 5.2.10.

Proposition 5.2.11. *Let (P, Q, R) be a triple as in Theorem 5.2.10. Then any linear combination*

$$\alpha P(x) - \beta(1-x)^{L+N+1} Q(x) - \gamma x^{L+M+1} R(x)$$

with complex coefficients is either identically 0, in which case $\alpha = \beta = \gamma$, or is a polynomial with only simple roots outside $\{0, 1, \infty\}$.

Proof: Consider the rational function

$$\varphi(x) := \frac{x^{L+M+1} R(x)}{P(x)} = 1 - \frac{(1-x)^{L+N+1} Q(x)}{P(x)}$$

and the associated covering $\varphi : \mathbb{P}^1 \to \mathbb{P}^1$.

By Theorem 5.2.10, P, Q, R have exact degree L, M, N and do not vanish at 0 or 1. Now let e_x be the ramification index of φ at the point x. Then

$$\begin{aligned} e_0 &= L + M + 1 && \text{(because } P(0) \neq 0,\ R(0) \neq 0) \\ e_1 &= L + N + 1 && \text{(because } P(1) \neq 0,\ Q(1) \neq 0) \\ e_\infty &= M + N + 1 && \text{(because } \deg(P) = L,\ \deg(R) = N) \end{aligned}$$

and in any case $e_x \geq 1$ and $\deg(\varphi) \leq L + M + N + 1$. Therefore, Hurwitz's theorem B.4.6 yields

$$\begin{aligned} -2 &= \deg(\varphi) \cdot (-2) + (e_0 - 1) + (e_1 - 1) + (e_\infty - 1) + \sum_{x \notin \{0,1,\infty\}} (e_x - 1) \\ &= -2 + 2(L + M + N + 1 - \deg(\varphi)) + \sum_{x \notin \{0,1,\infty\}} (e_x - 1) \\ &\geq -2. \end{aligned}$$

Equality must hold, hence $\deg(\varphi) = L + M + N + 1$ (thus P, Q, R are pairwise coprime) and φ is unramified outside $0, 1, \infty$.

Let $\lambda \in \mathbb{C} \setminus \{0, 1\}$ be with

$$\alpha P(x) - \beta(1-x)^{L+M+1}Q(x) - \gamma x^{L+M+1}R(x) = (x - \lambda)^2 S(x)$$

for some polynomial $S(x)$. By identity (5.2) on page 130, we get

$$(\alpha - \beta)P(x) - (\gamma - \beta)x^{L+M+1}R(x) = (x - \lambda)^2 S(x).$$

Note that $P(\lambda) \neq 0$, because $P(x)$ and $R(x)$ have no common zeros. Dividing by $P(x)$, we find

$$-(\gamma - \beta)\varphi(x) = (x - \lambda)^2 \frac{S(x)}{P(x)} - (\alpha - \beta).$$

Since φ is unramified at λ and $P(\lambda) \neq 0$, it follows that $\gamma = \beta$. Hence λ is a multiple root of $(\alpha - \beta)P(x)$. Since $P(\lambda) \neq 0$, we conclude that $\alpha = \beta$. □

Proof of Theorem 5.2.1. Preliminary lemmas: We need two lemmas. In what follows, a, b, x_1, $x_2, \dots$ will denote algebraic numbers.

Lemma 5.2.12. *Suppose that* $ax_1 + bx_2 = c$, $a'x_1 + b'x_2 = c'$ *and* $ab' \neq a'b$. *Then we have*

$$h(\mathbf{x}) \leq \log 2 + h((a : b : c)) + h((a' : b' : c')).$$

Proof: By Cramer's rule, we have

$$x_1 = \frac{cb' - c'b}{ab' - a'b}, \qquad x_2 = \frac{ac' - a'c}{ab' - a'b}.$$

Hence

$$\begin{aligned} h(\mathbf{x}) &= h((ab' - a'b : cb' - c'b : ac' - a'c)) \\ &= \sum_v \log \max(|ab' - a'b|_v, |cb' - c'b|_v, |ac' - a'c|_v) \\ &\leq \log 2 + \sum_v \log \max(|a|_v, |b|_v, |c|_v) + \sum_v \log \max(|a'|_v, |b'|_v, |c'|_v) \\ &= \log 2 + h((a : b : c)) + h((a' : b' : c')). \end{aligned}$$

□

Corollary 5.2.13. *Suppose that* $x_1 + x_2 = 1$ *and* $y_1 + y_2 = 1$, *with non-zero* x_1, x_2, y_1, y_2, *and* $\mathbf{x} \neq \mathbf{y}$. *Then*

$$h(\mathbf{x}) \leq \log 2 + h(\mathbf{y}\mathbf{x}^{-1}).$$

Proof: Use Lemma 5.2.12 with $a = 1$, $b = 1$ and $a' = y_1/x_1$, $b' = y_2/x_2$. □

The next lemma is the key to the proof.

Lemma 5.2.14. *Suppose that* $x_1 + x_2 = 1$, $y_1 + y_2 = 1$, *with non-zero* x_1, x_2, y_1, y_2. *Let* $n \geq 2$ *be an integer. Then*

$$h(\mathbf{x}) \leq \kappa + \frac{1}{n-1} h(\mathbf{y}\mathbf{x}^{-2n})$$

for an absolute constant κ.

Remark 5.2.15. We may take $\kappa = \log 42$.

Proof: Let $L, M, N \geq 1$ be positive integers. By Theorem 5.2.10, we have

$$a\,x_1^{L+M} + b\,x_2^{L+N} = c \tag{5.6}$$

with

$$a := x_1 R_{L,M,N}(x_1), \quad b := x_2 Q_{L,M,N}(x_1), \quad c := P_{L,M,N}(x_1).$$

Another, and obvious, relation is

$$a' x_1^{L+M} + b' x_2^{L+N} = 1 \tag{5.7}$$

with

$$a' := y_1 x_1^{-L-M}, \quad b' := y_2 x_2^{-L-N}.$$

Now we define a condition $C(L, M, N)$ by

$$ab' \neq a'b. \tag{$C(L,M,N)$}$$

We claim that either $C(L, M, N)$ or $C(L-1, M, N)$ holds. Suppose $C(L, M, N)$ does not hold. This is the same as saying that

$$f_{L,M,N}(x) := x^{L+M+1} R_{L,M,N}(x) - \frac{y_1}{y_2} (1 - x)^{L+N+1} Q_{L,M,N}(x)$$

vanishes at $x = x_1$. By Proposition 5.2.11, x_1 must be a simple zero of $f_{L,M,N}(x)$, that is $f'_{L,M,N}(x_1) \neq 0$. Differentiating and using the last identities in Theorem 5.2.10, we see that

$$f_{L-1,M,N}(x_1) = -\frac{1}{L+M+N+1} f'_{L,M,X}(x_1) \neq 0,$$

proving our claim.

Therefore, either equations (5.6) and (5.7) are linearly independent, or the same equations, but now with parameters $(L-1, M, N)$ in place of (L, M, N), are linearly independent; this second alternative is the same as saying that

$$a'' x_1^{L+M} + b'' x_2^{L+N} = c'' \tag{5.8}$$

with

$$a'' := R_{L-1,M,N}(x_1), \quad b'' := Q_{L-1,M,N}(x_1), \quad c'' := P_{L-1,M,N}(x_1),$$

and equation (5.7) are linearly independent.

Now we specialize $(L, M, N) = (n, n, n)$. Suppose first that equations (5.6) and (5.7) are linearly independent. Then Lemma 5.2.12 shows that

$$2n\, h(\mathbf{x}) = h((1 : x_1^{2n} : x_2^{2n})) \leq \log 2 + h((a : b : c)) + h((a' : b' : 1)). \tag{5.9}$$

Let us write P, Q, R for $P_{L,M,N}$, $Q_{L,M,N}$, $R_{L,M,N}$. In order to estimate $h((a : b : c))$, we note that the formulas for P, Q, R in Theorem 5.2.10, together with the equations $x_1 + x_2 = 1$ and $1 - 1/x_1 = -x_2/x_1$, give

$$\log|Q(x_1)|_v \leq \log^+ \left| \binom{L+M+N+1}{M} \right|_v + M \log^+ |x_1|_v$$

$$\log|R(x_1)|_v \leq \log^+ \left| \binom{L+M+N+1}{N} \right|_v + N \log^+ |x_2|_v$$

$$\log|P(x_1)|_v \leq \log^+ \left| \binom{L+M+N+1}{L} \right|_v + L \max(\log|x_1|_v, \log|x_2|_v).$$

In the special case $(L, M, N) = (n, n, n)$ we consider here, this gives

$$\begin{aligned} h((a : b : c)) &\leq \log \binom{3n+1}{n} + \sum_v \max((n+1)\log^+ |x_1|_v, (n+1)\log^+ |x_2|_v) \\ &= \log \binom{3n+1}{n} + (n+1) h(\mathbf{x}). \end{aligned}$$

By definition

$$h((a' : b' : 1)) = h(\mathbf{y} x^{-2n}),$$

and, in view of (5.9), we deduce

$$2n\,h(\mathbf{x}) \le \log 2 + \log \binom{3n+1}{n} + (n+1)h(\mathbf{x}) + h(\mathbf{y}x^{-2n}),$$

hence

$$h(\mathbf{x}) \le \frac{1}{n-1}\log\left(2\binom{3n+1}{n}\right) + \frac{1}{n-1}\,h(\mathbf{y}x^{-2n}). \tag{5.10}$$

If instead equations (5.6) and (5.7) on page 134 are linearly dependent, equations (5.7) and (5.8) on page 135 must be linearly independent. The same calculation as before now shows that

$$h(\mathbf{x}) \le \frac{1}{n}\log\left(2\binom{3n}{n}\right) + \frac{1}{n}\,h(\mathbf{y}x^{-2n}),$$

which is better than (5.10). Thus (5.10) holds in any case.

The maximum of $\frac{1}{n-1}\log\left(2\binom{3n+1}{n}\right)$ occurs for $n = 2$ and equals $\log 42$. This proves the lemma. □

5.2.16. *Continuation of the proof of Theorem 5.2.1:* Let Γ be a finitely generated subgroup of $\overline{\mathbb{Q}}^\times \times \overline{\mathbb{Q}}^\times$, of rank r. Let Γ_{tors} be its subgroup of torsion elements. Then $\Gamma/\Gamma_{\text{tors}}$ is a free abelian group of rank r, which we may identify with $\mathbb{Z}^r$. Let $\mathcal{Z}$ be the set of solutions of $x + y = 1$ in Γ and let $\mathcal{Z}_0$ be its image in $\mathbb{Z}^r$ under the projection $\Gamma \longrightarrow \mathbb{Z}^r$. We claim that

$$|\mathcal{Z}| \le 2\,|\mathcal{Z}_0|\,. \tag{5.11}$$

Indeed, elements of $\mathcal{Z}$ with same image in $\mathcal{Z}_0$ can be written as $(a\varepsilon, b\zeta)$ with a, b fixed and ε and ζ roots of unity. Consider the triangle in the complex plane with vertices at 0, $a\varepsilon$ and 1. Then the equation $a\varepsilon + b\zeta = 1$ shows that its sides have length 1, $|a|$, $|b|$.

There are at most two such triangles (intersect a circle of radius $|a|$ and centre 0 and a circle of radius $|b|$ and centre 1), showing that the projection of $\mathcal{Z}$ onto $\mathcal{Z}_0$ is at most two-to-one.

We define a norm $\|\ \|$ on $\mathbb{R}^r$ as follows. Let $\mathbf{x} = (x_1, x_2) \in \Gamma$ be any representative of $\mathbf{u} \in \mathbb{Z}^r \cong \Gamma/\Gamma_{\text{tors}}$, and set

$$\|\mathbf{u}\| = \widehat{h}(\mathbf{x}) = h(x_1) + h(x_2);$$

this is well defined because changing x_1 or x_2 by a root of unity does not change their height. Next, we extend this to $\mathbb{Q}^r$ by setting $\|\lambda\mathbf{u}\| = |\lambda|\cdot\|\mathbf{u}\|$, which is consistent with the definition of $\|\ \|$ because $h(x^\lambda) = |\lambda|\cdot h(x)$ for $\lambda \in \mathbb{Q}$. Finally, we extend this to $\mathbb{R}^r$ by continuity.

The triangle inequality $\|\mathbf{u}+\mathbf{v}\| \le \|\mathbf{u}\| + \|\mathbf{v}\|$ is clear from the properties of the standard height, and we want to show that it is a norm. This requires a little proof.

We know that $\| \ \|$ is positive on $\mathbb{Q}^r \setminus \mathbf{0}$, hence not negative on $\mathbb{R}^r$, but it is not yet clear that it remains positive on $\mathbb{R}^r \setminus \mathbb{Q}^r$. Indeed, this is not a general fact as can be seen from the example $\|(u,v)\| = |u - \alpha v|$ with a real irrational α.

The argument that $\| \ \|$ is a norm is due to Cassels and runs as follows.

Consider the subspace $V_0 := \{\mathbf{u} \in \mathbb{R}^r \mid \|u\| = 0\}$. Then $\| \ \|$ induces a norm on V/V_0 by $\|\mathbf{u} + V_0\| = \|\mathbf{u}\|$. By orthogonal projection with respect to the euclidean structure on $\mathbb{R}^r$, we identify $W := V/V_0$ with $V_0^\perp$. It is clear that $B_t := \{\mathbf{u} \in \mathbb{R}^r \mid \|\mathbf{u}\| \leq t\}$ is a closed, convex, symmetric set. If $\|\mathbf{u}\|$ were not a norm on $\mathbb{R}^r$, the set B_t would be a cylinder over $\{\mathbf{w} \in W \mid \|\mathbf{w}\| \leq t\}$.

Since all norms on W are equivalent, Minkowski's first theorem in C.2.19 would give infinitely many lattice points in B_t, and hence infinitely many elements $\mathbf{x} = (x_1, x_2) \in \Gamma$ with $\widehat{h}(\mathbf{x}) \leq t$. Since Γ is finitely generated, this would contradict Northcott's theorem in 1.6.8. Hence $\| \ \|$ is a norm with associated ball B_t of radius t.

It is clear that

$$\frac{1}{2}\widehat{h}(\mathbf{x}) \leq \max(h(x_1), h(x_2)) \leq h(\mathbf{x}) \leq \widehat{h}(\mathbf{x}).$$

In view of this inequality, Lemma 5.2.14 shows that for $\mathbf{u}, \mathbf{v} \in \mathcal{Z}_0$ and any integer $n \geq 2$ we have

$$\|\mathbf{u}\| \leq 2\kappa + \frac{2}{n-1}\|\mathbf{v} - 2n\mathbf{u}\|. \tag{5.12}$$

In the same way, Corollary 5.2.13 shows that

$$\|\mathbf{u}\| \leq \log 4 + 2\|\mathbf{v} - \mathbf{u}\| \quad \text{if } \mathbf{u} \neq \mathbf{v}. \tag{5.13}$$

The idea behind the last two displayed inequalities is the following.

For a vector $\mathbf{u} \in \mathbb{R}^r$ let $\nu(\mathbf{u}) = \mathbf{u}/\|\mathbf{u}\|$ be the associated unit vector with respect to the norm $\| \ \|$. Suppose that the vectors $\nu(\mathbf{u})$ and $\nu(\mathbf{v})$ are nearly the same, so that $\mathbf{u}$ and $\mathbf{v}$ point about in the same direction. If $\|\mathbf{v}\|$ is much larger than $\|\mathbf{u}\|$, then we can find an integer n such that $\|\mathbf{v} - 2n\mathbf{u}\|$ is small compared with $n\|\mathbf{u}\|$, and now (5.12) can be used to get an upper bound for $\|\mathbf{u}\|$.

The details are quite simple. Let $\varepsilon > 0$ be a small positive constant and let $\mathbf{u}, \mathbf{v} \in \mathcal{Z}_0$ be two points with

$$\|\nu(\mathbf{v}) - \nu(\mathbf{u})\| \leq \varepsilon, \qquad \|\mathbf{v}\| \geq \|\mathbf{u}\|.$$

Let $n = \lfloor \|\mathbf{v}\|/(2\|\mathbf{u}\|) \rfloor$, so that $0 \le \|\mathbf{v}\| - 2n\|\mathbf{u}\| < 2\|\mathbf{u}\|$. If $n \ge 2$, then (5.12) gives

$$\begin{aligned}
\|\mathbf{u}\| &\le 2\kappa + \frac{2}{n-1}\|\mathbf{v} - 2n\mathbf{u}\| \\
&= 2\kappa + \frac{2}{n-1}\Big\| \|\mathbf{v}\| \cdot \nu(\mathbf{v}) - 2n\|\mathbf{u}\| \cdot \nu(\mathbf{u}) \Big\| \\
&\le 2\kappa + \frac{2}{n-1}(\|\mathbf{v}\| - 2n\|\mathbf{u}\|) + \frac{4n}{n-1}\|\mathbf{u}\| \cdot \|\nu(\mathbf{v}) - \nu(\mathbf{u})\| \\
&\le 2\kappa + \frac{4+4n\varepsilon}{n-1}\|\mathbf{u}\|.
\end{aligned}$$

We take $\varepsilon = \frac{1}{10}$ and note that $(4+4n\varepsilon)/(n-1) \le \frac{1}{2}$ if $n \ge 45$. In this case the above chain of inequalities yields $\|\mathbf{u}\| \le 2\kappa + \frac{1}{2}\|\mathbf{u}\|$ and $\|\mathbf{u}\| \le 4\kappa$. If instead $1 \le n < 45$, we note that $\|\mathbf{v}\| - 2n\|\mathbf{u}\| < 2\|\mathbf{u}\|$, hence $\|\mathbf{v}\| \le 90\|\mathbf{u}\|$.

We have shown:

Lemma 5.2.17. *Let* $\mathbf{u}, \mathbf{v} \in \mathcal{Z}_0$ *and suppose that* $\|\nu(\mathbf{v}) - \nu(\mathbf{u})\| \le \frac{1}{10}$ *and* $4\kappa < \|\mathbf{u}\| \le \|\mathbf{v}\|$. *Then*

$$\|\mathbf{u}\| \le \|\mathbf{v}\| \le 90\|\mathbf{u}\|.$$

5.2.18. *Conclusion of the proof of Theorem 5.2.1:* Let us call **large** a solution $\mathbf{x} \in \Gamma$ of $x_1 + x_2 = 1$ if $\widehat{h}(\mathbf{x}) = h(x_1) + h(x_2) \ge \max(4\kappa, 5)$ and **small** otherwise.

The counting of large solutions is done in two steps, first by providing an upper bound for the number of points $\mathbf{u} \in \mathcal{Z}_0$ such that $H \le \|\mathbf{u}\| \le AH$ and lying in a fixed cone

$$C(\varepsilon; \mathbf{a}) := \{\mathbf{w} \in \mathbb{R}^r \mid \|\nu(\mathbf{w}) - a\| \le \varepsilon\},$$

and then by covering all of $\mathbb{R}^r$ by means of finitely many cones $C(\varepsilon; \mathbf{a}_i)$.

For the first step we use (5.13). Suppose we have two distinct points $\mathbf{u}, \mathbf{v} \in \mathcal{Z}_0 \cap C(\varepsilon; \mathbf{a})$ with

$$\max(4\kappa, 5) < \|\mathbf{u}\| \le \|\mathbf{v}\| \le (1+\delta)\|\mathbf{u}\|.$$

Then (5.13) gives

$$\begin{aligned}
\|\mathbf{u}\| &\le \log 4 + 2\|\mathbf{v} - \mathbf{u}\| \\
&= \log 4 + 2\Big\| \|\mathbf{v}\| \cdot \nu(\mathbf{v}) - \|\mathbf{u}\| \cdot \nu(\mathbf{u}) \Big\| \\
&\le \log 4 + 2(\|\mathbf{v}\| - \|\mathbf{u}\|) + 2\|\mathbf{u}\| \cdot \|\nu(\mathbf{v}) - \nu(\mathbf{u})\| \\
&\le \log 4 + (2\delta + 4\varepsilon)\|\mathbf{u}\|.
\end{aligned}$$

If we take for example $\delta = \frac{1}{4}$ and $\varepsilon = \frac{1}{20}$, we obtain $\|\mathbf{u}\| \le (10/3)\log 4 < 5$, contradicting the assumption $\|\mathbf{u}\| \ge \max(\kappa, 5)$. Thus we have a **gap principle**

$$\|\mathbf{v}\| > \frac{5}{4}\|\mathbf{u}\|.$$

Suppose we have m large solutions in a cone $C(\frac{1}{20}\,;\mathbf{a})$, say $\mathbf{u}_i \in \mathcal{Z}_0 \cap C(\frac{1}{20}\,;\mathbf{a})$ with $\max(4\kappa, 5) < \|\mathbf{u}_1\| \le \|\mathbf{u}_2\| \le \cdots$. Then $\|\nu(\mathbf{u}_m) - \nu(\mathbf{u}_1)\| \le \frac{1}{10}$, and hence, by Lemma 5.2.17, we have $\|\mathbf{u}_m\| \le 90\,\|\mathbf{u}_1\|$. On the other hand, the preceding gap principle shows that $\|\mathbf{u}_m\| \ge (\frac{5}{4})^{m-1}\|\mathbf{u}_1\|$. Hence $m - 1 \le \log 90/\log(5/4) < 21$ and, by (5.11) on page 136, we cannot have more than 42 large solutions with image in any given cone $C(\frac{1}{20}\,;\mathbf{a})$.

We need one more lemma:

Lemma 5.2.19. *Let $\|\ \|$ be a norm on $\mathbb{R}^r$. Let E be a subset of the ball $B_t := \{\mathbf{x} \in \mathbb{R}^r \mid \|\mathbf{x}\| \le t\}$ of radius t. Then for any $\varepsilon > 0$, we can cover E with $(1 + 2t/\varepsilon)^r$ translates, all centred on the set E, of the ball B_ε.*

Proof: Indeed, consider a maximal set $\mathcal{E}$ of non-overlapping balls of radius $\varepsilon/2$ with centres on E. Since they are contained in a ball of radius $t + \varepsilon/2$ and they are disjoint, their number does not exceed the ratio of the volumes of $B_{t+\varepsilon/2}$ and $B_{\varepsilon/2}$, namely $(1 + 2t/\varepsilon)^r$.

On the other hand, doubling the radius we obtain a covering of E. Otherwise, if $\mathbf{x}^*$ is a point of E not covered in this way, the ball $\|\mathbf{x} - \mathbf{x}^*\| \le \varepsilon/2$, which is centred on E, would be disjoint from $\mathcal{E}$ and $\mathcal{E}$ would not be maximal. □

In our case, taking $\varepsilon = \frac{1}{20}$, we infer from Lemma 5.2.19 that we can cover all of $\mathbb{R}^r$ with not more than 41^r cones $C(\frac{1}{20}\,;a)$.

We have already shown that any such cone determines at most 42 large solutions and we conclude that the total number of large solutions does not exceed $42 \cdot 41^r$.

It remains to give a bound for the number of small solutions, and this is a consequence of Theorem 4.2.3. We apply this theorem with $d = 1$ and $n = 2$, and deduce that there are two constants $\gamma = \gamma(1,2) > 0$ and $N = N(1,2) < \infty$ such that, for any $a,\, b \in \overline{\mathbb{Q}}^\times$, we have at most N solutions $\mathbf{x} = (x_1, x_2) \in \overline{\mathbb{Q}}^\times \times \overline{\mathbb{Q}}^\times$ of

$$ax_1 + bx_2 = 1, \qquad \widehat{h}(\mathbf{x}) \le \gamma.$$

Let Γ' be the division group of Γ, namely

$$\Gamma' := \left\{\boldsymbol{\alpha} \in \overline{\mathbb{Q}}^\times \times \overline{\mathbb{Q}}^\times \mid \boldsymbol{\alpha}^n \in \Gamma,\ \text{for some } n \ge 1\right\}.$$

By Lemma 5.2.19 applied to $E = B_t \cap \mathbb{Q}^r$, there are $\lfloor (1 + 2t/\gamma)^r \rfloor$ translates of the ball B_γ, all centred at rational points and covering B_t.

This means that we can find points $(a_i, b_i) \in \Gamma'$, numbering not more than $\lfloor (1+2t/\gamma)^r \rfloor$, such that every $\mathbf{x} = (x_1, x_2) \in \Gamma$ with $x_1 + x_2 = 1$ and $\widehat{h}(\mathbf{x}) \leq t$ can be written, for some i, as $x_1 = a_i\xi$, $x_2 = b_i\eta$ with $a_i\xi + b_i\eta = 1$ and $h(\xi) + h(\eta) \leq \gamma$. Since there are at most N such (ξ, η), we deduce that the number of (x, y) in question does not exceed $N \cdot \lfloor (1+2t/\gamma)^r \rfloor$.

We can take $t = \max(4\kappa, 5)$. Hence the number of small solutions does not exceed $N \cdot \lfloor (1 + 2\max(4\kappa, 5)/\gamma)^r \rfloor$. Thus the total number of solutions does not exceed

$$42 \cdot 41^r + N \cdot \lfloor (1 + 2\max(4\kappa, 5)/\gamma)^r \rfloor .$$

This completes the proof of Theorem 5.2.1 if Γ is finitely generated. In the general case, Γ is the union of its finitely generated subgroups (of rank at most r). Since the above upper bound depends only on the rank, this proves the claim. □

5.3. Applications

The importance of the generalized unit equation stems from the fact that many diophantine problems can be reduced to it. In this section, we review some of the most interesting applications.

5.3.1. Let K be a number field, S a finite set of places of K containing all places at infinity, and let $O_{S,K}$ denote the ring of S-integers of K, and let $U_{S,K}$ be the group of units of $O_{S,K}$. Let also $F(x, y) \in O_{S,K}[x, y]$ be homogeneous of degree $r \geq 3$ with coefficients in $O_{S,K}$ and assume that F has at least three non-proportional linear factors in a factorization over $\overline{K}$.

The **Thue–Mahler equation** is the equation $F(x, y) \in U_{S,K}$, to be solved with $x, y \in O_{S,K}$.

In 1909, using a new method based on diophantine approximation, Thue proved that the equation $F(x, y) = m$, with $F(x, y) \in \mathbb{Z}[x, y]$ and with three non-proportional linear factors over $\mathbb{C}$ has only finitely many solutions in integers (see 6.2.1 for the argument). Through the work of Siegel and Mahler, this was extended to equations in number fields to be solved in S-integers and to the more general Thue–Mahler equation, with the proviso of considering equivalent two solutions differing only by multiplication by an S-unit. We have:

Theorem 5.3.2. *The number of equivalence classes of solutions $(x, y) \in O_{S,K}^2$ of the Thue–Mahler equation $F(x, y) \in U_{S,K}$ does not exceed $C_1 \cdot C_2^{12\binom{r}{3}|S|}$, where C_1 and C_2 are the constants introduced in Theorem 5.2.1.*

Remark 5.3.3. With a different method J.-H. Evertse [**109**] has obtained the improved bound $(5 \cdot 10^6 r)^{|S|}$, which shows a much better dependence on r. The example $x^r + a(x - y)(2x - y) \cdots (rx - y) = 1$, with solutions $(1, j)$ with

$j = 1, \ldots, r$ shows that already with $|S| = 1$ we may have r solutions of a Thue equation $F(x, y) = 1$ of degree r.

On the other hand, L. Caporaso, J. Harris, and B. Mazur [**56**], and later D. Abramovich [**2**], have obtained some evidence for the conjecture that the number of K-rational points on a curve C of genus at least 2, defined over a number field K, admits a bound depending only on the genus of the curve and the degree of the number field K; in particular, their conjecture implies that there should be a bound for the number of solutions of a Thue equation depending only on r and the degree of K.

Proof: The following argument, which reduces a Thue–Mahler equation to a unit equation, goes back to Siegel.

We may assume that $F(x, y) = a_0 x^r + a_1 x^{r-1} y + \cdots + a_r y^r$ has degree r in x and leading coefficient $a_0 = 1$. This step is not really essential to the proof, but makes things a little simpler. To verify this assertion, choose any solution (x_0, y_0) of the Thue–Mahler equation $F(x, y) \in U_{S,K}$. Let A be the matrix

$$A = \begin{pmatrix} a & b \\ c & d \end{pmatrix}$$

with

$$\begin{aligned} a &= F(x_0, y_0)^{-1}(a_0 x_0^{r-1} + a_1 x_0^{r-2} y_0 + \cdots + a_{r-1} y_0^{r-1}) \\ b &= F(x_0, y_0)^{-1} a_r y_0^{r-1} \\ c &= -y_0 \\ d &= x_0. \end{aligned}$$

Then $\det(A) = 1$ and A has entries in $O_{S,K}$. Therefore, the Thue–Mahler equation $F(x, y) \in U_{S,K}$ is equivalent to the other Thue–Mahler equation

$$G(x, y) := F(x_0, y_0)^{-1} F(dx - by, -cx + ay) \in U_{S,K}$$

with leading coefficient $G(1, 0) = 1$.

Let α_1, α_2, α_3 be three distinct roots of $F(x, 1) = 0$ over $\overline{K}$, and define:

(a) $K' = K(\alpha_1, \alpha_2, \alpha_3)$;

(b) S' the set of places of K' over S, $O_{S',K'}$ the ring of S'-integers of K', and $U_{S',K'}$ the group of units of $O_{S',K'}$.

The field K' has degree at most $r(r-1)(r-2)$ over K and $|S'| \leq r(r-1)(r-2)|S|$ (use Corollary 1.3.2). The group $U_{S',K'}$ has rank $|S'| - 1$ by Dirichlet's unit theorem (see Theorem 1.5.13). Now we define Γ to be the group of pairs (u, v) with $u, v \in U_{S',K'}$; it is clear that Γ has rank s not exceeding $2(|S'| - 1) \leq 2r(r-1)(r-2)|S| - 2$.

Since F has leading coefficient 1, all roots $\alpha_1, \ldots, \alpha_r$ of the equation $F(x, 1) = 0$ are integral over $O_{S,K}$. For a solution $(x, y) \in O_{S,K}^2$ of the Thue–Mahler equation, each factor $x - \alpha_i y$ is integral over $O_{S,K}$ and the product is an S-unit. Hence the factors $x - \alpha_i y$ are S'-units for $i = 1, 2, 3$.

On the other hand, the three linear forms $x - \alpha_i y$, $i = 1, 2, 3$, must be linearly dependent, and in fact a linear relation is

$$\frac{\alpha_2 - \alpha_3}{\alpha_2 - \alpha_1} \frac{x - \alpha_1 y}{x - \alpha_3 y} + \frac{\alpha_3 - \alpha_1}{\alpha_2 - \alpha_1} \frac{x - \alpha_2 y}{x - \alpha_3 y} = 1.$$

This is an equation of type $Au + Bv = 1$ with $(u, v) \in \Gamma$. We extend Γ to a new group Γ' of rank at most $s + 1$ by adding a new generator (A, B) and apply Theorem 5.2.1. Then the number of solutions (u, v) does not exceed $C_1 \cdot C_2^{s+1}$. Conversely, (u, v) determines (x, y) up to multiplication by a scalar, and it follows that (u, v) determines at most one equivalence class of solutions (x, y) of the Thue–Mahler equation $F(x, y) \in U_{S,K}$. □

5.3.4. Another equation which can be treated by similar methods is the **hyperelliptic equation**

$$by^2 = a_0 x^r + a_1 x^{r-1} + \cdots + a_r$$

with coefficients in $O_{S,K}$, $b \neq 0$, to be solved with $(x, y) \in O_{S,K}$. For its treatment, the reader is required to have some basic knowledge of algebraic number theory. We recall that the **class group** of a number field is the group of fractional ideals in K modulo the principal fractional ideals.

There is little loss in generality if we assume that $b = 1$ and that $f(x) := a_0 x^r + a_1 x^{r-1} + \cdots + a_r$ has no multiple roots. In fact, we can always write $bf(x) = F(x)H(x)^2$ with $F, H \in O_{S,K}[x]$ and $F(x)$ without multiple roots. This yields the equation

$$Y^2 = F(X),$$

where $Y = by/H(x)$ and $X = x$. Thus $Y \in K$ if $X \in O_{S,K}$ and since Y is integral over $O_{S,K}$, which is integrally closed, we see that $Y \in O_{S,K}$ too.

Theorem 5.3.5. *Let $f(x) = a_0 x^r + a_1 x^{r-1} + \cdots + a_r$ be a polynomial of degree $r \geq 3$ with no multiple roots and coefficients in $O_{S,K}$. Let $\omega(D_f)$ be the number of distinct prime ideals of K dividing the discriminant D_f of $f(x)$ which are not contained in S. Also let $\alpha_1, \alpha_2, \alpha_3$ be three roots of $f(x)$ and let*

$$K' = K(a_0^{\frac{r-1}{2}}, \alpha_1, \alpha_2, \alpha_3).$$

Finally, suppose that the Sylow 2-subgroup of the class group of the ring of integers $O_{K'}$ of the field K' is generated by ν elements. Then the equation

$$y^2 = f(x)$$

has at most $C_1 \cdot (2C_2)^{16[K':K](|S|+\omega(D_f))+16\nu}$ solutions $(x, y) \in O_{S,K}$ with $y \neq 0$, where C_1 and C_2 are the constants introduced in Theorem 5.2.1.

Proof: We use a method of Siegel to reduce the equation to a finite number of unit equations. We consider first the special case in which:

(a) the polynomial $f(x)$ is monic;

(b) $f(x)$ has three roots α_i, $i = 1, 2, 3$, in K;

(c) the discriminant D_f is a unit in $O_{S,K}$;

(d) the ring $O_{S,K}$ is a unique factorization domain.

We write $f(x) = (x - \alpha_1)(x - \alpha_2)(x - \alpha_3)h(x)$. By (a), (b), and Gauss's lemma in 1.6.3, we have $\alpha_1, \alpha_2, \alpha_3 \in O_{S,K}$ and $h(x) \in O_{S,K}[x]$.

Let $x \in O_{S,K}$ be a solution of $y^2 = f(x)$ with $f(x) \neq 0$. We claim that the principal ideals $[x - \alpha_1]$, $[x - \alpha_2]$, $[x - \alpha_3]$, $[h(x)]$ in $O_{S,K}$ are pairwise coprime. The first three ideals are coprime because $\alpha_i - \alpha_j \in [x - \alpha_i, x - \alpha_j]$ and $\alpha_i - \alpha_j$ divides D_f, which is a unit by assumption (c). A similar argument applies to $x - \alpha_i$ and $h(x)$, noting that by (a) all roots are integral over $O_{S,K}$ and working with a factorization of $h(x)$ in the splitting field of f.

Now the ideal equation

$$[x - \alpha_1]\,[x - \alpha_2]\,[x - \alpha_3]\,[h(x)] = [y]^2$$

shows that

$$[x - \alpha_i] = \mathfrak{y}_i^2, \qquad i = 1, 2, 3$$

for some ideal $\mathfrak{y}_i$ of $O_{S,K}$. By assumption (d), we conclude that the square root of $[x - \alpha_i]$ must be a principal ideal. Thus we can write

$$x - \alpha_i = u_i y_i^2$$

with $u_i \in U_{S,K}$ and $y_i \in O_{S,K}$, for $i = 1, 2, 3$. Note also that we can take u_1, u_2, and u_3 modulo squares. By Dirichlet's unit theorem in 1.5.13, the group of units $U_{S,K}$ of $O_{S,K}$ is the direct product of a cyclic torsion group and a free abelian group of rank $|S| - 1$. Therefore, we need not consider more than $8^{|S|}$ triples (u_1, u_2, u_3). Eliminating x from these equations we find

$$u_i y_i^2 - u_j y_j^2 = \alpha_j - \alpha_i \qquad \text{for} \quad i, j = 1, 2, 3 \quad \text{and} \quad i \neq j. \tag{5.14}$$

Let $F = K(\sqrt{u_1}, \sqrt{u_2}, \sqrt{u_3})$ and let S' be the set of places of F lying over S. In the field F, the equations (5.14) factorize as

$$(\sqrt{u_i}y_i - \sqrt{u_j}y_j)(\sqrt{u_i}y_i + \sqrt{u_j}y_j) = \alpha_j - \alpha_i$$

with $\sqrt{u_i}y_i \pm \sqrt{u_j}y_j \in O_{S',K'}$, while $\alpha_j - \alpha_i \in U_{S',K'}$ by (c). Thus $v_{ij} := \sqrt{u_i}y_i - \sqrt{u_j}y_j$ is a unit in $O_{S',K'}$ for $i, j = 1, 2, 3$ and $i \neq j$. On the other hand, we have identically

$$(v_{12}/v_{13}) + (v_{23}/v_{13}) = 1.$$

Now $(v_{12}/v_{13}, v_{23}/v_{13}) \in U_{S',K'} \times U_{S',K'}$, which has rank $2|S'|-2$. According to Theorem 5.2.1, this equation has at most $C_1 \cdot C_2^{2|S'|-2}$ solutions. Also $|S'| \leq [F : K]\,|S| \leq 8\,|S|$.

Hence let us fix $(v_{12}/v_{13}, v_{23}/v_{13})$, so that we can write $v_{ij} = c_{ij}w$ with the $\{c_{ij}\}$ having not more than $C_1 \cdot C_2^{2|S'|-2}$ possibilities. We have

$$\begin{aligned} c_{ij}^{-1}(\alpha_j - \alpha_i) &= c_{ij}^{-1}(u_i y_i^2 - u_j y_j^2) \\ &= c_{ij}^{-1}\left(u_i y_i^2 - (\sqrt{u_i} y_i - c_{ij} w)^2\right) \\ &= 2\sqrt{u_i} y_i w - c_{ij} w^2. \end{aligned}$$

For a given i we have two distinct values $j, k \neq i$ and $c_{ij} \neq c_{ik}$. Hence

$$c_{ij}^{-1}(\alpha_j - \alpha_i) - c_{ik}^{-1}(\alpha_k - \alpha_i) = (c_{ik} - c_{ij})w^2.$$

This determines w^2 uniquely and hence w up to sign. Once w is given, y_i and x are uniquely determined.

If we take into account the number of triples (u_1, u_2, u_3) and the number of sets $\{c_{ij}\}$, we conclude that the number of solutions of the equation $y^2 = f(x)$ in $O_{S,K}$ with $y \neq 0$ does not exceed

$$2C_1 \cdot 8^{|S|}(C_2)^{2|S'|-2} \leq C_1 \cdot (2C_2)^{16|S|}.$$

To complete the proof in the general case, we have only to enlarge the field K and the set S so that our assumptions (a) to (d) are verified.

For (a), it suffices to add $a_0^{\frac{r-1}{2}}$ to the field K.

For (b), it suffices to add α_i, $i = 1, 2, 3$, to the field K.

For (c), it suffices to add to S the set S_2 of places v for which $\mathrm{ord}_v(D_f) > 0$.

This gives us an extension K' of K (of degree $[K' : K] \leq 2r(r-1)(r-2)$) and a new set S_3, the places of K' lying over S and S_2. Thus we may assume that (a), (b), (c) are satisfied.

For (d), we use:

Proposition 5.3.6. *Let K be a number field. Then we can find a finite set of places S of K such that for any finite set of places $T \in M_K$ with $T \supset S$, the ring $O_{T,K}$ is a principal ideal domain and hence a unique factorization domain.*

Proof: The ring of integers O_K is not necessarily a unique factorization domain. Since the class group of a number field is finite (see e.g. **[172]**, Ch.V or **[162]**, Th. 2.7.1), there are ideals $\mathfrak{I}_1, \ldots, \mathfrak{I}_r$ in O_K forming a finite set of representatives for the class group of O_K. Let $\mathfrak{M}_1, \ldots, \mathfrak{M}_n$ be a finite set of prime ideals of O_K, containing all maximal ideals dividing at least one of the ideals $\mathfrak{I}_j$, $j = 1, \ldots, r$.

The set

$$M := \bigcap_{j=1}^{n} (O_K \setminus \mathfrak{M}_j)$$

is multiplicatively closed.

Let R be the localization of O_K in M. The ring R is again a Dedekind domain ([**156**],Th.10.4). In order to show that R is a unique factorization domain, it is enough to prove that any maximal ideal $\mathfrak{m}$ in R is principal. The maximal ideal $\mathfrak{M} := \mathfrak{m} \cap O_K$ of O_K generates $\mathfrak{m}$. Moreover, $\mathfrak{M}$ is equivalent to a product of ideals $\mathfrak{M}_j$, because they generate the class group of O_K. Therefore, $\mathfrak{m}$ is also equivalent to a product of ideals $R\mathfrak{M}_j$.

Since $R\mathfrak{M}_j = R$ for every j, we see that $\mathfrak{M}$ is generated by a single element. We conclude that R is a principal ideal domain. This proves what we want, with S the set of places determined by the prime ideals dividing the ideals $\mathfrak{I}_j$. □

This proves a slightly weaker version of the theorem, in which ν is the number of distinct prime ideals dividing a set of ideals which generate the class group of K. For the more precise statement, we need two observations. First, note that in place of (d) it suffices that the 2-primary part of the class group of O_K is trivial. Second, as shown by Landau in 1907, every ideal class contains prime ideals. This follows from the general form of Dirichlet's density theorem, a particular case of which states that the set of prime ideals $\mathfrak{p}$ of O_K in a given ideal class of the class group C_K is a set of positive natural density $1/|C_K|$ (see e.g. [**215**], Ch.VII, §2, Prop.7.10, Cor.4).

Hence in the proof of the preceding proposition we can take all ideals $\mathfrak{I}_j$ to be prime ideals. This also shows that we can take ν to be the cardinality of a set of generators of the 2-subgroup of the class group, completing the proof. □

Remark 5.3.7. The usefulness of Theorem 5.3.5 in applications is somewhat limited by the presence of the factor $(2C_2)^{16\nu}$ in the given bound.

At present, the only general estimate we have on ν is comparable with the logarithm of the full class number of K', and hence this factor is comparable with a power of the class number of K', which in turn is comparable with a power of the height $H(f)$ of the polynomial f. Really useful estimates would be of order $H(f)^{\varepsilon}$ for any fixed $\varepsilon > 0$. Note however that this will be the case if $f(x)$ has three roots in K and either r is odd or a_0 is a square in K, since then $K' = K$ and ν will be independent of f.

Remark 5.3.8. The so-called superelliptic equation $y^m = f(x)$ with $m \geq 3$ can be treated pretty much in the same way, with a reduction to a Thue–Mahler equation, of Fermat type, of degree m. We leave the details to the reader.

5.4. Effective methods

A discussion of the unit equation would not be complete without mentioning its effective solution obtained by Baker's method or by the so-called Thue–Siegel principle. In its simplest formulation, everything follows from:

Theorem 5.4.1. *Let K be a number field and Γ a finitely generated subgroup of $K^\times$. Let $0 < \varepsilon \leq 1$ and $v \in M_K$. There is an effectively computable function $C(K, \Gamma, v, \varepsilon)$ such that every solution of the diophantine inequality*

$$|1 - \gamma|_v \leq H(\gamma)^{-\varepsilon}, \qquad \gamma \in \Gamma$$

satisfies $h(\gamma) \leq C(K, \Gamma, v, \varepsilon)$.

Remark 5.4.2. The best bounds for $C(K, \Gamma, v, \varepsilon)$ are obtained *via* Baker's theory of linear forms in logarithms in many variables, as in A. Baker and G. Wüstholz **[16]** in the archimedean case and Kunrui Yu **[335]** in the non-archimedean case, see also Y. Bugeaud **[53]** and Y. Bugeaud and M. Laurent **[54]**.

A self-contained proof of Theorem 5.4.1, obtained with quite a different method (the so-called Thue–Siegel method) is in **[31]** and E. Bombieri and P.B. Cohen **[32]**. The special case $A = 1$ of the estimate in **[32]**, which holds in the non-archimedean case, yields the following completely explicit result.

We define $\rho(x) \geq e^5$ to be the solution of $\rho/(\log \rho)^5 = x$ if $x > e^5 5^{-5}$ and $\rho(x) = e^5$ otherwise and denote by $\xi_1, \ldots, \xi_t$ a set of generators of Γ/tors. Let K be a number field of degree d, v be a non-archimedean place of K dividing the rational prime p and with residue class degree f_v. For $D_v^* := \max(1, d/(f_v \log p))$, define $h'(x)$ to be the modified height $h'(x) = \max(h(x), 1/D_v^*)$ and $H'(x) = \exp(h'(x))$. Finally, let

$$C = 66 p^{f_v} (D_v^*)^6 \quad \text{and} \quad Q = (2t\rho(C/\varepsilon))^t \prod_{i=1}^{t} h'(\xi_i).$$

Then any solution $\gamma \in \Gamma$ of $|1 - \gamma|_v < H'(\gamma)^{-\varepsilon}$ has height bounded by

$$h(\gamma) \leq 16 p^{f_v} \rho(C/\varepsilon) Q \max\left(1, 4p^{f_v} Q\right).$$

5.4.3. It is easy to see how Theorem 5.4.1 can be used to solve effectively the unit equation $x + y = 1$ in the group $\Gamma = U_{S,K}$ of S-units of K. We give a quick sketch of the argument.

Since y is an S-integer, we have $\sum_{v \in S} \log^+ |y|_v = h(y)$ and also $\sum_{v \in S} \log |y|_v = 0$ by the product formula, because $\log |y|_v = 0$ for $v \notin S$. Therefore, there is $v \in S$ such that

$$\log |y|_v \leq -\frac{1}{|S|} h(y),$$

which is the same as

$$|1 - x|_v = |y|_v \leq H(y)^{-1/|S|}.$$

Moreover, it is clear that $H(x) = H(1 - y) \leq 2H(y)$ (use Proposition 1.5.15), hence

$$|1 - x|_v \leq 2^{1/|S|} H(x)^{-1/|S|}. \tag{5.15}$$

If $H(x) \leq 4$, we have the desired bound. If instead $H(x) > 4$, it is immediate from (5.15) that

$$|1 - x|_v \leq H(x)^{-1/(2|S|)},$$

hence in any case Theorem 5.4.1 yields

$$h(x) \leq \max\{\log 4, C(K, \Gamma, v, 1/(2|S|))\}.$$

Remark 5.4.4. Lang's proof of the theorem stated in the introduction is obtained by using Hurwitz's genus formula to show that, for large m, a component of the pull-back of the curve $C \subset \mathbb{G}_m^n$ by the isogeny $\mathbf{x} \mapsto \mathbf{x}^m$ has necessarily large genus. Then we conclude by an application of the well-known Siegel's finiteness theorem on integral points on curves. (For curves over a number field, see Theorem 7.3.9 and Remark 7.3.10. For the general case, see [**169**], Ch.8, Th.2.4.)

We give here a sketch of a different argument, for the case when the curve is defined over a number field, because it reduces the proof to the statement of Theorem 5.4.1 rather than the ineffective Siegel theorem. We prove:

Theorem 5.4.5. *Let C be a geometrically irreducible closed curve in $\mathbb{G}_m^n$, defined over a number field K, not a translate of a subtorus of $\mathbb{G}_m^n$, and let Γ be any finitely generated subgroup of $\mathbb{G}_m^n(\overline{K})$. Then $C \cap \Gamma$ is an effectively computable finite set.*

Proof: By using the projections $\mathbf{x} \mapsto (x_i, x_j)$ onto $\mathbb{G}_m^2$, we easily reduce the problem to the case $n = 2$ and where C is given by the equation $f(x, y) = 0$.

Since Γ is finitely generated, there is a number field L and a finite set $S \subset M_L$ containing all archimedean places such that $\Gamma \subset (O_{S,L}^\times)^2$. Replacing K by L and enlarging Γ, we may assume that $\Gamma = \Gamma_1 \times \Gamma_1$, where $\Gamma_1 \subset O_{S,K}^\times$ is finitely generated.

For any $(\alpha, \beta) \in \Gamma$, the affine multiplicative height satisfies

$$H((\alpha, \beta)) := \prod_{v \in M_K} \max(1, |\alpha|_v, |\beta_v|_v) \leq \max_{v \in S}(1, |\alpha|_v, |\beta|_v)^{|S|}. \tag{5.16}$$

Now we let (α, β) range over $C \cap \Gamma$ and we want to get an effective upper bound for the height. Replacing α by α^{-1} or β by β^{-1}, which does not affect the standard height $h(\alpha) + h(\beta)$, and replacing C by the image of the corresponding automorphism, we may asssume that the maximum in (5.16) is attained in $v \in S$ with $\min(|\alpha|_v, |\beta|_v) \geq 1$. Now consider the polynomial $f(x, y) = \sum a_{ij} x^i y^j$ and order terms as

$$|a_{pq} \alpha^p \beta^q|_v \geq |a_{rs} \alpha^r \beta^s|_v \geq \cdots$$

Since $f(\alpha, \beta) = 0$, the two largest terms must be of the same order of magnitude, hence

$$\begin{aligned} p \log|\alpha|_v + q \log|\beta|_v &= r \log|\alpha|_v + s \log|\beta|_v + O(1) \\ &\geq i \log|\alpha|_v + j \log|\beta|_v + O(1) \end{aligned} \tag{5.17}$$

for all monomials $x^i y^j$ appearing in $f(x, y)$. Note that we may restrict our attention to the set of $(\alpha, \beta) \in C \cap \Gamma$ having the above properties with respect to the fixed absolute value $v \in S$ and with $|a_{pq} \alpha^p \beta^q|_v \geq |a_{rs} \alpha^r \beta^s|_v$ as the largest terms in f for fixed $(p, q), (r, s)$. Here and in the following, the Landau (and Vinogradov) symbols are with respect to this set. In particular, (p, q) and (r, s) must be linearly independent if $H((\alpha, \beta))$ is large, because of (5.17) and $\min(\log|\alpha|_v, \log|\beta|_v) \geq 0$.

Consider now another monomial $x^i y^j$. We have $(i,j) = A_{ij} \cdot (p,q) + B_{ij} \cdot (r,s)$ for certain rational numbers A_{ij}, B_{ij}, with denominators dividing $D = |ps - qr| \geq 1$, hence $DA_{ij}, DB_{ij} \in \mathbb{Z}$. By (5.17), we deduce

$$\begin{aligned} p\log|\alpha|_v + q\log|\beta|_v &\geq i\log|\alpha|_v + j\log|\beta|_v + O(1) \\ &= (A_{ij}p + B_{ij}r)\log|\alpha|_v + (A_{ij}q + B_{ij}s)\log|\beta|_v + O(1) \\ &= (A_{ij} + B_{ij})(p\log|\alpha|_v + q\log|\beta|_v) + O(1). \end{aligned}$$

Therefore, if $H((\alpha,\beta))$ is large enough, we must have either $A_{ij} + B_{ij} = 1$ or $A_{ij} + B_{ij} \leq 1 - 1/D$. Now we define $\mathcal{I}$ to be the set of all pairs (i,j) such that $A_{ij} + B_{ij} = 1$. Thus we have for $(i,j) \in \mathcal{I}$ an equation

$$x^i y^j = (x^p y^q)(x^{r-p} y^{s-q})^{B_{ij}}. \tag{5.18}$$

If instead $(i,j) \notin \mathcal{I}$, then $A_{ij} + B_{ij} \leq 1 - 1/D$ implies that

$$\log|\alpha^i \beta^j|_v \leq (1 - 1/D)\log|\alpha^p \beta^q|_v + O(1). \tag{5.19}$$

We abbreviate $X = x^p y^q$, $Y = x^r y^s$ and note that $x^{Di} y^{Dj}$ are monomials in X, Y. Then we have by (5.18)

$$f(x^D, y^D) = X^D R(Y/X) + \sum_{(i,j) \notin \mathcal{I}} a_{ij} x^{Di} y^{Dj}, \tag{5.20}$$

where

$$R(t) = \sum_{(i,j) \in \mathcal{I}} a_{ij} t^{DB_{ij}}.$$

Now we specialize (ξ,η) with $(\xi^D,\eta^D) = (\alpha,\beta)$ and correspondingly write Ξ, H for the specializations of X and Y. For $(i,j) \notin \mathcal{I}$, the bound (5.19) yields

$$|\xi^{Di}\eta^{Dj}|_v = O(|\Xi|_v^{D-1}).$$

Since $f(\xi^D,\eta^D) = f(\alpha,\beta) = 0$, from (5.20) and (5.16), we find in case of $pq \neq 0$ that

$$|R(\mathsf{H}/\Xi)|_v \ll |\Xi|_v^{-1} \leq \max(|\alpha|_v, |\beta|_v)^{-1/D} \leq H((\alpha,\beta))^{-1/(|S|D)}.$$

A similar exponential bound follows easily from the definition of (p,q) if $pq = 0$. By (5.17) again, $|\mathsf{H}/\Xi|_v$ is bounded and bounded away from 0 and from the last displayed equation we see that

$$|1 - \zeta^{-1}\mathsf{H}/\Xi|_v \ll H((\alpha,\beta))^{-c}$$

for some root ζ of $R(t)$ and some $c > 0$.

Finally, $\zeta^{-1}\mathsf{H}/\Xi$ belongs to the finitely generated group Γ_2 obtained by adding ζ to the division group $\{P \in \mathbb{G}_m^2 \mid P^D \in \Gamma_1\}$ of order D of Γ_1. It is also clear that $H((\alpha,\beta))^{-c} \ll H(\zeta^{-1}\mathsf{H}/\Xi)^{-\kappa}$ for some $\kappa > 0$. Thus we may apply Theorem 5.4.1 to Γ_2 and conclude that $H(\alpha^{r-p}\beta^{s-q})$ is bounded. By Northcott's theorem in 1.6.8, $\alpha^{r-p}\beta^{s-q}$ belongs to a finite set, hence we have shown that (α,β) belongs to a finite union of effectively computable torus cosets in $\mathbb{G}_m^2$. Since C is geometrically irreducible and not a torus coset by hypothesis, the intersection of C with these torus cosets is finite and effectively computable. □

For comparison with this argument, see also the proof of Theorem 7.4.7.

5.5. Bibliographical notes

Special cases of the unit equation appear in the work of Siegel [**283**] and Mahler [**187**], and finiteness of the number of solutions is proved through a reduction to a finite set of Thue equations. The first general formulation in a geometric setting was done by S. Lang in 1960 [**166**]. S. Lang's conjectured extension [**167**] to the division group of a finitely generated group was later proved by P. Liardet [**182**].

Theorem 5.2.1 is the coronation of a long series of successive improvements in counting the number of solutions of unit equations. Our proof follows [**23**] quite closely. K.K. Choi has shown, in an unpublished note, that we can take $C_1 = 30$, $C_2 = 70$ in this theorem. The first such bound depending only on the rank was obtained, for the generalized S-unit equation in a number field, by J.-H. Evertse [**103**] in a paper which sparked much research in this area.

The argument in Example 5.2.4 is due to D. Zagier and simplifies a more precise calculation due to P. Erdős, C.L. Stewart, and R. Tijdeman [**101**], which leads to a better value of the constant c. Example 5.2.5 is due to J. Berstel and is mentioned in Beukers and Schlickewei [**23**], where it is shown that such an equation has at most 61 solutions. The remark in Example 5.2.6 on complex solutions of the unit equation in a cyclotomic field is due to H.W. Lenstra; this applies in a more general setting, notably CM-fields.

The reduction of a Thue equation to a unit equation is in [**283**], Zweiter Teil, §1. Siegel studies the unit equation by taking cosets of units modulo high powers, thereby reducing it to a finite set of equations $ax^r + by^r = c$, for which he had independently proved finiteness of the number of integral solutions using diophantine approximation methods.

Theorem 5.3.2 is due to Evertse [**103**]. Uniform polynomial bounds in r were independently obtained in [**30**] and, with a sharper result and better proof, in [**109**].

The reduction of a hyperelliptic equation to a unit equation is in a two-page paper [**282**] by C.L. Siegel in 1926, published under the pseudonym X.

6 ROTH'S THEOREM

6.1. Introduction

The Liouville inequality in 1.5.21, or its projective version in 2.8.21, while simple and useful, does not tell the real truth about how well we can approximate algebraic numbers by algebraic numbers in a fixed field K.

In 1909 A. Thue obtained the first improvement on Liouville's theorem about approximation of algebraic numbers by rational numbers. He proved the following result:

Thue's theorem: *Let α be a real algebraic number of degree $d \geq 3$ and let $\varepsilon > 0$. Then there are only finitely many rational numbers p/q, $q \geq 1$, such that*

$$\left|\alpha - \frac{p}{q}\right| \leq \frac{1}{q^{\frac{d}{2}+1+\varepsilon}}.$$

As a consequence of this theorem, Thue proved that the **Thue equation**

$$F(x,y) = m,$$

where $F \in \mathbb{Z}[x,y]$ is homogeneous of degree d with at least three non-proportional linear factors over $\mathbb{C}$, has only finitely many solutions in integers x, y, for every fixed non-zero integer m.

The main drawback of Thue's theorem is that it is ineffective, in the sense that no bound can be placed *a priori* on the height of the rational approximations p/q. Loosely speaking, this is due to the fact that no procedure is given to decide whether a solution exists with height above a given constant. Since Thue's method obtains information on solutions assuming that one solution is known, this leads in the end to ineffectivity.

On the other hand, many questions in number theory can be reduced to questions of diophantine approximation as above, so that the result of Thue, even with its inherent ineffectivity, definitely is of considerable importance.

Thue's theorem went through various successive improvements. First of all, aside from the fairly obvious extension to approximations in general number fields,

Siegel showed that the exponent $d/2+1$ can be replaced by

$$\min_{s\in\mathbb{N}}\left(s+\frac{d}{s+1}\right)<2\sqrt{d}\,.$$

The fact that this exponent is of order $o(d)$ rather than d turned out to be quite important in Siegel's proof of the finiteness of the number of integral points on a curve of genus $g\geq 1$, in treating the case $g=1$ (for the case $g\geq 2$, Siegel had to develop a corresponding method dealing with simultaneous approximations). A little later, Mahler developed the same method over the p-adic numbers, thereby obtaining as an application the finiteness of the number of solutions of a Thue equation in S-integers rather than ordinary integers; again, this had significant applications to other problems in number theory. However, all these extensions of Thue's theorem also suffer from the same problem of ineffectivity.

Thue's method depended on an auxiliary construction with polynomials in two variables. It was expected that a similar construction in m variables would yield further drastic improvements, and this was explored by Siegel and Schneider in the 1930s. They could not deal with one crucial point of the construction, namely the non-vanishing of the auxiliary polynomial when evaluated at a special point. It was only in 1955 that Roth was able to overcome this stumbling block, thereby obtaining the sharp exponent 2 in place of $d/2+1$ in Thue's theorem.

In Section 6.2, we start by proving finiteness of integer solutions for the Thue equation, then we formulate Roth's theorem for number fields with respect to a finite set of places and we sketch the proof in the case $K=\mathbb{Q}$ with respect to the ordinary absolute value.

In Section 6.3, we introduce the index of a polynomial as a measure for its vanishing in a given point. It is used together with Wronskian techniques to prove Roth's lemma, the crucial point for the non-vanishing of the auxiliary polynomial mentioned above. Section 6.4 is reserved for the proof of Roth's theorem.

In Section 6.5, we prove Vojta's generalization with moving targets, we give quantitative results for the number of exceptional good approximations, and we mention the Cugiani–Mahler theorem. This section provides us with additional information, which may be omitted in a first reading.

For this chapter, the reader should be familiar with the results about Siegel's lemma in Section 2.9. However, we do not need here the most sophisticated formulations, for example the dependence on the discriminant of the number field will be irrelevant and we may deduce Roth's theorem also from a relative version of the easier Corollary 2.9.2. Roth's theorem will be rather important in the sequel, it will be used and generalized in Chapters 7 and 14 and Roth's lemma is a tool in the proof of Schmidt's subspace theorem in Chapter 7 and in the proof of Mordell's conjecture in Chapter 11.

6.2. Roth's theorem

This section states Roth's theorem and gives a sketch of the proof.

6.2.1. In order to see how diophantine approximation can be applied to diophantine equations, we start with the argument that Thue's theorem implies finiteness of integer solutions of the Thue equation (see the introduction in 6.1 for the statements, and see Theorem 5.3.2 for a generalization and a quantitative result).

The argument is by contradiction. First, we assume F irreducible. We use the decomposition

$$F\left(\frac{x}{y},1\right) = a_d\left(\frac{x}{y}-\alpha_1\right)\cdots\left(\frac{x}{y}-\alpha_d\right) = \frac{m}{y^d}$$

into linear factors. If there are infinitely many integer solutions (x_n, y_n) of the Thue equation, then $|y_n| \to \infty$ and we may assume, by passing to a subsequence, that x_n/y_n tends to a zero α_j. As the other factors are bounded away from 0, we get infinitely many solutions of $|x/y - \alpha_j| \leq C|y|^{-d}$ for some constant $C > 0$. Since $d \geq 3$, this contradicts Thue's theorem.

In general, let $F_1, \ldots, F_r$ be the non-constant irreducible polynomials in $\mathbb{Z}[x,y]$ dividing F. By a linear change of coordinates, we may assume that y is not a divisor of F. Dirichlet's box principle gives finitely many divisors m_j of m such that the system of equations $F_j(x,y) = m_j$ $(j = 1, \ldots, r)$ has infinitely many solutions (x_n, y_n). The above argument shows that x_n/y_n approaches a zero of every $F_j(x,1)$ and hence $r = 1$. Since F has at least three different linear factors, we get $\deg(F_1) \geq 3$. Now the irreducible case considered above leads to a contradiction with the initial assumption of infinitely many solutions of $F_1(x,y) = m_1$.

6.2.2. Now we give Lang's general formulation of **Roth's theorem** over number fields; the same statement for the rational field $\mathbb{Q}$ belongs to D. Ridout.

We use the notions introduced in Chapter 1, 1.4.3: For a place v of a number field K, we denote by $|\ |_v$ the normalized absolute value (as in (1.6) on page 11) to get the product formula. As usual, we denote again by $|\ |_v$ its extension to the completion K_v. The absolute exponential height $H(\alpha) = e^{h(\alpha)}$ for an algebraic number is as defined in 1.5.7.

Theorem 6.2.3. *Let K be a number field with a finite set S of places. For each $v \in S$ let $\alpha_v \in K_v$ be K-algebraic. Let $\kappa > 2$. Then there are only finitely many $\beta \in K$ such that*

$$\prod_{v \in S} \min\left(1, |\beta - \alpha_v|_v\right) \leq H(\beta)^{-\kappa}.$$

The classical theorem of Roth is the special case $K = \mathbb{Q}$ and $S = \{\infty\}$, so that $|\ |_v$ is the ordinary absolute value in $\mathbb{R}$.

Remark 6.2.4. Theorem 6.2.3 is ineffective in the sense that the proof does not give an upper bound for $H(\beta)$. However, it does give an upper bound for the number of solutions β, see 6.5.3.

A refinement of Theorem 6.2.3, in which we allow α_v to vary with β, will be mentioned later in Section 6.5.

6.2.5. Theorem 6.2.3 makes sense also if we allow $\alpha_v = \infty$, just by replacing the meaningless $|\infty - \beta|_v$ by $|1/\beta|_v$. This can be seen by applying a linear transformation $T(x) = (ax+b)/(cx+d)$, $a,b,c,d \in \mathbb{Z}$, such that $T(\alpha_v)$ are all finite and applying the theorem with $T(\alpha_v)$ and $T(\beta)$. Since $H(T(\beta)) \gg\ll H(\beta)$, our claim follows.

A more elegant way of dealing with this consists in working on the projective line $\mathbb{P}^1$ rather than the affine line $\mathbb{A}^1$, replacing the affine v-adic distance $|\alpha_v - \beta|_v$ with the projective v-adic distance $\delta_v(\alpha_v, \beta)$ introduced in 2.8.16, and the height $H(\beta)$ with the exponential Arakelov height on $\mathbb{P}^1$.

6.2.6. Consider the special case $K = \mathbb{Q}$, $S = \{\infty, p\}$ with $\alpha_\infty = \infty$, $\alpha_p = \alpha$ an algebraic integer in $\mathbb{Q}_p$ and $\beta = n \in \mathbb{Z}$. Then $|1/n|_\infty = 1/H(n)$, therefore Theorem 6.2.3 implies that

$$|\alpha - n|_p < |n|^{-1-\varepsilon}$$

has only finitely many solutions in integers n, for every fixed $\varepsilon > 0$.

6.2.7. The following application of Ridout's form of Roth's theorem is due to K. Mahler [**185**]. Let $g(k)$ be the smallest number such that every positive integer is a sum of at most $g(k)$ positive integral kth powers. It is a classical theorem of Lagrange that $g(2) = 4$. Waring stated the empirical theorem that $g(3) = 9$, $g(4) = 19$, and so on, and Hilbert proved in general $g(k) < \infty$. As noted by J.A. Euler (son of Leonhard Euler, see L.E. Dickson [**89**], p.717), the number $\lfloor (\frac{3}{2})^k \rfloor 2^k - 1$ requires $\lfloor (\frac{3}{2})^k \rfloor - 1$ powers 2^k and $2^k - 1$ powers 1^k for its representation, making it clear that $g(k) \geq 2^k + \lfloor (\frac{3}{2})^k \rfloor - 2$.

It turned out, after the researches of Hardy, Littlewood, and Vinogradov, on the problem, that the number of kth powers needed to represent large integers was substantially less than the above lower bound for $g(k)$, thus reducing the problem of determining $g(k)$ to a finite calculation for every k. Through the work of Dickson, Pillai, Rubugunday, and Niven, this eventually led to a complete solution for $k \geq 6$ of the original **Waring's problem**, in which there were two possible answers, namely $g(k) = 2^k + \lfloor (\frac{3}{2})^k \rfloor - 2$ if $\lceil (\frac{3}{2})^k \rceil - (\frac{3}{2})^k \geq (\frac{3}{4})^k$, and another more complicated result otherwise (see W.J. Ellison [**99**] for a detailed account and references).

Let us apply Theorem 6.2.3 choosing $S = \{\infty, 2, 3\}$, $\alpha_\infty = 1$, $\alpha_2 = \infty$, $\alpha_3 = 0$, and $\beta = 3^k/(n \cdot 2^k)$ with $n = \lceil (\frac{3}{2})^k \rceil$. We have (using the remark in 6.2.5): $|\alpha_2 - \beta|_2 \leq 2^{-k}$, $|\alpha_3 - \beta|_3 = 3^{-k}|n|_3^{-1}$, and $H(\beta) \geq 3^k |n|_3$. After an easy simplification, we find that the inequality

$$|\alpha_\infty - \beta|_\infty = |1 - 3^k/(n \cdot 2^k)| < 2^k 3^k |n|_3 (3^k |n|_3)^{-2-\varepsilon}$$

has only finitely many solutions k. If we multiply by n, we verify *a fortiori* that

$$\left\lceil \left(\frac{3}{2}\right)^k \right\rceil - \left(\frac{3}{2}\right)^k \geq 3^{-\varepsilon k}$$

for all but finitely many positive integers k, for any fixed $\varepsilon > 0$. If we take $\varepsilon = \log(4/3)/\log 3$ and use the above solution of Waring's problem, we deduce Mahler's theorem that $g(k) = 2^k + \lfloor (\frac{3}{2})^k \rfloor - 2$ for all sufficiently large integers k. Due to the ineffectiveness of Roth's theorem, it remains an open problem to determine an effective k_0 such that this result holds for $k \geq k_0$.

6.2.8. The proof given here can be easily axiomatized to obtain the result for more general fields of characteristic 0, for example function fields of characteristic 0 (see [**169**], Ch.7). On the other hand, Roth's theorem does not hold in function fields of characteristic p. The following example is due to Mahler.

Let $K = k(t)$, where $k = \overline{\mathbb{F}}_p$ is an algebraic closure of the finite field $\mathbb{F}_p$ with p elements. Let $|\ |_v$ be the absolute value on K such that $|t|_v = c^{-1}$, with $c > 1$. The completion K_v can be identified with the field $K((t))$ of formal Laurent series

$$\sum_{m=h}^{\infty} a_m t^m, \qquad h \in \mathbb{Z}$$

in the uniformizing parameter t, with

$$\left| a_h t^h + a_{h+1} t^{h+1} + \cdots \right|_v = c^{-h}$$

provided $a_h \neq 0$.

Let $q = p^a$ and consider the Artin–Schreier equation

$$x^q - x + t = 0$$

and the associated finite separable extension $E = K[x]/(x^q - x + t)$ of degree q. There is an extension w of v to M_E such that $E_w = K_v$, with a solution α of

$$\alpha^q - \alpha + t = 0$$

given by

$$\alpha = t + t^q + t^{q^2} + t^{q^3} + \cdots.$$

If we take $\beta = t + t^q + \cdots + t^{q^m}$, we have $H(\beta) = c^{q^m}$ and $|\alpha - \beta|_v = c^{-q^{m+1}}$, whence

$$|\alpha - \beta|_v = H(\beta)^{-[E:K]}.$$

Therefore in this case we cannot get any sharpening over the obvious Liouville inequality and Roth's theorem does not hold as soon as $q \geq 3$.

6.2.9. We first give a sketch of the main steps in the proof of Roth's theorem in the simplest possible case, namely $K = \mathbb{Q}$ and $S = \{\infty\}$, so that $|\ |_v$ is the ordinary euclidean absolute value. There is only one α to worry about. Suppose we have infinitely many rational approximations p/q to α such that

$$\left|\alpha - \frac{p}{q}\right| \leq q^{-\kappa}.$$

Then, for any positive integer m and any large constant M, we can find m rational approximations to α, namely p_j/q_j, $j = 1, \ldots, m$, with $\log q_1 > L$ and also

$$\log q_{j+1} > M \log q_j, \qquad j = 1, \ldots, m-1.$$

Such a sequence of approximations will be said to be **(*L, M*)-independent.**

Step I: The auxiliary construction at the algebraic point.

We abbreviate $\mathbf{x} = (x_1, \ldots, x_m)$. Construct a polynomial $P(\mathbf{x}) \in \mathbb{Z}[\mathbf{x}]$ with partial degrees $d_1, \ldots, d_m$, vanishing to a (weighted) high order at $(\alpha, \ldots, \alpha)$.

The degrees d_j are chosen so that the quantities $d_j \log q_j$ are nearly the same, for $j = 1, \ldots, m$. Vanishing to high order means vanishing of ΔP at $(\alpha, \ldots, \alpha)$ with Δ any differential operator of order $(i_1, \ldots, i_m)$ with

$$\frac{i_1}{d_1} + \cdots + \frac{i_m}{d_m} < t.$$

Here t is a parameter, which we want to be large.

Carrying out this step is done by an application of the pigeon-hole principle, say Siegel's lemma. The number of equations is asymptotic to $V_m(t) d_1 \cdots d_m$, with $V_m(t)$ the volume of the region

$$\mathcal{V}_m(t) := \{\mathbf{x} \in \mathbb{R}^m \mid x_1 + \cdots + x_m \le t, \, 0 \le x_j \le 1 \text{ for } j = 1, \ldots, m\}.$$

The number of coefficients is asymptotic to $d_1 \cdots d_m$, and, if $[\mathbb{Q}(\alpha) : \mathbb{Q}]\, V_m(t) \le 1 - \delta$, we can find such a polynomial P with integer coefficients bounded by $C^{d_1 + \cdots + d_m}$, for a suitable constant C depending only on $H(\alpha)$, m, and δ.

Step II: Non-vanishing at the rational point.

Show that the polynomial P constructed in step I, or a suitable derivative of it of rather small order, does not vanish at the rational point $(p_1/q_1, \ldots, p_m/q_m)$, provided the points p_i/q_i are (L, M)-independent and L, M are sufficiently large.

This was the difficult step that, before Roth's work, we could do only for $m = 1$ or 2. This step uses in an essential way the hypothesis that the approximations are (L, M)-independent.

Step III: The upper bound.

Since P vanishes to high order at the algebraic points $(\alpha, \ldots, \alpha)$, the Taylor expansion at $(\alpha, \ldots, \alpha)$ shows that, if $|p_j/q_j - \alpha| < q_j^{-\kappa}$, then

$$|P(p_1/q_1, \ldots, p_m/q_m)| \le C'^{d_1 + \cdots + d_m} \max_j q_j^{-\kappa t d_j}$$

for a constant C' depending only on α, m, and δ.

Step IV: The Liouville lower bound.

Since P does not vanish at the rational point, it is bounded away from 0 as

$$q_1^{-d_1} \cdots q_m^{-d_m} \leq |P(p_1/q_1, \dots, p_m/q_m)| .$$

Step V: Comparison of the upper and lower bounds.

We have chosen

$$d_1 \log q_1 \sim \cdots \sim d_m \log q_m$$

and since C'^{d_j} is negligible with respect to $q_j^{d_j}$ (because $q_j > e^L$), comparison of the upper and lower bounds shows that

$$\kappa t \leq m + O(1/M);$$

the constant involved in the $O(\ )$ symbol depends on α, m, and δ.

Thus, as M tends to ∞, we find $\kappa \leq m/t$ provided $[\mathbb{Q}(\alpha) : \mathbb{Q}]\, V_m(t) \leq 1 - \delta$. A simple probability estimate shows that, if we choose $t = (\frac{1}{2} - \varepsilon)m$, then $V_m(t)$ tends to 0 as m increases, so that this choice of t is admissible for large m. This gives $\kappa \leq 1/(\frac{1}{2} - \varepsilon)$ leading to a contradiction for $\varepsilon > 0$ sufficiently small, and Roth's theorem follows.

6.3. Preliminary lemmas

In this section we prove several preliminary results needed for the proof of Roth's theorem. It will also be convenient to use the notation and formalism already developed in Chapter 1.

6.3.1. We abbreviate $\mathbf{x} = (x_1, \dots, x_m)$, $\boldsymbol{\alpha}_v = (\alpha_{v1}, \dots, \alpha_{vm})$ and write

$$\binom{\mathbf{m}}{\boldsymbol{\mu}} = \prod_{j=1}^{m} \binom{m_j}{\mu_j}$$

and

$$\partial_{\boldsymbol{\mu}} = \frac{1}{\mu_1! \cdots \mu_m!} \left(\frac{\partial}{\partial x_1}\right)^{\mu_1} \cdots \left(\frac{\partial}{\partial x_m}\right)^{\mu_m} .$$

We have

$$\partial_{\boldsymbol{\mu}} \mathbf{x}^{\mathbf{m}} = \binom{\mathbf{m}}{\boldsymbol{\mu}} \mathbf{x}^{\mathbf{m}-\boldsymbol{\mu}}. \tag{6.1}$$

6.3.2. We work with polynomials in $F[x_1, \dots, x_m]$ for a field F, vanishing to high order at a point. For $P \in F[x_1, \dots, x_m]$ and positive weights $\mathbf{d} = (d_1, \dots, d_m)$, we define the **index** of P at a point $\boldsymbol{\alpha} = (\alpha_1, \dots, \alpha_m)$ to be

$$\operatorname{ind}(P; \mathbf{d}; \boldsymbol{\alpha}) = \min_{\boldsymbol{\mu}} \left\{ \frac{\mu_1}{d_1} + \cdots + \frac{\mu_m}{d_m} \mid \partial_{\boldsymbol{\mu}} P(\boldsymbol{\alpha}) \neq 0 \right\} .$$

The following properties hold:

(a) $\operatorname{ind}(P+Q;\mathbf{d};\boldsymbol{\alpha}) \geq \min(\operatorname{ind}(P;\mathbf{d};\boldsymbol{\alpha}), \operatorname{ind}(Q;\mathbf{d};\boldsymbol{\alpha}))$;

(b) $\operatorname{ind}(PQ;\mathbf{d};\boldsymbol{\alpha}) = \operatorname{ind}(P;\mathbf{d};\boldsymbol{\alpha}) + \operatorname{ind}(Q;\mathbf{d};\boldsymbol{\alpha})$;

(c) $\operatorname{ind}(\partial_{\boldsymbol{\mu}} P;\mathbf{d};\boldsymbol{\alpha}) \geq \operatorname{ind}(P;\mathbf{d};\boldsymbol{\alpha}) - \dfrac{\mu_1}{d_1} - \dfrac{\mu_2}{d_2} - \cdots - \dfrac{\mu_m}{d_m}$.

By means of the Taylor expansion at α, we see easily that $\operatorname{ind}(P;\mathbf{d};\boldsymbol{\alpha}) = \infty$ if and only if $P = 0$. Together with (a) and (b), this says that the index is a valuation.

6.3.3. It is convenient to introduce the set

$$\mathcal{V}_m(t) := \{\mathbf{x} \mid x_1 + \cdots + x_m \leq t,\ \ 0 \leq x_j \leq 1\}$$

and its volume $V_m(t) = \operatorname{vol}(\mathcal{V}_m(t))$. For $\mathbf{t} \in \mathbb{R}^n_+$, we define

$$V_m(\mathbf{t}) := \sum_{i=1}^{n} V_m(t_i).$$

Lemma 6.3.4. *Let $\boldsymbol{\alpha}_i$, $i = 1, \ldots, n$, be points $\boldsymbol{\alpha}_i = (\alpha_{i1}, \ldots, \alpha_{im})$ with coordinates α_{ij} in a finite extension F/K of a number field K, of degree $r = [F : K]$. Suppose that $\mathbf{t} \in \mathbb{R}^n_+$ satisfies*

$$rV_m(\mathbf{t}) < 1.$$

Then, for all sufficiently large integers $d_1, \ldots, d_m$, there is $P \in K[x_1, \ldots, x_m]$, not identically 0 and with partial degrees at most $d_1, \ldots, d_m$, such that:

(a) *the index is bounded below by $\operatorname{ind}(P;\mathbf{d};\boldsymbol{\alpha}_i) \geq t_i$ for $i = 1, \ldots, n$;*

(b) *the height of P is bounded by*

$$h(P) \leq \frac{r}{1 - rV_m(\mathbf{t})} \sum_{i=1}^{n} \sum_{j=1}^{m} V_m(t_i)(h(\alpha_{ij}) + \log 2 + o(1))d_j$$

as $d_j \to \infty$ for $j = 1, \ldots, m$.

Proof: We abbreviate $I = (i_1, \ldots, i_m)$, $J = (j_1, \ldots, j_m)$, set $P(\mathbf{x}) = \sum p_J \mathbf{x}^J$ and consider the set of equations

$$\partial_I P(\boldsymbol{\alpha}_k) = 0 \qquad \text{for} \qquad \frac{i_1}{d_1} + \cdots + \frac{i_m}{d_m} < t_k; \quad k = 1, \ldots, n.$$

This is a linear system in the coefficients p_J of P, which we want to solve non-trivially in K; the coefficients of the linear system are in the field F, of degree $[F : K] = r$. The number N of unknowns is $N = (d_1 + 1) \cdots (d_m + 1) \sim d_1 \cdots d_m$ as all $d_j \to \infty$, while the number M of equations is asymptotically $M \sim V_m(\mathbf{t})\, d_1 \cdots d_m$, because the number of lattice points $(i_1/d_1, \ldots, i_m/d_m)$ in $\mathcal{V}_m(t)$ is asymptotic to $V_m(t)\, d_1 \cdots d_m$. In order to verify this, let Z be this number. If we associate to each lattice point $(i_1/d_1, \ldots, i_m/d_m)$ the parallelopiped $i_\nu/d_\nu \leq x_\nu \leq (i_\nu + 1)/d_\nu$, $\nu = 1, \ldots, m$, we immediately see that $V_m(t) d_1 \cdots d_m \leq Z$.

For an upper bound for Z, we note that, if $(i_1/d_1, \ldots, i_m/d_m)$ is a lattice point in $\mathcal{V}_m(t)$, then

$$\frac{i_1+1}{d_1} + \cdots + \frac{i_m+1}{d_m} \leq t + \frac{1}{d_1} + \cdots + \frac{1}{d_m}$$

and $i_\nu + 1 \leq d_\nu + 1$. It follows that, if we rescale $\mathcal{V}_m(t)$ by a factor $1 + \max(1, t^{-1})(1/d_1 + \cdots + 1/d_m)$, then the rescaled domain contains all parallelopipeds associated to lattice points in $\mathcal{V}_m(t)$. Hence

$$V_m(t)d_1 \cdots d_m \leq Z \leq V_m(t)\left(1 + \max(1, t^{-1})\left(\frac{1}{d_1} + \cdots + \frac{1}{d_m}\right)\right)^m d_1 \cdots d_m,$$

thus showing that $Z \sim V_m(t)d_1 \cdots d_m$ as $d_j \to \infty$.

In particular, we have $N > rM$ if $rV_m(\mathbf{t}) < 1$ and $d_j \to \infty$. By (6.1) on page 156, the matrix of coefficients has entries

$$\mathcal{A} = \left(\binom{J}{I}(\boldsymbol{\alpha}_k)^{J-I}\right)$$

with rows indexed by (I, k) and columns indexed by J.

We apply Siegel's lemma as given in Theorem 2.9.19. In our case, it suffices to produce only one small solution to our system of equations. Theorem 2.9.19 and 2.9.8 imply that, if $N > rM$, there is a solution $\mathbf{x}$ such that

$$H(\mathbf{x}) \leq |D_K|^{1/2r}\left(\sqrt{N}\prod_{(I,k)} H(\mathcal{A}_{(I,k)})\right)^{r/(N-rM)}, \tag{6.2}$$

where $\mathcal{A}_{(I,k)}$ is the (I, k)-th row of $\mathcal{A}$.

For fixed (I, k), we estimate the height of the corresponding row vector $\mathcal{A}_{(I,k)}$ of $\mathcal{A}$ as follows. The vector $\mathcal{A}_{(I,k)}$ has entries $\binom{J}{I}(\boldsymbol{\alpha}_k)^{J-I}$, and hence

$$H(\mathcal{A}_{(I,k)}) \leq \prod_{j=1}^{m}(2H(\alpha_{kj}))^{d_j}.$$

This bound is independent of I.

As noted before, for fixed k we have about $V_m(t_k)\, d_1 \cdots d_m$ possibilities for I. Thus the product of the heights of the rows of $\mathcal{A}$ is bounded by

$$\left(\prod_{k=1}^{n}\prod_{j=1}^{m}(d_j+1)^{V_m(t_k)/2}\left(2\,H(\alpha_{kj})\right)^{d_j V_m(t_k)}\right)^{(1+o(1))d_1\cdots d_m}.$$

The conclusion follows from (6.2), noting that the term $|D_K|^{1/2r}$ and the terms $(d_j+1)^{V_m(t_k)/2}$ are negligible with respect to $2^{d_j V_m(t_k)}$, so they do not affect the asymptotics in the final estimate. □

Lemma 6.3.5. *If* $0 \le \varepsilon \le \frac{1}{2}$, *then*

$$V_m\left(\left(\frac{1}{2}-\varepsilon\right)m\right) \le e^{-6m\varepsilon^2}.$$

Proof: We use a familiar method of probability theory. We set $\chi(x) = 1$ if $x < 0$ and 0 if $x > 0$. Since $\chi(x) < e^{-\lambda x}$ for every $\lambda > 0$, we have

$$\begin{aligned}
V_m\left(\left(\frac{1}{2}-\varepsilon\right)m\right) &= \int_{-\frac{1}{2}}^{\frac{1}{2}} \cdots \int_{-\frac{1}{2}}^{\frac{1}{2}} \chi(x_1+\cdots+x_m+m\varepsilon)\,\mathrm{d}x_1\cdots\mathrm{d}x_m \\
&\le \int_{-\frac{1}{2}}^{\frac{1}{2}} \cdots \int_{-\frac{1}{2}}^{\frac{1}{2}} e^{-\lambda(m\varepsilon+\sum x_j)}\,\mathrm{d}x_1\cdots\mathrm{d}x_m \\
&= \left(\int_{-\frac{1}{2}}^{\frac{1}{2}} e^{-\lambda(\varepsilon+x)}\,\mathrm{d}x\right)^m \\
&= \exp(-mU(\lambda))
\end{aligned}$$

with

$$U(\lambda) = \varepsilon\lambda - \log\left(\frac{\sinh(\lambda/2)}{\lambda/2}\right).$$

It is possible to show that this estimate is quite precise, in the sense that

$$\log\left(V_m\left(\left(\frac{1}{2}-\varepsilon\right)m\right)\right) = -m\max_{\lambda} U(\lambda) + O(\log m)$$

uniformly in ε. For our purposes, it suffices to note that

$$\sinh(u)/u = 1 + \frac{u^2}{3!} + \frac{u^4}{5!} + \cdots \le 1 + \frac{u^2}{6} + \frac{(u^2/6)^2}{2!} + \cdots = e^{u^2/6},$$

giving $\log(\sinh(u)/u) \le u^2/6$. If we choose $\lambda = 12\varepsilon$, we get what we want. □

6.3.6. The simplest way to achieve Step II in the proof of Roth's theorem is by means of **Roth's lemma:**

Lemma 6.3.7. *Let* $P(x_1,\ldots,x_m)$ *be a polynomial in* m *variables, with partial degrees at most* $d_1,\ldots,d_m$ *with* $d_i \ge 1$, *with coefficients in* $\overline{\mathbb{Q}}$ *and not identically* 0. *Let* $(\xi_1,\ldots,\xi_m) \in \overline{\mathbb{Q}}^m$ *and let* $0 < \sigma \le \frac{1}{2}$. *Suppose that:*

(a) *the weights* $d_1,\ldots,d_m$ *are rapidly decreasing, namely*

$$d_{j+1}/d_j \le \sigma;$$

(b) *the point* $(\xi_1,\ldots,\xi_m)$ *has components with large height, in the sense that*

$$\min_j d_j h(\xi_j) \ge \sigma^{-1}(h(P) + 4md_1).$$

Then

$$\operatorname{ind}(P;\mathbf{d};\boldsymbol{\xi}) \le 2m\,\sigma^{1/2^{m-1}}.$$

Remark 6.3.8. The constant $2m$ appearing in the conclusion of the theorem is not optimal but its actual value is of little importance here.

Proof: The proof is by induction on m. If the polynomial P is the product of two polynomials in disjoint sets of variables, it is easy to obtain an upper bound for the index if we have an upper bound for the index of the factors. The point is that, even if P does not split in such a fashion, a suitable generalized Wronskian of P does. If the degrees d_i form a rapidly decreasing sequence, we can get sufficient control on the order of derivatives involved (it is here that we use hypothesis (a)), and the induction works by splitting one variable at a time. The details are as follows.

6.3.9. If $m = 1$, the bound

$$\operatorname{ind}(P; d_1; \xi_1)\, d_1 h(\xi_1) \leq h(P) + d_1 \log 2$$

follows from Proposition 1.6.5 and the fact that $(x_1 - \xi_1)^{\operatorname{ind}(P) d_1}$ is a factor of P. This gives Roth's lemma for $m = 1$ with the better constants $\log 2$ and 1 in place of 4 and 2.

In order to perform the splitting of the Wronskian, we write P in the form

$$P = \sum_{j=0}^{s} f_j(x_1, \ldots, x_{m-1}) g_j(x_m),$$

where $s \leq d_m$ and where the f_j s, and similarly the g_j s, are linearly independent polynomials defined over $\overline{\mathbb{Q}}$.

We recall the **Wronskian criterion** for linear independence.

Proposition 6.3.10. *Let K be a field of characteristic 0 and let $x_1, \ldots, x_m$ be algebraically independent over K. Let $\varphi_j \in K[x_1, \ldots, x_m]$, $j = 1, \ldots, n$, be n polynomials. Then $\varphi_1, \ldots, \varphi_n$ are linearly independent over K if and only if some generalized Wronskian*

$$W_{\boldsymbol{\mu}_1, \ldots, \boldsymbol{\mu}_n}(x_1, \ldots, x_m) := \det \begin{pmatrix} \partial_{\boldsymbol{\mu}_1} \varphi_1 & \partial_{\boldsymbol{\mu}_1} \varphi_2 & \cdots & \partial_{\boldsymbol{\mu}_1} \varphi_n \\ \partial_{\boldsymbol{\mu}_2} \varphi_1 & \partial_{\boldsymbol{\mu}_2} \varphi_2 & \cdots & \partial_{\boldsymbol{\mu}_2} \varphi_n \\ \cdot & \cdot & \cdots & \cdot \\ \partial_{\boldsymbol{\mu}_n} \varphi_1 & \partial_{\boldsymbol{\mu}_n} \varphi_2 & \cdots & \partial_{\boldsymbol{\mu}_n} \varphi_n \end{pmatrix}$$

with $|\boldsymbol{\mu}_i| = \mu_{1i} + \mu_{2i} + \cdots + \mu_{mi} \leq i - 1$ not identically 0.

Proof: If $\varphi_1, \ldots, \varphi_n$ are linearly dependent over K, then all generalized Wronskians vanish. Indeed, let

$$c_1 \varphi_1 + c_2 \varphi_2 + \cdots + c_n \varphi_n = 0$$

be a linear dependence relation among the φ_j. If we apply the differential operators $\partial_{\boldsymbol{\mu}_i}$ $(i = 1, \ldots, n)$ to this relation, we obtain a homogeneous linear system in the coefficients c_j and its determinant must vanish, proving what we want.

If instead $\varphi_1, \ldots, \varphi_n$ are linearly independent over K, we proceed as follows. Consider the Kronecker substitution $(x_1, x_2, \ldots, x_m) \mapsto (t, t^d, \ldots, t^{d^{m-1}})$, where t is a new indeterminate. This maps monomials in $x_1, \ldots, x_m$ into powers of t and is injective on the set of monomials with partial degrees strictly less than d. It follows that, if d is larger than the partial degrees of the φ_js then $\varphi_1, \ldots, \varphi_n$ are linearly independent over K if and only if the polynomials

$$\Phi_j(t) := \varphi_j(t, t^d, \ldots, t^{d^{m-1}})$$

are linearly independent over K. By Wronski's well-known result, this is the case if and only if the Wronskian

$$W(t) := \det\left((\mathrm{d}/\mathrm{d}t)^{i-1}\Phi_j\right)_{i,j=1,\ldots,n} \tag{6.3}$$

is not identically 0. We have, for certain universal polynomials $a_{\boldsymbol{\mu},i}(t; d, m) \in \mathbb{Q}[t]$

$$(\mathrm{d}/\mathrm{d}t)^{i-1}\Phi_j = \sum_{|\boldsymbol{\mu}| \leq i-1} a_{\boldsymbol{\mu},i}(t; d, m)\, \partial_{\boldsymbol{\mu}}\varphi_j(t, \ldots, t^{d^{m-1}})$$

and substituting into (6.3) we see that $W(t)$ is a linear combination of generalized Wronskians $W_{\boldsymbol{\mu}_1,\ldots,\boldsymbol{\mu}_n}(t, t^d, \ldots, t^{d^{m-1}})$ with $|\boldsymbol{\mu}_i| \leq i-1$. Since $W(t)$ is not identically 0, some generalized Wronskian does not vanish identically. □

6.3.11. We return to the proof of 6.3.7. Because of linear independence, applying the above proposition shows that there are two Wronskians

$$U(x_1, \ldots, x_{m-1}) := \det(\partial_{\boldsymbol{\mu}_i} f_j)_{i,j=0,\ldots,s}$$

and

$$V(x_m) := \det(\partial_\nu g_j)_{\nu,j=0,\ldots,s},$$

which are not identically 0; here we have $\boldsymbol{\mu}_i = (\mu_{1i}, \ldots, \mu_{m-1,i})$ and $|\boldsymbol{\mu}_i| \leq s \leq d_m$. We multiply the two determinants and obtain

$$W(x_1, \ldots, x_m) := \det(\partial_{\boldsymbol{\mu}_i,\nu} P) = U(x_1, x_2, \ldots, x_{m-1})V(x_m).$$

Since $d_{j+1}/d_j \leq \frac{1}{2}$ for every j, we have

$$d_1 + \cdots + d_m \leq 2d_1.$$

The partial degrees of U and V are bounded by $((s+1)d_1, \ldots, (s+1)d_{m-1})$ and $(s+1)d_m$; we have

$$h(U) + h(V) = h(W) \leq (s+1)\big(h(P) + 4d_1\big). \tag{6.4}$$

Only the last display requires some explanation. We have $h(W) = h(U) + h(V)$ because U and V involve disjoint sets of variables, see Proposition 1.6.2. We estimate $h(W)$ as follows, by expanding the determinant (π runs over permutations), using the Laplace expansion and then applying Lemma 1.3.7, we get

$$h(W) \leq \sum_v \max_\pi \log \Big| \prod_{i=0}^{s} \partial_{\boldsymbol{\mu}_i,\pi(i)} P \Big|_v + \log((s+1)!).$$

Then Gauss's lemma in 1.6.3 and Gelfond's lemma in 1.6.11 lead to

$$h(W) \le \sum_v \max_\pi \sum_{i=0}^s \log \left|\partial_{\boldsymbol{\mu}_i,\pi(i)} P\right|_v$$
$$+ (s+1)(d_1 + d_2 + \cdots + d_m) \log 2 + \log((s+1)!) .$$

Now (6.1) on page 156 and $d_1 + \ldots + d_m \le 2d_1$ yield

$$\begin{aligned} h(W) &\le \sum_{i=0}^s (h(P) + (d_1 + d_2 + \cdots + d_m) \log 2) \\ &\quad + (s+1)\{(d_1 + d_2 + \cdots + d_m) \log 2 + \log(d_m + 1)\} \\ &< (s+1)\big(h(P) + 4d_1\big), \end{aligned}$$

where in the last step we also used $\log(d_m + 1) \le d_m \le \frac{1}{2} d_1$. This proves (6.4).

We obtain a lower bound for $\operatorname{ind}(W)$ by expanding the determinant for W, using properties 6.3.2 (a), (b), and (c) of the index to estimate from below the index of W in terms of the index of a typical term in the expansion.

In what follows, we abbreviate $\operatorname{ind}(\cdot)$ for $\operatorname{ind}(\cdot; \mathbf{d}; \boldsymbol{\xi})$. By 6.3.2 (c), we have the estimate

$$\begin{aligned} \operatorname{ind}(\partial_{\boldsymbol{\mu},\nu} P) &\ge \operatorname{ind}(P) - \frac{\mu_1}{d_1} - \cdots - \frac{\mu_{m-1}}{d_{m-1}} - \frac{\nu}{d_m} \\ &\ge \operatorname{ind}(P) - \frac{d_m}{d_{m-1}} - \frac{\nu}{d_m} \\ &\ge \operatorname{ind}(P) - \frac{\nu}{d_m} - \sigma. \end{aligned}$$

Moreover, since the index is never negative, we can improve this to

$$\operatorname{ind}(\partial_{\boldsymbol{\mu},\nu} P) \ge \max\left(\operatorname{ind}(P) - \frac{\nu}{d_m}, 0\right) - \varepsilon. \tag{6.5}$$

By 6.3.2 (a), (b), inequality (6.5), and expanding W by means of the Laplace expansion, we get (here π runs over permutations of $0, \ldots, s$)

$$\begin{aligned} \operatorname{ind}(W) &\ge \min_\pi \left(\sum_{i=0}^s \operatorname{ind}(\partial_{\boldsymbol{\mu}_i,\pi(i)} P) \right) \\ &\ge \min_\pi \sum_{i=0}^s \left(\max\left(\operatorname{ind}(P) - \frac{\pi(i)}{d_m}, 0 \right) - \sigma \right) \\ &= \sum_{i=0}^s \left(\max\left(\operatorname{ind}(P) - \frac{i}{d_m}, 0 \right) - \sigma \right) \\ &\ge (s+1) \min\left(\frac{1}{2} \operatorname{ind}(P), \frac{1}{2} \operatorname{ind}(P)^2 \right) - (s+1)\sigma. \end{aligned} \tag{6.6}$$

Here the last step comes from the easy inequality

$$\sum_{i=0}^{s} \max\left(t - \frac{i}{s}, 0\right) \geq (s+1) \min\left(\frac{1}{2}\, t, \frac{1}{2}\, t^2\right).$$

Next, we obtain an upper bound for $\mathrm{ind}(W)$ by noting that

$$\mathrm{ind}(W) = \mathrm{ind}(U) + \mathrm{ind}(V) \tag{6.7}$$

and using Roth's lemma inductively on the number of variables to estimate $\mathrm{ind}(U)$ and $\mathrm{ind}(V)$.

Suppose we have proved Roth's lemma for polynomials in $l < m$ variables. We apply the inductive assumption of Roth's lemma to U and V but with $(s+1)d_j$ in place of d_j. In view of the bounds obtained in (6.4) on page 161 for $h(U)$ and $h(V)$, the hypotheses of Roth's lemma are satisfied. Therefore, we obtain

$$\mathrm{ind}(U) \leq 2(m-1)(s+1)\,\sigma^{1/2^{m-2}}, \qquad \mathrm{ind}(V) \leq (s+1)\,\sigma$$

(use the better bound given in 6.3.9 for the case $m = 1$ when estimating $\mathrm{ind}(V)$). We insert these two estimates in (6.7) obtaining an upper bound for $\mathrm{ind}(W)$, and compare with the lower bound (6.6). This gives

$$\min\left(\mathrm{ind}(P), \mathrm{ind}(P)^2\right) \leq 4(m-1)\sigma^{1/2^{m-2}} + 4\sigma.$$

In any case $\mathrm{ind}(P) \leq m$, hence the preceding bound may be simplified to

$$\mathrm{ind}(P)^2 \leq 4m(m-1)\sigma^{1/2^{m-2}} + 4m\varepsilon \leq 4m^2\sigma^{1/2^{m-2}}.$$

This completes the induction and the proof of Roth's lemma. □

6.4. Proof of Roth's theorem

The general case of Roth's theorem is proved along similar lines as outlined in Section 6.2, but the presence of several places means that we can only compare approximations which have a similar behaviour at each place v. This creates additional complications.

We shall prove the following statement:

Theorem 6.4.1. *Let K be a number field with a finite set S of places. Let F be a finite-dimensional extension of K and, for $v \in S$, let $\alpha_v \in F$. We extend $|\ |_v$ to an absolute value $|\ |_{v,K}$ of F. Then for any $\kappa > 2$, there are only finitely many $\beta \in K$ such that*

$$\prod_{v \in S} \min(1, |\beta - \alpha_v|_{v,K}) \leq H(\beta)^{-\kappa}. \tag{6.8}$$

This statement implies Roth's theorem in 6.2.3 by embedding $\overline{K}$ into $\overline{K_v}$ for each $v \in S$, extending $| \ |_v$ to $\overline{K_v}$ and using the reduction $| \ |_{v,K}$ to the subfield $F = K(\{\alpha_v \mid v \in S\}) \subset \overline{K}$. In fact, Theorem 6.4.1 is equivalent to Roth's theorem; we leave the proof to the reader, since the converse implication will not be used here.

Proof of Theorem 6.4.1: We assume that (6.8) is satisfied for infinitely many $\beta \in K$ and obtain a contradiction at the end.

We need a reduction, due to Mahler, which allows us to restrict our considerations to approximations with similar behaviour at every place $v \in S$.

6.4.2. *Step 0: Approximation classes.*

We define

$$\Lambda(\beta) := \prod_{v \in S} \min\left(1, |\beta - \alpha_v|_{v,K}\right).$$

A $\beta \in K$ for which $\Lambda(\beta) < 1$ is said to be a **non-trivial approximation.** Consider, for a non-trivial β, the vector

$$\left(\log \min\left(1, |\beta - \alpha_v|_{v,K}\right) / \log \Lambda(\beta)\right)_{v \in S}.$$

This is a point in the $|S|$-dimensional unit cube lying on the hyperplane where the sum of the coordinates is 1. We partition this cube by means of a grid of semi-open subcubes of side $1/N$ where N is a positive integer, and classify β according to the subcube containing the corresponding vector. The set of non-trivial approximations β determining a same subcube is called an **approximation class**. The quantity $1/N$ is the **size** of the approximation class.

Let $\boldsymbol{\lambda} := (\lambda_v)_{v \in S}$ be the south-west corner of a subcube, namely the point

$$\boldsymbol{\lambda} = (\lfloor N x_v \rfloor / N)_{v \in S}$$

with $\boldsymbol{x}$ any point in the subcube. Then we denote by $Q(\boldsymbol{\lambda})$ the corresponding subcube and by $\mathcal{C}(\boldsymbol{\lambda}; N)$ the approximation class of size $1/N$ determined by $Q(\boldsymbol{\lambda})$.

For every $\beta \in \mathcal{C}(\boldsymbol{\lambda}; N)$ and $v \in S$, we have by definition

$$\Lambda(\beta)^{\lambda_v + \frac{1}{N}} < \min\left(1, |\beta - \alpha_v|_{v,K}\right) \leq \Lambda(\beta)^{\lambda_v}. \tag{6.9}$$

Note also that

$$1 - \frac{|S|}{N} \leq \sum_{v \in S} \lambda_v \leq 1, \tag{6.10}$$

because $Q(\boldsymbol{\lambda})$ always contains a point $\mathbf{x}$ with $\sum x_v = 1$ if $\mathcal{C}(\boldsymbol{\lambda}; N)$ is not empty.

Lemma 6.4.3. *The number of approximation classes of size* $1/N$ *determined by non-trivial approximations does not exceed*

$$\binom{N+|S|}{|S|} < 2^{N+|S|}.$$

Proof: Consider an approximation class $\mathcal{C}(\boldsymbol{\lambda}; N)$. Then $n_v := N\lambda_v$ is a non-negative integer and, by (6.10), we have

$$\sum_{v \in S} n_v \leq N.$$

The number of solutions of this inequality is $\binom{N+|S|}{|S|}$. □

6.4.4. *Choosing independent solutions.*

Since by hypothesis $\Lambda(\beta) \leq H(\beta)^{-\kappa}$ has infinitely many solutions, the preceding lemma shows that for any N there is an approximation class $\mathcal{C}(\boldsymbol{\lambda}; N)$ containing infinitely many such βs.

Let $\beta_1, \ldots, \beta_m$ be elements of K and let $M \geq 2$. We say that the β_js are (L, M)**-independent** if $h(\beta_1) \geq L$ and $h(\beta_{j+1}) \geq M\, h(\beta_j)$ for $j = 1, \ldots, n-1$.

By Northcott's theorem in 1.6.8, any infinite sequence in K contains an infinite subsequence of (L, M)-independent elements, therefore for any N and L, M we can find an infinite subsequence of (L, M)-independent elements belonging to a fixed approximation class $\mathcal{C}(\boldsymbol{\lambda}; N)$.

6.4.5. *Step I: The auxiliary polynomial.*

Let D be a large real number, which in the end will tend to ∞, and choose

$$d_j = \lfloor D/h(\beta_j) \rfloor, \qquad j = 1, \ldots, m.$$

Let $\mathbf{t} = (t_v)_{v \in S}$ with $t_v = (\frac{1}{2} - \varepsilon)m$ and let $\boldsymbol{\alpha}_v := (\alpha_v, \ldots, \alpha_v) \in F^m$, $\boldsymbol{\beta} = (\beta_1, \ldots, \beta_m) \in K^m$. We also abbreviate $r := [F : K]$.

We want to apply Lemma 6.3.4 to this situation. By Lemma 6.3.5, we have

$$rV_m(\mathbf{t}) = r\,|S|\,V_m\left(\left(\frac{1}{2} - \varepsilon\right)m\right) < r\,|S|\,e^{-6m\varepsilon^2} \leq \frac{1}{2}$$

provided

$$m > \frac{\log(2r\,|S|)}{6\varepsilon^2},$$

which we shall assume for the rest of this section. Thus the hypothesis of Lemma 6.3.4 is verified. We estimate (b) of Lemma 6.3.4 by noting that $rV_m(\mathbf{t})/(1 - rV_m(\mathbf{t})) \leq 1$. In this way we obtain a non-trivial polynomial P with coefficients in K, partial degrees at most $d_1, \ldots, d_m$, such that $\operatorname{ind}(P; \mathbf{d}; \boldsymbol{\alpha}_v) \geq (\frac{1}{2} - \varepsilon)m$ for $v \in S$, and

$$h(P) \leq \left\{ \sum_{v \in S} \sum_{j=1}^{m} (h(\alpha_v) + \log 2)/h(\beta_j) \right\} D + o(D) \tag{6.11}$$

as $D \to \infty$. If we define

$$C_1 := |S| \, (\max_{v \in S} h(\alpha_v) + \log 2), \tag{6.12}$$

we obtain for large D the bound

$$h(P) \leq 2C_1 D/L. \tag{6.13}$$

6.4.6. *Step II: Application of Roth's lemma.*

We would like P to have the additional property that $P(\boldsymbol{\beta}) \neq 0$. In order to do this, we apply Roth's lemma to show that the polynomial P so constructed does not vanish too much at $\boldsymbol{\beta}$ if $\boldsymbol{\beta}$ is (L, M)-independent and L, M are large, and then work with a suitable derivative of P rather than P itself. The details are as follows.

Let $0 < \sigma \leq \frac{1}{2}$. By Roth's lemma in 6.3.7, we have

$$\operatorname{ind}(P; \mathbf{d}; \boldsymbol{\beta}) \leq 2m \, \sigma^{1/2^{m-1}}$$

provided $d_{j+1}/d_j \leq \sigma$ and $d_j h(\beta_j) \geq \sigma^{-1}(h(P) + 4md_1)$.

In our case $d_j = \lfloor D/h(\beta_j) \rfloor \sim D/h(\beta_j)$ and $h(\beta_{j+1}) \geq M h(\beta_j)$, therefore the first condition $d_{j+1}/d_j \leq \sigma$ is verified if

$$M \geq 2\sigma^{-1}$$

and D is large enough, which we shall suppose. Similarly, using $d_j h(\beta_j) \sim D$, $d_1 \leq D/h(\beta_1) \leq D/L$ and (6.13) we see that the condition $d_j h(\beta_j) \geq \sigma^{-1}(h(P) + 4md_1)$ is verified for large D if

$$D \geq \sigma^{-1} \frac{2C_1 + 5m}{L} D,$$

which is so if $L \geq (2C_1 + 5m)\sigma^{-1}$.

We conclude that, if $M \geq 2\sigma^{-1}$, $L \geq (2C_1 + 5m)\sigma^{-1}$ and D is large enough, we have

$$\operatorname{ind}(P; \mathbf{d}; \boldsymbol{\beta}) \leq 2m \, \sigma^{1/2^{m-1}}.$$

We choose $\sigma = \varepsilon^{2^{m-1}}$ and deduce that there is $\boldsymbol{\mu}$ such that $\partial_{\boldsymbol{\mu}} P(\boldsymbol{\beta}) \neq 0$ and

$$\sum_{j=1}^{m} \frac{\mu_j}{d_j} \leq 2m\varepsilon.$$

Now recall that by construction we have $\operatorname{ind}(P; \mathbf{d}; \boldsymbol{\alpha}_v) \geq (\frac{1}{2} - \varepsilon)m$, for $v \in S$. Let $Q = \partial_{\boldsymbol{\mu}} P$. We have, with C_1 given by (6.12):

Lemma 6.4.7. *Suppose that $\beta_1, \ldots, \beta_m$ are (L, M)-independent with*

$$m > \log(2r\,|S|)/(6\varepsilon^2) \quad \text{and} \quad L \geq (2C_1 + 5m)\,\varepsilon^{-2^{m-1}}, \quad M \geq 2\varepsilon^{-2^{m-1}}.$$

Then, for every sufficiently large D, there is a polynomial $Q \in K[x_1, \ldots, x_m]$, with partial degrees at most $d_j = \lfloor D/h(\beta_j) \rfloor$, such that:

(a) $\operatorname{ind}(Q; \mathbf{d}; \boldsymbol{\alpha}_v) \geq \left(\frac{1}{2} - 3\varepsilon\right)m$ *for $v \in S$;*

(b) $Q(\boldsymbol{\beta}) \neq 0$;

(c) $h(Q) \leq 4C_1D/L$.

Proof: We have already verified statement (b), and 6.3.2 (c) implies (a). In order to prove (c), it suffices to note (use (6.1) on page 156) that the differential operator ∂_{μ} increases coefficients by not more that $2^{d_1+\cdots+d_m} < 4^{d_1}$. Then use (6.13). □

6.4.8. *Step III: The upper bound.*

We begin by giving an upper bound for $\log|Q(\boldsymbol{\beta})|_v$, for each place v. As usual, we abbreviate $\log^+ t = \max(0, \log t)$ for $t \geq 0$, and we also define

$$\varepsilon_v := \begin{cases} [K_v : \mathbb{Q}_v]/[K : \mathbb{Q}] & \text{if } v \text{ is archimedean} \\ 0 & \text{if } v \text{ is non-archimedean.} \end{cases}$$

If $v \notin S$, we estimate $\log|Q(\boldsymbol{\beta})|_v$ trivially, namely

$$\log|Q(\boldsymbol{\beta})|_v \leq \log|Q|_v + \sum_{j=1}^{m}(\log^+|\beta_j|_v + \varepsilon_v\, o(1))\, d_j. \tag{6.14}$$

Here the term $o(1)$ tends to 0 as $d_j \to \infty$.

If instead $v \in S$, we expand Q in Taylor series with center $\boldsymbol{\alpha}_v$, obtaining

$$Q(\boldsymbol{\beta}) = \sum \partial_{\mathbf{j}} Q(\boldsymbol{\alpha}_v)(\beta_1 - \alpha_v)^{j_1} \cdots (\beta_m - \alpha_v)^{j_m},$$

and estimate each term in the Taylor expansion. Now note that by Lemma 6.4.7 (a) we have

$$\partial_{\mathbf{j}} Q(\boldsymbol{\alpha}_v) = 0 \quad \text{if} \quad \frac{j_1}{d_1} + \cdots + \frac{j_m}{d_m} < \left(\frac{1}{2} - 3\varepsilon\right) m; \tag{6.15}$$

and also a direct estimate yields

$$\log|\partial_{\mathbf{k}} Q(\boldsymbol{\alpha}_v)|_{v,K} \leq \log|Q|_v + \sum_{j=1}^{m} \log^+|\alpha_v|_{v,K}(d_j - k_j) + \varepsilon_v(\log 2 + o(1))d_j. \tag{6.16}$$

We have the easily verified inequality

$$\log|a - b|_{v,K} \leq -\log^+ \frac{1}{|a - b|_{v,K}} + \log^+|a|_{v,K} + \log^+|b|_{v,K} + \varepsilon_v \log 2$$

(the first term on the right-hand side suffices if $|a-b|_{v,K} \le 1$, while the remaining terms take care of the case $|a-b|_{v,K} > 1$), hence using (6.16) we find

$$
\begin{aligned}
\log & \left| \partial_{\mathbf{k}} Q(\boldsymbol{\alpha}_v) \prod_{j=1}^{m} (\beta_j - \alpha_v)^{k_j} \right|_{v,K} \\
& \le -\sum_{j=1}^{m} k_j \log^+ \frac{1}{|\beta_j - \alpha_v|_{v,K}} + \log |Q|_v \\
& \quad + \sum_{j=1}^{m} (\log^+ |\beta_j|_v + \log^+ |\alpha_v|_{v,K} + (\log 4 + o(1))\varepsilon_v)\, d_j .
\end{aligned}
$$

Now we can estimate $\log \left|Q(\boldsymbol{\beta})\right|_v$

$$
\begin{aligned}
\log |Q(\boldsymbol{\beta})|_v &= \log \left| \sum \partial_{\mathbf{k}} Q(\boldsymbol{\alpha}_v) \prod_{j=1}^{m} (\beta_j - \alpha_v)^{k_j} \right|_{v,K} \\
&\le \max_{\mathbf{k}} \log \left| \partial_{\mathbf{k}} Q(\boldsymbol{\alpha}_v) \prod_{j=1}^{m} (\beta_j - \alpha_v)^{k_j} \right|_{v,K} + \varepsilon_v \sum_{j=1}^{m} \log(d_j + 1) \\
&\le -\min{}' \left(\sum_{j=1}^{m} k_j \log^+ \frac{1}{|\beta_j - \alpha_v|_{v,K}} \right) + \log |Q|_v \\
&\quad + \sum_{j=1}^{m} (\log^+ |\beta_j|_v + \log^+ |\alpha_v|_{v,K} + (\log 4 + o(1))\varepsilon_v)\, d_j , \qquad (6.17)
\end{aligned}
$$

where $\min'$ means that the minimum is taken over $(k_1, \dots, k_m)$ not as in (6.15), that is with

$$
\frac{k_1}{d_1} + \cdots + \frac{k_m}{d_m} \ge \left(\frac{1}{2} - 3\varepsilon \right) m. \qquad (6.18)
$$

We put together (6.14) and (6.17), note that $\sum_v \varepsilon_v = 1$, and find

$$
\begin{aligned}
\sum_{v \in M_K} \log \left|Q(\boldsymbol{\beta})\right|_v &\le -\sum_{v \in S} \min{}' \left(\sum_{j=1}^{m} k_j \log^+ \frac{1}{|\beta_j - \alpha_v|_{v,K}} \right) \\
&\quad + h(Q) + \sum_{j=1}^{m} \left(h(\beta_j) + \sum_{v \in S} \log^+ |\alpha_v|_{v,K} + \log 4 + o(1) \right) d_j .
\end{aligned}
$$

Lemma 6.4.7 (c) provides an upper bound for $h(Q)$, and we have already observed that $\sum d_j \le 2d_1 \le 2D/L + o(D/L)$. Also $h(\beta_j) d_j \sim D$ and $h(\beta_j) \ge L$. Hence the last displayed inequality simplifies to

$$\sum_{v\in M_K} \log|Q(\boldsymbol{\beta})|_v \leq -\sum_{v\in S} \min{}' \left(\sum_{j=1}^m k_j \log^+ \frac{1}{|\beta_j - \alpha_v|_{v,K}}\right) + \left(m + \frac{C_2}{L}\right) D + o(D) \tag{6.19}$$

with for example $C_2 = 4C_1 + 2\log 4 + 2|S| \max_{v\in S} \log^+ |\alpha_v|_{v,K}$.

It remains to estimate the minimum in the last displayed inequality. It is here that we use the fact that the approximations β_j are of similar type. Let $\mathcal{C}(\boldsymbol{\lambda}; N)$ be the approximation class of the approximations β_j. Recall that

$$\Lambda(\beta_j) := \prod_{v\in S} \min(1, |\beta_j - \alpha_v|_{v,K}).$$

By the hypothesis (6.8) on page 163 and by (6.9) on page 164, we have

$$\lambda_v\, \kappa\, h(\beta_j) \leq \lambda_v \log \frac{1}{\Lambda(\beta_j)} \leq \log^+ \frac{1}{|\beta_j - \alpha_v|_{v,K}};$$

therefore, using $h(\beta_j)d_j \sim D$, we get

$$\begin{aligned}\sum_{v\in S} \min{}' \left(\sum_{j=1}^m k_j \log^+ \frac{1}{|\beta_j - \alpha_v|_{v,K}}\right) &\geq \sum_{v\in S} \min{}' \left(\sum_{j=1}^m \lambda_v\, \kappa\, h(\beta_j)\, k_j\right) \\ &= \kappa \left(\sum_{v\in S} \lambda_v\right) \min{}' \left(\sum_{j=1}^m h(\beta_j)\, d_j\, \frac{k_j}{d_j}\right) \\ &\sim D\kappa \left(\sum_{v\in S} \lambda_v\right) \min{}' \left(\sum_{j=1}^m \frac{k_j}{d_j}\right).\end{aligned}$$

Now (6.10) on page 164 gives $\sum \lambda_v \geq 1 - |S|/N$ and (6.18) gives

$$\min{}' \left(\sum k_j/d_j\right) \geq \left(\frac{1}{2} - 3\varepsilon\right) m.$$

We substitute in the last displayed inequality and find

$$\sum_{v\in S} \min{}' \left(\sum_{j=1}^m k_j \log^+ \frac{1}{|\beta_j - \alpha_v|_{v,K}}\right) \geq \kappa \left(1 - \frac{|S|}{N}\right) \left(\frac{1}{2} - 3\varepsilon\right) mD + o(D). \tag{6.20}$$

Finally, we substitute (6.20) into (6.19), obtaining the desired majorization

$$\sum_{v\in M_K} \log|Q(\boldsymbol{\beta})|_v \leq -\kappa \left(1 - \frac{|S|}{N}\right) \left(\frac{1}{2} - 3\varepsilon\right) mD + \left(m + \frac{C_2}{L}\right) D + o(D). \tag{6.21}$$

6.4.9. *Step IV: The lower bound.*

Since $Q(\boldsymbol{\beta}) \neq 0$ is in K, the lower bound for $Q(\boldsymbol{\beta})$ is given in a most elegant way by the product formula

$$\sum_{v \in M_K} \log |Q(\boldsymbol{\beta})|_v = 0 .$$

6.4.10. *Step V: Comparison of the upper and lower bounds.*

Comparison of the upper bound (6.21) and the product formula in 6.4.9 yields (after division by mD and letting $D \to \infty$)

$$-\kappa\left(1 - \frac{|S|}{N}\right)\left(\frac{1}{2} - 3\varepsilon\right) + 1 + \frac{C_2}{mL} \geq 0 ,$$

which (assuming $\frac{1}{2} > 3\,\varepsilon$) we may rewrite as

$$\kappa \leq \left(1 + \frac{C_2}{L}\right)\left(1 - \frac{|S|}{N}\right)^{-1}\left(\frac{1}{2} - 3\varepsilon\right)^{-1} . \tag{6.22}$$

The right-hand side of this inequality tends to 2 as $\varepsilon \to 0$, $N \to \infty$ and $L \to \infty$, contradicting $\kappa > 2$.

For the reader's convenience, we state in which order these various parameters need to be chosen. First, note that the constants C_1 and C_2 depend only on the given data $K, F, (\alpha_v)_{v \in S}$ in Theorem 6.4.1. Arguing by contradiction, we assume that there are infinitely many solutions β of (6.8) on page 163 for some $\kappa > 2$. We choose $\varepsilon > 0$ so small that $(\frac{1}{2} - 3\varepsilon)^{-1} < \kappa$. Then we fix m, L, M so that the assumptions of Lemma 6.4.7 are satisfied. Moreover, we may assume that L, N are so large that (6.22) above is not satisfied. There is one approximation class of size $1/N$ containing infinitely many solutions β of (6.8) on page 163, and in particular containing m (L, M)-independent solutions $\beta_1, \ldots, \beta_m$. Once all these data have been fixed the new parameter D is introduced, assumed very large, and the polynomial Q satisfying (a) to (c) of Lemma 6.4.7 is constructed. This leads to the asymptotic inequality (6.21) and, letting $D \to \infty$, to (6.22), which is the desired contradiction. □

6.5. Further results

We describe here some additional results which complement Roth's theorem.

6.5.1. As noted by Vojta, it is possible to give a refinement of Theorem 6.2.3 in which we allow the point $\boldsymbol{\alpha}$ to vary with β; this is referred to as **Roth's theorem with moving targets** (because the target $\boldsymbol{\alpha}$ of the approximation β is allowed to change with β). We have:

Theorem 6.5.2. *Let K be a number field and let S be a finite subset of M_K. Let F be a finite extension of K and for each $v \in S$, extend $|\ |_v$ to an absolute value $|\ |_{v,K}$ of F. Let $\kappa > 2$. Then we cannot have an infinite sequence of points $(\boldsymbol{\alpha}_j, \beta_j)$ such that:*

(a) $\quad \boldsymbol{\alpha}_j = (\alpha_{vj})_{v\in S} \in F^{|S|}, \qquad \beta_j \in K;$

(b) $\quad 1 + \sum_{v\in S} h(\alpha_{vj}) = o(h(\beta_j)) \quad$ *as* $j \to \infty$;

(c) $\quad \prod_{v\in S} \min(1, |\beta_j - \alpha_{vj}|_{v,K}) \leq H(\beta_j)^{-\kappa}.$

Proof: We replace $\boldsymbol{\alpha}$ by $\boldsymbol{\alpha}_j$ and define $\boldsymbol{\alpha}_v = (\alpha_{v1}, \ldots, \alpha_{vm})$ in the preceding arguments. Now (6.11) on page 166 will hold with the new upper bound

$$\left\{ \sum_{v\in S} \sum_{j=1}^{m} (h(\alpha_{vj}) + \log 2)/h(\beta_j) \right\} D + o(D).$$

By our assumption (b) and by going to a subsequence, we may assume that

$$\sum_{v\in S} \sum_{j=1}^{m} (h(\alpha_{vj}) + \log 2)/h(\beta_j) \leq m/L$$

and the proof goes through as before. □

6.5.3. Theorem 6.2.3 and Theorem 6.5.2 can be made quantitative in the sense that we can bound the number of solutions β in Theorem 6.2.3 and the length of the sequence $(\boldsymbol{\alpha}_j, \beta_j)$ in Theorem 6.5.2; in the case of Theorem 6.5.2, we must replace the $o(h(\beta_j))$ appearing in condition (b) by an explicit quantity $\delta(\kappa)h(\beta_j)$, where $\delta(\kappa) \to 0$ sufficiently fast as $\kappa \to 2$.

The proof of quantitative statements of this type utilizes in an essential way a **strong gap principle**, which provides exponential growth for the sequence of heights of solutions in a fixed approximation class. A simple statement is as follows:

Theorem 6.5.4. *Let K be a number field and S be a finite subset of M_K. Let F be a finite-dimensional extension of K and, for $v \in S$, let $\alpha_v \in F$. We extend $|\ |_v$ to an absolute value $|\ |_{v,K}$ of F. Let β, $\beta' \in K$ be different elements in a same approximation class of size $1/N$ as defined in 6.4.2, such that $\Lambda(\beta) \leq H(\beta)^{-\kappa}$, $\Lambda(\beta') \leq H(\beta')^{-\kappa}$ and $h(\beta') \geq h(\beta)$. Then*

$$h(\beta') \geq -\log 4 + \left(\left(1 - \frac{|S|}{N} \right) \kappa - 1 \right) h(\beta).$$

Remark 6.5.5. If N is so large that $c := \left(1 - \frac{|S|}{N}\right)\kappa - 1 > 1$, we obtain $h(\beta') \geq c\,h(\beta) - \log 4$. Thus the sequence of logarithmic heights of solutions grows at least in geometric progression, whence the name 'strong gap principle'.

Proof: For $v \in S$, we have $|\alpha_v - \beta'|_{v,K} < 1$ if and only if $|\alpha_v - \beta|_{v,K} < 1$ (since β, β' are in the same approximation class). Passing from S to $S' = \{v \in S \mid |\alpha_v - \beta|_{v,K} < 1\}$, the quantities in the theorem do not change and the lower bound gets even better. So we may assume that $|\alpha_v - \beta|_{v,K} < 1$ and $|\alpha_v - \beta'|_{v,K} < 1$ for $v \in S$. By (6.9) on page 164, we have for $v \in S$

$$\begin{aligned}\log|\beta - \beta'|_v &= \log|(\alpha_v - \beta') - (\alpha_v - \beta)|_v \\ &\leq \max(\log|\alpha_v - \beta'|_{v,K}, \log|\alpha_v - \beta|_{v,K}) + \varepsilon_v \log 2 \\ &\leq -\min\big(\kappa\lambda_v h(\beta'), \kappa\lambda_v h(\beta)\big) + \varepsilon_v \log 2 \\ &= -\kappa\lambda_v h(\beta) + \varepsilon_v \log 2.\end{aligned}$$

Now we sum over $v \in S$ and using (6.10) on page 164, we find

$$\sum_{v\in S} \log|\beta - \beta'|_v \leq -\left(1 - |S|N\right)\kappa h(\beta) + \log 2.$$

Finally, the fundamental inequality (1.8) on page 20 and Proposition 1.5.15 give (note that $\beta \neq \beta'$)

$$\sum_{v\in S} \log|\beta - \beta'|_v \geq -h(\beta - \beta') \geq -h(\beta) - h(\beta') - \log 2$$

and the result follows from the last two displayed inequalities. □

Lemma 6.5.6. *Let K be a number field, S a finite subset of M_K and F a finite extension of K. Let $\kappa > 2$ and let $\alpha_v \in F$ for $v \in S$. Let N be an integer so large that $c := \left(1 - \frac{|S|}{N}\right)\kappa - 1 > 1$ and let $X > \log 16/(c-1)$ and $A > 1$. Then the inequality*

$$\prod_{v\in S} \min(1, |\beta - \alpha_v|_{v,K}) \leq H(\beta)^{-\kappa}$$

has at most $\lceil \log A / \log \frac{c+1}{2} \rceil \binom{N+|S|}{|S|}$ solutions β with $h(\beta) \in (X, AX]$.

Proof: The interval $(X, AX]$ is contained in the union of $\lceil \log A / \log \frac{c+1}{2} \rceil$ intervals, each of type $((\frac{c+1}{2})^k X, (\frac{c+1}{2})^{k+1} X]$.

By the pigeon-hole principle, if there were more than $\lceil \log A / \log \frac{c+1}{2} \rceil \binom{N+|S|}{|S|}$ approximations β with height $h(\beta) \in (X, AX]$, we would find an interval of type $(Y, \frac{c+1}{2} Y]$ with $Y > (\log 16)/(c-1)$, and at least $\binom{N+|S|}{|S|}$ approximations β, such that $h(\beta) \in (Y, \frac{c+1}{2} Y]$. Now, using Lemma 6.4.3, another application of the pigeon-hole principle would give two elements β and β' with $h(\beta), h(\beta') \in (Y, cY - \log 4]$ and in a same approximation class $\mathcal{C}(\boldsymbol{\lambda}; N)$. This contradicts Remark 6.5.5. □

6.5.7. It is now clear how to obtain a bound for the number of solutions of the inequality in Theorem 6.4.1. Let $\varepsilon > 0$ be such that $(\frac{1}{2} - 3\varepsilon)^{-1} < \kappa$, let $m = \lceil \log(2r|S|)/(6\varepsilon^2) \rceil$, $L \geq m(C_1 + 5)\varepsilon^{-2^m - 1}$ and $M = 2\varepsilon^{-2^m - 1}$ (as required in Lemma 6.4.7). We choose L, N so large that we contradict (6.22) on page 170. Then we cannot find m solutions β_j $(j = 1, \ldots, m)$ that are (L, M)-independent and belong to a fixed approximation class $\mathcal{C}(\boldsymbol{\lambda}; N)$.

Consider all solutions $\beta \in K$ of

$$\prod_{v\in S} \min(1, |\beta - \alpha_v|_{v,K}) \leq H(\beta)^{-\kappa}$$

in a fixed approximation class $\mathcal{C}(\boldsymbol{\lambda}; N)$ and group them in subsets $\mathcal{N}_0, \mathcal{N}_1, \ldots, \mathcal{N}_j, \ldots$ as follows.

The first subset $\mathcal{N}_0$ contains all such solutions with $h(\beta) \leq L$.

The other sets $\mathcal{N}_j$ are defined inductively: Suppose $\mathcal{N}_l$ is given for $l \leq j$; then $\mathcal{N}_{j+1}$ is obtained by first taking a solution β_{j+1} with smallest height and not in any $\mathcal{N}_l$ with $l \leq j$, and then defining $\mathcal{N}_{j+1}$ as the set of solutions β with $h(\beta_{j+1}) \leq h(\beta) < Mh(\beta_{j+1})$. If such a β_{j+1} does not exist, then $\mathcal{N}_l$ is empty for $l > j$.

The sequence $\beta_1, \beta_2, \ldots$ of solutions so constructed is (L, M)-independent and belongs to a fixed approximation class $\mathcal{C}(\boldsymbol{\lambda}; N)$. Since by hypothesis (6.22) on page 170 is not verified, and since (6.22) on page 170 was obtained on the assumption that we had m (L, M)-independent solutions with m, L, M satisfying the hypothesis of Lemma 6.4.7, we see that this sequence has less than m elements. It follows that all solutions in $\mathcal{C}(\boldsymbol{\lambda}; N)$ lie in the union of the sets $\mathcal{N}_i$, $i = 0, \ldots, m'$ for some $m' < m$. By the proof of Lemma 6.5.6, we see that each $\mathcal{N}_j$ contains at most $\lceil \log M / \log \frac{c+1}{2} \rceil$ elements. Moreover, Lemma 6.4.3 shows that the number of approximation classes does not exceed $\binom{N+|S|}{|S|}$.

Now we subdivide solutions into:

(a) **very small solutions**, namely those with

$$h(\beta) \leq \log 16/(c-1) \text{ with } c = \left(1 - \frac{|S|}{N}\right)\kappa - 1 > 1;$$

(b) **small solutions**, those with $\log 16/(c-1) < h(\beta) \leq L$;

(c) **large solutions**, those with $h(\beta) > L$.

A bound for the number of very small solutions can be obtained by Northcott's theorem in 1.6.8.

A bound for the number of small solutions can be obtained by appealing to Lemma 6.5.6 for the interval $(\log 16/(c-1), L]$, obtaining

$$\lceil \log L / \log \frac{c+1}{2} \rceil \binom{N+|S|}{|S|}.$$

Note also that in this case we automatically have a bound for the height.

We cannot give a bound for the height of large solutions, but the above considerations show that their number does not exceed

$$m \lceil \log M / \log \frac{c+1}{2} \rceil \binom{N+|S|}{|S|}.$$

In order to get an idea of the size of this bound, we set $\delta = \min(\kappa - 2, 1)$. Then the above quantity is majorized by

$$(2r|S|)^{c_1 \delta^{-2}} (c_2/\delta)^{|S|} \tag{6.23}$$

for certain absolute constants c_1, c_2. For $|S| = 1$, this is due to H. Davenport and K.F. Roth **[79]** with the bound $\exp(70r^2\delta^{-2})$.

6.5.8. The bound for the number of large solutions obtained by the above method is rather large as $\delta \to 0$. It is possible to improve drastically this bound by replacing Roth's lemma by more sophisticated results, notably Dyson's lemma in several variables or Faltings's product theorem (see Theorem 7.6.4). For example, an improved bound obtained using Dyson's lemma is

$$(\log(2r|S|/\delta))^{c_3}(c_4/\delta)^{|S|+4}$$

for some absolute (and not very large) constants c_3, c_4. This is superior to (6.23) above.

6.5.9. No refinement of Roth's theorem is known replacing $\kappa > 2$ by $2 + f(h(\beta))$, where $f(t) \to 0$ as $t \to \infty$. Heuristic arguments based on the analogy between diophantine approximation and Nevanlinna theory suggest that $f(t) = (1 + o(1))(\log t)/t$ should be admissible for this purpose (see Remark 13.2.25 and S. Lang and H. Trotter [**175**]).

On the other hand, something can still be said if the function $f(t)$ tends to 0 sufficiently slowly as $t \to \infty$. We have the **Cugiani–Mahler theorem** which we state without proof (see E Bombieri and A.J. van der Poorten [**36**] for details):

Theorem 6.5.10. *Let K be a number field and let S be a finite subset of M_K. For each $v \in S$ let $\alpha_v \in F \cap K_v$, where F is a finite extension of K of degree r. Let also*

$$f(t) = 6\sqrt{\log r}\,\sqrt[4]{\frac{\log\log(t + \log 4)}{\log(t + \log 4)}}.$$

Let (β_j) be the sequence of solutions in K of

$$\prod_{v \in S} \min(1, |\beta_j - \alpha_v|_{v,K}) < (4H(\beta_j))^{-2-f(h(\beta_j))},$$

ordered by strictly increasing height. Then either the sequence (β_j) is finite or

$$\limsup_{j\to\infty} \frac{h(\beta_{j+1})}{h(\beta_j)} = \infty.$$

6.6. Bibliographical notes

For further information about diophantine approximation in positive characteristic, see M. Kim, D.S. Thakur, and J.F. Voloch [**161**].

The problem of finding a refined Roth's theorem as in 6.5.9 is still open. The first partial result was obtained in 1958 by Cugiani for approximations in $\mathbb{Q}$, with

$$f(t) = C(r)/\sqrt{\log\log t}$$

for a suitable function $C(r)$ of the degree r of the algebraic number α (he also gives a value for $C(r)$). A little later M. Cugiani [**74**] extended this result to approximations in an arbitrary number field K. After the appearance of Cugiani's paper for the case of $K = \mathbb{Q}$, K. Mahler [**186**], Appendix B, independently extended the earlier result of Cugiani to the case of a general number field K. Nowadays, this is referred to as the Cugiani–Mahler theorem.

More recently, alternative ways to Roth's method for dealing with the difficult point of the non-vanishing of the polynomial at the special point have been obtained by H. Esnault and E. Viehweg [**102**] in their m-dimensional version of Dyson's lemma, and by G. Faltings [**114**] with his product theorem (see Theorem 7.6.4) and its quantitative versions due independently to J.-H. Evertse [**106**], R.G. Ferretti [**119**], and G. Rémond [**241**]. We have not considered in this chapter these very important improvements of Roth's lemma and we have limited ourselves to a rather elementary and explicit treatment along Roth's original line.

Roth's theorem with moving targets is proved in P. Vojta [**315**] and later M. Ru and P. Vojta [**250**] extended the theorem to a version of Schmidt's subspace theorem with moving targets. Vojta's argument, inspired by N. Steinmetz's paper on Nevanlinna's second theorem with moving targets [**290**] (see also the Bibliographical Notes in Chapter 13), obtains the result as a consequence of Schmidt's subspace theorem (see Theorem 7.2.2). The simple direct proof outlined here appears to be new.

The improved bound in 6.5.8 for the case $|S| = 1$, with explicit values of c_3, c_4, see [**36**], Th.2 and the Note at the end of the proof. A similar result was independently obtained by H. Luckhardt [**184**] combining Dyson's lemma with methods of mathematical logic. This was extended to several places by R. Gross [**132**].

The improved function $f(t)$ in Theorem 6.5.10 given here is obtained in [**36**], Th.3. The improvements mentioned in 6.5.8 and 6.5.10 stem from replacing Roth's lemma with the deeper Dyson's lemma in several variables of Esnault and Viehweg, *loc. cit.*; the versions of Faltings's product theorem mentioned above would suffice as well. Except for this point, the structure of the proofs follows rather closely the arguments in [**79**] and [**74**].

7 THE SUBSPACE THEOREM

7.1. Introduction

This chapter deals with Schmidt's far-reaching extension of Roth's theorem to systems of inequalities in linear forms. This is not a routine generalization; entirely new difficulties appear in the course of the proof, which Schmidt resolved by introducing new ideas from Minkowski's geometry of numbers.

In the case of systems of inequalities, it is possible to have infinitely many solutions. However, even then a finiteness theorem still holds, in the sense that solutions are contained in finitely many proper linear subspaces of the ambient space. This paves the way for applying induction arguments.

As is the case for Thue's and Roth's theorems, again Schmidt's subspace theorem is ineffective in the sense that no bound can be placed *a priori* on the height of the finitely many linear spaces which contain the solutions. At any rate, it remains a very flexible tool with wide applicability in many questions; the reader will find some unusual applications of the subspace theorem in this chapter.

It is also possible, as for Roth's theorem, to give an effective bound for the number of linear spaces containing the solutions. This requires rather sophisticated methods beyond the scope of this book and will not be done here.

An extension of Schmidt's theorem with a formulation allowing a finite set of places, entirely analogous to Ridout's and Lang's generalizations of Roth's theorem, was later obtained by Schlickewei. This is quite important in applications.

Section 7.2 contains several equivalent formulations of the subspace theorem. Then in Section 7.3 we consider applications of the subspace theorem related to diophantine approximation and we give an alternative proof of Siegel's theorem on integral points, due to Corvaja and Zannier. In Section 7.4 we apply the subspace theorem to the generalized unit equation. As a consequence we give Schmidt's solution of the norm-form equation, Laurent's solution of the Lang conjecture for tori and a nice result of Corvaja and Zannier from elementary number theory.

The proof of the subspace theorem, in the general form obtained by Schlickewei, is given in full in Section 7.5. It is quite involved, first developing the "Roth machinery" and then using new ideas of Schmidt, with further simplifications by Evertse.

In the last two sections, we describe, without proofs, further important results. We begin with Faltings's product theorem in Section 7.6, which is a very important geometric alternative to Roth's lemma. In Section 7.7 we describe a deep extension of the subspace theorem obtained by Faltings and Wüstholz, dealing with inequalities determined by forms of arbitrary degree; the finiteness statement becomes that the solutions are contained in a proper algebraic subvariety of the ambient space. Their method, quite different from Schmidt's, uses deep tools from algebraic geometry such as the theory of semi-stable bundles, Faltings's product theorem and an induction argument. Somewhat surprisingly, it was later shown by Evertse and Ferretti that the Faltings–Wüstholz theorem can also be obtained directly by an ingenious application of the subspace theorem.

In this chapter, the reader is assumed to be familiar with Chapter 6. For Section 7.4 it is also helpful to know the basic results of Chapters 3 and 5.

7.2. The subspace theorem

We begin with Schlickewei's and Evertse's formulation [**108**] of **Schmidt's subspace theorem** over number fields, in a projective setting.

7.2.1. We follow the notions introduced in Chapter 1 and used in Chapter 6: For a place v on a number field K, we denote by $|\ |_v$ the normalized absolute value (as in (1.6) on page 11) to get the product formula. As usual, we denote again by $|\ |_v$ its extension to the completion K_v. The multiplicative height $H(\alpha) = e^{h(\alpha)}$ for an algebraic number is as defined in 1.5.7. The multiplicative projective height $H(\mathbf{x})$ on algebraic points of $\mathbb{P}^n_{\overline{\mathbb{Q}}}$ is defined as in 1.5.4. For $\mathbf{x} \in K^{n+1}$, let $|\mathbf{x}|_v := \max_i |x_i|_v$.

Theorem 7.2.2. *Let K be a number field, $S \subset M_K$ a finite set of places, $n \in \mathbb{N}$ and $\varepsilon > 0$. For every $v \in S$, let $\{L_{v0}, \ldots, L_{vn}\}$ be a linearly independent set of linear forms in the variables $X_0, \ldots, X_n$ with K-algebraic coefficients in K_v. Then there are finitely many hyperplanes $T_1, \ldots, T_h$ of $\mathbb{P}^n_K$ such that the set of solutions $\mathbf{x} \in \mathbb{P}^n_K(K)$ of*

$$\prod_{v \in S} \prod_{i=0}^{n} \frac{|L_{vi}(\mathbf{x})|_v}{|\mathbf{x}|_v} < H(\mathbf{x})^{-n-1-\varepsilon}$$

is contained in $T_1 \cup \cdots \cup T_h$.

Remark 7.2.3. The condition that L_{vi} has K-algebraic coefficients in K_v, rather than in some finite extension of K_v, is not restrictive. We may even reduce everything to the case where the coefficients are in K, by the following argument. As in Theorem 6.4.1, we may replace the completions K_v by a field of definition F

for all forms L_{vi} using suitable extensions of $| \ |_v$ to the number field F. We may assume that F is a finite Galois extension of K. Then use $L_{ui} = L_{vi}$ for the place $u \in M_F$ corresponding to the chosen extension of v to F. For the other places $w|v$ ($w \in M_F$), there is $\sigma \in \mathrm{Gal}(F/K)$ with $w = u \circ \sigma^{-1}$ (Corollary 1.3.5) and we set $L_{wi} := \sigma(L_{vi})$. The subspace theorem for F and this set of linear forms implies the subspace theorem in the relative setting, because the left-hand sides are the same for $\mathbf{x} \in K^{n+1}$.

Note that the field of definition of the subspaces can always be assumed to be K, since we may replace T_j by the linear span of all solutions $\mathbf{x} \in \mathbb{P}^n_{\overline{K}}(K)$ contained in T_j.

7.2.4. A very important refinement of the theorem above is a quantitative version in which we also control the number of hyperplanes needed to contain all solutions. The first result of this type was obtained by Schmidt and the best bounds are in J.-H. Evertse [**108**] and J.-H. Evertse and H.P. Schlickewei [**111**]. These bounds are quite sharp and uniform with respect to the number field K, which is essential in many applications, an example is Theorem 7.4.1. Here we will limit ourselves to the qualitative statement above, referring to the original papers [**108**], [**111**], for statements and proofs of the quantitative versions.

The proof will be given in Section 7.5. The following affine version of the subspace theorem is worth noting. We denote by $O_{S,K}$ the ring of S-integers (see 1.5.10).

Corollary 7.2.5. *Let K, S, L_{vi} be as before with S containing all archimedean places and let $\varepsilon > 0$. Then there are finitely many linear subspaces $T_1, \ldots, T_h$ of K^{n+1} such that the set of S-integral solutions $\mathbf{x} \in O_{S,K}^{n+1} \setminus \{\mathbf{0}\}$ of*

$$\prod_{v \in S} \prod_{i=0}^{n} |L_{vi}(\mathbf{x})|_v < H(\mathbf{x})^{-\varepsilon}$$

is contained in $T_1 \cup \cdots \cup T_h$.

Proof: Since $\mathbf{x} \in O_{S,K}^{n+1} \setminus \{\mathbf{0}\}$, we have

$$H(\mathbf{x}) = \prod_{v \in M_K} |\mathbf{x}|_v \leq \prod_{v \in S} |\mathbf{x}|_v.$$

Hence we see that every S-integral solution in Corollary 7.2.5 induces a projective solution in Theorem 7.2.2. We conclude that Corollary 7.2.5 follows from Theorem 7.2.2. □

Theorem 7.2.6. *The affine subspace theorem as in Corollary 7.2.5 implies the subspace theorem as in Theorem 7.2.2.*

Proof: Since the subspace theorem for a set of places S implies *a fortiori* the subspace theorem for a subset $S' \subset S$, we may assume that S is so large that it contains all the archimedean places and also that $O_{S,K}$ is a principal ideal domain (see Proposition 5.3.6). Now let $\mathbf{x} \in K^{n+1} \setminus \{\mathbf{0}\}$ verify the projective diophantine inequality in the statement of Theorem 7.2.2 and let $\mathfrak{X}$ be the fractional ideal $\mathfrak{X} = \sum x_i O_{S,K}$ of the ring $O_{S,K}$. Then, if $v \notin S$, we have

$$|\mathbf{x}|_v = \max_i |x_i|_v = \max_{\xi \in \mathfrak{X}} |\xi|_v. \tag{7.1}$$

On the other hand, we have $\mathfrak{X} = (\delta)$ for some $\delta \in K^\times$. Therefore, from equation (7.1) we infer that

$$\max_i |x_i|_v = |\delta|_v$$

for every place $v \notin S$. Hence, setting $\mathbf{x}' := \delta^{-1}\mathbf{x}$, we have

$$|\mathbf{x}'|_v = \max_i |x_i'|_v = 1$$

for $v \notin S$, whence

$$\prod_{v \in S} |\mathbf{x}'|_v = H(\mathbf{x}') = H(\mathbf{x})$$

and also $\mathbf{x}' \in O_{S,K}^{n+1} \setminus \{\mathbf{0}\}$. Now $\mathbf{x}'$ is an affine solution of the inequality in Corollary 7.2.5. This shows the equivalence of the projective and affine diophantine inequalities in Theorem 7.2.2 and Corollary 7.2.5, completing the proof. □

Example 7.2.7. Roth's theorem in 6.2.3 is the special case $n = 1$ of the subspace theorem. In the form given in this book, it states that, given K and S as above and given K-algebraic $\alpha_v \in K_v$ for $v \in S$, the inequality

$$\prod_{v \in S} \min(1, |\beta - \alpha_v|_v) < H(\beta)^{-2-\varepsilon} \tag{7.2}$$

has only finitely many solutions $\beta \in K$. Note that, by splitting the solutions of (7.2) into finitely many subsets according to the places $v \in S$ for which $|\beta - \alpha_v|_v < 1$, we see immediately that (7.2) is equivalent to the statement that the solutions to

$$\prod_{v \in S^*} |\beta - \alpha_v|_v < H(\beta)^{-2-\varepsilon}, \quad |\beta - \alpha_v|_v < 1 \quad (v \in S^*) \tag{7.3}$$

form a finite set, for S^* any subset of S.

Let $L_{v0} = X_0$, $L_{v1} = X_1 - \alpha_v X_0$, and take $\mathbf{x} = (1 : \beta) \in \mathbb{P}^1(K)$. Then

$$\frac{|L_{v0}(\mathbf{x})|_v}{|\mathbf{x}|_v} \frac{|L_{v1}(\mathbf{x})|_v}{|\mathbf{x}|_v} = |\beta - \alpha_v|_v \max(1, |\beta|_v)^{-2}.$$

Now if $|\beta - \alpha_v|_v < 1$ we have $|\beta|_v < 1 + |\alpha_v|_v$ and $\max(1, |\beta|_v)$ is bounded above independently of β. Therefore, if (7.3) holds, we also have

$$\prod_{v \in S^*} \frac{|L_{v0}(\mathbf{x})|_v}{|\mathbf{x}|_v} \frac{|L_{v1}(\mathbf{x})|_v}{|\mathbf{x}|_v} < CH(\mathbf{x})^{-2-\varepsilon}$$

for some constant C. Now by Northcott's theorem in 1.6.8, we have $C < H(\mathbf{x})^{\varepsilon/2} = H(\beta)^{\varepsilon/2}$ except for finitely many $\beta \in K$. Thus we may apply the subspace theorem with $n = 1$ and $\varepsilon/2$ in place of ε and conclude that the solutions β of (7.3) form a finite set. Roth's theorem follows.

Next we give a stronger formulation of the subspace theorem due to P. Vojta [**307**] and which is of importance in some applications.

Definition 7.2.8. Let F be a field. A set $\{L_1, \ldots, L_m\}$ of linear forms in the ring $F[X_0, \ldots, X_n]$ is said to be in ***general position*** if any subset of cardinality not exceeding $n+1$ is linearly independent over F.

Theorem 7.2.9. *Let K be a number field and $S \subset M_K$ be a finite subset of places of K. For every $v \in S$, let $\{L_{v0}, \ldots, L_{vm_v}\}$ be a set of linear forms in $K_v[X_0, \ldots, X_n]$ in general position and with K-algebraic coefficients. Let $\varepsilon > 0$. Then there are finitely many hyperplanes $T_1, \ldots, T_h$ of $\mathbb{P}^n_K$ such that the set of projective solutions $\mathbf{x} \in \mathbb{P}^n_K(K)$ of*

$$\prod_{v \in S} \prod_{i=0}^{m_v} \frac{|L_{vi}(\mathbf{x})|_v}{|\mathbf{x}|_v} < H(\mathbf{x})^{-n-1-\varepsilon}$$

is contained in $T_1 \cup \cdots \cup T_h$.

Proof: We note first that we may assume $m_v \geq n$. This is clear by extending the set of linear forms $L_{0v}, \ldots, L_{m_v v}$ with suitable standard coordinates to a basis of the space of linear forms in case of $m_v < n$. Next, we reduce to the case $m_v = n$. By partitioning the set of solutions $\mathbf{x}$ into finitely many classes, we may assume after a permutation of the forms L_{vi} that

$$|L_{v0}(\mathbf{x})|_v \leq |L_{v1}(\mathbf{x})|_v \leq \cdots \leq |L_{vm_v}(\mathbf{x})|_v.$$

We keep only the forms L_{vi} with $i \leq n$ and remove the forms with $i > n$, obtaining a new set $\widetilde{L}_{vi}$ of linear forms with $\widetilde{L}_{vi} = L_{vi}$ for $i \leq n$. The forms L_{vi} with $i \leq n$ form a basis of the space of all linear forms, hence

$$|\mathbf{x}|_v \ll \max_{i \leq n} |L_{vi}(\mathbf{x})|_v = |L_{vn}(\mathbf{x})|_v \leq |L_{vi}(\mathbf{x})|_v$$

for $i > n$. Therefore, on each subset of solutions we have

$$\prod_{v \in S} \prod_{i=0}^{n} \frac{|\widetilde{L}_{vi}(\mathbf{x})|_v}{|\mathbf{x}|_v} \leq C \prod_{v \in S} \prod_{i=0}^{m_v} \frac{|L_{vi}(\mathbf{x})|_v}{|\mathbf{x}|_v} < C\,H(\mathbf{x})^{-n-1-\varepsilon}$$

for some constant C independent of $\mathbf{x}$. As in Example 7.2.7, the constant C can be removed by passing to a smaller ε, hence the result we want follows from the subspace theorem in 7.2.2 applied to the set of forms $\{\widetilde{L}_{v0}, \ldots, \widetilde{L}_{vn}\}$, $v \in S$. □

7.3. Applications

There are several consequences of the subspace theorem in the theory of diophantine approximation of algebraic numbers, and we examine here a few of them. At the end of this section, we present a proof of Siegel's theorem on integral points relying on the subspace theorem. For this, the reader should be familiar with the theory of algebraic curves as provided by Appendix A.13.

7.3.1. The first important application of the subspace theorem is to show that algebraic numbers behave, to a great extent, like random numbers as far as diophantine approximation is concerned.

A well-known theorem of Dirichlet states that, if $1, \alpha_1, \ldots, \alpha_n$ are real numbers, then for every positive integer N there is a point $\mathbf{x} \in \mathbb{Z}^{n+1} \setminus \{\mathbf{0}\}$ with $\max |x_i| \leq N$, $i = 1, \ldots, n$, such that

$$|x_0 + \alpha_1 x_1 + \cdots + \alpha_n x_n| \leq N^{-n}.$$

The easy proof is obtained applying the pigeon-hole principle to

$$\{\alpha_1 x_1 + \cdots + \alpha_n x_n \pmod 1 \mid x_i = 0, \ldots, N\},$$

or by geometry of numbers applying Minkowski's first theorem in C.2.19 to the symmetric convex body of volume 2^{n+1} given by

$$|X_0 + \alpha_1 X_1 + \cdots + \alpha_n X_n| \leq N^{-n}, \quad |X_i| \leq N,\ i = 1, \ldots, n.$$

The following result, which is equivalent to Roth's original theorem if $n = 1$, shows that this result is essentially sharp for linear forms with real algebraic coefficients.

Theorem 7.3.2. *Let $\alpha_0, \ldots, \alpha_n$ be algebraic numbers. Then for every $\varepsilon > 0$ the inequality*

$$0 < |\alpha_0 x_0 + \cdots + \alpha_n x_n| \leq H(\mathbf{x})^{-n-\varepsilon} \tag{7.4}$$

has only finitely many solutions $\mathbf{x} \in \mathbb{Z}^{n+1}$.

Remark 7.3.3. The non-vanishing of the linear combination of the algebraic numbers $\alpha_0, \ldots, \alpha_n$ is essential here. The statement of the theorem in this form makes it applicable without assuming a linear independence condition, which is useful in certain cases.

Remark 7.3.4. The preceding theorem can be easily extended to systems of homogeneous linear diophantine inequalities, the simplest example being

$$|x_0| \leq N, \quad |\alpha_i x_0 - x_i| \leq N^{-1/n-\varepsilon} (i = 1, \ldots, n).$$

The solubility of this for real α_i, $\varepsilon = 0$, and arbitrary N is a classical result about simultaneous diophantine approximation, easily proved by Minkowski's first theorem. For the general case, see W.M. Schmidt [**268**], Ch. VI, §2.

Proof of Theorem 7.3.2: The proof is by induction on n. If $n = 0$, there is nothing to prove. We may assume that $\alpha_i \neq 0$ for every i. We apply the subspace theorem in 7.2.2 and Remark 7.2.3 with $K = \mathbb{Q}$, $v = \infty$, $L_{0v} = \alpha_0 X_0 + \cdots + \alpha_n X_n$, $L_{vi} = X_i$, $i = 1, \ldots, n$, obtaining that the set $\mathcal{X}$ of solutions is contained in finitely many rational linear subspaces $T_i \subset \mathbb{P}^n_{\mathbb{Q}}$. Let T be one of these subspaces. Then it provides a linear relation

$$\sum_{j=0}^{n} A_j X_j = 0,$$

which we may assume to hold on the set $\mathcal{X}$ (by partitioning). Without loss of generality, we may suppose $A_n \neq 0$. It follows that

$$0 < |\alpha_0 x_0 + \cdots + \alpha_n x_n| = |\beta_0 x_0 + \cdots + \beta_{n-1} x_{n-1}|$$

with $\beta_i = \alpha_i - \alpha_n A_i / A_n$, and *a fortiori*

$$0 < |\beta_0 x_0 + \cdots + \beta_{n-1} x_{n-1}| \leq H(\mathbf{x})^{-n-\varepsilon} \leq H(\mathbf{x}')^{-n+1-\varepsilon}$$

with $\mathbf{x}' = (x_0, \ldots, x_{n-1})$. Now the induction hypothesis applies and we obtain finitely many possibilities for $\mathbf{x}'$. Moreover, since $\alpha_n \neq 0$, given $\mathbf{x}'$ there are only finitely many possibilities for x_n, and there are only finitely many possibilities for the subspace T. Hence the set $\mathcal{X}$ is finite. □

The following corollary is due to Schmidt, [**268**], Ch.VIII, Th.9A.

Corollary 7.3.5. *Let $\alpha \in \mathbb{C}$ be algebraic, let d be a positive integer, and let $\varepsilon > 0$. Then there are only finitely many complex algebraic numbers ξ of degree at most d such that*

$$|\alpha - \xi| \leq H(f_\xi)^{-d-1-\varepsilon},$$

where $H(f_\xi)$ is the height of the minimal polynomial of ξ.

Proof: We may assume that ξ is not a conjugate of α. Let

$$f_\xi(X) = x_d X^d + \cdots + x_0$$

be the minimal polynomial of ξ over $\mathbb{Z}$ and with coprime coefficients. Then $f_\xi(\alpha) \neq 0$ and using the mean-value theorem we get

$$\begin{aligned} 0 < |x_0 + x_1\alpha + \cdots + x_d\alpha^d| &= |f_\xi(\alpha)| \\ &\ll_\alpha |\alpha - \xi|\, H(f_\xi) \\ &\leq H(f_\xi)^{-d-\varepsilon} = H(\mathbf{x})^{-d-\varepsilon}. \end{aligned}$$

By the preceding theorem, there are only finitely many solutions $\mathbf{x}$ to this inequality (by a familiar argument already used at the end of Example 7.2.7, the constant involved in the $\ll_\alpha$ symbol is irrelevant here). □

Remark 7.3.6. Note that $H(f_\xi) \gg\ll H(\xi)^{\deg(\xi)}$, which follows from Proposition 1.6.6 and Lemma 1.6.7.

7.3.7. It is an interesting open problem to find the best exponent κ_d for which the inequality

$$|\alpha - \xi| \leq cH(f_\xi)^{-\kappa_d+\varepsilon}$$

has infinitely many real algebraic solutions ξ of degree at most d, for every fixed $\varepsilon > 0$ and every real α not an algebraic number of degree at most d, for some constant c depending on α and ε. If $d = 1$ and α is irrational, Dirichlet's theorem shows that $\kappa_1 = 2$ (even with $\varepsilon = 0$). If $d = 2$ and α is not rational or a quadratic irrational, a difficult theorem of H. Davenport and W.M. Schmidt [**80**] shows that $\kappa_2 = 3$ (even with $\varepsilon = 0$). However, their method breaks down if $d \geq 3$ and it remains completely unclear what the correct result should be.

In this connexion, it is instructive to consider the not unrelated problem of approximating real numbers by algebraic integers of degree at most d. The first non-trivial case is $d = 2$, where we can easily show (see H. Davenport and W.M. Schmidt [**81**]) that the inequality

$$|\alpha - \xi| \leq cH(f_\xi)^{-2}$$

has infinitely many solutions in algebraic integers ξ of degree at most 2 for some constant c depending on α, provided α is irrational. The next case $d = 3$ turned out to be quite interesting. Davenport and Schmidt proved that, if α is not algebraic of degree at most 2, then the inequality

$$|\alpha - \xi| \ll_\alpha H(f_\xi)^{-(3+\sqrt{5})/2}$$

has infinitely many solutions with ξ an algebraic integer of degree at most 3. The exponent $(3 + \sqrt{5})/2$ looked somewhat strange and the authors commented, "We have no reason to think that the exponents in these theorems are best possible."

Thus it was a great surprise when D. Roy [**246**] constructed a real transcendental number α and a constant $c > 0$ such that

$$|\alpha - \xi| \geq cH(f_\xi)^{-(3+\sqrt{5})/2}$$

for every algebraic integer ξ of degree at most 3.

7.3.8. We conclude this section with an application of Corollary 7.2.5, which illustrates its power. It is an alternative direct proof, due to P. Corvaja and U. Zannier [**72**], of a famous theorem of Siegel on integral points on curves. The standard proof uses diophantine approximation (see 14.3.5) on the Jacobian and Roth's theorem (see [**277**], §7.3, for details).

Let C be a geometrically irreducible affine curve over a number field K and let S be a finite set of places containing the archimedean places. We assume that C is given as a closed subvariety of $\mathbb{A}^n_K$. An S**-integral point** of C is a point of C with $O_{K,S}$-integral coordinates. This notion depends on the embedding of C into affine space.

Let $\pi : \widetilde{C}_{\text{aff}} \to C$ be the normalization of C and we extend the affine curve $\widetilde{C}_{\text{aff}}$ to a smooth projective curve $\widetilde{C}$, which is unique up to isomorphism (see A.13.2, A.13.3). The points in $\widetilde{C} \setminus \widetilde{C}_{\text{aff}}$ are called the **points of C at ∞**.

Now we are ready to state **Siegel's theorem on integral points.**

Theorem 7.3.9. *If C has at least three distinct points at ∞, then C has only finitely many S-integral points.*

Remark 7.3.10. The usual version of Siegel's theorem is stronger, requiring the condition of at least three distinct point at ∞ only if $\widetilde{C}$ has genus 0. We briefly sketch how it may be deduced from Theorem 7.3.9.

We may assume C smooth, as we will show at the beginning of the proof of Theorem 7.3.9. If $\widetilde{C}$ has positive genus g, we may take, after replacing K by a larger number field, an unramified covering of $\widetilde{C}$ to obtain a new curve $\widetilde{C}'$ with at least three distinct points at ∞ with C' the inverse image of C. In order to show existence, we note that the first homology group $H_1(\widetilde{C}_{\mathrm{an}}, \mathbb{Z})$ is a free abelian group of rank $2g$. By standard techniques from the theory of covering spaces (see Section 12.3), we may easily construct a normal subgroup of the fundamental group of finite index ≥ 3 and hence a finite unramified covering of $\widetilde{C}_{\mathrm{an}}$ of degree ≥ 3. By Theorem 12.3.12, the covering is defined over a finite extension of K and leads to our desired unramified covering $\pi : \widetilde{C}' \to \widetilde{C}$ of degree ≥ 3; hence C' has at least 3 points at ∞.

Moreover, the Chevalley–Weil theorem as in Theorem 10.3.11 shows that rational points on C lift to rational points over a finite extension K' of K. By enlarging S, we may assume that π extends to an unramified finite morphism over $O_{S,K}$ and the valuative criterion of properness ([**148**], Th. II.4.7) shows that S-integral points on C lift to S'-integral points on C' for S' the set of places of K' lying over S. Thus we can apply the above theorem to C' to deal with arbitrary curves of positive genus.

Proof of Theorem 7.3.9: To begin with, we show that, by increasing the set S, we may assume that C is normal and therefore smooth.

Let R be the ring of regular functions on C. By a base change to a larger number field, we may assume that the K-rational points of C lift to K-rational points of the normalization (using birationality from A.13.2). The S-integral points on C are those at which all coordinate functions $x_1, \ldots, x_n$ take values in $O_{S,K}$. Now let $f \in K(C)$ be integral over R. Then f satisfies some equation

$$f^N + \sum_{j=1}^{N} p_j(\mathbf{x}) f^{N-j} = 0,$$

where $p_j(\mathbf{X}) \in K[\mathbf{X}]$. By enlarging S, we may assume that $p_j(\mathbf{X}) \in O_{S,K}[\mathbf{X}]$. Since $O_{S,K}$ is integrally closed in K, we see that f continues to take values in $O_{S,K}$ at the S-integral points of C. Thus adding f to the coordinate ring R preserves S-integrality (possibly by enlarging S). Therefore, by enlarging S we may replace R by its integral closure, which is what we needed to prove.

We denote by $\widetilde{C}$ the associated smooth projective curve. Clearly, we may assume now that $C \subset \widetilde{C}$. Let $Q_1, \ldots, Q_r$ be the distinct points at ∞ on C, which we may assume to be K-rational by enlarging K again. Let N be a large integer to be chosen later and let V be the space of rational functions φ in $K(C)^\times$ such that $\mathrm{div}(\varphi) \geq -N([Q_1] + \cdots + [Q_r])$, that is with poles of order at most N at the points Q_i. We include also $\varphi = 0$ in V. By the Riemann–Roch theorem in A.13.5, V is a K-vector space of dimension

$$d = \dim(V) \geq Nr + 1 - g, \tag{7.5}$$

where g is the genus of $\widetilde{C}$. Let $\boldsymbol{\varphi} = \{\varphi_1, \dots, \varphi_d\}$ be a basis of V. Since φ_j is regular on C we may assume, by enlarging S if needed, that $\varphi_j(P) \in O_{S,K}$ whenever P is an S-integral point in $C(K)$.

Now let $(P_\nu)_{\nu \in \mathbb{N}}$ be a sequence of S-integral points on C. Let $v \in M_K$. Since $\widetilde{C}$ is projective, $\widetilde{C}(K_v)$ is compact for the v-adic topology (use Example 2.6.4); hence, going to a subsequence if needed, we may assume that for $v \in S$ the points P_ν converge v-adically to a point $P^v \in \widetilde{C}(K_v)$. We now write $S = S' \cup S''$, where S' is the set of places v for which $P^v \in \{Q_1, \dots, Q_r\}$.

For the other places $v \in S''$, we note that $|\varphi_j(P_\nu)|_v$ is uniformly bounded, because the points P^v are in C and the functions φ_j are regular on C.

For a place $v \in S'$, we define a filtration

$$V = W_{v1} \supset W_{v2} \supset \dots$$

by setting

$$W_{vj} = \{\varphi \in V \mid \mathrm{ord}_{P^v}(\varphi) \geq j - 1 - N\}.$$

We have $\dim(W_{vj}/W_{v,j+1}) \leq 1$ (as in Lemma 8.10.9), hence $\dim(W_{vj}) \geq d - j + 1$. We now pick a basis for W_{vd} and complete it successively to bases of the vector spaces $W_{v,d-1}, W_{v,d-2}, \dots, W_{v1}$. This gives us a basis $w_{v1}, \dots, w_{vd}$ of V such that

$$\mathrm{ord}_{P^v}(w_{vj}) \geq j - 1 - N \quad \text{for} \quad j = 1, \dots, d.$$

Expressing the w_{vj} in terms of the basis $\boldsymbol{\varphi}$ of V, we obtain d independent linear forms $L_{vj}(\boldsymbol{\varphi})$ defined over K (because $v \in S'$ and $P^v \in \widetilde{C}(K)$) with

$$\mathrm{ord}_{P^v}(L_{vj}(\boldsymbol{\varphi})) \geq j - 1 - N \quad \text{for} \quad j = 1, \dots, d. \tag{7.6}$$

Finally, we define independent linear forms for $v \in S''$ by setting $L_{vj}(\boldsymbol{\varphi}) = \varphi_j$, for $j = 1, \dots, d$.

For $v \in S'$, let λ_v be a local parameter at P^v, i.e. $\mathrm{ord}_{P^v}(\lambda_v) = 1$ (recall that $P^v \in \widetilde{C}(K)$ because of the definition of S'). Then, since the sequence $(P_\nu)_{\nu \in \mathbb{N}}$ converges v-adically to P^v for $v \in S'$, we deduce from (7.6) that

$$|L_{vj}(\boldsymbol{\varphi}(P_\nu))|_v \ll |\lambda_v(P_\nu)|_v^{j-1-N} \quad \text{for } v \in S', j = 1, \dots, d \text{ and } \nu \text{ sufficiently large.}$$

For $v \in S''$, we have that $|L_{vj}(\varphi(P_\nu))|_v$ is bounded, because of the definition of S''. Therefore, by multiplying these inequalities we get

$$\prod_{v\in S}\prod_{j=1}^{d} |L_{vj}(\varphi(P_\nu))|_v \ll \left(\prod_{v\in S'} |\lambda_v(P_\nu)|_v\right)^{d(d-2N-1)/2} \tag{7.7}$$

for ν sufficiently large. On the other hand, the $\varphi_j(P_\nu)$ are S-integers; hence $|\varphi_j(P_\nu)|_v \leq 1$ for $v \notin S$, and also $|\varphi_j(P_\nu)|_v$ is bounded for $v \in S''$. For $v \in S'$ and ν sufficiently large, we have $|\varphi_j(P_\nu)|_v \ll |\lambda_v(P_\nu)|_v^{-N}$ because the maximum order of pole of φ_j at P^v is at most N. This shows that

$$H((\varphi_1(P_\nu) : \cdots : \varphi_d(P_\nu))) \ll \left(\prod_{v\in S'} |\lambda_v(P_\nu)|_v\right)^{-N}. \tag{7.8}$$

By combining (7.7) and (7.8), we infer that

$$\prod_{v\in S}\prod_{j=1}^{d} |L_{vj}(\varphi(P_\nu))|_v \ll H((\varphi_1(P_\nu) : \cdots : \varphi_d(P_\nu)))^{-\frac{d(d-2N-1)}{2N}}$$

always for ν sufficiently large. Since, by hypothesis $r \geq 3$, inequality (7.5) yields $d - 2N - 1 \geq (rN + 1 - g) - (2N + 1) > 0$ for $N > g$. Moreover, we may assume that $H((\varphi_1(P_\nu) : \cdots : \varphi_d(P_\nu))) \to \infty$ because otherwise the ratios $\varphi_j(P_\nu)/\varphi_1(P_\nu)$ would belong to a finite set independent of ν by Northcott's theorem from 1.6.8. Hence the points P_ν would belong to a finite set, which is what we want to prove. Thus we may apply the subspace theorem as in Corollary 7.2.5 with any fixed $0 < \varepsilon < d(d-2N-1)/(2N)$, concluding that the S-integer points $(\varphi_1(P_\nu), \ldots, \varphi_d(P_\nu))$ lie in a finite union of linear subspaces of K^d. Since the functions φ_j are linearly independent on the curve $\widetilde{C}$, these linear subspaces cannot contain $\varphi(C)$ (at least if there are infinitely many distinct points P_ν) and must have a finite intersection with $\varphi(C)$. Hence, in any case, the points P_ν belong to a finite set. □

7.4. The generalized unit equation

In Chapter 5 we examined in some detail the unit equation $x + y = 1$ and its applications, in particular to the finiteness of solutions of significant classes of diophantine equations in two variables. The following theorem of J.-H. Evertse, H.P. Schlickewei and W.M. Schmidt [**112**] is an extension of Theorem 5.2.1 to several variables. We will not prove this theorem but we state it here because of its importance.

Theorem 7.4.1. *Let K be a field of characteristic 0, let $K^\times$ be its multiplicative subgroup, and let Γ be a subgroup of $(K^\times)^n$ of finite $\mathbb{Q}$-rank r and let $a_1, \ldots, a_n \in K^\times$. Let $\mathcal{X}$ be the set of solutions of*

$$a_1x_1 + \cdots + a_nx_n = 1$$

such that $(x_1, \ldots, x_n) \subset \Gamma$ *and no proper subsum of* $a_1x_1 + \cdots + a_nx_n$ *vanishes. Then* $\mathcal{X}$ *is a finite set of cardinality at most*

$$|\mathcal{X}| \leq e^{(6n)^{3n}(r+1)}.$$

A remarkable feature of this result is the uniformity of the bound for the number of solutions, which, as in Theorem 5.2.1, is simply exponential in the rank of Γ and independendent of the field K. This theorem is quite difficult to prove. We prove now a weaker, but still very useful, version of Theorem 7.4.1, due independently to Schlickewei and van der Poorten (see the bibliography in [**260**]) and J.-H. Evertse [**104**].

Theorem 7.4.2. *Let* K *be a number field and let* S *be a finite set of places of* K *containing all archimedean places, with group of* S*-units* $U_{S,K}$. *Let* $\mathcal{X}$ *be the set of solutions of*

$$x_1 + \cdots + x_n = 1$$

such that $(x_1, \ldots, x_n) \in (U_{S,K})^n$ *and no proper subsum of* $x_1 + \cdots + x_n$ *vanishes. Then* $\mathcal{X}$ *is a finite set.*

Corollary 7.4.3. *Let* $\mathcal{X}$ *be the set of solutions of*

$$x_1 + \cdots + x_n = 1$$

such that $(x_1, \ldots, x_n) \in (U_{S,K})^n$. *Then there is a finite set* $\mathcal{F} \subset U_{S,K}$ *such that every* $\mathbf{x} \in \mathcal{X}$ *has at least one coordinate in* $\mathcal{F}$.

Proof of corollary: Clear by induction on n and Theorem 7.4.2. □

Proof of theorem: We follow U. Zannier [**337**]. The proof is by induction on n, the case $n = 1$ being obvious.

We say that a solution $\mathbf{x} \in \mathcal{X}$ of the S-unit equation $x_1 + \cdots + x_n = 1$ is non-degenerate if no proper subsum of $x_1 + \cdots + x_n$ vanishes. We partition $\mathcal{X}$ into finitely many subsets according to the set of indices j_v $(v \in S)$ such that

$$j_v = \min\{j \mid |x_j|_v = \max_i |x_i|_v\}$$

and then it suffices to prove the result for each subset.

Let us fix one of these subsets, say $\mathcal{X}'$, and define linear forms L_{vj} by $L_{vj} = X_j$ if $j \neq j_v$ and $L_{vj_v} = X_1 + \cdots + X_n$. Since $L_{vj_v}(\mathbf{x}) = x_1 + \cdots + x_n = 1$ and since $|x_{j_v}|_v = |\mathbf{x}|_v$, we have for $v \in S$

$$\prod_{j=1}^{n} |L_{vj}(\mathbf{x})|_v = \prod_{j \neq j_v} |x_j|_v = |\mathbf{x}|_v^{-1} \prod_{j=1}^{n} |x_j|_v.$$

On the other hand, we have for every j

$$\prod_{v \in S} |x_j|_v = 1$$

because $x_j \in U_{S,K}$ (use the product formula in 1.4.3). Therefore, from the last two displayed equations, we infer that

$$\prod_{v \in S} \prod_{j=1}^{n} |L_{vj}(\mathbf{x})|_v = \prod_{v \in S} |\mathbf{x}|_v^{-1}.$$

Moreover, we have

$$H(\mathbf{x}) = \prod_{v \in M_K} |\mathbf{x}|_v = \prod_{v \in S} |\mathbf{x}|_v$$

because each coordinate x_j is an S-unit, and we conclude that

$$\prod_{v \in S} \prod_{j=1}^{n} |L_{vj}(\mathbf{x})|_v = H(\mathbf{x})^{-1}.$$

Thus we can apply Corollary 7.2.5 with $\varepsilon = 1$ and obtain that the solutions $\mathbf{x} \in \mathcal{X}'$ lie in a finite union $\mathcal{S}$ of proper linear subspaces of K^n. Now we partition $\mathcal{X}'$ into finitely many subsets, such that in a typical subset $\mathcal{X}''$ we have a relation $\sum a_j x_j = 0$. We may suppose, after a permutation of coordinates, that $a_n \neq 0$. Eliminating x_n and using the equation $x_1 + \cdots + x_n = 1$, we then obtain an equation

$$b_1 x_1 + \cdots + b_{n-1} x_{n-1} = 1$$

with $b_j \in K$, not all 0. By removing vanishing subsums from the above relation, we end up with a new relation $\sum_{i \in I} b_i x_i = 1$, where no proper subsum of $\sum_{i \in I} b_i x_i$ vanishes. Thus by partitioning once again $\mathcal{X}''$ into finitely many subsets, it suffices to deal with one such relation. We enlarge S to a new finite set S', where each b_i $(i \in I)$ is an S'-unit, and we obtain a non-degenerate solution $\mathbf{y}$ in S'-units, $y_i = b_i x_i$, of the equation

$$\sum_{i \in I} y_i = 1.$$

Since $|I| \leq n - 1$, the induction hypothesis shows that the $b_i x_i$ $(i \in I)$, and hence the coordinates x_i $(i \in I)$ themselves belong to a finite set. Thus we have proved that the inductive hypothesis implies that there is a finite set Φ such that any solution $\mathbf{x} \in \mathcal{X}$ has at least one coordinate in Φ. Let $x_{i_0} = c$, $c \in \Phi$, be one of these relations. Then $c \neq 1$ because $\mathbf{x}$ is a non-degenerate solution. By enlarging S to a new set S'' so that $1 - c$ becomes an S''-unit and setting $z_i = (1 - c)^{-1} x_i$, we have that $z_i = (1 - c)^{-1} x_i$ $(i \neq i_0)$ is an S''-unit and yields a non-degenerate solution of

$$\sum_{i \neq i_0} z_i = 1.$$

Thus we can apply induction again to conclude that all remaining coordinates x_i also belong to a finite set. □

7.4.4. Theorem 7.4.2 is a very powerful tool and we devote the next few paragraphs to some of its consequences. The first application is to the so-called norm-form equation, a generalization of the Thue equation (for irreducible polynomials). For simplicity, we shall consider here only norm-form equations over $\mathbb{Z}$.

Let $\omega_1, \ldots, \omega_n \in \overline{\mathbb{Q}}$ and let $K = \mathbb{Q}(\omega_1, \ldots, \omega_n)$, $d = [K : \mathbb{Q}]$. Let $L(X)$ be the linear form

$$L(\mathbf{X}) = \sum_{j=1}^{n} \omega_j X_j,$$

let $\mathcal{S}$ be the set of distinct embeddings $\sigma : K \to \mathbb{C}$, and define

$$L^\sigma(\mathbf{X}) = \sum_{j=1}^{n} \sigma(\omega_j) X_j.$$

The **norm-form** associated to $L(\mathbf{X})$ is

$$N_{K/\mathbb{Q}}(L(\mathbf{X})) = \prod_{\sigma \in \mathcal{S}} L^\sigma(\mathbf{X}).$$

It is a homogeneous polynomial of degree d in the n variables $X_1, \ldots, X_n$, with rational coefficients. The corresponding **norm-form equation** is the equation $N_{K/\mathbb{Q}}(L(\mathbf{X})) = c$ with $c \in \mathbb{Q}^\times$, to be solved with integral $\mathbf{x} \in \mathbb{Z}^n$.

A more intrinsic way of looking at norm-form equations is as follows. Define

$$\mathfrak{M} = L(\mathbb{Z}^n) = \sum_{j=1}^{n} \omega_j \mathbb{Z}.$$

Then $\mathfrak{M}$ is a free $\mathbb{Z}$-module in K of rank $\leq d$. Hence we may just as well consider the equation $N_{K/\mathbb{Q}}(\mu) = c$ for $\mu \in \mathfrak{M}$, where $\mathfrak{M}$ is an arbitrary finitely generated $\mathbb{Z}$-module in K. Moreover, we may always assume that $\omega_1, \ldots, \omega_n$ are linearly independent over $\mathbb{Z}$.

We first recall some results from algebraic number theory. A $\mathbb{Z}$-submodule $\mathfrak{M}$ of K of rank $n = d$ is called a **full module**. An **order** is a full module of K, which is also a subring of K containing 1. We denote by $\mathfrak{O}^{\times,+}$ the elements of $\mathfrak{O}$ of norm 1. By A.I. Borevich and I.R. Shafarevich [**42**], Ch.II, §2, Th.4, it is a subgroup of the group of units $\mathfrak{O}^\times$, clearly of index 1 or 2. Since the ring $\mathfrak{O}$ is finitely generated as a $\mathbb{Z}$-module, $\mathfrak{O}$ is contained in O_K. We conclude that every order has finite index in the maximal order O_K.

Dirichlet's unit theorem extends to orders $\mathfrak{O}$ saying that $\mathfrak{O}^\times$ is a finitely generated abelian group of rank $r + s - 1$, where r (resp. s) is the number of real (resp. complex) places of K ([**42**], Ch.II, §4, Th.5).

An obvious instance in which a norm-form equation may have infinitely many solutions is when $\mathfrak{M}$ is a full module. Then the set of $m \in K$ such that $m\mathfrak{M} \subset \mathfrak{M}$

is an order $\mathfrak{O}$ in K ([**42**], Ch.II, §2, Th.3). Since by definition $\mathfrak{O}\mathfrak{M} = \mathfrak{M}$, if the norm-form equation $N_{K/\mathbb{Q}}(\mu) = c$ has a solution μ_0, it is immediate that $\mu_0\mathfrak{O}^{\times,+}$ is also a family of solutions, necessarily infinite if K is not $\mathbb{Q}$ or imaginary quadratic (by the generalization of Dirichlet's unit theorem).

The same phenomenon may occur, even if $n < d$, if there are a proper subfield $K' \subset K$, a full module $\mathfrak{M}'$ of K' with associated order $\mathfrak{O}'$, and an element $\alpha \in K^\times$ such that $\alpha\mathfrak{M}' \subset \mathfrak{M}$. If $\mu' \in \mathfrak{M}'$, we have $\alpha\mu' \in \mathfrak{M}$ and

$$N_{K/\mathbb{Q}}(\alpha\mu') = N_{K/\mathbb{Q}}(\alpha) N_{K'/\mathbb{Q}}(\mu')^{[K:K']},$$

hence if there is a solution $\mu_0 = \alpha\mu_0'$ with $\mu_0' \in M'$, we see that $\mu_0(\mathfrak{O}')^{\times,+}$ is a set of solutions if $[K : K']$ is odd, and $\mu_0(\mathfrak{O}')^\times$ (thus allowing for units of norm -1) if $[K : K']$ is even. Again, we get an infinite set of solutions if K' is not $\mathbb{Q}$ or imaginary quadratic. If $[K : K']$ is even and $(\mathfrak{O}')^{\times,+}$ is of index 2 in $(\mathfrak{O}')^\times$, the set $\mu_0(\mathfrak{O}')^\times$ splits as

$$\mu_0(\mathfrak{O}')^\times = \mu_0(\mathfrak{O}')^{\times,+} \cup \mu_0\eta(\mathfrak{O}')^{\times,+}$$

with $\eta \in (\mathfrak{O}')^\times$ of norm $N_{K'/\mathbb{Q}}(\eta) = -1$.

A set of solutions $\mu_0(\mathfrak{O}')^{\times,+}$ or $\mu_0\eta(\mathfrak{O}')^{\times,+}$ (which may occur only if $[K : K']$ is even) obtained by such a procedure is a **family of solutions** associated to a pair $(\mathfrak{M}', \mu_0)$, where $\mathfrak{M}'$ is a full module in a subfield $K' \subset K$ with associated order $\mathfrak{O}'$ and where μ_0 is a solution of the norm-form equation with $\mu_0 \in \alpha\mathfrak{M}'$ for a suitable $\alpha \in K^\times$ with $\alpha\mathfrak{M}' \subset \mathfrak{M}$.

Example 7.4.5. A typical example is $L(\mathbf{X}) = X_1 + \sqrt{2}X_2 + \sqrt{3}X_3$, $K = \mathbb{Q}(\sqrt{2}, \sqrt{3})$. Then the associated module $\mathfrak{M} = \mathbb{Z} + \sqrt{2}\mathbb{Z} + \sqrt{3}\mathbb{Z}$ has rank 3 and is not full in K. We are looking for an infinite family of solutions associated to $(\mathfrak{M}', \mu_0)$. The corresponding subfield K' has to be $\mathbb{Q}(\sqrt{6})$, $\mathbb{Q}(\sqrt{3})$, or $\mathbb{Q}(\sqrt{2})$. Clearly, we have

$$K'\mu_0 \subset \mathbb{Q}\mathfrak{M} = \mathbb{Q} + \sqrt{2}\mathbb{Q} + \sqrt{3}\mathbb{Q}$$

for such a family of solutions. We conclude that the family of solutions is contained in

$$\mathfrak{M}^{K'} = \sqrt{2}\mathbb{Z} + \sqrt{3}\mathbb{Z},\ \mathbb{Z} + \sqrt{3}\mathbb{Z} \text{ or } \mathbb{Z} + \sqrt{2}\mathbb{Z}.$$

The second and the third module are already full modules in the corresponding K' and for the first we have the full module $\mathfrak{M}' = 2\mathbb{Z} + \sqrt{6}\mathbb{Z}$ in $\mathbb{Q}(\sqrt{6})$ such that $\frac{1}{\sqrt{2}}\mathfrak{M}' = \mathfrak{M}^{K'}$. The restrictions of the norm-form equation $N_{K/\mathbb{Q}}(\mathbf{x}) = c$ to the modules $\mathfrak{M}^{K'}$ have the form

$$(2X_2^2 - 3X_3^2)^2 = c, \quad (X_1^2 - 3X_3^2)^2 = c, \quad (X_1^2 - 2X_2^2)^2 = c.$$

In order to have an infinite family of solutions, it is necessary that $c = d^2$ for some $d \in \mathbb{Q}^\times$ and every solution of one of the Pell equations

$$2X_2^2 - 3X_3^2 = \pm d, \quad X_1^2 - 3X_3^2 = \pm d, \quad X_1^2 - 2X_2^2 = \pm d$$

induces an infinite family of solutions of the original norm-form equation. (We recall that a **Pell equation** is of the form $X^2 - aY^2 = b$ for some square free integer a and any $b \in \mathbb{Z} \setminus \{0\}$. Since $2X_2^2 - 3X_3^2 = 6(X_2 + X_3)^2 - (2X_2 + 3X_3)^2$, the first equation is also of this type. The procedure to solve Pell's equation effectively is well known, we refer to [**42**], Ch.II, §7.)

The natural conjecture is then that the solutions of a non-degenerate norm-form equation $N_{K/\mathbb{Q}}(L(\mathbf{X})) = c$ consist of finitely many families of solutions. This was answered in the affirmative by W.M. Schmidt [**267**] in 1972, as a consequence of his subspace theorem.

Theorem 7.4.6. *Let $\mathfrak{M}$ be a finitely generated $\mathbb{Z}$-module in K. Then the set of solutions of the norm-form equation $N_{K/\mathbb{Q}}(\mu) = c$, $\mu \in \mathfrak{M}$, consists of the union of finitely many families of solutions.*

Proof: The proof is by verifying the theorem first if $\mathfrak{M}$ is a full module in K (which is automatically the case if $d = 1$) and then proceeding by induction on the degree $d = [K : \mathbb{Q}]$.

Suppose first that $\text{rank}(\mathfrak{M}) = d$, hence $\mathfrak{M}$ is a full module in K with associated order $\mathfrak{O}$. Every solution μ_0 determines a family of solutions $\mu_0 \mathfrak{O}^{\times,+}$ and to prove the theorem we need to show that the number of distinct families so obtained is finite. To see this, note that the group U of units of $\mathfrak{O}$ of norm 1 has finite index in the group of all units of K (by the Dirichlet unit theorem for orders). A classical result of Minkowski shows that for any $\xi \in K$ there is a unit η such that

$$|\sigma(\xi\eta)| \leq C \left|N_{K/\mathbb{Q}}(\xi)\right|^{1/d} \tag{7.9}$$

for σ running over all embeddings $\sigma : K \to \mathbb{C}$, for some constant C depending only on K. A direct simple proof runs as follows (cf. [**215**], Ch.3, Lemma 3.5). Replacing ξ by $n\xi$ for suitable $n \in \mathbb{Z}$, we may assume that ξ is a non-zero algebraic integer. Let $\omega_1, \ldots, \omega_d$ be an integral basis of O_K and let $B = \lfloor |N_{K/\mathbb{Q}}(\xi)^{1/d}| \rfloor$. Consider all numbers $\omega_1 x_1 + \cdots + \omega_d x_d$ with integer x_i and $0 \leq x_i \leq B$. There are $(B+1)^d > |N_{K/\mathbb{Q}}(\xi)|$ such numbers, while there are $|N_{K/\mathbb{Q}}(\xi)|$ residue classes $(\text{mod}\, \xi O_K)$. Therefore, by the pigeon-hole principle there are two such numbers in a same residue class with their difference equal to $\alpha\xi$ for some $\alpha \in O_K \setminus \{0\}$. Then $|\sigma(\alpha\xi)| \leq d \max_i |\sigma(\omega_i)|\, B$ and it follows that $N_{K/\mathbb{Q}}(\alpha)$ is bounded independently of ξ. Thus there are only finitely many possibilities for α up to multiplication by units, i.e. $\alpha = \beta\eta$, where β belongs to a finite set and η is a unit. Then $|\sigma(\eta\xi)| \leq d \max_{i\beta} |\sigma(\beta^{-1}\omega_i)|\, B$, proving what we want.

Therefore, since U has finite index in the group of all units, the same conclusion as in (7.9) holds with $\eta \in U$, provided we replace C by a larger constant which may depend on the order $\mathfrak{O}$. Moreover, there is an integer $b > 0$ such that $b\mathfrak{M} \subset O_K$. We conclude that in any given family of solutions there is an element μ_0 with height bounded in terms of $N_{K/\mathbb{Q}}(\mu_0) = c$ and b (using (7.9) at archimedean places and $|\mu_0|_v \leq |b|_v^{-1}$ for non-archimedean places). By Northcott's theorem in 1.6.8, the number of families is finite, as asserted. This proves the theorem if $\mathfrak{M}$ is a full module in K.

Suppose now that $n = \text{rank}(\mathfrak{M}) < d$. Let F be a Galois closure of K with Galois group G and let H be the subgroup of G fixing K. Then we identify $\{\sigma : K \hookrightarrow F\}$ with a set $\mathcal{S}$ of representatives in G of G/H.

Since $\mathfrak{M}$ has rank $n < d = |\mathcal{S}|$ as a $\mathbb{Z}$-module and $\mathcal{S}$ acts trivially on $\mathbb{Z}$, there is a relation of linear dependence in F

$$\sum_{\sigma \in \mathcal{S}} a_\sigma \sigma(\mu) = 0 \quad (\mu \in \mathfrak{M}).$$

Some coefficients a_σ may be 0 and we select a relation

$$\sum_{\sigma \in \mathcal{S}'} a_\sigma \sigma(\mu) = 0 \quad (\mu \in \mathfrak{M}) \tag{7.10}$$

for a subset $\mathcal{S}' \subset \mathcal{S}$, where now $a_\sigma \neq 0$.

After these preliminaries, we take a sufficiently large finite set $S \subset M_F$ containing the archimedean places such that $\sigma(\mathfrak{M}) \subset O_{S,F}$ for every σ and moreover c and the coefficients a_σ $(\sigma \in \mathcal{S}')$ are S-units. Then the equation $N_{K/\mathbb{Q}}(\mu) = c$ implies that $\sigma(\mu) \in O_{S,F}^\times$ for every σ and we conclude that

$$a_\sigma \sigma(\mu) \in O_{S,F}^\times \quad (\sigma \in \mathcal{S}').$$

Therefore, by applying Corollary 7.4.3 to (7.10), we infer that there is a finite set $\mathcal{F}$ such that any solution μ in our given class satisfies

$$\frac{\tau(\mu)}{\sigma(\mu)} = f \, \frac{a_\sigma}{a_\tau} \tag{7.11}$$

for a suitable choice (depending on μ) of $f \in \mathcal{F}$ and $\sigma, \tau \in \mathcal{S}'$, $\sigma \neq \tau$.

This gives us a further subdivision of solutions into finitely many subclasses in which f, σ, and τ remain fixed. Let us fix such a subclass $\mathcal{X}'$ and an element $\alpha \in \mathcal{X}'$. Then by (7.11) we have

$$\tau(\alpha^{-1}\mu) = \sigma(\alpha^{-1}\mu)$$

for every $\mu \in \mathcal{X}'$. By applying σ^{-1} and setting $g = \sigma^{-1}\tau$, we obtain

$$g(\alpha^{-1}\mu) = \alpha^{-1}\mu$$

for a certain $g \notin H$. It follows from this that $\alpha^{-1}\mu$ lies in a proper subfield K' of K, namely the fixed field of the group $\langle H, g \rangle$ generated by H and g. Therefore, if $\mathfrak{M}'$ is the $\mathbb{Z}$-module generated by the elements $\alpha^{-1}\mu$ $(\mu \in \mathcal{X}')$, we have that $\mathfrak{M}' \subset K'$ and $\alpha\mathfrak{M}' \subset \mathfrak{M}$. For $\mu' \in \mathfrak{M}'$, we have

$$N_{K/\mathbb{Q}}(\alpha\mu') = N_{K/\mathbb{Q}}(\alpha) N_{K'/\mathbb{Q}}(\mu')^{[K:K']},$$

and the norm-form equation $N_{K/\mathbb{Q}}(\alpha\mu') = c$ is equivalent to the new norm-form equation $N_{K'/\mathbb{Q}}(\mu') = \varepsilon$, with $\varepsilon = 1$ if $[K : K']$ is odd and $\varepsilon = \pm 1$ otherwise. Since K' is a proper subfield of K, the inductive hypothesis applies to K' and $\mathcal{X}'$ is indeed the union of finitely many families of solutions. Since there are only finitely many subclasses $\mathcal{X}'$ to consider, the result follows. $\square$

The following theorem is due to M. Laurent [**178**].

Theorem 7.4.7. *Let Γ be a finitely generated subgroup of $(\overline{\mathbb{Q}}^\times)^n$ and let Σ be any subset of Γ. Then the Zariski closure of Σ in $\mathbb{G}_m^n$ is a finite union of translates of algebraic subgroups of $\mathbb{G}_m^n$.*

Remark 7.4.8. In his paper Laurent proves the stronger statement, previously conjectured by Lang, in which the subgroup Γ is a subgroup of $(\mathbb{C}^\times)^n$ of finite $\mathbb{Q}$-rank. This was obtained from the above theorem by using additional arguments from Kummer theory. A proof of the stronger result is also immediate if in the argument below we use Theorem 7.4.1, rather than Corollary 7.4.3.

Proof of Theorem 7.4.7: Let V be an irreducible component of the Zariski closure of Σ and let $f_1(\mathbf{X}) = \cdots = f_r(\mathbf{X}) = 0$ be a set of polynomials defining V set theoretically. Let

$$f(\mathbf{X}) = \sum_{i=0}^{I} c_i \mathbf{X}^{\mathbf{a}_i} \quad (c_i \neq 0)$$

be one of these polynomials f_i. We take $S \subset M_K$ to be a finite set of places such that $\Gamma \subset (U_{S,K})^n$ (it suffices to do so for a set of generators) and also $c_i \in U_{S,K}$ for every i, and apply Corollary 7.4.3 to the equation

$$\sum_{i=1}^{I} \left(-\frac{c_i}{c_0}\right) \mathbf{g}^{\mathbf{a}_i - \mathbf{a}_0} = 1 \quad (\mathbf{g} \in V \cap \Sigma).$$

Then we conclude that there are $\lambda_1, \ldots, \lambda_J \in U_{S,K}$ such that

$$V \cap \Sigma \subset \bigcup_{i=1}^{I} \bigcup_{j=1}^{J} \{\mathbf{X}^{\mathbf{a}_i} - \lambda_j \mathbf{X}^{\mathbf{a}_0} = 0\}.$$

Therefore, passing to the Zariski closures, we infer that

$$V \subset \bigcup_{i=1}^{I} \bigcup_{j=1}^{J} \{\mathbf{X}^{\mathbf{a}_i} - \lambda_j \mathbf{X}^{\mathbf{a}_0} = 0\}.$$

Hence there are $i, j \geq 1$ such that $\mathbf{X}^{\mathbf{a}_i} - \lambda_j \mathbf{X}^{\mathbf{a}_0} = 0$ on V, because V is irreducible. Now we can eliminate the monomial $\mathbf{X}^{\mathbf{a}_0}$ from $f(\mathbf{X})$, obtaining a polynomial $\widetilde{f}(\mathbf{X})$ in which the number of monomials has decreased by at least 1 and we may replace $f = 0$ by $\mathbf{X}^{\mathbf{a}_i} - \lambda_j \mathbf{X}^{\mathbf{a}_0} = \widetilde{f} = 0$ in the set of defining equations for V. Proceeding in this way, we see that V is defined by binomial equations of type $\mathbf{X}^{\mathbf{a}} - \lambda \mathbf{X}^{\mathbf{b}} = 0$; hence V is a translate of an algebraic subgroup of $\mathbb{G}_m^n$. □

7.4.9. Our final application of the subspace theorem is a theorem in arithmetic, due to Corvaja and Zannier [**73**], see also [**337**], p.50.

Theorem 7.4.10. *Let $S \subset M_{\mathbb{Q}}$ be a finite set of places including the place at ∞, and let $\varepsilon > 0$. Then the set Σ of integer pairs (u, v) with prime factors only from S and such that*

$$\mathrm{GCD}(u-1, v-1) \geq \max(|u|, |v|)^{\varepsilon}$$

is contained in the union of a finite set and finitely many proper algebraic subgroups of $\mathbb{G}_m^2$. In particular, the subset of pairs $(u, v) \in \Sigma$ with u and v multiplicatively independent S-units is a finite set.

As a special case, we have:

Corollary 7.4.11. *Let $a > b > c \geq 1$ be positive integers and let $P(a, b, c)$ be the greatest prime factor of $(ab+1)(ac+1)$. Then $P(a, b, c) \to \infty$ as $a \to \infty$.*

Proof of corollary: We argue by contradiction. Suppose there is an infinite set of triples (a, b, c) with $P(a, b, c) \leq P$. Let S be the set of all primes $p \leq P$, including also ∞ and set $u = ab+1$, $v = ac+1$. Then u, v are S-units, $u > v$ and $\mathrm{GCD}(u-1, v-1) \geq a$.

Thus Theorem 7.4.10 applies and we deduce that there are infinitely many u, v verifying an equation $u^m = v^n$ (see Corollary 3.2.15). Since $u > v$ are positive integers, we may assume that $1 \leq m < n$ are positive and coprime, whence $u = t^n$, $v = t^m$ for some integer t. Now $\mathrm{GCD}((t^m-1)/(t-1), (t^n-1)/(t-1)) = 1$ (otherwise there is a prime p such that the reduction of both polynomials $(\mathrm{mod}\ p)$ has the zero $\bar{t}$, which is impossible for coprime m, n). Therefore, we have

$$a \leq \mathrm{GCD}(u-1, v-1) = t - 1 < u^{1/n} \leq (ab+1)^{1/2} < a$$

because $b < a$. This is a contradiction, concluding the proof. □

Proof of Theorem 7.4.10: We may assume that S contains at least one non-archimedean place, $|v| \geq |u|$ and $v \neq 1$. Let d be the denominator of the fraction $(u-1)/(v-1)$ in its lowest terms, thus $d \leq 2|v|^{1-\varepsilon}$ because $\mathrm{GCD}(u-1, v-1) \geq \max(|u|, |v|)^{\varepsilon}$ by hypothesis. We define $z_j \in \mathbb{Q}$ and $c_j \in \mathbb{Z}$ for $j \in \mathbb{N}$ by

$$z_j := u^{j-1} \frac{u-1}{v-1} = \frac{c_j}{d}.$$

We have the approximation

$$\frac{1}{v-1} = \sum_{r=1}^{h} \frac{1}{v^r} + O\left(|v|^{-h-1}\right)$$

whence, multiplying by $u^{j-1}(u-1)$, we obtain

$$\left| z_j + \sum_{r=1}^{h} \frac{u^{j-1}}{v^r} - \sum_{r=1}^{h} \frac{u^j}{v^r} \right| = O\left(|u|^j |v|^{-h-1} \right). \tag{7.12}$$

We shall view such an inequality as providing a small value of a linear form in independent variables corresponding to z_j, u^{j-1}/v^r, u^j/v^r and apply the affine version Corollary 7.2.5 of the subspace theorem to a suitable set of linear forms, which will include the linear forms associated to the inequalities (7.12).

Let k be a positive integer to be chosen later and let $n = k+(k+1)h$. We consider linear forms $L_{\nu i}$, indexed by $i = 1, \ldots, n$ and $\nu \in S$, in variables $W_1, \ldots, W_k$ and Y_{jr}, $j = 0, \ldots, k$, $r = 1, \ldots, h$, defined as follows. It will be notationally convenient to define $\mathbf{X}$ to be the vector

$$\mathbf{X} = (X_1, \ldots, X_n) := (W_1, \ldots, W_k, Y_{01}, \ldots, Y_{0h}, \ldots, Y_{k1}, \ldots, Y_{kh}).$$

For $i = 1, \ldots, k$ and $\nu = \infty$ we set

$$L_{\infty i} := W_i + \sum_{r=1}^{h} Y_{i-1,r} - \sum_{r=1}^{h} Y_{ir}.$$

For $(\nu, i) \notin \{(\infty, 1), \ldots, (\infty, k)\}$ we define instead

$$L_{\nu i} := X_i.$$

Obviously, for each $\nu \in S$ the linear forms $L_{\nu 1}, \ldots, L_{\nu n}$ are linearly independent. Now define for a pair $(u, v) \in \Sigma$ the point $\mathbf{x} = (x_1, \ldots, x_n)$ to be

$$\mathbf{x} := dv^h \left(z_1, \ldots, z_k, v^{-1}, \ldots, v^{-h}, uv^{-1}, \ldots, uv^{-h}, \ldots, u^k v^{-1}, \ldots, u^k v^{-h} \right).$$

Then, for $i > k$, we have $x_i = du^a v^b$, for suitable integers a and b, hence x_i equals d times an S-unit. Since $L_{\nu i} = x_i$, we easily deduce that

$$\prod_{\nu \in S} |L_{\nu i}(\mathbf{x})|_\nu \le d \qquad \text{for} \quad i > k.$$

Therefore, we have

$$\begin{aligned} \prod_{\nu \in S} \prod_{i=1}^{n} |L_{\nu i}(\mathbf{x})|_\nu &\le d^{n-k} \prod_{\nu \in S} \prod_{i=1}^{k} |L_{\nu i}(\mathbf{x})|_\nu \\ &= d^{n-k} \left(\prod_{i=1}^{k} |L_{\infty i}(\mathbf{x})|_\infty \right) \prod_{\nu \in S \setminus \{\infty\}} \prod_{i=1}^{k} |x_i|_\nu. \end{aligned} \tag{7.13}$$

For $i \le k$, we have $x_i = dv^h z_i \in v^h \mathbb{Z}$, whence

$$\prod_{\nu \in S \setminus \{\infty\}} |x_i|_\nu \le |v|^{-h}. \tag{7.14}$$

Moreover, by (7.12) we have $|L_{\infty i}(\mathbf{x})| = O\left(d|u|^i|v|^{-1}\right)$ for $i = 1, \ldots, k$. Putting this estimate and (7.14) in (7.13), we get

$$\prod_{\nu \in S} \prod_{i=1}^{n} |L_{\nu i}(\mathbf{x})|_\nu = O\left(d^n |u|^{k(k+1)/2} |v|^{-hk-k}\right).$$

Recalling that $d \leq 2|v|^{1-\varepsilon}$ and $|u| \leq |v|$, from the last displayed inequality we obtain

$$\prod_{\nu \in S} \prod_{i=1}^{n} |L_{\nu i}(\mathbf{x})|_\nu = O\left(|v|^{h+k(k+1)/2-\varepsilon n}\right). \tag{7.15}$$

Finally, each x_i is an integer and we have $\max |x_i| \leq 2d|v|^{h+k} < 4|v|^{h+k+1}$, hence $H(\mathbf{x}) \leq 4|v|^{h+k+1}$. In view of (7.15), we conclude that

$$\prod_{\nu \in S} \prod_{i=1}^{n} |L_{\nu i}(\mathbf{x})|_\nu \ll_{h,k} H(\mathbf{x})^{-\delta}$$

with $\delta = (\varepsilon n - h - k(k+1)/2)/(h+k+1)$ provided that $\delta > 0$. If we take for example $k \geq 2/\varepsilon$ and $h \geq k^2 + 1$, it is clear that we get $\delta > 0$. The subspace theorem in Corollary 7.2.5 now proves that the vectors $\mathbf{x}$ all lie on a certain finite union of subspaces of $\mathbb{Q}^n$. Let $\sum a_i x_i = 0$ be one of these subspaces. If we substitute the value of x_i in such an equation, we find a non-trivial equation of type

$$\frac{f(u)}{v-1} + \frac{g(u,v)}{v^h} = 0,$$

for some polynomials f and g. The rational function $f(X)/(Y-1) + g(X,Y)/Y^h$ cannot vanish identically on $\mathbb{G}_m^2$, otherwise $Y - 1$ would divide $f(X)$, yielding $f = 0$, and then $g = 0$ too, a contradiction. Thus the points (u, v), which belong to a finitely generated subgroup of $\mathbb{G}_m^2$, are also on a finite union of curves of type $Y^h f(X) + g(X,Y)(Y-1) = 0$. By Laurent's theorem in 7.4.7, we conclude that the set Σ lies in a finite union of translates of algebraic subgroups of $\mathbb{G}_m^2$.

In order to complete the proof of the theorem, it suffices to show that, if a translate of such an algebraic subgroup contains infinitely many points in Σ, then it is already a subgroup. Let $X^m Y^n = c$ be an equation defining such a translate (use Corollary 3.2.15). Let $(u, v) \in \Sigma$ and write $g = GCD(u-1, v-1)$. Then from $u^m v^n = c$ we must have $c \equiv 1 \pmod{g}$. Since by hypothesis $g \geq \max(|u|, |v|)^\varepsilon$, it is now clear that, if such a translate contains infinitely many points of Σ, we must have $c = 1$. This proves the first conclusion of the theorem.

For the second conclusion, it is obvious that, if u and v are multiplicatively independent, then the pair (u, v) does not lie in any proper algebraic subgroup of $\mathbb{G}_m^2$ (use Corollary 3.2.15), and the result follows from the first part of the theorem. □

7.5. Proof of the subspace theorem

We follow here the classical proof of Schmidt, with important modifications introduced by H.P. Schlickewei [**261**] and J.-H. Evertse [**108**] in order to cover the case of arbitrary number fields and allow a finite set of places.

7.5.1. By Theorem 7.2.6, it is sufficient to prove the subspace theorem in its affine form Corollary 7.2.5. The proof is by contradiction.

The first step in the proof consists in following, as far as possible, the blueprint provided by the proof of Roth's theorem. Here a major new difficulty appears, namely the non-vanishing of the auxiliary construction cannot be done at a single point and it requires the consideration of n independent points. This gives only a rather weak version of the subspace theorem.

The conclusion of the proof is to show how to go from n independent points to one point only. This is where we need new ideas. Schmidt's original method consisted in applying the weaker form of the subspace theorem to a new set of linear forms, obtained by taking exterior products of the given linear forms, and using geometry of numbers to deduce the strong version of the subspace theorem. This part of Schmidt's proof has been substantially simplified by Evertse [**108**] and we will follow his exposition here, with some further simplifications, because our goal is only the easier qualitative form of the subspace theorem. As formally stated in the next article, we will not keep track of constants depending only on K, S and the linear forms L_{vi}. This will allow substantial simplifications in the exposition.

7.5.2. Before starting the proof of the subspace theorem, we need some notation. We write

$$\mathbf{x}_h = (x_{h0}, \ldots, x_{hn}) \quad (h = 1, \ldots, m), \qquad \mathbf{X} = (\mathbf{x}_1, \ldots, \mathbf{x}_m)$$

and similarly for linear forms and vectors of linear forms

$$\mathbf{L}_v(\mathbf{x}) = (L_{v0}(\mathbf{x}), \ldots, L_{vn}(\mathbf{x})) \quad (v \in S).$$

If $\mathbf{i} = (i_0, \ldots, i_n)$ is an $(n+1)$-tuple of non-negative integers, we also write

$$|\mathbf{i}| = i_0 + \cdots + i_n, \quad \mathbf{x}^{\mathbf{i}} = x_0^{i_0} \ldots x_n^{i_n}, \quad \mathbf{i}! = i_0! \cdots i_n!.$$

We will also need to use some elementary exterior algebra and our notation will be as follows. Let V be a vector space over a field F, of dimension $n+1$, with basis e_i $(i = 0, \ldots, n)$. For $k = 1, \ldots, n$, we equip the exterior power $\wedge^k V$ with the standard basis

$$e_{i_1} \wedge \cdots \wedge e_{i_k}$$

with $i_1 < i_2 < \cdots < i_k$, in lexicographic order.

We extend the standard scalar product $\mathbf{x} \cdot \mathbf{y} = \sum_{j=0}^{n} x_j y_j$ on V to $\wedge^k V$ by Laplace's identity

$$(\mathbf{x}_1 \wedge \cdots \wedge \mathbf{x}_k) \cdot (\mathbf{y}_1 \wedge \cdots \wedge \mathbf{y}_k) = \det((\mathbf{x}_i \cdot \mathbf{y}_j))_{i,j=1,\ldots,k} \tag{7.16}$$

and by multilinearity. Obviously, this is just the standard scalar product on $\wedge^k V$ with respect to the standard basis $(e_{i_1} \wedge \cdots \wedge e_{i_k})_{i_1 < \cdots < i_k}$.

Similarly as in Riemannian geometry, we define a K-linear $*$-operator on the exterior algebra $\wedge V := \bigoplus_k \wedge^k V$ by setting

$$(e_{i_1} \wedge \cdots \wedge e_{i_k})^* := (-1)^{\operatorname{sign}\pi} e_{j_0} \wedge \cdots \wedge e_{j_{n-k}},$$

where $\pi = (j_0, \ldots, j_{n-k}, i_1, \ldots, i_k)$ is a permutation of $\{0, \ldots, n\}$. For $\omega \in \wedge^k V$, we have

$$\omega^{**} = (-1)^{k(n+1-k)} \omega.$$

We use it only for $k = 1$ and $k = n$, where it is trivial. Note also that $*$ is an isometry with respect to the scalar product. We have the Laplace expansion

$$\mathbf{x}_0 \cdot (\mathbf{x}_1 \wedge \cdots \wedge \mathbf{x}_n)^* = \det(\mathbf{x}_0, \ldots, \mathbf{x}_n), \tag{7.17}$$

which we will use several times.

A similar notation will be used with regards to linear forms. If $L(\mathbf{x}) = \mathbf{a} \cdot \mathbf{x} = a_0 x_0 + \cdots + a_n x_n$, then operations on L will be understood as operating on the vector $\mathbf{a}$.

Constants $C_0, C_1, \ldots$ denote unspecified constants depending only on the given forms L_{vi} and K (hence in particular on n, S and a field F of definition for the linear forms), but we will not make any special effort in giving explicit values for them. We shall also use the Vinogradov symbol $\ll$ and the Landau symbol $O(\cdot)$ in the same way, namely up to constants depending only on the given forms L_{vi} and K. However, all estimates will be uniform in the parameters ε, m, and the weights d_i.

7.5.3. *Step 0: Approximation classes and approximation domains.*

We denote by F a finite extension of K, which is a field of definition for all forms L_{vi} ($v \in S$, $i = 0, \ldots, n$) (recall that they have algebraic K_v-coefficients) and set $r = [F : K]$. As in Theorem 6.4.1, let $|\ |_{v,K}$ be an extension of the absolute value $|\ |_v$ to F such that $|L_v(\mathbf{x})|_v = |L_v(\mathbf{x})|_{v,K}$ for every $\mathbf{x} \in K^{n+1}$. In order to simplify the exposition, the reader could assume that $F = K$ (see Remark 7.2.3). We will work however in full generality, with the forms L_{vi} having coefficients in $F \cap K_v$.

The following step is not really necessary for the proof but it leads to cleaner estimates. Since we can always enlarge the set S, we may and shall assume that S contains the archimedean places and that the ring $O_{S,K}$ is a principal ideal domain (see Proposition 5.3.6) and, by the proof of Theorem 7.2.6, it suffices to deal only with S-integral vectors $\mathbf{x}$ satisfying the additional condition

$$\prod_{v \in S} |\mathbf{x}|_v = H(\mathbf{x}).$$

Such vectors will be called **primitive**.

Thus we begin by assuming that there is an infinite set of S-integral primitive solutions of $\mathbf{x} \in O_{S,K}^{n+1} \setminus \{\mathbf{0}\}$ of

$$\prod_{v \in S} \prod_{i=0}^{n} |L_{vi}(\mathbf{x})|_{v,K} < H(\mathbf{x})^{-\varepsilon}. \tag{7.18}$$

If $\mathbf{x} = (x_0, \ldots, x_n) \in O_{S,K}^{n+1}$ is primitive, clearly $u\mathbf{x}$ is again primitive whenever u is an S-unit. Moreover, if $\mathbf{x}$ is a solution of the **basic inequality** (7.18), then $u\mathbf{x}$ is again a solution. Such solutions form an equivalence class. The following lemma shows that in every equivalence class there is an element with small affine height, as defined in 1.5.7. In order to distinguish it from the projective height, we set

$$h_{\text{aff}}(\mathbf{x}) := h((1, \mathbf{x})).$$

Lemma 7.5.4. *There is a positive constant C_0, depending only on S and K, with the following property. Let $\mathbf{x} = (x_0, \ldots, x_n) \in O_{S,K}^{n+1}$ be a primitive point. Then there is an S-unit u in $O_{S,K}$ such that*

$$h(\mathbf{x}) \leq h_{\text{aff}}(u\mathbf{x}) \leq h(\mathbf{x}) + C_0.$$

Proof: The inequality $h(\mathbf{x}) \leq h_{\text{aff}}(u\mathbf{x})$ is obvious because $h(u\mathbf{x}) = h(\mathbf{x})$.

Now suppose for example that $x_0 \neq 0$. Since x_0 is a non-zero S-integer, the product formula shows that

$$\sum_{v \in S} \log |x_0|_v = -\sum_{v \notin S} \log |x_0|_v \geq 0.$$

By Dirichlet's unit theorem (see Theorem 1.5.13), the points $(\log |u|_v)_{v \in S}$, $u \in U_{S,K}$, form a lattice Λ in the real subspace $\sum_{v \in S} t_v = 0$ of $\mathbb{R}^{|S|}$. Therefore, there is an S-unit u such that $\log |ux_0|_v = \log |u|_v + \log |x_0|_v \geq -C$ for $v \in S$, with C a positive constant depending only on the lattice Λ. Since $u\mathbf{x}$ is primitive, we deduce that

$$\begin{aligned} h(u\mathbf{x}) &= \sum_{v \in S} \log |u\mathbf{x}|_v \\ &\geq \sum_{v \in S} \{-C + \max_{i=0,\ldots,n} \log^+ |ux_i|_v\} \\ &= -|S|\, C + h_{\text{aff}}(u\mathbf{x}). \end{aligned}$$

The lemma follows by taking $C_0 = |S|\, C$. □

Corollary 7.5.5. *Every equivalence class of primitive solutions of the basic inequality* (7.18) *contains an element* $\mathbf{x}$ *such that for every* $v \in S$ *we have*

$$h_{\mathrm{aff}}(\mathbf{L}_v(\mathbf{x})) \le h(\mathbf{x}) + C_1.$$

In particular, we have for $L_{vi}(\mathbf{x}) \neq 0$

$$\big| \log |L_{vi}(\mathbf{x})|_{v,K} \big| \le r h(\mathbf{x}) + rC_1. \tag{7.19}$$

Proof: Since the forms L_{vi} $(i = 0, \dots, n)$ are linearly independent, we have for $w \in M_F$

$$\big| \log |\mathbf{L}_v(\mathbf{x})|_w - \log |\mathbf{x}|_w \big| \le \gamma_w,$$

with γ_w depending only on the forms L_{vi} and equal to 0 up to finitely many w. We conclude that $h_{\mathrm{aff}}(\mathbf{L}_v(\mathbf{x}))$ and $h_{\mathrm{aff}}(\mathbf{x})$ differ only by a bounded quantity and an application of Lemma 7.5.4 yields the first claim. Now let w be the place of F with $w|v$. Then Proposition 1.2.7 shows that

$$|\ |_{v,K} = |\ |_w^{[F:K]/[F_w:K_v]}$$

and the claim follows from the fundamental inequality in (1.8) on page 20 and $[F:K] = r$. □

7.5.6. As in the proof of Roth's theorem, it is necessary to consider only solutions $\mathbf{x}$ for which the factors $|L_{vi}(\mathbf{x})|_{v,K}$, for each (v,i), have a similar behaviour compared with $H(\mathbf{x})$. In other words, we want

$$\left(h(\mathbf{x})^{-1} \log |L_{vi}(\mathbf{x})|_{v,K}\right)_{v \in S, i=0,\dots,n} \tag{7.20}$$

to be nearly constant along our set of primitive solutions. If $L_{vi}(\mathbf{x}) = 0$ for some (v,i), then the solution $\mathbf{x}$ lies in the hyperplane defined by the linear form L_{vi}. Hence such solutions satisfy the conclusion of the theorem to be proved. Thus, by the preceding corollary, we may and shall assume that for all (v,i) we have $L_{vi}(\mathbf{x}) \neq 0$ and that (7.19) holds.

The next step consists in splitting solutions into finitely many approximation classes, as it was done in the proof of Roth's theorem. Since we are not interested here in counting the number of approximation classes, it will suffice to note that, given any infinite set of solutions of inequality (7.18), and given $N > 0$ (which will be taken very large), then by (7.19) there exists a cube in $\mathbb{R}^{(n+1)|S|}$ of edge size $1/N$, with north-east corner at a point $(c_{vi}) \in [-2r, 2r]^{(n+1)|S|}$, containing (7.20) for an infinite subset of the given set of primitive solutions. (The point of the constant $2r$ is to swallow the contribution of rC_1 in (7.19) as soon as $h(\mathbf{x})$ is large enough.)

Thus we have an infinite **primitive approximation class** of primitive solutions of inequality (7.18), consisting of solutions satisfying

$$\left(c_{vi} - \frac{1}{N}\right) h(\mathbf{x}) \le \log |L_{vi}(\mathbf{x})|_{v,K} \le c_{vi} h(\mathbf{x}) \tag{7.21}$$

for each pair (v, i). By this inequality and (7.18) on page 199, we necessarily have

$$\sum_{v \in S} \sum_{i=0}^{n} c_{vi} \leq -\varepsilon + (n+1)|S|/N < -\frac{\varepsilon}{2} \tag{7.22}$$

provided $N > 2(n+1)|S|/\varepsilon$, which we shall suppose henceforth.

Next, we are going to apply geometry of numbers in the context of Sections C.1 and C.2. We recall that $K_{\mathbb{A}}$ denotes the adele ring of K. Let R_v be the discrete valuation ring in the completion K_v. For $v \in S$ and any $Q \geq 1$, we define the **v-adic approximation domain** $\Pi_v(Q)$ of level Q to be the parallelopiped in K_v^{n+1} given by

$$\Pi_v(Q) := \left\{ \boldsymbol{\xi}_v \in K_v^{n+1} \mid |L_{vi}(\boldsymbol{\xi}_v)|_{v,K} \leq Q^{c_{vi}} \, \forall i = 0, \ldots, n \right\}.$$

For archimedean v, this is a compact convex symmetric subset of $K_{\mathbb{A}}^{n+1}$. For non-archimedean v, it is easy to see that $\Pi_v(Q)$ is a K_v-lattice. The **approximation domain** is defined in $K_{\mathbb{A}}^{n+1}$ by

$$\Pi(Q) = \prod_{v \in S} \Pi_v(Q) \times \prod_{v \notin S} R_v.$$

Then $\Pi(Q)$ is similar as the domains considered in Minkowski's second theorem (see Theorem C.2.11). The only difference is that the archimedean factors $\Pi_v(Q)$ are closed instead of open. With the same definition for the successive minima $\lambda_1, \ldots, \lambda_{n+1}$ of $\Pi(Q)$ as before, it is trivial to check that Minkowski's second theorem also holds for closed domains as $\Pi(Q)$. Recall that we use the volume with respect to the following measures: Let β_v and β be the Haar measures on K_v and $K_{\mathbb{A}}$ respectively, as defined and normalized in C.1.9, and denote by β_v^{n+1} and β^{n+1} the corresponding measures on K_v^{n+1} and $K_{\mathbb{A}}^{n+1}$.

Lemma 7.5.7. *Let $\Delta_v = \det(L_{vi})$ be the determinant of the linear forms $L_{v0}, \ldots, L_{vn}$ and let p be the rational prime with $v|p$. Then with $d = [K : \mathbb{Q}]$ we have:*

(a) *If $v \notin S$, then*

$$\beta_v^{n+1}(\Pi_v(Q)) = \beta_v(R_v)^{n+1} = \left| D_{K_v/\mathbb{Q}_p} \right|_p^{(n+1)/2}.$$

(b) *If $v \in S$ is not archimedean, then*

$$\beta_v^{n+1}(\Pi_v(Q)) = \left| D_{K_v/\mathbb{Q}_p} \right|_p^{(n+1)/2} |\Delta_v|_{v,K}^{-d} \, Q^{d \sum_i c_{vi}}.$$

(c) *If $K_v = \mathbb{R}$, then*

$$\beta_v^{n+1}(\Pi_v(Q)) = 2^{n+1} |\Delta_v|_{v,K}^{-d} \, Q^{d \sum_i c_{vi}}.$$

(d) *If $K_v = \mathbb{C}$, then*

$$\beta_v^{n+1}(\Pi_v(Q)) = (2\pi)^{n+1} |\Delta_v|_{v,K}^{-d} \, Q^{d \sum_i c_{vi}}.$$

(e) *Let r, s be the number of real and complex places. Then*

$$\beta^{n+1}(\Pi(Q)) = 2^{r(n+1)}(2\pi)^{s(n+1)}|D_{K/\mathbb{Q}}|^{-(n+1)/2}\left(\prod_{v\in S}|\Delta_v|_{v,K}^{-d}\right) Q^{d\sum_{vi} c_{vi}}.$$

Proof: Statement (a) is obvious from the definition of β_v and, similarly, we would get (b), (c), (d) if $\Pi_v(Q)$ is the cuboid $\{|\xi_{vi}|_v \leq Q^{c_{vi}} \mid i = 0, \ldots, n\}$. In general, a transformation of coordinates $\xi_{vi} = L_{vi}(\boldsymbol{\xi}_v')$ is necessary to get $\Pi_v(Q)$ from the cuboid. Then the volume changes by a factor $\|\det(L_{vi})\|_{v,K}^{-1} = |\Delta_v|_{v,K}^{-d}$ (use C.1.3) proving (b)–(d). Finally (e) follows from (C.4) on page 606. □

Corollary 7.5.8. *The volume of an approximation domain is*

$$\beta^{n+1}(\Pi(Q)) \ll Q^{-d\varepsilon/2}.$$

Proof: Clear from the preceding lemma and (7.22). □

We summarize the results obtained so far as:

Lemma 7.5.9. *Suppose the affine subspace theorem in 7.2.5 is false for $\varepsilon > 0$. Let $d = [K : \mathbb{Q}]$, let F be a finite extension of K, which is a field of definition for all forms L_{vi}, and define $r = [F : K]$. Then there are real numbers c_{vi} ($v \in S$, $i = 0, \ldots, n$) with $|c_{iv}| \leq 2r$ and*

$$\sum_{v\in S}\sum_{i=0}^{n} c_{iv} \leq -\frac{\varepsilon}{2},$$

and an infinite set $\mathcal{X}$ of points of K^{n+1}, with the following properties:

(a) *Distinct subsets of n elements of $\mathcal{X}$ span distinct linear subspaces of K^{n+1}.*

(b) *Let ι be the diagonal embedding $\iota : K^{n+1} \longrightarrow K_{\mathbb{A}}^{n+1}$. Then for $\mathbf{x} \in \mathcal{X}$ we have*

$$\iota(\mathbf{x}) \in \Pi(H(\mathbf{x})).$$

(c) $\beta^{n+1}(\Pi(Q)) \ll Q^{-d\varepsilon/2}$.

Proof: Since we assume the falsity of the subspace theorem, there is an infinite set of primitive solutions of inequality (7.18) on page 199 verifying statement (a). Noting that any infinite subset of this set continues to verify (a) and that there must be an approximation class containing such an infinite subset, we have (b) for $\mathbf{x}$ in this approximation class in view of (7.21) on page 200. Finally, we have already verified (c) in Corollary 7.5.8. □

7.5.10. We shall show that given $\varepsilon > 0$ and the real numbers c_{vi} the conclusion of the lemma cannot hold, thereby proving the subspace theorem. The proof is articulated in two separate parts.

The first part is a natural extension of the Roth machinery to our higher-dimensional situation. However, to make it work we need to assume an additional condition, namely that $\Pi(Q)$ contains n linearly independent points $\iota(\mathbf{y}_i)$ $(i = 1, \ldots, n)$. Then we show that the conclusion of Lemma 7.5.9 cannot hold. Note that if $n = 1$ this condition is automatically satisfied (we take $\mathbf{y}_1 = \mathbf{x}$), but if $n \geq 2$ this appears to be a very restrictive condition.

The second part of the proof is quite different in nature and uses geometry of numbers to show, starting with any set of linear forms L_{vi} for which the conclusion of Lemma 7.5.9 holds, that there exists a new set of linear forms for which the same conclusion still holds and moreover the additional condition of existence of independent points is also verified. Since this contradicts the result in the first part, the subspace theorem follows.

Before stating the precise result to be proved in the first part of the proof, we need a definition and two simple results. We assume that the subspace theorem is false, hence Lemma 7.5.9 is in force for the set $\mathcal{X}$. Let $\Pi(Q)$ be the approximation domain of level Q associated to the set $\{c_{vi}\}$.

Definition 7.5.11. *The K-vector space $V(Q)$ is the linear subspace of K^{n+1} spanned by the points $\mathbf{x} \in K^{n+1}$ such that $\iota(\mathbf{x}) \in \Pi(Q)$. The **rank** of $\Pi(Q)$ is the dimension of the K-vector space $V(Q)$.*

Lemma 7.5.12. *Let $\mathcal{X} = \{\mathbf{x}_\nu\}$ and define $Q_\nu = H(\mathbf{x}_\nu)$. Then for all but finitely many $\mathbf{x}_\nu \in \mathcal{X}$ we have*

$$1 \leq \operatorname{rank}(\Pi(Q_\nu)) \leq n.$$

Proof: By Lemma 7.5.9, we have $\iota(\mathbf{x}_\nu) \in \Pi(Q_\nu)$. Thus $V(Q_\nu)$ is not the 0 space and $\operatorname{rank}(\Pi(Q_\nu)) \geq 1$.

In order to prove $\operatorname{rank}(\Pi(Q_\nu)) \leq n$, we use geometry of numbers. Let $\lambda_1, \ldots, \lambda_{n+1}$ be the successive minima of $\Pi(Q_\nu)$ and let $\iota(\mathbf{x}^{(i)})$ be linearly independent points defining the successive minima λ_i. Then $V(Q_\nu)$ is the space spanned by the points $\mathbf{x}^{(i)}$ such that $\lambda_i \leq 1$ and

$$\operatorname{rank}(\Pi(Q_\nu)) = \max_i\{i \mid \lambda_i \leq 1\}.$$

On the other hand, the lower bound in Minkowski's second theorem (see Theorem C.2.11) shows that

$$\lambda_{n+1}^{n+1} \geq \lambda_1 \ldots \lambda_{n+1} \gg \beta^{n+1}(\Pi(Q_\nu))^{-1/d},$$

hence Corollary 7.5.8 yields $\lambda_{n+1} \gg Q_\nu^{\varepsilon/(2n+2)}$ leading to $\lambda_{n+1} > 1$ for large Q_ν. By Northcott's theorem in 2.4.9, we get the claim. □

For the following main result of the first part of the proof, we no longer assume the falsity of the subspace theorem. We will return to the indirect proof only in the second part, starting in Step VIII.

Theorem 7.5.13. *Let S be a finite set of places on K containing the archimedean places. For $v \in S$, let $L_{v0}, \ldots, L_{vn}$ be independent linear forms on K^{n+1} with* *coefficients in a finite extension F/K. Let $c_{vi} \in \mathbb{R}$ $(v \in S,\ i = 0, \ldots, n)$ with $\sum_{vi} c_{vi} < 0$ defining an approximation domain $\Pi(Q)$ of level Q as in 7.5.6. Then $\{V(Q) \mid Q \geq 1,\ \mathrm{rank}(\Pi(Q)) = n\}$ is a finite set.*

We break the proof of this theorem into several steps, trying to imitate the arguments given in Chapter 6 for Roth's theorem. We fix $0 < \varepsilon \leq 1$ with $\sum_{vi} c_{vi} \leq -\varepsilon/2$. The cardinality of the above set of subspaces may be expressed in terms of $K, S, L_{vi}, \varepsilon$ (see **[108]**, Th.C), but we restrict our attention to the qualitative result.

7.5.14. *Step I: The auxiliary polynomial.*

In the proof of Roth's theorem, we start by constructing a polynomial in several variables vanishing to high weighted order at points $(\alpha_v, \ldots, \alpha_v) \in K_v^m$ with $v \in S$. We begin here in a similar way, but here it proves to be convenient to work with multihomogeneous polynomials. So we need to develop some notation first.

For an m-tuple of positive integers $\mathbf{d} = (d_1, \ldots, d_m)$, we denote by $\mathbf{P}(\mathbf{d})$ the K-vector space of multihomogeneous polynomials $P(\mathbf{x}_1, \ldots, \mathbf{x}_m)$ of degree d_h in the block of variables $\mathbf{x}_h$. By Example A.6.13, we can identify

$$\mathbf{P}(\mathbf{d}) = \Gamma(\mathbb{P}, O_{\mathbb{P}}(\mathbf{d})), \quad \mathbb{P} := \underbrace{\mathbb{P}_K^n \times \cdots \times \mathbb{P}_K^n}_{m \quad \text{factors}}.$$

Let $\mathbf{I} = (\mathbf{i}_1, \ldots, \mathbf{i}_m)$ with $\mathbf{i}_h = (i_{h0}, \ldots, i_{hn})$. Using 6.3.1, we define

$$\partial_{\mathbf{I}} := \frac{1}{\mathbf{i}_1!} \cdots \frac{1}{\mathbf{i}_m!} \left(\frac{\partial}{\partial \mathbf{x}_1}\right)^{\mathbf{i}_1} \cdots \left(\frac{\partial}{\partial \mathbf{x}_m}\right)^{\mathbf{i}_m} = \partial_{\mathbf{i}_1} \cdots \partial_{\mathbf{i}_m}.$$

Similarly as in the homogeneous case, the normalizations yield that

$$f(\mathbf{x}) = \sum_{\mathbf{I}} \partial_{\mathbf{I}} f(\mathbf{0}) \mathbf{x}_1^{\mathbf{i}_1} \cdots \mathbf{x}_m{}^{\mathbf{i}_m}$$

for any polynomial or more generally any power series f.

Let $\mathbf{L}(\mathbf{x}) = (L_0(\mathbf{x}), \ldots, L_n(\mathbf{x}))$ be linearly independent linear forms over F. Then, given $P \in \mathbf{P}(\mathbf{d})$ and the differential operator operator $\partial_{\mathbf{I}}$, we can write

$$\partial_{\mathbf{I}} P(\mathbf{x}_1, \ldots, \mathbf{x}_m) = \sum_{\mathbf{J}} a(\mathbf{L}; \mathbf{J}; \mathbf{I})\, \mathbf{L}(\mathbf{x}_1)^{\mathbf{j}_1} \cdots \mathbf{L}(\mathbf{x}_m)^{\mathbf{j}_m}$$

with $\mathbf{j}_h = (j_{h0}, \ldots, j_{hn})$, $\mathbf{J} = (\mathbf{j}_1, \ldots, \mathbf{j}_m)$, and $a(\mathbf{L}; \mathbf{J}; \mathbf{I}) \in F$. By homogeneity, we also have $|\mathbf{j}_h| = d_h - |\mathbf{i}_h|$ for $h = 1, \ldots, m$.

If $\mathbf{V} = (\mathbf{v}_1, \ldots, \mathbf{v}_m)$, we will write $\mathbf{V}_i$ for the vector

$$\mathbf{V}_i = (v_{1i}, \ldots, v_{mi}).$$

Also, it will prove to be convenient to use the notation

$$(\mathbf{V}/\mathbf{d}) := \frac{|\mathbf{v}_1|}{d_1} + \cdots + \frac{|\mathbf{v}_m|}{d_m}.$$

Lemma 7.5.15. *Let* $0 < \eta \le 2/(n+1)$, $r := [F : K]$ *and*

$$m \ge \frac{4}{(n+1)(n+2)\eta^2} \log(2(n+1)r|S|).$$

Then there are constants C_2, C_3 *depending only on the forms* L_{vi} *and* K *such that for* $d_1, \ldots, d_m$ *sufficiently large, there is a non-zero polynomial* $P \in \mathbf{P}(\mathbf{d})$ *with*

$$h(P) \le C_2\,|\mathbf{d}|, \qquad h((a(\mathbf{L}_v; \mathbf{J}, \mathbf{I}))) \le C_3\,|\mathbf{d}| \qquad (v \in S).$$

Moreover, $a(\mathbf{L}_v; \mathbf{J}; \mathbf{I}) = 0$ *for* $v \in S$*, whenever* $\mathbf{J}$ *and* $\mathbf{I}$ *satisfy*

$$(\mathbf{J}_i/\mathbf{d}) \le \frac{m}{n+1} - 2m\eta \quad \text{or} \quad (\mathbf{J}_i/\mathbf{d}) \ge \frac{m}{n+1} + 2nm\eta$$

for some $i = 0, \ldots, n$ *and*

$$(\mathbf{I}/\mathbf{d}) \le m\eta.$$

Here and in the following, the lower bound for $d_1, \ldots, d_m$ may depend on the given data including η, m, and so on.

Proof: We give here a proof which parallels the argument used in Lemma 6.3.4 in the construction of the auxiliary polynomial. We are interested here only in asymptotics for $d_1 \to \infty, \ldots, d_m \to \infty$ for fixed m.

The dimension of the vector space $\mathbf{P}(\mathbf{d})$ is the number of m-tuples $\mathbf{J} = (\mathbf{j}_1, \ldots, \mathbf{j}_m)$ with non-negative integer components such that $(|\mathbf{j}_1|, \ldots, |\mathbf{j}_m|) = \mathbf{d}$.

Since prescribing j_{hi} for $i = 0, \ldots, n-1$ with the sum not exceeding d_h determines j_{hn}, we have $\dim(\mathbf{P}(\mathbf{d})) \sim d_1 \ldots d_m V_0$ with

$$V_0 = \prod_{h=1}^{m} \int_0^1 \cdots \int_0^1 \chi_{[0,1]}\left(\sum_{i=0}^{n-1} x_{hi}\right) dx_{h0} \cdots dx_{h,n-1}$$

and $\chi_{[a,b]}$ the characteristic function of the interval $[a, b]$. The integral equals $(n!)^{-1}$ and we obtain

$$V_0 = (n!)^{-m}. \tag{7.23}$$

Let us fix $v \in S$ and compute, for $\eta > 0$, an asymptotic upper bound for the number of linear conditions (with coefficients in F) imposed on an element P of $\mathbf{P}(\mathbf{d})$ by the vanishing of the coefficients $a(\mathbf{L}_v; \mathbf{J}; \mathbf{0})$ whenever

$$(\mathbf{J}_i/\mathbf{d}) \le \frac{m}{n+1} - m\eta \tag{7.24}$$

for some fixed i, for example $i = 0$. The number of such m-tuples $\mathbf{J}$ is asymptotic to $d_1 \cdots d_m V$ with

$$V = \int_0^1 \cdots \int_0^1 \chi_{[0, \frac{m}{n+1} - m\eta]}\left(\sum_{h=1}^{m} x_{h0}\right) \left\{\prod_{h=1}^{m} \chi_{[0,1-x_{h0}]}\left(\sum_{i=1}^{n-1} x_{hi}\right)\right\} \prod_{h=1}^{m} \prod_{i=0}^{n-1} dx_{hi}.$$

Therefore, we have

$$V = \int_0^1 \cdots \int_0^1 \chi_{[0, \frac{m}{n+1} - m\eta]} \left(\sum_{h=1}^m x_{h0} \right) \prod_{h=1}^m \frac{(1 - x_{h0})^{n-1}}{(n-1)!} \, dx_{10} \cdots dx_{m0}.$$

In order to obtain a good upper bound for V, we proceed as in Lemma 6.3.5 noting the majorization

$$\chi_{[a,b]}(x) \le e^{\lambda \cdot (b-x)},$$

valid for $\lambda \ge 0$. This decouples the variables x_{h0} and we get

$$V \le \left(e^{\lambda/(n+1) - \lambda\eta} \int_0^1 e^{-\lambda x} \frac{(1-x)^{n-1}}{(n-1)!} \, dx \right)^m.$$

for any $\lambda \ge 0$. Suppose now that $0 < \lambda \le n + 4$. By expanding the exponential into a MacLaurin series and integrating term by term we obtain

$$\begin{aligned} \int_0^1 e^{-\lambda x} \frac{(1-x)^{n-1}}{(n-1)!} \, dx &= \sum_{k=0}^\infty \frac{(-\lambda)^k}{(n+k)!} \\ &\le \frac{1}{n!} \left\{ 1 - \frac{\lambda}{n+1} + \frac{1}{(n+1)(n+2)} \lambda^2 \right\} \\ &< \frac{1}{n!} \exp\left(-\frac{\lambda}{n+1} + \frac{1}{(n+1)(n+2)} \lambda^2 \right). \end{aligned}$$

Hence from the last two displayed equations we conclude that

$$V < \frac{1}{(n!)^m} e^{-m\lambda\eta + m\lambda^2 \frac{1}{(n+1)(n+2)}}$$

provided $0 < \lambda \le n + 4$. If we choose for example $\lambda = \eta(n+1)(n+2)/2$ with $\eta \le 2/(n+1)$, we get

$$V < \frac{1}{(n!)^m} e^{-(n+1)(n+2)\eta^2 m/4}. \tag{7.25}$$

Now we apply the relative Siegel lemma in 2.9.19 to find a non-zero P with small coefficients satisfying (7.24) for some $i = 0, \ldots, n$. The calculation is entirely analogous to the proof of Lemma 6.3.4. The final result is that, if

$$r(n+1)|S| \, V/V_0 \le \frac{1}{2}, \tag{7.26}$$

then $h(P) \le C_2' |\mathbf{d}|$, for $d_1, \ldots, d_m$ sufficiently large and with C_2' depending only on the forms L_{vi} and K.

Applying a differential operator $\partial_{\mathbf{I}}$ to P increases the height by not more than $(\log 2)|\mathbf{d}|$, proving $h(\partial_{\mathbf{I}} P) \le C_2 |\mathbf{d}|$. Then a bound $h((a(\mathbf{L}_v; \mathbf{J}; \mathbf{I}))) \le C_3 |\mathbf{d}|$ is obtained looking at $(\partial_{\mathbf{I}} P)(\mathbf{M}_v(\mathbf{x}))$, where $\mathbf{M}_v \circ \mathbf{L}_v(\mathbf{x}) = \mathbf{x}$.

We note the vanishing of $a(\mathbf{L}_v;\mathbf{J};\mathbf{I})$ whenever

$$(\mathbf{J}_i/\mathbf{d}) \leq m/(n+1) - 2m\eta$$

for some $i = 0,\ldots,n$ and for every $\mathbf{I}$ with $(\mathbf{I}/\mathbf{d}) \leq m\eta$.

For $(\mathbf{I}/\mathbf{d}) \leq m\eta$, the vanishing of $a(\mathbf{L}_v;\mathbf{J};\mathbf{I})$ in the complementary range

$$(\mathbf{J}_i/\mathbf{d}) \geq m/(n+1) + 2nm\eta \quad \text{for some } i$$

is automatic, because $a(P;\mathbf{J};\mathbf{I}) \neq 0$ only if $(\mathbf{J}_i/\mathbf{d}) > m/(n+1) - 2m\eta$ for every i. In view of the condition $|\mathbf{j}_h| \leq d_h$, this implies that $\mathbf{J}$ verifies

$$(\mathbf{J}_{i_0}/\mathbf{d}) < \sum_{i=0}^{n}(\mathbf{J}_i/\mathbf{d}) - \left(\frac{m}{n+1} - 2m\eta\right) n \leq m - \frac{mn}{n+1} + 2mn\eta = \frac{m}{n+1} + 2mn\eta$$

for every i_0, hence $\mathbf{J}$ is outside of the complementary range defined above.

It remains to verify condition (7.26). By (7.23) on page 205 and (7.25) this is is verified as soon as m is so large that

$$r(n+1)|S|e^{-(n+1)(n+2)\eta^2 m/4} \leq \frac{1}{2},$$

which is our initial assumption in the lemma. □

7.5.16. *Step II: A generalization of Roth's lemma.*

In the higher-dimensional setting of the subspace theorem, the non-vanishing of the auxiliary polynomial cannot be obtained at a single point $(\mathbf{x}_1,\ldots,\mathbf{x}_m)$ and has to be replaced by the vanishing on a product $V_1 \times \cdots \times V_m$, where the factors V_h are K-vector subspaces of K^{n+1}, all of dimension n. Accordingly, the notion of index of a multihomogeneous polynomial has to be changed as follows.

Let $M_1,\ldots,M_m \in \overline{\mathbb{Q}}[x_0,\ldots,x_n]$ be non-zero linear forms, let $\mathbf{M} = (M_1,\ldots,M_m)$, and let $\mathbf{d} = (d_1,\ldots,d_m)$ be an m-tuple of positive numbers.

Definition 7.5.17. *Denote by $\mathfrak{I}(t;\mathbf{d};\mathbf{M})$ the ideal in $\overline{\mathbb{Q}}[\mathbf{x}_1,\ldots,\mathbf{x}_m]$ generated by all monomials $M_1(\mathbf{x}_1)^{j_1}\cdots M_m(\mathbf{x}_m)^{j_m}$ with*

$$(\mathbf{j}/\mathbf{d}) \geq t.$$

Then for a multihomogeneous polynomial P, the index of P with respect to $\mathbf{M}$ and weights $\mathbf{d}$ is

$$\mathrm{ind}(P;\mathbf{d};\mathbf{M}) := \sup\{t \geq 0 \mid P \in \mathfrak{I}(t;\mathbf{d};\mathbf{M})\}.$$

7.5.18. For $n=1$, $M_j = x_1 - \alpha_j x_0 (j=1,\ldots,m)$ and $Q := P(1,x_1,\ldots,1,x_m)$, it is clear that $\mathrm{ind}(P;\mathbf{d};\mathbf{M})$ is the same as $\mathrm{ind}(Q;\mathbf{d};\boldsymbol{\alpha})$ defined in 6.3.2. Clearly, the properties from 6.3.2 extend to the multihomogeneous case.

Now we can state the **generalized Roth's lemma:**

Lemma 7.5.19. *Let $P(\mathbf{x}_1,\dots,\mathbf{x}_m) \in \overline{\mathbb{Q}}[\mathbf{x}_1,\dots,\mathbf{x}_m]$ be a multihomogeneous polynomial, not identically 0, with partial degrees at most $d_1,\dots,d_m$. Let $\mathbf{M} = (M_1,\dots,M_m)$ be non-zero linear forms in $\overline{\mathbb{Q}}[x_0,\dots,x_n]$, and let $0 < \sigma \le \frac{1}{2}$. Suppose that:*

(a) *the degrees $d_1,\dots,d_m$ are rapidly decreasing, namely*

$$d_{j+1}/d_j \le \sigma \qquad (j = 1,\dots,m-1);$$

(b) *the linear forms M_j have large height, namely*

$$\min_j d_j h(M_j) \ge n\sigma^{-1}(h(P) + 4md_1).$$

Then

$$\operatorname{ind}(P;\mathbf{d};\mathbf{M}) \le 2m\sigma^{1/2^{m-1}}.$$

Proof: The idea is to specialize, in each group of variables $\mathbf{x}_h$, all variables except two to 0 (which we may relabel as (x_{h0}, x_{h1})) and apply Roth's lemma from 6.3.7. However, we must make sure that the specialized polynomial is not identically 0 and also we must be able to compare index and heights before and after specialization.

We may assume $n \ge 2$ since if $n = 1$ this is simply Roth's lemma in a homogeneous setting. Let F be a number field containing the coefficients of P and $\mathbf{M}$. Let $\mathbf{b} = (b_0,\dots,b_n) \in F^{n+1} \setminus \{\mathbf{0}\}$ and suppose for simplicity that $b_0 \ne 0$. Then

$$h(\mathbf{b}) \le \sum_{v \in M_F} \sum_{i=1}^{n} \log(\max(|b_0|_v, |b_i|_v))$$
$$\le n \max_{i=1,\dots,n} h((b_0, b_i)).$$

Thus, after relabeling the variables $(x_{j0},\dots,x_{jn})$ we may assume that the linear forms

$$M_j(\mathbf{x}_j) = b_{j0}x_{j0} + \cdots + b_{jn}x_{jn}$$

specialize under $x_{ji} = 0$ $(i = 2,\dots,n)$ to

$$\widetilde{M}_j(\mathbf{x}_j) = b_{j0}x_{j0} + b_{j1}x_{j1}$$

with

$$h(M_j) \ge h(\widetilde{M}_j) \ge \frac{1}{n} h(M_j).$$

Since $b_{j0} \ne 0$, we can write uniquely

$$P = \sum_{\mathbf{j}} c(\mathbf{j}) M_1(\mathbf{x}_1)^{j_1} \cdots M_m(\mathbf{x}_m)^{j_m} q_{\mathbf{j}}(\mathbf{x}'_1,\dots,\mathbf{x}'_m), \tag{7.27}$$

where the coefficients $q_{\mathbf{j}}$ are polynomials in the variables $\mathbf{x}'_j = (x_{j1},\dots,x_{jm})$. This makes it plain that in computing the index of P we may restrict ourselves to decompositions as in (7.27).

After removing from P the highest factor x_{12}^k dividing it, we specialize $x_{12}=0$, obtaining a new polynomial P^*, not identically 0. Since the coefficients of P^* are a subset of the set of coefficients of P, we certainly have

$$h(P^*) \le h(P).$$

Moreover, by the uniqueness of the decomposition (7.27), if x_{12}^k divides P, then every, q_{j} in (7.27) is divisible by x_{12}^k, and it follows that

$$\mathrm{ind}(P^*; \mathbf{M}|_{x_{12}=0}; \mathbf{d}) = \mathrm{ind}(P; \mathbf{M}; \mathbf{d}).$$

Proceeding step-by-step in this way, we eventually arrive at a multihomogeneous polynomial $\widetilde{P}(\widetilde{\mathbf{x}}_1, \ldots, \widetilde{\mathbf{x}}_m)$, in the variables $\widetilde{\mathbf{x}}_j = (x_{j0}, x_{j1})$, not identically 0, with multidegree $\widetilde{\mathbf{r}} \le \mathbf{d}$ componentwise, such that

$$h(\widetilde{P}) \le h(P), \quad \mathrm{ind}(\widetilde{P}; \widetilde{\mathbf{M}}; \mathbf{d}) = \mathrm{ind}(P; \mathbf{M}; \mathbf{d}), \quad h(\widetilde{M}_j) \ge h(M_j)/n$$

for $j = 1, \ldots, m$.

We apply Roth's lemma in 6.3.7 to the polynomial $\widetilde{P}$, with $\widetilde{\mathbf{M}}$ in place of $\boldsymbol{\xi}$ (this is due to the fact that we work here in a homogeneous setting). Since $d_j h(\widetilde{M}_j) \ge n^{-1} d_j h(M_j)$, and $h(\widetilde{P}) \le h(P)$, condition (b) of Roth's lemma is verified as soon as

$$\min_j d_j h(M_j) \ge n\sigma^{-1}(h(P) + 4md_1).$$

Therefore

$$\mathrm{ind}(P; \mathbf{M}; \mathbf{d}) = \mathrm{ind}(\widetilde{P}; \widetilde{\mathbf{M}}; \mathbf{d}) \le 2m\sigma^{1/2^{m-1}}. \qquad \square$$

7.5.20. *Step III: The height of $V(Q)$.*

As in Section 2.8, we define the height of a non-zero subspace V of $\overline{\mathbb{Q}}$ by $h(V) := h(b_1 \wedge \cdots \wedge b_k)$, where $b_1, \ldots, b_k$ is a basis of V. In the Roth case $n = 1$ we have $\dim(V(Q)) = 1$ and $V(Q)$ is generated by $\mathbf{x}$, hence

$$h(V(Q)) = h(\mathbf{x}) = \log(Q).$$

The goal of this key step is to show that a result of comparable strength still holds if $n \ge 2$.

Lemma 7.5.21. *Let* $\mathrm{rank}(\Pi(Q)) = n$, *let* $c_{\max} = \max_{vi} c_{vi}$, *and suppose that* $\log(Q) \ge C_4\,\varepsilon^{-1}$. *There is a linear space W, independent of $\Pi(Q)$ and ε, such that either $V(Q) = W$ or*

$$(4r\,|S|)^{-1}\varepsilon\,\log(Q) - C_5 \le h(V(Q)) \le nc_{\max}\,|S|\,\log(Q) + C_6.$$

Proof: Let $\mathbf{y}^{(1)}, \ldots, \mathbf{y}^{(n)}$ be a basis of $V(Q)$ with $\iota(\mathbf{y}^{(i)}) \in \Pi(Q)$. Then $V(Q)$ is given by an equation

$$\mathbf{y}^{(1)} \wedge \cdots \wedge \mathbf{y}^{(n)} \wedge \mathbf{x} = 0. \tag{7.28}$$

Hence

$$h(V(Q)) = h(\mathbf{y}^{(1)} \wedge \cdots \wedge \mathbf{y}^{(n)}) \le \sum_{i=1}^{n} h(\mathbf{y}^{(i)}) + C_6'. \tag{7.29}$$

We also have for $v \in S$, using $\iota(\mathbf{y}^{(i)}) \in \Pi(Q)$ and that the form L_{vi} $(i = 0, \ldots, n)$ are linearly independent

$$|\mathbf{y}^{(i)}|_{v,K} \ll |\mathbf{L}_v(\mathbf{y}^{(i)})|_{v,K} \le \max_j Q^{c_{vj}} \le Q^{c_{\max}},$$

while $|\mathbf{y}^{(i)}|_v \le 1$ for $v \notin S$. In view of (7.29), we get the upper bound for $h(V(Q))$.

The proof of the lower bound is more intricate. For the linear forms

$$\widehat{L}_{vk} = (L_{v0} \wedge \cdots \wedge L_{v,k-1} \wedge L_{v,k+1} \wedge \cdots \wedge L_{vn})^* \quad (k = 0, \ldots, n)$$

we set

$$D_{vk} = \widehat{L}_{vk}((\mathbf{y}^{(1)} \wedge \cdots \wedge \mathbf{y}^{(n)})^*).$$

Since the $*$-operator is an isometry and by Laplace's identity (7.16) on page 198 we have

$$D_{vk} = \det\Big(L_{vi}(\mathbf{y}^{(j)}) \Big)_{i \in \{0,\ldots,n\} \setminus \{k\}; j \in \{1,\ldots,n\}}.$$

Therefore, expanding the determinant and using $|L_{vi}(\mathbf{y}^{(j)})|_{v,K} \le Q^{c_{vi}}$, we find

$$\begin{aligned} |D_{vk}|_{v,K} &\le \max(1, |n!|_v) \max_{\pi} \prod_{i \ne k} |L_{vi}(\mathbf{y}^{\pi(i)})|_{v,K} \\ &\le \max(1, |n!|_v)\, Q^{-c_{vk} + \sum_i c_{vi}}, \end{aligned} \tag{7.30}$$

where π ranges over all bijective mappings $\pi : \{0, \ldots, n\} \setminus \{k\} \to \{1, \ldots, n\}$.

Now suppose for the time being that there is an $|S|$-tuple $\{i_v\}$ such that

$$\sum_{v \in S} c_{vi_v} \ge -\frac{\varepsilon}{4}, \qquad D_{vi_v} \ne 0 \quad (v \in S). \tag{7.31}$$

Then by (7.30) we get

$$0 < \prod_{v \in S} |D_{vi_v}|_{v,K} \le n!\, Q^{-\sum_v c_{vi_v} + \sum_{vi} c_{vi}} \le n!\, Q^{\frac{\varepsilon}{4} - \frac{\varepsilon}{2}} = n!\, Q^{-\frac{\varepsilon}{4}}. \tag{7.32}$$

Since $0 \ne D_{vi_v} \in F$, the fundamental inequality in (1.8) on page 20 yields

$$\begin{aligned} \sum_{v \in S} \log |D_{vi_v}|_{v,K} &\ge -r \sum_{v \in S} h(D_{vi_v}) \\ &= -r \sum_{v \in S} h(\widehat{L}_{vi_v}((\mathbf{y}^{(1)} \wedge \cdots \wedge \mathbf{y}^{(n)})^*)) \\ &\ge -r\, |S|\, h(V(Q)) - C_7. \end{aligned} \tag{7.33}$$

By (7.32) and (7.33), we get

$$-r\,|S|\,h(V(Q)) - C_7 \le -\frac{\varepsilon}{4}\,\log(Q) + \log(n!),$$

thereby proving the lower bound estimate for $h(V(Q))$.

It remains to verify the preceding assumption on the existence of the $|S|$-tuple $\{i_v\}$. Hence let us suppose that this is not the case. Let

$$\mathcal{I}_v = \{i \in \{0,\dots,n\} \mid D_{vi} \ne 0\}.$$

The system of equations

$$\widehat{\mathbf{L}}_{vi}(\mathbf{y}) = 0 \qquad (i \in \{0,\dots,n\} \setminus \mathcal{I}_v)$$

has the non-trivial solution $\mathbf{y} = (\mathbf{y}^{(1)} \wedge \cdots \wedge \mathbf{y}^{(n)})^* \in K^{n+1} \setminus \{\mathbf{0}\}$. However, this system of equations depends only on the linear forms L_{vi} and not on $\Pi(Q)$.

Hence let us fix a non-trivial solution $\mathbf{w} \in K^{n+1} \setminus \{\mathbf{0}\}$ of this system (if we deal with an empty system, any non-zero vector in K^{n+1} will work here) and let W be the K-vector space $W = \{\mathbf{x} \in K^{n+1} \mid \mathbf{w}\cdot\mathbf{x} = 0\}$. We will show that, if $\log(Q)$ is sufficiently large, then $V(Q) \subset W$, hence $V(Q) = W$ because they have the same dimension n, which is the alternative conclusion of the lemma.

In order to prove this, we deduce from Laplace's expansion (7.17) on page 198 the identity

$$\mathbf{x}\cdot\mathbf{w} = \det(\mathbf{L}_v)^{-1}\sum_{i=0}^{n}(-1)^i L_{vi}(\mathbf{x})\widehat{L}_{vi}(\mathbf{w})$$

for any $\mathbf{x}$ and $\mathbf{w}$. Moreover, in our case we have $\widehat{L}_{vi}(\mathbf{w}) = 0$ for $i \notin \mathcal{I}_v$, hence

$$\mathbf{x}\cdot\mathbf{w} = \det(\mathbf{L}_v)^{-1}\sum_{i\in\mathcal{I}_v}(-1)^i L_{vi}(\mathbf{x})\widehat{L}_{vi}(\mathbf{w}). \tag{7.34}$$

Now for each $v \in S$ choose j_v such that

$$c_{vj_v} = \max_{i\in\mathcal{I}_v} c_{vi}$$

and note that, if $\mathbf{x} \in V(Q) \setminus W$ with $\iota(\mathbf{x}) \in \Pi(Q)$, then we have

$$|L_{vi}(\mathbf{x})|_{v,K} \le Q^{c_{vi}} \le Q^{c_{vj_v}} \quad (i \in \mathcal{I}_v).$$

Thus for $v \in S$ and every such $\mathbf{x}$ we have a bound

$$|\mathbf{x}\cdot\mathbf{w}|_v \ll Q^{c_{vj_v}}. \tag{7.35}$$

On the other hand, since $j_v \in \mathcal{I}_v$ our assumption of the non-existence of a good $|S|$-tuple as in (7.31) shows that

$$\sum_{v\in S} c_{vj_v} < -\varepsilon/4.$$

Then, by the product formula, (7.34), (7.35), and the fundamental inequality (1.8) on page 20, we find

$$1 = \prod_{v \in M_K} |\mathbf{x} \cdot \mathbf{w}|_v = \prod_{v \in S} |\mathbf{x} \cdot \mathbf{w}|_v \prod_{v \notin S} |\mathbf{x} \cdot \mathbf{w}|_v \leq \prod_{v \in S} |\mathbf{x} \cdot \mathbf{w}|_v \prod_{v \notin S} |\mathbf{w}|_v$$
$$\ll Q^{\sum_{v \in S} c_{vj_v}} \ll Q^{-\varepsilon/4}.$$

Therefore, $\log(Q) \ll \varepsilon^{-1}$ contradicting, for large C_4, the hypothesis $\log(Q) \geq C_4\varepsilon^{-1}$ of the lemma. □

7.5.22. *Step IV: Application of the generalized Roth's lemma.*

Here we combine Lemma 7.5.15 and the generalized Roth's lemma from 7.5.19, obtaining a polynomial vanishing to high order at $\mathbf{L}_v$, $v \in S$, but not identically 0 when restricted to $V_1 \times \cdots \times V_m$, where the V_h s are suitable linear subspaces of K^{n+1}.

We choose parameters as follows

$$m \geq \frac{4}{(n+1)(n+2)\eta^2} \log(2(n+1)[F:K]\,|S|),$$
$$\sigma = (\eta/4)^{2^{m-1}}, \quad \eta \leq \frac{1}{n+1}.$$

We prove Theorem 7.5.13 by contradiction. Then there is a sequence $Q_\nu \geq 1$ with $\operatorname{rank}(\Pi_v(Q_\nu)) = n$ and $V(Q_\nu) \neq V(Q_\mu)$ for $\mu \neq \nu$. We may assume that $V(Q_\nu)$ omits the exceptional subspace W from Lemma 7.5.21. Going to a subsequence (again denoted by Q_ν), we may assume that $\log(Q_\nu) \to \infty$ at an arbitrarily fast rate. Indeed, if this were not the case, then $V(Q_\nu)$ would have bounded height and, by Northcott's theorem in 2.4.9, there would be only finitely many spaces $V(Q_\nu)$. Hence we may and shall assume that

$$\log(Q_1) \geq C,$$
$$\log(Q_{j+1}) \geq 2\sigma^{-1}\log(Q_j) \tag{7.36}$$

for every j and any given constant C, which may depend on the given parameters.

Since $\operatorname{rank}(\Pi(Q_\nu)) = n$, the vector space $V(Q_\nu)$ is defined by a single equation

$$b_{\nu 0}x_0 + \cdots + b_{\nu n}x_n = 0$$

and we denote by M_ν the associated linear form, thus (7.28) on page 209 shows

$$h(V(Q_j)) = h(M_j)$$

for every j. Now we take $d_j = \lfloor D/\log Q_j \rfloor$ $(j = 1, \ldots, m)$ and apply Lemma 7.5.15. This gives us, for large D, a certain non-zero polynomial P of multidegree $\mathbf{d}$. We claim that P and $\mathbf{M} = (M_1, \ldots, M_m)$ satisfy the hypotheses of the generalized Roth lemma in 7.5.19.

In order to verify (a) and (b) of Lemma 7.5.19 we appeal to Lemma 7.5.21.

Condition (a), in view of our choice of the weights d_j, follows from (7.36) for D sufficiently large, which we will assume from now on.

For condition (b), we note first that, if $\log(Q_1) \geq C_4\varepsilon^{-1}$, we have by Lemma 7.5.21 the estimate

$$\begin{aligned} d_j h(M_j) &= d_j h(V(Q_j)) \\ &\geq \lfloor D/\log(Q_j)\rfloor((4r\,|S|)^{-1}\varepsilon\,\log(Q_j) - C_5) \geq C_8\varepsilon D. \end{aligned}$$

On the other hand, the bound for $h(P)$ from Lemma 7.5.15 and the just verified $d_{j+1} \leq \sigma d_j$ yield

$$\begin{aligned} n\sigma^{-1}(h(P) + 4md_1) &\leq n\sigma^{-1}(C_2 + 4m)(D/\log(Q_1))(1 + \sigma + \cdots + \sigma^{m-1}) \\ &\ll \sigma^{-1}mD/\log(Q_1), \end{aligned}$$

which is negligible with respect to εD if $\log(Q_1) \geq C_9(\varepsilon\sigma)^{-1}m$ with C_9 large enough. This proves that condition b) is satisfied for large $\log(Q_1)$.

Therefore, Lemma 7.5.19 yields

$$\mathrm{ind}(P; \mathbf{d}; \mathbf{M}) \leq m\eta/2.$$

It follows that there is $\mathbf{I} = (\mathbf{i}_1, \ldots, \mathbf{i}_m)$ with

$$(\mathbf{I}/\mathbf{d}) \leq m\eta/2$$

such that $\partial_{\mathbf{I}} P$ does not vanish identically on the product space

$$V(Q_1) \times \cdots \times V(Q_m).$$

7.5.23. *Step V: Non-vanishing at a small point of* $V(Q_1) \times \cdots \times V(Q_m)$.

What we really want is the non vanishing of a derivative of P at a point $\mathbf{Y} = (\mathbf{y}_1, \ldots, \mathbf{y}_m)$ with $\mathbf{y}_h$ of small height, say comparable with Q_h. The next easy lemma allows us to do so from the information we have gathered so far.

Lemma 7.5.24. *Let k be field of characteristic 0 and $x_1, \ldots, x_N$ algebraically independent over k. Let $f(x_1, \ldots, x_N) \in k[x_1, \ldots, x_N]$ be a non-zero polynomial of degree at most e_j in x_j, and let $B > 0$. Then there are rational integers z_j and i_j $(j = 1, \ldots, N)$, with*

$$|z_j| \leq B, \qquad 0 \leq i_j \leq e_j/B$$

for $j = 1, \ldots, N$, such that for $\mathbf{i} = (i_1, \ldots, i_N)$ we have

$$\partial_{\mathbf{i}} f(z_1, \ldots, z_N) \neq 0.$$

Proof: By induction on N. If $N = 1$, it suffices to note that f cannot be divisible by $\{\prod_{|b|\leq B}(x_1 - b)\}^{\lfloor e_1/B\rfloor + 1}$, which has degree strictly greater than e_1. The general case is a straightforward induction. □

The general idea in applying this lemma is to consider many points which are linear combinations of a basis $\mathbf{y}^{(i)}$ with small coefficients and show that the auxiliary polynomial we have constructed cannot vanish at every such point. However, we must be sure that the height of the polynomial evaluated at the point does not increase more that $O(|\mathbf{d}|)$. This means that the size B of the coefficients must be kept bounded and then we would have too few points at our disposal to meet our goal. On the other hand, applying a differential operator $\partial_{\mathbf{I}}$ only increases the height by $O(|\mathbf{d}|)$, and $\partial_{\mathbf{i}} f$ cannot vanish at a point for every $\mathbf{i}$ unless f is identically 0. Thus, if we vary not only the choice of the point but also vary the polynomial by applying a differential operator, we can prove what we want without increasing the height too much in the process.

The details are as follows. Let $\mathbf{y}_h^{(1)}, \ldots, \mathbf{y}_h^{(n)} \in K^{n+1}$ be linearly independent points with $\iota(\mathbf{y}_h^{(l)}) \in \Pi(Q_h)$, which is possible because $\Pi(Q_h)$ has rank n. We write $\mathbf{z}_h = (z_{h1}, \ldots, z_{hn})$, $\mathbf{Z} = (\mathbf{z}_1, \ldots, \mathbf{z}_m)$, and $\mathbf{Y}_h = (\mathbf{y}_h^{(1)}, \ldots, \mathbf{y}_h^{(n)})$. Then the polynomial

$$R(\mathbf{Z}) := (\partial_{\mathbf{I}} P)(\mathbf{z}_1 \cdot \mathbf{Y}_1, \ldots, \mathbf{z}_m \cdot \mathbf{Y}_m)$$

is not identically 0 because $\partial_{\mathbf{I}} P$ does not vanish identically on $V(Q_1) \times \cdots \times V(Q_m)$. Clearly, $R(\mathbf{Z})$ has degree at most d_h in the block of variables $\mathbf{z}_h$. Hence by the previous lemma there are a point $\mathbf{Z}' \in \mathbb{Z}^{mn}$, and rational integers j_{hl} $(h = 1, \ldots, m,\ l = 1, \ldots, n)$, such that

$$|z'_{hl}| \le B, \qquad 0 \le j_{hl} \le d_h/B$$

and

$$(\partial_{\mathbf{J}} R)(\mathbf{Z}') \neq 0.$$

Since $\partial_{\mathbf{J}} R$ is a linear combination of derivatives $\partial_{\mathbf{I}'} P$ evaluated at the point

$$\mathbf{X}' = (\mathbf{x}'_1, \ldots, \mathbf{x}'_m) := \left(\sum_{l=1}^{n} z'_{1l} \cdot \mathbf{y}_1^{(l)}, \ldots, \sum_{l=1}^{n} z'_{ml} \cdot \mathbf{y}_m^{(l)} \right),$$

there is a differential operator $\partial_{\mathbf{I}'}$ with $\mathbf{I}' = \mathbf{I} + \mathbf{I}^*$ for some $\mathbf{I}^* = (\mathbf{i}_1^*, \ldots, \mathbf{i}_m^*)$ with $|\mathbf{i}_h^*| \le n d_h / B$, such that

$$(\partial_{\mathbf{I}'} P)(\mathbf{X}') \neq 0.$$

Hence $(\mathbf{I}'/\mathbf{d}) \le \frac{m\eta}{2} + \frac{mn}{B}$ and if we choose $B = 2n/\eta$ we see that

$$(\mathbf{I}'/\mathbf{d}) \le m\eta. \tag{7.37}$$

We write $T(\mathbf{X}) := \partial_{\mathbf{I}'} P(\mathbf{X})$.

Lemma 7.5.25. *Let m and $0 < \eta \le 1/(n+1)$ be given, with*

$$m \ge \frac{4}{(n+1)(n+2)\eta^2} \log(2(n+1)[F : K]\,|S|).$$

Let $\sigma = (\eta/4)^{2^{m-1}}$ *and let* Q_h $(h = 1, \dots, m)$ *be such that* $\mathrm{rank}(\Pi(Q_h)) = n$ *and, for a certain constant* C_{10},

$$\log(Q_1) \geq C_{10}(\varepsilon\sigma)^{-1}m, \quad \log(Q_{h+1}) \geq 2\sigma^{-1}\log(Q_h) \qquad (h = 1, \dots, m-1).$$

For $h = 1, \dots, m$, *let* $\mathbf{y}_h^{(l)}$ $(l = 1, \dots, n)$ *be a basis of* $V(Q_h)$ *such that* $\iota(\mathbf{y}_h^{(l)}) \in \Pi(Q_h)$. *Then there are a non-zero multihomogeneous polynomial* $T(\mathbf{X}) \in K[\mathbf{X}]$, *of multidegree majorized by* $\mathbf{d}$ *componentwise, and rational integers* z_{hl} *with*

$$|z_{hl}| \leq 2n/\eta$$

with the following properties. Let $\mathbf{x}'_h = \sum_{l=1}^{n} z_{hl}\mathbf{y}_h^{(l)}$. *Then:*

(a) $T(\mathbf{X}') \neq 0$.

(b) $h(T) \ll d_1$.

(c) *If* $v \in S$ *and* $T(\mathbf{X}) = \sum_{\mathbf{J}} a(\mathbf{L}_v; \mathbf{J})\mathbf{L}_v(\mathbf{x}_1)^{\mathbf{j}_1} \dots \mathbf{L}_v(\mathbf{x}_m)^{\mathbf{j}_m}$, *then*

$$a(\mathbf{L}_v; \mathbf{J}) = 0 \quad \textit{unless} \quad \left|(\mathbf{J}_i/\mathbf{d}) - \frac{m}{n+1}\right| \leq 2nm\eta \quad \textit{for every } i.$$

(d) $h((a(\mathbf{L}_v; \mathbf{J}))) \ll d_1$ *for every* $v \in S$.

Proof: Statement (a) follows from the construction of $\mathbf{X}'$. Also, (b) and (d) follow from $h(P) \ll |\mathbf{d}|$ (see Lemma 7.5.15) and $|\mathbf{d}| \ll d_1$. Finally, we note that $a(\mathbf{L}_v; \mathbf{J})$ is non-zero if and only if $a(\mathbf{L}_v; \mathbf{J}; \mathbf{I}')$ is non-zero. Hence (c) follows from (7.37) and Lemma 7.5.15. □

7.5.26. *Step VI: Conclusion of the proof of Theorem 7.5.13.*

The proof of Theorem 7.5.13 is now easy and follows the blueprint of the proof of Roth's theorem. On the one hand, we have by the product formula and $T(\mathbf{X}') \neq 0$ that

$$\sum_{v \in M_K} \log |T(\mathbf{X}')|_v = 0.$$

On the other hand, $|\mathbf{x}'_h|_v \leq 1$ for $v \notin S$ and hence

$$\sum_{v \in M_K} \log |T(\mathbf{X}')|_v \leq \sum_{v \in S} \log |T(\mathbf{X}')|_v + \sum_{v \notin S} \log |T|_v.$$

By a change of coordinates, it is easy to check that

$$\sum_{v \in S} \max_{\mathbf{J}} |a(\mathbf{L}_v; \mathbf{J})|_{v,K} = \sum_{v \in S} \log |T|_v + O(d_1)$$

and hence the last three displayed formulas and Lemma 7.5.25 lead to

$$0 \leq \sum_{v \in S} \log \max_{a(\mathbf{L}_v;\mathbf{J}) \neq 0} |\mathbf{L}_v(\mathbf{x}'_1)^{\mathbf{j}_1} \cdots \mathbf{L}_v(\mathbf{x}'_m)^{\mathbf{j}_m}|_{v,K} + O(d_1).$$

For $v \in S$ and $a(\mathbf{L}_v; \mathbf{J}) \neq 0$, using

$$d_h \log(Q_h) = D + o(D), \quad |z_{hl}| \leq 2n/\eta, \quad |L_{vi}(\mathbf{y}_h^{(l)})|_{v,K} \leq Q_h^{c_{vi}},$$

and Lemma 7.5.25 (c), we have for D sufficiently large

$$\begin{aligned}
\log \left|\mathbf{L}_v(\mathbf{x}'_1)^{\mathbf{j}_1} \cdots \mathbf{L}_v(\mathbf{x}'_m)^{\mathbf{j}_m}\right|_{v,K} &= \sum_{h=1}^{m}\sum_{i=0}^{n} j_{hi} \log |L_{vi}(\mathbf{x}'_h)|_{v,K} \\
&\leq \sum_{h=1}^{m}\sum_{i=0}^{n} j_{hi}\{O(1) + \max_{hl} \log |z_{hl}|_v + c_{vi} \log(Q_h)\} \\
&= \sum_{h=1}^{m}\sum_{i=0}^{n} \frac{j_{hi}}{d_h} c_{vi} d_h \log(Q_h) + O(\log(1/\eta)\, d_1) \\
&= D \sum_{i=0}^{n} c_{vi}(\mathbf{J}_i/\mathbf{d}) + O(\log(1/\eta)\, D/\log(Q_1)) \\
&= \left(\sum_{i=0}^{n} c_{vi}\right) \frac{m}{n+1} D + O(m\eta\, D) + O(\log(1/\eta)\, D/\log(Q_1)).
\end{aligned}$$

Therefore, putting together the last two sequences of inequalities and recalling the definition of ε at the beginning of the proof of Theorem 7.5.13, we deduce

$$\begin{aligned}
0 &\leq \left(\sum_{v \in S}\sum_{i=0}^{n} c_{vi}\right) \frac{m}{n+1} D + O(m\eta\, D) + O(\log(1/\eta)\, D/\log(Q_1)) \\
&\leq -\frac{\varepsilon/2}{n+1} m\, D + O(m\eta D) + O(\log(1/\eta)\, D/\log(Q_1)).
\end{aligned}$$

Hence, dividing by mD, we conclude that

$$0 \leq -\frac{\varepsilon/2}{n+1} + O(\eta) + O\left(\frac{\log(1/\eta)}{m \log(Q_1)}\right).$$

Since we are allowed to take Q_1 arbitrarily large, we get $0 \leq -\varepsilon/(2n+2) + O(\eta)$, a contradiction if η is positive and small. This completes the proof of Theorem 7.5.13. □

7.5.27. *Step VII: A general strategy and Evertse's lemma.*

In order to provide some motivation for what follows, we begin by describing in a special case Schmidt's original strategy for the proof of the subspace theorem.

Consider the simplest case, namely $n = 2$, $K = \mathbb{Q}$, $S = \{\infty\}$ and three linear forms L_1, L_2, L_3. Let $\lambda_1, \lambda_2, \lambda_3$ be the three successive minima of $\Pi(Q_\nu)$. If $\lambda_1 \leq \lambda_2 \leq 1$, the rank is 2 and Theorem 7.5.13 can be applied. If instead $\lambda_1 \leq 1 < \lambda_2 \leq \lambda_3$, the rank is 1. By Minkowski's second theorem, we have

$$\beta^3(\Pi(Q_\nu))^{-1} \ll \lambda_1\lambda_2\lambda_3 \ll \beta^3(\Pi(Q_\nu))^{-1}.$$

The volume of $\Pi(Q_\nu)$ is relatively small, namely

$$\beta^3(\Pi(Q_\nu)) \ll Q_\nu^{-\varepsilon/2}.$$

Consider now the new linear forms $L_{ij} = L_i \wedge L_j$ in the space $V = \mathbb{R}^3 \wedge \mathbb{R}^3 \cong \mathbb{R}^3$. There is a naturally associated parallelopiped $\Pi_2(Q_\nu)$ in V, determined by the forms L_{ij} and the set of exponents $c_{ij} = c_i + c_j$.

Let $\mathbf{x}^{(1)}, \mathbf{x}^{(2)}, \mathbf{x}^{(3)}$ determine the successive minima of $\Pi(Q_\nu)$. Then the vectors $\mathbf{x}^{(ij)} = \mathbf{x}^{(i)} \wedge \mathbf{x}^{(j)}$ in V are linearly independent and

$$|L_{ij}(\mathbf{x}^{(pq)})| \ll \lambda_p \lambda_q Q_\nu^{c_{ij}}.$$

From this, it is plain that the successive minima $\lambda_{1,2} \le \lambda_{1,3} \le \lambda_{2,3}$ of $\Pi_2(Q)$ (it is convenient to use this indexing here) are majorized by

$$\lambda_{p,q} \ll \lambda_p \lambda_q.$$

Since $\beta^3(\Pi_2(Q_\nu))$ is of the same order as $\beta^3(\Pi(Q_\nu))^2$, Minkowski's second theorem shows that

$$(\lambda_1\lambda_2\lambda_3)^2 \ll \prod \lambda_{p,q} \ll (\lambda_1\lambda_2\lambda_3)^2.$$

This estimate, in conjunction with the above upper bound for $\lambda_{p,q}$, yields

$$\lambda_p\lambda_q \ll \lambda_{p,q} \ll \lambda_p\lambda_q.$$

This is a special case of a general theorem by Mahler on the so-called compound convex bodies.

If $\lambda_2\beta^3(\Pi(Q_\nu)) \ge C$ for some sufficiently large positive constant C, Minkowski's second theorem yields $\lambda_{1,3} \le 1$ for Q_ν sufficiently large. Then Theorem 7.5.13 will be applicable to $\Pi_2(Q_\nu)$, with the conclusion that the points $\mathbf{x}^{(p)} \wedge \mathbf{x}^{(q)}$ belong to finitely many linear subspaces of $\mathbb{R}^3 \wedge \mathbb{R}^3$. From this, we can deduce that the points $\mathbf{x}^{(1)}$ also belong to finitely many linear subspaces of $\mathbb{R}^3$. This will prove what we want except in the relatively narrow range $1 < \lambda_2 < C\beta^3(\Pi(Q_\nu))^{-1}$.

One more idea is needed to complete the proof. In order to deal with the remaining range, Schmidt deforms the parallelopiped $\Pi_2(Q_\nu)$ by stretching it but keeping fixed the point $\mathbf{x}^{(1)}$, controlling the change in the successive minima using a lemma of Davenport. The new successive minima can then be brought in a range to which we can apply Theorem 7.5.13, concluding the argument.

7.5.28. A clever simplification of Schmidt's original proof was later found by Evertse, by means of a simple lemma bypassing Schmidt's use of Mahler's results on successive minima of compound convex bodies, as well as Davenport's lemma on successive minima of parallelopipeds after stretching.

Evertse's lemma, given here in a slightly simplified form paying no attention to numerical constants, is the following:

Lemma 7.5.29. *Let K be a number field, S a finite set of places of K containing all the archimedean places and let $\mathbf{x}^{(1)},\dots,\mathbf{x}^{(n+1)}$ be a basis of a K-vector space V. Further, for $v \in S$, let $L_{v0},\dots,L_{vn}$ be linearly independent linear forms on V with coefficients in K_v, and let μ_{vj} be real numbers such that*

$$0 < \mu_{v1} \le \mu_{v2} \le \cdots \le \mu_{v,n+1}, \qquad \left| L_{vk}(\mathbf{x}^{(j)}) \right|_v \le \mu_{vj}$$

for all k, j. Then there are vectors

$$\mathbf{v}^{(1)} = \mathbf{x}^{(1)}, \quad \mathbf{v}^{(i)} = \sum_{j=1}^{i-1} \xi_{ij}\mathbf{x}^{(j)} + \mathbf{x}^{(i)}$$

with $\xi_{ij} \in O_{S,K}$ $(1 \le j < i \le n+1)$, bijective maps $\pi_v : \{1,\dots,n+1\} \to \{0.\dots,n\}$, and a positive constant C such that for $v \in S$ and $i,j = 1,\dots,n+1$ it holds

$$\left| L_{v,\pi_v(i)}(\mathbf{v}^{(j)}) \right|_v \le \begin{cases} C\min(\mu_{vi},\mu_{vj}) & \text{if } v \text{ is archimedean,} \\ \min(\mu_{vi},\mu_{vj}) & \text{if } v \text{ is not archimedean.} \end{cases}$$

The constant C depends only on K, S, and the set of linear forms L_{vi}.

Proof: By induction on n, the case $n = 0$ being trivial. Now suppose $n \ge 1$ and that the lemma holds for $n-1$ in place of n.

Let us fix $v \in S$ and let V' be the K-vector subspace of V with basis $\mathbf{x}^{(i)}$ $(i = 1,\dots,n)$. Then the linear system

$$\sum_{k=0}^{n} L_{vk}(\mathbf{x}^{(j)})\alpha_{vk} = 0, \quad (j = 1,\dots,n)$$

has a non-trivial solution $(\alpha_{v0},\dots,\alpha_{vn}) \in K_v^{n+1}$. We define $\pi_v(n+1)$ to be any index for which

$$\left|\alpha_{v,\pi_v(n+1)}\right|_v = \max_i |\alpha_{vi}|_v.$$

Obviously, $\alpha_{v,\pi_v(n+1)} \ne 0$. Hence, setting

$$\beta_{vi} = -\alpha_{vi}/\alpha_{v,\pi_v(n+1)},$$

we have

$$L_{v,\pi_v(n+1)}(\mathbf{x}^{(j)}) = \sum_{k \ne \pi_v(n+1)} \beta_{vk}L_{vk}(\mathbf{x}^{(j)}), \quad |\beta_{vk}|_v \le 1. \tag{7.38}$$

The restrictions of the forms L_{vi} to V' yield a set of linear forms of rank n and the restriction of $L_{v,\pi_v(n+1)}$ to V' is linearly dependent on the restrictions of the remaining linear forms. Hence the restrictions to V' of the linear forms L_{vk},

$k \neq \pi_v(n+1)$, are linearly independent over K_v. By the induction hypothesis, there are bijective maps π_v from $\{1,\ldots,n\}$ to $\{0,\ldots,n\}\backslash\pi_v(n+1)$, and vectors

$$\mathbf{v}^{(1)} = \mathbf{x}^{(1)}, \quad \mathbf{v}^{(i)} = \sum_{j=1}^{i-1} \xi_{ij}\mathbf{x}^{(j)} + \mathbf{x}^{(i)}$$

for $i = 1,\ldots,n$, with $\xi_{ij} \in O_{S,K}$, such that for $i,j = 1,\ldots,n$ and all $v \in S$ we have

$$\left|L_{v,\pi_v(i)}(\mathbf{v}^{(j)})\right|_v \leq \begin{cases} C'\min(\mu_{vi},\mu_{vj}) & \text{if } v \text{ is archimedean,} \\ \min(\mu_{vi},\mu_{vj}) & \text{if } v \text{ is not archimedean.} \end{cases} \tag{7.39}$$

By definition of μ_{vj} and (7.38), this estimate continues to hold if $i = n+1$ and $j = 1,\ldots,n$. Thus to complete the proof it suffices to show there are $\xi'_{n+1,j} \in O_{S,K}$ such that, setting

$$\mathbf{v}^{(n+1)} = \sum_{j=1}^{n} \xi'_{n+1,j}\mathbf{v}^{(j)} + \mathbf{x}^{(n+1)},$$

we have for $i = 1,\ldots,n$ and $v \in S$

$$\left|L_{v,\pi_v(i)}(\mathbf{v}^{(n+1)})\right|_v \leq \begin{cases} C\mu_{vi} & \text{if } v \text{ is archimedean,} \\ \mu_{vi} & \text{if } v \text{ is not archimedean.} \end{cases}$$

Indeed, for $i = n+1$ this follows as above from (7.38).

Since the linear forms $L_{v,\pi_v(i)}$, $i \neq n+1$ are linearly independent on V', which is also generated by $\mathbf{v}^{(1)},\ldots,\mathbf{v}^{(n)}$, we can solve the linear system

$$L_{v,\pi_v(i)}(\mathbf{x}^{(n+1)}) = \sum_{j=1}^{n} \gamma_{vj} L_{v,\pi_v(i)}(\mathbf{v}^{(j)}) \quad (i \neq n+1) \tag{7.40}$$

with $\gamma_{vj} \in K_v$.

Now note that $O_{S,K}$ is a lattice in $\prod_{v\in S} K_v$, meaning that it is a discrete subgroup with compact quotient. This follows as in the proof of Proposition C.2.6, which is the special case, where S is the set of archimedean places. Hence the lattice has a bounded fundamental domain and thus there is a vector $(\xi'_{n+1,1},\ldots,\xi'_{n+1,n}) \in O_{S,K}^n$ such that for $v \in S$ we have

$$|\xi'_{n+1,j} + \gamma_{vj}|_v \leq A_v,$$

with A_v bounded only in terms of K and S. Moreover, we may assume that $A_v = 1$ for non-archimedean v by the following simple argument. There is a non-zero $m \in O_{S,K}$, depending only on K and S, such that $|m|_v \leq \min(1, A_v^{-1})$ for $v \nmid \infty$ $(v \in S)$ (for example, a sufficiently high power of the product of the rational primes p with $v|p$). Then, applying the argument to $m^{-1}\gamma_{vi}$ and multiplying by

m afterwards, we obtain the result with $A_v = 1$ for $v \not| \infty$ and A_v replaced by $|m|_v A_v$ at the infinite places. Hence, setting

$$\mathbf{v}^{(n+1)} = \mathbf{x}^{(n+1)} + \sum_{j=1}^{n} \xi'_{n+1,j} \mathbf{v}^{(j)}$$

we infer from (7.40) and the inductive step (7.39) the estimate

$$\begin{aligned} \left| L_{v,\pi_v(i)}(\mathbf{v}^{(n+1)}) \right|_v &= \left| L_{v,\pi_v(i)}(\mathbf{x}^{(n+1)}) + \sum_{j=1}^{n} \xi'_{n+1,j} L_{v,\pi_v(i)}(\mathbf{v}^{(j)}) \right|_v \\ &= \left| \sum_{j=1}^{n} (\gamma_{vj} + \xi'_{n+1,j}) L_{v,\pi_v(i)}(\mathbf{v}^{(j)}) \right|_v \\ &\le \begin{cases} |n|_v A_v C' \, \mu_{vi} & \text{if } v \text{ is archimedean,} \\ \mu_{vi} & \text{if } v \text{ is not archimedean,} \end{cases} \end{aligned}$$

for $i = 1, \dots, n$. □

7.5.30. *Step VIII: Application of the Grassmann algebra.*

Let $\Pi(Q)$ be an approximation domain associated to the forms L_{vi} and parameters c_{vi} as in Step 0, and suppose that $R := \mathrm{rank}(\Pi(Q))$ is such that $1 \le R \le n$. Then R is determined by $\lambda_R \le 1 < \lambda_{R+1}$ and, as already noted in the proof of Lemma 7.5.12, we have

$$\lambda_{n+1} \gg Q^{\varepsilon/(2n+2)}.$$

Now we define k to be the smallest integer in the interval $[R, n]$ such that the quotient λ_k/λ_{k+1} is minimal. Since

$$\frac{\lambda_k}{\lambda_{k+1}} \le \left(\prod_{j=R}^{n} \frac{\lambda_j}{\lambda_{j+1}} \right)^{\frac{1}{n+1-R}} = \left(\frac{\lambda_R}{\lambda_{n+1}} \right)^{\frac{1}{n+1-R}} \le \lambda_{n+1}^{-\frac{1}{n+1-R}},$$

we have

$$\frac{\lambda_k}{\lambda_{k+1}} \ll Q^{-\varepsilon/\{2(n+1)n\}}. \tag{7.41}$$

As usual, we write

$$\varepsilon_v = \begin{cases} [K_v : \mathbb{R}]/[K : \mathbb{Q}] & \text{if } v \text{ is archimedean,} \\ 0 & \text{otherwise.} \end{cases}$$

Then we define

$$\mu_{vj} = \lambda_j^{\varepsilon_v} \quad (j = 1, \dots, n+1).$$

Since $\sum_{v|\infty} \varepsilon_v = 1$ (Corollary 1.3.2), we have

$$\prod_{v|\infty} \mu_{vj} = \lambda_j, \quad \mu_{vj} = 1 \quad \text{if } v\nmid\infty,\ v \in S. \tag{7.42}$$

Let $\mathbf{x}^{(j)} \in O_{S,K}^{n+1}$ $(j = 1, \ldots, n+1)$ be linearly independent points such that $\iota(\mathbf{x}^{(j)})$ determine the successive minima of $\Pi(Q)$. Then

$$\left|L_{vi}(\mathbf{x}^{(j)})\right|_{v,K} \leq \mu_{vj} Q^{c_{vi}}$$

for $v \in S$, $i = 0, \ldots, n$ and $j = 1, \ldots, n+1$ (recall that the absolute values are normalized so that for $v|\infty$ we have $|ta|_v = t^{\varepsilon_v}|a|_v$ for $t > 0$). We apply Evertse's lemma to this situation and infer that there are $\mathbf{v}^{(i)} \in O_{S,K}^{n+1}$ $(i = 1, \ldots, n+1)$ such that

$$\mathbf{v}^{(1)} = \mathbf{x}^{(1)}, \quad \mathbf{v}^{(i)} = \sum_{j=1}^{i-1} \xi_{ij} \mathbf{x}^{(j)} + \mathbf{x}^{(i)}$$

and for each $v \in S$ there is a bijection $\pi_v : \{1, \ldots, n+1\} \to \{0, \ldots, n\}$ such that

$$\left|L_{v,\pi_v(i)}(\mathbf{v}^{(j)})\right|_{v,K} \leq \begin{cases} C_{11} \min(\mu_{vi}, \mu_{vj})\, Q^{c_{v,\pi_v(i)}} & \text{if } v|\infty, \\ Q^{c_{v,\pi_v(i)}} & \text{if } v\nmid\infty, \end{cases} \tag{7.43}$$

for $i, j = 1, \ldots, n+1$.

Now we pass to the Grassmann algebra of order $n+1-k$, where k, as defined at the beginning of this article, verifies (7.41). We abbreviate

$$\begin{aligned} \mathbf{i} &= (i_1, \ldots, i_{n+1-k}), \quad \text{where} \quad \{i_1 < i_2 < \cdots < i_{n+1-k}\}, \\ L_{v,\pi_v(\mathbf{i})} &= L_{v,\pi_v(i_1)} \wedge L_{v,\pi_v(i_2)} \wedge \cdots \wedge L_{v,\pi_v(i_{n+1-k})}, \\ \mathbf{v}^{(\mathbf{i})} &= \mathbf{v}^{(i_1)} \wedge \mathbf{v}^{(i_2)} \wedge \cdots \wedge \mathbf{v}^{(i_{n+1-k})}, \\ c_{v,\pi_v(\mathbf{i})} &= c_{v,\pi_v(i_1)} + c_{v,\pi_v(i_2)} + \cdots + c_{v,\pi_v(i_{n+1-k})}, \\ \lambda_{\mathbf{i}} &= \prod_{\nu=1}^{n+1-k} \lambda_{i_\nu}, \qquad \mu_{v\mathbf{i}} = \prod_{\nu=1}^{n+1-k} \mu_{v,i_\nu}. \end{aligned}$$

The linear forms $L_{v,\pi_v(\mathbf{i})}$ are linearly independent and so are the points $\mathbf{v}^{(\mathbf{i})}$.

By (7.43) and the Laplace identity (7.16) on page 198, it is immediate that, for some constant C_{12}, we have for every $\mathbf{i}$ and $\mathbf{j}$ and $v \in S$

$$\left|L_{v,\pi_v(\mathbf{i})}(\mathbf{v}^{(\mathbf{j})})\right|_{v,K} \leq \begin{cases} C_{12} \mu_{v\mathbf{i}}\, Q^{c_{v,\pi_v(\mathbf{i})}} & \text{if } v|\infty, \\ Q^{c_{v,\pi_v(\mathbf{i})}} & \text{if } v\nmid\infty. \end{cases} \tag{7.44}$$

Moreover, if $\mathbf{i} = (k+1, k+2, \ldots, n+1)$ but $\mathbf{j} \neq (k+1, k+2, \ldots, n+1)$ Evertse's lemma shows that we can do a little better, namely

$$\left| L_{v,\pi_v(\mathbf{i})}(\mathbf{v}^{(\mathbf{j})}) \right|_{v,K} \leq C_{12} \cdot (\mu_{vk}/\mu_{v,k+1}) \cdot \mu_{v\mathbf{i}}\, Q^{c_{v,\pi_v(\mathbf{i})}} \quad \text{if } v|\infty, \tag{7.45}$$

because in this case $j_1 \leq k$, hence

$$\mu_{v\mathbf{j}} \leq \mu_{vk}\mu_{v,k+2} \cdots \mu_{v,n+1} = (\mu_{vk}/\mu_{v,k+1}) \cdot \mu_{v\mathbf{i}}.$$

Now we can prove:

Lemma 7.5.31. *Let $\mathcal{S}(Q)$ be the symmetric convex domain in $\wedge^{n+1-k}(K_{\mathbb{A}}^{n+1})$ defined in obvious notation by*

$$\begin{aligned}
\left|L_{v,\pi_v(\mathbf{i})}(\mathbf{X})\right|_{v,K} &\leq C_{12}\mu_{v\mathbf{i}}\, Q^{c_{v,\pi_v(\mathbf{i})}} && v|\infty,\ \mathbf{i} \neq (k+1,\ldots,n+1),\\
\left|L_{v,\pi_v(\mathbf{i})}(\mathbf{X})\right|_{v,K} &\leq C_{12} \cdot (\mu_{vk}/\mu_{v,k+1}) \cdot \mu_{v\mathbf{i}}\, Q^{c_{v,\pi_v(\mathbf{i})}} && v|\infty,\ \mathbf{i} = (k+1,\ldots,n+1),\\
\left|L_{v,\pi_v(\mathbf{i})}(\mathbf{X})\right|_{v,K} &\leq Q^{c_{v,\pi_v(\mathbf{i})}} && v \in S,\ v\not|\infty,\\
|\mathbf{X}|_v &\leq 1 && v \notin S.
\end{aligned}$$

Let $\lambda_j(\mathcal{S}(Q))$ be the successive minima of $\mathcal{S}(Q)$. Then we have

$$\begin{aligned}
\lambda_j(\mathcal{S}(Q)) &\leq 1 && \text{if } j < \binom{n+1}{k},\\
\lambda_j(\mathcal{S}(Q)) &> C_{13} Q^{\varepsilon/\{2n(n+1)\}} && \text{if } j = \binom{n+1}{k}.
\end{aligned}$$

Proof: The first part of the thesis of the lemma (namely, for $j < \binom{n+1}{k}$) is obvious from (7.44) and (7.45), since they provide independent points in $\mathcal{S}(Q)$. For the second part, we apply Minkowski's second theorem, which gives

$$1 \ll \left(\prod_{j=1}^{\binom{n+1}{k}} \lambda_j(\mathcal{S}(Q)) \right) \beta^{\binom{n+1}{k}}(\mathcal{S}(Q))^{1/d} \ll 1, \tag{7.46}$$

where the constants involved depending only on K, S, and n. By Lemma 7.5.7 and (7.42)

$$\begin{aligned}
\beta^{\binom{n+1}{k}}(\mathcal{S}(Q))^{1/d} &\ll \prod_{v \in S} \left\{ \frac{\mu_{vk}}{\mu_{v,k+1}} \prod_{\mathbf{i}} \mu_{v\mathbf{i}}\, Q^{c_{v,\pi_v(\mathbf{i})}} \right\}\\
&\ll \frac{\lambda_k}{\lambda_{k+1}} \left(\prod_{\mathbf{i}} \lambda_{\mathbf{i}} \right) Q^{\sum_{v\mathbf{i}} c_{v,\pi_v(\mathbf{i})}}\\
&= \frac{\lambda_k}{\lambda_{k+1}} \left(\lambda_1 \cdots \lambda_{n+1} \right)^{\binom{n}{k}} Q^{\binom{n}{k} \sum_{vi} c_{v,\pi_v(i)}}.
\end{aligned}$$

By Minkowski's second theorem, Lemma 7.5.7, and (7.41) on page 220 this may be bounded by

$$\ll \frac{\lambda_k}{\lambda_{k+1}}\left(\lambda_1\cdots\lambda_{n+1}\beta^{n+1}(\Pi(Q))^{1/d}\right)^{\binom{n}{k}} \ll \frac{\lambda_k}{\lambda_{k+1}} \ll Q^{-\varepsilon/\{2n(n+1)\}}.$$

We have already noticed that $\lambda_j(\mathcal{S}(Q)) \leq 1$ if $j < \binom{n+1}{k+1}$, hence the second part of the thesis follows from the left-hand side of (7.46). □

7.5.32. *Step IX: Proof of the subspace theorem.*

We apply Theorem 7.5.13 as follows. Let (Q_ν) be a unbounded family such that we have approximation domains $\Pi(Q_\nu)$ as in Step 0 with $\mathrm{rank}(\Pi(Q_\nu)) = R$, where $1 \leq R < n$ (the case $R = n$ being already covered by Theorem 7.5.13).

By going to a subfamily, we may assume that the parameter k defined at the beginning of Step VIII (see 7.5.30) is constant along this family.

With $\mu_{v\mathbf{i}}$ as in Step VIII (relative now to $Q = Q_\nu$), we note that there is a constant $C_{14} > 0$ such that

$$Q_\nu^{-C_{14}} \leq \mu_{v\mathbf{i}} \leq Q_\nu^{C_{14}}$$

for every $v|\infty$ and every $\mathbf{i}$ (of course, $\mu_{v\mathbf{i}} = 1$ if $v\not|\infty$). The argument is the same as for the proof of Corollary 7.5.5. Clearly, it suffices to verify the corresponding result for the successive minima of $\Pi(Q_\nu)$. By 7.5.6 and Lemma 7.5.7, we have

$$\beta^{n+1}(\Pi(Q_\nu)) \gg Q_\nu^{d\sum_{vi} c_{vi}} \geq Q_\nu^{-2rd(n+1)|S|}$$

and hence, by Minkowski's second theorem, it suffices to obtain a lower bound for the first minimum. Since the linear forms $L_{v0}, \ldots, L_{vn}$ are linearly independent, we have for every $v \in S$ and $\mathbf{x} \in K^{n+1} \setminus \{\mathbf{0}\}$ with $\iota(\mathbf{x}) \in \Pi(Q_\nu)$

$$\log|\mathbf{x}|_v \leq \max_i \log|L_{vi}(\mathbf{x})|_{v,K} + C_{15} \leq c_{vi}\log(Q_\nu) + C_{15}$$

for some constant C_{15}. Since $c_{vi} \leq 2r$, from this it follows $h_{\mathrm{aff}}(x_i) \ll \log(Q_\nu)$ and

$$-C_{16}\log(Q_\nu) \leq \log\max_i |L_{vi}(\mathbf{x})|_{v,K}$$

for large Q_ν (use (1.8) on page 20). If we apply this with $\mathbf{x}$ such that $\iota(\mathbf{x})$ determines the first successive minimum of $\Pi(Q_\nu)$ and with $v|\infty$, then the right-hand side is bounded by $\log|\lambda_1|_v + c_{vi}\log(Q_\nu)$, proving what we want.

Once this observation has been made, we proceed as we did in defining approximation classes and, going once more to a subfamily, we may assume that, given any small positive number $\gamma > 0$, there are bounded real numbers $d_{v\mathbf{i}}$ $(v|\infty)$ such that

$$-\gamma\varepsilon_v + d_{v\mathbf{i}} < \log(C_{12}\mu_{v\mathbf{i}})/\log(Q_\nu) \leq d_{v\mathbf{i}} \quad \text{if } \mathbf{i} \neq (k+1, \ldots, n+1)$$

and

$$-\gamma\varepsilon_v + d_{v\mathbf{i}} < \log(C_{12}(\mu_{vk}/\mu_{v,k+1})\mu_{v\mathbf{i}})/\log(Q_\nu) \le d_{v\mathbf{i}} \quad \text{if } \mathbf{i} = (k+1,\ldots,n+1).$$

If $v \in S$ and $v \not| \infty$, we set $d_{v\mathbf{i}} = 0$.

If $\Pi_k(Q_\nu)$ is the parallelopiped in $\wedge^{n+1-k}(K_{\mathbb{A}}^{n+1})$ defined by

$$\begin{aligned} \left|L_{v,\pi_v(\mathbf{i})}(\mathbf{X})\right|_{v,K} &\le Q_\nu^{c_{v,\pi_v(\mathbf{i})}+d_{v\mathbf{i}}} && (v \in S) \\ \left|\mathbf{X}\right|_{v,K} &\le 1 && (v \notin S), \end{aligned}$$

then it is obvious that $\mathcal{S}(Q_\nu) \subset \Pi_k(Q_\nu)$ and in particular $\lambda_j(\mathcal{S}(Q_\nu)) \ge \lambda_j(\Pi_k(Q_\nu))$ for every j. On the other hand, the volume of $\Pi_k(Q_\nu)$ does not increase too much if γ is small, in fact Lemma 7.5.7 yields

$$\beta^{\binom{n+1}{k}}(\Pi_k(Q_\nu))^{1/d} \ll \beta^{\binom{n+1}{k}}(\mathcal{S}(Q_\nu))^{1/d}\, Q_\nu^{\binom{n+1}{k}\gamma}.$$

Therefore, if we take for example $\gamma = \varepsilon/\{4n(n+1)\binom{n+1}{k}\}$, by Minkowski's second theorem as in the proof of Lemma 7.5.31, we still find, for large Q_ν, that

$$\begin{aligned} \lambda_j(\Pi_k(Q_\nu)) &\le 1 && \text{if } j < \binom{n+1}{k}, \\ \lambda_j(\Pi_k(Q_\nu)) &> C_{17} Q_\nu^{\varepsilon/\{4n(n+1)\}} && \text{if } j = \binom{n+1}{k}. \end{aligned}$$

Hence

$$\operatorname{rank}(\Pi_k(Q_\nu)) = \binom{n+1}{k} - 1$$

as soon as Q_ν is sufficiently large, which we may suppose.

We have

$$\begin{aligned} \sum_{v\in S}\sum_{\mathbf{i}} d_{v\mathbf{i}} &\le \sum_{v|\infty}\left(\frac{1}{\log(Q_\nu)}\left(\log\frac{\mu_{vk}}{\mu_{v,k+1}} + \sum_{\mathbf{i}}\log(C_{12}\mu_{v\mathbf{i}})\right) + \gamma\varepsilon_v\right) \\ &= \frac{1}{\log(Q_\nu)}\left(\log\frac{\lambda_k}{\lambda_{k+1}} + \log\left(\prod_{\mathbf{i}}\lambda_{\mathbf{i}}\right) + O(1)\right) + \gamma \\ &= \frac{1}{\log(Q_\nu)}\left(\log\frac{\lambda_k}{\lambda_{k+1}} + \binom{n}{k}\log(\lambda_1\cdots\lambda_{n+1}) + O(1)\right) + \gamma. \end{aligned}$$

Therefore, using (7.41) on page 220 and again Minkowski's second main theorem together with Lemma 7.5.7, we get

$$\sum_{v\in S}\sum_{\mathbf{i}} d_{v\mathbf{i}} \le -\frac{\varepsilon}{2(n+1)n} - \binom{n}{k}\sum_{vi} c_{vi} + \gamma + \frac{O(1)}{\log(Q_\nu)}.$$

Now we note that

$$\sum_{\mathbf{i}} \sum_{v \in S} \left(c_{v,\pi_v(\mathbf{i})} + d_{v\mathbf{i}}\right) = \sum_{vi} c_{vi} + \sum_{vi} d_{v\mathbf{i}} = \binom{n}{k} \sum_{vi} c_{vi} + \sum_{v\mathbf{i}} d_{v\mathbf{i}}.$$

Thus the above and our choice of γ gives

$$\sum_{\mathbf{i}} \sum_{v \in S} \left(c_{v,\pi_v(\mathbf{i})} + d_{v\mathbf{i}}\right) \leq -\frac{\varepsilon}{2(n+1)n} \left(1 - \frac{1}{2\binom{n+1}{k}}\right) + \frac{O(1)}{\log(Q_\nu)},$$

which is negative for Q_ν sufficiently large.

Now we are able to apply Theorem 7.5.13 to this situation, concluding that the vector spaces $V(\Pi_k(Q_\nu))$ form a finite set.

Let $\mathbf{v}^{(\mathbf{i})} = \mathbf{v}^{(i_1)} \wedge \cdots \wedge \mathbf{v}^{(i_{n+1-k})}$ be the points constructed in 7.5.30. Obviously, they are linearly independent over K. Those with $\mathbf{i} \neq (k+1, \ldots, n+1)$ belong to $V(\Pi_k(Q_\nu))$, hence form a basis of $V(\Pi_k(Q_\nu))$.

Lemma 7.5.33. *Let V be a vector space over K of dimension $n+1$ and let W be a subspace of dimension k. Let $\mathbf{x}^{(1)}, \ldots, \mathbf{x}^{(k)}$ be a basis of W and extend it to a basis $\mathbf{x}^{(1)}, \ldots, \mathbf{x}^{(n+1)}$ of V. Let W' denote the subspace of $\wedge^{n+1-k}V$ generated by $\left(\mathbf{x}^{(\mathbf{i})}\right)_{\mathbf{i} \neq (k+1,\ldots,n+1)}$. Then W' is independent of the choice of the basis and is uniquely determined by W.*

Proof: Recall that $\wedge^{n+1-k}V$ is the dual of $\wedge^k V$ with respect to the non-degenerate pairing $\langle \omega, \omega' \rangle = \omega \wedge \omega'$.

Since W' is the annihilator $W^\perp$ of $\wedge^k W = K\mathbf{x}^{(1)} \wedge \cdots \wedge \mathbf{x}^{(k)}$ with respect to this pairing, W' is well defined by W. Since the Grassmann coordinate $\wedge^k W = (W')^\perp$ determines W uniquely, we get the claim. □

Now we can finish the proof of the subspace theorem.

We argue by contradiction, hence we get an infinite set $\mathcal{X} = \{\mathbf{x}_\nu\}$ as in Lemma 7.5.9. By Northcott's theorem in 2.4.9, $Q_\nu := H(\mathbf{x}_\nu)$ is unbounded. By Lemma 7.5.12 , we may assume that $\operatorname{rank}(\Pi_v(Q_\nu)) \leq n$. If the rank is equal to n for infinitely many Q_ν, then Theorem 7.5.13 contradicts statement (a) of Lemma 7.5.9. So we may assume that $\operatorname{rank}(\Pi_v(Q_\nu)) < n$ for all ν and we may apply our above considerations in Step IX.

By Lemma 7.5.33, the K-vector space $V(\Pi_k(Q_\nu))$ determines uniquely the K-vector space $W_k(Q_\nu)$ of dimension k generated by $\mathbf{v}^{(i)}$ $(i = 1, \ldots, k)$. Since the spaces $V(\Pi_k(Q_\nu))$ form a finite set, the associated spaces $W_k(Q_\nu)$ also form a finite set. Since $k \geq \operatorname{rank}(\Pi(Q_\nu))$, we conclude that $\mathbf{x}_\nu$ is a linear combination of the vectors in K^{n+1}, denoted by $\mathbf{x}^{(1)}, \ldots, \mathbf{x}^{(k)}$, determining the first k successive minima . By construction, this holds also for $\mathbf{v}^{(1)}, \ldots, \mathbf{v}^{(k)}$ and hence the points $\mathbf{x}_\nu$ belong to finitely many proper subspaces of K^{n+1}, contradicting statement (a) of Lemma 7.5.9. Since the hypothesis of Lemma 7.5.9 was the falsity of the subspace theorem, the subspace theorem must hold.

7.6. Further results: the product theorem

In this section we give a quick review, without proofs, of important progress in this area.

7.6.1. In his landmark paper [114], G. Faltings introduced a completely new approach to study the index of a multihomogeneous polynomial at a point. We describe now the simplest version of Faltings's basic result, the product theorem.

We work in a product

$$\mathbb{P}_K := \mathbb{P}_K^{n_1} \times \cdots \times \mathbb{P}_K^{n_m}$$

of projective spaces, over an algebraically closed field K of characteristic 0 and with sections

$$f \in \Gamma(\mathbb{P}, O_{\mathbb{P}}(d_1, \ldots, d_m))$$

associated to the ample line bundle $O_{\mathbb{P}}(d_1, \ldots, d_m)$ (here the degrees $d_1, \ldots, d_m$ are positive integers). Recall that f may be identified with a multihomogeneous polynomial, homogeneous of degree d_h in the variables $\mathbf{x}_h = (x_{h0}, \ldots, x_{hn_h})$ (see Example A.6.13).

7.6.2. For $x \in \mathbb{P}_K$, we define the **index** of f in x with respect to the weights $\mathbf{d}$ by

$$\operatorname{ind}(f; \mathbf{d}; x) := \min\{(\mathbf{I}/\mathbf{d}) \mid \partial_{\mathbf{I}} f(x) \neq 0\},$$

where $\mathbf{I} = (\mathbf{i}_1, \ldots, \mathbf{i}_m)$ with $\mathbf{i}_h$ ranging over $\mathbb{N}^{n_h}$ for $h = 1, \ldots, m$ and where $(\mathbf{I}/\mathbf{d})$ and $\partial_{\mathbf{I}}$ are defined as in 7.5.14.

This notion extends the definition of the index in 6.3.2 in the following way. For a polynomial $F \in K[x_1, \ldots, x_m]$ with partial degrees at most d_h, we may consider the multihomogenization

$$f(x_{10}, x_{11}; \ldots; x_{m0}, x_{m1}) := x_{10}^{d_1} \cdots x_{m0}^{d_m} F(x_{11}/x_{10}, \ldots, x_{m1}/x_{m0})$$

of multidegree $\mathbf{d}$. By passing from $x \in \mathbb{A}_K^n$ to the multiprojective space $(\mathbb{P}_K^1)^m$, it is clear that the index of F in x with respect to the weight $\mathbf{d}$ as defined in 6.3.2 is the same as $\operatorname{ind}(f; \mathbf{d}; x)$.

7.6.3. Let $\sigma \geq 0$. Faltings's product theorem gives information on the geometry of the set Z_σ of $\mathbb{P}_K$ on which $\operatorname{ind}(f; \mathbf{d}; x) \geq \sigma$. Since Z_σ is the zero set of the multihomogeneous polynomials $\partial_{\mathbf{I}} f$ $((\mathbf{I}/\mathbf{d}) < \sigma)$, it is a closed subvariety of $\mathbb{P}_K$.

Now we can state a simple version of Faltings's **product theorem.**

Theorem 7.6.4. *Let K be an algebraically closed field of characteristic 0. Let $m, n_1, \ldots,$ n_m be positive integers, $\mathbb{P} = \mathbb{P}^{n_1} \times \cdots \times \mathbb{P}^{n_m}$. Then, for every $\varepsilon > 0$, there is $C > 0$, depending on ε and $m, n_1, \ldots, n_m$, with the following property. Suppose:*

(a) *$d_1 > \cdots > d_m$ are rapidly decreasing positive integers, namely $d_h/d_{h+1} \geq C$ for $h = 1, \ldots, m-1$.*

(b) *$f \in \Gamma(\mathbb{P}_K, O_{\mathbb{P}}(d_1, \ldots, d_m)) \setminus \{0\}$.*

(c) *For some* $\sigma \geq 0$, Z *is an irreducible component of both* Z_σ *and* $Z_{\sigma+\varepsilon}$ *(with respect to the weight vector* $\mathbf{d}$*).*

Then:

(i) Z *is a product of closed subvarieties* Z_i *of* $\mathbb{P}^{n_i}$*, i.e.* $Z = Z_1 \times \cdots \times Z_m$.

(ii) *The degrees* $\deg(Z_i)$ *are bounded in terms of* ε *and* $n_1 + \cdots + n_m$ *only.*

(iii) *If* $K = \overline{\mathbb{Q}}$*, the varieties* Z_i *admit presentations* $\mathbf{p}_i$ *in the sense of Section 2.5 such that*

$$d_1 h(\mathbf{p}_1) + \cdots + d_m h(\mathbf{p}_m) \ll h(f) + d_1,$$

where the implied constant depends on ε *and* $n_1, \ldots, n_m$.

We will not prove here this important result, referring to Faltings's paper **[114]**, to the article of M. van der Put **[303]** for a simple proof of (i) and (ii), and to the versions with explicit good constants in Evertse **[106]**, Ferretti **[119]**, and Rémond **[241]**.

Remark 7.6.5. Part (iii) of the thesis of this theorem is best stated taking for $h(Z_i)$ a more intrinsic notion of height, rather than the hand-made height through presentations. This is done by Faltings in **[114]**, by defining the height as an intersection number of arithmetic cycles in $\mathbb{P}^{n_i}$. Another definition is by taking the height of the Chow point defining Z_i; this second definition is equivalent to Faltings's, up to a simple uniformly bounded error term (see J.-B. Bost, H. Gillet, C. Soulé **[45]**, Sec.4.3).

7.6.6. The product theorem is used as follows. Let N be an integer $N > \dim(\mathbb{P})$ and let $\sigma > 0$. Assume that $\operatorname{ind}(f; \mathbf{d}; x) \geq \sigma$ at some $x \in \mathbb{P}$. Then there exist a chain

$$\mathbb{P} \neq Z_1 \supset Z_2 \supset \cdots \supset Z_N \ni x$$

with Z_i an irreducible component of $Z_{i\sigma/N}$. Since each Z_i is irreducible, the dimension drops every time we have $Z_i \neq Z_{i+1}$ and it follows from $N > \dim(\mathbb{P})$ that $Z_i = Z_{i+1}$ for some i. Taking $\varepsilon = \sigma/N$, we may apply the product theorem and deduce that Z_i is a product variety $Z_i = Z_{i1} \times \cdots \times Z_{im}$. Obviously, we must have $Z_{i\mu} \neq \mathbb{P}^{n_\mu}$ for some μ and the study of the vanishing of f on Z_i is reduced, by projecting to some linear subspace of $\mathbb{P}^{n_\mu}$, to the study of the vanishing in a multiprojective space with smaller dimension than that of $\mathbb{P}$. Then we apply induction.

It turns out that this inductive procedure is much more efficient that Roth's method using Wronskians, where we lose a square root every time we increase m by 1. For example, proving Roth's lemma using the product theorem allows us to replace $\sigma^{1/2^{m-1}}$ by the much better $\sigma^{1/m}$ (with minor changes for the other constants, see **[106]**, Th.3). In quantitative results, this has the effect of replacing doubly exponential bounds by simply exponential bounds.

7.7. The absolute subspace theorem and the Faltings–Wüstholz theorem

It is possible to obtain quantitative versions of the subspace theorem, in which we control the number of subspaces containing all solutions of height exceeding a certain bound, in the same way as was the case with the Davenport–Roth refinement of Roth's theorem (see (6.23) on page 173, 6.5.8). Evertse and Schlickewei [**111**] have obtained a remarkably strong result of this type, which in view of its strong uniformity with respect to its dependence on the field K and the set of places S has proved to be a powerful tool in applications (Theorem 7.4.1 is an example).

7.7.1. Let K be a number field with a finite set S of places. For $v \in S$, let $L_{v0}, \dots, L_{vn}$ be linearly independent linear forms in $x_0, \dots, x_n$ with coefficients in $\overline{K}$. We assume that the coefficients of the linear forms $L_{v0}, \dots, L_{vn}$ are contained in a field extension of K of degree at most D and that $H_{\mathrm{Ar}}(L_{iv}) \leq H$. We denote by $|\ |_{v,K}$ an extension of $|\ |_v$ to $\overline{K}$.

Now we can state the **absolute subspace theorem** of Evertse and Schlickewei (see [**111**], Th.3.1, for explicit constants and a proof).

Theorem 7.7.2. *Let $\varepsilon > 0$. Then there are proper linear subspaces $T_1, \dots, T_h$ of $\mathbb{P}^n_{\overline{K}}$ with h bounded in terms of n, $|S|$, D, and ε, with the property that the set of solutions $\mathbf{x} \in \mathbb{P}^n(\overline{K})$ of*

$$\prod_{v \in S} \prod_{i=0}^{n} \max_{\sigma \in \mathrm{Gal}(\overline{K}/K)} \frac{|L_{vi}(\sigma\mathbf{x})|_{v,K}}{|\sigma\mathbf{x}|_{v,K}} \leq H(\mathbf{x})^{-n-1-\varepsilon} \prod_{v \in S} |\det(L_{vi})|_{v,K}$$

and with $H(\mathbf{x})$ bounded below by a constant given in terms of n, ε, and H, is contained in $T_1 \cup \cdots \cup T_h$.

7.7.3. It is a natural question to obtain an extension of the subspace theorem in which linear forms are replaced by homogeneous forms of higher degree, and more generally by functions measuring a distance from an arbitrary algebraic variety. An answer to this question was obtained by G. Faltings and G. Wüstholz in 1994, in an innovative paper [**117**], which, as a special case, also gave a completely new proof of the subspace theorem.

We describe their result, beginning with the linear case. Let $K \subset F$ be number fields and let S be a finite set of places of F. For each $w \in S$, we fix a finite index set I_w, non-zero linear forms $L_{wi} \in F[x_0, \dots, x_n]$ and $c_{wi} \geq 0$ $(i \in I_w)$.

Let $V = \Gamma(\mathbb{P}^n_K, O_{\mathbb{P}^n}(1))$ and $V_F = V \otimes_K F$. For every $w \in S$ and any positive real number p, we consider the subspace of V_F generated by the linear forms for which $c_{wi} \geq p$. In this way we get a finite chain of subspaces

$$V_L = W^0 \supsetneq W^1 \supsetneq \cdots \supsetneq W^e \supsetneq W^{e+1} = 0$$

of V_L. Let p_{wj} be the minimum of the c_{wi} when i runs over the indices given by the generators of W^j, for $j = 0, \dots, e$. We also put $p_{w0} = 0$ if W^0 is not generated by the forms L_{wi}.

For a subspace W of V, we define an invariant

$$\mu(W) := \sum_{w \in S} \sum_{j=0}^{e} p_{wj} \frac{\dim((W^j \cap W_F)/(W^{j+1} \cap W_F))}{\dim(W)}.$$

The set of filtrations so obtained on V is **jointly semistable**, if for each non-zero proper subspace $W \subset V$ we have $\mu(W) \leq \mu(V)$.

Then the **Faltings–Wüstholz theorem** is as follows. We state it with the normalizations used in this book; for the proof and for details of the following remarks, we refer to [**117**].

Theorem 7.7.4. *Assume that the linear forms* L_{wi} $(w \in S, i \in I_w)$ *define a jointly semistable filtration on* V *and that* $\mu(V) > 1$. *Then the number of points* $\mathbf{x} \in \mathbb{P}^n(K)$ *with*

$$\frac{|L_{wi}(\mathbf{x})|_w}{|\mathbf{x}|_w} < H(\mathbf{x})^{-c_{wi}} \qquad (w \in S, \ i \in I_w)$$

is finite.

7.7.5. A more general theorem where the linear forms L_{wi} need not be jointly semistable is then obtained by considering the first non-trivial step W in the Harder–Narasimhan filtration of V. This is the unique subspace W of V characterized by the property that $(\mu(W), \dim(W))$ is maximal with respect to the lexicographic order.

Let $\mathbb{P}^*(V)$ denote the projective space of one-dimensional quotient spaces of V. The conclusion now is that if $\mu(W) > 1$ then there are only finitely many $\mathbf{x} \in \mathbb{P}^*(V)(K) \setminus \mathbb{P}^*(V/W)$ such that

$$\frac{|L_{wi}(\mathbf{x})|_w}{|\mathbf{x}|_w} < H(\mathbf{x})^{-c_{wi}} \qquad (w \in S, \ i \in I_w).$$

It is not difficult to deduce from this Schmidt's subspace theorem.

7.7.6. The Faltings–Wüstholz theorem can be applied to the study of a system of inequalities

$$|f_{wi}(\mathbf{x})|_w < H(\mathbf{x})^{-c_{wi}} \qquad (w \in S, \ i \in I_w),$$

where now $f_{wi} \in F[x_0, \ldots, x_n]$ are homogeneous forms of any degree. One may assume that they have all the same degree r and then one associates to this the corresponding linear forms obtained by a Segre embedding $\mathbb{P}^n \to \mathbb{P}^N$ using all monomials of degree r, see 1.5.14. Although the results obtained in this way probably are not optimal, they are usually stronger than those obtained by a straightforward application of Schmidt's subspace theorem.

7.7.7. The computation of the invariants $\mu(V)$ and the verification of the semistability condition is not easy. R.G. Ferretti [**120**] has considered more generally replacing the forms f_{wi}, which define hypersurfaces in $\mathbb{P}^n$, by projective subvarieties of $\mathbb{P}^n$, and has shown how to compute the associated invariants using the Chow form associated to a subvariety of $\mathbb{P}^n$.

Finally, in an interesting paper J.-H. Evertse and R.G. Ferretti [**110**] have been able to combine this point of view with the absolute subspace theorem in 7.7.2, obtaining a rather strong absolute version of the general Faltings–Wüstholz theorem. A new idea in their paper is the use of a more general type of Segre embedding, which is chosen in an optimal way so as to produce the best exponents. As a consequence, they obtain the Faltings–Wüstholz theorem as a consequence of the original Schmidt's subspace theorem and of their analysis of generalized Segre embeddings.

7.8. Bibliographical notes

The exposition in Sections 7.2 and 7.3 follows to a large extent material gleaned from J.-H. Evertse's expository paper [**105**]. The presentation of Siegel's theorem and Section 7.4 follow closely Zannier [**337**], Ch.II–IV.

The proof of the subspace theorem in the present form is the result of many years of step-by-step progress. The first step towards it was W.M. Schmidt's paper [**264**] of 1967, in which he solved the problem in the case $n = 2$ and $K = \mathbb{Q}$. In that paper the role of geometry of numbers emerges clearly through the use of Mahler's theorems on successive minima of polar bodies. However, the extension to the general case required control of all successive minima and this was done only in 1970 in [**266**], when Schmidt introduced the tool of the Grassmann algebra. This was followed by Schlickewei's generalization with several places and a general number field K.

A new direction began with Schmidt's extension of the Davenport–Roth theorem to the multidimensional case in [**269**]. This line of research culminated in the absolute subspace theorem 7.7.2 of Evertse and Schlickewei [**111**]. The remarkable uniformity of their result with respect to fields of definition and the set of places S has proven to be essential in applications. Essential ingredients in the proof of the absolute subspace theorem are an absolute version of geometry of numbers (with a corresponding absolute Siegel lemma) found by Roy and Thunder [**247**], [**248**], a precise gap principle for the sequence of solutions, and a precise quantitative version of Faltings's product theorem.

Vojta in [**307**] gives a succint account of the subspace theorem stressing certain analogies with the work of L. Ahlfors [**7**] on meromorphic curves. A version of the subspace theorem allowing "moving targets," similar to Theorem 6.5.2 in Roth's case, is in Ru and Vojta [**250**].

The paper by Faltings and Wüstholz [**117**] gave a deep geometric extension of the subspace theorem, using new methods quite independent of Schmidt's. The main tools are Faltings's fundamental product theorem and the introduction of the Harder–Narasimhan filtration in order to be able to apply probabilistic methods for the construction of the auxiliary polynomial.

The proof of Theorem 7.5.13 is patterned after Evertse in [**108**], with several simplifications because we do not keep track of constants. The rest of the proof is modeled after Evertse's treatment of the rational case, see [**105**]. For Faltings's product theorem, we also recommend the illuminating review of [**117**] by J.-H. Evertse in [**107**].

8 ABELIAN VARIETIES

8.1. Introduction

This chapter contains fundamental preparatory material on abelian varieties and Jacobians of algebraic curves. Abelian varieties are defined as complete, geometrically irreducible, and geometrically reduced group varieties. The main properties of abelian varieties are obtained using the seesaw principle, the theorem of the cube, and the theorem of the square.

Classically, the theorem of the cube is proved directly using the theory of the Jacobian of a curve, then deducing the existence of the Picard variety, and the theorem of the square. On the other hand, once we have the Picard variety, it is easy to deduce both the theorem of the square and the theorem of the cube. Thus our philosophy will be to take for granted the existence of the Picard variety, borrowed as a result about representable functors from algebraic geometry, and deduce from this the basic theorems we need.

Section 8.2 contains preliminary material on group varieties, limited however to what is needed to develop the theory of abelian varieties. Section 8.3 deals with elliptic curves, including the well-known method for obtaining a Weierstrass model *via* the Riemann–Roch theorem.

The next three sections deal with the Picard variety and the theorems of the square and the cube. Section 8.7 studies the basic isogeny defined by multiplication by n. Section 8.8 contains the characterization of odd elements in the Picard variety. Section 8.9 studies the factorization of an abelian variety into simple abelian varieties and proves Poincaré's complete reducibility theorem. Section 8.10 deals with the construction of the Jacobian of a curve and gives the main properties of the theta divisor, needed for the proof of the Mordell conjecture (Faltings's theorem).

In order to read this chapter, the reader should have some knowledge of algebraic geometry as provided by Appendix A. At the end of almost every section, we give a complex analytic description of the results. These expositions are rather sketchy and will not be used anywhere else in the book, but they might be useful for readers with a background in complex analysis (see also Section A.14).

8.2. Group varieties

Let K be a field and $\overline{K}$ an algebraic closure of K. We assume that all occurring varieties and morphisms are defined over K.

After the basic definitions, we state and prove the constancy lemma. If a morphism φ defined on a product $X \times Y$ is constant on one fibre $X \times \{y_0\}$, then completeness of X implies that φ is constant on every $X \times \{y\}$. Abelian varieties are complete group varieties. As applications of the constancy lemma, we get the striking facts that abelian varieties are commutative and that every morphism of abelian varieties is a translation of a homomorphism.

Another important tool is the use of translations to prove that a generic property of a group variety holds everywhere. In this way, we will show that abelian varieties are smooth, that the dimension formula and other properties hold for homomorphisms and that the tangent bundle is trivial. Another important result is that a rational map to an abelian variety is a morphism at all smooth points. Finally, complex abelian varieties are biholomorphic to complex tori equipped with positive definite Riemann forms.

In this section, the reader is assumed to be familiar with complete varieties (see Section A.6) and with the concept of smoothness (see Section A.7).

Definition 8.2.1. *A variety G with morphisms*

$$m : G \times G \longrightarrow G, \qquad (x, y) \longmapsto xy \quad (\textit{multiplication}),$$

$$\iota : G \longrightarrow G, \qquad x \longmapsto x^{-1} \quad (\textit{inverse}),$$

and with an element $\varepsilon \in G(K)$ is called a ***group variety*** *(over K) if $G(\overline{K})$ is a group with multiplication, inverse, identity induced by m, ι, ε.*

If G_1, G_2 are group varieties with multiplications m_1, m_2, then a morphism $\varphi : G_1 \to G_2$ with $\varphi \circ m_1 = m_2 \circ (\varphi \times \varphi)$ is called a ***homomorphism of group varieties****. If there is also a homomorphism $\psi : G_2 \to G_1$ such that $\psi \circ \varphi$ and $\varphi \circ \psi$ are both the identity, then φ is called an* ***isomorphism of group varieties****. If $G_1 = G_2$, then a homomorphism (resp. isomorphism) is called an* ***endomorphism*** *(resp.* ***automorphism****) as usual.*

A closed subvariety of G, whose $\overline{K}$-rational points form a subgroup of $G(\overline{K})$, is a group variety. We say that it is a ***closed subgroup*** *of G.*

Example 8.2.2. The linear tori studied in Chapter 3 are commutative affine group varieties. In this chapter, we want to perform a similar study for the following objects:

Definition 8.2.3. *An* ***abelian variety*** *is a geometrically irreducible and geometrically reduced complete group variety.*

A ***homomorphism of abelian varieties*** *is nothing else other than a homomorphism of group varieties. An* ***abelian subvariety*** *B of an abelian variety A is a geometrically irreducible and geometrically reduced closed subgroup of A. As B is closed in A, it is again an abelian variety.*

Example 8.2.4. Let M_n denote the set of $n \times n$-matrices with entries in $\overline{K}$. The identification of $\mathbb{A}^{n^2}(\overline{K})$ with M_n, together with addition, makes the latter into an irreducible affine group variety over K. The determinant is a morphism $\det : M_n \to \mathbb{A}^1_K$, so we have an affine open irreducible subvariety $GL(n)_K$, defined as the complement of the vanishing locus of the determinant (cf. Example A.3.12). It is immediately seen that $GL(n)_K$ and matrix multiplication form a group variety. All closed subgroups, as for example the special linear group

$$SL(n)_K := \{a \in GL(n)_K \mid \det(a) = 1\}$$

or the upper triangular matrices, are affine group varieties.

Remark 8.2.5. We mention some general facts about the structure of group varieties not used in the book. Every affine group variety is isomorphic to a closed subgroup of $GL(n)$ (cf. W.C. Waterhouse [**323**], 3.4). We will not consider here the theory of affine group varieties and we refer the reader to the literature.

Let G be an irreducible group variety over a perfect field K. By a theorem of Chevalley (cf. S. Bosch, W. Lütkebohmert, and M. Raynaud [**44**], Th.9.2.1), there is a smallest irreducible affine closed subgroup H and an abelian variety A such that we have an exact sequence

$$0 \longrightarrow H \longrightarrow G \longrightarrow A \longrightarrow 0.$$

To study general group varieties we have to understand both affine group varieties and abelian varieties. Since the trivial group variety $\mathbb{A}^0_K$ is the only complete geometrically irreducible affine variety (see A.6.15 (d)), no other affine group variety is an abelian variety.

Next, we come to the **constancy lemma**:

Lemma 8.2.6. *Let X, Y, Z be varieties such that X is complete and both X and Y are geometrically irreducible. If $f : X \times Y \to Z$ is a morphism such that*

$$f(X \times \{y_0\}) = \{z_0\}$$

for some $y_0 \in Y$ and $z_0 \in Z$, then $f(X \times \{y\})$ is a point for every $y \in Y$.

Proof: By base change, we may assume that K is algebraically closed. Let U be an open affine neighborhood of z_0. The image C of $f^{-1}(Z \backslash U)$ by the projection $X \times Y \to Y$ is closed because X is complete. Then $V := Y \backslash C$ is an open neighborhood of y_0 and, for any $y \in V$, we have a morphism $X \to U$, given by $x \mapsto f(x, y)$. Since X is complete and irreducible and U is affine, the morphism has to be constant for any $y \in V$, with image $f(x_0, y)$ choice of a point $x_0 \in X$ (use A.6.15).

Now note that

$$S := \{y \in Y \mid |f(X \times \{y\})| = 1\} = \bigcap_{x_1, x_2 \in X} \{y \in Y \mid f(x_1, y) = f(x_2, y)\}$$

is closed in Y. Since it contains the non-empty open subset V of Y and since Y is irreducible, we conclude that $S = Y$, proving our claim. □

Example 8.2.7. The affine line $\mathbb{A}^1_K$ is not a complete variety, because $xy = 1$ is a closed subvariety of $\mathbb{A}^1_K \times \mathbb{A}^1_K$, while its projection on the second factor is $\mathbb{A}^1_K \backslash \{0\}$, which is not closed in $\mathbb{A}^1_K$. Now consider the morphism $f : \mathbb{A}^1_K \times \mathbb{A}^1_K \to \mathbb{A}^1_K$, given by $(x, y) \mapsto xy$. Then f satisfies the hypothesis of the constancy lemma in 8.2.6, namely

$$f(\mathbb{A}^1_K \times \{0\}) = \{0\}.$$

This shows that the constancy lemma in 8.2.6 does not hold for non-complete X.

Corollary 8.2.8. *Let X, Y be geometrically irreducible varieties with at least one K-rational point. We assume that X is complete. A morphism $f : X \times Y \longrightarrow G$ of a product into a group variety G factorizes as $f(x, y) = g(x)h(y)$, for suitable morphisms $g : X \longrightarrow G$ and $h : Y \longrightarrow G$.*

Proof: We choose a point $y_0 \in Y(K)$ and define $g : X \longrightarrow G$ by $g(x) = f(x, y_0)$. The morphism $F : X \times Y \longrightarrow G$ defined by $F(x,y)g(x)^{-1}f(x,y)$ satisfies $F(X \times \{y_0\}) = \{\varepsilon\}$, where ε is the identity of G. Now the constancy lemma in 8.2.6 shows that $F(X \times \{y\})$ is a point, say $h(y)$, for every $y \in Y$, and $f(x, y) = g(x)h(y)$. In order to verify that h is a K-morphism, note that $h = f(x_0, \cdot)g(x_0)^{-1}$ for any $x_0 \in X(K)$. □

Corollary 8.2.9. *Let $\varphi : A \to G$ be a morphism of the abelian variety A into the group variety G. Then the map*

$$\psi : A \longrightarrow G, \qquad a \longmapsto \varphi(a)\varphi(\varepsilon_A)^{-1}$$

is a homomorphism of group varieties.

Proof: Apply the constancy lemma (see Lemma 8.2.6) with $f : A \times A \to G$, given by

$$(x, y) \mapsto \psi(x)\psi(y)\psi(xy)^{-1},$$

and with y_0 and z_0 the identity elements $\varepsilon_A, \varepsilon_G$ of A and G. We conclude that the restriction of f to $A \times \{y\}$ is a constant map for every y. Since $f(\{\varepsilon_A\} \times A) = \{\varepsilon_G\}$, we deduce that f is constant, with image the identity of G. □

Corollary 8.2.10. *An abelian variety is commutative.*

Proof: By Corollary 8.2.9, the inverse map ι is a homomorphism. This is equivalent to commutativity. □

8.2.11. This allows us to change conventions. From now on, we write an abelian variety additively, hence

$$m(x, y) = x + y,$$
$$\iota(x) = -x,$$

and the identity is denoted by 0. For $a \in A$, the morphism $\tau_a(x) := x + a$ is called **translation** by a.

For $n \in \mathbb{Z}$, we denote by $[n]$ the endomorphism of A, which is multiplication by n. The kernel of $[n]$ is denoted by $A[n]$. It is the torsion subgroup of A. We will also use these notations for any abelian group.

Proposition 8.2.12. *A geometrically reduced group variety is smooth.*

Proof: By base change and using A.7.14, we may assume K algebraically closed. By A.7.16, there is an open dense smooth subset U. As above, we can define left- and right-translation by a point of the group variety. They are automorphisms and so the left-translation of U is also smooth. If we vary the left-translations, then we get an open cover of the group variety, proving the claim. □

Proposition 8.2.13. *For a group variety G over K, the following conditions are equivalent:*

(a) *G is connected;*

(b) *G is geometrically connected (i.e. $G_{\overline{K}}$ is connected);*

(c) *G is irreducible;*

(d) *G is geometrically irreducible.*

In particular, a connected complete geometrically reduced group variety over K is an abelian variety.

Proof: First, we note that a K-variety with at least one K-rational point is connected if and only if it is geometrically connected (use A. Grothendieck **[137]**, Prop.4.5.13). This proves equivalence of (a) and (b). Every irreducible variety is connected. So it remains to prove that (b) implies (d). We may assume that K is algebraically closed and G connected. By Proposition 8.2.12, G is smooth and therefore G is the disjoint union of its irreducible components (cf. A.7.14). We conclude that G is irreducible. □

8.2.14. Let $\varphi : G \longrightarrow H$ be a homomorphism of group varieties. Then the image $\mathrm{im}(\varphi)$ is a closed subgroup of H (**[85]**, Ch.II, §5, Prop.5.1).

8.2.15. There are more problems in handling $\ker(\varphi)$. The main difficulty is that, if we insist in defining $\ker(\varphi)$ as a group variety, then we lose the main formalism associated to group homomorphisms. This is due to the appearance of nilpotent elements in $\ker(\varphi)$, i.e. $\ker(\varphi)$ need not be reduced. For example, the support of the kernel of the Frobenius homomorphism $x \mapsto x^p$ on the multiplicative group $\mathbb{G}_m$ over $\mathbb{F}_p$ contains only 1, and therefore cannot be distinguished from the kernel of the identity map. These difficulties disappear in the context of group schemes. For details, we refer to I.R. Shafarevich **[280]**, Ch.V, 4.2. The natural way is to define $\ker(\varphi)$ as the cartesian product $G \times_H \operatorname{Spec}(K)$ in the category of schemes with respect to the Cartesian diagram

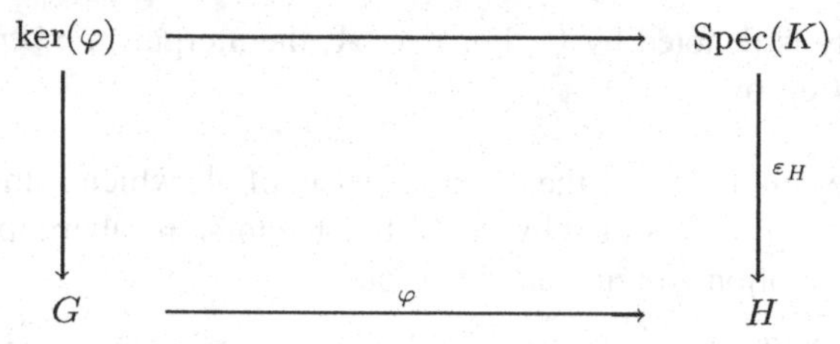

where ε_H is the map of $\operatorname{Spec}(K)$ to the neutral element of H. Then $\ker(\varphi)$ is a closed subscheme such that its $\overline{K}$-rational points form a group. On the other hand, $\ker(\varphi)$ need not be reduced and so is not necessarily a group variety. Working with group schemes is therefore the natural way of overcoming these obstacles, leading to a coherent theory. The more elementary classical theory in the framework of group varieties is adequate only for separable maps. On the other hand, a famous result of Cartier states that every group scheme in characteristic 0 is reduced, i.e. is a group variety (see **[85]**, Ch.II, §6, no.1).

Since the fact that $\ker(\varphi)$ is in general only a group scheme plays no role in this text, we only consider

$$\ker(\varphi) := \{x \in G(\overline{K}) \mid \varphi(x) = \varepsilon_H\}$$

as a closed subgroup of G, unless specified otherwise.

Next, we consider the **dimension theorem of group varieties.**

Theorem 8.2.16. *Let $\varphi : G \to H$ be a surjective homomorphism of irreducible group varieties. Then*

$$\dim(G) = \dim(H) + \dim(\ker(\varphi)).$$

Proof: Since all fibres are isomorphic to $\ker(\varphi)$, this follows from the dimension theorem of varieties (cf. A.12.1). □

Proposition 8.2.17. *Let $\varphi : G \to H$ be a surjective homomorphism of irreducible group varieties. Then φ is flat. Moreover, if $\dim(G) = \dim(H)$, then φ is finite and $|\ker(\varphi)|$ is equal to the separable degree of the field extension $K(G)$ over $K(H)$.*

Proof: By generic flatness (cf. A.12.13), there is an open dense subset U of G such that $\varphi|_U$ is flat. Of course, any translate of U is as good as U itself. Assuming for a moment K algebraically closed, we may cover G by translates of U. This proves flatness of φ. If K is not algebraically closed, then we perform base change to $\overline{K}$. However, we have to work in the category of schemes to ensure that a morphism over K is flat if and only if its base change to $\overline{K}$ is flat (cf. **[137]**, Prop.2.5.1).

If $\dim(G) = \dim(H)$, then there is an open dense subset U' of H such that φ induces a finite map $U := \varphi^{-1}(U') \to U'$ whose fibres have cardinality equal to the separable degree of $K(G)$ over $K(H)$ (cf. A.12.9). Also, this cardinality equals $|\ker(\varphi)|$. Again, we assume that K is algebraically closed to cover G by translates of U proving finiteness of φ. If K is not algebraically closed, then we use base change to $\overline{K}$ and the fact that a morphism over K is finite if and only if its base change to $\overline{K}$ is finite (cf. **[137]**, Prop.2.7.1). □

8.2.18. A **rational curve** is a curve birational to $\mathbb{P}^1_K$. A variety X is called **rationally connected** if any two points in $X(\overline{K})$ may be connected by a rational curve over $\overline{K}$. It follows from the constancy lemma in 8.2.6 that abelian varieties do not contain rational curves (cf. Corollary 8.2.20 below). In particular, a morphism $X \longrightarrow A$ of X into an abelian variety A contracts the rational curves of X to points. It follows that any morphism of a rationally connected variety, such as projective space $\mathbb{P}^n$, into an abelian variety is constant.

Proposition 8.2.19. *Any morphism $f : \mathbb{P}^1_K \longrightarrow G$ of the projective line into a group variety is constant.*

Proof: Let $(x_0 : x_1)$ be homogeneous coordinates on $\mathbb{P}^1_K$. The map $s : \mathbb{P}^1_K \times \mathbb{A}^1_K \longrightarrow \mathbb{P}^1_K$ given by $s((x_0 : x_1), y) = (x_0 : (x_1 + x_0 y))$ is a morphism. Now let $f : \mathbb{P}^1_K \longrightarrow G$ be a morphism of $\mathbb{P}^1_K$ into the group variety G. We apply Corollary 8.2.8 to the composition

$$\mathbb{P}^1_K \times \mathbb{A}^1_K \xrightarrow{s} \mathbb{P}^1_K \xrightarrow{f} G$$

and obtain that $f \circ s$ factorizes as $f(s(x,y)) = g(x)h(y)$ for two suitable morphisms $g : \mathbb{P}^1_K \longrightarrow G$, $h : \mathbb{A}^1_K \longrightarrow G$.

We set first $y = 0$, note that $s(x,0) = x$ to get $g(x) = f(x)h(0)^{-1}$. Thus

$$f(s(x,y)) = f(x)h(0)^{-1}h(y).$$

Next we set $x = \infty$, note that $s(\infty, y) = \infty$ to get

$$f(\infty) = f(\infty)h(0)^{-1}h(y).$$

This shows that $h(y) = h(0)$, so h is a constant map and $f(s(x,y)) = f(x)$. Finally we set $x = 0$, note that $s(0,y) = y$, and find $f(y) = f(0)$. □

Corollary 8.2.20. *Let U be an open set of the projective line $\mathbb{P}^1_K$. Then any morphism $f : U \longrightarrow A$ of U into an abelian variety is constant.*

Proof: By the valuative criterion of properness (cf. A.11.10), f extends to a morphism from $\mathbb{P}^1_K$ to A. By Proposition 8.2.19, we get the claim. □

Theorem 8.2.21. *Let $\varphi : X \dashrightarrow G$ be a rational map of a smooth variety X to a group variety G and let $U_{\max}$ be the domain of φ. Then every irreducible component of $X \setminus U_{\max}$ is of codimension 1.*

Proof: By base change and A.11.9, we may assume K algebraically closed. By A.7.14, we may assume that X is irreducible. Consider the rational map $\Phi : X \times X \dashrightarrow G$, given by $\Phi(x, y) := \varphi(x)\varphi(y)^{-1}$. It is clear that the restriction of Φ to the diagonal Δ is constant equal to the identity ε of G. First, we prove that φ is defined in $x \in X$ if and only if Φ is defined in (x, x). If φ is defined in x, then Φ is obviously defined in (x, x) and $\Phi(x, x) = \varepsilon$. This proves also that Φ is defined and constant ε on an open dense subset of the diagonal. Conversely, we assume that Φ is defined in (x, x). The above shows that $\Phi(x, x) = \varepsilon$. Since Φ is defined in an open neighbourhood W of (x, x), there is an open neighbourhood V of x such that $\{x\} \times V \subset W$. Let $y \in U_{\max} \cap V$. Then we define $\varphi(x) := \Phi(x, y)\varphi(y)$. This defines a morphism even in an open neighbourhood of x and agrees with the given φ on $U_{\max}$. This proves $x \in U_{\max}$.

By A.11.8, Φ is defined in (x, x) if and only if $f \mapsto f \circ \Phi$ maps $\mathcal{O}_{G,\varepsilon}$ to $\mathcal{O}_{X\times X,(x,x)}$. The latter is equivalent to the condition that for every $f \in \mathcal{O}_{G,\varepsilon} \setminus \{0\}$, the poles of $\operatorname{div}(f \circ \Phi)$ omit (x, x) (cf. A.8.21). Given $f \in \mathcal{O}_{G,\varepsilon}$, let P_f be the divisor of poles of $f \circ \Phi$. We have seen above that Φ is defined on an open subset of the diagonal, hence every irreducible component of $\Delta \cap P_f$ has codimension 1 in Δ (cf. A.8.24). We conclude that the complement of $U_{\max}$ is equal to the union of all $\Delta \cap P_f$ projected to X. This proves the claim. □

Corollary 8.2.22. *A rational map from a smooth variety to an abelian variety is a morphism.*

Proof: Let $\varphi : X \dashrightarrow A$ be a rational map with domain $U_{\max}$. By the valuative criterion of properness (cf. A.11.10), $X \setminus U_{\max}$ has codimension at least 2. Then $U_{\max} = X$ follows by appealing to Theorem 8.2.21. □

Remark 8.2.23. Next, we prove that the differential of multiplication on a group variety is given by addition. This is quite easy to see for a complex abelian variety given analytically by V/Λ, where Λ is a lattice in the complex vector space V (see 8.2.27 below). Then V may be seen as the tangent space at 0 and the differential of addition on V/Λ is just addition on V.

Proposition 8.2.24. *Let $m : G \times G \longrightarrow G$ be multiplication of a smooth group variety G. Then the differential of m at ε is the map $T_{G,\varepsilon} \oplus T_{G,\varepsilon} \longrightarrow T_{G,\varepsilon}$ given by addition of tangent vectors.*

Proof: Note that $T_{G\times G,(\varepsilon,\varepsilon)} = T_{G,\varepsilon} \oplus T_{G,\varepsilon}$ (cf. A.7.17). For $\partial \in T_{G,\varepsilon}$, we have

$$\mathrm{d}m(\partial, 0) = \mathrm{d}m \circ \mathrm{d}\iota(\partial),$$

where $\iota : G \to G \times G, g \mapsto (g, \varepsilon)$. Since $\mathrm{d}m \circ \mathrm{d}\iota = \mathrm{d}(m \circ \iota)$ is the identity map, we conclude that $\mathrm{d}m(\partial, 0) = \partial$. In the same way, we prove $\mathrm{d}m(0, \partial) = \partial$. By linearity of $\mathrm{d}m$, this gives the claim. □

Corollary 8.2.25. *Let G be a smooth group variety and for $n \in \mathbb{Z}$, let $[n] : G \longrightarrow G$ be the morphism $x \mapsto x^n$. Then the differential of $[n]$ at ε is the endomorphism of $T_{G,\varepsilon}$ given by multiplying tangent vectors with n.*

Proof: For $\partial \in T_{G,\varepsilon}$ and for the diagonal morphism $\Delta : G \times G \to G$, we have

$$\mathrm{d}[2](\partial) = \mathrm{d}(m \circ \Delta)(\partial) = \mathrm{d}m \circ \mathrm{d}\Delta(\partial) = \mathrm{d}m(\partial, \partial) = 2\partial$$

proving the case $n = 2$. The general case follows by induction with the same argument. □

Proposition 8.2.26. *Let G be an irreducible smooth group variety. Then the tangent bundle T_G on G is a trivial vector bundle of rank equal to* $\dim(G)$.

Proof: Let $\partial_\varepsilon \in T_{G,\varepsilon}$. By translation, we extend ∂_ε to a vector field ∂ on G. More precisely, let $\tau_x(y) := yx$ be right translation on G and let $\partial_x(f) := \partial_\varepsilon(f \circ \tau_x)$ for any $x \in G$ and $f \in \mathcal{O}_{G,x}$. Standard arguments for derivatives show that ∂ is a vector field on G. Clearly, linearly independent tangent vectors in ε extend to vector fields, which are linearly independent in every fibre. This proves the claim. □

8.2.27. We review here the analytic description of an abelian variety. For details and proofs, we refer to D. Mumford **[212]**, Ch.I. Let A be an abelian variety over $\mathbb{C}$ endowed with its complex analytic structure (cf. A.14). Then A is a compact complex manifold. Let $V := T_0A$. The kernel of $\exp : V \to A$ is a lattice Λ in V and we get an isomorphism $V/\Lambda \xrightarrow{\sim} A$ of complex Lie groups. By Proposition 8.2.26, we may identify V with $H^0(A, T_A) = H^0(A, \Omega^1_A)^*$, i.e. with the dual of the space of global holomorphic 1-forms. Then we may identify the first homology group $H_1(A, \mathbb{Z})$ with Λ by using the period map $\omega \mapsto \int_\gamma \omega$ for $\gamma \in H_1(A, \mathbb{Z})$.

Conversely, let Λ be a lattice in a finite-dimensional complex vector space V. Then the compact complex manifold $T := V/\Lambda$ is called a **complex torus**. A hermitian form H on V is called a **Riemann form** for T if $E := \Im H$ is an integer valued alternating bilinear form on Λ. Note that an alternating bilinear form E on Λ with values in $\mathbb{Z}$ is the imaginary part of a Riemann form if and only if $E(iv, iw) = E(v, w)$ for all $v, w \in V$. We get the Riemann form by $H(v, w) := E(iv, w) + iE(v, w)$. The torus T is an abelian variety if and only if there is a positive definite Riemann form for T. Moreover, this is equivalent for T to be the complex space associated to a complex algebraic variety.

To give an idea of the Riemann form, let L be a line bundle on the complex torus T. Here in these analytical remarks, this always means a holomorphic line bundle. Then the cohomology group $H^k(T, \mathbb{Z})$ can be identified with the group of alternating $\mathbb{Z}$-valued k-forms on Λ. This is clear for $k = 1$ and follows from the isomorphism $\Lambda^k H^1(T, \mathbb{Z}) \xrightarrow{\sim} H^k(T, \mathbb{Z})$ induced by cup product. It is easy to see that multiplication by i is an isometry with respect to the alternating bilinear form corresponding to the Chern class $c_1(L) \in H^2(T, \mathbb{Z})$ and so it may be viewed as the imaginary part of a unique Riemann form H. Then H is positive definite if and only if L is ample. Note that two line bundles have the same Chern class (and hence the same Riemann form) if and only if they are algebraically equivalent (use A.14.8).

An analytic homomorphism $\varphi : T = V/\Lambda \to T' = V'/\Lambda'$ of complex tori is the quotient map of the linear map $d\varphi : V \to V'$ of tangent spaces using that $d\varphi(\Lambda) \subset \Lambda'$ is induced from

$$\Lambda = H_1(T, \mathbb{Z}) \to \Lambda' = H_1(T', \mathbb{Z}), \quad \gamma \mapsto \varphi \circ \gamma.$$

Conversely, every linear map $\psi : V \to V'$ with $\psi(\Lambda) \subset \Lambda'$ induces an analytic homomorphism $\varphi : T \to T'$ with $d\varphi = \psi$.

8.3. Elliptic curves

By Proposition 8.2.26, the cotangent bundle of an abelian variety over the field K is trivial. Thus an abelian variety of dimension 1 has genus 1, i.e. is an elliptic curve. In this section we prove the converse statement, namely that an elliptic curve has a group structure and is an abelian variety. By Corollary 8.2.9, this group structure is unique up to translations.

Elliptic curves are a major tool in arithmetic and play a role also in other parts of mathematics. The interested reader may consult the monographs by D. Husemöller [**155**] or J. Silverman [**284**] for a deeper study of the subject. We assume the reader to be familiar with the theory of algebraic curves as provided by Section A.13.

Definition 8.3.1. *An **elliptic curve** over K is a geometrically irreducible smooth projective curve E of genus $g(E) = 1$ defined over K, **equipped with a rational point** $P_0 \in E(K)$.*

8.3.2. Note that geometrically irreducible is the same as irreducible by A.7.14. Let E be an elliptic curve over K and let D be a divisor on $E_{\overline{K}}$ of degree $\deg(D) > 0$. The space of global sections $\Gamma(E_{\overline{K}}, O(D))$ may be realized as the subspace

$$\overline{L}(D) := \{f \in \overline{K}(E_{\overline{K}})^{\times} \mid \operatorname{div}(f) \geq -D\} \cup \{0\}$$

in $\overline{K}(E_{\overline{K}})$, using the homomorphism $s \mapsto s/s_D$. By the Riemann–Roch theorem in A.13.5, we have

$$\dim_{\overline{K}} \overline{L}(D) = \deg(D), \tag{8.1}$$

hence the corresponding linear system $|D_{\overline{K}}|$ has dimension $\deg(D) - 1$. It follows that two distinct points on E never are rationally equivalent over $\overline{K}$.

Let us fix a base point $P_0 \in E(K)$. For two points $P_1, P_2 \in E(\overline{K})$, let $D := [P_1] + [P_2] - [P_0]$. Thus $\deg(D) = 1$ and $\overline{L}(D)$ is one-dimensional, generated by a function f, unique up to multiplication by a scalar. By construction, this function has a pole divisor majorized by $[P_1] + [P_2]$. If $P_0 \notin \{P_1, P_2\}$, then f has pole divisor $[P_1] + [P_2]$ and vanishes at P_0 and at exactly one other point P_3, which is the unique point rationally equivalent to $[P_1] + [P_2] - [P_0]$. This makes sense even if P_1 or P_2 equals P_0. Thus we get a well-defined composition law on E by $(P_1, P_2) \longmapsto P_1 + P_2 := P_3$.

We should distinguish carefully between addition of points P_1, P_2 on E and of the corresponding divisors $[P_1], [P_2]$. Remembering that $\operatorname{Pic}^0(E_{\overline{K}})$ is the group of rational equivalence classes of divisors of degree 0 (see A.9.40), we get an additive map

$$E \longrightarrow \mathrm{Pic}^0(E_{\overline{K}}), \qquad P \longmapsto \mathrm{cl}([P]-[P_0]).$$

Formula (8.1) shows easily that this map is bijective. We shall give in 8.3.4 an equation for E as a plane cubic curve, defined over K. This will lead in 8.3.6 to a geometric interpretation of the group operations and in particular the inverse map will be shown to be a morphism. In 8.3.7, we deduce explicit formulas for our composition law in terms of the affine coordinates of the plane cubic curve. They will be postulated in the addition law 8.3.8. We shall use them in 8.3.9 to prove that the composition law is a morphism. This will prove:

Proposition 8.3.3. *If the group structure on an elliptic curve E over K with base point $P_0 \in E(K)$ is given by the bijective map*

$$E \longrightarrow \mathrm{Pic}^0(E_{\overline{K}}), \qquad P \longmapsto \mathrm{cl}([P]-[P_0]),$$

then E is an abelian variety defined over K.

8.3.4. We give here the classical argument showing that E has a model given by a smooth cubic curve.

In the same way as in 8.3.2, we realize $\Gamma(E, O(D))$ explicitly by

$$L(D) := \{f \in K(E)^\times \mid \mathrm{div}(f) \geq -D\} \cup \{0\}$$

for any divisor D on E. If $\deg(D) > 0$, then the Riemann–Roch theorem in A.13.5 again shows that $L(D)$ has dimension $\deg(D)$.

We have an ascending chain of K-vector spaces

$$L([P_0]) \subset L(2[P_0]) \subset L(3[P_0]) \subset L(4[P_0]) \subset L(5[P_0]) \subset L(6[P_0])$$

and the jth member has dimension j.

Obviously, 1 is a basis of $L([P_0])$. Since P_0 is defined over K, there are $x, y \in K(E)$ such that $1, x$ is a basis of $L(2[P_0])$ and $1, x, y$ is a basis of $L(3[P_0])$. Looking at the order of pole at P_0, it is clear that $1, x, y, x^2$ is a basis of $L(4[P_0])$ and $1, x, y, x^2, xy$ is a basis of $L(5[P_0])$. Moreover, $x^3, y^2 \in L(6[P_0])$. This gives us seven elements $1, x, y, x^2, xy, x^3, y^2$ spanning $L(6[P_0])$. They must be linearly dependent over K and we obtain a linear relation

$$c_0 + c_1x + c_2y + c_3x^2 + c_4xy + c_5x^3 + c_6y^2 = 0$$

with coefficients in K. By the above, c_5 and c_6 are different from zero, so that we may normalize $c_5 = -1$. If we divide by c_6^3 and replace x by x/c_6 and y by y/c_6^2, we get a relation of the form

$$y^2 + a_1xy + a_3y = x^3 + a_2x^2 + a_4x + a_6 \tag{8.2}$$

with $a_i \in K$.

Since $\deg(3[P_0]) = 3 = 2g(E) + 1$, the divisor $3[P_0]$ is very ample (cf. A.13.7). Hence the basis of $L(3[P_0])$ corresponding to 1, x, y induces a closed embedding of E into $\mathbb{P}^2_K$ (cf. Remark A.6.11). We know by (8.2) that the image of E is contained in the projective curve with **Weierstrass equation**

$$x_0x_2^2 + a_1x_0x_1x_2 + a_3x_0^2x_2 = x_1^3 + a_2x_0x_1^2 + a_4x_0^2x_1 + a_6x_0^3$$

in the homogeneous coordinates $(x_0 : x_1 : x_2)$ of $\mathbb{P}^2_K$.

It is an easy matter to prove that the curve defined above is geometrically irreducible, hence it gives a projective model of E as a smooth plane cubic curve. Note also that the rational functions $x = x_1/x_0$ and $y = x_2/x_0$ are nothing else than the two functions x, y defined before, hence the affine form (8.2) of the Weierstrass equation describes the affine curve $E \cap \{x_0 \neq 0\}$. The only point of E outside this part is the point $(0 : 0 : 1) \in \mathbb{P}^2_K$, corresponding to $P_0 \in E$. It is easily seen that, in this model, P_0 is an inflexion point of E.

Remark 8.3.5. If $\mathrm{char}(K) \neq 2$, then replacing y by $\frac{1}{2}(y - a_1x - a_3)$ leads to a Weierstrass equation with $a_1 = a_3 = 0$. Then the Jacobi criterion shows that a Weierstrass equation describes a smooth curve C in $\mathbb{P}^2_K$ if and only if the discriminant of the cubic polynomial $x^3 + a_2x^2 + a_4x + a_6$ is not zero (see Proposition 10.2.3 for the argument). By the genus formula

$$g(C) = \frac{1}{2}\,(\deg(C) - 1)(\deg(C) - 2),$$

for a smooth plane curve (cf. A.13.4), this is an elliptic curve. If $\mathrm{char}(K) \neq 3$, then a further linear transformation leads to the well-known Weierstrass normal form

$$y^2 = 4x^3 - g_2x - g_3$$

of the elliptic curve. For generalizations and details, see **[284]**.

8.3.6. Now we go back to arbitrary characteristic. We describe more explicitly the group structure of the abelian group E, beginning by proving that the inverse operation is a morphism.

Consider the rational equivalence relation

$$[P_1] + [P_2] + [P_3] \sim 3[P_0] \tag{8.3}$$

on $E_{\overline{K}}$. This relation is equivalent to the geometric statement that the points P_1, P_2, P_3 are the three intersection points, counted with multiplicity, of a straight line with E. We verify this as follows. The lines in $\mathbb{P}^2_{\overline{K}}$ are just the divisors of the global sections of $O_{\mathbb{P}^2_{\overline{K}}}(1)$ and, by construction, the restriction of this line bundle to E is isomorphic to $O(3[P_0])$. First, we assume that $[P_1] + [P_2] + [P_3] \sim 3[P_0]$. Then it exists $s' \in \Gamma(E_{\overline{K}}, O(3[P_0]))$ with $\mathrm{div}(s') = [P_1] + [P_2] + [P_3]$ (cf. A.8.21). By construction of the embedding $E \hookrightarrow \mathbb{P}^2_{\overline{K}}$, there is $s \in \Gamma(\mathbb{P}^2_{\overline{K}}, O_{\mathbb{P}^2_{\overline{K}}}(1))$ with $s' = s|_E$. Then the line $\ell = \mathrm{div}(s)$ is the line through the three points P_i . Indeed, by definition of proper intersection product (cf. A.9.20), we have

$$\ell.E = \operatorname{div}(s|_E) = \operatorname{div}(s') = [P_1] + [P_2] + [P_3].$$

The converse statement is proved in the same way by reversing the previous argument.

The zero element of E is the point $P_0 = (0 : 0 : 1)$.

The inverse $P_2 := -P_1$ of a point $P_1 \in E$ is characterized by the rational equivalence relation $[P_1] + [P_2] \sim 2[P_0]$, which can be rewritten as the special case

$$[P_0] + [P_1] + [P_2] \sim 3[P_0]$$

of (8.3).

It follows that P_0, P_1, P_2 are on a straight line and in fact, noting that $P_0 = (0 : 0 : 1)$, we see that, if $P_1 \neq P_0$, then P_2 is the residual finite intersection of E with the vertical line in the (x, y)-plane going through P_1. If (x_1, y_1) are the affine coordinates of P_1, then, using (8.2) on page 241, the affine coordinates (x_2, y_2) of P_2 are given by

$$x_2 = x_1,$$
$$y_2 = -a_1 x_1 - a_3 - y_1.$$

Thus the inverse map is an automorphism of the affine part of E defined over K. On the other hand, a rational map of a smooth projective curve is always a morphism (cf. A.11.10). We conclude that the above restriction extends to an automorphism of E. This requires that 0 is mapped to 0, hence the inverse map is a morphism on E defined over K.

8.3.7. We study here addition on the elliptic curve. By 8.3.6, it is enough to construct

$$P_3 = -(P_1 + P_2).$$

The point P_3 is characterized by the rational equivalence relation (8.3). As we have seen in 8.3.6, P_3 is the third intersection point of the line ℓ through P_1 and P_2 with E, taking this line to be the tangent line to E at P_1 if $P_1 = P_2$.

If $P_1 \neq P_0$ and $P_2 \notin \{P_0, -P_1\}$, then the third intersection point of the line through P_1, P_2 with E is contained in the (x, y)-plane. Let $y = ax + b$ be the equation for this line. We eliminate y in (8.2) on page 241 obtaining a cubic equation for x, with two known solutions x_1, x_2. This equation has the form

$$x^3 - (a^2 + a_1 a - a_2)x^2 + \text{ terms of lower degree } = 0.$$

The third solution x_3 is determined by the trace $x_1 + x_2 + x_3 = a^2 + a_1 a - a_2$. Since $P_1 + P_2 = -P_3$, applying the inverse as in 8.3.6 we obtain:

Proposition 8.3.8 (Addition Law). *Let E be the elliptic curve in normal form*

$$y^2 + a_1xy + a_3y = x^3 + a_2x^2 + a_4x + a_6.$$

Then the origin O of the group E is the unique point at infinity and the group law $+$ is defined as follows. Let $P_1 = (x_1, y_1)$, $P_2 = (x_2, y_2)$ be two finite points on E and set

$$a = \frac{y_2 - y_1}{x_2 - x_1} \qquad \text{if } x_1 \neq x_2,$$

$$a = \frac{3x_1^2 + 2a_2x_1 + a_4 - a_1y_1}{2y_1 + a_1x_1 + a_3} \qquad \text{if } x_1 = x_2,$$

$$b = y_1 - ax_1.$$

Then:

(a) *The inverse of P_1 is given by $-P_1 = (x_1, -a_1x_1 - a_3 - y_1)$.*

(b) *If $x_2 = x_1$ and $y_2 = -a_1x_1 - a_3 - y_1$, then $P_1 + P_2 = O$.*

(c) *Otherwise, we have $P_1 + P_2 = (a^2 + a_1a - a_2 - x_1 - x_2,\ -(a + a_1)(a^2 + a_1a - a_2 - x_1 - x_2) - a_3 - b)$.*

The addition law and the associative law for an elliptic curve in Weierstrass form can be seen visually as in the following picture:

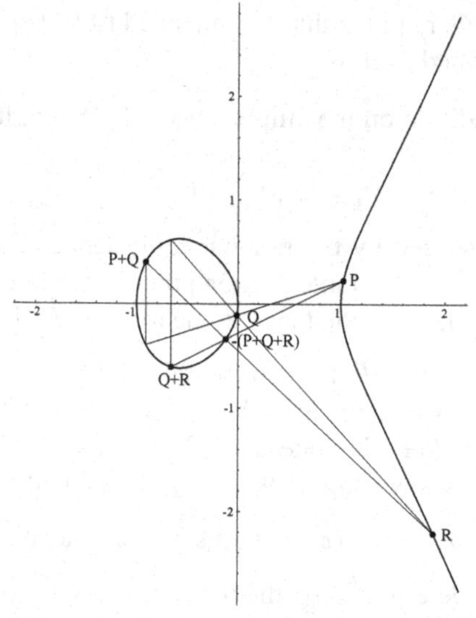

The addition law and the associative law on $y^2 = x^3 - x$

8.3.9. The addition law shows that addition is a rational map. In order to finish the proof of Proposition 8.3.3, it remains to show that $+$ is a morphism. By A.11.9, we may assume that K is algebraically closed. In a first step, we prove that translation τ_Q by $Q \in E$ is a morphism. We may assume $Q \neq O$. By the formulas in Proposition 8.3.8, τ_Q is a rational map which restricts to a morphism $E \setminus \{O, Q, -Q\} \to E \setminus \{Q, O, Q+Q\}$. Since every rational map between projective smooth curves extends to a morphism (cf. A.11.10), we get a morphism $\tau'_Q : E \to E$, which agrees with τ_Q on $E \setminus \{O, Q, -Q\}$. It remains to prove that $\tau_Q = \tau'_Q$. For $R \in E$, we get $\tau'_Q \circ \tau'_R = \tau'_{Q+R}$. In particular, every τ'_Q is an isomorphism with inverse τ'_{-Q}. We conclude that τ'_Q maps $\{O, Q, -Q\}$ onto $\{Q, Q+Q, O\}$. For any $R \notin \{O, Q, -Q, Q+Q, -Q-Q\}$, we have

$$\tau'_R(\tau'_Q(Q)) = \tau'_{Q+R}(Q) = \tau'_Q(\tau'_R(Q)) = \tau'_Q(Q+R) = Q+Q+R.$$

This excludes $\tau'_Q(Q) = Q$ immediately. On the other hand, we know $\tau'_R(O) \in \{O, R, R+R\}$. Hence $\tau'_Q(Q) = O$ is only possible if $Q+Q = O$. This proves

$$\tau'_Q(Q) = Q+Q = \tau_Q(Q).$$

The equation

$$\tau'_Q(-Q) = O = \tau_Q(-Q)$$

is proved in a similar fashion. Thus, using that τ'_Q is a bijection, we conclude that $\tau'_Q(O) = Q = \tau_Q(O)$. We have handled all exceptions, thereby proving that $\tau_Q = \tau'_Q$.

Next, we prove that addition is a morphism. The formulas in 8.3.8 show that addition is a rational map m, which is a morphism outside of

$$Z := \{(P,P) \mid P \in E\} \cup \{(P,-P) \mid P \in E\} \cup (E \times \{O\}) \cup (\{O\} \times E).$$

For $(P, Q) \in Z$, there are $R, S \in E$ such that $(P+R, Q+S) \notin Z$. Since translations are morphisms by our above considerations, we see that

$$\tau_{-P-Q} \circ m \circ (\tau_R \times \tau_S)$$

is a morphism in a neighbourhood of (P, Q) and agrees with $+$ everywhere. This proves that $+$ is a morphism. □

8.3.10. Complex analytically, an elliptic curve is biholomorphic to $\mathbb{C}/\Lambda$ for a lattice Λ in $\mathbb{C}$ (cf. 8.2.27). In dimension 1 the converse is true, i.e. every one-dimensional complex torus is biholomorphic to an abelian variety: The imaginary part of a Riemann form must be an integer multiple of the alternating bilinear form E_0 given in the following way. Let λ_1, λ_2 be a positively oriented $\mathbb{Z}$-basis of Λ; then E_0 is characterized by

$$v \wedge w = E_0(v, w)\lambda_1 \wedge \lambda_2 \quad (v, w \in \mathbb{C}).$$

A Riemann form is positive definite if and only if its imaginary part is a negative multiple of E_0. Note however that in higher dimensions a complex torus need not be an abelian variety (in fact, this is the case for a general complex torus).

The description of the elliptic curve determined by $\mathbb{C}/\Lambda$ is done quite explicitly by means of the **Weierstrass $\wp$- function** associated to the lattice Λ, namely

$$\wp(z) := \frac{1}{z^2} + \sum_{\omega \in \Lambda \setminus \{0\}} \left\{ \frac{1}{(z-\omega)^2} - \frac{1}{\omega^2} \right\}.$$

It is a Λ-periodic meromorphic function on $\mathbb{C}$ and has double periods at the lattice points. It satisfies the first-order differential equation

$$\wp'(z)^2 = 4\wp(z)^3 - g_2\wp(z) - g_3,$$

where the coefficients g_2, g_3 are given by

$$g_2 := 60 \sum_{\omega \in \Lambda \setminus \{0\}} \frac{1}{\omega^4}, \quad g_3 := 140 \sum_{\omega \in \Lambda \setminus \{0\}} \frac{1}{\omega^6}.$$

The coefficients g_2, g_3 are examples of Eisenstein series.

The map $z \mapsto (\wp(z), \wp'(z))$ is biholomorphic from $\mathbb{C}/\Lambda$ onto the elliptic curve with affine Weierstrass equation $y^2 = 4x^3 - g_2x - g_3$. This map is also an isomorphism of groups. For further details, we refer to [**284**].

8.4. The Picard variety

Elliptic curves are the only standard explicit examples of abelian varieties, because higher-dimensional abelian varieties can be defined only by means of a very large number of equations, and little can be understood on abelian varieties by looking directly at these equations.

In this respect, the cubic model of an elliptic curve is rather special and not representative of the general situation. However, abelian varieties are ubiquitous in algebraic geometry and they occur most naturally, through the Picard variety, in the parametrization of families of divisor classes on a variety. For singular varieties, it is better to use the Picard group instead of divisor classes (cf. Section A.8 and A.9.18).

This section is devoted to fundamental facts about Picard varieties. Here, the reader is assumed to be familiar with the basic properties of the Picard group as provided by A.5.16, with Section A.8 about divisors and with the concept of algebraic equivalence for line bundles (see end of Section A.9). We fix the ground field K and an algebraic closure $\overline{K}$.

8.4.1. If $\varphi : X \longrightarrow Y$ is a morphism of varieties over K and $y \in Y$, then the fibre of φ over y is denoted by X_y. The pull-back of $\mathbf{c} \in \operatorname{Pic}(X)$ to the fibre X_y is denoted by $\mathbf{c}_y$. It is an element of $\operatorname{Pic}(X_y)$. Note that X_y and $\mathbf{c}_y$ are only defined over $K(y)$. Often, we identify X with X_y using the map $x \mapsto (x, y)$. Of course, this is only defined over $K(y)$.

In the following, we consider $\mathbf{c} \in \mathrm{Pic}(X \times Y)$ and the fibres with respect to the projections p_1, p_2 onto the factors. For $x \in X, y \in Y$, we have

$$\mathbf{c}_y = \mathbf{c}|_{X \times \{y\}} \in \mathrm{Pic}(X_{K(y)}), \quad \mathbf{c}_x = \mathbf{c}|_{\{x\} \times Y} \in \mathrm{Pic}(Y_{K(x)}).$$

The next result is called the **seesaw principle**:

Theorem 8.4.2. *Let X be a geometrically irreducible smooth complete variety over K and Y an irreducible smooth variety over K. Let $\mathbf{c} \in \mathrm{Pic}(X \times Y)$ and suppose that there is a dense open subset U of Y such that $\mathbf{c}_y = \mathbf{0}$ for all $y \in U$. Then $\mathbf{c}$ is equal to the pull-back of an element of $\mathrm{Pic}(Y)$ by p_2.*

Proof: By the semicontinuity theorem (see [**148**], Th.III.12.8 for the projective case and A. Grothendieck [**136**], 7.7, or [**212**], II, §5, for the proper case), there is an open dense subset V of Y with $H^0(X \times V, \pm\mathbf{c}) \neq 0$ and thus we can find corresponding non-zero global sections s_+, s_-. Now $s_+ \otimes s_-$ is a regular function on $X \times V$. Since X is complete and geometrically irreducible, this regular function has to be constant on every fibre over V (use A.6.15). By passing to a smaller V, we may assume that $\mathrm{div}(s_+ \otimes s_-) = 0$ on $X \times V$. Therefore $\mathrm{div}(s_\pm) = 0$ and hence the restriction of $\mathbf{c}$ to $X \times V$ is trivial (cf. A.8.18). Here and in what follows, we use the fact that on a smooth variety we may identify Cartier- and Weil-divisors (cf. A.8.21).

Let s be an invertible meromorphic section of $\mathbf{c}$. Then there is a rational function $f \in K(X \times Y)^\times$ such that $\mathrm{div}\,(f)$ and $\mathrm{div}\,(s)$ are equal on $X \times V$. Therefore, their difference $\mathrm{div}\,(f) - \mathrm{div}\,(s)$ is supported in $X \times Z$, where Z is a closed subvariety of codimension 1 in Y. If $Z_1, \dots, Z_r$ are the irreducible components of Z, then $X \times Z_1, \dots, X \times Z_r$ are the irreducible components of $X \times Z$ (cf. A.4.11). Therefore the divisor $\mathrm{div}\,(f) - \mathrm{div}\,(s)$ is a linear combination of the $X \times Z_i$ $(i = 1, \dots, r)$, i.e. a pull-back of a divisor on Y. This proves the claim. □

Remark 8.4.3. The seesaw principle holds even without smoothness assumptions (see [**212**], II, §5). We often use the seesaw principle in the following form.

Corollary 8.4.4. *Let X, Y be smooth varieties over K and assume that Y is irreducible and that X is complete and geometrically irreducible. Let $\mathbf{c} \in \mathrm{Pic}(X \times Y)$ with $\mathbf{c}_y = \mathbf{0}$ for all y in an open dense subset of Y and with $\mathbf{c}_x = \mathbf{0}$ for some $x \in X(K)$. Then $\mathbf{c} = \mathbf{0}$.*

Proof: By Theorem 8.4.2, we have $\mathbf{c} = p_2^*\mathbf{c}'$ for some $\mathbf{c}' \in \mathrm{Pic}(Y)$. Now consider the closed embedding $\iota_x : Y \longrightarrow X \times Y$, $y \mapsto (x, y)$, mapping Y isomorphically onto the fibre over x. Since $p_2 \circ \iota_x$ is the identity map on Y, we get

$$\mathbf{c}' = \iota_x^* p_2^* \mathbf{c}' = \mathbf{c}_x = \mathbf{0}.$$

This also proves that $\mathbf{c} = \mathbf{0}$. □

Corollary 8.4.5. *Let A be an abelian variety over K, let p_i be the ith projection $A \times A$ onto A, and let m be addition as usual. The following conditions are equivalent for* $\mathbf{c} \in \mathrm{Pic}(A)$*:*

(a) $m^*(\mathbf{c}) = p_1^*(\mathbf{c}) + p_2^*(\mathbf{c})$;

(b) $\tau_a^*(\mathbf{c}) = \mathbf{c}$ *for all* $a \in A$.

If (a) and (b) are satisfied, then $[-1]^*(\mathbf{c}) = -\mathbf{c}$.

Note that identity (b) takes place in $\mathrm{Pic}(A_{K(a)})$ after identifying A with $A \times \{a\}$ as usual.

Proof: The equivalence is a consequence of

$$(m^*(\mathbf{c}) - p_1^*(\mathbf{c}) - p_2^*(\mathbf{c}))\,|_{A \times \{a\}} = \tau_a^*(\mathbf{c}) - \mathbf{c}$$

and the seesaw principle from 8.4.4. If we pull-back equation (a) by the morphism

$$A \longrightarrow A \times A, \qquad a \longmapsto (a, -a),$$

then we get $[-1]^*(\mathbf{c}) = -\mathbf{c}$. □

8.4.6. Let X be an irreducible smooth complete variety defined over a field K. Recall that $\mathrm{Pic}^0(X)$ denotes the subgroup of $\mathrm{Pic}(X)$ consisting of those classes represented by line bundles algebraically equivalent to the trivial line bundle (cf. A.9.35). A fundamental result is that the group $\mathrm{Pic}^0(X)$ is canonically an abelian variety. In what follows, we would like to describe this algebraic structure on $\mathrm{Pic}^0(X)$.

We assume $X(K)$ not empty and we fix a point P_0 on X, which will play the role of a base point. By A.7.14, we know that X is geometrically irreducible. Let T be an irreducible variety. We say that $\mathbf{c} \in \mathrm{Pic}(X \times T)$ is a **subfamily of $\mathbf{Pic^0(X)}$ parametrized by $\mathbf{T}$** if:

(a) $\mathbf{c}_t \in \mathrm{Pic}^0(X_{K(t)})$ for any $t \in T$;

(b) $\mathbf{c}_{P_0} = \mathbf{0} \in \mathrm{Pic}(T)$.

By the seesaw principle in 8.4.2 and Remark 8.4.3, $\mathbf{c}$ is uniquely determined by the family $(\mathbf{c}_t)_{t \in T}$ and by condition (b).

The following fundamental result due to Poincaré gives the existence of a universal family $\mathbf{p}$ and parameter space B, such that any subfamily $\mathbf{c}$ parametrized by any T is obtained by an appropriate pull-back of $\mathbf{p}$. We denote by id_X the identity map of X.

Theorem 8.4.7. *There is a subfamily* $\mathbf{p}$ *of* $\mathrm{Pic}^0(X)$*, parametrized by an irreducible smooth complete variety B, with the following universal property. For* *any subfamily* $\mathbf{c}$ *of* $\mathrm{Pic}^0(X)$*, parametrized by an irreducible variety* T*, there is a unique morphism* $\varphi : T \to B$ *with* $(\mathrm{id}_X \times \varphi)^*(\mathbf{p}) = \mathbf{c}$.

B is called the **Picard variety** of X and $\mathbf{p}$ is the **Poincaré class.** If $(B', \mathbf{p}')$ is another such pair, then there are morphisms $\varphi : B' \to B$, $\varphi' : B \to B'$ such that $\mathbf{p}' = (\mathrm{id}_X \times \varphi)^*(\mathbf{p})$ and $\mathbf{p} = (\mathrm{id}_X \times \varphi')^*(\mathbf{p}')$. Since $(\mathrm{id}_X \times (\varphi \circ \varphi'))^*(\mathbf{p}) = \mathbf{p}$, we conclude $\varphi \circ \varphi' = \mathrm{id}_B$ by uniqueness. Interchanging the role of $(B, \mathbf{p})$ and $(B', \mathbf{p}')$, we notice that φ is an isomorphism. In this sense, the pair $(B, \mathbf{p})$ is uniquely determined.

8.4.8. The proof of Theorem 8.4.7 is beyond the scope of this book. We advise the reader to accept this fundamental but difficult result of algebraic geometry. For those with a solid background in algebraic geometry, we give some references to deduce the theorem from the existence of the Picard scheme. By a theorem of Murre and Oort, $\mathrm{Pic}(X)$ is representable by a scheme $Pic(X)$ for any proper scheme X over K (cf. **[44]**, Th.8.2.3). Then $Pic(X)$ with this scheme structure is called the **Picard scheme** of X. If X is a smooth irreducible variety over K with base point $P_0 \in X(K)$, there is $\widetilde{\mathbf{p}} \in \mathrm{Pic}(X \times Pic(X))$ with $\widetilde{\mathbf{p}}_{P_0} = \mathbf{0}$ satisfying the following universal property: For any scheme T over K and any $\mathbf{c} \in \mathrm{Pic}(X \times T)$ with $\mathbf{c}_{P_0} = \mathbf{0}$, there is a unique morphism $\varphi : T \to Pic(X)$ with $(!\mathrm{id}_X \times \varphi)^*(\widetilde{\mathbf{p}}) = \mathbf{c}$ (cf. **[44]**, Prop.8.2.4, and use **[136]**, Prop.7.8.6, to check the assumptions).

By A. Grothendieck **[133]**, Th.2.1 and Cor.3.2, the connected component B of $Pic(X)$ containing $\mathbf{0}$ together with its induced reduced scheme structure is a smooth complete variety over K. Note that the Picard scheme itself need not to be smooth if $\mathrm{char}(K) \neq 0$. By the universal property of $\widetilde{\mathbf{p}}$, we easily deduce that the formation of the Picard scheme $Pic(X)$ and the class $\widetilde{\mathbf{p}}$ is compatible with base change to field extensions of K. Since B is an irreducible smooth variety containing $\mathbf{0}$, we conclude that the restriction of $\widetilde{\mathbf{p}}$ to $X \times B$ is a subfamily of $\mathrm{Pic}^0(X)$ parametrized by B.

We claim that the restriction $\mathbf{p}$ of $\widetilde{\mathbf{p}}$ to B satisfies the hypothesis of our theorem. It is enough to show that for a subfamily $\mathbf{c}$ of $\mathrm{Pic}^0(X)$ parametrized by an irreducible variety T, the morphism $\varphi : T \to Pic(X)$ from the universal property of $\widetilde{\mathbf{p}}$ factors through the open and closed subscheme B of $Pic(X)$. Since T and B are both irreducible, it is enough to show that $\varphi(t) \in B(\overline{K})$ for some $t \in T(\overline{K})$. By the universal property again, the point $\varphi(t)$ is characterized by $\mathbf{c}_t = \mathbf{p}_{\varphi(t)} \in \mathrm{Pic}^0(X_{K(t)})$. By definition of algebraic equivalence, there is a subfamily $\mathbf{d}$ of $\mathrm{Pic}^0(X_{K(t)})$ parametrized by an irreducible smooth variety S over $K(t)$ such that $\mathbf{d}_{s_1} = \mathbf{0}$ and $\mathbf{d}_{s_2} = \mathbf{c}_t$ for some $s_1, s_2 \in T(K(t))$. Now, by the universal property and compatibility with base change, there is a morphism $\psi : S \to Pic(X)_{K(t)}$ of schemes over K with $(\mathrm{id}_X \times \psi)^*(\mathbf{p}_{K(t)}) = \mathbf{d}$. Then $\psi(s_1) = \mathbf{0}$ and $\psi(s_2) = \mathbf{c}_t$ in $Pic(X)(K(t))$. Therefore, ψ maps S into B and hence $\mathbf{c}_t \in B(\overline{K})$.

8.4.9. The following result will show that the F-rational points of the Picard variety may be identified with $\mathrm{Pic}^0(X_F)$ for any extension F/K. In particular, the set of points of the Picard variety corresponds to $\mathrm{Pic}^0(X_{\overline{K}})$. **We denote the Picard variety of X cursively by $Pic^0(X)$** to distinguish from the subgroup $\mathrm{Pic}^0(X)$ of $\mathrm{Pic}(X)$.

In the classical geometric setting of an algebraically closed field this caution is not necessary. However, taking for example an elliptic curve over a field which is not algebraically closed, it is easy to choose a divisor of degree 0 which is not invariant under $\mathrm{Gal}(\overline{K}/K)$ and hence not defined over K, making it clear that the Picard variety has more points than $\mathrm{Pic}^0(X)$.

Let $\mathbf{p}$ be the Poincaré class on $X \times Pic^0(X)$. As a byproduct of 8.4.8, we obtain the following result. If the reader has not followed the remarks there, he should accept the corollary as well and skip its proof.

Corollary 8.4.10. *Let F be an extension field of K:*

(a) *By base change, we have* $\mathrm{Pic}(X) \subset \mathrm{Pic}(X_F)$.

(b) $Pic^0(X_F) = Pic^0(X)_F$ *and its Poincaré class is obtained from* $\mathbf{p}$ *by base change to* F.

(c) $Pic^0(X)(F) = \mathrm{Pic}^0(X_F)$ *by identifying b with* $\mathbf{p}_b$.

Proof: We have seen that the formation of the Picard scheme $Pic(X)$ is compatible with base change. This proves (a) immediately. Because $\mathrm{Pic}^0(X)$ is smooth with K-rational point $\mathbf{0}$, we know that $Pic^0(X)_F$ is connected (use A.7.14), thus equal to the connected component $Pic^0(X_F)$ of $Pic(X)$ endowed with its induced reduced scheme structure. By 8.4.8, also the formation of the universal class $\widetilde{\mathbf{p}}$ on $X \times Pic(X)$ is compatible with base change. As the Poincaré class is the restriction of $\widetilde{\mathbf{p}}$, this proves (b) completely. Finally, (c) follows immediately from (b). □

Remark 8.4.11. By the seesaw principle as in Corollaries 8.4.4 and 8.4.10 c), the Poincaré class $\mathbf{p}$ is uniquely characterized by the conditions:

(a) $\mathbf{p}_\mathbf{c} = \mathbf{c}$ for any $\mathbf{c} \in Pic^0(X)$;

(b) $\mathbf{p}_{P_0} = \mathbf{0}$.

Note that, in the situation of Theorem 8.4.7, the morphism φ is given by

$$\varphi(t) = \mathbf{c}_t = \mathbf{p}_{\varphi(t)}.$$

This is clear by restriction of $(\mathrm{id}_X \times \varphi)^*(\mathbf{p}) = \mathbf{c}$ to the fibre $X \times \{t\}$ and then using the rule $(f \circ g)^* = g^* \circ f^*$ to show that

$$\mathbf{c}_t = (\mathrm{id}_X \times \varphi)^*(\mathbf{p})|_{X \times \{t\}}(\mathrm{id}_X \times \varphi(t))^*(\mathbf{p}) = \mathbf{p}_{\varphi(t)} = \varphi(t).$$

Such arguments will be often used in the sequel to deduce identities in the Picard group.

Theorem 8.4.12. *Together with its canonical group structure induced by tensor product of line bundles, $Pic^0(X)$ is an abelian variety over K.*

Proof: It is enough to show that $B := Pic^0(X)$ is a group variety (using B smooth and Remark 8.2.13). Let p_1, p_2 be the canonical projections of $X \times B \times B$

onto $X \times B$. For $\mathbf{c} := p_1^*(\mathbf{p}) + p_2^*(\mathbf{p})$ and $\mathbf{a}, \mathbf{b} \in B$, the restriction of $\mathbf{c}$ to the fibre $X \times \{\mathbf{a}\} \times \{\mathbf{b}\}$ is equal to $\mathbf{a} + \mathbf{b}$ (identifying $X \times \{\mathbf{a}\} \times \{\mathbf{b}\}$ with X). In order to see this, note that the restriction of $p_1^*(\mathbf{p})$ is equal to the restriction of $\mathbf{p}$ to $X \times \{\mathbf{a}\}$ and then use Remark 8.4.11. Since $\mathbf{c}$ is a subfamily of $\mathrm{Pic}^0(X)$ parametrized by $B \times B$, there is a unique morphism $m : B \times B \to B$ with $(\mathrm{id}_X \times m)^*(\mathbf{p}) = \mathbf{c}$. By Remark 8.4.11, we obtain

$$m(\mathbf{a}, \mathbf{b}) = \mathbf{p}_{m(\mathbf{a},\mathbf{b})} = \mathbf{c}_{(\mathbf{a},\mathbf{b})} = \mathbf{a} + \mathbf{b}$$

and so addition is a morphism. Let $\iota : B \to B$ be the unique morphism with $(\mathrm{id}_X \times \iota)^*(\mathbf{p}) = -\mathbf{p}$. We get similarly

$$\iota(\mathbf{b}) = \mathbf{p}_{\iota(\mathbf{b})} = -\mathbf{p}_{\mathbf{b}} = -\mathbf{b}$$

and so the inverse is also a morphism. □

We summarize our results in:

Theorem 8.4.13. *Let X be an irreducible smooth complete variety over K and let $P_0 \in X(K)$ be a base point of X. Then the group $\mathrm{Pic}^0(X_{\overline{K}})$ has a unique structure as an abelian variety over K, called the Picard variety and denoted by $Pic^0(X)$, with the properties:*

(a) *There is $\mathbf{p} \in \mathrm{Pic}(X \times Pic^0(X))$ such that $\mathbf{p}_{\mathbf{b}} = \mathbf{b}$ for $\mathbf{b} \in Pic^0(X)$ and $\mathbf{p}_{P_0}$ is trivial.*

(b) *For any subfamily $\mathbf{c}$ of $\mathrm{Pic}^0(X)$ parametrized by an irreducible variety T over K, the set-theoretic map*

$$T \longrightarrow Pic^0(X), \qquad t \longmapsto \mathbf{c}_t$$

is actually a morphism over K.

The uniquely determined class $\mathbf{p}$ is called the Poincaré class.

8.4.14. Let X' be also a complete smooth variety over K with base point $P_0' \in X'(K)$ and let $\varphi : X \to X'$ be a morphism such that $\varphi(P_0) = P_0'$. Then the map

$$\widehat{\varphi} : Pic^0(X') \longrightarrow Pic^0(X), \qquad \mathbf{c}' \longmapsto \varphi^*(\mathbf{c}')$$

is a homomorphism of abelian varieties called the **dual map** of φ.

To prove this, we remark first that the pull-back of the Poincaré class $\mathbf{p}'$ of X' to $X \times Pic^0(X')$ under $\varphi \times \mathrm{id}_{Pic^0(X')}$ is a subfamily of $\mathrm{Pic}^0(X)$ parametrized by $Pic^0(X')$. For $\mathbf{c}' \in Pic^0(X')$, the restriction of that pull-back to $X \times \{\mathbf{c}'\}$ is equal to $\varphi^*(\mathbf{c}')$ by Theorem 8.4.13 (a). We conclude by Theorem 8.4.13 (b) that the dual map is a morphism. Corollary 8.2.9 shows that it is a homomorphism. Actually, it is characterized by

$$(\mathrm{id}_X \times \widehat{\varphi})^*(\mathbf{p}) = (\varphi \times \mathrm{id}_{Pic^0(X')})^*(\mathbf{p}').$$

8.4.15. At the end, we describe the situation complex analytically. For details, we refer to P. Griffiths and J. Harris [**130**], pp.326–332. Let X be an irreducible proper smooth complex variety endowed with its structure as a compact connected complex manifold (cf. A.14). The considerations hold more generally for a connected compact Kähler manifold. As in Example A.10.10, the transition functions $(g_{\alpha\beta})$ of a line bundle L on X may be viewed as a Čech-cocycle with values in $\mathcal{O}_X^\times$ and we may identify $\mathrm{Pic}(X)$ with $H^1(X, \mathcal{O}_X^\times)$. The exponential map induces a short exact sequence

$$0 \longrightarrow \mathbb{Z}_X \longrightarrow \mathcal{O}_X \xrightarrow{\exp} \mathcal{O}_X^\times \longrightarrow 0,$$

where $\mathbb{Z}_X$ is the sheaf associated to the constant presheaf $\mathbb{Z}$ on X. The beginning of the associated long exact cohomology sequence is

$$0 \longrightarrow \mathbb{Z} \longrightarrow \mathbb{C} \longrightarrow \mathbb{C}^\times \longrightarrow H^1(X, \mathbb{Z}) \longrightarrow$$
$$\longrightarrow H^1(X, \mathcal{O}_X) \longrightarrow H^1(X, \mathcal{O}_X^\times) \xrightarrow{c_1} H^2(X, \mathbb{Z}) \longrightarrow \cdots \quad .$$

The map c_1 gives the ***Chern class*** of line bundles. A line bundle is algebraically equivalent to 0 if and only if its Chern class is 0 (cf. [**130**], p.462). If we use the canonical isomorphism

$$H^1(X, \mathcal{O}_X) \cong H^{0,1}(X)$$

arising from the Dolbeault complex, we conclude that the Picard variety is biholomorphic to the complex torus $H^{0,1}(X)/H^1(X, \mathbb{Z})$.

8.5. The theorem of the square and the dual abelian variety

In Section 8.4 we have defined the Picard variety $Pic^0(X)$ of an irreducible smooth complete variety X with K-rational base point. Now X is always an abelian variety A over the field K and the base point is the origin. Then the associated Picard variety is called the dual abelian variety denoted by $\widehat{A}$.

The theorem of the square says that for any $\mathbf{c} \in \mathrm{Pic}(A)$ the point $\varphi_{\mathbf{c}}(a) := \tau_a^*(\mathbf{c}) - \mathbf{c}$ is in $\widehat{A}$ and additive in $a \in A$. Over $\mathbb{C}$, it is quite clear that the translated $\tau_a^*(\mathbf{c})$ is algebraically equivalent to $\mathbf{c}$ using a path from 0 to a for the deformation. In the special case of an elliptic curve E with origin P_0 and divisor D we have

$$\tau_P^*(D) \sim D - [P] + [P_0]$$

and the theorem of the square is evident from $[P] - [P_0]$ algebraically equivalent to 0 and $[P+Q] \sim [P] + [Q] - [P_0]$. The theorem of the square in the general case will be obtained from our results about the Picard variety.

As a consequence of the theorem of the square, we will prove that an abelian variety is always projective. If $\mathbf{c}$ is ample, we will see that $\varphi_{\mathbf{c}}$ is surjective and has finite kernel, thus $\widehat{A}$ has the same dimension as A.

At the end, we will mention a direct construction of the dual abelian variety, biduality and the complex analytic situation. These considerations are not essential for the sequel of the book.

Theorem 8.5.1. *Let* $\mathbf{c} \in \mathrm{Pic}(A)$ *and* $a \in A$ *. Then* $\varphi_{\mathbf{c}}(a) := \tau_a^*(\mathbf{c}) - \mathbf{c} \in Pic^0(A)(K(a))$ *and* $\varphi_{\mathbf{c}} : A \to Pic^0(A)$ *is a homomorphism of abelian varieties over* K.

Proof: Let p_i be the ith projection of $A \times A$ onto A and consider

$$\mathbf{c}' := m^*(\mathbf{c}) - p_1^*(\mathbf{c}) - p_2^*(\mathbf{c})$$

on $A \times A$. We already remarked in the proof of Corollary 8.4.5 that

$$\mathbf{c}'|_{A\times\{a\}} = \tau_a^*(\mathbf{c}) - \mathbf{c}$$

for $a \in A$. Thus $\varphi_{\mathbf{c}}(a) \in \mathrm{Pic}^0(A_{K(a)}) = Pic^0(A)(K(a))$ by the definition of algebraic equivalence and Corollary 8.4.10. Since $\mathbf{c}'|_{\{0\}\times A} = \mathbf{0}$, $\mathbf{c}'$ is a subfamily of $\mathrm{Pic}^0(A)$ parametrized by A. Theorem 8.4.13 shows that $\varphi_{\mathbf{c}}$ is morphism of varieties defined over K. Since $\varphi_{\mathbf{c}}(0)$ is trivial, the map is a homomorphism of abelian varieties (Corollary 8.2.9). □

The next statement is the **theorem of the square:**

Theorem 8.5.2. *For $a, b \in A$, we have*

$$\tau_{a+b}^*(\mathbf{c}) + \mathbf{c} = \tau_a^*(\mathbf{c}) + \tau_b^*(\mathbf{c}).$$

Proof: Immediate from Theorem 8.5.1, by substracting $2\mathbf{c}$ on both sides. □

Theorem 8.5.3. *Let $\mathbf{b} \in \mathrm{Pic}(A)$ such that $\varphi_{\mathbf{b}} = 0$. Then for any ample $\mathbf{c} \in \mathrm{Pic}(A)$, there is some $a \in A$ with*

$$\mathbf{b} = \tau_a^*(\mathbf{c}) - \mathbf{c}.$$

See **[212]**, II.8, Th.1 for a cohomological proof of this key fact.

Remark 8.5.4. The kernel of $\varphi_{\mathbf{c}}$ gives much information about $\mathbf{c}$. If $\mathbf{c}$ is ample, then the kernel is finite. We will prove a partial converse of this statement, which we will use in Section 8.10 in the proof that the Θ-divisor on the Jacobian is ample. On the other hand, $\ker(\varphi_{\mathbf{c}}) = A$ if $\mathbf{c} \in \mathrm{Pic}^0(A)$. These statements about the kernel will be proved next and afterwards we shall sketch a construction of the dual abelian variety.

Proposition 8.5.5. *A class $\mathbf{c} \in \mathrm{Pic}(A)$ is ample if and only if $\ker(\varphi_{\mathbf{c}})$ is finite and in addition $H^0(A, n\mathbf{c}) \neq 0$ for some positive integer n.*

Proof: Assume that $\mathbf{c}$ is ample. Let B be the connected component of the closed subgroup $\ker(\varphi_{\mathbf{c}})$ containing 0. For any $b \in B$, we have

$$\tau_b^*(\mathbf{c}) = \mathbf{c},$$

and hence

$$[-1]^*(\mathbf{c}|_B) = -\mathbf{c}|_B$$

by Corollary 8.4.5. Since

$$\mathbf{0}_B = \mathbf{c}|_B + [-1]^*(\mathbf{c}|_B)$$

is ample, B has to be the trivial abelian subvariety $\{0\}$ (use A.6.15). Therefore $\ker(\varphi_{\mathbf{c}})$ is finite. Choose n so large that $n\mathbf{c}$ is very ample. This gives $H^0(A, n\mathbf{c}) \neq 0$.

In the other direction, we may assume that $H^0(A, \mathbf{c}) \neq 0$, i.e. there is an effective divisor D such that $O(D)$ is in the class of $\mathbf{c}$. The next lemma shows that $\mathbf{c}$ is ample. □

Lemma 8.5.6. *Let D be an effective divisor on A and suppose that the subgroup $\{a \in A \mid \tau_a^*(D) = D\}$ is finite. Then D is ample on A.*

Proof: Note that D is ample if and only if $D_{\overline{K}}$ is ample on $A_{\overline{K}}$ (cf. Remark A.6.12); hence we may assume K algebraically closed. The proof then proceeds by proving first that the linear system $|2D|$ is base-point free and defines a morphism φ of A into some projective space. Then we prove that φ is a finite morphism and the conclusion comes by pull-back. The details are as follows.

Let $a, b \in A$. If b is in the support of the effective divisor

$$E_a := \tau_a^*(D) + \tau_{-a}^*(D),$$

then $b + a$ or $b - a$ is in the support of D. (The reader need not worry about pull-back of divisors because it is done with respect to an isomorphism. In this case the pull-back is just given by the inverse images of the components, keeping the multiplicities.) For any given $b \in A$ we can always find a point $a \notin (D - b) \cup (b - D)$, i.e. $b \notin \mathrm{supp}(E_a)$. Then by the theorem of the square in 8.5.2 the effective divisor E_a is an element of $|2D|$. Thus the linear system $|2D|$ is base-point free hence, by A.6.8, it defines a morphism $\varphi : A \longrightarrow \mathbb{P}_K^n$.

The morphism φ is proper (cf. A.6.15 (b), (e)). Let F be an irreducible component of any fibre. All elements of $|2D|$ are pull-backs of hyperplanes by the definition of φ. Now for any $a \in A$ either F is contained in the support of E_a or $F \cap \mathrm{supp}(E_a) = \emptyset$, hence we can find $a \in A$ such that F and the support of E_a are disjoint, namely

$$a \notin \mathrm{supp}(D) - F.$$

Let Z be an irreducible component of D. Then $Z - F$ is an irreducible closed subset of A not containing a. We conclude that $Z - F$ is of codimension 1. Now note that for any $b \in F$ we have

$$Z - F = Z - b,$$

whence it follows that Z is invariant by translations in $F - F$. Therefore, the same is true for D instead of Z. By assumption, this is only possible for $\dim(F) = 0$ and we conclude that φ has finite fibres. Thus, since φ is a proper morphism, it must also be finite (cf. A.12.4). Finally, recalling that the pull-back of an ample class by a finite morphism is again ample (see A.12.7), we see that $2D$ is ample. □

Corollary 8.5.7. *An abelian variety is projective.*

Proof: Let U be an affine open subset of A containing 0. We may assume that $\dim(A) \geq 1$. Let $Z_1, \ldots, Z_r$ be the irreducible components of $A \setminus U$. Enlarging them, we may assume that $Z_1, \ldots, Z_r$ are prime divisors. In order to see this note that the complement of a divisor in an affine smooth variety is affine (use A.2.10). Setting

$$D := \sum_{i=1}^{r} Z_i,$$

the subgroup $B := \{a \in A \mid \tau_a^*(D) = D\}$ is closed and for $b \in B$, we have $U + b = U$. Since $0 \in U$, we have

$$B \subset U.$$

As a complete variety, B must be finite (see A.6.15 (d)). Lemma 8.5.6 shows that D is ample, hence A is projective. □

8.5.8. The situation is even better than as described above. On an abelian variety, any ample divisor D is rationally equivalent to an effective divisor. As we have seen in the proof of Lemma 8.5.6, the linear system $|2D|$ is base-point free. In the case of elliptic curves, it is obvious that $3D$ is very ample. The striking fact is that this remains true in general. We will not need these results in this book and we only refer to **[212]** for a proof.

Proposition 8.5.9. *For* $\mathbf{b} \in \mathrm{Pic}(A)$, *the following statements are equivalent:*

(a) $\mathbf{b} \in \mathrm{Pic}^0(A)$.

(b) $\ker(\varphi_{\mathbf{b}}) = A$.

(c) *For every ample* $\mathbf{c} \in \mathrm{Pic}(A)$, *there is* $a \in A$ *such that* $\mathbf{b} = \tau_a^*(\mathbf{c}) - \mathbf{c}$.

(d) *There is an ample* $\mathbf{c} \in \mathrm{Pic}(A)$ *such that* $\mathbf{b} = \tau_a^*(\mathbf{c}) - \mathbf{c}$ *for some* $a \in A$.

Proof: First, we prove first that (a) $\Rightarrow$ (b). By Corollary 8.4.10, we may assume that K is algebraically closed. Let

$$\varphi : A \times Pic^0(A) \longrightarrow Pic^0(A)$$

be the map given by

$$(a, \mathbf{b}) \longmapsto \tau_a^*(\mathbf{b}).$$

We will prove below that it is a morphism. For $T := A \times Pic^0(A)$, consider

$$\mathbf{c} := (m \times \mathrm{id}_{Pic^0(A)})^*(\mathbf{p}) \in \mathrm{Pic}(A \times T),$$

where m denotes the addition morphism as usual. Note that the restriction of $m \times \mathrm{id}_{Pic^0(A)}$ to $A \times \{a\} \times \{\mathbf{b}\}$ is given by $\tau_a \times \{\mathbf{b}\}$, by identifying $A \times \{a\} \times \{\mathbf{b}\}$ with A. By Remark 8.4.11 and the rule $(f \circ g)^* = g^* \circ f^*$, we get

$$\mathbf{c}\big|_{A \times \{a\} \times \{\mathbf{b}\}} = \tau_a^*(\mathbf{b})$$

and similarly

$$\mathbf{c}\big|_{\{0\} \times T} = \mathbf{p}.$$

Let us denote by p_2 the projection of $A \times T$ onto T. The subfamily $\mathbf{c} - p_2^*(\mathbf{p})$ of $\mathrm{Pic}^0(A)$ parametrized by T induces a morphism $T \to Pic^0(A)$, which is equal to φ (cf. Theorem 8.4.13). Since $\varphi(A \times \{\mathbf{0}\}) = \mathbf{0}$, the constancy lemma in 8.2.6 shows $\tau_a^*(\mathbf{b}) = \mathbf{b}$ for all $a \in A$. This proves the claim.

The implication (b) $\Rightarrow$ (c) is Theorem 8.5.3. The existence of an ample class is assured by Corollary 8.5.7 and so (d) is a consequence of (c). The implication (d) $\Rightarrow$ (a) is part of Theorem 8.5.1. $\square$

Definition 8.5.10. *The Picard variety* $Pic^0(A)$ *is called the* ***dual abelian variety*** *of A and will be denoted by* $\widehat{A}$.

Corollary 8.5.11. *The dual abelian variety* $\widehat{A}$ *has the same dimension as* A.

Proof: There is an ample $\mathbf{c} \in \mathrm{Pic}(A)$ by Corollary 8.5.7. By Proposition 8.5.9 and Corollary 8.4.10, the homomorphism $\varphi_{\mathbf{c}} : A \to \widehat{A}$ is surjective. By Proposition 8.5.5, the kernel is finite, therefore the dimension theorem in 8.2.16 proves the claim. □

8.5.12. For K algebraically closed, we sketch a construction of the dual abelian variety $\widehat{A}$ independently of the existence of the Picard scheme. For details, see [**212**], Ch.III, §13. First, the theorem of the square is proved, i.e. $\varphi_{\mathbf{c}}$ is a homomorphism of abstract groups for any $\mathbf{c} \in \mathrm{Pic}(A)$. By Corollary 8.5.7, there is an ample $\mathbf{c} \in \mathrm{Pic}(A)$. Let B be the connected component of $\ker(\varphi_{\mathbf{c}})$. We can prove that $\ker(\varphi_{\mathbf{c}})$ is closed and hence B is an abelian variety, $\mathbf{c}|_B$ is ample and $\tau_b^*(\mathbf{c}) = \mathbf{c}$ for any $b \in B$. Corollary 8.4.5 shows that $[-1]^*(\mathbf{c})|_B = -\mathbf{c}|_B$ and since [-1] is an automorphism, this is ample. Therefore

$$\mathbf{0}_B = \mathbf{c}|_B + (-\mathbf{c})|_B$$

is ample (cf. A.6.10 (c)) and so $B = \{0\}$ (see A.6.15).

Next, we show that $\varphi_{\mathbf{c}}$ is surjective. Thus the dual abelian variety is the quotient of A by a finite group scheme with underlying space $\ker(\varphi_{\mathbf{c}})$. If the characteristic of K is zero, then this group scheme is the closed subgroup on $\ker(\varphi_{\mathbf{c}})$.

As suggested by the name, we have the following fact (needed only in the example below):

Theorem 8.5.13. *Let* $\mathbf{p}$ *be the Poincaré class of* A. *Then* A *is the dual abelian variety of* $Pic^0(A)$ *and the Poincaré class of* $Pic^0(A)$ *is the pull-back of* $\mathbf{p}$ *by the isomorphism*

$$Pic^0(A) \times A \longrightarrow A \times Pic^0(A), \quad (\mathbf{b}, a) \mapsto (a, \mathbf{b}).$$

For a proof, see [**212**], Ch.III, §13.

Example 8.5.14. If E is an elliptic curve with origin P_0, then we have seen in 8.3 that $P \mapsto \mathrm{cl}([P] - [P_0])$ is an isomorphism $E \to Pic^0(E)$ of groups. It is also a morphism of varieties as it is induced by the subfamily $\mathrm{cl}(\Delta - \{P_0\} \times C - C \times \{P_0\})$ of $\mathrm{Pic}^0(E)$ parametrized by E, where Δ is the diagonal. Applying the same consideration to the dual elliptic curve $\widehat{E} = Pic(E)$ and using Theorem 8.5.13, we conclude that E is canonically isomorphic to $\widehat{E}$ as an abelian variety. Thus E is self dual with Poincaré class on $E \times E$ induced by $\mathrm{cl}(\Delta - \{P_0\} \times C - C \times \{P_0\})$.

8.5.15. Here we assume that $K = \mathbb{C}$. Let $T = V/\Lambda$ be a complex torus. By 8.2.27, every complex abelian variety is biholomorphic to a complex torus. We describe the results of this section from the point of view of complex analysis. For proofs and further details, we refer to [**212**], Ch.I and especially II.9.

The pull-back of a line bundle L on T to V with respect to the quotient map $\pi : V \longrightarrow T$ is trivial as every line bundle on V is trivial (Cousin's theorem) and L may be obtained as the quotient of $V \times \mathbb{C}$ by an action of Λ of the form

$$T_\lambda(v, z) = (v + \lambda, e_\lambda(v) \cdot z) \quad (\lambda \in \Lambda, v \in V). \tag{8.4}$$

The $e_\lambda(z) \in \mathbb{C}^\times$ is obtained by the canonical isomorphism of the fibres of π^*L over v and $v + \lambda$.

The ***theorem of Appell–Humbert*** says that the trivialization $V \times \mathbb{C}$ of $\pi^* L$ may be normalized in the following way: Let H be the Riemann form on V with imaginary part E induced by the Chern class of L (cf. 8.2.27). Then there is a unique map $\alpha : \Lambda \to \mathbb{T} := \{z \in \mathbb{C} \mid |z| = 1\}$ with

$$\alpha(\lambda_1 + \lambda_2) = e^{i\pi E(\lambda_1,\lambda_2)}\alpha(\lambda_1)\alpha(\lambda_2) \quad (\lambda_1, \lambda_2 \in \Lambda) \tag{8.5}$$

such that L is isomorphic to the quotient of $V \times \mathbb{C}$ by the action of Λ given by

$$e_\lambda(v) := \alpha(\lambda)e^{\pi H(v,\lambda)+\frac{\pi}{2}H(\lambda,\lambda)} \tag{8.6}$$

in (8.4). Conversely, every Riemann form H and every $\alpha : \Lambda \to \mathbb{T}$ with (8.5) give rise to a line bundle L on V given as the quotient of $V \times \mathbb{C}$ by the action of Λ induced by (8.6) in (8.4), and H is the Riemann form associated to $c_1(L)$. We have seen in 8.4.15 that $\mathrm{Pic}^0(T)$ is the kernel of the Chern class map $c_1 : \mathrm{Pic}(T) \to H^2(T, \mathbb{Z})$. By 8.2.27, we may identify $H^2(T, \mathbb{Z})$ with the alternating $\mathbb{Z}$-valued bilinear forms on Λ and $c_1(L)$ corresponds to E. Hence $\mathrm{cl}(L) \in \mathrm{Pic}^0(T)$ if and only if its Riemann form $H = 0$. In this case, identity (8.5) shows that α is a homomorphism of groups. Hence we may identify $\mathrm{Hom}(\mathbb{T}, \mathbb{C}^\times)$ with $\mathrm{Pic}^0(T)$.

Let $a \in T$ and let $L' := \tau_a^* L$. Clearly, L and L' have the same Chern class and hence H is also the Riemann form of L'. For any $v \in V$ with $\pi(v) = a$, we compute easily that

$$\gamma_v(\lambda) := \alpha'(\lambda)/\alpha(\lambda) = e^{2\pi i E(v,\lambda)} \quad (\lambda \in \Lambda).$$

Note that $\gamma_v \in \mathrm{Hom}(\Lambda, \mathbb{C}^\times)$ represents $\varphi_L(a) := \tau_a^* L \otimes L^{-1}$. Obviously, we have $\gamma_{v+w} = \gamma_v \gamma_w$ proving the theorem of the square.

If L is ample, then T is an abelian variety A, and H is positive definite (cf. 8.2.27). Hence every homomorphism $\Lambda \to \mathbb{R}$ has the form $E(\cdot, v)$ for suitable $v \in V$. We conclude that every homomorphism $\alpha : \Lambda \to \mathbb{T}$ has the form $\alpha = \exp(2\pi i E(\cdot, v))$. Hence the line bundle in $\mathrm{Pic}^0(A)$ corresponding to α is isomorphic to $\varphi_L(a) = \tau_a^* L \otimes L^{-1}$ for $a = \pi(v)$ proving Theorem 8.5.3.

For any line bundle L on T, we have $\ker(\varphi_L) = \Lambda^\perp/\Lambda$, where

$$\Lambda^\perp := \{v \in V \mid E(v, \Lambda) \subset \mathbb{Z}\}.$$

By Proposition C.2.4, $\ker(\varphi_L)$ is finite if and only if $\Lambda^\perp$ is a lattice. The latter is easily shown to be equivalent for H to be not degenerate (i.e. the matrix of H has full rank). In particular, if L is ample, then $\ker(\varphi_L)$ is finite. On the other hand, we have seen that $\mathrm{cl}(L) \in \mathrm{Pic}^0(T)$ if and only if $E = 0$ (using that H is determined by E). Clearly, this is equivalent to $\Lambda^\perp = V$, i.e. to $\ker(\varphi_L) = T$.

8.6. The theorem of the cube

We first deduce some elementary facts for involutions and quadratic functions on an abelian group and then prove the theorem of the cube. The theorem of the cube states that the pull-back of a fixed line bundle on an abelian variety is a quadratic function in the morphism. It will be deduced from the theorem of the square. In fact, both statements are easily seen to be equivalent. The theorem of the cube will be a fundamental tool in the study of canonical heights on abelian varieties given in the next chapter. We conclude this section by sketching a complex analytic proof of the theorem of the cube using theta functions.

8.6.1. Let M be an abelian group with an involution *, i.e. a linear map

$$M \longrightarrow M, \qquad x \longmapsto x^*$$

with

$$(x^*)^* = x$$

for any $x \in M$. An element x of M is called **even** (resp. **odd**), if $x^* = x$ (resp. $x^* = -x$). The even (resp. odd) elements form a subgroup of M. The intersection of these two subgroups is equal to the 2-torsion points of M. For $x \in M$, we are looking for a decomposition $x = x_+ + x_-$ into an even part x_+ and an odd part x_-. Such x_+ (resp. x_-) is determined up to 2-torsion.

Lemma 8.6.2. *Let $x \in M$. Then $2x$ has a decomposition into an even part and an odd part. If the subgroup of odd elements is divisible by 2, then x has also such a decomposition.*

Proof: The decomposition $2x = (x + x^*) + (x - x^*)$ does the job. Divisibility by 2 means that we have an odd z with $2z = x - x^*$. Let $x_+ := x - z$ and $x_- := z$. We have to show that x_+ is even. This follows from

$$(x_+)^* = x^* - z^* = (x - 2z) + z = x_+,$$

completing the proof. □

Example 8.6.3. Let A be an abelian variety (or more generally any group variety) over K. Then we consider the involution $\mathbf{c} \mapsto [-1]^*\mathbf{c}$ on the abelian group $\mathrm{Pic}(A)$. Hence a line bundle is called **even** (resp. **odd**) if we have $[-1]^*L \cong L$ (resp. $[-1]^*L \cong L^{\otimes(-1)}$).

Proposition 8.6.4. *On every abelian variety, there is an even very ample line bundle.*

Proof: By Corollary 8.5.7, there is a very ample $\mathbf{c} \in \mathrm{Pic}(A)$. Then $\mathbf{c} + [-1]^*\mathbf{c}$ is very ample (cf. A.6.10 (c)) and even. □

8.6.5. Let $q : M \to N$ be a set-theoretic map of abelian groups. If the function

$$b : M \times M \longrightarrow N, \qquad (x, y) \longmapsto q(x + y) - q(x) - q(y)$$

is bilinear, then q is called a **quadratic function** with **associated bilinear form** b. Obviously, b is symmetric. A **quadratic form** is a quadratic function which is homogeneous of degree 2 with respect to multiplication by integers. The quadratic functions form an abelian group. Moreover, we have an involution given by

$$q^*(x) := q(-x).$$

By Lemma 8.6.2 and its proof, we have a canonical decomposition of $2q$ into an even element Q and an odd element L given by

$$Q(x) := q(x) + q(-x)$$

and

$$L(x) := q(x) - q(-x).$$

The even Q is called the **associated quadratic form** of q. Note that $q(0) = 0$ and hence $b(x, -x) = -Q(x)$. We get $Q(x) = b(x, x)$ and so $Q(x)$ is homogeneous of degree 2. Similarly, L is called the **associated linear form** of q. An easy calculation shows that L is linear.

Lemma 8.6.6. *Let $q : M \to N$ be a quadratic function and let $n \in \mathbb{Z}$. Then*

$$q(nx) = \frac{n^2 + n}{2} q(x) + \frac{n^2 - n}{2} q(-x)$$

for all $x \in M$.

Proof: We proceed by induction on $|n|$. The result is clear for $|n| \leq 1$. Let b be the bilinear form associated to q. Then

$$0 = b(nx, x) + b(nx, -x)$$

gives

$$0 = q((n+1)x) - 2q(nx) + q((n-1)x) - q(x) - q^*(x).$$

By the induction hypothesis, we get easily the claim. □

Corollary 8.6.7. *A quadratic function is even (resp. odd) if and only if it is a quadratic form (resp. homogeneous of degree 1).*

Proof: Immediate from Lemma 8.6.6. □

Example 8.6.8. Let $M := (\mathbb{Z}/2\mathbb{Z})^2$ and $N := \mathbb{Z}/2\mathbb{Z}$. Consider the function $q : M \to N$ given by $q(\mathbf{x}) = 0$ if and only if $\mathbf{x} = \mathbf{0}$. Then q is an odd quadratic function which is *not* linear.

8.6.9. For $k \in \mathbb{N} \setminus \{0\}$ and $I \subset \{1, \ldots, k\}$, the homomorphism

$$S_I : M^k \longrightarrow M$$

is defined by

$$S_I(x_1, \ldots, x_k) := \sum_{i \in I} x_i.$$

In particular, $s_\emptyset$ is identically 0.

Lemma 8.6.10. *Let $q : M \to N$ be a quadratic function and let k be an integer. If $k \geq 3$, then we have for $\mathbf{x} \in M^k$:*

$$\sum_{I \subset \{1,\ldots,k\}} (-1)^{|I|} q(S_I(\mathbf{x})) = 0.$$

Proof: We proceed by induction on k. If we formulate the bilinearity of b in terms of q, then we get

$$q(x+y+z) - q(x+z) - q(x+y) - q(y+z) + q(x) + q(y) + q(z) = 0.$$

This proves the claim for $k = 3$. For $k > 3$, we obtain

$$\begin{aligned}
&\sum_{I \subset \{1,\ldots,k\}} (-1)^{|I|} q(S_I(\mathbf{x})) \\
&\quad = \sum_{I \subset \{1,\ldots,k-1\}} (-1)^{|I|} q(S_I(\mathbf{x})) - \sum_{J \subset \{1,\ldots,k-1\}} (-1)^{|J|} q(S_J(\mathbf{x}) + x_k) \\
&\quad = - \sum_{J \subset \{1,\ldots,k-1\}} (-1)^{|J|} b(S_J(\mathbf{x}), x_k) - \sum_{J \subset \{1,\ldots,k-1\}} (-1)^{|J|} q(x_k).
\end{aligned}$$

The first sum equals 0, by induction applied to the linear and hence quadratic function $b(\cdot, x_k)$. The second sum is 0 by carrying out the identity $(1-1)^{k-1} = 0$. □

We apply the above to the abelian group $M = \mathrm{Mor}(X, A)$ of morphisms from X to A and to the abelian group $N = \mathrm{Pic}(X)$.

Theorem 8.6.11. *Let X be a variety over the field K and let A be an abelian variety over K with $\mathbf{c} \in \mathrm{Pic}(A)$. Then the map of* $\mathrm{Mor}(X, A)$ *into* $\mathrm{Pic}(X)$ *given by $\varphi \mapsto \varphi^*(\mathbf{c})$ is quadratic.*

Let $k \geq 3$ and $X = A^k$ with ith projection p_i onto A. For $I \subset \{1, \ldots, k\}$, we have

$$S_I(p_1, \ldots, p_k) = \sum_{i \in I} p_i.$$

Lemma 8.6.10 shows that the theorem implies

$$\sum_{I \subset \{1,\ldots,k\}} (-1)^{|I|} S_I(p_1, \ldots, p_k)^*(\mathbf{c}) = \mathbf{0}.$$

For $k = 3$, this equation

$$\sum_{I \subset \{1,2,3\}} (-1)^{|I|} \left(\sum_{i \in I} p_i \right)^* (\mathbf{c}) = \mathbf{0} \tag{8.7}$$

is called the **theorem of the cube.**

Proof of Theorem 8.6.11: Let $\varphi_1, \varphi_2, \varphi_3 \in \mathrm{Mor}(X, A)$ and let

$$\Phi : X \longrightarrow A^3, \quad x \longmapsto (\varphi_1(x), \varphi_2(x), \varphi_3(x)).$$

We pull-back equation (8.7) to X using Φ, getting bilinearity as in the proof of Lemma 8.6.10. So it is enough to prove the theorem of the cube. Let $\mathbf{c}'$ be the left-hand side of (8.7). For $a, b, c \in A$, we observe

$$\mathbf{c}'|_{\{a\}\times\{b\}\times A} = \tau_{a+b}^*(\mathbf{c}) - \tau_a^*(\mathbf{c}) - \tau_b^*(\mathbf{c}) + \mathbf{c}$$

and this is zero by the theorem of the square (see Theorem 8.5.2). In the same way, the restriction of $\mathbf{c}$ to $\{a\} \times A \times \{c\}$ is trivial. By the seesaw principle (see Corollary 8.4.4), $\mathbf{c}|_{\{a\}\times A\times A}$ is trivial for any $a \in A$. Now $\mathbf{c}|_{A\times\{b\}\times\{c\}}$ is also trivial and, by the seesaw principle again, $\mathbf{c}$ is trivial. □

Remark 8.6.12. We proved the theorem of the cube using the theorem of the square. The latter was an immediate consequence of the existence of the Picard variety. Another way is to prove directly the so-called **theorem of the cube for varieties** (for a proof, see **[212]**, Ch.II, §6):

Let X, Y, Z be geometrically irreducible and geometrically reduced varieties over K with X, Y complete and let $\mathbf{c} \in \mathrm{Pic}(X \times Y \times Z)$. If there are $x_0 \in X(K)$, $y_0 \in Y(K)$ and $z_0 \in Z(K)$ such that the restrictions of $\mathbf{c}$ to $\{x_0\} \times Y \times Z$, $X \times \{y_0\} \times Z$ and $X \times Y \times \{z_0\}$ are all zero, then $\mathbf{c} = \mathbf{0}$.

The theorem of the cube is an immediate consequence of the theorem of the cube for varieties (use the above for $\mathbf{c}'$).

8.6.13. First, we give a short introduction into theta functions (for details, cf. S. Lang **[168]**, Ch.VI, Ch.VIII) and then we use them to prove the theorem of the cube complex analytically. Let Λ be a lattice in the complex vector space V and let $L : V \times \Lambda \to \mathbb{C}$, $J : \Lambda \to \mathbb{C}$ be maps with $L(v, \lambda)$ linear in $v \in V$. A holomorphic (resp. meromorphic) ***theta function*** for Λ of type (L, J) is a holomorphic (resp. meromorphic) function $\theta : V \to \mathbb{C}$ satisfying

$$\theta(v + \lambda) = \exp(2\pi i L(v, \lambda) + 2\pi i J(\lambda)) \cdot \theta(v) \quad (v \in V, \lambda \in \Lambda).$$

It is easy to see that $E(\lambda, \mu) := L(\lambda, \mu) - L(\mu, \lambda) \quad (\lambda, \mu \in \Lambda)$ is the imaginary part of a Riemann form H. If b is a symmetric bilinear form on V, l is a linear form on V, and $c \in \mathbb{C}$, then

$$\theta(v) := \exp(2\pi i b(v, v) + 2\pi i l(v) + c)$$

is a theta function for Λ of type $L(v, \lambda) := 2b(v, \lambda)$, $J(\lambda) := l(\lambda) + b(\lambda, \lambda)$. These are exactly the meromorphic theta functions with trivial divisor and thus they are called ***trivial theta functions***.

A ***normalized*** holomorphic (resp. meromorphic) ***theta function*** with Riemann form H and quadratic character α (i.e. a function $\alpha : \Lambda \to \mathbb{T}$ satisfying (8.5) on page 257 is a holomorphic (resp. meromorphic) function $\theta : V \to \mathbb{C}$ with

$$\theta(v + \lambda) = \alpha(\lambda) \cdot \exp(\pi H(v, \lambda) + \frac{\pi}{2} H(\lambda, \lambda)) \cdot \theta(v) \tag{8.8}$$

for all $v \in V, \lambda \in \Lambda$. Note that it is a theta function and H is indeed the Riemann form constructed above. For a theta function, there is always a normalized theta function

$\widetilde{\theta}$ associated to θ such that $\theta/\widetilde{\theta}$ is a trivial theta function. Note that $\widetilde{\theta}$ is unique up to a multiplicative non-zero constant.

For example, let Z be a symmetric complex $g \times g$ matrix with $\Im Z$ positive definite and let Λ be the lattice $\mathbb{Z}^g + Z \cdot \mathbb{Z}^g$ in $\mathbb{C}^g$. Then the Riemann bilinear relations (cf. **[130]**, p.306) say that this is equivalent for V/Λ to be an abelian variety. Then

$$H(\mathbf{v}, \mathbf{w}) := \mathbf{v}^t (\Im Z)^{-1} \overline{\mathbf{w}} \tag{8.9}$$

is a positive definite Riemann form for Λ. The ***Riemann theta function*** for Λ is the entire function defined by the Fourier series

$$\theta(\mathbf{v}, Z) = \sum_{\mathbf{k} \in \mathbb{Z}^g} \exp(\pi i \mathbf{k}^t Z \mathbf{k} + 2\pi i \mathbf{k}^t \mathbf{v}).$$

It is an easy matter to verify that $\theta(\cdot, Z)$ is a theta function for Λ of type

$$L(\mathbf{v}, \mathbf{m} + Z\mathbf{n}) = -\mathbf{n}^t \mathbf{v}, \quad J(\mathbf{m} + Z\mathbf{n}) = -\frac{1}{2} \mathbf{n}^t Z \mathbf{n}.$$

The associated Riemann form is indeed the H from (8.9). The normalized theta function associated to $\theta(\cdot, Z)$ is

$$\widetilde{\theta}(\mathbf{v}, Z) = \exp\left(\frac{\pi i}{2} \mathbf{v}^t (\Im Z)^{-1} \mathbf{v}\right) \cdot \theta(\mathbf{v})$$

with quadratic character $\alpha(\mathbf{m} + Z\mathbf{n}) = \exp(\pi i \mathbf{m}^t \mathbf{n})$.

Let L be a line bundle on the complex torus $T := V/\Lambda$. By the theorem of Appell–Humbert (cf. 8.5.15), the trivialization $V \times \mathbb{C}$ of $\pi^* L$ may be normalized by a Riemann form H and a quadratic character $\alpha : \Lambda \to \mathbb{T}$. Then a global (resp. meromorphic) section of L lifts to a normalized holomorphic (resp. meromorphic) theta function for Λ. This induces a one-to-one correspondence between global (resp. meromorphic) sections of L and normalized holomorphic (resp. meromorphic) theta functions with Hermitian form H and quadratic character α. If L is ample or equivalently H positive definite (cf. 8.2.27), then V/Λ is an abelian variety A and the **Riemann–Roch theorem for abelian varieties** states that

$$\dim(\Gamma(A, L)) = \sqrt{|\det E(\lambda_j, \lambda_k)|},$$

where (λ_j) is a $\mathbb{Z}$-basis for the lattice Λ.

We conclude this section by proving the theorem of the cube complex analytically. Let $A = V/\Lambda$ be an abelian variety together with a line bundle L. We choose a trivialization $V \times \mathbb{C}$ of $\pi^* L$ as in the theorem of Appell–Humbert. Since A is algebraic, there is a non-zero meromorphic section of L inducing a normalized meromorphic theta function θ with Riemann form H and quadratic character α. Then we consider the meromorphic function

$$f(v_1, v_2, v_3) := \frac{\theta(v_1 + v_2 + v_3)\theta(v_1)\theta(v_2)\theta(v_3)}{\theta(v_1 + v_2)\theta(v_1 + v_3)\theta(v_2 + v_3)}$$

on $V \times V \times V$. It corresponds to a meromorphic section of

$$\bigotimes_{I \subset \{1,2,3\}} \left(\sum_{i \in I} p_i\right)^* \left(L^{(-1)^{|I|}}\right).$$

For $\lambda_1, \lambda_2, \lambda_3 \in \Lambda$, the identity (8.8) implies immediately the equation

$$f(v_1 + \lambda_1, v_2 + \lambda_2, v_3 + \lambda_3) = f(v_1, v_2, v_3)$$

using the facts that α is a homomorphism and H is bilinear. We conclude that the automorphy factor $e_{(\lambda_1,\lambda_2,\lambda_3)}$ of the relevant line bundle above is trivial and hence this line bundle is also trivial. This proves the theorem of the cube complex analytically.

8.7. The isogeny multiplication by n

Let A be an abelian variety over the field K. We assume that the reader is familiar with the basic notions and results of Section A.12. In this section we study the endomorphism $[n] : A \longrightarrow A$ given by multiplication by $n \in \mathbb{Z}$ on the abelian variety A. The main result is Proposition 8.7.2 dealing with its kernel $A[n]$. This endomorphism plays a fundamental role in the study of abelian varieties, both geometrically and from the arithmetic point of view. The arithmetic significance of this isogeny is made clear by the construction of the Néron–Tate height in the next chapter and by its role in the proof of the Mordell–Weil theorem, which will be treated in Chapter 10.

Over the complex numbers, it is a classical fact that

$$A[n] \cong (\mathbb{Z}/n\mathbb{Z})^{2\dim(A)}.$$

In fact, we may identify the complex manifold A with $\mathbb{C}^d/\Lambda$, where Λ is a lattice and d is the complex dimension of A; multiplication by n on A is induced by ordinary multiplication by n on $\mathbb{C}^d$. If we forget the complex structure, we get an isomorphism with the $2d$-dimensional real manifold $(\mathbb{R}/\mathbb{Z})^{2d}$ with multiplication by n induced by the usual multiplication by n on $\mathbb{R}^{2d}$, and the result becomes obvious. If $\mathrm{char}(K)$ is coprime to n, then we will see below that the situation is the same.

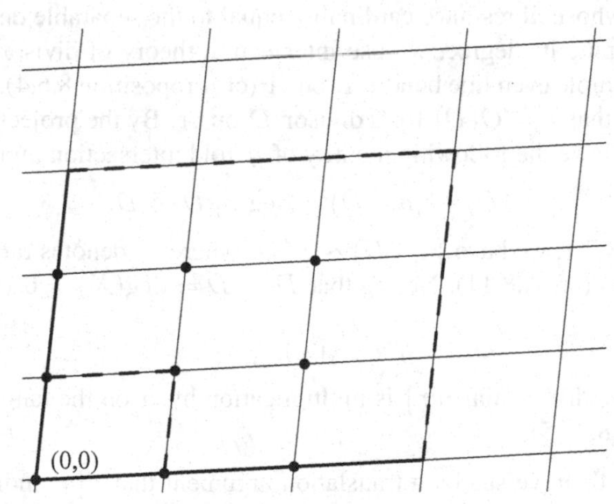

The isogeny $[3]$ on $\mathbb{C}/\Lambda$ and its nine points kernel, marked with dots

However, all this changes quite drastically for $\mathrm{char}(K)|n$. Then the map $[p]$ is always inseparable and its reduced kernel $A[p]$ has cardinality p^a, with a an integer $0 \leq a \leq \dim(A)$. All values of a in this range may occur (the case $a = 0$ is the so-called supersingular case). As it is mentioned in 8.2.15, it is better to give the kernel of $[p]$ the structure of a (non-reduced) group scheme, but this will not be treated here.

Proposition 8.7.1. *Let* $\mathbf{c} \in \mathrm{Pic}(A)$ *and* $n \in \mathbb{Z}$. *Then*

$$[n]^*(\mathbf{c}) = \frac{n^2+n}{2}\mathbf{c} + \frac{n^2-n}{2}[-1]^*\mathbf{c}.$$

In particular, we have $[n]^*(\mathbf{c}) = n^2\mathbf{c}$ *if* $\mathbf{c}$ *is even and* $[n]^*(\mathbf{c}) = -\mathbf{c}$ *if* $\mathbf{c}$ *is odd.*

Proof: By Theorem 8.6.11, the function

$$q : \mathbb{Z} \longrightarrow \mathrm{Pic}(A), \qquad n \longmapsto [n]^*\mathbf{c}$$

is quadratic. Then the claim follows from Lemma 8.6.6 □

Proposition 8.7.2. *Let* $n \in \mathbb{Z}\setminus\{0\}$. *Then* $[n]$ *is a finite flat surjective morphism of degree* $n^{2\dim(A)}$. *The separable degree of* $[n]$ *equals the number of points of any fibre. If* $\mathrm{char}(K) \not| n$, *then* $[n]$ *is an étale morphism and*

$$A[n] \cong (\mathbb{Z}/n\mathbb{Z})^{2\dim(A)}.$$

If $p = \mathrm{char}(K)$ *divides* n, *then* $[n]$ *is not separable.*

Proof: Let g be the dimension of A. By Corollary 8.5.7, there is an ample $\mathbf{c} \in \mathrm{Pic}(A)$. The restriction of $[n]^*(\mathbf{c})$ to $A[n]$ is trivial. Since $[-1]$ is an automorphism, $[-1]^*(\mathbf{c})$ is also ample. Proposition 8.7.1 shows that $[n]^*(\mathbf{c})$ is ample and so is the restriction to $A[n]$. Therefore, $A[n]$ has to be finite. Now the dimension theorem from 8.2.16 and Proposition 8.2.17 show that $[n]$ is a surjective finite flat morphism, whose fibres have cardinality equal to the separable degree of $[n]$. In order to compute its degree, we use intersection theory of divisors (see A.9). There is a very ample even line bundle L on A (cf. Proposition 8.6.4). By A.9.18, we may assume that $L = O(D)$ for a divisor D on A. By the projection formula (cf. A.9.24), we have the following identity of g-fold intersection numbers

$$[n]^*(D)\cdots[n]^*(D) = \deg[n]\; D\cdots D.$$

By Proposition 8.7.1, we have $[n]^*(D) \sim n^2 D$, where $\sim$ denotes rational equivalence of divisors (cf. A.8.11). Noting that $D\cdots D = \deg(X) \neq 0$ (see A.9.33), we deduce

$$n^{2g} = \deg[n].$$

We know that the differential $\mathrm{d}[n]$ is multiplication by n on the tangent space at 0 (cf. Corollary 8.2.25).

If $\mathrm{char}(K) \not| n$, then we see by a translation argument that $\mathrm{d}[n]$ induces an isomorphism on tangent spaces. By Proposition B.3.5, the morphism $[m]$ is étale and hence separable (cf. A.12.19). We have seen that the number of points of $A[n]$ is equal to the separable degree of $[n]$. This means $|A[n]| = n^{2g}$. For any $m|n$, it follows that the subgroup $A[m]$ of $A[n]$ has m^{2g} elements. By the theory of finite abelian groups, $A[n]$ is isomorphic to $(\mathbb{Z}/n\mathbb{Z})^{2g}$.

If $p = \mathrm{char}(K)|n$, then the differential $\mathrm{d}[n]$ vanishes at 0. Hence $\mathrm{d}[n]$ vanishes everywhere by a translation argument. Since a separable dominant morphism is generically étale (cf. A.12.19), Proposition B.3.5 shows that $[n]$ is not separable. □

8.8. Characterization of odd elements in the Picard group

Let A be an abelian variety. All varieties are assumed to be defined over the field K.

Recall from Example 8.6.3 that we have a canonical involution of $\mathrm{Pic}(A)$ defining even and odd elements. First, we prove that $\mathrm{Pic}^0(A)$ is a divisible subgroup and thus we get a decomposition into even and odd parts on the Picard group. Another application is the beautiful result that the classes in the Picard group algebraically equivalent to 0 are precisely the odd classes. Note that this is quite evident for an elliptic curve E with origin P_0. Then a divisor class D is algebraically equivalent to 0 if and only if $\deg(D) = 0$. On the other hand, 8.3.6 yields $[-1]^*D \sim 2\deg(D)[P_0] - D$, proving that D is odd if and only if $\deg(D) = 0$. At the end, we will prove that the Poincaré class of an abelian variety is even.

Proposition 8.8.1. *If* $\mathbf{c} \in \mathrm{Pic}(A)$ *and* $r \in \mathbb{Z} \setminus \{0\}$ *with* $r\mathbf{c} \in \mathrm{Pic}^0(A)$, *then* $\mathbf{c} \in \mathrm{Pic}^0(A)$.

Proof: Obviously, we have $r\varphi_{\mathbf{c}} = \varphi_{r\mathbf{c}}$ and the latter is equal to zero by Proposition 8.5.9 (a), (b). Theorem 8.5.1 tells us that $\varphi_{\mathbf{c}}$ is a homomorphism of abelian varieties and so $\varphi_{\mathbf{c}} = 0$ (Proposition 8.7.2). Using once more Proposition 8.5.9 (a), (b), this implies that $\mathbf{c} \in \mathrm{Pic}^0(A)$. □

Corollary 8.8.2. *Let* $\mathbf{c} \in \mathrm{Pic}(A)$. *Then there are an odd element* $\mathbf{c}_-$ *and an even element* $\mathbf{c}_+$ *of* $\mathrm{Pic}(A)$ *such that* $\mathbf{c} = \mathbf{c}_- + \mathbf{c}_+$. *The element* $\mathbf{c}_-$ *is determined only up to* 2*-torsion elements in* $\mathrm{Pic}(A)$.

Proof: This follows from 8.6.1, Lemma 8.6.2 and Proposition 8.8.1. □

Theorem 8.8.3. *If* $\mathbf{c} \in \mathrm{Pic}(A)$, *then* $[-1]^*\mathbf{c} - \mathbf{c} \in \mathrm{Pic}^0(A)$. *Moreover, the following statements are equivalent:*

(a) $\mathbf{c}$ *is odd.*

(b) *For any variety* X, *the map* $\mathrm{Mor}(X, A) \longrightarrow \mathrm{Pic}(A)$, *given by* $\varphi \longmapsto \varphi^*(\mathbf{c})$, *is linear.*

(c) *If* p_i *is the* i*th projection of* $A \times A$ *onto* A, *then*

$$(p_1 + p_2)^*(\mathbf{c}) = p_1^*(\mathbf{c}) + p_2^*(\mathbf{c}).$$

(d) $\tau_a^*(\mathbf{c}) = \mathbf{c}$ *for all* $a \in A$.

(e) $\mathbf{c} \in \mathrm{Pic}^0(A)$.

(f) *For all ample* $\mathbf{c}' \in \mathrm{Pic}(A)$, *there is an* $a \in A$ *such that* $\mathbf{c} = \tau_a^*(\mathbf{c}') - \mathbf{c}'$.

(g) *There is an ample* $\mathbf{c}' \in \mathrm{Pic}(A)$ *such that* $\mathbf{c} = \tau_a^*(\mathbf{c}') - \mathbf{c}'$ *for some* $a \in A$.

Proof: The equivalence of (d), (e), (f), and (g) is Proposition 8.5.9. By Corollary 8.4.5, (c) and (d) are equivalent. In order to show that (c) $\Rightarrow$ (b), we choose $\varphi_1, \varphi_2 \in \mathrm{Mor}(X, A)$ and define φ to be the morphism of X into $A \times A$ given by $\varphi(x) := (\varphi_1(x), \varphi_2(x))$. Pulling back identity (c) by φ, we obtain

$$(\varphi_1 + \varphi_2)^*(\mathbf{c}) = \varphi_1^*(\mathbf{c}) + \varphi_2^*(\mathbf{c}),$$

which is statement (b). We note that (b) $\Rightarrow$ (a) is trivial.

Next, we show that $[-1]^*(\mathbf{c}) - \mathbf{c} \in \mathrm{Pic}^0(A)$. For $a \in A$, we have $[-1] \circ \tau_a = \tau_{-a} \circ [-1]$ proving

$$\tau_a^*([-1]^*(\mathbf{c})) - [-1]^*(\mathbf{c}) = [-1]^*(\tau_{-a}^*(\mathbf{c}) - \mathbf{c}). \tag{8.10}$$

Since $\tau_{-a}^*(\mathbf{c}) - \mathbf{c} \in \mathrm{Pic}^0(A)$ (Theorem 8.5.1), (8.10) is equal to $\mathbf{c} - \tau_{-a}^*(\mathbf{c})$ by the implication (e) $\Rightarrow$ (a). By the theorem of the square in 8.5.2, the latter equals $\tau_a^*(\mathbf{c}) - \mathbf{c}$. So we have proved

$$\tau_a^*([-1]^*(\mathbf{c})) - [-1]^*(\mathbf{c}) = \tau_a^*(\mathbf{c}) - \mathbf{c}.$$

The implication (d) $\Rightarrow$ (e) shows $[-1]^*\mathbf{c} - \mathbf{c} \in \mathrm{Pic}^0(A)$.

Finally, we prove (a) $\Rightarrow$ (e). Let $\mathbf{c}$ be an odd element of $\mathrm{Pic}^0(A)$. Then

$$-2\mathbf{c} = [-1]^*(\mathbf{c}) - \mathbf{c} \in \mathrm{Pic}^0(A)$$

and so $\mathbf{c} \in \mathrm{Pic}^0(A)$ (Proposition 8.8.1). □

Theorem 8.8.4. *Let* $\widehat{A}$ *be the dual abelian variety with corresponding Poincaré class* $\mathbf{p} \in \mathrm{Pic}(A \times \widehat{A})$. *Then* $\mathbf{p}$ *is even.*

Proof: Let $\mathbf{b} \in \widehat{A}$. By Remark 8.4.11 and Theorem 8.8.3, we have

$$([-1]^*(\mathbf{p}))\big|_{A \times \{\mathbf{b}\}} = [-1]^*(\mathbf{p}|_{A \times \{-\mathbf{b}\}})[-1]^*(-\mathbf{b}) = \mathbf{b}.$$

Since

$$([-1]^*(\mathbf{p}))\big|_{\{0\} \times B} = [-1]^*(\mathbf{p}|_{\{0\} \times B}) = \mathbf{0},$$

we get $[-1]^*(\mathbf{p}) = \mathbf{p}$ by Remark 8.4.11. □

8.9. Decomposition into simple abelian varieties

We assume in this section that A is an abelian variety over the field K.

Recall from algebra that a module is called completely reducible if it is a direct sum of irreducible modules. Abelian varieties have a similar property due to Poincaré where the decomposition into a product of simple abelian varieties is now up to isogenies. The algebraic proof is based on the concept of dual abelian varieties from Section 8.5. The analytic proof sketched at the end replaces duality by the use of a positive Riemann form. As an application of Poincaré's complete reducibility theorem and our description of the kernel of multiplication by an integer from the previous section, we deduce that the group of homomorphisms is torsion free.

This section is only a side line in our book and not used anywhere else, but is of major importance in the theory of abelian varieties. Our presentation is just the beginning of a wonderful theory representing homomorphisms on the Tate modules (cf. **[212]**, Ch.IV).

Definition 8.9.1. *A surjective homomorphism of abelian varieties of the same dimension is called an* ***isogeny****.*

8.9.2. By the dimension theorem in 8.2.16, a surjective homomorphism is an isogeny if and only if the kernel is finite. If $\mathbf{c} \in \mathrm{Pic}(A)$ is ample, then $\varphi_{\mathbf{c}}$ is an isogeny by Corollary 8.5.11 and its proof. For $n \in \mathbb{Z} \setminus \{0\}$, it follows from Proposition 8.7.2 that $[n]$ is an isogeny.

The next result is called **Poincaré's complete reducibility theorem.**

Theorem 8.9.3. *Let B an abelian subvariety of A. Then there is an abelian subvariety C of A such that the restriction of addition gives an isogeny*

$$m : B \times C \longrightarrow A.$$

Proof: By Corollary 8.5.7, there is an ample $\mathbf{c} \in \mathrm{Pic}(A)$. Denote by $\iota : B \hookrightarrow A$ the inclusion and let $\widehat{\iota} : \widehat{A} \to \widehat{B}$ be the dual map, as in 8.4.14. By definition, we have

$$\widehat{\iota} \circ \varphi_{\mathbf{c}}|_B = \varphi_{\iota^*(\mathbf{c})}.$$

Since $\iota^*(\mathbf{c})$ is ample on B, by what was remarked in 8.9.2 we have that $\varphi_{\iota^*(\mathbf{c})}$ is an isogeny, hence $\varphi_{\iota^*(\mathbf{c})}$ has finite kernel. Let C be the kernel of $\widehat{\iota} \circ \varphi_{\mathbf{c}}$. The above shows that $B \cap C$ is finite, whence $m : B \times C \to A$ has finite kernel. The dimension theorem in 8.2.16 applied to $\widehat{\iota} \circ \varphi_{\mathbf{c}}$ gives

$$\dim(C) + \dim(\widehat{B}) = \dim(\widehat{A})$$

(remember that $\widehat{\iota} \circ \varphi$ is onto because its restriction to B is an isogeny). By Corollary 8.5.11, we have $\dim(\widehat{A}) = \dim(A)$ and $\dim(\widehat{B}) = \dim(B)$. Therefore $m : B \times C \to A$ is a homomorphism of abelian varieties of the same dimension, with finite kernel. By the dimension theorem again, we conclude that the homomorphism is onto A, i.e. an isogeny. □

Definition 8.9.4. *An abelian variety $B \neq \{0\}$ is called* ***simple*** *if $\{0\}$ and B are its only abelian subvarieties.*

Corollary 8.9.5. *There are simple abelian subvarieties $B_1, \ldots, B_r$ of A such that addition gives an isogeny $m : B_1 \times \cdots \times B_r \longrightarrow A$.*

Proof: Proceed by induction on the dimension and apply Theorem 8.9.3. □

Proposition 8.9.6. *Let A_1, A_2 be abelian varieties over K and let $\mathrm{Hom}(A_1, A_2)$ be the set of homomorphisms from A_1 to A_2. Then, with respect to addition of homomorphisms, the group $\mathrm{Hom}(A_1, A_2)$ is a torsion-free abelian group.*

Proof: Let us assume that we have $m \in \mathbb{Z} \setminus \{0\}$ and $\varphi \in \mathrm{Hom}(A_1, A_2)$ such that $[m] \circ \varphi = 0$. Let l be a prime different from $\mathrm{char}(K)$ with $l \not| \, m$. For $r \in \mathbb{N}$, the restriction of φ induces a homomorphism $A_1[l^r] \to A_2[l^r]$. Proposition 8.7.2 shows that

$$A_i[l^r] \cong (\mathbb{Z}/l^r\mathbb{Z})^{2\dim(A_i)}.$$

Since $[m] \circ \varphi = 0$ and $l \not| \, m$, we conclude

$$\varphi|_{A_1[l^r]} = 0.$$

Now let B be a simple abelian subvariety of A_1. The above applied to B instead of A_1 shows that φ vanishes on infinitely many points of B. Therefore the restriction of φ to B is zero (take the closure of the vanishing points and use that B is simple). Corollary 8.9.5 shows that $\varphi = 0$. □

Remark 8.9.7. We can even prove that $\mathrm{Hom}(A_1, A_2)$ is a free abelian group of rank smaller or equal to $4\dim(A_1)\dim(A_2)$. It can also be shown that, for an isogeny $\varphi : A_1 \to A_2$, there is an isogeny $\psi : A_2 \to A_1$ such that $\psi \circ \varphi$ is equal to multiplication with some non-zero integer. With this in mind, we show that the factors in Corollary 8.9.5 are uniquely determined up to isogeny and renumbering. For the proof of these results, we refer to **[212]**, Ch.IV, §19, or to **[204]**, §12.

8.9.8. Complex analytically, A may be identified with a complex torus V/Λ with a positive definite Riemann form H on V for Λ (cf. 8.2.27). An abelian subvariety B of A is given by a subspace W of V such that $\Lambda \cap W$ is a lattice in W. Then $B = W/(\Lambda \cap W)$.

Now Poincaré's complete reducibility theorem may be proved complex analytically in the following way. Let $W^\perp$ be the orthogonal complement with respect to H. As H is not degenerate (i.e. its matrix has full rank), the same holds for $E := \Im H$. We may view E as a non-degenerate alternating bilinear form on the $\mathbb{Q}$-vector space $\Lambda \otimes \mathbb{Q}$ and hence the orthogonal complement of $\Lambda \cap W$ in $\Lambda \otimes \mathbb{Q}$ with respect to E has dimension equal to $2(\dim(V) - \dim(W))$. It is easily checked that the orthogonal complement is spanned by $\Lambda \cap W^\perp$, where $\perp$ is first meant with respect to E. Now for a complex subspace of V, the orthogonal complements with respect to E and H are the same. We conclude that $\Lambda \cap W^\perp$ is a lattice in $W^\perp$. Let C be the abelian subvariety $W^\perp/(\Lambda \cap W^\perp)$. As $(\Lambda \cap W) \oplus (\Lambda \cap W^\perp)$ is a sublattice of Λ (using H positive definite), we get an isogeny $B \times C \to A$. The kernel has cardinality equal to the index of the lattices.

8.10. Curves and Jacobians

In this section, we assume that the reader is familiar with the Riemann–Roch theorem of curves from A.13.5 and with related cohomological arguments provided by Section A.10. Let C be an irreducible smooth projective curve over a field K of genus $g \geq 1$ with base point $P_0 \in C(K)$. By A.7.14, the existence of a K-rational point ensures that C is geometrically irreducible.

Definition 8.10.1. *The Picard variety of C is called the **Jacobian variety** of C.*

We denote the Jacobian by J. By A.9.40, J is equal as a group to the rational equivalence classes of divisors of degree 0 on $C_{\overline{K}}$. For every intermediate field $K \subset L \subset \overline{K}$, Corollary 8.4.10 shows that the L-rational points of J may be identified with the rational equivalence classes defined over L. The definition also makes sense for $g = 0$, but then the Jacobian variety is $\{0\}$.

8.10.2. The goal of this section is to deduce the properties of the Jacobian variety from our previous results for Picard varieties. However, this is not the historical approach, where Weil and Chow constructed the Jacobian directly from products of C. We first sketch Weil's approach in 8.10.3 and then Chow's approach in 8.10.4. As both constructions are more elementary than the use of the Picard variety (involving some machinery of algebraic geometry), we should note that they do not give immediately the universal property of the Picard variety. In the logic of this book, our developement of the theory does not follow the historical approach.

In 8.10.5, we briefly describe the well-known analytic construction of the Jacobian making it clear that the holomorphic differential forms on C and J are the 'same'. This result holds generally for any base field K, but we skip the algebraic proof here because it is best proved in the language of schemes (see J.S. Milne [**205**]). As an immediate consequence, we shall see that the Jacobian is g-dimensional.

On J, there is an important divisor Θ given as the $(g-1)$-fold sum of the image of C. The goal is to prove that Θ is ample and that the isogeny φ_Θ (cf. 8.5.1) is an isomorphism, i.e. the Jacobian is self-dual. These facts will have important diophantine consequences in the following chapters because we can endow $J(\overline{K})$ with a canonical inner product which, at least in the number field case, allows us to work with an euclidean norm to count rational points.

After some introductory lemmas for divisors on curves, which follow from the Riemann–Roch theorem, we prove in Proposition 8.10.13 that C may be seen as a closed subvariety of J. Then we study in 8.10.14–8.10.21 the interrelation between the Poincaré classes of C and J leading to the self-duality of J and to the ampleness of Θ in Theorem 8.10.22. In 8.10.23, we proceed complex analytically to construct a canonical Riemann form for the Jacobian and we give Riemann's theorem relating the theta divisor with Riemann's theta function.

8.10.3. The first construction goes back to Jacobi, in the analytic setting. The construction over fields of arbitrary characteristic was obtained by A. Weil, as sketched below.

We assume that K is algebraically closed. Let S_r be the symmetric group with r elements. The group S_r acts on C^r by permuting factors. By considering symmetric functions, we show that $C^{(r)} := C^r/S_r$ is a smooth variety of dimension r, which clearly parametrizes the effective divisors on C of degree r.

Now consider the special case $r = g$. By the Riemann–Roch theorem (see A.13.6), if D is any divisor of degree g, the linear system $|D|$ has dimension $\dim H^1(C, \mathcal{O}(D)) \geq 0$ and therefore is not empty. Generally, its dimension will be 0 (see Lemma 8.10.10 below for details); if the dimension is larger than 0, D is called a **special divisor**. Note that, by Roch's part of the Riemann–Roch theorem (the duality), D is special if and only if the linear system $|K_C - D|$, where K_C denotes a canonical divisor of C, is not empty. For example, if $g = 2$ we see that D is special if and only if it is a canonical divisor. The study of special divisors for curves of high genus is highly non-trivial, and we refer for example to E. Arbarello, M. Cornalba, P. Griffiths, and J. Harris [**13**] for the reader who wants to learn more on this topic. We denote by $U_{(g)}$ the Zariski open subset of $C^{(g)}$ of non-special effective divisors of degree g (see Lemma 8.10.11 below).

Recall that P_0 is the base point on C and let D_1, D_2 be two points on $U_{(g)}$. The divisor $D_1 + D_2 - gP_0$ has degree g and if it were not special it would give a well-defined third point D_3 on $U_{(g)}$, namely the unique effective divisor in $|D_1 + D_2 - gP_0|$. We show next that there is an open dense subset $V_{(g)}$ of $U_{(g)}$ such that the addition law $D_1 \oplus D_2 = D_3$ induces a well-defined morphism

$$V_{(g)} \times V_{(g)} \longrightarrow U_{(g)}.$$

This is in fact a so-called birational group law on $C^{(g)}$, where the sum D_3 of $D_1, D_2 \in V_{(g)}$ is given by the rational equivalence

$$D_1 + D_2 - gP_0 \sim D_3 \geq 0.$$

Now Weil gives a general process which attaches to every birational group law on a smooth variety X a group variety G and a birational map φ of X into G with $\varphi(ab) = \varphi(a)\varphi(b)$ whenever the left-hand side is defined. Moreover, G is uniquely defined up to canonical isomorphisms. So, in our case, we get a commutative group variety J. It can be identified with $\mathrm{Pic}^0(C)$ in such a way that the birational map $C^{(g)} \dashrightarrow J$ is equal to the morphism $j_{(g)} : C^{(g)} \to \mathrm{Pic}^0(C)$ given by

$$j_{(g)}(P_1, \ldots, P_g) = \mathrm{cl}([P_1] + \cdots + [P_g] - g[P_0]).$$

Since $C^{(g)}$ is complete, the same is true for J (see A.6.15 (c)) and so J is an abelian variety of dimension g.

It is a point of historical significance that this last part of the construction required, and in fact motivated, the concept of abstract variety in the sense of Weil. It still remained to see that the Jacobian so constructed was indeed a projective variety.

8.10.4. Somewhat later Chow gave a different construction which obtained J directly as a projective variety. Again, we assume for simplicity K algebraically closed. Chow bypassed the difficulty created by special divisors by constructing first not the Jacobian itself, but rather a certain projective bundle over it. The idea is the following.

Consider $C^{(n)}$ with $n \geq 2g-1$. Let D be an effective divisor of degree n; then the Serre duality in A.10.29 gives

$$\dim H^1(C, \mathcal{O}(D)) = 0,$$

therefore $|D|$ is a projective space of dimension precisely $n-g$, which is a subvariety of $C^{(n)}$. Conversely, every closed subvariety of $C^{(n)}$ isomorphic to $\mathbb{P}^{n-g}$ is equal to a linear system $|D|$ of an effective divisor D of degree n. Just note that any line in $C^{(n)}$ joins rationally equivalent divisors by the very definition of rational equivalence. Thus the variety parametrizing the projective subspaces of $C^{(n)}$ of dimension $n-g$ may be identified with the rational equivalence classes of effective divisors on C of degree n. By the Riemann–Roch theorem, every divisor of degree n is rationally equivalent to an effective divisor. By means of the map $D \mapsto D - n[P_0]$, the above parameter variety may be identified with $J = Pic^0(C)$.

In fact, the subvarieties of a projective variety X of fixed degree and fixed dimension form an algebraic family (Chow, Van der Waerden, Cayley); constructive proofs can be obtained via elimination theory and the theory of Cayley–Chow form. This suffices to give J the structure of a projective variety and the only problem is to show that it is smooth, which Chow proved by a direct calculation.

8.10.5. In the context of complex geometry there is a more familiar construction of the Jacobian variety (for details consult **[130]**).

Let γ be a 1-cycle on C. Then $\int_\gamma \omega$ is a linear functional on the holomorphic 1-forms on C, whose value depends only on the homology class of γ in $H_1(C, \mathbb{Z})$. Thus we obtain a homomorphism

$$H_1(C, \mathbb{Z}) \longrightarrow H^0(C, \Omega^1_C)^*$$

of the homology group $H_1(C, \mathbb{Z})$ into the dual $H^0(C, \Omega^1_C)^*$ of the space of holomorphic 1-forms on C. This embeds $H_1(C, \mathbb{Z})$ as a lattice in $H^0(C, \Omega^1_C)^*$. Then the complex torus

$$J = H^0(C, \Omega^1_C)^* \,/\, H_1(C, \mathbb{Z})$$

realizes the Jacobian variety complex analytically (cf. 8.10.23 for details). We have an embedding

$$j\colon C \longrightarrow J, \quad P \longmapsto \int_{\gamma_p},$$

where γ_p is any path connecting the base point P_0 with P. The value $j(P)$ is independent of the choice of the path. Independently of the choice of the base point P_0, we have a homomorphism

$$\operatorname{Pic}^0(C) \longrightarrow J, \quad \sum_{i=1}^{n}([P_i] - [Q_i]) \longmapsto \sum_{i=1}^{n}(j(P_i) - j(Q_i)).$$

Abel's theorem gives the injectivity and the **Jacobi inversion theorem** the surjectivity of this homomorphism. There is a natural isomorphism of $H^0(J, \Omega^1_J)$ onto the dual of the tangent space $T_{J,0}$ (use Proposition 8.2.26). Pull-back induces an isomorphism

$$H^0(J, \Omega^1_J) \xrightarrow{\simeq} H^0(C, \Omega^1_C). \tag{8.11}$$

Remark 8.10.6. This isomorphism holds for any base field K. More precisely, let $J = Pic^0(C)$ be the Jacobian of C and let us consider the map $j : C \to J$, given by $P \mapsto \mathrm{cl}([P] - [P_0])$. It follows from the theory of Picard varieties that j is a morphism of varieties over K. In fact, let Δ be the diagonal in $C \times C$ and let p_1, p_2 be the projections of $C \times C$ onto C, then $\mathrm{cl}(\Delta) - p_1^*\mathrm{cl}([P_0]) - p_2^*\mathrm{cl}([P_0])$ is a subfamily of $\mathrm{Pic}^0(C)$ parametrized by C. By Theorem 8.4.13, we conclude that j is a morphism. Then j is the $j_{(1)}$ in Weil's approach 8.10.3 and j is the same as the analytic j in 8.10.5. It can be proved that j^* gives the isomorphism (8.11) (use [**44**], Th.8.4.1 and Prop. 8.4.2, or [**205**], Prop.2.2).

Corollary 8.10.7. *The Jacobian variety of C has dimension g.*

Proof: By Proposition 8.2.26, the tangent bundle T_J is a trivial vector bundle of rank $\dim(J)$. By duality, the same holds for the cotangent bundle. Now recall that the only regular functions on an irreducible complete variety over an algebraically closed field are the constants (use A.6.15 (b), (d)). Since J is geometrically reduced, compatibility of cohomology and base change holds (see A.10.28). This shows that $H^0(J, \mathcal{O}_J) = K$ and hence

$$\dim H^0(J, \Omega_J^1) = \dim(J) \cdot \dim H^0(J, \mathcal{O}_J) = \dim(J).$$

The claim follows using the isomorphism (8.11). □

Remark 8.10.8. An important role on the Jacobian variety J is played the **theta divisor**

$$\Theta := \underbrace{j(C) + \cdots + j(C)}_{g-1}.$$

We shall show below that Θ is indeed a divisor on J. In order to deduce elementary properties of j and of the theta divisor, we need the following three lemmas. For a divisor D and a line bundle L, it is convenient to use the notation $L(D) := L \otimes O(D)$ and similarly for the sheaf of sections.

Lemma 8.10.9. *Assume that the ground field K is algebraically closed. Let L be a line bundle on C. Then for every $P \in C$, we have*

$$\dim \Gamma(C, L(-[P])) \geq \dim \Gamma(C, L) - 1.$$

Equality holds if and only if P is not a base point of L.

Proof: Let s_P be the canonical global section of $O([P])$ corresponding to the divisor $[P]$ on C. Then we identify $\Gamma(C, L(-[P]))$ with a subspace of $\Gamma(C, L)$ using the injective map $s \mapsto s \otimes s_P$. Clearly, $\Gamma(C, L(-[P]))$ is the kernel of the evaluation map

$$\Gamma(C, L) \longrightarrow K, \quad s \mapsto s(P)$$

at P. This map is surjective if and only if P is not a base point of L (cf. A.5.20). A fancier way to deduce this is the use of the skyscraper sheaf $\mathcal{K}_P$ at P (see Example A.10.20). Taking global sections in the short exact sequence

$$0 \longrightarrow \mathcal{L}(-[P]) \longrightarrow \mathcal{L} \longrightarrow \mathcal{K}_P \longrightarrow 0,$$

where $\mathcal{L}$ is the sheaf of sections of L, we get the same result. Finally, using the dimension formula from linear algebra, we get the claim. □

Lemma 8.10.10. *Let L be a line bundle on C and let $r \in \{1, \ldots, \dim \Gamma(C, L)\}$. Then there is a dense open subset U of C^r such that*

$$\dim \Gamma\left(C_{\overline{K}}, L_{\overline{K}}\left(-\sum_{j=1}^{r}[P_j]\right)\right) = \dim \Gamma(C, L) - r$$

for all $(P_1, \ldots, P_r) \in U$.

Proof: A repeated application of Lemma 8.10.9 gives us a sequence of distinct points $P_1, \ldots, P_r$ such that

$$\dim \Gamma\left(C_{\overline{K}}, L_{\overline{K}}\left(-\sum_{j=1}^{r}[P_j]\right)\right) = \dim H^0(C, L) - r.$$

On the right-hand side, we have used that forming cohomology groups is compatible with base change (see A.10.28). Fix a basis $s_1, \ldots, s_n$ of $\Gamma(C, L)$ and let $Q_1, \ldots, Q_r$ be distinct points of C. We have an exact sequence

$$0 \longrightarrow \mathcal{L}_{\overline{K}}\left(-\sum_{j=1}^{r}[Q_j]\right) \longrightarrow \mathcal{L}_{\overline{K}} \longrightarrow \bigoplus_{i=1}^{r} \mathcal{K}_{Q_j} \longrightarrow 0$$

of sheaves on $C_{\overline{K}}$, where $\mathcal{L}$ is the sheaf of sections of L and $\mathcal{K}_{Q_j}$ is the skyscraper sheaf at Q_j. The first part of the long exact cohomology sequence (see A.10.22) reads as

$$0 \longrightarrow H^0\left(C_{\overline{K}}, \mathcal{L}_{\overline{K}}\left(-\sum_{i=1}^{r}[Q_j]\right)\right) \longrightarrow H^0(C_{\overline{K}}, \mathcal{L}_{\overline{K}}) \longrightarrow \overline{K}^r$$

The last map is evaluation at $Q_1, \ldots, Q_r$. It is surjective if and only if

$$\det\big(s_i(Q_j)\big)_{i \in I; j=1,\ldots,r} \neq 0$$

for some $I \subset \{1, \ldots, n\}$ of cardinality r. Let U be the set of points in C^r satisfying this last condition. Obviously, U is an open subset of C^r. By the equivalence above, the points in U satisfy the conclusion of the lemma. Finally, we have seen at the beginning of the proof that $(P_1, \ldots, P_r) \in U$. Hence U is not empty and is an open dense subset of the irreducible smooth variety C^r (cf. A.4.11). □

Lemma 8.10.11. *Let $r \in \{1, \ldots, g\}$. Then*

$$U_r := \left\{(P_1, \ldots, P_r) \in C^r \mid \forall i \neq j \Rightarrow P_i \neq P_j,\ \dim \Gamma\left(C_{\overline{K}}, O\left(\sum_{j=1}^{r}[P_j]\right)\right) = 1\right\}$$

is an open dense subset of C^r.

Proof: By Lemma 8.10.10 and its proof applied to $\mathcal{L} = \Omega^1_C$, the set of $(P_1, \ldots, P_r)$ with

$$\dim H^0\left(C_{\overline{K}}, \Omega^1_{C_{\overline{K}}}\left(-\sum_{j=1}^r [P_j]\right)\right) = g - r$$

and $P_i \neq P_j$ for all $i \neq j$ is an open dense subset of C^r. By the Riemann–Roch theorem in A.13.5, this set is U_r. □

Remark 8.10.12. For any $r \in \mathbb{N}$, we have a map

$$j_r : C^r \longrightarrow J, \quad (P_1, \ldots, P_r) \mapsto \mathrm{cl}\left(\sum_{j=1}^r [P_j] - r[P_0]\right).$$

Since $j = j_1$ and addition on J are both morphisms (see Remark 8.10.6), we easily deduce that j_r is a morphism. Note that its image is closed because C^r is complete (see A.6.15 (b)). Let $a := j_r(P_1, \ldots, P_r)$, then the fibre over a is

$$j_r^{-1}(a) = \left\{(Q_1, \ldots, Q_r) \in C^r \mid \sum_{j=1}^r [Q_j] \sim \sum_{j=1}^r [P_j]\right\}.$$

Let $r \in \{1, \ldots, g\}$ and $(P_1, \ldots, P_r) \in U_r$ (as introduced in Lemma 8.10.11). Then the fibre over a is obtained by permuting the entries, namely

$$j_r^{-1}(j_r(P_1, \ldots, P_r)) = \{(P_{\pi(1)}, \ldots, P_{\pi(r)}) \mid \pi \in S_r\}.$$

By the dimension theorem in A.12.1, we conclude that $\dim j_r(C^r) = r$. In particular, Corollary 8.10.7 implies that the morphism j_g is surjective. Moreover, $\Theta = j_{g-1}(C^{g-1})$ is indeed a divisor.

Proposition 8.10.13. *The map $j : C \to J, P \mapsto \mathrm{cl}([P] - [P_0])$, is a closed embedding.*

Proof: We may assume that K is algebraically closed (cf. A.6.12). Since $g \geq 1$, two points of $C_{\overline{K}}$ are rationally equivalent if and only if they are equal (use the Riemann–Roch theorem or give a direct proof). Hence j is one-to-one. We claim that $\mathrm{d}j$ induces an injective map between tangent spaces. In order to prove this, it is enough to show that the dual is surjective between cotangential spaces. We have seen that for any $a \in J$, a cotangential vector in a extends canonically to a global section of Ω^1_J (Proposition 8.2.26). By the isomorphism (8.11) on page 271, it is enough to show that the evaluation map $\Gamma(C, \Omega^1_C) \to T^*_{C,P}$ is surjective for $P \in C$. Note that the kernel of the evaluation map is $\Gamma(C, \Omega^1_C(-[P]))$. By injectivity of j, we know that $\Gamma(C, O([P]))$ has dimension 1. By the Riemann–Roch theorem in A.13.5, we conclude that the kernel is $g - 1$ dimensional and hence the evaluation map is surjective.

Since J is a projective variety (Corollary 8.5.7), we have a closed embedding $J \to \mathbb{P}^n_K$. In order to prove that j is a closed embedding, we have to show that the linear system corresponding to the induced map $C \to \mathbb{P}^n_K$ separates points and tangent vectors ([148], Remark II.7.8). The first (resp. second) condition follows by injectivity of j (resp $\mathrm{d}j$). □

8.10.14. As a divisor on J, we also consider

$$\Theta^- := [-1]^*\Theta = \underbrace{-j(C) - \cdots - j(C)}_{g-1}.$$

In $\mathrm{Pic}(J)$, we use $\boldsymbol{\theta} := \mathrm{cl}(O(\Theta))$ and $\boldsymbol{\theta}^- := [-1]^*\boldsymbol{\theta}$. For $a \in J$, we set $j_a := \tau_{-a} \circ j$, i.e. j_a is the map $C \to J$ given by $j_a(P) = j(P) - a$.

The pull-back of a divisor D' with respect to a morphism $\varphi : X \to X'$ of irreducible smooth varieties over K is well defined as a divisor if $\varphi(X)$ is not contained in the support of D' (see A.8.26). In this case, viewing D' as a Cartier divisor on X' locally given on U'_α by a rational function f'_α, the pull-back $\varphi^*(D')$ is given on $\varphi^{-1}(U'_\alpha)$ by the rational function $f'_\alpha \circ \varphi$. Note that $\varphi^*(D')$ is well defined in $CH^1(X)$ for *any* divisor D' on X' (see A.9.18). If φ is an isomorphism (as $[-1]$ is in the cases above) and if D' is a prime divisor, then $\varphi^*(D') = \varphi^{-1}(D')$.

Proposition 8.10.15. *Assume K algebraically closed. For all $(P_1, \ldots, P_g) \in C^g$, we have the rational equivalence relation*

$$\sum_{i=1}^{g} [P_i] \sim j_a^*(\Theta^-)$$

of divisors on C, where $a := j_g(P_1, \ldots, P_g)$.

If K is not algebraically closed, then Corollary 8.4.10 (a) shows that the rational equivalence holds over any field of definition for $P_1, \ldots, P_g$.

Proof: The idea is to show that the intersection of $j(C)$ and $\Theta^- + a$ is transverse for generic $(P_1, \ldots, P_g)$. Then the proposition will follow in the generic case from our first step below. An application of the theorem of the square will lead to the general case.

First step: For $(P_1, \ldots, P_g) \in C^g$ with $\dim \Gamma(C, O(\sum_{i=1}^g [P_i])) = 1$, we have

$$(\Theta^- + a) \cap j(C) = \{j(P_1), \ldots, j(P_g)\}.$$

To prove the first step, we note that for $Q \in C$, we have $j(Q) \in \Theta^- + a$ if and only if there are $Q_1, \ldots, Q_{g-1} \in C$ such that

$$[Q] - [P_0] \sim (g-1)[P_0] - \sum_{i=1}^{g-1} [Q_i] + \sum_{i=1}^{g} [P_i] - g[P_0].$$

This is equivalent to

$$[Q] + \sum_{i=1}^{g-1} [Q_i] \sim \sum_{i=1}^{g} [P_i].$$

By assumption, the linear system $|\sum_{i=1}^g [P_i]|$ consists only of $\sum_{i=1}^g [P_i]$. This proves immediately the first step.

Second step: For $1 \leq r \leq g$ and $(P_1, \ldots, P_r) \in U_r$ (defined in Lemma 8.10.11), the differential $\mathrm{d}j_r : T_{C^r,(P_1,\ldots,P_r)} \to T_{J,j_r(P_1,\ldots,P_r)}$ is injective.

As in the proof of Proposition 8.10.13, it is enough to show that the dual map is surjective. The kernel of the dual map

$$T^*_{J,j_r(P_1,\ldots,P_r)} \longrightarrow T^*_{C^r,(P_1,\ldots,P_r)} \cong \bigoplus_{i=1}^r T^*_{C,P_i}$$

may be identified again with

$$\bigcap_{i=1}^r \Gamma(C, \Omega^1_C([P_i]) = \Gamma(C, \Omega^1_C(-[P_1] - \cdots - [P_r])$$

using (8.11) on page 271 and using that regular functions on C are constant (see A.6.15 (b), (d)). By the Riemann–Roch theorem in A.13.5 (cf. proof of Lemma 8.10.11), we conclude that the kernel has dimension $g - r$. Since $\dim(J) = g$ (Corollary 8.10.7), this proves the second claim.

Third step: For generic $(P_1, \ldots, P_g) \in C^g$, the intersection of $\Theta^- + a$ and $j(C)$ is transverse.

For the definition of transverse intersection, see Example A.9.22. Generic means that it holds in an open dense subset of C^g. Thus we may assume $(P_1, \ldots, P_g) \in U_g$. From the first step, we get

$$\{j(P_1), \ldots, j(P_g)\} = (\Theta^- + a) \cap j(C).$$

We have to omit the singular part of $\Theta^- + a$, i.e. we do not want $j(P_i) \in \Theta^-_{\text{sing}} + a$ for any $i \in \{1, \ldots, g\}$. This incidence is equivalent to

$$j(P_1) + \cdots + j(P_{i-1}) + j(P_{i+1}) + \cdots + j(P_g) \in \Theta_{\text{sing}}.$$

Clearly, these are closed conditions of codimension ≥ 1 on C^g. So we may assume that $j(C)$ intersects $\Theta^- + a$ only in the smooth locus. It remains to show that the tangent spaces $T_{j(C),j(P_i)}$ and $T_{\Theta^- + a, j(P_i)}$ span $T_{J,j(P_i)}$. Without loss of generality, we may assume $i = 1$.

We consider the morphism

$$\varphi : C^g \to J, \quad (P_1, \ldots, P_g) \mapsto j(P_1) - j(P_2) - \cdots - j(P_g).$$

For $\mathbf{P} := (P_1, \ldots, P_g) \in U_g$, we claim that $\mathrm{d}\varphi_{\mathbf{P}}$ is injective. Again, it is enough to show surjectivity of the dual map

$$(\delta\varphi)_{\mathbf{P}} : T^*_{J,\varphi(\mathbf{P})} \longrightarrow T^*_{C^g,\mathbf{P}} \cong \bigoplus_{i=1}^g T^*_{C,P_i}.$$

We may identify the left-hand side with $\Gamma(C,\Omega^1_C)$ (*via* the isomorphism (8.11) on page 271 and Proposition 8.2.26). For $\omega\in\Gamma(C,\Omega^1_C)$, we have $(\delta j)_P(\omega)=\omega(P)$ and Corollary 8.2.25 leads easily to $\delta(-j)_P=-\omega(P)$ for any $P\in C$. This shows

$$(\delta\varphi)_{\mathbf{P}}(\omega)=(\omega(P_1),-\omega(P_2),\ldots,-\omega(P_g))\in\bigoplus_{i=1}^{g}T^*_{C,P_i}$$

and, as in the second step, we deduce that $(\mathrm{d}\varphi)_{\mathbf{P}}$ is one-to-one. By dimensionality reasons, $(\mathrm{d}\varphi)_{\mathbf{P}}$ is an isomorphism for all $\mathbf{P}\in U_g$.

We note that $(\mathrm{d}j)_{P_1}:T_{C,P_1}\xrightarrow{\sim}T_{j(C),j(P_1)}$ by Proposition 8.10.13. As the claim has to be proved only generically, we may assume that $(P_2,\ldots,P_g)\in U_{g-1}$. The second step shows that

$$\mathrm{d}(-j_{g-1}):T_{C^{g-1},(P_2,\ldots,P_g)}\xrightarrow{\sim}T_{\Theta^-,-j_{g-1}(P_2,\ldots,P_g)}.$$

We have also used that we map to the smooth locus of Θ^-. We apply $(\mathrm{d}\varphi)_{\mathbf{P}}$ to the canonical direct sum decomposition

$$T_{C^g,(P_1,\ldots,P_g)}\cong T_{C,P_1}\oplus T_{C^{g-1},(P_2,\ldots,P_g)}$$

from A.7.17. These four isomorphisms lead to the interior direct sum decomposition

$$T_{J,\varphi(\mathbf{P})}=T_{j(C)-j_{g-1}(P_2,\ldots,P_g),\varphi(\mathbf{P})}\oplus T_{\Theta^-+j(P_1),\varphi(\mathbf{P})}.$$

The third claim now comes by applying the isomorphism $\mathrm{d}\tau_{j_{g-1}(P_2,\ldots,P_g)}$.

Fourth step: Proof of the proposition for generic $(P_1,\ldots,P_g)\in C^g$.

We choose a generic $(P_1,\ldots,P_g)$ on C^g. By the third step, we know that the intersection of Θ^-+a and $j(C)$ is transverse, hence all components in the intersection product have multiplicity one (Example A.9.22). By the first step, we get

$$(\Theta^-+a).j(C)=\sum_{i=1}^{g}j(P_i).$$

If we identify C with $j(C)$, then it follows from the definitions that $j_a^*(\Theta^-)$ corresponds to the above intersection product. This proves the fourth step.

Fifth step: Proof of the proposition for all $(P_1,\ldots,P_g)\in C^g$.

We note first that for two dense open subsets U,V of J, we have $J=U-V$. This follows easily from the fact that the intersection of U and $a+V$ is not empty for all $a\in J$. By the fourth step and since j_g is a surjective closed morphism, there is an open dense subset U of J such that

$$\sum_{i=1}^{g}[P_i]\sim j_a^*(\Theta^-)$$

for all $P_1, \ldots, P_g \in C$ with $a = j_g(P_1, \ldots, P_g) \in U$. Now let $(P_1, \ldots, P_g)$ be *any* point in C^g. Since $J = U + U - U$, there are $x, y, z \in U$ with $a = x + y - z$. The map j_g is surjective (cf. Remark 8.10.12), hence we have $(Q_1, \ldots, Q_g) \in C^g$ mapping to x. Let $D_x := \sum_{i=1}^g [Q_i]$. We define divisors D_y and D_z on C in a similar way. The theorem of the square from 8.5.2 shows that

$$\tau_{z-x-y}^*(\Theta^-) \sim \tau_{-x}^*(\Theta^-) + \tau_{-y}(\Theta^-) - \tau_{-z}^*(\Theta^-).$$

This proves

$$j_a^*(\Theta^-) \sim j^* \circ \tau_{z-x-y}^*(\Theta^-) \sim j_x^*(\Theta^-) + j_y^*(\Theta^-) - j_z^*(\Theta^-).$$

Since the claim is true for x, y, z, we conclude

$$j_a^*(\Theta^-) \sim D_x + D_y - D_z.$$

The relation

$$j_g(P_1, \ldots, P_g) = a = x + y - z$$

implies

$$\sum_{i=1}^g [P_i] \sim D_x + D_y - D_z$$

proving the proposition. □

Corollary 8.10.16. *For all* $(P_1, \ldots, P_g) \in U_g$ *and* $a := j_g(P_1, \ldots, P_g)$, *we have*

$$\sum_{i=1}^g [P_i] = j_a^*(\Theta^-)$$

as an identity of divisors.

Proof: We may assume, by base change, that K is algebraically closed. By Proposition 8.10.15, we have

$$\sum_{i=1}^g [P_i] \sim j_a^*(\Theta^-).$$

Both sides are effective divisors on C. By assumption, the linear system $|\sum_{i=1}^g [P_i]|$ is zero-dimensional proving the claim. □

Corollary 8.10.17. *For* $\mathbf{a} \in J = Pic^0(C)$, *we have*

$$j_{\mathbf{a}}^*(\boldsymbol{\theta}^-) - j^*(\boldsymbol{\theta}^-) = \mathbf{a}.$$

Proof: By base change and Corollary 8.4.10, we may assume K algebraically closed. Then the claim follows from surjectivity of j_g (cf. Remark 8.10.12) and Proposition 8.10.15. □

There are two Poincaré classes in the context of Jacobians. One is the Poincaré class $\mathbf{p}_C \in \mathrm{Pic}(C \times J)$, the other is the Poincaré class $\mathbf{p}_J \in \mathrm{Pic}(J \times \widehat{J})$, where $\widehat{J}$ is the dual abelian variety of J. In the final part of this section, we shall study relations between these Poincaré classes and the class $\boldsymbol{\theta} \in \mathrm{Pic}(J)$ of the theta divisor. This will lead us to the self-duality of the Jacobian.

Proposition 8.10.18. *Let Δ denote the diagonal in $C \times C$. Then*

$$(\mathrm{id}_C \times j)^*(\mathbf{p}_C) = \mathrm{cl}(O(\Delta - C \times \{P_0\} - \{P_0\} \times C)).$$

Proof: By the characterization of the Poincaré class in Remark 8.4.11, we get for $P \in C$:

$$(\mathrm{id}_C \times j)^*(\mathbf{p}_C)\big|_{C\times\{P\}} \mathbf{p}_C\big|_{C\times\{j(P)\}} = \mathrm{cl}(O([P] - [P_0])).$$

Since the restriction of $O(\Delta - C \times \{P_0\} - \{P_0\} \times C)$ to $C \times \{P\}$ is in the same class, we get the claim by the seesaw principle in Corollary 8.4.4 (noting that the restrictions to $\{P_0\} \times C$ of both classes are $\mathbf{0}$). $\square$

Proposition 8.10.19. *Let $m : J \times J \to J$ be addition and let p_1, p_2 be the projections of $J \times J$ onto the corresponding factor. For $\mathbf{c} := m^*\boldsymbol{\theta}^- - p_1^*\boldsymbol{\theta}^- - p_2^*\boldsymbol{\theta}^- \in \mathrm{Pic}(J \times J)$, we have*

$$(j \times \mathrm{id}_J)^*(\mathbf{c}) = -\mathbf{p}_C.$$

Proof: For $\mathbf{a} \in J = Pic^0(C)$, we have $(p_C)_{\mathbf{a}} = \mathbf{a}$. By Corollary 8.10.17, we conclude that

$$(\mathbf{p}_C)_{\mathbf{a}} = j^* \circ \tau_{-\mathbf{a}}^*(\boldsymbol{\theta}^-) - j^*\boldsymbol{\theta}^-.$$

We have already seen in the proof of Corollary 8.4.5 that

$$\mathbf{c}\big|_{J\times\{\mathbf{a}\}} = \tau_{\mathbf{a}}^*(\boldsymbol{\theta}^-) - \boldsymbol{\theta}^-.$$

By pull-back of the identity to C, we obtain

$$((j \times \mathrm{id}_J)^*(\mathbf{c}))_{\mathbf{a}} = j^* \circ \tau_{\mathbf{a}}^*(\boldsymbol{\theta}^-) - j^*\boldsymbol{\theta}^-.$$

This is equal to $(\mathbf{p}_C)_{-\mathbf{a}} = -\mathbf{a} = -(\mathbf{p}_C)_{\mathbf{a}}$. Since the restriction of $\mathbf{c}$ to $\{0\} \times J$ is $\mathbf{0}$, we get

$$(j \times \mathrm{id}_J)^*(\mathbf{c})\big|_{\{P_0\}\times J} = \mathbf{0}.$$

As $\mathbf{p}_C|_{\{P_0\}\times C} = \mathbf{0}$ as well, the claim follows from the seesaw principle (see Corollary 8.4.4). $\square$

Proposition 8.10.20. *Let $\varphi_{\boldsymbol{\theta}}, \varphi_{\boldsymbol{\theta}^-}$ be the morphisms $J \to \widehat{J}$ introduced in 8.5.1 and let $\mathbf{c} := m^*\boldsymbol{\theta}^- - p_1^*\boldsymbol{\theta}^- - p_2^*\boldsymbol{\theta}^- \in \mathrm{Pic}(J \times J)$, as in Proposition 8.10.19. Then*

$$(\mathrm{id}_J \times \varphi_{\boldsymbol{\theta}^-})^*(\mathbf{p}_J) = (\mathrm{id}_J \times \varphi_{\boldsymbol{\theta}})^*(\mathbf{p}_J) = \mathbf{c}.$$

Proof: Let $a \in J$. Then

$$\varphi_{\boldsymbol{\theta}^-}(a) = \tau_a^*(\boldsymbol{\theta}^-) - \boldsymbol{\theta}^-$$

by definition. By Theorem 8.4.13, we get

$$(\mathrm{id}_J \times \varphi_{\boldsymbol{\theta}^-})^*(\mathbf{p}_J)\big|_{J\times\{a\}} \mathbf{p}_J\big|_{J\times\{\varphi_{\boldsymbol{\theta}^-}(a)\}} = \tau_a^*(\boldsymbol{\theta}^-) - \boldsymbol{\theta}^-.$$

The latter is equal to $\mathbf{c}\big|_{J\times\{a\}}$ (see proof of Proposition 8.10.19) and so we obtain, by the seesaw principle in 8.4.4, that

$$(\mathrm{id}_J \times \varphi_{\boldsymbol{\theta}^-})^*(\mathbf{p}_J) = \mathbf{c}.$$

Replacing $\boldsymbol{\theta}^-$ by $\boldsymbol{\theta}$, we get in the same way

$$(\mathrm{id}_J \times \varphi_{\boldsymbol{\theta}})^* (\mathbf{p}_J) = m^*\boldsymbol{\theta} - p_1^*\boldsymbol{\theta} - p_2^*\boldsymbol{\theta}.$$

If we apply $[-1]^*$, then the left-hand side of the equation does not change, because $\varphi_{\boldsymbol{\theta}}$ is a homomorphism (see Theorem 8.5.1) and $\mathbf{p_J}$ is even (see Theorem 8.8.4). Instead, the pull-back of the right-hand side is equal to $\mathbf{c}$. Our assertion follows. □

8.10.21. We summarize here our findings.

Given a curve C of genus $g \geq 1$ with base point $P_0 \in C(K)$, there is a natural embedding j of C into the Jacobian variety. By 8.5.1, we have a dual homomorphism $\widehat{\jmath} \colon \widehat{J} \to J$. The theta divisor is defined by

$$\Theta := \underbrace{j(C) + \cdots + j(C)}_{g-1 \text{ times}}$$

and the corresponding class in $\mathrm{Pic}(J)$ is denoted by $\boldsymbol{\theta}$. Let $\boldsymbol{\theta}^- := [-1]^*\boldsymbol{\theta}$ and let $\varphi_{\boldsymbol{\theta}} : J \to \widehat{J}$ be the natural morphism introduced in Theorem 8.5.1. There are three canonical morphisms from $J \times J$ onto J, namely addition m, first projection p_1 and second projection p_2. The pull-back of the Poincaré class $\mathbf{p}_J \in \mathrm{Pic}(J \times \widehat{J})$ by $\mathrm{id}_J \times \varphi_{\boldsymbol{\theta}}$ is equal to the class

$$\mathbf{c} := m^*\boldsymbol{\theta}^- - p_1^*\boldsymbol{\theta}^- - p_2^*\boldsymbol{\theta}^-$$

and it follows from the proof of Proposition 8.10.20 that

$$\mathbf{c} = m^*\boldsymbol{\theta} - p_1^*\boldsymbol{\theta} - p_2^*\boldsymbol{\theta}.$$

The next theorem shows that $\varphi_{\boldsymbol{\theta}}$ is an isomorphism. We can identify J and $\widehat{J}$ by this natural isomorphism, so it makes sense to consider J as self-dual and $\mathbf{c}$ as the Poincaré class of J.

Theorem 8.10.22. *The map $\varphi_{\boldsymbol{\theta}}$ is an isomorphism of J onto $\widehat{J}$ whose inverse is $-\widehat{\jmath}$. Moreover, $\boldsymbol{\theta}$ is ample.*

Proof: We first show $-\widehat{\jmath} \circ \varphi_{\boldsymbol{\theta}} = \mathrm{id}_J$. For $\mathbf{a} \in J$, we have

$$-\widehat{\jmath} \circ \varphi_{\boldsymbol{\theta}}(\mathbf{a}) = -j^* \left(\tau_{\mathbf{a}}^*(\boldsymbol{\theta}) - \boldsymbol{\theta}\right).$$

Since $\tau_{\mathbf{a}}^*(\boldsymbol{\theta}) - \boldsymbol{\theta} \in \widehat{J} = Pic^0(J)$ is odd (Theorem 8.8.3), we may replace $-j^*$ on the right-hand side by $j^* \circ [-1]^*$, leading to

$$\widehat{\jmath} \circ \varphi_{\boldsymbol{\theta}}(\mathbf{a}) = j^* \circ \tau_{-\mathbf{a}}^*(\boldsymbol{\theta}^-) - j^*(\boldsymbol{\theta}^-) = j_{\mathbf{a}}^*(\boldsymbol{\theta}^-) - j^*(\boldsymbol{\theta}^-) = \mathbf{a}$$

by Corollary 8.10.17. We get $-\widehat{\jmath} \circ \varphi_{\boldsymbol{\theta}} = \mathrm{id}_J$. In particular, $\varphi_{\boldsymbol{\theta}}$ is injective. Since J and $\widehat{J}$ have the same dimension (Corollary 8.5.11), the dimension theorem in 8.2.16 proves that $\varphi_{\boldsymbol{\theta}}$ is also surjective proving the first claim. By Proposition 8.5.5, $\boldsymbol{\theta}$ has to be ample. □

To complete the picture we show that Θ^- is a translation of Θ. Let K_C be an effective canonical divisor of C. It can be viewed as a point of $C^{(2g-2)}$ (cf. 8.10.3). The image $\mathbf{k} := j_{(2g-2)}(K_C) \in J$ is independent of the choice of K_C.

Proposition 8.10.23. *With the above notation, we have*

$$\Theta^- = \tau_{\mathbf{k}}^*(\Theta).$$

Proof: Let $a \in \Theta$, then there is an effective divisor $D \in C^{(g-1)}$ mapping to a. By the Riemann–Roch theorem in A.13.5

$$\dim \Gamma(C, O(D)) - \dim \Gamma(C, \Omega^1(-D)) = 0.$$

Thus there is an effective divisor D' of degree $g-1$ such that $D + D'$ is a canonical divisor, say K_C. Therefore

$$j_{(2g-2)}(K_C) - j_{(g-1)}(D') = j_{(g-1)}(D) = a.$$

This proves

$$\Theta^- \supset \tau_{\mathbf{k}}^*(\Theta).$$

Both are prime divisors and so they are equal. □

8.10.24. We end this chapter by interpreting the situation complex analytically (for details, cf. [**130**]). We claimed in 8.10.5 that the Jacobian variety J is given by the complex torus

$$H^0(C, \Omega_C^1)^* / H_1(C, \mathbb{Z}). \tag{8.12}$$

Note first that we have a complex isomorphism $H^{0,1}(C) \xrightarrow{\sim} H^0(X, \Omega^1)^*$ given by mapping the $(0,1)$-form ρ to the linear functional $\int \cdot \wedge \rho$. By Serre duality in A.10.29 and 8.4.15, we conclude that (8.12) is indeed the Jacobian variety.

We give a direct argument that (8.12) is a complex abelian variety. The intersection number induces a canonical alternating $\mathbb{Z}$-valued bilinear form E on $H_1(C, \mathbb{Z})$ given by $E(\gamma_1, \gamma_2) := \gamma_2 \cdot \gamma_1$. It is enough to show that E is the imaginary part of a positive definite Hermitian form on $H^{0,1}(C)$. By Poincaré duality, we have an isomorphism $H_1(C, \mathbb{R}) \xrightarrow{\sim} H^1_{\mathrm{dR}}(C, \mathbb{R})$ by mapping the cycle γ to η_γ characterized by

$$\int_\gamma \eta = \int_C \eta \wedge \eta_\gamma \quad (\eta \in H^1_{\mathrm{dR}}(C, \mathbb{R})).$$

The lattice $H_1(C, \mathbb{Z})$ is realized in $H^{0,1}(C)$ by mapping γ to the projection $\eta_\gamma^{0,1}$ of η_γ with respect to the Hodge decomposition. Now recall that intersection products of cycles correspond via Poincaré duality to wedge products of forms (cf. [**130**], p.59), i.e.

$$\gamma \cdot \delta = \int_C \eta_\gamma \wedge \eta_\delta.$$

Then it is easily seen that E is the imaginary part of the positive definite Hermitian form

$$H(\rho, \mu) = -2i \int \rho \wedge \overline{\mu} \tag{8.13}$$

on $H^{0,1}(C)$.

Topologically, the compact Riemann surface C is characterized by its g holes and there is a basis $\gamma_1, \ldots, \gamma_{2g}$ of $H_1(C, \mathbb{Z})$ such that the intersection matrix has the form

$$(\gamma_j \cdot \gamma_k) = \begin{pmatrix} 0 & I_g \\ -I_g & 0 \end{pmatrix}, \tag{8.14}$$

where I_g is the $g \times g$-identity matrix (to see this, choose $\gamma_1, \ldots, \gamma_g$ "around" the holes and $\gamma_{g+1}, \ldots, \gamma_{2g}$ "through" the holes). We choose a basis $\omega_1, \ldots, \omega_g$ of $H^0(C, \Omega^1_C)$ such that the period matrix has the form

$$\left(\int_{\gamma_k} \omega_j \right)_{j=1,\ldots,g;\, k=1,\ldots,2g} = (I_g\ Z).$$

Then $\omega_1, \ldots, \omega_g$ give rise to complex coordinates on the torus J, which we identify with $\mathbb{C}^g/\Lambda$ such that $\gamma_1, \ldots, \gamma_g$ correspond to the the standard basis. Since H is a positive definite Riemann form, we easily deduce that Z is a symmetric $g \times g$-matrix with $\Im Z$ positive definite. We have $\Lambda = \mathbb{Z}^g + Z \cdot \mathbb{Z}^g$ as in the example of 8.6.13. Moreover, the Riemann form in (8.13) equals the one in (8.9) on page 262. By (8.14) and 8.6.13 again, we deduce that the Riemann theta function

$$\theta(\mathbf{v}, Z) = \sum_{\mathbf{k} \in \mathbb{Z}^g} \exp(\pi i \mathbf{k}^t Z \mathbf{k} + 2\pi i \mathbf{k}^t \mathbf{v})$$

corresponds to a global section s of a line bundle L on J, with Riemann form H and with $\dim \Gamma(J, L) = 1$. ***Riemann's theorem*** (cf. [**130**], p.338) states that the theta divisor Θ from 8.10.8 is equal to $\operatorname{div}(s)$ up to translation.

8.11. Bibliographical notes

The presentation of the material is as in the classical treatises of S. Lang [**165**] and Mumford [**212**], to which our exposition owes a great deal. We have also used the modern survey articles of J.S. Milne [**204**], [**205**] at several places. The reader is interested in going deeper into the complex analytic theory of abelian varieties may consult Griffiths and Harris [**130**].

The theory of abelian varieties over an arbitrary field was initiated by A. Weil [**327**], motivated by his proof of the Riemann hypothesis, where we also find his construction of the Jacobian. Chow's construction is given in W.L. [**67**].

An important part of the theory, which we have left aside, is the study of ℓ-adic representations and Tate groups, for which we refer to Mumford [**212**].

On another point, while Weierstrass models for elliptic curves are very useful and simple to study, it is very difficult to describe abelian varieties of dimension > 1 by homogeneous equations. This leads to a deep study of theta functions for which the reader may consult D. Mumford's Tata lectures [**214**].

9 NÉRON–TATE HEIGHTS

9.1. Introduction

We have already seen the advantages of Weil's normalized height $h(x)$ on $\mathbb{G}_m$ compared with more naive definitions: It is homogeneous of degree 1, it is not negative, and torsion points on $\mathbb{G}_m$ are precisely the points of height 0 (Kronecker's theorem in 1.5.9). In particular, the height defines a distance function on $\mathbb{G}_m(\overline{\mathbb{Q}})/\mathrm{tors}$. The heights associated to divisors of varieties studied in Chapter 2 retain similar properties only if we consider them up to bounded functions. Working, as Weil did, only in the associated equivalence class is formally pleasing because of its functorial properties, but the price paid is that Weil's equivalence relation on heights is too coarse for some of the most important applications, such as the proof of Faltings's theorem.

It was a fundamental discovery of Néron that Weil's equivalence classes of heights associated to divisors on abelian varieties contain a unique representative with all the nice functorial properties of Weil's equivalence classes. Néron's original construction gave much more, namely a 'best model' (the Néron model) for abelian varieties over a number field, and a decomposition of the height as a sum of local heights, much in the same way as it happens for the height on $\mathbb{G}_m$. The Néron heights obtained in this way can be decomposed as a sum of a quadratic function and a linear function, which is an extremely useful property in applications.

For many applications, it is enough to have the global Néron height. An elementary proof of the existence of a normalized height associated to a divisor class on an abelian variety was then found by Tate. This turns out to be sufficient for our purposes in this book.

In Sections 9.2 and 9.3 we construct the Néron–Tate heights using Tate's limit argument and study the associated bilinear form. In Section 9.4 we make a deeper study of the Néron–Tate height on Jacobians, in particular the height associated to the Poincaré class, which will be needed in the proof of Faltings's theorem, given in Chapter 11. We prove Mumford's formula for the height h_Δ associated to the diagonal Δ on the product $C \times C$ of a curve C with itself, and then deduce Mumford's gap principle for rational points on a curve.

In Section 9.5, which is based on Section 2.7, we describe Néron's canonical local heights. This complements the picture and will not be used anywhere else in this book. In Section 9.6, we use the theory of heights to deduce and extend a result of Sprindžuk on the distribution of poles and, as an application, we give a generalization of an old theorem of Runge leading in turn to a new proof of Hilbert's irreducibility theorem, based on the theory of local heights. This section depends on additional material requiring more knowledge from the reader and may be skipped in a first reading.

Chapters 2 (at least Sections 2.2–2.4) and 8 are the prerequisites for reading this chapter.

9.2. Néron–Tate heights

Let K be a field with product formula and let A be an abelian variety over K. All varieties and morphisms are assumed to be defined over K.

9.2.1. Let X be a complete variety over K. By Theorem 2.3.8, we have the height homomorphism

$$\mathbf{h} : \mathrm{Pic}(X) \longrightarrow \mathbb{R}^{X(\overline{K})}/O(1),$$

which associates to $\mathbf{c}$ the equivalence class of heights $\mathbf{h_c}$.

In general, there is no canonical height function associated to $\mathbf{c} \in \mathrm{Pic}(X)$. They are only determined up to bounded functions. But on an abelian variety, there is a canonical choice $\widehat{h}_\mathbf{c}$ of a height function in any class $\mathbf{h_c}$ characterized by good behaviour with respect to the group operation. The reader is assumed to be familiar with the notions and results from Section 8.6.

By the theorem of the cube in 8.6.11, for every $\mathbf{c} \in \mathrm{Pic}(A)$ we have a quadratic function

$$\mathrm{Mor}(X, A) \longrightarrow \mathrm{Pic}(X), \qquad \varphi \longmapsto \varphi^*(\mathbf{c}). \tag{9.1}$$

Note that the decomposition $\mathbf{c} = \mathbf{c}_+ + \mathbf{c}_-$ of $\mathbf{c}$ into an even part $\mathbf{c}_+$ and an odd part $\mathbf{c}_-$ (Corollary 8.8.2) gives a decomposition of our quadratic function into a quadratic form $\varphi \mapsto \varphi^*(\mathbf{c}_+)$; hence with the homogeneity property

$$(n\varphi)^*(\mathbf{c}_+) = n^2\varphi^*(\mathbf{c}_+)$$

(see Proposition 8.7.1), and into a linear form $\varphi \mapsto \varphi^*(\mathbf{c}_-)$ (see Theorem 8.8.3). The composite of the height homomorphism and the quadratic function in (9.1) is a quadratic function

$$q : \mathrm{Mor}(X, A) \longrightarrow \mathbb{R}^{X(\overline{K})}/O(1), \quad \varphi \longmapsto \mathbf{h}_{\varphi^*(\mathbf{c})}.$$

We conclude that $q = q_+ + q_-$ for the quadratic form $q_+(\varphi) := \mathbf{h}_{\varphi^*(\mathbf{c}_+)}$ and the linear form $q_-(\varphi) := \mathbf{h}_{\varphi^*(\mathbf{c}_-)}$. Since 2 is invertible in the abelian group

$\mathbb{R}^{X(\overline{K})}/O(1)$, this decomposition is unique, in contrast to $\mathbf{c} = \mathbf{c}_+ + \mathbf{c}_-$, which is unique only up to 2-torsion in $\mathrm{Pic}(X)$. In terms of 8.6.5, $2q_+$ (resp. $2q_-$) is the associated quadratic (resp. linear) form of q.

9.2.2. Let $h_{\mathbf{c}_\pm}$ be an arbitrary height function in the class $\mathbf{h}_{\mathbf{c}_\pm}$. For any integer n, we have $n^2\mathbf{h}_{\mathbf{c}_+} = \mathbf{h}_{[n]^*(\mathbf{c}_+)}$ and $n\mathbf{h}_{\mathbf{c}_-} = \mathbf{h}_{[n]^*(\mathbf{c}_-)}$. By Theorem 2.3.8, there is a constant $C(n)$ such that for every $a \in A$

$$|h_{\mathbf{c}_+}(na) - n^2 h_{\mathbf{c}_+}(a)| \le C(n)$$

and

$$|h_{\mathbf{c}_-}(na) - n\, h_{\mathbf{c}_-}(a)| \le C(n).$$

These conditions serve to choose a canonical height function. We can do it in the following abstract situation.

9.2.3. Let $\mathcal{N}$ be a multiplicatively closed subset of $\mathbb{R}$ (resp. $\mathbb{R}_+$) acting on a set S by means of a map such that $n(mx) = (nm)x$ for $x \in S$. A function $h : S \longrightarrow \mathbb{R}$ is **quasi-homogeneous** of degree $d \in \mathbb{N}$ (resp. $d \in \mathbb{R}_+$) for $\mathcal{N}$ if for $n \in \mathcal{N}$ there is a positive constant $C(n)$ such that

$$|h(nx) - n^d h(x)| \le C(n) \quad \text{for every } x \in S, \tag{9.2}$$

and is **homogeneous** of degree d for $\mathcal{N}$ if $h(nx) = n^d h(x)$.

The example we have in mind is the following: $S = A(\overline{K})$, $h = h_{\mathbf{c}}$, $\mathcal{N} = \mathbb{Z}$ and the action of n is multiplication by n in the abelian group $A(\overline{K})$. Then we have seen in 9.2.2 that $h_{\mathbf{c}_+}$ and $h_{\mathbf{c}_-}$ are quasi-homogeneous of degree 2 and 1.

Lemma 9.2.4. *Let $\mathcal{N}$ act on the set S as before and let $h : S \longrightarrow \mathbb{R}$ be quasi-homogeneous of degree $d > 0$. If $\mathcal{N}$ has an element of absolute value > 1, then there is a unique homogeneous function $\widehat{h} : S \longrightarrow \mathbb{R}$ of degree d for $\mathcal{N}$ such that $\widehat{h} - h$ is bounded.*

Proof: Assume for a moment that a homogeneous function $\widehat{h}$ of degree d for $\mathcal{N}$ exists, with $h - \widehat{h}$ bounded. Then for $x \in S$ and $n \in \mathcal{N}$, we have

$$\widehat{h}(x) = \lim_{|n|\to\infty} n^{-d}\widehat{h}(nx) = \lim_{|n|\to\infty} n^{-d}h(nx),$$

because $h - \widehat{h}$ is bounded. This proves uniqueness and gives us an idea of how to show existence. Apparently, in order for this argument to succeed, we need $C(n) = o(n^d)$, a condition we do not want to impose *a priori*. On the other hand, noting that $h(mnx) = h(m(nx))$ allows us to get control of $C(mn)$ in terms of $C(m)$ and $C(n)$. This is enough for proving the existence of the limit if we stay with a suitable subsequence, and this suffices for the proof. The details are as follows.

Let us fix $m \in \mathcal{N}$, $m > 1$. For a positive integer r, estimate (9.2) with $n = m$ and $m^{r-1}x$ in place of x, gives

$$|h(m^r x) - m^d h(m^{r-1}x)| \leq C(m),$$

whence

$$\begin{aligned} |h(m^r x) - m^{rd} h(x)| &= \left| \sum_{i=1}^{r} m^{d(i-1)} h(m^{r-i+1}x) - m^{di} h(m^{r-i}x) \right| \\ &\leq \sum_{i=1}^{r} m^{d(i-1)} \cdot \left| h(m^{r-i+1}x) - m^d h(m^{r-i}x) \right| \\ &\leq \frac{m^{dr} - 1}{m^d - 1} C(m). \end{aligned}$$

(To be precise, we should use always x instead of $m^0 \cdot x$.) Replacing x by $m^s x$ for any $s \in \mathbb{N}$, we get

$$|h(m^{r+s}x) - m^{rd} h(m^s x)| \leq \frac{m^{dr} - 1}{m^d - 1} \cdot C(m)$$

and we conclude that

$$|m^{-(r+s)d} h(m^{r+s}x) - m^{-sd} h(m^s x)| \leq \frac{C(m)}{(m^d - 1)m^{ds}} \tag{9.3}$$

for every $r, s \in \mathbb{N}$. This shows that

$$\left(m^{-sd} h(m^s x)\right)_{s \in \mathbb{N}}$$

is a Cauchy sequence, and we denote by $\widehat{h}(x)$ its limit. Using (9.3) for $s = 0$, $r \to \infty$, we get

$$|\widehat{h}(x) - h(x)| \leq \frac{C(m)}{m^d - 1}.$$

If we use again (9.2) with $m^s x$ in place of x and $n \in \mathcal{N}$, we obtain

$$\begin{aligned} \widehat{h}(nx) &= \lim_{s \to \infty} m^{-sd} \left(h(m^s nx) - n^d h(m^s x) + n^d h(m^s x) \right) \\ &= n^d \, \widehat{h}(x). \end{aligned}$$

We have proved the existence of a homogeneous function $\widehat{h}$ of degree d for $\mathcal{N}$, such that $\widehat{h} - h$ remains bounded. □

If we combine 9.2.1–9.2.4, then we obtain canonical global height functions associated to every class of $\mathrm{Pic}(A)$. This procedure is called **Tate's limit argument**.

Corollary 9.2.5. *Let* $\mathbf{c} \in \mathrm{Pic}(A)$ *and let* $\mathbf{c} = \mathbf{c}_+ + \mathbf{c}_-$ *be a decomposition into an even part* $\mathbf{c}_+$ *and an odd part* $\mathbf{c}_-$. *Then the classes* $\mathbf{h}_{\mathbf{c}_\pm}$ *are independent of the choice of the decomposition. There is a unique homogeneous height function* $\widehat{h}_{\mathbf{c}_\pm}$ *in the class* $\mathbf{h}_{\mathbf{c}_\pm}$, *of degree* 2 *in the* $+$ *case and degree* 1 *in the* $-$ *case.*

Definition 9.2.6. *The height function* $\widehat{h}_{\mathbf{c}} := \widehat{h}_{\mathbf{c}_+} + \widehat{h}_{\mathbf{c}_-}$ *is called the* **Néron–Tate height** *associated to* $\mathbf{c}$.

All the formalism in Chapter 2 involving heights on abelian varieties is now true for Néron–Tate heights as exact equations, and not just up to bounded functions. More precisely:

Theorem 9.2.7. *The Néron–Tate heights on abelian varieties have the following properties:*

(a) *The map*

$$\widehat{h} : \mathrm{Pic}(A) \longrightarrow \mathbb{R}^{A(\overline{K})}, \qquad \mathbf{c} \longmapsto \widehat{h}_{\mathbf{c}}$$

is a homomorphism.

(b) *If* $\varphi : A \longrightarrow B$ *is a homomorphism of abelian varieties, then*

$$\widehat{h}_{\varphi^*(\mathbf{c})} = \widehat{h}_{\mathbf{c}} \circ \varphi$$

for any $\mathbf{c} \in \mathrm{Pic}(B)$.

(c) *Let* $\mathbf{c} \in \mathrm{Pic}(A)$ *be even. If* $\mathbf{c}$ *is base-point free or ample, then* $\widehat{h}_{\mathbf{c}} \geq 0$.

Proof: For (a), we have to prove the identity

$$\widehat{h}_{\mathbf{c}} + \widehat{h}_{\mathbf{c}'} = \widehat{h}_{\mathbf{c}+\mathbf{c}'}$$

for $\mathbf{c}, \mathbf{c}' \in \mathrm{Pic}(A)$, while for (b) we need

$$\widehat{h}_{\varphi^*(\mathbf{c})} = \widehat{h}_{\mathbf{c}} \circ \varphi$$

for $\mathbf{c} \in \mathrm{Pic}(B)$.

It is enough to prove them for odd or even $\mathbf{c}$. Both sides of either identity are then homogeneous and in the same class, by Theorem 2.3.8. Corollary 9.2.5 shows that they are equal. This proves (a) and (b).

To prove (c), we may assume that $\mathbf{c}$ is base-point free. For if $\mathbf{c}$ is ample, then there is an $m \in \mathbb{N}$ such that $m\mathbf{c}$ is very ample. In particular, $m\mathbf{c}$ is base-point free and by (a), we have $m\widehat{h}_{\mathbf{c}} = \widehat{h}_{m\mathbf{c}}$. So let $\mathbf{c}$ be base-point free and even. There is a morphism φ of A into some projective space such that $\varphi^* O_{\mathbb{P}^n_K}(1)$ is in the class of $\mathbf{c}$ (see A.6.8) and hence h_φ is in the class $\mathbf{h}_{\mathbf{c}}$ (see 2.4.5). We have seen in the proof of Lemma 9.2.4 that

$$\widehat{h}_{\mathbf{c}}(a) \lim_{n \to \infty} n^{-2} h_\varphi(na)$$

for any $a \in A$. Since h_φ is a non-negative function, the same holds for $\widehat{h}_{\mathbf{c}}$. □

Theorem 9.2.8. *The Néron–Tate height* $\widehat{h}_{\mathbf{c}}$ *is the unique quadratic function in the class* $\mathbf{h}_{\mathbf{c}}$. *Moreover,* $2\widehat{h}_{\mathbf{c}_+}$ *is the associated quadratic form and* $2\widehat{h}_{\mathbf{c}_-}$ *is the associated linear form.*

Proof: First, we note that the function

$$b(a, a') := \widehat{h}_{\mathbf{c}}(a + a') - \widehat{h}_{\mathbf{c}}(a) - \widehat{h}_{\mathbf{c}}(a')$$

is bilinear in a, a'. This follows easily from the theorem of the cube in 8.6.11, using Theorem 9.2.7. The associated quadratic and linear forms are given by

$$\widehat{h}_{\mathbf{c}}(a) \pm \widehat{h}_{\mathbf{c}}(-a) = \widehat{h}_{\mathbf{c} \pm [-1]^* \mathbf{c}}(a) = 2\widehat{h}_{\mathbf{c}_{\pm}}(a),$$

again by Theorem 9.2.7.

It remains to prove uniqueness. By definition, the quadratic function is determined up to bounded functions. Hence the same is true for the associated quadratic (or linear) form. Corollary 9.2.5 shows that they are unique. □

Remark 9.2.9. As observed by several authors, the Tate construction given before is not limited to abelian varieties. Let X be a projective variety over K and let $\varphi : X \to X$ be a morphism over K. Assume that we have $\mathbf{c} \in \operatorname{Pic}(X)$ and $k, l \in \mathbb{Z}, |k| > |l|$, such that

$$l\varphi^*(\mathbf{c}) = k\mathbf{c}. \tag{9.4}$$

Then we claim that there is a unique height function $\widehat{h}_\varphi$ in the class $\widehat{\mathbf{h}}_{\mathbf{c}}$ such that

$$l\widehat{h}_\varphi(\varphi(x)) = k\widehat{h}_\varphi(x).$$

As a special case, the Néron–Tate height on an abelian variety for an even or odd class is obtained by taking $\varphi = [m]$ for some $m \in \mathbb{N}, m \geq 2$.

In order to prove the claim, we may assume $l \neq 0$, because if $l = 0$ clearly we have $\widehat{h}_\varphi = 0$. We consider the semigroup $\mathcal{N} := \{\lambda^r \mid r \in \mathbb{N}\}$, where $\lambda := k/l$. Then $\mathcal{N}$ acts on the set $S := X$ by means of $\lambda^r \cdot x := \varphi^r(x)$, where $\varphi^r = \varphi \circ \varphi \circ \cdots \circ \varphi$ is the composition of φ with itself r times. Then (9.4) proves that a height function $h_{\mathbf{c}}$ is quasi-homogeneous of degree k/l (use Theorem 2.3.8). By Lemma 9.2.4, we get the claim for

$$\widehat{h}_\varphi(x) := \lim_{r \to \infty} \lambda^{-r} h_{\mathbf{c}}(\varphi^r(x)).$$

The same proof as for Theorem 9.2.8(b) shows that $\widehat{h}_\varphi$ is a non-negative function if $\mathbf{c}$ is base-point free or ample.

Theorem 9.2.10. *If K is a number field and $\mathbf{c}$ is ample, then $\widehat{h}_\varphi(x) = 0$ if and only if x is preperiodic, i.e. the sequence x, $\varphi(x)$, $\varphi^2(x), \ldots$ contains only finitely many distinct points.*

Proof: If $l = 0$, then a positive multiple of $\mathbf{c}$ is very ample and $\mathbf{0}$ (use A.6.10). Thus X is finite (use A.6.15) and the claim follows from $\widehat{h}_\varphi = 0$. So we may assume $l \neq 0$.

Suppose that x is preperiodic. Then the sequence $(\varphi^r(x))$ contains only finitely many distinct elements, making it plain that

$$\widehat{h}_\varphi(x) = \lim_{r \to \infty} \lambda^{-r} h_{\mathbf{c}}(\varphi^r(x)) = 0.$$

Suppose that $\widehat{h}_\varphi(x) = 0$. Then

$$\widehat{h}_\varphi(\varphi^r(x)) = \lambda^r \widehat{h}_\varphi(x) = 0$$

for every r, therefore

$$\left|h_{\mathbf{c}}(\varphi^r(x))\right| \leq \left|\widehat{h}_\varphi(\varphi^r(x))\right| + C(\varphi) = C(\varphi)$$

is bounded for every r. On the other hand, we have $\varphi^r(x) \in K(x)$. Hence the points $\varphi^r(x)$ have degree at most $[K(x) : \mathbb{Q}]$ and bounded height. By Northcott's theorem in 2.4.9 they form a finite set and x is preperiodic for φ. □

Example 9.2.11. The following special case is quite interesting, because of its connexions with dynamical systems.

Let f be a non-constant rational function on $\mathbb{P}^1_K$. We assume that f, as a morphism $f : \mathbb{P}^1_K \to \mathbb{P}^1_K$, has degree $n \geq 2$. We deduce the relation

$$\mathrm{cl}(f^* O_{\mathbb{P}^1}(1)) = n \cdot \mathrm{cl}(O_{\mathbb{P}^1}(1)) \in \mathrm{Pic}(\mathbb{P}^1_K) \cong \mathbb{Z}$$

and Remark 9.2.9 shows that there is a unique real-valued function $\widehat{h}_f$ on $\mathbb{P}^1_K$ with:

(a) $\left|\widehat{h}_f(x) - h(x)\right| \leq C(f)$ for some constant $C(f)$ independent of x;

(b) $\widehat{h}_f(f(x)) = n\widehat{h}_f(x)$.

Remark 9.2.12. Note that Theorem 9.2.10 extends Kronecker's theorem from 1.5.9. Indeed, if $f(x) = x^n$ in the situation of Example 9.2.11, we have $\widehat{h}_f(x) = h(x)$, where h is the Weil height. The preperiodic points of f are 0, ∞ and the roots of unity.

In the next section, we shall prove a counterpart of Kronecker's theorem for abelian varieties, namely Theorem 9.3.5, proved along similar lines.

9.3. The associated bilinear form

Let A be an abelian variety over a field K with product formula.

9.3.1. Let M be an abelian group and let b be a real-valued symmetric bilinear form on M. We have in mind the example $M = A(\overline{K})$ and a certain bilinear form associated to a Néron–Tate height. The kernel of b is the abelian group

$$N := \{x \in M \mid b(x, y) = 0 \text{ for every } y \in M\}.$$

Then b induces a symmetric bilinear form $\overline{b}$ on $\overline{M} := M/N$ and the kernel of $\overline{b}$ is zero. Since b is real valued, $\overline{M}$ is torsion free and all torsion elements of M are contained in N. We conclude that

$$\overline{M} \longrightarrow \overline{M}_{\mathbb{R}} := \overline{M} \otimes_{\mathbb{Z}} \mathbb{R}, \qquad \overline{m} \longmapsto \overline{m} \otimes 1$$

is injective. Let $\overline{M}'$ be a finitely generated subgroup of $\overline{M}$. The restriction of $\overline{b}$ to the free abelian group $\overline{M}'$ extends uniquely to a bilinear form $\overline{b}'$ on $\overline{M}'_{\mathbb{R}}$. Let $\overline{M}'_{\mathbb{Q}} = \overline{M}' \otimes_{\mathbb{Z}} \mathbb{Q}$. An easy argument shows that $\overline{M}'_{\mathbb{Q}} \subset \overline{M}_{\mathbb{Q}}$ and so $\overline{M}'_{\mathbb{R}} \subset \overline{M}_{\mathbb{R}}$. Since $\overline{M}_{\mathbb{R}}$ is the union of all $\overline{M}'$ and the bilinear forms $\overline{b}'$ coincide on overlappings by uniqueness, we have a unique extension of $\overline{b}$ to a bilinear form $b_{\mathbb{R}}$ on $\overline{M}_{\mathbb{R}}$.

9.3.2. We would like that the bilinear form $b_{\mathbb{R}}(x,y)$ determines a scalar product and an associated norm $\|x\|^2 = b_{\mathbb{R}}(x,x)$ on $\overline{M}_{\mathbb{R}}$.

To this end, it is of course necessary that $\overline{b}(x,x) > 0$ for every $x \in \overline{M} \setminus \{0\}$. Suppose this is the case. By clearing denominators, $\overline{b}(x,x) > 0$ for $x \in \overline{M}_{\mathbb{Q}} \setminus \{0\}$; therefore, by continuity we also have $b_{\mathbb{R}}(x,x) \geq 0$ for $x \in \overline{M}_{\mathbb{R}} \setminus \{0\}$. Note however that this is not enough for $b_{\mathbb{R}}$ to be positive definite, as is seen in the following example: If α is a transcendental number in $\mathbb{R}$, then the quadratic form in $\mathbb{R}^2$ given by $q(\mathbf{x}) := (x_1 - \alpha x_2)^2$ is positive semidefinite. We have $q(\alpha, 1) = 0$, but $q(\mathbf{x}) > 0$ whenever $\mathbf{x} \in \overline{\mathbb{Q}}^2 \setminus \{0\}$, because α is transcendental.

Thus some care is needed if we want to obtain a scalar product from the bilinear form b.

Lemma 9.3.3. *With the notation and assumptions of 9.3.1, the bilinear form $b_{\mathbb{R}}$ is positive definite if and only if for every finitely generated subgroup $\overline{M}'$ of $\overline{M}$ and for every $C > 0$ the set*

$$\{x \in \overline{M} \mid b_{\mathbb{R}}(x,x) \leq C\}$$

is finite.

Proof: We may assume that $\overline{M}$ is finitely generated. Since $\overline{M}$ is torsion-free, it is a lattice in $\overline{M}_{\mathbb{R}}$. If $b_{\mathbb{R}}$ is a scalar product, then there are only finitely many lattice points in a bounded set (Proposition C.2.4). This proves the result in one direction.

Conversely, assume that $b_{\mathbb{R}}$ is not positive definite. We may assume that $b_{\mathbb{R}}$ is positive semidefinite. Otherwise, the set $\{x \in \overline{M} \mid b_{\mathbb{R}}(x,x) \leq C\}$ is clearly infinite. There is a $y \in \overline{M}_{\mathbb{R}} \setminus \{0\}$ such that $b_{\mathbb{R}}(y,y) = 0$. For $b_{\mathbb{R}}$ positive semidefinite, the Cauchy–Schwarz inequality is valid. Therefore, y is in the kernel of $b_{\mathbb{R}}$. By construction, the restriction of $b_{\mathbb{R}}$ to $\overline{M} \times \overline{M}$ has trivial kernel and hence $y \notin \overline{M}_{\mathbb{Q}}$.

Choose a basis $x_1, \ldots, x_r$ of $\overline{M}$. It is also a basis of $\overline{M}_{\mathbb{R}}$. For any $n \in \mathbb{N}$, there is a $y_n \in \overline{M}$ such that the coordinates of $y_n - ny$ are in the interval $[0,1]$. The elements $y_n - ny$ are contained in the compact cube

$$\left\{ \sum_{i=1}^{r} \alpha_i x_i \mid 0 \leq \alpha_i \leq 1 \right\},$$

while on the other hand

$$b(y_n, y_n) = b(y_n - ny,\ y_n - ny).$$

Since $b_{\mathbb{R}}$ is continuous, it is bounded on that cube, say by C. Since $y \notin \overline{M}_{\mathbb{Q}}$, the set $\{y_n \mid n \in \mathbb{N}\}$ is infinite and contained in

$$\{x \in \overline{M} \mid b_{\mathbb{R}}(x,x) \leq C\}.$$

This proves the lemma. □

9.3.4. Now we apply these considerations to $M = A(\overline{K})$. The reader should be familiar with our notions about quadratic functions introduced in 8.6.5. Let $\mathbf{c} \in \mathrm{Pic}(A)$ and let b be the bilinear form associated to the quadratic function $\widehat{h}_{\mathbf{c}}$. The associated quadratic form and hence b itself depend only on $\mathbf{c}_+$ by Theorem 9.2.8. Hence we may assume that $\mathbf{c}$ is even. In view of 9.3.1, we get a symmetric bilinear form $b_{\mathbb{R}}$ on $\overline{M}_{\mathbb{R}}$.

Assume that $\mathbf{c}$ is also ample. Then $\widehat{h}_{\mathbf{c}}$ is a non-negative function by Theorem 9.2.7 and hence $b_{\mathbb{R}}$ is positive semidefinite as in 9.3.2. There are now two problems: First, we would like to know the kernel N of b and, second, we would like to have at our disposal the necessary and sufficient condition of Lemma 9.3.3 for a scalar product. The latter is satisfied if there are only finitely many L-rational points of bounded height relative to $\mathbf{c}$ for any finite field extension L/K. By Northcott's theorem in 2.4.9, this holds for a number field.

Now we assume that the condition of Lemma 9.3.3 holds. Our goal is to determine N. Let $x \in A$ be a point with $\widehat{h}_{\mathbf{c}}(x) = 0$. Then for every integer n we have $\widehat{h}_{\mathbf{c}}(nx)n^2\widehat{h}_{\mathbf{c}}(x) = 0$, hence the set $\{nx \mid n \in \mathbb{Z}\}$ is finite. By the pigeon-hole principle, there will be two distinct integers m, n such that $nx = mx$; hence x is a torsion point. We have already seen in 9.3.1 that the torsion elements of M are contained in the kernel N, and hence N is the torsion subgroup of $M = A(\overline{K})$.

Theorem 9.3.5. *Let K be a number field and let $\mathbf{c}$ be ample and even. Then $\widehat{h}_{\mathbf{c}}$ vanishes exactly on the torsion subgroup of $A(\overline{K})$. Moreover, there is a unique scalar product $\langle\ ,\ \rangle$ on the abelian group $A(\overline{K}) \otimes_{\mathbb{Z}} \mathbb{R}$ such that*

$$\widehat{h}_{\mathbf{c}}(x) = \langle x \otimes 1, x \otimes 1 \rangle$$

for every $x \in A(\overline{K})$.

Proof: This follows from the preceding discussion in 9.3.4, because $\overline{M}_{\mathbb{R}} = \overline{M} \otimes_{\mathbb{Z}} \mathbb{R}$ is canonically isomorphic to $M \otimes_{\mathbb{Z}} \mathbb{R}$, as N is the torsion subgroup of $M = A(\overline{K})$. □

In the next result, we relate b to the Néron–Tate height of the Poincaré class. We use the definitions and notations of 8.4–8.6.

Proposition 9.3.6. *Let $\mathbf{c} \in \mathrm{Pic}(A)$ and let b be the symmetric bilinear form associated to $\widehat{h}_{\mathbf{c}}$. Moreover, let $\boldsymbol{\delta} \in \mathrm{Pic}(A \times \widehat{A})$ be the Poincaré class of A and let $\varphi_{\mathbf{c}} : A \longrightarrow \widehat{A}$ be the homomorphism of Theorem 8.5.1. Then*

$$b(a, a') = \widehat{h}_{\boldsymbol{\delta}}(a, \varphi_{\mathbf{c}}(a'))$$

for every $a, a' \in A(\overline{K})$.

Proof: By definition

$$b(a,a') = \widehat{h}_{\mathbf{c}}(a+a') - \widehat{h}_{\mathbf{c}}(a) - \widehat{h}_{\mathbf{c}}(a')$$
$$= \widehat{h}_{\mathbf{c}} \circ \tau_{a'}(a) - \widehat{h}_{\mathbf{c}}(a) - \widehat{h}_{\mathbf{c}}(a').$$

For the moment, let us keep a' fixed and view the above as functions of a. By Theorem 2.3.8, we conclude that

$$\widehat{h}_{\mathbf{c}} + b(\cdot, a')$$

is a representative in the class $\mathbf{h}_{\tau_{a'}^*(\mathbf{c})}$. Then the representative above is a quadratic function too, being a sum of a quadratic function and a linear form. Now Theorem 9.2.8 shows that

$$\widehat{h}_{\tau_{a'}^*(\mathbf{c})}(a) = \widehat{h}_{\mathbf{c}}(a) + b(a,a')$$

and hence by Theorem 9.2.7 we get

$$b(a,a') = \widehat{h}_{\varphi_{\mathbf{c}}(a')}(a).$$

It follows that it is enough to prove

$$\widehat{h}_{\mathbf{c}'} \widehat{h}_{\boldsymbol{\delta}}(\cdot, \mathbf{c}') \tag{9.5}$$

for $\mathbf{c}' \in \widehat{A} = Pic^0(A)$.

On the other hand, the point $\mathbf{c}'$ is the pull-back of $\boldsymbol{\delta}$ to $A \times \{\mathbf{c}'\}$ (see Theorem 8.4.13), hence (9.5) holds up to a bounded function on $A(\overline{K})$ (by Theorem 2.3.8). In order to get equality, by Theorem 9.2.8 it is enough to show that $\widehat{h}_{\boldsymbol{\delta}}$ is bilinear. If $a \in A$, then applying Theorem 9.2.7 to the homomorphism $\varphi : A \longrightarrow A \times \widehat{A}$ given by $\varphi(a) = (a, 0)$ and using $\varphi^*(\boldsymbol{\delta}) = \mathbf{0}$ (Theorem 8.4.13), we get

$$\widehat{h}_{\boldsymbol{\delta}}(a, 0) = \widehat{h}_{\varphi^*(\boldsymbol{\delta})}(a) = 0.$$

In the same way we verify that

$$\widehat{h}_{\boldsymbol{\delta}}(0, a') = 0$$

for $a' \in \widehat{A}$. We conclude that the bilinear form associated to the quadratic function $\widehat{h}_{\boldsymbol{\delta}}$, evaluated at $((a,0),\ (0,a'))$, is equal to $\widehat{h}_{\boldsymbol{\delta}}(a,a')$. This proves bilinearity. □

From the proof, we also obtain

Corollary 9.3.7. *With the notation of the proof of Proposition 9.3.6, it holds*

$$\widehat{h}_{\tau_{a'}^*(\mathbf{c})}(a) = \widehat{h}_{\mathbf{c}}(a) + b(a,a')$$

and

$$\widehat{h}_{\mathbf{c}'} = \widehat{h}_{\boldsymbol{\delta}}(\cdot, \mathbf{c}').$$

Corollary 9.3.8. *Let* $\mathbf{c}' \in \mathrm{Pic}^0(A)$ *and let* $\mathbf{c} \in \mathrm{Pic}(A)$ *be an even ample class. Then*

$$\widehat{h}_{\mathbf{c}'} = O(\widehat{h}_{\mathbf{c}}^{1/2}).$$

Proof: Let $a \in A(\overline{K})$. Corollary 9.3.7 shows that

$$\widehat{h}_{\mathbf{c}'}(a)\widehat{h}_{\boldsymbol{\delta}}(a, \mathbf{c}').$$

By 8.9.2, there is an element a' of A such that $\mathbf{c}' = \varphi_{\mathbf{c}}(a')$. Let b be the bilinear form associated to $\widehat{h}_{\mathbf{c}}$. Then, applying Proposition 9.3.6, we conclude that

$$\widehat{h}_{\mathbf{c}'}(a) = b(a, a').$$

We have seen in 9.3.4 that b induces a symmetric bilinear form on $A(\overline{K}) \otimes_{\mathbb{Z}} \mathbb{R}$, which is positive semidefinite because $\widehat{h}_{\mathbf{c}}$ is a non-negative function (use 9.3.2 and Theorem 9.2.7). So we can apply the Cauchy–Schwarz inequality to get

$$|\widehat{h}_{\mathbf{c}'}(a)|^2 \leq b(a,a) \cdot b(a', a') = 4\widehat{h}_{\mathbf{c}}(a) \cdot \widehat{h}_{\mathbf{c}}(\varphi_{\mathbf{c}}(a'))\,.$$

This proves the claim. □

Remark 9.3.9. Let X be a projective smooth variety and fix an ample class $\mathbf{c} \in \mathrm{Pic}(X)$. If $\mathbf{c}' \in \mathrm{Pic}^0(X)$, then

$$h_{\mathbf{c}'} = O(|h_{\mathbf{c}}|^{1/2} + 1).$$

The proof involves the basic functorial properties of Picard varieties from Section 8.4 to reduce the claim to the case of abelian varieties, where it follows from Corollary 9.3.8. By base change, we may assume that X has a base point $P_0 \in X(K)$. Let $A := Pic^0(X)$ and let $\mathbf{p}_X, \mathbf{p}_A$ be the Poincaré classes of X and A. The pull-back of $\mathbf{p}_X$ with respect to the isomorphism $A \times X \to X \times A$, obtained by switching the entries, is denoted by $\mathbf{p}_X^t$. Then $\mathbf{p}_X^t$ is a subfamily of $\mathrm{Pic}^0(A)$ parametrized by X, which follows from $(\mathbf{p}_X)_{P_0} = \mathbf{0}, (\mathbf{p}_X)_{\mathrm{o}} = \mathbf{0}$ (see Remark 8.4.11) and from the definition of algebraic equivalence. By Theorem 8.4.7 applied to A instead of X, there is a unique morphism $\varphi : X \to \widehat{A}$ with $(\mathrm{id}_A \times \varphi)^*(\mathbf{p}_A) = \mathbf{p}_X^t$. We may view $\mathbf{c}'$ as point $a \in A(K)$. For $\mathbf{c}'' := (\mathbf{p}_A)_a \in \mathrm{Pic}^0(\widehat{A})$, we get

$$\varphi^*(\mathbf{c}'') = (\mathrm{id}_A \times \varphi)^* \circ \left(\mathbf{p}_A|_{\{a\}\times\widehat{A}}\right) = \mathbf{p}_X^t|_{\{a\}\times X} = \mathbf{c}'\,,$$

where the last step was by Remark 8.4.11. Hence $h_{\mathbf{c}'} = h_{\mathbf{c}''} \circ \varphi + O(1)$ by Theorem 2.3.8.

Let $\widehat{\mathbf{c}}$ be an ample even line bundle on $\widehat{A}$, whose existence is guaranteed by Proposition 8.6.4. Then there is $n \in \mathbb{N}$ such that $n\mathbf{c} - \varphi^*(\widehat{\mathbf{c}})$ is base-point free (see A.6.10). Hence $\widehat{h}_{\widehat{\mathbf{c}}} \circ \varphi = O(|h_{\mathbf{c}}| + 1)$ by Theorem 2.3.8 and Proposition 2.3.9. We conclude that it is enough to prove the claim for $\mathbf{c}'', \widehat{\mathbf{c}}$ on the abelian variety $\widehat{A}$ instead of $\mathbf{c}', \mathbf{c}$ and this is just Corollary 9.3.8.

Corollary 9.3.10. *Let X be a projective smooth variety over K, let $\mathbf{c} \in \mathrm{Pic}(X)$ be ample and let $\mathbf{c}' \in \mathrm{Pic}(X)$ be algebraically equivalent to $\mathbf{c}$. Then*

$$h_{\mathbf{c}'} = h_{\mathbf{c}} + O(|h_{\mathbf{c}}|^{1/2} + 1).$$

9.4. Néron–Tate heights on Jacobians

Let C be an irreducible smooth projective curve of genus $g > 0$ over a field K with product formula. By base change, we may assume that C has a K-rational base point P_0 (see 1.5.22). We denote the Jacobian of C by J and identify J with its dual $\widehat{J}$ as in 8.10.21 and Theorem 8.10.22. Then the Poincaré class $\boldsymbol{\delta}$ corresponds to

$$\boldsymbol{\delta} = m^*\boldsymbol{\theta} - p_1^*\boldsymbol{\theta} - p_2^*\boldsymbol{\theta} \in \mathrm{Pic}^0(J \times J),$$

where $\boldsymbol{\theta}$ is the theta divisor 8.10.8, m is the sum homomorphism and p_i is the projection of $A \times A$ onto the ith factor.

Proposition 9.4.1. *The Néron–Tate height* $\widehat{h}_{\boldsymbol{\delta}} : J(\overline{K}) \times J(\overline{K}) \longrightarrow \mathbb{R}$ *is a symmetric positive semidefinite bilinear form.*

Proof: Using the above identification of J and $\widehat{J}$ by $\varphi_{\boldsymbol{\theta}}$, Proposition 9.3.6 asserts that $\widehat{h}_{\boldsymbol{\delta}}$ is the symmetric bilinear form associated to the quadratic function $\widehat{h}_{\boldsymbol{\theta}}$.

Let $\Delta : J \longrightarrow J \times J$ be the diagonal homomorphism. Proposition 8.7.1 shows $[2]^*\boldsymbol{\theta} = 3\boldsymbol{\theta} + \boldsymbol{\theta}^-$ where $\boldsymbol{\theta}^- = [-1]^*\boldsymbol{\theta}$ as usual. Hence

$$\begin{aligned}\Delta^*\boldsymbol{\delta} &= (m \circ \Delta)^*\boldsymbol{\theta} - (p_1 \circ \Delta)^*\boldsymbol{\theta} - (p_2 \circ \Delta)^*\boldsymbol{\theta} \\ &= [2]^*\boldsymbol{\theta} - 2\boldsymbol{\theta} \\ &= \boldsymbol{\theta} + \boldsymbol{\theta}^-.\end{aligned}$$

For $a \in J(\overline{K})$, we conclude

$$\widehat{h}_{\boldsymbol{\delta}}(a, a) = \widehat{h}_{\boldsymbol{\theta}+\boldsymbol{\theta}^-}(a)$$

by Theorem 9.2.7 again. Since $\boldsymbol{\theta}$ is ample (see Theorem 8.10.22), $\boldsymbol{\theta}^- = [-1]^*\boldsymbol{\theta}$ is ample. Therefore, $\boldsymbol{\theta} + \boldsymbol{\theta}^-$ is an ample even class and the corresponding Néron–Tate height is a non-negative function (Theorem 9.2.7 (c)), proving the claim. □

9.4.2. In the light of Proposition 9.4.1, we will use the following notation for $a, a' \in J(\overline{K})$

$$\begin{aligned}\langle a, a' \rangle &:= h_{\boldsymbol{\delta}}(a, a'), \\ |a| &:= \widehat{h}_{\boldsymbol{\delta}}(a, a)^{1/2} = \widehat{h}_{\boldsymbol{\theta}+\boldsymbol{\theta}^-}(a)^{1/2}.\end{aligned}$$

The symmetric positive semidefinite bilinear form $\langle\ ,\ \rangle$ is called the **canonical form** of J. For a divisor D, it is also convenient to use the notation h_D for a height function associated to the isomorphim class of $O(D)$.

Proposition 9.4.3. *Let Δ be the diagonal in $C \times C$ and let $j : C \longrightarrow J$ be the natural embedding from 8.10.8. Then* ***Mumford's formula*** *for $P, Q \in C(\overline{K})$ is*

$$h_\Delta(P,Q) = \frac{1}{2g}\,|j(P)|^2 + \frac{1}{2g}\,|j(Q)|^2 - \langle j(P),\, j(Q)\rangle$$
$$- \frac{1}{2g}\,\widehat{h}_{\boldsymbol{\theta}-\boldsymbol{\theta}^-}(j(P)) - \frac{1}{2g}\,\widehat{h}_{\boldsymbol{\theta}-\boldsymbol{\theta}^-}(j(Q)) + O(1).$$

Proof: We abbreviate $z = j(P)$, $w = j(Q)$. By what was proved in 8.10.18–8.10.21, we get

$$(j \times j)^*(\boldsymbol{\delta}) = \mathrm{cl}(O(C \times \{P_0\} + \{P_0\} \times C - \Delta)).$$

By Theorem 2.3.8, we get

$$h_\Delta(P,Q) = h_{C\times\{P_0\}}(P,Q) + h_{\{P_0\}\times C}(P,Q) - \langle z, w\rangle + O(1). \tag{9.6}$$

By Theorem 2.3.8 again, we have

$$h_{C\times\{P_0\}}(P,Q) = h_{p_2^*([P_0])}(P,Q) = h_{[P_0]}(Q) + O(1) \tag{9.7}$$

and similarly

$$h_{\{P_0\}\times C}(P,Q) = h_{[P_0]}(P) + O(1). \tag{9.8}$$

Now Proposition 8.10.15 shows that $g[P_0]$ is in the class $j^*(\boldsymbol{\theta}^-)$ and Theorem 2.3.8 implies

$$h_{[P_0]}(P) = \frac{1}{g}\,\widehat{h}_{\boldsymbol{\theta}^-}(z) + O(1)$$
$$= \frac{1}{2g}|z|^2 - \frac{1}{2g}\,\widehat{h}_{\boldsymbol{\theta}-\boldsymbol{\theta}^-}(z) + O(1).$$

Substituting this in (9.7), (9.8) and putting the results into (9.6), we get Mumford's formula. □

Remark 9.4.4. Since $\boldsymbol{\theta} - \boldsymbol{\theta}^-$ is an odd class and $\boldsymbol{\theta} + \boldsymbol{\theta}^-$ is an even ample class, we get by Corollary 9.3.8

$$h_\Delta(P,Q) = \frac{1}{2g}|j(P)|^2 + \frac{1}{2g}|j(Q)|^2 - \langle j(P), j(Q)\rangle + O(|j(P)| + |j(Q)| + 1).$$

As shown by Mumford, this formula has some rather interesting consequences for curves of genus $g \geq 2$.

Proposition 9.4.5. *Assume that C has genus $g \geq 2$ and let $\cos\alpha \in (\frac{1}{g}, 1)$, $\varepsilon > 0$. Then there is a constant $B = B(C, P_0, \varepsilon) > 0$ such that for any pair $(P, Q) \in C(\overline{K})$ one of the following four possibilities occurs:*

(a) $P = Q$;

(b) $\langle j(P), j(Q)\rangle < (\cos\alpha) \cdot |j(P)| \cdot |j(Q)|$;

(c) $\min(|j(P)|, |j(Q)|) \leq B$;

(d) $(2g\,\cos\alpha - 1 - \varepsilon)\min(|j(P)|, |j(Q)|) \leq \max(|j(P)|, |j(Q)|)$.

Proof: We may assume that (a), (b) do not hold. Then we need to prove that either (c) or (d) hold. Again, we abbreviate $z = j(P)$, $w = j(Q)$ and assume $|z| \geq |w|$. By Remark 9.4.4, we have

$$\langle z, w\rangle + h_\Delta(P, Q) = \frac{1}{2g}\,|z|^2 + \frac{1}{2g}\,|w|^2 + O(|z| + 1).$$

Since $P \neq Q$ and Δ is an effective divisor, we may assume by Proposition 2.3.9 that $h_\Delta(P, Q) \geq 0$. Using also the negation of (b), we conclude

$$(\cos\alpha)\,|z|\,|w| \leq \frac{1}{2g}\,|z|^2 + \frac{1}{2g}\,|w|^2 + O(|z| + 1).$$

We may also assume that $|z| \geq 1$. We set $r := |z|\,/\,|w|$ and find from the preceding inequality that

$$\cos\alpha \leq \frac{1}{2g}\left(r + \frac{1}{r}\right) + O\left(\frac{1}{|w|}\right).$$

We multiply the last inequality by $2g$, note that $1/r \leq 1$ and find

$$2g\,\cos\alpha - 1 - O\left(\frac{1}{|w|}\right) \leq r = \frac{|z|}{|w|}.$$

If we choose B sufficiently large, then either (c) or (d) hold. □

Corollary 9.4.6. *With the notation of Proposition 9.4.5, let P, Q be points on C. Then, if $j(P) - j(Q)$ is in the kernel of $\langle\ ,\ \rangle$, either $P = Q$ or $|j(P)| = |j(Q)| \leq B$. In particular, if $j(P) - j(Q)$ is a torsion point and $|j(Q)| > B$, then $P = Q$.*

Proof: It is an immediate consequence of the preceding Proposition 9.4.5. □

9.4.7. Our next goal is to count rational points of C (still assuming $g \geq 2$), using a procedure similar to that used in the proof of Theorem 5.2.1. As in Section 9.3, the canonical bilinear form $b = \widehat{h}_\delta$ extends to a symmetric positive semidefinite bilinear form $\widetilde{b}$ on $J(\overline{K}) \otimes_{\mathbb{Z}} \mathbb{R}$. Let $N_{\mathbb{R}}$ be its kernel; then $\widetilde{b}$ induces a scalar product on $E := J(\overline{K}) \otimes_{\mathbb{Z}} \mathbb{R}/N_{\mathbb{R}}$, again denoted by $\langle\ ,\ \rangle$. By Corollary 9.4.6 above, the map

$$i : C(\overline{K}) \to E, \qquad P \mapsto j(P) \otimes 1 + N_{\mathbb{R}},$$

is one-to-one on the subset of points P such that $|j(P)| > B$.

Definition 9.4.8. *A point $P \in C(\overline{K})$ is called* **small** *if $|j(P)| \leq B$, otherwise it is called* **large**.

9.4.9. We may choose (and fix) $0 < \alpha < \pi/2$ and $\varepsilon > 0$ such that $\cos\alpha \in (\frac{1}{g}, 1)$ is such that $\lambda := 2g\cos\alpha - 1 - \varepsilon > 1$.

In the euclidean space E, we have the following geometric interpretation of Proposition 9.4.5. If P, Q are different large points such that $i(P)$ and $i(Q)$ include an angle $\leq \alpha$ and if $|j(P)| \leq |j(Q)|$, then $\lambda|j(P)| \leq |j(Q)|$. This shows that we have gaps between points on C pointing in approximatively the same direction.

Let us consider the cone

$$T := \{x \in E \mid \langle x, a\rangle \geq \cos(\alpha/2) \cdot |x| \cdot |a|\}$$

with center 0, angle $\alpha/2$ and axis through $a \in E$. We order the large points in $C(\overline{K})$ mapping to T in a sequence Q_0, Q_1, $\ldots$ such that

$$B < |j(Q_0)| \leq |j(Q_1)| \leq |j(Q_2)| \leq \ldots.$$

The above shows $|j(Q_k)| \geq \lambda^k |j(Q_0)|$ for every k. For $H > B$, let $n_T(H)$ be the number of large points $Q \in C(\overline{K})$ mapping to T with $|j(Q)| \leq H$. We get

$$n_T(H) \leq \left\lceil \frac{\log(H/B)}{\log\lambda} \right\rceil.$$

9.4.10. The above bound for $n_T(H)$ is uniform with respect to T (for fixed angle) and yields a counting of all large $\overline{K}$-rational points of C mapping to T. This can be used, in some circumstances, to count all large points in $C(K)$ with bounded height. To this end, it is necessary to assume that $J(K)$ is a finitely generated group. Another possibility consists in fixing *a priori* a finitely generated subgroup Γ of $J(\overline{K})$, and consider only the subset of large points $P \in C(\overline{K})$ for which $j(P) \in \Gamma$. The question whether we can take $J(K)$ for such a group Γ can then be examined independently. As we shall see later in the chapter dedicated to the Mordell–Weil theorem, $J(K)$ is finitely generated if K is a number field or K is a function field over a finite field.

Thus we fix a subgroup Γ of $J(\overline{K})$ of finite rank $r := \operatorname{rank}_{\mathbb{Q}}(\Gamma)$, where the rank is the maximum number of $\mathbb{Z}$-linearly independent elements in Γ.

We associate to Γ the finite-dimensional real vector subspace E_Γ spanned by the image of Γ in E. It is clear that $\dim(E_\Gamma) \leq r$.

For $x \in E \setminus \{0\}$, we set $\nu(x) = x/|x|$. Then ν maps cones to spherical caps and we get a bound for the minimal number of cones needed to cover E_Γ by Lemma 5.2.19. From the proof of Lemma 5.2.19, we deduce immediately the following result leading to a little sharpening of the bound for the number of large points.

Lemma 9.4.11. *Let $\|\ \|$ be a norm on $\mathbb{R}^r$ and let $\rho > 0$. If $\mathbf{x}_1, \ldots, \mathbf{x}_n \in \mathbb{R}^r$ have norm 1 and if $\|\mathbf{x}_i - \mathbf{x}_j\| > \rho$, then $n \leq (1 + 2/\rho)^r$.*

Proposition 9.4.12. *For $\rho = 2\sin(\alpha/2)$ and $r = \operatorname{rank}_{\mathbb{Q}}(\Gamma)$, the number $n_\Gamma(H)$ of large points $Q \in C(\overline{K})$ with $j(Q) \in \Gamma$ and $|j(Q)| \leq H$ does not exceed*

$$n_\Gamma(H) \leq \left\lceil \frac{\log(H/B)}{\log \lambda} \right\rceil \cdot \lfloor (1+2/\rho)^r \rfloor.$$

In particular, $n_\Gamma(H) \ll \log H$.

Proof: For any $k \in \mathbb{N}$, we count the number of $Q \in C(\overline{K})$ with $\lambda^k B < |j(Q)| \leq \lambda^{k+1} B$. By 9.4.9, the angle between two such points Q, Q' is $> \alpha$. We conclude that $|\nu(Q) - \nu(Q')| > \rho$ for $\rho := 2\sin(\alpha/2)$. By Lemma 9.4.11, there are at most $(1+2/\rho)^r$ such points. Now the interval $(B, H]$ may be covered by $\lceil \log(H/B)) \rceil$ such intervals, proving the claim. □

9.4.13. Still assuming $g \geq 2$, we may choose $\alpha = \pi/6$ and $\varepsilon > 0$ such that $\lambda = 2g\cos\alpha - 1 - \varepsilon \geq 2$ and $\rho = 2\sin(\pi/12) = \sqrt{2 - \sqrt{3}} > \frac{1}{2}$. With this choice of parameters, we may summarize our findings about **Mumford's gap principle:**

Theorem 9.4.14. *Let C be an irreducible smooth projective curve of genus $g \geq 2$ over K with base point $P_0 \in C(K)$ leading to a closed embedding j of C into the Jacobian variety J and let Γ be a subgroup of J of finite $\mathbb{Q}$-rank r. Then there is a constant $B > 0$ depending only on C and P_0 with the following properties:*

(a) *If we choose any cone T in E with center 0 and angle $\alpha/2$ (see 9.4.9) and if we order $\{Q \in C \mid j(Q) \in \Gamma,\ |j(Q)| > B,\ i(Q) \in T\}$ by increasing norm, then $|j(Q_{n+1})| \geq 2|j(Q_n)|$ for every $n \in \mathbb{N}$.*

(b) *For $H > B$, the number $n_\Gamma(H)$ of points $Q \in C$ with $j(Q) \in \Gamma, B < |j(Q)| \leq H$ is bounded by*

$$n_\Gamma(H) \leq \left\lceil \frac{\log(H/B)}{\log 2} \right\rceil \cdot 5^r.$$

(c) *In particular, $n_\Gamma(2H) - n_\Gamma(H) \leq 2 \cdot 5^r$.*

The question of bounding the number of small points requires different considerations. In Chapter 5 we used a uniform version of Zhang's theorem, as at the end of the proof of Theorem 5.2.1. Another way of dealing with the problem is to apply a finiteness result, as in Northcott's theorem in 2.4.9. This approach leads to the following Northcott condition already used in 4.5.1:

Definition 9.4.15. *The field K satisfies the* ***Northcott property (N)*** *if $\{\alpha \in K \mid h(\alpha) \leq R\}$ is finite for every $R > 0$.*

Proposition 9.4.16. *If K satisfies (N), then for any projective variety X over K and any ample class $\mathbf{c} \in \operatorname{Pic}(X)$, the set $\{x \in X(K) \mid h_{\mathbf{c}}(x) \leq H\}$ is finite for every $H > 0$.*

Proof: Since $h_{\mathfrak{c}}$ is determined up to bounded functions, this property is well defined. Now the proof is the same as our deduction of Northcott's theorem in 2.4.9 from 1.6.8. □

By definition, $|\ |^2$ is a height function with respect to an ample class, hence:

Proposition 9.4.17. *The Northcott property (N) for a field K implies that the number of small K-rational points of C is finite.*

Remark 9.4.18. If K satisfies (N), then we see as in 9.3.4 that the kernel of $\langle\ ,\ \rangle$ on $J(K)$ is the torsion subgroup of $J(K)$. By Lemma 9.3.3, $\langle\ ,\ \rangle$ is a scalar product on $E_{J(K)} = J(K) \otimes_{\mathbb{Z}} \mathbb{R}$. Thus, if Γ is a subgroup of $J(K)$, we get $\dim(E_\Gamma) = \operatorname{rank}_{\mathbb{Q}}(\Gamma)$.

9.4.19. Faltings' theorem in 11.1.1 says that $n_{J(K)}(H)$ is in fact bounded whenever K is a number field. On the other hand, there are fields K satisfying (N) for which $J(K)$ is finitely generated and Mumford's bound $n_{J(K)}(H) \ll \log H$ in Proposition 9.4.12 is best possible. The following example is due to Serre **[277]**, p.80.

Example 9.4.20. Let K be a finite field with q elements and let X be an irreducible smooth projective curve. There is a natural set $M_{K(X)}$ of absolute values on the function field $K(X)$, satisfying the product formula (see 1.4.6–1.4.9). A point $P \in \mathbb{P}^n(K(X))$ corresponds to a rational map $\varphi : X \dashrightarrow \mathbb{P}^n_K$ defined over K. But φ is always defined in codimension 1 (by the valuative criterion of properness and smoothness of X, see A.11.10), hence is a morphism. To φ we attach the line bundle $L = \varphi^* O_{\mathbb{P}^n}(1)$ generated by the global sections $s_j := \varphi^*(x_j)$, $0 \leq j \leq n$. Example 2.4.11 shows that $h(P) = \deg_L(X)$.

We claim that the field $K(X)$ has the Northcott property (N). To prove that, let $H > 0$ and $f \in K(X)$ with $h(f) \leq H$. By Example 1.5.23, we have

$$h(f) = -\sum_Z \deg(Z) \min(0, \operatorname{ord}_Z(f)),$$

where Z ranges over all prime divisors of X. It is easy to see that over a finite field, only finitely many prime divisors of bounded degree can occur. We conclude that the number of pole divisors of all rational functions f with $h(f) \leq H$ is finite. By A.8.22, f is determined up to a locally constant function by its pole divisor. Since $\overline{K} \cap K(X)$ is a finite field, we get the claim.

Let C be a geometrically irreducible smooth projective curve of genus $g \geq 2$ over K. We may assume that X is geometrically irreducible and that there is a point $P \in C(K(X)) \setminus C(K)$. This can certainly be achieved by a finite base extension of $K(X)$ (extending K also), which is again a function field of an irreducible normal projective curve by Lemma 1.4.10. The points of a normal curve are regular (see A.8.5) and the finite field K is perfect, hence the curve has to be smooth again (see A.7.12 and A.7.13).

Denote the Jacobian of C by J. By Remark 10.6.5, $J(K(X))$ is a finitely generated abelian group. Thus the assumptions of Proposition 9.4.12 are satisfied.

We recall the definition of the Frobenius map over K. For a variety Y over K, it is the K-morphism $F : Y \to Y$, given, in affine coordinates $\mathbf{x}$ on a chart, by $\mathbf{x} \mapsto \mathbf{x}^q$. For a line bundle L over Y, we have $F^*(L) \cong L^{\otimes q}$, which is easily proved by considering transition functions. Moreover, for every K-morphism ψ, we have $F \circ \psi = \psi \circ F$. In particular, $F : J \to J$ is a homomorphism of abelian varieties, defined over K. By Theorem 9.2.7, we get

$$|F(z)|^2 = \widehat{h}_{\theta+\theta^-}(F(z)) = \widehat{h}_{F^*(\theta+\theta^-)}(z) = q|z|^2 \tag{9.9}$$

for any $z \in J(K(X))$. Note here that both the Jacobian and the theta divisor are compatible with base change to $K(X)$ (use Corollary 8.4.10).

By (9.9), for $P_n := F^n(P) \in C(K(X))$, we have $|j(P_n)|^2 = q^n|j(P)|^2$. Since $P \notin C(K)$, the points P_n are pairwise different. Here we have used that the $K(X)$-rational fixed points of a Frobenius power have coordinates in a finite field extension of K and hence in K (as K is algebraically closed in $K(X)$ by A.4.11). Moreover, we note that $|j(P)| \neq 0$ (which is clear from Remark 9.4.18 because $j(P) \notin J(\overline{K})$) and hence is not a torsion point (Proposition 8.7.2). This gives $n_{J(K(X))}(H) \gg \log H$ as desired.

Example 9.4.21. The following example gives an explicit construction of a curve for which we have $n_\Gamma(H) \gg \log H$. The reader will verify that it is really not different in nature from the preceding example.

Let K be a finite field with q elements, let n be a positive integer and let $a_{ij}(t) \in K(t)$, $i, j = 0, 1, 2$, be such that

$$\sum_{i,j=0}^{2} a_{ij}(t) x_i x_j^{q^n} = 0$$

defines a smooth plane curve C of degree $q^n + 1$ in $\mathbb{P}^2_{K(t)}$ with homogeneous coordinates $(x_0 : x_1 : x_2)$. By the Jacobian criterion in A.7.15, this happens if $\det(a_{ij}(t))$ is not identically 0. The curve C has genus $g(C) = q^n(q^n - 1)/2$ (see A.13.4) and in particular $g(C) \geq 2$ if $q^n \geq 3$. Let P be a point on C rational over $K(t)$. It is easy to see that the tangent line to C at P has intersection multiplicity q^n with C at P; therefore, since C has degree $q^n + 1$, it intersects C residually in a single point P', again rational over $K(t)$. Thus, starting with a point P_0, we obtain by the above tangent process a sequence of points $P_1, P_2, \ldots$ all defined over $K(t)$. In general, the sequence obtained in this way is infinite and we have

$$m \ll \log(h(P_m)) \ll m.$$

An explicit example is the curve (now in affine coordinates x, y)

$$x^{q+1} + txy^q - y^{q+1} = 1.$$

The associated equation $\alpha^{q+1} + t\alpha - 1 = 0$ has a solution with a formal continued fraction expansion

$$\alpha = [0; t, t^q, t^{q^2}, t^{q^3}, \ldots].$$

It is now easy to see, using classical elementary properties of continued fractions (see for example G.H. Hardy and E.M. Wright [**145**] or O. Perron [**233**]), that, if we set $(x_{-1}, y_{-1}) = (1, 0)$, $(x_0, y_0) = (0, 1)$, $(x_1, y_1) = (1, t)$, and in general

$$x_{m+1} = t^{q^m} x_m + x_{m-1}, \qquad y_{m+1} = t^{q^m} y_m + y_{m-1},$$

then

$$[0; t, t^2, \ldots, t^{q^{m-1}}] = \frac{x_m}{y_m}$$

and

$$x_m^{q+1} + tx_m y_m^q - y_m^{q+1} = (-1)^{m+1}.$$

Thus we obtain a polynomial solution for every odd m, and in fact for any m if we are in characteristic 2. Moreover, it is the same as the geometric construction above, i.e. for $P_0 = (1,0)$, we get $P_m = (x_{2m-1}, y_{2m-1})$ for all $m \in \mathbb{N}$. We leave the details to the reader.

This gives an example of an affine curve (in fact, a Thue equation) over $K[t]$, of genus $q(q-1)/2$, with infinitely many integral points over $K[t]$, occurring precisely at the rate predicted by Mumford's bound.

9.5. The Néron symbol

We give here Néron's decomposition of the Néron–Tate height into canonical local heights. This will lead to the Néron symbol on arbitrary smooth complete varieties, namely a pairing between a divisor and a disjoint zero-dimensional cycle, both assumed to be algebraically equivalent to 0. This section is based on Section 2.7 and provides additional material not essential to the rest of the book. Instead of working with local heights, we adopt the equivalent concept of metrized line bundles promoted by Arakelov theory with much success. Examples are given at the end and in particular we will describe quite explicitly the theory when applied to Tate's elliptic curves.

Let K be a field with a given set of places M_K. For every place $v \in M_K$, we fix an absolute value $|\ |_v$ on K in the equivalence class determined by v. We assume that $\{v \in M_K \mid |\alpha|_v \neq 1\}$ is finite for every $\alpha \in K \setminus \{0\}$. We also fix an embedding $K \subset \overline{K}$ and denote by M a set of places of $\overline{K}$ such that every $u \in M$ restricts to a $v \in M_K$. We denote by $|\ |_u$ the unique extension of $|\ |_v$ to an absolute value in the equivalence class of u.

9.5.1. Let L be a line bundle on a complete K-variety X. For the moment, we fix a place u, which is the reference for boundedness and metrics. Let $\|\ \|, \|\ \|'$ be locally bounded metrics on L. By Proposition 2.6.6, the norm of the constant section 1 with respect to the metric $\|\ \|/\|\ \|'$ of O_X is a bounded function ρ and we set

$$d(\|\ \|, \|\ \|') := \sup_{x \in X} |\log \rho(x)|.$$

Obviously, d is a distance function on the space of locally bounded metrics on L. The existence of such metrics was shown in Proposition 2.7.5. By choosing one such metric $\|\ \|'$, the map $\|\ \| \mapsto \log(\|\ \|, \|\ \|')$ is a non-canonical isometry onto the Banach space of bounded real functions on X with supremum norm.

Lemma 9.5.2. *The distance d on the space of locally bounded metrics of L satisfies:*

(a) *If $n \in \mathbb{Z}$, then $d(\|\ \|^{\otimes n}, \|\ \|'^{\otimes n}) = |n| \cdot d(\|\ \|, \|\ \|')$.*

(b) *If $\varphi : X' \to X$ is a morphism, then $d(\varphi^*\|\ \|, \varphi^*\|\ \|') \leq d(\|\ \|, \|\ \|')$.*

Proof: These properties follow immediately from the definitions. □

9.5.3. We pick up the case of a dynamical system as in 9.2.9. Let $\varphi : X \to X$ be a morphism of a complete K-variety X. We fix $k, l \in \mathbb{Z}$, $|k| > |l|$, and we consider a line bundle L on X with an isomorphism

$$\theta : \varphi^*(L)^{\otimes l} \xrightarrow{\sim} L^{\otimes k}. \tag{9.10}$$

Theorem 9.5.4. *There is a unique locally bounded M-metric $(\| \ \|_u)_{u \in M}$ on L satisfying*

$$\| \ \|_u^{\otimes k} \circ \theta = (\varphi^* \| \ \|_u)^{\otimes l}. \tag{9.11}$$

Proof: Let us fix $u \in M$ as in 9.5.1 and let B be the Banach space of locally bounded u-metrics on L. Note that an u-metric $\| \ \|$ on $L^{\otimes k}$ induces a metric $\| \ \|^{1/k}$ on L characterized by $\|s(x)\|^{1/k} = \|s^{\otimes k}(x)\|^{1/k}$ for every local section s of L. We consider the map

$$\Phi : B \longrightarrow B, \quad \| \ \| \mapsto \left((\varphi^* \| \ \|)^{\otimes l} \circ \theta^{-1} \right)^{1/k}.$$

By Lemma 9.5.2, Φ is a contraction with factor $\lambda = |l/k| < 1$. Banach's fixed point theorem in 11.5.15 yields that Φ has a unique fixed point, meaning that there is a unique locally bounded u-metric $\| \ \|_u$ on L satisfying (9.11). To prove that $(\| \ \|_u)_{u \in M}$ is a locally bounded M-metric, we choose any locally bounded M-metric $(\| \ \|_{0,u})_{u \in M}$ of L. Then the proof of Banach's fixed point theorem gives $\| \ \|_u = \lim_{k \to \infty} \Phi^k(\| \ \|_{0,u})$ and a detailed analysis of the occuring estimates proves

$$d(\| \ \|_{0,u}, \| \ \|_u) \leq \frac{1}{1-\lambda} \cdot d(\| \ \|_{0,u}, \Phi(\| \ \|_{0,u})).$$

The right-hand side is M-bounded. We conclude that $(\| \ \|_u / \| \ \|_{0,u})_{u \in M}$ is a locally bounded M-metric and hence the same is true for $(\| \ \|_u)_{u \in M}$. □

Remark 9.5.5. We denote the M-metric in Theorem 9.5.4 by $(\| \ \|_{\theta,u})_{u \in M}$. If θ' is another isomorphism in (9.10), then $\theta' \circ \theta^{-1}$ is an automorphism of $L^{\otimes k}$ given by multiplication with a nowhere vanishing regular function a. By A.6.15, a is constant on every irreducible component of $X_{\overline{K}}$. From (9.11), we get

$$|a|_u^{-1} \cdot (\| \ \|_{\theta,u}^{\otimes k} \circ \theta') = (\varphi^* \| \ \|_{\theta,u})^{\otimes l}.$$

Uniqueness in Theorem 9.5.4 implies $\| \ \|_{\theta',u} = |a|_u^{\frac{1}{l-k}} \| \ \|_{\theta,u}$.

9.5.6. Usually, we cannot hope for a canonical isomorphism θ. We clarify the role of θ by introducing rigidifications.

Let X be an irreducible smooth complete variety over K with base point $P_0 \in X(K)$. A **rigidification** of a line bundle L on X is the choice of an element $\nu \in L_{P_0}(K)$. An isomorphism of rigidified line bundles is an isomorphism of the underlying line bundles mapping the rigidification of the first to the rigidification of the latter.

The **tensor product** of two rigidified line bundles is defined by

$$(L_1, \nu_1) \otimes (L_2, \nu_2) = (L_1 \otimes L_2, \nu_1 \otimes \nu_2).$$

The **pull-back** with respect to a base point preserving morphism $\psi : X' \to X$ of irreducible smooth varieties over K is defined by

$$\psi^*(L, \nu) = (\psi^*(L), \psi^*(\nu))$$

using the canonical homomorphism $\psi^* : L_{P_0} \to \psi^*(L)_{P_0'}$. The advantage of rigidifications is that it makes relations in the Picard group concrete. If $(L, \nu), (L, \nu')$ are rigidified line bundles with $L \cong L'$, then there is a unique isomorphism $\theta : L \xrightarrow{\sim} L'$ with $\theta(\nu) = \nu'$. Existence is obvious and as any other isomorphism is given by multiplication with a nowhere vanishing regular function λ, uniqueness is clear from the fact that λ has to be constant (use that X is geometrically irreducible (cf. A.7.14) and then completeness in A.6.15). By A.4.11, we get $\lambda \in K$.

Recall that on an abelian variety, we always use 0 as a base point. The next result gives us **canonical M-metrics** on rigidified line bundles of an abelian variety.

Theorem 9.5.7. *For every rigidified line bundle (L, ν) of an abelian variety A over K, there is a unique locally bounded M-metric $(\| \ \|_{\nu,u})_{u \in M}$ satisfying the properties:*

(a) *An isomorphism of rigidified line bundles is an isometry.*

(b) $\| \ \|_{\nu_1,u} \otimes \| \ \|_{\nu_2,u} = \| \ \|_{\nu_1 \otimes \nu_2,u} \quad (u \in M)$.

(c) *With the rigidification 1 of O_A, it holds* $\| \ \|_{1,u} = | \ |_u \quad (u \in M)$.

(d) *If $\psi : A' \to A$ is a homomorphism of abelian varieties over K, then*

$$\psi^* \| \ \|_{\nu,u} = \| \ \|_{\psi^*\nu,u}.$$

Proof: First, we prove that the theorem holds for all even line bundles. We fix an integer $m \in \mathbb{Z}$, $|m| \geq 2$. By Proposition 8.7.1 and 9.5.6, there is a unique isomorphism

$$\theta : [m]^*(L, \nu) \xrightarrow{\sim} (L, \nu)^{\otimes m^2} \tag{9.12}$$

of rigidified line bundles. We apply Theorem 9.5.4 with $\varphi = [m]$ to get a unique locally bounded M-metric $(\| \ \|_{\nu,u})_{u \in M}$ on L satisfying

$$\| \ \|_{\nu,u}^{\otimes m^2} \circ \theta = [m]^* \| \ \|_{\nu,u} \quad (u \in M). \tag{9.13}$$

Using elementary properties of line bundles, it is easy to show that (a)–(d) hold.

To prove uniqueness, assume that every rigidified even line bundle (L, ν) on A is endowed with a locally bounded M-metric $(\| \ \|_{\nu,u})_{u \in M}$ satisfying (a)–(d). For any integer m, $|m| \geq 2$, property (d) implies

$$[m]^* \| \ \|_{\nu,u} = \| \ \|_{[m]^*\nu,u} \quad (u \in M). \tag{9.14}$$

Using θ as in (9.12) and (a), (b), we get

$$\| \ \|_{[m]^*\nu,u} = \| \ \|_{\nu^{\otimes m^2},u} \circ \theta = \| \ \|_{\nu,u}^{\otimes m^2} \circ \theta. \tag{9.15}$$

Note that (9.14) and (9.15) give identity (9.13), which characterize the metric uniquely. This proves uniqueness.

Similarly, we show existence and uniqueness for odd line bundles replacing the tensor power m^2 by m. An arbitrary line bundle L is isomorphic to the tensor product of an even line bundle L_+ and an odd line bundle L_-. We use this factorization, which is unique up to 2-torsion in the Picard group (Corollary 8.8.2), to get well-defined canonical M-metrics $(\| \ \|_{\nu,u})_{u \in M}$ on L with the required properties. □

Remark 9.5.8. The theorem gives canonical metrics for rigidified line bundles on all abelian varieties simultaneously. We have seen in the proof that uniqueness already follows for all rigidified line bundles on a given abelian variety from (a), (b) and the special case $\psi = [m]$ of (d), where $m \in \mathbb{Z}$, $|m| \geq 2$ is fixed.

Remark 9.5.9. Note that $\|\nu\|_{\nu,u} = 1$. This follows from uniqueness because the locally bounded M-metrics $(\|\nu\|_{\nu,u}^{-1} \| \; \|_{\nu,u})_{u \in M}$ also satisfy (a)–(d) of Theorem 9.5.7. If ν' is another rigidification on L, then

$$\| \; \|_{\nu',u} = |\nu/\nu'|_u \cdot \| \; \|_{\nu,u} \quad (u \in M).$$

Hence the M-metric $(\| \; \|_{\nu,u})_{u \in M}$ is canonically determined by L up to $(|a|_u)_{u \in M}$ for some $a \in K^\times$. The corresponding **canonical local height** will be canonically determined by the divisor up to $(\log |a|_u)_{u \in M}$ for some $a \in K^\times$.

We define an **M-constant** to be a family $(\gamma_v)_{v \in M}$ of real numbers with $\gamma_v \neq 0$ only for finitely many $v \in M$. Hence the canonical local height above is determined by the divisor up to an M-constant.

9.5.10. In order to eliminate this indeterminacy, we introduce first some notation. Let f be a non-zero rational function on a smooth irreducible variety X. Recall that $Z_0(X_{\overline{K}})$ denotes the group of zero-dimensional cycles on the base change $X_{\overline{K}}$. We denote here by $[P]$ the cycle associated to $P \in X$. For $Z = \sum n_j P_j \in Z_0(X_{\overline{K}})$ with $\operatorname{supp}(Z) \cap \operatorname{supp}(\operatorname{div}(f)) = \emptyset$, we define

$$f(Z) = \prod_j f(P_j)^{n_j} \in \overline{K}^\times.$$

If $\lambda_{\widehat{D}}$ is a local height with respect to a Néron divisor $\widehat{D}$ on X, it is natural to extend it additively to zero-dimensional cycles, i.e.

$$\lambda_{\widehat{D}}(Z, u) := \sum_j n_j \lambda_{\widehat{D}}(P_j, u)$$

for all $Z = \sum_j n_j P_j \in Z_0(X_{\overline{K}})$ with $\operatorname{supp}(Z) \cap \operatorname{supp}(D) = \emptyset$ and $u \in M$.

By $B_0(X)$, we denote the subgroup of $Z_0(X_{\overline{K}})$, which is the kernel of the degree map. In the next result, we will make use of the fact that the indeterminacy of the canonical local heights from Remark 9.5.9 cancels by restriction to $B_0(X)$.

We also make use of the pull-back τ_a^* of cycles with respect to translation. Pull-back of cycles is defined with respect to flat morphisms, but here for an isomorphism it is simply the inverse of push-forward, meaning that $\tau_a^*(Y) = Y - a$ for any prime cycle.

Theorem 9.5.11. *Let A be an abelian variety over K. For $u \in M$, a divisor D on A and $Z \in B_0(A)$ with $\operatorname{supp}(D) \cap \operatorname{supp}(Z) = \emptyset$, there is a pairing $(D, Z)_u \in \mathbb{R}$ called the* **Néron symbol**, *which is uniquely characterized by the following properties:*

(a) *If D, D' are divisors with $(\operatorname{supp}(D) \cup \operatorname{supp}(D')) \cap \operatorname{supp}(Z) = \emptyset$, then*
$$(D + D', Z)_u = (D, Z)_u + (D', Z)_u.$$

(b) *If $Z, Z' \in B_0(A)$ with $\operatorname{supp}(D) \cap (\operatorname{supp}(Z) \cup \operatorname{supp}(Z'))$, then*
$$(D, Z + Z')_u = (D, Z)_u + (D, Z')_u.$$

(c) *If* $f \in K(A)^\times$ *with* $\mathrm{supp}(\mathrm{div}(f)) \cap \mathrm{supp}(Z) = \emptyset$, *then*

$$(\mathrm{div}(f), Z)_u = -\log|f(Z)|_u.$$

(d) *If* $a \in A$, *then*

$$(\tau_a^*(D), \tau_a^*(Z))_u = (D, Z)_u.$$

(e) *For every* $P_0 \in A \setminus \mathrm{supp}(D)$, *the function*

$$(A \setminus \mathrm{supp}(D)) \times M \longrightarrow \mathbb{R}, \quad (P, u) \mapsto (D, [P] - [P_0])_u$$

is locally M*-bounded.*

Proof: We endow $O(D)$ with a rigidification ν. Then Theorem 9.5.7 gives a canonical locally bounded M-metric $\| \ \|_{\nu,u}$ on $O(D)$ and we denote the corresponding Néron divisor by $\widehat{D}^\nu$. It gives rise to a local height $\lambda_{\widehat{D}^\nu}$ on A, which we extend to $Z_0(A_{\overline{K}})$ as in 9.5.10. For $Z \in B_0(A)$ with $\mathrm{supp}(D) \cap \mathrm{supp}(Z) = \emptyset$, we set

$$(D, Z)_u := \lambda_{\widehat{D}^\nu}(Z, u).$$

We have seen in Remark 9.5.9 that $\lambda_{\widehat{D}^\nu}$ depends on the rigidification ν by an additive M-constant. Since $\deg(Z) = 0$, we conclude that $(D, Z)_u$ does *not* depend on the rigidification.

Properties (a), (c), (e) follow immediately from Theorem 9.5.7, and (b) is obvious from the definition of the local height on $Z_0(A_{\overline{K}})$. In order to prove (d), let ν, ν_a be rigidifications on $O(D)$ and $O(\tau_a^* D)$. Lemma 9.5.13 below shows that $\tau_a^* \| \ \|_{\nu,u} = C_u \cdot \| \ \|_{\nu_a,u}$ for a constant $C_u > 0$. As we have seen above, M-constants do not influence the pairing and (d) follows from

$$(\tau_a^*(D), \tau_a^*(Z))_u = \lambda_{\tau_a^* \widehat{D}^\nu}(\tau_a^*(Z), u) = \lambda_{\widehat{D}^\nu}((\tau_a)_* \circ \tau_a^*(Z), u) = \lambda_{\widehat{D}^\nu}(Z, u).$$

To prove uniqueness, let $(\ , \ )'$ be another pairing with properties (a)–(e). We consider

$$[D, Z]_u := (D, Z)_u - (D, Z)'_u.$$

By (c), it depends only on the rational equivalence class $\mathrm{cl}(D)$ of D. Now we need the following **moving lemma:**

Lemma 9.5.12. *Given a Cartier divisor* D *and finitely many points* $P_1, \dots, P_g$ *on a projective variety over an infinite field, there is always a Cartier divisor* D' *such that* $D - D'$ *is a principal Cartier divisor with*

$$\{P_1, \dots, P_g\} \cap \mathrm{supp}(D') = \emptyset.$$

Proof: Every Cartier divisor on a projective variety is, up to a principal Cartier divisor, equal to the difference of two very ample divisors (see A.6.10(a), A.8.16). So we may assume D very ample giving rise to a closed embedding ι into projective space. Then K infinite allows us to choose a hyperplane H omitting $P_1, \dots, P_g$ and we get the claim for $D' := \iota^*(H)$. □

Continuation of the proof of Theorem 9.5.11: Note first that in the whole section we have implicitly assumed that K is infinite. Otherwise, no place on K would exist. The moving lemma allows us to define the pairing $[D, Z]_u$ for all divisors D and all $Z \in B_0(A)$ without assuming $\mathrm{supp}(D) \cap \mathrm{supp}(Z) = \emptyset$. By the theorem of the square in 8.5.2, we have

$$m(\tau_a^*(D) - D) \sim \tau_{ma}^*(D) - D$$

for every $m \in \mathbb{Z}$ and hence (d) proves

$$m[\tau_a^*(D) - D, Z]_u = [D, (\tau_{ma})_*(Z) - Z]_u. \tag{9.16}$$

For a moment, we assume that $Z = [P] - [P_0]$ for $P, P_0 \in A \setminus \operatorname{supp}(D)$. By (e), the right-hand side of (9.16) is a locally M-bounded function of $b = ma$ on the open subset

$$U_D := \{b \in A \mid \operatorname{supp}(D) \cap \{P + b, P_0 + b\} = \emptyset\}.$$

By the moving lemma, we may replace D by finitely many divisors $D' \sim D$ such that the open subsets $U_{D'}$ cover A. Independence of D' and Remark 2.6.14 prove that the right-hand side of (9.16) is a locally M-bounded function of $b = ma$ on the whole A. Since A is an M-bounded set (Proposition 2.6.17), we conclude that the left-hand side of (9.16) is an M-bounded function of m and a. Letting $m \to \infty$ and taking into account (a), we deduce that

$$[\tau_a^*(D), Z]_u = [D, Z]_u.$$

By linearity (b) of the pairing in Z, this identity holds for all $Z \in B_0(A)$. From (d), we deduce

$$[D, (\tau_a)_*(Z) - Z]_u = 0.$$

Thus $[D, \cdot\,]_u$ vanishes on the subgroup of $B_0(A)$ generated by the cycles of the form $(\tau_a)_*(Z) - Z$ for varying $a \in A$ and $Z \in B_0(A)$. By induction, it is easily seen that this subgroup is equal to the kernel of the homomorphism

$$S : B_0(A) \longrightarrow A, \quad \sum_j m_j[P_j] \mapsto \sum_j m_j P_j.$$

In order to prove uniqueness, it is enough to show that $[D, [P] - [0]]_u = 0$ for every $P \in A$. For $m \in \mathbb{Z}$, the cycle

$$m\,([P] - [0]) - ([mP] - [0])$$

has degree 0 and is in the kernel of S. By our considerations above and (b), we get

$$m[D, [P] - [0]]_u = [D, [mP] - [0]]_u. \tag{9.17}$$

With the same arguments as for (9.16), we deduce that the right-hand side of (9.17) is an M-bounded function of $Q = mP$. Hence the left-hand side of (9.17) is an M-bounded function of m and P and as $m \to \infty$ this is possible only if $[D, [P] - [P_0]]_u = 0$. This proves uniqueness. □

Lemma 9.5.13. *Let (L, ν) be a rigidified line bundle on the abelian variety A over K, let $a \in A$ and let ν_a be a rigidification on $\tau_a^*(L)$. For $u \in M$, there is a constant $C_u > 0$ such that*

$$\tau_a^* \| \ \|_{\nu,u} = C_u \cdot \| \ \|_{\nu_a,u}.$$

Proof: In order to compare the two metrics, we consider as in 9.5.1 the function ρ on A given as the norm of 1 with respect to the metric $\tau_a^*\| \ \|_{\nu,u} / \| \ \|_{\nu_a,u}$ on O_A. We have to prove that

$$\omega_{L,a}(P) := \log \rho(P) - \log \rho(0)$$

vanishes identically in $P \in A$. By Remark 9.5.9, it is clear that $\omega_{L,a}$ does not depend on the choice of the rigidifications ν, ν_a. From Theorem 9.5.7, we easily deduce the following properties:

(a) $\omega_{L,a}$ depends only on the isomorphism class of L.

(b) $\omega_{L1\otimes L2,a} = \omega_{L1,a} + \omega_{L2,a}$.

(c) If $\varphi : A' \to A$ is a homomorphism of abelian varieties and $a' \in A'$, then

$$\omega_{\varphi^* L,a'} = \omega_{L,\varphi(a')} \circ \varphi.$$

The main point is to prove that $\omega_{L,a}$ is also linear in a. The theorem of the cube in 8.6.11 gives us a unique isomorphism

$$\bigotimes_{I \subset \{1,2,3\}} \left(\sum_{i \in I} p_i \right)^* (L, \nu)^{(-1)^{|I|}} \xrightarrow{\sim} (O_{A^3}, 1) \tag{9.18}$$

of rigidified line bundles on A^3.

Now replacing (L, ν) by the metrized line bundle $\overline{L}^\nu := (L, \| \ \|_{\nu,u})$ and $(O_{A^3}, 1)$ by $\overline{O}_{A^3}$ with the trivial metric, we see that this is an isometry by Theorem 9.5.7. Then by pull-back with respect to the morphism $P \mapsto (P, a, a')$, we get an isometry

$$\tau_{a+a'}^*(\overline{L}^\nu) \otimes \tau_a^*(\overline{L}^\nu)^{-1} \otimes \tau_{a'}^*(\overline{L}^\nu)^{-1} \otimes \overline{L}^\nu \xrightarrow{\sim} \overline{O}_{A^3}^C \tag{9.19}$$

in the theorem of the square. Here, $\overline{O}_{A^3}^C$ is endowed with a constant metric $\|1\|_C := C$ for some constant $C > 0$. The constant arises from the fact that the trivial contributions from the theorem of the cube may have non-trivial metrics. By Theorem 9.5.7, a similar isometry as in (9.19) holds, provided we replace the metrics by the canonical ones on $\tau_a^*(L), \tau_{a'}^*(L)$ and $\tau_{a+a'}^*(L)$ with respect to some rigidifications. This proves immediately

(d) $\omega_{L,a+a'} = \omega_{L,a} + \omega_{L,a'}$

because the constants do not influence (d).

In order to prove that $\omega_{L,a}$ vanishes identically, it is enough to consider separately the even and odd cases. Suppose first that L is even. Then we from $[m]^*L \cong L^{\otimes m^2}$ we infer that

$$m^2 \omega_{L,a} = \omega_{[m]^*L,a} = \omega_{L,ma} \circ [m] = m \cdot \omega_{L,a} \circ [m].$$

Because the metrics in question are bounded, it is clear that $\omega_{L,a}$ is a bounded function and, letting $m \to \infty$, we conclude that $\omega_{L,a} = 0$.

If instead L is odd, then there is a unique isomorphism

$$(p_1 + p_2)^*(L, \nu) \otimes p_1^*(L, \nu)^{-1} \otimes p_2^*(L, \nu)^{-1}$$

of rigidified line bundles (Theorem 8.8.3). By pull-back with respect to the morphism $P \mapsto P + a$, we get an isometry

$$\tau_a^* \overline{L}^\nu \otimes (\overline{L}^\nu)^{-1} \xrightarrow{\sim} \overline{O}_A^C$$

as above. This proves directly that $\omega_{L,a} = 0$. □

Corollary 9.5.14. *If K satisfies the product formula and L is a line bundle on A, then the Néron–Tate height $\widehat{h}_L$ is equal to the global height associated to $(L, \| \ \|_\nu)$ (see 2.7.17), where $\| \ \|_\nu$ is the canonical M-metric on L with respect to any rigidification ν.*

Proof: By (9.18) and Proposition 2.7.18, we see that $h_{(L, \| \ \|_\nu)}$ is a quadratic function in the height class $\mathbf{h}_L$. By Theorem 9.2.8, we get the claim. □

Remark 9.5.15. It is obvious that the Néron symbol is compatible with extension of the base field. So it makes sense to work over an algebraic closure $\overline{K}$. Note however that the Néron symbol does not extend to arbitrary complete smooth varieties X over K, this happens only if D belongs to the group $B^1(X)$ of divisors algebraically equivalent to zero.

Theorem 9.5.16. *Let X be an irreducible smooth complete variety over $\overline{K}$ with base point P_0. For every line bundle L on X algebraically equivalent to 0 and rigidification ν, there is a unique locally bounded M-metric $(\| \ \|_{\nu,u})_{u\in M}$ on L satisfying the following properties:*

(a) *An isomorphism of rigidified line bundles is an isometry.*

(b) $\| \ \|_{\nu_1,u} \otimes \| \ \|_{\nu_2,u} = \| \ \|_{\nu_1 \otimes \nu_2,u} \quad (u \in M)$.

(c) *On O_X with rigidification 1, it holds $\| \ \|_{1,u} = | \ |_u \quad (u \in M)$.*

(d) *If $\varphi : X' \to X$ is a morphism of irreducible smooth complete varieties over $\overline{K}$ mapping the base point of X' to the base point of X, then*

$$\varphi^* \| \ \|_{\nu,u} = \| \ \|_{\varphi^*\nu,u}.$$

Proof: Let $A := Pic^0(X)$ and $\widehat{A} := Pic^0(A)$. By the theory of Picard varieties (Theorem 8.4.7), there is a unique morphism $\psi : X \to \widehat{A}$ with $(\mathrm{id}_A \times \psi)^* \mathbf{p}_A = \mathbf{p}_X^t$, where $\mathbf{p}_X^t$ is the pull-back of $\mathbf{p}_X$ to $A \times X$ with respect to $(a, P) \mapsto (P, a)$. For $\mathbf{c} = \mathrm{cl}(L) \in A$, we have

$$\psi^*(\mathbf{p}_A|_{\{\mathbf{c}\}\times\widehat{A}}) = \mathbf{p}_X|_{X\times\{\mathbf{c}\}} = \mathbf{c}.$$

By Theorem 9.5.7, the rigidified line bundles in the class $\mathbf{p}_A|_{\{\mathbf{c}\}\times\widehat{A}}$ have canonical metrics and we use pull-back with respect to ψ to get a well-defined locally bounded M-metric $(\| \ \|_{\nu,u})_{u\in M}$ on L. From the properties of Picard varieties developed in Section 8.4, we easily deduce (a)–(d). By applying Theorem 9.5.7 to odd line bundles, as we have done in its proof, and noting in addition that odd is the same as algebraically equivalent to 0 (Theorem 8.8.3), we get uniqueness by construction. □

Theorem 9.5.17. *Let X be an irreducible smooth complete variety over $\overline{K}$ and let $u \in M$. For $D \in B^1(X)$ and $Z \in B_0(X)$ with $\mathrm{supp}(D) \cap \mathrm{supp}(Z) = \emptyset$, there is a Néron symbol $(D, Z)_u \in \mathbb{R}$ uniquely determined by the following properties:*

(a) *$(D, Z)_u$ is bilinear in D and Z.*

(b) *If $f \in K(X)^\times$ with $\mathrm{supp}(\mathrm{div}(f)) \cap \mathrm{supp}(Z) = \emptyset$, then*

$$(\mathrm{div}(f), Z)_u = -\log|f(Z)|_u.$$

(c) *If $\varphi : X' \to X$ is a morphism of irreducible smooth complete varieties over K and $Z' \in B_0(X')$ with $\varphi(\mathrm{supp}(Z')) \cap \mathrm{supp}(D) = \emptyset$, then*

$$(\varphi^*(D), Z')_u = (D, \varphi_*(Z'))_u.$$

(d) *For $P_0 \in X \setminus \mathrm{supp}(D)$, the function $P \mapsto (D, [P] - [P_0])_u$ is locally M-bounded on $(X \setminus \mathrm{supp}(D)) \times M$.*

Proof: We choose a base point $P_0 \in X$ and we endow $O(D)$ with a rigidification ν. By Theorem 9.5.16, we get a canonical Néron divisor $\widehat{D}^\nu = (D, \| \ \|_\nu)$ and we set

$$(D, Z)_u = \lambda_{\widehat{D}^\nu}(Z, u).$$

Since $\deg(Z) = 0$, it is clear that $(D, Z)_u$ does not depend on the choices of base point and rigidification. Now we easily deduce (a)–(d) from Theorem 9.5.16.

For uniqueness, we proceed as in the proof of Theorem 9.5.11. We get a pairing $[D, Z]_u$ with properties (a),(c), and (d), which is defined for all $D \in B^1(X), Z \in B_0(X)$ and which depends only on the rational equivalence class of D. By Theorem 9.5.11, the restriction of the Néron symbol to abelian varieties must be unique, i.e. $[D, Z]_u = 0$ for abelian varieties. In general, we have seen in the proof of Theorem 9.5.16 that D is rationally equivalent to the pull-back of a divisor algebraically equivalent to 0 on the dual abelian variety of $Pic^0(X)$. We conclude from (c) and the above that $[D, Z]_u = 0$. This proves uniqueness. □

9.5.18. Let X, X' be irreducible smooth complete varieties over $\overline{K}$. We consider a divisor E on $X \times X'$ called a **correspondence**. For $P \in X$ such that $\{P\} \times X' \not\subset \mathrm{supp}(E)$ (resp. $P' \in X'$ with $X \times \{P'\} \not\subset \mathrm{supp}(E)$), we define a divisor on X' (resp. X) by

$$E(P) := (p_2)_*(E.(\{P\} \times X'))$$

and

$${}^tE(P') = (p_1)_*(E.(X \times \{P'\})),$$

where p_1, p_2 are the projections of $X \times X'$ onto the factors and where we use the proper intersection product from A.9.20. By linearity, we extend these operations to zero-dimensional cycles. Almost by definition, the resulting divisor has to be algebraically equivalent to 0 if the zero-dimensional cycle has degree 0.

Theorem 9.5.19. *Let $Z \in B_0(X), Z' \in B_0(X')$ with* $\mathrm{supp}(Z) \times \mathrm{supp}(Z')$ *disjoint from* $\mathrm{supp}(E)$. *Then the* **reciprocity law**

$$(E(Z), Z')_u = ({}^tE(Z'), Z)_u$$

holds for every $u \in M$.

Proof: We deal first with the case where X and X' are abelian varieties. Let P_j be an irreducible component of Z. For the canonical meromorphic section s_E of $O(E)$, we have

$$E(P_j) = (p_2)_* \left(\mathrm{div}(s_E|_{\{P_j\} \times X'})\right).$$

Let $(\| \ \|_{j,u})_{u \in M}$ be the canonical M-metric of $O(E(P_j))$ with respect to a rigidification. If $Z = \sum m_j[P_j]$ and $Z' = \sum m'_k[P'_k]$, then the proof of Theorem 9.5.17 shows that

$$(E(Z), Z')_u = -\sum_{j,k} m_j m'_k \log \|s_{E(P_j)}(P'_k)\|_{j,u}. \tag{9.20}$$

By Theorem 9.5.7 and Lemma 9.5.13, the pull-back of a canonical metric $\| \ \|_{E,u}$ on $O(E)$ with respect to the morphism $P' \mapsto (P_j, P')$ is equal to $\| \ \|_{j,u}$ times a positive constant C_u. The latter does not influence (9.20) because $\deg(Z') = 0$, thus we obtain

$$(E(Z), Z')_u = -\sum_{j,k} m_j m'_k \log \|s_E(P_j, P'_k)\|_{E,u}.$$

This is completely symmetric in Z and Z', proving the case of abelian varieties.

The general case can be reduced to the previous special case of abelian varieties by appealing to the theory of the Picard variety. First, we choose base points $P_0 \in X$ and $P'_0 \in X'$ such that $\{P_0\} \times \mathrm{supp}(Z')$ and $\mathrm{supp}(Z) \times \{P'_0\}$ are both disjoint from $\mathrm{supp}(E)$. The proper intersection product $E.(X \times \{P'_0\})$ (resp. $E.(\{P_0\} \times X')$) induces a well-defined divisor Y on X (resp. Y' on X'). We define the divisors $Y \times X'$ and $X \times Y'$ on $X \times X'$ by linearity in the components. Both have support disjoint from $\mathrm{supp}(Z) \times \mathrm{supp}(Z')$. The class $\mathbf{c}$ of $E - Y \times X' - X \times Y'$ in $\mathrm{Pic}(X \times X')$ is a subfamily of $\mathrm{Pic}^0(X)$ parametrized by X'. By Theorem 8.4.7, there is a unique morphism $\varphi' : X' \to A := Pic^0(X)$ with $(\mathrm{id}_X \times \varphi')^*(\mathbf{p}_X) = \mathbf{c}$.

On the other hand, there is a unique morphism $\varphi : X \to \widehat{A}$ with $(\mathrm{id}_A \times \varphi)^*(\mathbf{p}_A) = \mathbf{p}_X^t$. Therefore, we conclude that

$$(\varphi \times \varphi')^*(\mathbf{p}_A^t) = \mathbf{c}.$$

By the moving lemma in 9.5.12, there is a divisor Γ on $\widehat{A} \times A$ with class $\mathbf{p}_A^t \in \mathrm{Pic}(\widehat{A} \times A)$ and with support disjoint from $\varphi(\mathrm{supp}(Z)) \times \varphi'(\mathrm{supp}(Z'))$. Then $E_\Gamma := (\varphi \times \varphi')^*(\Gamma)$ is well defined as a divisor (see A.8.26). By construction, there is a rational function f on $X \times X'$ with

$$E = E_\Gamma + Y \times X' + X \times Y' + \mathrm{div}(f),$$

and the support of $\mathrm{div}(f)$ is disjoint from $\mathrm{supp}(Z) \times \mathrm{supp}(Z')$. It is easily seen from 9.5.18 that

$$(Y \times X')(Z) = 0, \quad (X \times Y')(Z) = 0$$

and

$$\mathrm{div}(f)(Z) = \mathrm{div}(f(Z, \cdot)).$$

From these identities and the similar ones for Z', we deduce the claim for the correspondences $Y \times X', X \times Y'$ and $\mathrm{div}(f)$. Hence we may assume that $E = E_\Gamma$. We claim that

$$\varphi'^*(\Gamma(\varphi_* Z)) = E(Z), \quad \varphi^*({}^t\Gamma(\varphi'_* Z')) = {}^tE(Z') \tag{9.21}$$

hold. By Theorem 9.5.17(c) and the case of abelian varieties considered above, this proves immediately the reciprocity law.

It remains to prove the two identities in (9.21). We begin by proving the first identity. By linearity, we may assume that $Z = [P] \in X$. The projection formula for proper intersection products (see **[125]**, Prop.2.3) yields

$$E(P) = (p_2)_*(E.(\{P\} \times \mathrm{id}_{X'})_*(X')) = (\{P\} \times \mathrm{id}_{X'})^* E$$

and a similar identity for $\Gamma(\varphi(P))$. We deduce

$$\begin{aligned} (\varphi')^*(\Gamma(\varphi(P))) &= (\varphi')^*(\{\varphi(P)\} \times \mathrm{id}_A)^*(\Gamma) \\ &= (\{P\} \times \mathrm{id}_{X'})^*(\varphi \times \varphi')^*(\Gamma) \\ &= E(P), \end{aligned}$$

proving what we want. The argument for the second identity is essentially the same. □

Corollary 9.5.20. *For an irreducible smooth projective curve C over $\overline{K}$ and $u \in M$, there is a unique pairing $(D, D')_u \in \mathbb{R}$, well defined for divisors D, D' of degree 0 with $\mathrm{supp}(D) \cap \mathrm{supp}(D') = \emptyset$, satisfying the following properties:*

(a) *It is bilinear.*

(b) *If $f \in K(C)^\times$, then $(\mathrm{div}(f), D')_u = -\log |f(D')|_u$.*

(c) *It is symmetric, i.e.* $(D, D')_u = (D', D)_u$.

(d) *For* $P_0 \notin \mathrm{supp}(D)$, *the map* $(P, u) \mapsto (D, P - P_0)_u$ *is locally* M*-bounded on* $(C \setminus \mathrm{supp}(D)) \times M$.

Proof: For existence, we use the Néron symbol. Clearly, it satisfies (a), (b), (d). The diagonal Δ is a correspondence on $C \times C$ and $\Delta(D) = D$, ${}^t\Delta(D') = D'$. By the reciprocity law in 9.5.19, we get

$$(D, D')_u = (\Delta(D), D')_u = ({}^t\Delta(D'), D)_u = (D', D)_u$$

proving (c).

To prove uniqueness, we consider again the difference $[\ ,\ ]_u$ of two such pairings. It is clear that $[D, D']_u$ depends only on the rational equivalence classes of D and D'. Let $m \in \mathbb{Z}$ and $P, P_0 \in C$. By the Riemann–Roch theorem in A.13.5, there is an effective divisor D'_+ with

$$D'_+ - g[P_0] \sim m([P] - [P_0]).$$

Because D'_+ is an effective divisor of degree g, property (d) implies that

$$[D, m([P] - [P_0])]_u = [D, D'_+ - g[P_0]]_u$$

is a bounded function of (P, m). In fact, we have to use the moving lemma as in the proof of Theorem 9.5.11 to get boundedness on the whole C. Using (a) and letting $m \to \infty$, we get $[D, [P] - [P_0]]_u = 0$ proving the claim. $\square$

Example 9.5.21. We consider the special case $K = \mathbb{C}$ with the usual absolute value $|\ |_u$. Let L be a line bundle on the complex abelian variety A. We claim that a metric $\|\ \|$ on L is a canonical metric with respect to a suitable rigidification if and only if $\|\ \|$ is a C^∞-metric with harmonic first Chern form. For details about the differential geometric tools, we refer to **[130]**.

To prove the claim, note first that the first Chern form determines the metric up to multiplication with a positive constant. For a divisor D on A, there is a unique harmonic representative ω_D of the first Chern class of $O(D)$ and we choose a C^∞-metric $\|\ \|'$ on $O(D)$ with first Chern form ω_D (**[130]**, p.148). If $Z = \sum_{j=1}^r m_j[P_j]$ is a zero-dimensional cycle of degree 0 with $\mathrm{supp}(D) \cap \mathrm{supp}(Z) = \emptyset$, then we set

$$(D, Z)_u := -\sum_{j=1}^r m_j \log \|s_D(P_j)\|'.$$

Using $\deg(Z) = 0$, it is clear that the pairing does not depend on the choice of $\|\ \|'$. Obviously, the pairing fulfills properties (a)–(c) of Theorem 9.5.11. Using the fact that the harmonic forms on a complex torus are the same as the translation invariant forms, we immediately deduce (d) as well. Finally, (e) follows from smoothness of the function $P \mapsto (D, [P] - [P_0])_u$ on $A \setminus \mathrm{supp}(D)$. We conclude that $(\ ,\ )_u$ is the Néron symbol. Let $\|\ \|$ be the canonical metric of $O(D)$ with respect to a rigidification. From the proof of Theorem 9.5.11, it is clear that

$$(D, [P] - [0])_u = \log \|s_D(0)\| - \log \|s_D(P)\|$$

for every $P \notin \mathrm{supp}(D)$. This shows easily that $\|\ \| = \|\ \|'$ up to multiplication with a positive constant.

Example 9.5.22. Now let K be a field with a complete discrete absolute value $|\ |_v$ normalized by $\log |K^\times|_v = \mathbb{Z}$. It has a unique extension to an absolute value $|\ |_u$ of $\overline{K}$ (see Proposition 1.2.7). Let A be an abelian variety over K with good reduction in v, i.e. there is a proper smooth scheme $\overline{A}$ over the discrete valuation ring R_v with $\overline{A}_K = A$ called the **Néron model**. Note also that $\overline{A}$ is a group scheme unique up to isomorphism (cf. 10.3.9).

We claim that a metric $\|\ \|$ on a line bundle L of A is a canonical metric with respect to some rigidification if and only if there is a line bundle $\mathcal{L}$ on $\overline{A}$ with $L = \mathcal{L}_K$ and $\|\ \| = \|\ \|_{\mathcal{L}}$ (Example 2.7.20).

For the proof, we note first that the restriction map $\mathcal{L} \mapsto \mathcal{L}_K$ gives an isomorphism from $\mathrm{Pic}(\overline{A})$ onto $\mathrm{Pic}(A)$. The inverse is induced by the map $D \mapsto \overline{D}$ of divisors taking Zariski closures of components. To see this, use that the special fibre of $\overline{A}$ over the residue field is irreducible and therefore every divisor supported in the special fibre is rationally equivalent to 0.

For $L = \mathcal{L}_K$ with rigidification ν, we set

$$\|\ \|'_{\nu,u} := \|\nu\|_{\mathcal{L}}^{-1} \cdot \|\ \|_{\mathcal{L}}.$$

It is easily checked that our metric $\|\ \|'_{\nu,u}$ on L satisfies (a)–(d) of Theorem 9.5.7, where we allow in (d) only endomorphisms of A. Because it is also bounded (Example 2.7.20), Remark 9.5.8 yields that the metric $\|\ \|'_{\nu,u}$ is the canonical metric $\|\ \|_{\nu,u}$ from Theorem 9.5.7. This proves our claim.

Remark 9.5.23. For an archimedean place u, we may always reduce to Example 9.5.21 by base change to $\mathbb{C}$. Hence canonical metrics $\|\ \|_u$ are always C^∞.

For a non-archimedean place $u \in M$, canonical metrics $\|\ \|_u$ are always continuous with respect to the u-topology and satisfy $\|L(\overline{K})\|_u = |\overline{K}|_u$. This follows from Tate's limit argument in the proof of Theorem 9.5.4 and Remark 2.7.6. In the case of good reduction in a discrete valuation, Example 9.5.22 shows that even $\|L(K)\|_u = |K|_u$ holds. However, the following example will show that this is not in general the case. Note also that $|\overline{K}|_u = \bigcup_{n\in\mathbb{N}\setminus\{0\}} |K|_u^{1/n}$ (see Lemma 11.5.2).

Example 9.5.24. Every complex elliptic curve is biholomorphic to a complex torus $\mathbb{C}/(\mathbb{Z}+\mathbb{Z}\tau)$, $\Im\tau > 0$. The map $\zeta = \exp(2\pi i z)$ gives an analytic group isomorphism to the **Tate uniformization** $\mathbb{C}^\times/q^{\mathbb{Z}}, q := e^{2\pi i\tau}$. As in 8.6.13, we consider the theta function

$$\theta(z,\tau) = \sum_{n=-\infty}^{\infty} e^{\pi i n^2\tau + 2\pi i n z}.$$

It transforms into

$$\theta(\zeta, q) = \sum_{n=-\infty}^{\infty} \tilde{q}^{n^2}\zeta^n, \tag{9.22}$$

where $\tilde{q} = e^{\pi i\tau}$. As an infinite product, it has the form

$$\theta(\zeta,q) = \prod_{n=1}^{\infty}(1-\tilde{q}^{2n}) \cdot \prod_{n=1}^{\infty}(1+\tilde{q}^{2n-1}\zeta) \cdot \prod_{n=1}^{\infty}(1+\tilde{q}^{2n-1}\zeta^{-1}),$$

which shows that $\theta(\cdot, q)$ has only simple zeros and they are precisely in $\zeta = -\tilde{q} \cdot q^{\mathbb{Z}}$ (for details, see K. Chandrasekharan [**62**], Ch.V, Th.6). Let $p : \mathbb{C}^\times \to \mathbb{C}^\times / q^{\mathbb{Z}}$ be the quotient map and let P be the 2-torsion point $p(-\tilde{q})$. Then $p^* O([P])$ is trivial and may be identified with $\mathbb{C}^\times \times \mathbb{C}$ such that the canonical global section s_P of $O([P])$ pulls back to $\zeta \mapsto (\zeta, \theta(\zeta, q))$ (use 8.6.13).

Conversely, $O([P])$ may be obtained as the quotient of the trivial bundle by a $\mathbb{Z}$-action

$$\mathbb{Z} \times (\mathbb{C}^\times \times \mathbb{C}) \longrightarrow (\mathbb{C}^\times \times \mathbb{C}), \quad (k; \zeta, v) \mapsto (q^k \zeta, e_k(\zeta, q) \cdot v).$$

To determine the cocycle $e_k(\zeta, q)$, note that

$$\theta(q^k \zeta, q) = \sum_{n=-\infty}^{\infty} \tilde{q}^{n^2+2kn} \zeta^n = \tilde{q}^{-k^2} \sum_{n=-\infty}^{\infty} \tilde{q}^{(n+k)^2} \zeta^n = \tilde{q}^{-k^2} \zeta^{-k} \theta(\zeta, q)$$

and hence $e_k(\zeta, q) = \tilde{q}^{-k^2} \zeta^{-k}$. Note that $O([P])$ gets rigidified by choosing ν corresponding to 1 in the fibre over 0 of the trivial bundle. For $m \in \mathbb{Z}$, $[m]^* O([P])$ is given by the cocycle

$$e_{km}(\zeta^m, q) = \tilde{q}^{-k^2 m^2} \zeta^{-km^2} = e_k(\zeta, q)^{m^2}.$$

The right-hand side is the cocycle of $O(m^2[P])$, hence $O([P])$ is even and we have also checked $[m]^* O([P]) \cong O(m^2[P]$. By our rigidifications, the isomorphism is uniquely determined. To determine the canonical metric $\| \ \|_\nu$, we compute the positive function $\mu := \|1\|_\nu$. In fact, 1 is a section of $p^* O([P]) = \mathbb{C}^\times \times \mathbb{C}$, but, locally, it may be also viewed as a section of $O([P])$ determining the metric completely. Compatibility of the fibres over ζ and $q^k \zeta$ leads to

$$\mu(q^k \zeta) = |e_{-k}(q^k \zeta, q)| \cdot \mu(\zeta). \tag{9.23}$$

On the other hand, $[m]^* \| \ \|_\nu = \| \ \|_\nu^{\otimes m^2}$ gives

$$\mu(\zeta^m) = \mu(\zeta)^{m^2}.$$

To find the canonical metric, we may start with any metric, i.e. with a μ_0 satisfying (9.23). For $|q| \leq |\zeta| \leq 1$, we choose $\mu_0(\zeta) = |\zeta|^{1/2}$ and we extend μ_0 by (9.23) to $\mathbb{C}^\times$ leading to a continuous function. Then the proof of Theorem 9.5.4 gives

$$\mu(\zeta) = \lim_{m \to \infty} \mu_0(\zeta^m)^{1/m^2}.$$

For $m \in \mathbb{Z}$, we choose $k(m) \in \mathbb{Z}$ with $\zeta_m := \zeta^m / q^{k(m)}$ satisfying $|q| < |\zeta_m| \leq 1$. Then (9.23) leads to

$$\mu(\zeta) = \lim_{m \to \infty} |e_{-k(m)}(\zeta^m, q)|^{1/m^2} = \lim_{m \to \infty} |q|^{-\frac{k(m)^2}{2m^2}} |\zeta|^{\frac{k(m)}{m}}.$$

We have $k(m) = \log |\zeta| / \log |q| + O(1)$, thus proving

$$\mu(\zeta) = |q|^{-\frac{1}{2}(\log|\zeta|/\log|q|)^2} \cdot |\zeta|^{\log|\zeta|/\log|q|} = |\zeta|^{\frac{\log|\zeta|}{2\log|q|}}. \tag{9.24}$$

It is easy to check that μ satisfies (9.23) and therefore (9.24) holds for all $\zeta \in \mathbb{C}^\times$.

We owe to Tate the counterpart of these considerations for the non-archimedean case. Let K be a field endowed with a complete (discrete) absolute value $| \ |$. For every $|q| < 1$, the analytic torus $\overline{K}^\times/q^{\mathbb{Z}}$ is isomorphic as an analytic group to **Tate's elliptic curve** E_q given by

$$y^2 + xy = x^3 + a_4 x + a_6,$$

where a_4, a_6 are convergent power series in q given by

$$a_4 = -\sum_{n=1}^{\infty} n^3 q^n/(1-q^n), \quad a_6 = -\frac{1}{12}\sum_{n=1}^{\infty}(7n^5 + 5n^3)q^n/(1-q^n).$$

The isomorphism $\overline{K}^\times/q^{\mathbb{Z}} \to E_q$ is given by

$$x(\zeta, q) = \sum_{n=-\infty}^{\infty} q^n\zeta/(1-q^n\zeta)^2 - 2\sum_{n=1}^{\infty} nq^n/(1-q^n)$$

$$y(\zeta, q) = \sum_{n=-\infty}^{\infty} q^{2n}\zeta^2/(1-q^n\zeta)^3 + \sum_{n=1}^{\infty} nq^n/(1-q^n)$$

and it is also an isomorphism of $\mathrm{Gal}(\overline{K}/K)$-modules. Note that the reduction of E_q is given by $y^2 + xy = x^3$, a case called **split multiplicative reduction**. In fact, exactly the elliptic curves with split multiplicative reduction have such a Tate uniformization. For details of all these facts, we refer to P. Roquette [**244**] or J.H. Silverman [**286**], Ch.5, Th.3.1, Th.5.3). For a fixed root $\tilde{q} = \sqrt{q}$, we define θ by (9.22) on page 312 and all considerations above remain valid. In particular, the canonical metric $\| \ \|_v$ is determined by (9.24).

For example, we assume that $| \ |$ is normalized by $\log|K^\times| = \mathbb{Z}$ and that $q = \pi^{2k}$ for a local parameter π. For $\tilde{q} := \pi^k$, the 2-torsion point $P = p(-\tilde{q})$ is K-rational and hence $L = O([P])$ is defined over K. Let $Q := p(\pi)$ and let L_Q be the fibre over Q. Then (9.24) leads to

$$\log \|L_Q(K)\|_v = \frac{1}{4k} + \mathbb{Z}.$$

9.6. Hilbert's irreducibility theorem

A famous theorem of Hilbert asserts that, if $f(x,y) \in K[x,y]$ is an irreducible polynomial over a number field K, then for infinitely many $\xi \in K$ the polynomial $f(x,\xi)$, obtained by specializing y to ξ, is also irreducible over K. Such a field is called **Hilbertian**.

In this section we give a proof of this theorem using the theory of heights developed here and in Chapter 2.

9.6.1. For the reader's convenience, we recall here from Chapter 2 some definitions and simple facts on local heights.

Let D be a Cartier divisor on a projective variety X defined over a number field K and let $\mathcal{D}$ be a presentation of D as defined in 2.2.1

$$\mathcal{D} = (s_D; L_+, \mathbf{s}; L_-, \mathbf{t}),$$

where $O(D) = L_+ \otimes L_-^{-1}$, L_+ and L_- are base-point free, $\mathbf{s}$ and $\mathbf{t}$ are generating global sections of L_+ and L_- respectively, and s_D is the meromorphic section of $O(D)$ defining the Cartier divisor D.

As always, we normalize absolute values on number fields as in 1.3.6. This normalization is not possible for absolute values on an algebraic closure $\overline{K}$ of the number field K. Let M be the set of places on $\overline{K}$. For $u \in M$, we normalize $|\ |_u$ so that if its restriction to K is equivalent to $|\ |_v$, $v \in M_K$, and if p is the restriction of v to $M_{\mathbb{Q}}$, then $|\ |_u$ restricted to $\mathbb{Q}$ is $|\ |_p$. In what follows, we reserve the notation u to indicate absolute values $|\ |_u$ with $u \in M$ and v, w to indicate absolute values in finite subextensions of $\overline{K}/K$.

For $u \in M$, we have associated local heights (see Section 2.2) on X given by

$$\lambda_{\mathcal{D}}(P,u) = \log \max_k \min_l \left| \frac{s_k}{s_D t_l}(P) \right|_u .$$

If F is a finite subextension of $\overline{K}/K$ with $F \supset K(P)$ and $|\ |_u$ restricts on F to the equivalent $|\ |_w$ with $w \in M_F$, then

$$\lambda_{\mathcal{D}}(P,w) = \frac{[F_w : \mathbb{Q}_p]}{[F : \mathbb{Q}]} \lambda_{\mathcal{D}}(P,u)$$

and for $v \in M_K$ the sum

$$\sum_{w|v,\ w \in M_F} \lambda_{\mathcal{D}}(P,w)$$

does not depend of the choice of F, as in Lemma 1.3.7.

For notational simplicity, in what follows we abbreviate

$$\varepsilon'_w = \frac{[F_w : \mathbb{Q}_p]}{[F : \mathbb{Q}]} .$$

We have seen in Theorem 2.2.11 and Remark 2.2.13 (see also Theorem 2.7.14 for such a result in a more general context) that, if $\lambda_{\mathcal{D}}$ and $\lambda_{\mathcal{D}'}$ are two local heights attached to different presentations of the same Cartier divisor D, then for $u|v$ with $v \in M_K$ we have

$$|\lambda_{\mathcal{D}}(\cdot,u) - \lambda_{\mathcal{D}'}(\cdot,u)| \le \gamma_v$$

for some constant $\gamma_v = \max_m \log^+ |p_m|_v$ for finitely many $p_m \in K$, depending only on the geometric data of the presentations. In particular, $\gamma_v = 0$ for all but finitely many $v \in M_K$.

For F a finite subextension of $\overline{K}/K$ such that $F \supset K(P)$, the sum

$$h_{\mathcal{D}}(P) := \sum_{w \in M_F} \lambda_{\mathcal{D}}(P,w)$$

is independent of the choice of F and is called the global Weil height attached to $\mathcal{D}$. By Theorem 2.3.6, it is uniquely defined as a quasi-function independent of the presentation, i.e. up to a uniformly bounded quantity.

We state now the main result of this section.

Theorem 9.6.2. *Let C be a smooth irreducible projective curve defined over a number field K and let $f : C \to \mathbb{P}^1$ be a surjective rational function defined over K. Suppose $Q \in f^{-1}(\infty)$ is a pole of f and let $\mathcal{Q}$ be a presentation of $[Q]$, with a corresponding family of local heights $\lambda_{\mathcal{Q}}(\cdot,u)$, $u \in M$.*

Then there is a family of real numbers $(\Delta_v)_{v\in M_K}$ with $\Delta_v \geq 0$ and $\Delta_v = 0$ for all but finitely many $v \in M_K$, depending only on C, f and the presentation $\mathcal{Q}$, with the following property.

For $P \in C(\overline{K}) \setminus f^{-1}(\infty)$ and any finite subextension F of $\overline{K}/K$ with $F \supset K(P,Q)$, we have

$$\sum_{v\in M_K} \sum_{\substack{w\in M_F,\ w|v \\ \lambda_{\mathcal{Q}}(P,w)>\varepsilon'_w \Delta_v}} \log^+ |f(P)|_w = -\frac{\operatorname{ord}_Q(f)}{\deg(f)} h(f(P)) + O(\sqrt{h(f(P))} + 1).$$

Remark 9.6.3. In intuitive terms, if $|f(P)|_v > e^{\Delta_v}$, then P is "close" to a pole Q of f. The meaning of the theorem is that, as v varies, each pole of f is approached by P with a probability proportional to the order of the pole.

To prove this theorem, we need first the following result:

Lemma 9.6.4. *Let λ, λ' be local heights relative to divisors D, D' with disjoint support, say given by presentations of D and D'. Then there is a family of real numbers $c_v \geq 0$ and with $c_v = 0$ for all but finitely many $v \in M_K$, depending only on the presentations, such that for $u|v \in M_K$ the following bound holds*

$$\min\left(|\lambda(P,u)|, |\lambda'(P,u)|\right) \leq c_v.$$

The lemma may be easily generalized to a complete variety and to a family of local heights relative to Néron divisors assuming that the intersection of all supports is empty.

Proof of lemma: Let us consider the open subsets $U := C \setminus \operatorname{supp}(D)$ and $U' := C \setminus \operatorname{supp}(D')$. Since $U \cup U' = C$ and C is M-bounded (see Proposition 2.6.17), there are M-bounded families $(E_u)_{u\in M}$ and $(E'_u)_{u\in M}$ of subsets of U and U', respectively, with $C(\overline{K}) = E_u \cup E'_u$ (see 2.6.3, 2.6.14). We note that s_D is a nowhere vanishing section over U. Since $\lambda(P,u) = -\log \|s_D(P)\|_u$ for a locally bounded metric on $O(D)$ (Proposition 2.7.11), we conclude that

$$\sup_{u\in M, u|v} \sup_{P\in E_u} |\lambda(P,u)|$$

is finite for all $v \in M_K$ and 0 up to finitely many. Working with a presentation, this may also be deduced directly from another subdivision of U and E_u using the remarks 2.6.3, 2.6.14. Similarly, λ' is locally M-bounded on $(E'_u)_{u\in M}$ and the claim follows. □

Proof of Theorem 9.6.2: By finite base change, we may assume that all poles of f are K-rational and that C is geometrically irreducible. Indeed, there is a finite Galois extension E/K such that the irreducible components of $C_{\overline{K}}$ are given by conjugates $(B^\sigma)_{\sigma\in R}$ for a geometrically irreducible curve B over E with $Q \in B(E)$ (see Example A.4.15). Then the theorem for $Q \in B$ proves the theorem for $Q \in C$ as λ_Q is M-bounded on the conjugates $B^\sigma \neq B$ (by Lemma 9.6.4).

Since the sum and pull-back of local heights remain a local height relative to suitable presentations (see 2.2.4 and 2.2.6), we see that

$$\log^+ |f(P)|_u + \operatorname{ord}_Q(f)\lambda_{\mathcal{Q}}(P,u)$$

is a local height relative to the effective divisor

$$D := f^*[\infty] + \mathrm{ord}_Q(f)[Q] = - \sum_{Q' \in f^{-1}(\infty) \setminus \{Q\}} \mathrm{ord}_{Q'}(f)[Q'].$$

The preceding lemma applied to D and $D' := -\mathrm{ord}_Q(f)[Q]$ shows that for $u|v \in M_K$ we also have

$$\min\left(-\mathrm{ord}_Q(f)\lambda_Q(P,u), \log^+|f(P)|_u + \mathrm{ord}_Q(f)\lambda_Q(P,u)\right) \leq c_v.$$

For $\Delta_v := -c_v/\mathrm{ord}_Q(f)$ and for a finite subextension F of $\overline{K}/K$ with $F \supset K(P)$, this leads to

$$\begin{aligned} \sum_{v \in M_K} \sum_{\substack{w \in M_F,\ w|v \\ \lambda_Q(P,w) > \varepsilon'_w \Delta_v}} & \log^+|f(P)|_w \\ &= -\mathrm{ord}_Q(f) \sum_{v \in M_K} \sum_{\substack{w \in M_F,\ w|v \\ \lambda_Q(P,w) > \varepsilon'_w \Delta_v}} \lambda_Q(P,w) + O(1) \qquad (9.25) \\ &= -\mathrm{ord}_Q(f)\, h_Q(P) + O(1). \end{aligned}$$

Since $f^*[\infty] - \deg(f)[Q] \in \mathrm{Pic}^0(C)$ (see A.9.40 and Corollary 8.4.10), Corollary 9.3.10 yields

$$h_Q(P) = \frac{1}{\deg(f)} h(f(P)) + O\left(\sqrt{h(f(P))} + 1\right).$$

If we insert this in (9.25), we get the claim. □

9.6.5. In 1887, C. Runge ([**255**], p.432) proved that, if $G(x,y)$ is an irreducible polynomial in $\mathbb{Z}[x,y]$ of degree d and if the homogeneous part $G_d(x,y)$ of degree d of G is not proportional to a power of an irreducible polynomial in $\mathbb{Z}[x,y]$, then $G(x,y)$ has only finitely many zeros $(x,y) \in \mathbb{Z} \times \mathbb{Q}$. Runge used this theorem to give a more precise criterion for G to have only finitely many integer zeros (see [**255**], p.434). Runge's method depends on the construction of rational approximations to algebraic functions of one variable and is quite explicit and applicable in practice.

Here we give an extension of **Runge's theorem**, as a consequence of Theorem 9.6.2 and Lemma 9.6.4.

Theorem 9.6.6. *Let C be a smooth irreducible projective curve defined over a number field K. Let $f : C \to \mathbb{P}^1$ be a surjective rational function on C defined over K.*

Let $P \in C$ such that $f(P) \in O_{K(f(P)),S}$ for a finite subset $S \subset M_{K(f(P))}$ containing all the archimedean places and satisfying the basic condition

$$[K(P):K] < |S|^{-1} \frac{\deg(f)}{\max_Q |\mathrm{ord}_Q(f)|\,[K(Q):K]}, \qquad (9.26)$$

where the maximum is taken over all poles of f. Then the set of such points P is finite and effectively computable.

Remark 9.6.7. It is a simple exercise, which we leave to the reader, to deduce Runge's theorem from Theorem 9.6.6 by taking $f = x$, $K = \mathbb{Q}$ and $S = \{\infty\}$.

Proof of theorem: For $u \in M$, $\log^+ |f(P)|_u$ is a local height relative to

$$f^*[\infty] = - \sum_{Q \in f^{-1}(\infty)} \mathrm{ord}_Q(f)[Q].$$

By Theorem 2.2.11 and Remark 2.2.13, there are constants $c_v \geq 0$, with $c_v = 0$ for all but finitely many $v \in M_K$ such that

$$\left| \log^+ |f(P)|_u + \sum_{Q \in f^{-1}(\infty)} \mathrm{ord}_Q(f) \lambda_{\mathcal{Q}}(P, u) \right| \leq c_v \tag{9.27}$$

for $u|v$. Since $\eta := f(P) \in O_{K(\eta),S}$, there is $v_\eta \in S$ with

$$\log^+ |\eta|_{v_\eta} \geq |S|^{-1} h(\eta). \tag{9.28}$$

Let F be as usual a sufficiently large finite Galois extension of $\overline{K}/K$, which is a field of definition for all points in $f^{-1}(\eta)$ and $f^{-1}(\infty)$. Fix once for all $w \in M_F$ and $u \in M$ with $u|w|v_\eta|v \in M_K$. By (9.27) and (9.28), we get

$$\left| \sum_{Q \in f^{-1}(\infty)} \mathrm{ord}_Q(f) \lambda_{\mathcal{Q}}(P, u) \right| \geq (\varepsilon'_{v_\eta})^{-1} |S|^{-1} h(\eta) - c_v.$$

Let $(\Delta_v)_{v \in M_K}$ be a family as in Theorem 9.6.2 working for all presentations $\mathcal{Q}$. We see that there is $Q \in f^{-1}(\infty)$ such that

$$\lambda_{\mathcal{Q}}(P, w) > \varepsilon'_w \Delta_v \tag{9.29}$$

as soon as

$$h(\eta) > \varepsilon'_{v_\eta} |S| \left(c_v + \deg(f) \Delta_v \right).$$

The right-hand side is bounded by an effective constant independent of the choice of P, S, and v_η (use (9.26)), hence (9.29) holds as soon $h(\eta)$ is sufficiently large, which we shall suppose henceforth.

Now let σ be an automorphism of F/K. We write w', Q' for $w \circ \sigma$ and σQ, and note that we may choose the presentations of the poles of f and associated local heights so that they are compatible by the action of σ. Since the action of $\mathrm{Gal}(F/K)$ on the places of F extending v is transitive (see Corollary 1.3.5), we have $\varepsilon'_w = \varepsilon'_{w'}$ and

$$\lambda_{\mathcal{Q}'}(\sigma P, w') \lambda_{\mathcal{Q}}(P, w) > \varepsilon'_{w'} \Delta_v.$$

It is now clear that, if $\{P_1, \ldots, P_r\}$, $\{Q_1, \ldots, Q_s\}$ are a full set of conjugates of P and Q over K, then

$$\sum_{i=1}^{r} \sum_{j=1}^{s} \sum_{\substack{w' \in M_F,\ w'|v \\ \lambda_{\mathcal{Q}_j}(P_i, w') > \varepsilon'_{w'} \Delta_v}} \log^+ |\eta|_{w'}$$

$$\geq \sum_{w' \in M_F,\ w'|v} \log^+ |\eta|_{w'} \geq \log^+ |\eta|_{v_\eta},$$

where the last step comes from Lemma 1.3.7. Noting that $\mathrm{ord}_{Q_j}(f) = \mathrm{ord}_Q(f)$ for every j and applying Theorem 9.6.2, we conclude with (9.28) that

$$|S|^{-1} h(\eta) \leq rs \frac{|\mathrm{ord}_Q(f)|}{\deg(f)} h(\eta) + O(\sqrt{h(\eta)} + 1).$$

By Galois theory, $r = [K(P) : K]$ and $s = [K(Q) : K]$, but since we have no information about which pole Q is involved here we must take the maximum of $s\,|\mathrm{ord}_Q(f)|$ over all poles of f. We conclude that $h(\eta)$ is bounded by an effective constant and the theorem follows from Northcott's theorem in 1.6.8. □

9.6.8. Theorem 9.6.6 can be applied directly to obtain irreducibility results about the fibres of f. For example, if $K = \mathbb{Q}$, $S = \{\infty\}$, and all poles of f are simple and defined over $\mathbb{Q}$, we get $[\mathbb{Q}(P) : \mathbb{Q}] = \deg(f)$ for $P \in f^{-1}(n)$, for all but finitely many integers $n \in \mathbb{Z}$. Hence there are $\deg(f)$ conjugates of P proving the irreducibility of the fibre.

Note that some condition about K and the poles of f must be imposed if we want to get such a finiteness result. The example of the curve with affine equation $y = x^2$ and $f = y$ shows that that the finiteness result does not hold if we omit the condition of simplicity of the poles at ∞. The example of the curve with affine equation $x^2 - 2y^2 = 1$ and $f = y$, with the associated Pell equation, shows that the finiteness result does not hold even if f has simple poles at ∞ defined over an extension of $\mathbb{Q}$.

The reader will perceive the connexion of such a finiteness theorem with Siegel's finiteness theorem on integral points on curves as in Theorem 7.3.9, and indeed a proof of the finiteness result can be obtained by using the full force of Siegel's theorem. However, Siegel's theorem is ineffective. It is therefore of some interest that Theorem 9.6.6 is strong enough to prove a general effective version of **Hilbert's irreducibility theorem.**

Theorem 9.6.9. *Let C be a smooth irreducible projective curve defined over a number field K and let $f : C \to \mathbb{P}^1$ be a surjective rational function on C, also defined over K. Then for all $n \in \mathbb{N}$ except for a set of natural density 0, the divisor $f^*[n]$ is a prime divisor over K.*

Remark 9.6.10. Recall that a subset $M \subset \mathbb{N}$ is called of **natural density** ρ if the following limit exists and

$$\rho = \lim_{x \to \infty} \frac{|\{n \in M \mid n \leq x\}|}{|\{n \in \mathbb{N} \mid n \leq x\}|}.$$

As will be clear from the proof, this theorem is effective in the following sense. There are effectively computable quantities r, $\kappa(B, r)$, an effectively computable polynomial $P(x_1, \ldots, x_{r-1}) \in K[x_1, \ldots, x_{r-1}] \setminus \{0\}$, depending on the set of ramification points of f, such that, if $1 \leq b_i \leq B$ with $b_i \in \mathbb{N}$, $n \geq \kappa(B, r)$, and $P(b_1, \ldots, b_{r-1}) \neq 0$, then at least one of the divisors $f^*[n]$, $f^*[n + b_i]$, $i = 1, \ldots, r-1$, is a prime divisor over K. In particular, the least $n \in \mathbb{N}$ such that $f^*[n]$ is a prime divisor over K is effectively computable.

Dèbes and Zannier [**88**] gave a proof of Hilbert's irreducibility theorem relying on G-functions. They used a clever trick by considering generic fractional linear transforms to omit ramification. Here, we put this in the context of algebraic geometry by working on an auxiliary curve to which we can apply directly Theorem 9.6.6. Before we come to the proof of Theorem 9.6.9, we need a couple of lemmas.

Lemma 9.6.11. *For $i = 0, \ldots, N$, let C_i be a smooth geometrically irreducible curve over a field F of characteristic 0 and let $f_i : C_i \to \mathbb{P}^1_F$ be a surjective rational function* *of degree* d_i. *We consider the fibre product* $g : \mathcal{C} \to \mathbb{P}^1_F \times \mathbb{A}^N_F$ *of the morphisms*

$$g_i : C_i \times \mathbb{A}^N_F \to \mathbb{P}^1_F \times \mathbb{A}^N_F, \quad (x_i, \mathbf{t}) \mapsto (f_i(x_i) - t_i, \mathbf{t}) \quad (i = 0, \ldots, N),$$

where we set $t_0 := 0$. *Let* R_i *be the set of points in* $\mathbb{P}^1_F \setminus \{\infty\}$ *over which* f_i *ramifies. Then the following properties hold:*

(a) g *is a flat finite morphism of degree* $d_0 \cdots d_N$;

(b) $\mathcal{C}$ *is an irreducible variety over* F;

(c) *All fibres* $\mathcal{C}_\mathbf{t}$ *of* $\mathcal{C}$ *over* $\mathbb{A}^N_F$ *with respect to* $p_2 \circ g$ *are geometrically connected projective curves;*

(d) *If* $\mathbf{t} \in \mathbb{A}^N_F$ *with* $t_i - t_j \notin R_i - R_j$ *for all* $i \neq j$ *in* $\{0, \ldots, N\}$ *(such* $\mathbf{t}$ *form an open dense subset), then a singular point* $(x_0, \ldots, x_N) \in \mathcal{C}_\mathbf{t}$ *has two components* x_i, x_j, *which are ramified over* ∞ *with respect to* f_i *and* f_j, *respectively.*

Proof of lemma: A surjective morphism of an irreducible variety onto a smooth curve is always flat ([**148**], Prop.III.9.7). So f_i is flat and g_i is obtained by base change and composition with an isomorphism. Hence every g_i and also the fibre product g are flat ([**148**], Prop.III.9.2), at least if we understand the fibre product in the sense of schemes. We will show below that $\mathcal{C}$ is indeed a variety.

Every g_i is finite (same reason as for flatness) and hence the fibre product g is finite. Thus the fibres are finite morphisms $g_\mathbf{t} : \mathcal{C}_\mathbf{t} \to \mathbb{P}^1_{F(\mathbf{t})}$ for every $\mathbf{t} \in \mathbb{A}^N_F$, proving that $\mathcal{C}_\mathbf{t}$ is a projective curve (see A.12.7).

Let $\mathbf{t}$ be a point as in (d). Since every R_i is finite, it is clear that such $\mathbf{t}$s form an open dense subset of $\mathbb{A}^N_F$. Let $\mathbf{x} = (x_0, \ldots, x_N) \in \mathcal{C}_\mathbf{t}$ with at most one x_i ramified over ∞. Even if the x_i lie over finite points, our assumption on $\mathbf{t}$ implies that at most one x_i is ramified.

In any case, the fibre product of the morphisms g_j omitting the ramified g_i is smooth at the point $(x_0, \ldots, x_{i-1}, x_{i+1}, \ldots, x_N)$, whence $\mathbf{x}$ is a smooth point of the map $\mathcal{C} \to C_i \times \mathbb{A}^N_F$ obtained by base change ([**148**], Prop.III.10.1(b)). Since C_i is a smooth curve, we conclude that the second projection p_2 of $C_i \times \mathbb{A}^N_F$ is a smooth morphism (again by base change) and hence $\mathbf{x}$ is a smooth point of the composition $\mathcal{C} \to \mathbb{A}^N_F$ ([**148**], Prop.III.10.1(c)). In particular, it is a smooth point of $\mathcal{C}_\mathbf{t}$, proving (d).

Since $p_2 \circ g$ is open as a flat morphism (see A.12.13), such points are dense in $\mathcal{C}$ (for varying $\mathbf{t}$) and hence the fibre product $\mathcal{C}$ (in the sense of schemes) has an open dense part, which is reduced. We conclude that $\mathcal{C}$ is a variety.

Let $\mathbf{t} \in \mathbb{A}^N_F$ be still as in (d). By flatness of g, every irreducible component of $\mathcal{C}_\mathbf{t}$ covers $\mathbb{P}^1_{F(\mathbf{t})}$. In order to show that $\mathcal{C}_\mathbf{t}$ is geometrically irreducible, it is enough to show that $\mathcal{C}'_\mathbf{t} := g_\mathbf{t}^{-1}(\mathbb{P}^1_F \setminus \{\infty\})$ is geometrically irreducible. By (d), we know that the latter is a smooth variety and so it is enough to show that $\mathcal{C}'_\mathbf{t}$ is geometrically connected (see A.7.14).

To prove this, we may argue complex analytically. If F is a number field (as in our application), this is obvious by using an embedding of F into $\mathbb{C}$ (see Section A.14). In general, we use the Lefschetz principle: $C'_{\mathbf{t}}$ is defined over a finitely generated subfield of F, which may be embedded into $\mathbb{C}$.

Let $R_{\mathbf{t}} := \bigcup_{i=0}^{N} (R_i - t_i)$. Then the same arguments as in the proof of (d) show that $g_{\mathbf{t}} : (C'_{\mathbf{t}})_{\mathrm{an}} \to \mathbb{P}^1_{\mathrm{an}} \setminus \{\infty\}$ is a finite ramified covering as in 12.3.8, and outside of $g_{\mathbf{t}}^{-1}(R_{\mathbf{t}})$ we get a topological covering (see 12.3.9) of degree $d_0 \cdots d_N$ (every fibre has exactly $d_0 \cdots d_N$ points by definition of the fibre product). We choose $\mathbf{x} = (x_0, \ldots, x_N) \in (C'_{\mathbf{t}})_{\mathrm{an}}$ with $y = g(\mathbf{x}) \notin R_{\mathbf{t}}$. To prove that $(C'_{\mathbf{t}})_{\mathrm{an}}$ is connected, it is enough to show that there is a path connecting $\mathbf{x}$ with any other fibre point over y. We may just change one entry at the time and, for notational simplicity, we change x_0 to $x'_0 \in g_0^{-1}(y)$.

The fundamental group $\pi_1(\mathbb{P}^1_{\mathrm{an}} \setminus (R_0 \cup \{\infty\}), y)$ is generated by the loops $(\sigma_z)_{z \in R_0}$, where σ_z is the loop starting in the base point y, then passing to a small neighbourhood of z, turning in a small positive circle around z and returning the same way back to y. This is a consequence of van Kampen's theorem and is proved in the case $|R_0| = 2$ in 12.6.1; the general case is by induction. We may assume that σ_z omits $R_{\mathbf{t}}$ and ∞.

By the theory of covering spaces, $\pi_1(\mathbb{P}^1_{\mathrm{an}} \setminus (R_0 \cup \{\infty\}), y)$ operates transitively on the fibre $g_0^{-1}(y)$ (see 12.3.5), hence there is $\gamma \in \pi_1(\mathbb{P}^1_{\mathrm{an}} \setminus (R_0 \cup \{\infty\}), y)$ such that the lift γ_0 to $(C_0)_{\mathrm{an}}$ with starting point x_0 ends in x'_0. Now we may assume that γ is a product of σ_zs (repetition possible). Then γ does not pass through $R_{\mathbf{t}} \cup \{\infty\}$ and we choose a g_i-lift γ_i in $(C_i)_{\mathrm{an}}$ with starting point x_i. It is enough to show that the lift of σ_z ends in x_i for $i = 1, \ldots, N$. Since $(C_i)_{\mathrm{an}}$ is a topological covering over a neighbourhood of z (as g_i does not ramify over z, otherwise $t_i - t_0 = t_i \in R_i - R_0$), we conclude that the lift of σ_z has x_i as end point (consider first the turn around z). Therefore, the path $\gamma = (\gamma_0, \ldots, \gamma_N)$ starts in $\mathbf{x}$ and ends in $(x'_0, x_1, \ldots, x_N)$. This proves that $C_{\mathbf{t}}$ is geometrically irreducible for every $\mathbf{t}$ as in (d).

Since we may do this for the generic point ξ of $\mathbb{A}_F^N$, we conclude that the generic fibre C_ξ of $C \to \mathbb{A}_F^N$ is a geometrically irreducible curve over $F(\mathbb{A}_F^N)$. We conclude that C is irreducible and hence we get (b). By Zariski's main theorem (note that $F(\mathbb{A}_F^N)$ is algebraically closed in the function field of C by geometric irreducibility (see A.4.11) and hence we may apply **[136]**, Cor.4.3.10), we conclude that the fibres $C_{\mathbf{t}}$ are geometrically connected for every $\mathbf{t} \in \mathbb{A}_F^N$, proving (c).

In the course of the proof, we have also seen that the generic fibre of g has degree $d_0 \cdots d_N$, thus proving the remaining part of (a). Here, we use the invariance of the degree with respect to base change. □

Remark 9.6.12. Alternatively, we give a more geometric way of deducing geometric connectedness from (d) without using the theory of covering spaces. As before, it is enough to show that $C_{\mathbf{t}}$ is geometrically connected for any $\mathbf{t}$ as in (d). We may assume F algebraically closed.

By induction on N and passing to normalizations, it suffices to consider the case $N = 1$. For $t \notin R_1 - R_0$, the curve

$$f_1(x_1) - f_0(x_0) = t$$

in $C_0 \times C_1$ is C_t. We consider the rational map

$$F : C_0 \times C_1 \dashrightarrow \mathbb{P}^1_F, \quad (x_0, x_1) \mapsto f_1(x_1) - f_0(x_0),$$

which is a morphism outside $\mathcal{Q} := f_0^{-1}(\infty) \times f_1^{-1}(\infty)$. By a suitable sequence of blowing-ups in points over $\mathcal{Q}$, we replace $C_0 \times C_1$ by a smooth surface X with a blowing-up morphism $\phi : X \to C_0 \times C_1$ such that the rational map $\widetilde{F} := F \circ \phi$ is a morphism ([**148**], Example II.7.17.3, Cor.V.5.4). Then $\widetilde{F}$ has a factorization

$$X \xrightarrow{\psi} \Gamma \xrightarrow{h} \mathbb{P}^1_F,$$

with ψ a proper surjective morphism with connected fibres and h a finite morphism (by the Stein factorization in A.12.8). Using the universal property of normalizations (see Lemma 1.4.10), we may replace Γ by its normalization, hence we may assume that Γ is a smooth irreducible projective curve.

In order to complete the proof, we need to show that h is an isomorphism, because then the fibre X_t is connected and hence the same is true for $\mathcal{C}_t = \phi(X_t)$.

Suppose this is not the case. Then Hurwitz's theorem in B.4.6 yields easily that h is ramified at least over two points of $\mathbb{P}^1_F$. We choose $z \in \Gamma$ with $h(z) \neq \infty$ such that h ramifies in z. The fibre $\psi^{-1}(z)$ is a curve in X. The complement of the exceptional divisor $E := \phi^{-1}(\mathcal{Q})$ is isomorphic to $(C_0 \times C_1) \setminus \mathcal{Q}$, hence we may identify $x \in \psi^{-1}(z) \setminus E$ with $(x_0, x_1) \in (C_0 \times C_1) \setminus \mathcal{Q}$. Since $f_1(x_1) - f_0(x_0) = h(\psi(x)) = h(z)$, the chain rule yields

$$(\partial f_0 / \partial \xi_0)(x_0) = (\partial f_1 / \partial \xi_1)(x_1) = 0 \tag{9.30}$$

for local parameters ξ_0, ξ_1 at x_0, x_1 and since the set of points $(x_0, x_1) \in C_0 \times C_1$ satisfying (9.30) is finite, we conclude that $\psi^{-1}(z) \subset E$.

On the other hand, recall that the intersection quadratic form determined by the components of a complete contractable curve E on a smooth surface is negative definite (see D. Mumford [**210**], p.6). All points of Γ are algebraically equivalent (see A.9.40) and hence the fibres of ψ are numerically equivalent, meaning in particular that the self-intersection of $\psi^{-1}(z)$ in X is 0 (see A.9.38). This is a contradiction, completing the proof.

9.6.13. In order to deal with the poles of g_t on the fibre $\mathcal{C}_t$, we generalize the situation of Lemma 9.6.11 a little bit.

Let $B, C_0, \dots, C_N$ be smooth irreducible projective curves over F and let $g_j : C_j \to B$ be surjective morphisms for $j = 0, \dots, N$. Then similarly as in Lemma 9.6.11, it follows that the fibre product $C_0 \times_B \cdots \times_B C_N$ is a projective curve. We assume for simplicity that it is geometrically irreducible and we denote by C' its normalization. Let $g_j' : C' \to C_j$ and $g' : C' \to B$ be the canonical morphisms induced by the normalization morphism.

Let $Q \in C'$ and let $Q_j := g_j'(Q)$. We denote the multiplicity of Q in the fibre with respect to g' by $m_Q(g')$. Similarly, m_j denotes the multiplicity of Q_j in the g_j-fibre. With π_j a local parameter at Q_j, we have

$$g_j = u_j \pi_j^{m_j}$$

for some unit $u_j \in \mathcal{O}_{C_j, Q_j}$. We define L_j to be the Galois closure of the field $F(Q_j, u_j(Q_j)^{1/m_j})$; clearly, it is independent of the choice of the uniformizing parameter π_j. Similarly, we define L_Q with respect to a decomposition of g' at Q, always taking the Galois closure in a fixed algebraic closure $\overline{F}$.

Lemma 9.6.14. *Under the hypothesis of 9.6.13, $m_Q(g')$ is the least common multiple of the multiplicities $m_0, \dots, m_N$ and L_Q is contained in the compositum of $L_0, \dots, L_N$ in $\overline{F}$.*

Proof: It is enough to prove the claim for $N = 1$, the general case is easily obtained by induction. By base change to $\mathbf{Q} = (Q_0, Q_1)$, we may assume Q_0, Q_1 to be F-rational.

We would like to argue complex analytically to work in a chart around $\mathbf{Q} \in C_0 \times C_1$ with coordinates (z_0, z_1), where the fibre product is given by the equation $u_0 z_0^{m_0} = u_1 z_1^{m_1}$. This would not change the relevant multiplicities and we could resolve the singularity at $(0,0)$ in the chart to get the normalization over $\mathbf{Q}$. However, to also get our claim about the fields L_j, the argument should be carried out over F. This will be done in the framework of formal geometry (see **[148]**, Section II.9), meaning that we will work over an infinitesimal neighbourhood of $\mathbf{Q}$ considering formal power series with coefficients in F instead of convergent complex power series. We will use that the residue field and the local parameter of the $\mathfrak{m}$-adic completion of a local ring remain the same as for the original local ring, hence the multiplicities and $L_j, L_{\mathbf{Q}}$ may be calculated in an infinitesimal neighbourhood.

The formal completion of $X := C_0 \times_B \cdots \times_B C_N$ along $\mathbf{Q}$ is $\mathrm{Spec}(\widehat{\mathcal{O}}_{X,\mathbf{Q}})$, obtained as the $\mathfrak{m}_{\mathbf{Q}}$-adic completion of the local ring $\mathcal{O}_{X,\mathbf{Q}}$ (see **[148]**, Example II.9.3). Note that $\widehat{\mathcal{O}}_{C_j,Q_j}$ is regular (**[148]**, Th.I.5.4.A) and hence is isomorphic to the ring of the formal power series $F[[\pi_j]]$ by the Cohen structure theorem (**[148]**, Th.I.5.5A). This yields easily

$$\widehat{\mathcal{O}}_{X,\mathbf{Q}} \cong F[[\pi_0, \pi_1]]/\langle u_0 \pi_0^{m_0} - u_1 \pi_1^{m_1} \rangle.$$

There is a finite succession of blowing-ups in points over $\mathbf{Q}$ realizing the normalization C' in a neighbourhood over $\mathbf{Q}$. This may be performed formal analytically, which also follows from the fact that the integral closure of $\widehat{\mathcal{O}}_{X,\mathbf{Q}}$ in its ring of fractions is isomorphic to the completion of the integral closure of $\mathcal{O}_{X,\mathbf{Q}}$ (see O. Zariski and P. Samuel **[338]**, Ch.VIII, Th.33). Hence we may replace X by the affine curve in $\mathbb{A}_F^2$ given by

$$A_0 x_0^{m_0} = A_1 x_1^{m_1},$$

where $A_j := u_j(Q_j)$. Indeed, they have the same infinitesimal neighbourhood at $\mathbf{Q}$, respectively at $(0,0)$. In order to verify this it suffices to look at the power series expansion of $(1+z)^{1/m_i}$ in $F[[x]]$ for any $z \in xF[[x]]$. Thus we may replace u_j by A_j by passing to a new local parameter, which may be identified with x_j.

We have a singularity at $(0,0)$ if and only if $m_i \geq 2$ for $i = 0, 1$. This is a singularity of a very simple type, which can be resolved locally in a single step by a local monoidal transformation. We may assume $2 \leq m_0 \leq m_1$. Let $k = \mathrm{GCD}(m_0, m_1)$ and write $m_i = k\mu_i$, $i = 0, 1$. Then there are positive integers a_i such that $\mu_0 a_0 - \mu_1 a_1 = 1$ and we define y_0, y_1 by

$$x_0 = y_0^{\mu_1} y_1^{a_0}, \quad x_1 = y_0^{\mu_0} y_1^{a_1},$$

hence

$$y_0 = x_0^{-a_1} x_1^{a_0}, \quad y_1 = x_0^{\mu_0} x_1^{-\mu_1}.$$

The strict transform of X by this transformation has a local equation

$$A_0 y_1^k = A_1.$$

Therefore, we get a resolution of the singularity at $(0,0)$ into k non-singular points located at the points $(0,y_1)$ with $y_1^k = A_1/A_0$. Now Q corresponds to one of the points $(0,y_1)$, where $\pi := y_0$ is a local parameter. Using this correspondence, we see that

$$g' = u_0 x_0^{m_0} = u_0 y_1^{a_0 m_0} \pi^{m_0 \mu_1}.$$

Hence $m_Q(g') = k\mu_0\mu_1$ proving the first claim. Moreover

$$u_0(Q) y_1^{a_0 m_0}(Q) = A_0 (A_1/A_0)^{\mu_0 a_0} = A_0^{-\mu_1 a_1} A_1^{\mu_0 a_0}.$$

Noting that

$$(A_0^{-a_1\mu_1} A_1^{a_0\mu_0})^{1/(k\mu_0\mu_1)} = A_0^{-a_1/m_0} A_1^{a_0/m_1}$$

we see that $L_Q \subset L_{Q_0} L_{Q_1}$, proving the second claim. □

9.6.15. Let $\mathbf{t} \in \mathbb{A}_F^N$ be as in Lemma 9.6.11(d) and assume in addition that $\mathbf{t}$ is F-rational. Let $C_{\mathbf{t}}'$ be the normalization of $C_{\mathbf{t}}$. Our goal is to apply Runge's theorem to the canonical morphism $f_{\mathbf{t}} : C_{\mathbf{t}}' \to \mathbb{P}_F^1$ given as the composition of $g_{\mathbf{t}}$ with the normalization morphism. So we need an estimate for $-\mathrm{ord}_Q(f_{\mathbf{t}})[F(Q):F]$, where Q is any pole of $f_{\mathbf{t}}$.

Lemma 9.6.16. *Suppose the hypothesis of 9.6.15 holds and let* $d := \deg(f)$. *Then*

$$-\mathrm{ord}_Q(f_{\mathbf{t}})[F(Q):F] \le d^d.$$

Proof: We will apply Lemma 9.6.14 for $C_j := C$, $B := \mathbb{P}_F^1$, $g_j := g_{j,\mathbf{t}}$ and hence $C' = C_{\mathbf{t}}'$. We note first that the canonical images $Q_j \in C_j$ of Q are all poles of f.

Let us choose $J \subset \{0,\dots,N\}$ such that, for every $i = 0,\dots,N$, there is exactly one $j \in J$ with Q_j conjugate to Q_i. For conjugates, we refer to Example A.4.14. Note that $-\mathrm{ord}_{Q_i}(g_i)$ is equal to the multiplicity m_i of the pole Q_i with respect to f. Similarly, the field L_i defined in 9.6.13 is the same with respect to f and with respect to g_i, since f and g_i differ only by a constant. Obviously, m_i and L_i do not depend on the conjugation class of Q_i. Now Lemma 9.6.14 yields

$$-\mathrm{ord}_Q(f_{\mathbf{t}})[F(Q):F] \le \prod_{j\in J} (m_j [L_j:F]).$$

Using the Euler φ-function and $d_i := [F(Q_i):F]$, we easily deduce

$$[L_i:F] \le d_i! m_i \varphi(m_i) \le d_i! m_i \max(m_i - 1, 1).$$

The displayed inequalities lead to

$$-\mathrm{ord}_Q(f_{\mathbf{t}})[F(Q):F] \le \prod_{j\in J} \left(d_j! m_j^2 \max(m_j-1,1)\right).$$

On the other hand, Example 1.4.12 shows

$$\sum_{j\in J} m_j d_j \le d.$$

By an easy induction on d, this gives

$$\prod_{j\in J} \left(d_j! m_j^2 \max(m_j-1,1)\right) \le d^d,$$

completing the proof. □

Proof of Theorem 9.6.9: We first reduce the proof to the case of a geometrically irreducible curve. The irreducible components C_i of $C_{\overline{K}}$ are defined over a finite Galois extension L

over K and they are all conjugate with respect to $\mathrm{Gal}(L/K)$. If Theorem 9.6.9 is known for the restriction $f_i : C_i \to \mathbb{P}^1_L$ of f, then $f_i^*[n]$ is a prime divisor over L, for $n \in \mathbb{N}$ outside a set of natural density 0. But then $f^*[n] = \sum_i f_i^*[n]$ has to be a prime divisor over K by considering the Galois action. So we may assume C geometrically irreducible.

For $r \geq 2$, let $g : \mathcal{C} \to \mathbb{P}^1_K \times \mathbb{A}^{r-1}_K$ be the fibre product considered in Lemma 9.6.11 with $C_i := C$ and $f_i := f$ for $i = 0, \dots, N := r-1$. Let R be the set of points in $\mathbb{P}^1_K \setminus \{\infty\}$ over which f ramifies and define $t_0 := 0$. We set

$$P(t_1, \dots, t_{r-1}) := \prod_{z \in R-R} \prod_{i=0}^{r-2} \prod_{j=i+1}^{r-1} (t_i - t_j - z),$$

hence the points $\mathbf{t}$ considered in Lemma 9.6.11(d) form the Zariski open dense subset $U := \{P \neq 0\}$.

Clearly, P is invariant under conjugation and hence defined over K. Note also that the proof of Lemma 9.6.11 shows that the fibre $g_{\mathbf{t}} : \mathcal{C}_{\mathbf{t}} \to \mathbb{P}^1_{K(\mathbf{t})}$ has degree d^r. To get an irreducible smooth projective curve, we will replace $\mathcal{C}_{\mathbf{t}}$ by its normalization denoted by $C'_{\mathbf{t}}$. By Lemma 9.6.11(d), they are isomorphic over $\mathbb{P}^1_{K(\mathbf{t})} \setminus \{\infty\}$. The induced morphism $f_{\mathbf{t}} : C'_{\mathbf{t}} \to \mathbb{P}^1_{K(\mathbf{t})}$ has also degree d^r since the normalization morphism is birational (see A.13.2).

Let $B \in \mathbb{N}$ and let $\mathcal{B}$ denote the box

$$\mathcal{B} = \{\mathbf{t} \mid 1 \leq t_i \leq B, i = 1, \dots, r-1\}.$$

The number of points $\mathbf{b} = (b_1, \dots, b_{r-1}) \in \mathbb{N}^{r-1}$ in the box $\mathcal{B}$ is B^{r-1}, while trivially the number of those which verify the further condition $P(\mathbf{b}) = 0$ is at most cB^{r-2}, for some constant c depending only on P. Now we apply Theorem 9.6.6 to $f_{\mathbf{b}} : C'_{\mathbf{b}} \to \mathbb{P}^1_K$, for $\mathbf{b}$ in the set

$$\mathcal{H}(\mathcal{B}) := \{\mathbf{b} \in \mathcal{B} \mid P(\mathbf{b}) \neq 0\}.$$

By Lemma 9.6.15, we have

$$\max_Q |\mathrm{ord}_Q(f_{\mathbf{b}})| \, [K(Q) : K] \leq d^d, \tag{9.31}$$

where $d := \deg(f)$ and Q ranges over the poles of $f_{\mathbf{b}}$. Now

$$f_{\mathbf{b}}^{-1}(n) = \{(P_0, \dots, P_{r-1}) \mid P_i \in f^{-1}(n + b_i), \, i = 0, \dots, r-1\}$$

(we set $b_0 = 0$). Obviously

$$[K(P_0, \dots, P_{r-1}) : K] \leq \prod_{i=0}^{r-1} [K(P_i) : K].$$

Note also that $\deg(f_{\mathbf{b}}) = d^r$. Therefore, using (9.31) and the last displayed inequality, Theorem 9.6.6 applied with S the set of archimedean places of K yields

$$\prod_{i=0}^{r-1} [K(P_i) : K] \geq [K : \mathbb{Q}]^{-1} d^{r-d} \tag{9.32}$$

for all $n \geq \kappa(B, r)$, where $\kappa(B, r)$ may depend on B, r, and the geometric data.

If $f^*[n + b_i]$ is not a prime divisor over K, then it has an irreducible component Y with $[K(Y) : K] \leq d/2$ (see Example 1.4.12) and hence every $P_i \in f^{-1}(n + b_i)$ satisfies

$[K(P_i):K] \leq d/2$. Thus by (9.32) the number s of indices i for which $f^*[n+b_i]$ is not a prime divisor over K satisfies

$$(d/2)^s d^{r-s} \geq [K:\mathbb{Q}]^{-1} d^{r-d},$$

hence $[K:\mathbb{Q}]d^d \geq 2^s$ and finally

$$s \leq (d \log d + \log([K:\mathbb{Q}]))/\log 2.$$

Now let $n \geq \kappa(B,r)$ and suppose that there are at least ρB integers $m \in (n, B+n]$ such that $f^*[m]$ is not a prime divisor over K. Choosing entries of the form $b = m - n$, we get at least $(\rho B)^{r-1} - cB^{r-2}$ points $\mathbf{b} \in \mathcal{H}(\mathcal{B})$. We choose $r > (d \log d + \log([K:\mathbb{Q}]))/\log 2$. Then the above shows that no such points $\mathbf{b} \in \mathcal{H}(\mathcal{B})$ exist, hence

$$\rho \leq (c/B)^{1/(r-1)}.$$

For B sufficiently large, this yields zero density for the set of integers m with $f^*[m]$ not a prime divisor. □

Remark 9.6.17. If we replace ε'_w by $[T_w : K_v]/[F:K]$ according to our normalizations in 1.3.12 then Theorem 9.6.2 and its proof hold for any field K with product formula.

Moreover, in Runge's theorem, we can still say that the height of such points is effectively bounded by a constant c. In the proof, some care is needed in case of a finite characteristic. For F, we choose a sufficiently large finite normal extension and the argument shows that we may replace $[K(P):K]$ and $[K(Q):K]$ in the basic condition of Theorem 9.6.6 by their separable degrees.

We claim that the same arguments as in the proof of Theorem 9.6.9 show that a field K with product formula and of characteristic 0 is Hilbertian.

To see this, we choose a non-archimedean $v \in M_K$ and $\alpha \in K$ such that $\log |\alpha|_v$ is above the bound c in Runge's theorem. Every $n \in \mathbb{N}$ coprime to the residue characteristic of v is a unit in R_v and hence $|\alpha|_v = |\alpha+n|_v \geq e^c$. We conclude that $h(\alpha+n) \geq c$. We apply Runge's theorem in the proof of Theorem 9.6.9 to the fibres $f_{\mathbf{b}}^{-1}(\alpha+n)$ and with S such that α is an S-unit. In the same way as in the number field case, we deduce that the set $n \in \mathbb{N}$ coprime to $\mathrm{char}(k(v))$ and with $f^*[\alpha+n]$ not a prime divisor over K has natural density 0 in $\mathbb{N}$.

In fact, R. Weissauer [**329**] proved by different methods from model theory that every field with product formula is Hilbertian.

9.7. Bibliographical notes

Large parts of this chapter are borrowed from the books of Lang [**169**] and Serre [**277**]. The additional remarks 9.2.9–9.2.12 on dynamical systems and further information may be found in G.S. Call and J.H. Silverman [**55**], J.H. Silverman [**285**]. The results from Section 9.4 are from D. Mumford [**211**]. Finally, Section 9.5 is due to Néron [**218**], see also [**169**], Ch.11.

As mentioned in the introduction, Néron obtained in [**217**] a best model for abelian varieties A over the quotient field of a discrete valuation ring R. This Néron model is a smooth group scheme over R with generic fibre A. It is proper over R if and only if A has good reduction. A detailed account for Néron models may be found in the book of S. Bosch, W. Lütkebohmert, and M. Raynaud [**44**].

In [**218**], an interpretation of the Néron symbol in terms of intersection multiplicities on the Néron model is given as in Example 9.5.22. For the singular case, we refer to [**169**], Ch.11, §5. Over the complex numbers, Néron [**218**] has also expressed the Néron symbol in terms of theta functions. For explicit formulas of Néron's canonical local height on elliptic curves, we refer to Silverman [**286**], Ch.VI, §3, §4.

Beilinson and Bloch have generalized the Néron pairing to a pairing $(Y, Z)_u$ for disjoint cycles Y and Z algebraically equivalent to 0 on the smooth complete variety X satisfying $\dim(Y) + \dim(Z) = \dim(X) + 1$ (see A. Beilinson [**20**], S. Bloch [**27**], where there are further generalizations and related conjectures). For a general approach of canonical local and global heights of subvarieties with respect to line bundles, the reader is referred to [**141**].

The literature on Hilbert's irreducibility theorem is quite ample and we refer to Lang [**169**], Ch.9 for a treatment of rather general cases. A critical analysis of both early and modern works on the subject is in Schinzel's monograph on polynomials [**259**]. Very general results were obtained by Weissauer [**329**] using methods of non-standard analysis and logic; proofs on classical lines were given by M. Fried [**124**].

Theorem 9.6.2 is due essentially to V.G. Sprindžuk [**289**], who used quite different techniques related to the theory of G-functions. The proof given here is in [**28**], with the corrections provided by P. Dèbes [**86**]. See also P. Dèbes [**87**] and P. Dèbes and U. Zannier [**88**]. The proof of Runge's theorem 9.6.6 follows an argument in [**28**]. The proof of Theorem 9.6.9 is a modification of [**88**] avoiding the theory of G-functions and using translations, rather than the general Möbius transformations of [**88**] and [**124**], in order to deal with Hilbert subsets of $\mathbb{N}$.

10 THE MORDELL–WEIL THEOREM

10.1. Introduction

The main content of this chapter is the proof of the Mordell–Weil theorem, namely the finite generation of the group of rational points of an abelian variety defined over a number field.

The finiteness of the rank of the group of rational points on an elliptic curve E defined over $\mathbb{Q}$ was proved by L.J. Mordell in his celebrated paper [**207**]. Mordell worked with the elliptic curve given by a quartic equation $y^2 = a_0x^4 + \ldots + a_4$ and used its parametrization by means of Jacobi elliptic functions and theta functions. It was by no means obvious at the time how to extend this result to elliptic curves over number fields and to abelian varieties, and this was done by Weil in his famous thesis [**324**].

A. Weil [**325**] also realized that, in the case of elliptic curves, it was somewhat simpler to work with a Weierstrass model rather than the quartic equation used by Mordell, replacing the addition and duplication formulas of elliptic functions used by Mordell by rational functions on the curve; since then this has become the standard elementary approach to the Mordell–Weil theorem for elliptic curves over a field.

The basic structure of the proof, which remains unchanged until now, is in two stages. The first stage consists in proving the so-called weak Mordell–Weil theorem, namely the finiteness of $A(K)/\phi A(K)$ for some non-trivial isogeny ϕ of the abelian variety A, usually taken to be $[m]$, namely multiplication by an integer $m \geq 2$. In the second stage, we use a Fermat descent argument to complete the proof. The explicit approach by Mordell using elliptic functions, even with Weil's simplifications, is not practical enough to be carried out explicitly on elliptic curves when dealing with multiplication by a general m, let alone on abelian varieties, which we do not quite know how to describe by means of a useful explicit set of equations. Thus in the general case we follow a more abstract point of view, culminating in the systematic use of Galois cohomology in the proof.

In this chapter, we shall follow a mid-course, beginning in Section 10.2 with the naive proof of the weak Mordell–Weil theorem for an elliptic curve over a number field.

Section 10.3 contains a detailed proof of the important Chevalley–Weil theorem on unramified morphisms, with its standard application (Corollary 10.3.13) that, for an abelian variety A over a number field K, the extension $K(\frac{1}{m}A(K))/K$ is finite.

This leads in Section 10.4 to an alternative proof of the weak Mordell–Weil theorem through the Kummer pairing, and its standard interpretation in Galois cohomology is the content of Section 10.5. We give also an extension of the weak Mordell–Weil theorem suitable to function fields of transcendence degree 1. The final short Section 10.6 concludes the proof of the Mordell–Weil theorem by means of the Fermat descent.

The Mordell–Weil theorem as proved here is ineffective. Indeed, as yet no general method is known for finding generators for the Mordell–Weil group $A(K)$ of an abelian variety A over a number field K. The ineffectiveness arises from the fact that no procedure is known for finding representatives in $A(K)$ of the finite group $A(K)/\phi A(K)$, due to our inability to decide whether a homogeneous space for A has a rational point over K and, if so, to find an algorithm to produce such a point. The question appears to be extraordinarily deep, even in the case of elliptic curves over $\mathbb{Q}$, and represents one of the most interesting open diophantine problems at the time of writing this book.

For Section 10.2, the reader is assumed to be familiar with the basics of elliptic curves presented in 8.3. For the other sections, we recommend first reading the whole of Chapter 8. The proof of the Mordell–Weil theorem in 10.6 uses also the fundamental properties of the Néron–Tate height from 9.2 and 9.3.

10.2. The weak Mordell–Weil theorem for elliptic curves

In this section we prove the finiteness of $E(K)/2E(K)$, for an elliptic curve E over a field K of characteristic $\mathrm{char}(K) \neq 2$.

10.2.1. We have seen in 8.3.4 that E may be viewed as a plane curve in $\mathbb{P}^2_K$, given in standard affine coordinates by

$$y^2 + a_1xy + a_3y = x^3 + a_2x^2 + a_4x + a_6$$

for some $a_i \in K$. Replacing y by $y - \frac{1}{2}(a_1x + a_3)$ (which is allowed because $\mathrm{char}(K) \neq 2$), we may assume that $a_1 = a_3 = 0$. Therefore, after this simplification, the affine part of E has equation

$$y^2 = (x - \alpha_1)\,(x - \alpha_2)\,(x - \alpha_3)$$

with $\alpha_i \in \overline{K}$, $i = 1, 2, 3$.

10.2.2. The intersection of E with the line $\mathbb{P}^2_K \setminus \mathbb{A}^2_K$ in $\mathbb{P}^2_K$ is a divisor $3O$, and the point $O = (0:0:1)$ is an inflexion point of E, which is taken as the identity of the group multiplication. The affine part of E is simply $E \setminus \{O\}$, and in what follows it will prove to be notationally convenient to write a point $P \in E \setminus \{O\}$ as $P = (x, y)$ in terms of its affine coordinates.

Proposition 10.2.3. *Let $\alpha_1, \alpha_2, \alpha_3 \in \overline{K}$ and $f(x) := (x-\alpha_1)(x-\alpha_2)(x-\alpha_3)$. Then the plane curve X in $\mathbb{P}^2_{\overline{K}}$, given in affine coordinates by $y^2 = f(x)$, is an elliptic curve over $\overline{K}$ if and only if the discriminant*

$$D_f := \prod_{i \neq j} (\alpha_i - \alpha_j)$$

of f is not 0.

Proof: It is easy to show that X is irreducible. By the Jacobi criterion A.7.15, $(x, y) \in X$ is a singular point if and only if

$$\frac{\partial}{\partial x}(y^2 - f(x)) = 0 \quad \textit{and} \quad \frac{\partial}{\partial y}(y^2 - f(x)) = 0,$$

hence if and only if $f'(x) = 0$ and $2y = 0$. But $y = 0$ is equivalent to $f(x) = 0$, while f and f' have a common zero if and only if the discriminant D_f of f vanishes.

In the affine neighborhood $\{y \neq 0\}$ of the point at infinity O, a defining equation for X is given by the polynomial

$$g(z, x) := z - (x - z\alpha_1)(x - z\alpha_2)(x - z\alpha_3)$$

and since $\frac{\partial g}{\partial z}(0, 0) = 1$, the point O is always smooth.

We conclude that X is smooth if and only if $D_f \neq 0$. In this case, the genus formula for plane curves in A.13.4 yields that X is an elliptic curve. □

Under the assumptions of 10.2.1, we describe now the morphism [2] explicitly. Let (x, y) be the standard affine coordinates of E. We have a group structure on E given by the zero element O at infinity and we have the geometric description of addition and inverse given in 8.3.6–8.3.7.

Let $P = (x_0, y_0) \in E(\overline{K})$; then $-2P$ is equal to the third intersection point of the tangent at P with E. If $2P = -2P = O$, this tangent is vertical, proving:

Proposition 10.2.4. *The group $E[2]$ of 2-torsion points of E consists of the identity element O and the points $(\alpha_i, 0)$, $i = 1, 2, 3$, of order 2.*

10.2.5. Let $P = (x_0, y_0) \in E(\overline{K})$ and suppose that P is not a 2-torsion point. The tangent line at P has equation

$$y = ax + b \tag{10.1}$$

with a, b determined as in 8.3.8, namely $a = f'(x_0)/(2y_0)$ and $b = y_0 - ax_0$. In order to determine the x-coordinate of the third intersection point $-2P$, we eliminate y from (10.1) and the equation $y^2 = f(x)$, obtaining the equation

$$(ax+b)^2 - (x-\alpha_1)(x-\alpha_2)(x-\alpha_3) = 0. \tag{10.2}$$

The polynomial on the left has a zero at x_0 of multiplicity at least 2 (it is 3 if P is a torsion point of order 3 on E), which accounts for two solutions. The third solution x_1 is the x-coordinate of $-2P$. Hence factoring the left-hand side of (10.2) into linear factors yields the identity

$$(ax+b)^2 - (x-\alpha_1)(x-\alpha_2)(x-\alpha_3) = -(x-x_0)^2(x-x_1)$$

of cubic polynomials in x. We specialize x to α_i, and find

$$(a\alpha_i + b)^2 = -(\alpha_i - x_0)^2(\alpha_i - x_1)$$

and

$$x_1 - \alpha_i = \left(\frac{a\alpha_i + b}{x_0 - \alpha_i}\right)^2 \tag{10.3}$$

for $i = 1,2,3$. In affine coordinates, by (10.1), we have $2P = (x_1, -ax_1 - b)$.

Suppose now that $\alpha_i \in K$ for $i = 1,2,3$. Then the preceding equation (10.3) shows that $x_1 - \alpha_i$ is a square in K, for $i = 1,2,3$. The following result gives the converse to this statement.

Lemma 10.2.6. *Under the hypotheses of 10.2.1, suppose that $\alpha_1, \alpha_2, \alpha_3 \in K$. Let (x_1, y_1) be the affine coordinates of a point $Q \in E(K)$, $Q \neq O$. Then $Q \in 2E(K)$ if and only if $x_1 - \alpha_i$ is a square in K for $i = 1,2,3$.*

Proof: If $Q \in 2E(K)$, then $x_1 - \alpha_i$ is a square in K by (10.3). On the other hand, let $x_1 - \alpha_i = u_i^2$, where $u_i \in K$ is determined up to sign. Consider the system of equations

$$u_i = \frac{a\alpha_i + b}{c - \alpha_i}, \qquad i = 1,2,3$$

in the unknowns a, b, c. We substitute $\alpha_i = x_1 - u_i^2$, getting after clearing denominators the three equations

$$\begin{aligned}
(ax_1 + b) - u_1 c - u_1^2 a &= -(x_1 - u_1^2)u_1 \\
(ax_1 + b) - u_2 c - u_2^2 a &= -(x_1 - u_2^2)u_2 \\
(ax_1 + b) - u_3 c - u_3^2 a &= -(x_1 - u_3^2)u_3.
\end{aligned}$$

We view this as an inhomogeneous linear system in the unknowns $ax_1 + b$, $-c$, $-a$. The determinant is the Vandermonde determinant $(u_2 - u_1)(u_3 - u_1)(u_3 - u_2)$, which is a factor of

$$(u_2^2 - u_1^2)(u_3^2 - u_1^2)(u_3^2 - u_2^2) = (\alpha_1 - \alpha_2)(\alpha_1 - \alpha_3)(\alpha_2 - \alpha_3) = \sqrt{D_f},$$

hence is not 0 by Proposition 10.2.3.

Since x_1 and u_1, u_2, u_3 are in K, solving the system by Cramer's rule shows that $ax_1 + b, -c, -a$ are in K. Since the morphism [2] is surjective, we know *a priori* that there is a point $P = (x_0, y_0) \in E(\overline{K})$ such that $Q = 2P$. For a suitable choice of $u_1, u_2, u_3 \in K$, we conclude from (10.3) that $x_0 = c$ and $a = f'(x_0)/(2y_0)$, $b = y_0 - ax_0$ as in 10.2.5. Hence P is K-rational. This proves the claim. □

Remark 10.2.7. The numbers u_i are determined only up to sign, hence we have eight possible choices for the triple (u_1, u_2, u_3). The corresponding rational points P give the four division points for which $2P$ has first coordinate x_1. Since Q and $-Q$ have the same first coordinate x_1 and are distinct, the eight solutions we have found are the eight division points for which $2P = \pm Q$.

10.2.8. Next, we consider addition. Let $P_1, P_2, P_3 \in E(K)$ such that $P_1 + P_2 + P_3 = O$ and $P_i \neq O$, $i = 1, 2, 3$. Let $y = ax + b$ be the line through P_1, P_2, P_3 and let (x_i, y_i) be the affine coordinates of the points P_i. Equation (10.2) has roots x_1, x_2, x_3, giving

$$(ax + b)^2 - (x - \alpha_1)(x - \alpha_2)(x - \alpha_3) = (x_1 - x)(x_2 - x)(x_3 - x). \quad (10.4)$$

As in the proof of (10.3) we set $x = \alpha_i$ and get

$$(a\alpha_i + b)^2 = (x_1 - \alpha_i)(x_2 - \alpha_i)(x_3 - \alpha_i) \quad (10.5)$$

for $i = 1, 2, 3$.

This gives us evidence for a group homomorphism $\varphi_i : E(K) \longrightarrow K^\times/K^{\times 2}$ given by $(x, y) \longmapsto x - \alpha_i \mod K^{\times 2}$, where $K^{\times 2}$ denotes the squares in $K^\times = K \setminus \{0\}$. However, this is defined only for $x \neq \alpha_i$. If we proceed as before but with $P_1 = (\alpha_i, 0)$, then differentiating (10.4) at the point α_i yields the equation

$$-(\alpha_i - \alpha_j)(\alpha_i - \alpha_k) = -(x_2 - \alpha_i)(x_3 - \alpha_i), \quad (10.6)$$

where j, k are the remaining two indices. Now, as we will verify in a moment, we obtain a homomorphism by setting

$$\varphi = (\varphi_1, \varphi_2, \varphi_3) : E(K) \longrightarrow (K^\times/K^{\times 2})^3$$

with

$$\varphi_i(P) := \begin{cases} 1 & \text{if } P = O, \\ x - \alpha_i \mod K^{\times 2} & \text{if } P = (x, y), x \neq \alpha_i, \\ (\alpha_i - \alpha_j)(\alpha_i - \alpha_k) \mod K^{\times 2} & \text{if } P = (\alpha_i, 0), \end{cases}$$

where again j, k denote the two other indices.

Lemma 10.2.9. *The map $\varphi : E(K) \longrightarrow (K^\times/K^{\times 2})^3$ is a group homomorphism with kernel $2E(K)$.*

Proof: By (10.5), (10.6), and commutativity we have

$$\varphi_i(P_1)\varphi_i(P_2)\varphi_i(P_3) = 1 \tag{10.7}$$

provided the points P_1, P_2, P_3 are different from O and at most one is equal to the 2-torsion point $(\alpha_i, 0)$.

If $P_1 = O$, then $P_3 = -P_2$ and P_2 and P_3 have the same x-coordinate, and (10.7) becomes obvious. It remains for consideration the case in which $P_i = (\alpha_i, 0)$ for $i = 1, 2, 3$. In this case (10.7) follows immediately from definition.

Since $\varphi_i(P) = \varphi_i(-P) = 1/\varphi_i(P)$, it follows that φ_i is a group homomorphism. By Lemma 10.2.6, $\ker(\varphi) = 2E(K)$. □

Proposition 10.2.10. *Let R be a unique factorization domain with quotient field K. Assume that* $\mathrm{char}(K) \neq 2$ *and that the group of units $R^\times$ in R is finitely generated. Let E be the elliptic curve given by*

$$y^2 = (x - \alpha_1)(x - \alpha_2)(x - \alpha_3),$$

where $\alpha_1, \alpha_2, \alpha_3$ are distinct elements of R. Then

$$|E(K)/2E(K)| \leq 4^r \cdot 2^{\sum_{i<j} \omega(\alpha_j - \alpha_i)},$$

where $\omega(\alpha)$ denotes the number of distinct prime factors of $\alpha \in R \setminus \{0\}$ and where r is the dimension of the $\mathbb{F}_2$-vector space $R^\times / R^{\times 2}$.

Proof: By Lemma 10.2.9, we have a homomorphism $\varphi : E(K) \longrightarrow (K^\times / K^{\times 2})^3$ with kernel $2E(K)$. We need to estimate the cardinality of the image.

Let S be a set of representatives of the primes of R and let $P \in E(K) \setminus \{O\}$ with affine coordinates (x, y). For $i = 1, 2, 3$, there are $b_i \in K$, $u_i \in R^\times$ and a_i a product of distinct primes of S such that

$$x - \alpha_i = b_i^2 u_i a_i. \tag{10.8}$$

It follows from

$$y^2 = (x - \alpha_1)(x - \alpha_2)(x - \alpha_3)$$

that the primes of the denominator of x occur with even multiplicity. Therefore, a_i is coprime to the denominator of b_i and, substituting (10.8) into the last equation, we see that $a_1 a_2 a_3$ and $u_1 u_2 u_3$ are squares in R. Hence there are $c_1, c_2, c_3 \in R$, pairwise coprime and product of distinct primes of S, such that

$$a_1 = c_2 c_3, \quad a_2 = c_1 c_3, \quad a_3 = c_1 c_2.$$

Let π be a prime of S dividing c_i. Since a_j and the denominator of b_j are coprime, π divides the numerator of $x - \alpha_j$ for $j \neq i$ and it follows that π divides $\alpha_j - \alpha_k$, where j, k are the other two indices. Therefore, the number of possibilities for π is bounded by $\omega(\alpha_j - \alpha_k)$ and there at most

$$2^{\sum_{i<j} \omega(\alpha_j - \alpha_i)}$$

such triples (c_1, c_2, c_3).

The image of $R^\times$ in $K^\times/K^{\times 2}$ is isomorphic to $\mathbb{F}_2^r$, restricting the range of u_i to 2^r possibilities. We also know that $u_1u_2u_3$ is a square, hence the number of triples (u_1, u_2, u_3) is bounded by 4^r.

Finally, it is easily seen that the points of $\varphi(E[2])$ may be represented by $(u_1c_2c_3, u_2c_3c_1, u_3c_1c_2)$ for admissible choices of u_i and c_j. □

The following corollary is a special case of the **weak Mordell–Weil theorem** for elliptic curves:

Corollary 10.2.11. *Let E be an elliptic curve over a number field K with 2-torsion also defined over K. Then $E(K)/2E(K)$ is finite.*

Proof: The ring of integers O_K is not necessarily a unique factorization domain. However, by Proposition 5.3.6, we can find a finite set of places S in K such that for any finite set of places $T \in M_K$ with $T \supset S$, the ring R of T-integers in K is a unique factorization domain. Its group of units $R^\times$ is finitely generated by Dirichlet's unit theorem in 1.5.13. We have seen in 10.2.1 and Proposition 10.2.4 that E is K-isomorphic to an elliptic curve of the form required in Proposition 10.2.10, because we can always enlarge the ring R so as to ensure that every $\alpha_i \in R$. The result now follows from Proposition 10.2.10. □

10.2.12. If the 2-torsion of E is not defined over the number field K, we can still prove the finiteness of $E(K)/2E(K)$ by an additional argument. Of course, a proof is obtained by performing a base change to a finite extension L of K over which the 2-torsion becomes rational, then prove the finiteness of $E(L)/2E(L)$, then prove that $E(L)$ is a finitely generated abelian group by Fermat descent, and conclude, by general results about abelian groups, that the subgroup $E(K)$ is also finitely generated of rank not exceeding the rank of $E(L)$.

Here we give a simple-minded direct proof of this result.

Lemma 10.2.13. *Let A be an abelian variety defined over a field K and let L be a finite separable extension of K. Let m be a positive integer and suppose that $A(L)/mA(L)$ is a finite group. Then $A(K)/mA(K)$ is a finite group.*

Proof: Let $d = [L : K]$ and let δ be a positive integer such that we can write $d = d_0d_1$ with $d_0 | m^{\delta-1}$ and $\mathrm{GCD}(d_1, m) = 1$. Let also $\mathcal{E}$ be a set of representatives of $A(L)/mA(L)$ in $A(L)$.

The group $A(L)/m^\delta A(L)$ is again a finite group, since a set of representatives for it is contained in the finite set

$$\mathcal{E}(\delta) = \mathcal{E} + m\mathcal{E} + \ldots + m^{\delta-1}\mathcal{E}.$$

Let F be the Galois closure of L over K, let $G = \mathrm{Gal}(F/K)$, let H be the subgroup of G of index d fixing L, and denote by R a full set of representatives for the left-cosets of H in G.

Let $x \in A(K) \subset A(L)$. Then we have

$$x - m^\delta y \in \mathcal{E}(\delta)$$

for some $y \in A(L)$. We apply the automorphisms $\sigma \in R$ to this equation and deduce

$$dx - m^\delta z \in \mathcal{E}'(\delta),$$

where $z := \sum_\sigma \sigma y$ and

$$\mathcal{E}'(\delta) := \left(\sum_{\sigma \in R} \sigma\right) \mathcal{E}(\delta).$$

Clearly, $z \in A(K)$ because any element $\tau \in \mathrm{Gal}(F/K)$ permutes the left-cosets of H. Since d_0 divides $m^{\delta-1}$, we may divide by d_0, getting

$$d_1 x - m\,(m^{\delta-1}/d_0) z \in A(K) \cap \frac{1}{d_0}\mathcal{E}'(\delta)$$

and $A(K) \cap \frac{1}{d_0}\mathcal{E}'(\delta)$ is still a finite set (use Proposition 8.7.2).

Finally, since d_1 and m are coprime, the euclidean algorithm produces integers u and v such that $d_1 u - mv = 1$. After multiplication by u, it follows that

$$x - m((m^{\delta-1}/d_0)uz - vx) \in A(K) \cap \frac{1}{d_0} u\mathcal{E}'(\delta). \qquad \square$$

Another proof of this lemma more in line with the Kummer theory will be given in Lemma 10.4.3. If we combine the lemma with Corollary 10.2.11, we conclude with the **weak Mordell–Weil theorem** for elliptic curves.

Theorem 10.2.14. *Let E be an elliptic curve defined over a number field K. Then $E(K)/2E(K)$ is finite.*

10.3. The Chevalley–Weil theorem

The reader is assumed to be familiar with the concepts of bounded sets from Section 2.6 and ramification from Section A.12 and Appendix B. To fix the ideas, let us consider for the moment an unramified finite morphism $\varphi : Y \to X$ of projective varieties over a field K. For $P \in Y, Q := \varphi(P)$ and a discrete valuation v of K, we may ask the question whether the field extension $K(P)/K(Q)$ is unramified at all places over v. This is not true in general, but the local form of the theorem of Chevalley–Weil states that the discriminant of the completions of $K(P)/K(Q)$ divides an element of the valuation ring independent of P.

This occupies us in the first part of this section, and then we globalize the statement to families of places. In particular, if K is a number field, then we will see that $K(P)/K(Q)$ is unramified over all but finitely many places v.

Together with Hermite's theorem, this will lead us to the finiteness of $K(\frac{1}{m}A(K))/K$ for an abelian variety A over K and $m \in \mathbb{Z} \setminus \{0\}$.

10.3.1. Let K be a field with a non-archimedean absolute value $|\ |$ on the algebraic closure $\overline{K} \supset K$. We will first prove a local version of the Chevalley–Weil theorem with respect to this absolute value, under fairly general conditions. We need to introduce first some notation:

Recall that, for $Q \in X(\overline{K})$, the residue field $K(Q)$ is the intermediate field of $\overline{K}/K$ generated by the coordinates of Q (in any affine chart). Let R_Q be the valuation ring of the restriction of $|\ |$ to $K(Q)$. The valuation ring of the restriction of $|\ |$ to K is denoted by R, and completions with respect to $|\ |$ will be denoted by $\widehat{\ }$.

Let $\varphi : Y \longrightarrow X$ be a morphism of K-varieties and let $P \in Y(\overline{K})$ with image $Q \in X(\overline{K})$. Since φ is defined over K, we have $K(Q) \subset K(P)$ and $R_Q = K(Q) \cap R_P$.

We define the **local discriminant**

$$\widehat{\mathfrak{d}}_{P/Q} := \{\det(Tr_{\widehat{K(P)}/\widehat{K(Q)}}(a_i b_j)) \mid a_1, \ldots, a_{\widehat{d}}, b_1, \ldots, b_{\widehat{d}} \in \widehat{R}_P\},$$

where we have abbreviated $\widehat{d}$ for the local degree $\widehat{d}_P := [\widehat{K(P)} : \widehat{K(Q)}]$. If R is a discrete valuation ring, then this agrees with the discriminant $\mathfrak{d}_{\widehat{R}_P/\widehat{R}_Q}$ from B.1.14.

Lemma 10.3.2. *The discriminant $\widehat{\mathfrak{d}}_{P/Q}$ is an ideal in $\widehat{R}_Q$.*

Proof: Replacing a_1 by λa_1 for any $\lambda \in \widehat{R}_Q$, it is evident that $\widehat{R}_Q \widehat{\mathfrak{d}}_{P/Q} \subset \widehat{\mathfrak{d}}_{P/Q}$. To prove the claim, it is enough to show that the trace of any $a \in \widehat{R}_P$ is in $\widehat{R}_Q$. This follows from the fact that a subset S of a valuation ring R is an ideal if and only if $RS \subset S$. Let $f(t) = t^m + a_{m-1}t^{m-1} + \cdots + a_0$ be the minimal polynomial of a over $\widehat{K(Q)}$. By transitivity of traces, it is clear that $Tr_{\widehat{K(P)}/\widehat{K(Q)}}(a)$ is a $\mathbb{Z}$-multiple of a_{m-1}. Completeness of $\widehat{K(Q)}$ ensures that all conjugates of a have the same absolute value (use Proposition 1.2.7) and hence $f(t) \in \widehat{R}_Q[t]$. This proves $Tr_{\widehat{K(P)}/\widehat{K(Q)}}(a) \in \widehat{R}_Q$ and the claim. □

Now we state the **local Chevalley–Weil theorem**. Briefly, it means that for an unramified morphism, the extension $K(P)/K(Q)$ cannot be too ramified.

Proposition 10.3.3. *Let us fix an embedding of the field K into $\overline{K}$ and let $|\ |$ be a non-archimedean absolute value on $\overline{K}$. Let $\varphi : Y \longrightarrow X$ be a finite unramified morphism of K-varieties and E be a bounded set in X. Then there is $\alpha \in R \setminus \{0\}$ such that $\alpha \in \widehat{\mathfrak{d}}_{P/Q}$ whenever $P \in Y(\overline{K})$, with $Q := \varphi(P) \in E$.*

Proof: It is known that an unramified morphism is locally equal to a closed embedding followed by a standard étale morphism (see A.12.17). This means that for every $y \in Y$ there is an affine neighborhood V of y and an affine neighborhood U of $\varphi(y)$ such that $\varphi|_V = \psi \circ i$, where $i : V \to W$ is a closed embedding and $\psi : W \to U$ is a standard étale morphism. To recall the latter, let $A := K[U]$. Then there is a monic polynomial $f(t) \in A[t]$ and an affine variety W' with coordinate ring $K[W'] = A[t]/fA[t]$ such that W is an open subset of W' and the formal derivative $f'(t)$ is a unit on W.

Since the finite morphism φ is proper (see A.12.4), Proposition 2.6.6 shows that $\varphi^{-1}(E)$ is bounded in Y. There are finitely many U_j, V_j of the above form covering X and Y, respectively. By Remark 2.6.3, there is a decomposition of $\varphi^{-1}(E)$ into bounded sets E_j' of V_j. Hence it is enough to prove the following claim:

Let $\psi : W \to U$ be a standard étale morphism as above and let E' be a bounded subset of W. Then there is $\alpha \in R \setminus \{0\}$ such that $\alpha \in \widehat{\mathfrak{d}}_{P/\varphi(P)}$ for all $P \in E'$.

Let

$$f(t) = t^d + a_1 t^{d-1} + \cdots + a_d, \quad a_i \in A.$$

By boundedness, there is $a \in R \setminus \{0\}$ such that

$$\max_{i=1,\ldots,d} \sup_{P \in E'} |a_i(\varphi(P))| \leq |a|^{-1}.$$

Note that, if the restriction of $|\ |$ to K is trivial, then $|\ |$ is trivial on $\overline{K}$ (use Proposition 1.2.7). Let $P \in E'$ and let $Q := \varphi(P)$. We have a natural surjection of $K[W]$ onto the residue field $K(P)$. Moreover, we have a commutative diagram

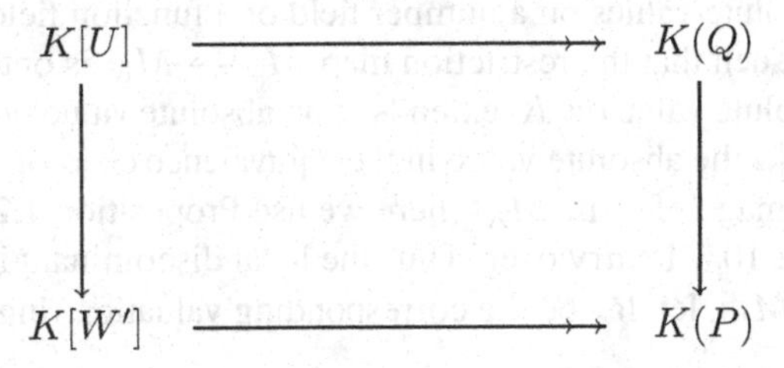

of natural ring homomorphisms. Therefore, $K(P)$ is generated by the image t_P of t as a $K(Q)$-algebra. Since t_P is a zero of the polynomial

$$f_Q(t) := t^d + a_1(Q)t^{d-1} + \cdots + a_d(Q),$$

the element $\xi := at_P \in K(P)$ is a zero of

$$g_Q(t) := t^d + aa_1(Q)t^{d-1} + \cdots + a^d a_d(Q).$$

By the choice of a, we have $g_Q(t) \in R_Q[t]$ and

$$|\xi|^d \leq \max_{i=1,\ldots,d} \left(|a|^i \, |a_i(Q)| \, |\xi|^{d-i}\right) \leq \max\left(1, \, |\xi|^{d-1}\right).$$

It follows that $|\xi| \leq 1$, i.e. $\xi \in R_P$. Let g_ξ be the minimal polynomial of ξ over $\widehat{K(Q)}$. It is a monic polynomial and there is a monic polynomial $h \in \widehat{K(Q)}[t]$ such that $g_Q = g_\xi h$. By Gauss's lemma (see Lemma 1.6.3), we have $g_\xi, h \in \widehat{R}_Q[t]$. The elements

$$1, \xi, \dots, \xi^{\widehat{d}-1}$$

form a basis of the $\widehat{K(Q)}$-vector space $\widehat{K(P)}$, which is contained in $\widehat{R}_P$. It follows that the discriminant $D_{g_\xi} \in \widehat{\mathfrak{d}}_{P/Q}$. By B.1.13 and Proposition 1.2.7, we have

$$|D_{g_\xi}| = |N_{\widehat{K(P)}/\widehat{K(Q)}}(g'_\xi(\xi))| = |g'_\xi(\xi)|^{\widehat{d}}. \tag{10.9}$$

For derivatives, the following identities hold

$$g'_\xi(\xi)h(\xi) = g'_Q(\xi) = a^{d-1} f'_Q(t_P) = a^{d-1} f'(P). \tag{10.10}$$

By assumption, f' is a unit on W and hence $(f')^{-1}$ is bounded on E'. Using (10.9) and (10.10), the same holds for $D_{g_\xi}^{-1}$ and hence there is an $\alpha \in R \setminus \{0\}$ with $|D_{g_\xi}| \geq |\alpha|$. Using $D_{g_\xi} \in \widehat{\mathfrak{d}}_{P/Q}$, we get the claim. □

10.3.4. In order to give the global version of the Chevalley–Weil theorem, we need an extension of the above to several absolute values. Let M_K be a set of non-archimedean places on K.

For every $v \in M_K$, we choose an absolute value $|\ |_v$ in the equivalence class v. All the considerations below are independent of these choices. We assume that, for every $\alpha \in K \setminus \{0\}$, the set $\{v \in M_K \mid |\alpha|_v \neq 1\}$ is finite. For example, think of the standard absolute values on a number field or a function field. Let M be a set of places on $\overline{K}$ such that the restriction map $M \longrightarrow M_K$ is onto. By Proposition 1.2.8, every absolute value on K extends to an absolute value on $\overline{K}$. For $u \in M$, we denote by $|\ |_u$ the absolute value in the equivalence class of u extending $|\ |_v$, where v is the image of u in M_K (here we use Proposition 1.2.3). For $u \in M$, the definitions of 10.3.1 carry over. Thus the local discriminant is now denoted by $\widehat{\mathfrak{d}}^u_{P/Q}$. For $v \in M_K$, let R_v be the corresponding valuation ring in K.

The **global Chevalley–Weil theorem** has the following form:

Theorem 10.3.5. *Under the hypothesis above, assume that $\varphi : Y \longrightarrow X$ is a finite unramified morphism of K-varieties and that $(E^u)_{u \in M}$ is an M-bounded family in X. Then for any $v \in M_K$, there is a non-zero $\alpha_v \in R_v$ such that $\alpha_v \in \widehat{\mathfrak{d}}^u_{P/Q}$ whenever $u \in M$ with $u|v$ and $P \in Y(\overline{K})$ with $Q := \varphi(P) \in E^u$. Moreover, we can choose $\alpha_v = 1$ for all but finitely many $v \in M_K$.*

Proof: Use the proof of Proposition 10.3.3 and keep track of the dependence on u. Note, by the definition of boundedness, that we may choose a and α depending only on v (and not on the particular choice of $u|v$). The details are left to the reader. □

The **global Chevalley–Weil theorem for discrete valuations** may be stated in terms of more familiar discriminants:

Corollary 10.3.6. *Let M_K be a set of discrete valuations of the field K, satisfying the finiteness condition of 10.3.4. Let $\varphi : Y \longrightarrow X$ be an unramified finite K-morphism of complete K-varieties. Then there are a finite subset S of M_K and for any $v \in S$ a non-zero element α_v of the maximal ideal in R_v, such that for any $P \in Y(\overline{K})$, $Q := \varphi(P)$ and any place w_0 of $K(Q)$ with $w_0|v$, the following statements hold:*

(a) *$K(P)/K(Q)$ is unramified over w_0 if $v \notin S$;*

(b) *$\alpha_v \in \mathfrak{d}_{P/Q}^{w_0}$ if $v \in S$.*

Here $\mathfrak{d}_{P/Q}^{w_0}$ is the discriminant of $\overline{R}_{w_0}$ over R_{w_0}, where R_{w_0} is the valuation ring of w_0 in $K(Q)$ and $\overline{R}_{w_0}$ is its integral closure in $K(P)$.

Proof: To prove the claim, we may assume Y irreducible and φ surjective (since φ is a closed map, A.12.4). By Proposition 2.6.17, $X(\overline{K})$ is M-bounded in X, where M is the set of places on $\overline{K}$ extending those of M_K. Now the result follows from the remarks below, Theorem 10.3.5 and the decomposition

$$\mathfrak{d}_{P/Q}^{w_0}\widehat{R}_{w_0} = \prod_w \widehat{\mathfrak{d}}_{P/Q}^{w}$$

of B.1.21, where w ranges over all places in $M_{K(P)}$ with $w|w_0$. Note that the extension $K(P)/K(Q)$ is separable because φ is unramified. The number of places w lying over w_0 is bounded by $[K(P) : K(Q)]$ (see Remark 1.3.3) and this is uniformly bounded (by the maximum of the $d = \deg(f)$ occuring in the proof of Proposition 10.3.3). Finally note that $K(P)/K(Q)$ is unramified over w_0 if and only if $1 \in \mathfrak{d}_{P/Q}^{w_0}$, see B.2.13. □

Now we can state the global version of the **Chevalley–Weil theorem for number fields.**

Theorem 10.3.7. *Let K be a number field and let $\varphi : Y \longrightarrow X$ be an unramified finite morphism of K-varieties. If X is complete, then there is a non-zero $\alpha \in O_K$ such that for any $P \in Y(\overline{K})$ and $Q := \varphi(P)$ the discriminant $\mathfrak{d}_{P/Q}$ of $O_{K(P)}$ over $O_{K(Q)}$ contains α.*

Proof: By B.1.20, we have in the notation of Corollary 10.3.6

$$\mathfrak{d}_{P/Q} = \prod_{w_0} \left(\mathfrak{d}_{P/Q}^{w_0} \cap O_{K(Q)}\right),$$

where w_0 ranges over all non-archimedean places of $K(Q)$. Now we apply Corollary 10.3.6 with M_K the set of non-archimedean places of K. If $v \in S$, then we may choose $\alpha_v \in O_K$. Note that in a Dedekind domain the product or intersection of coprime ideals is the same. Since $\mathfrak{d}_{P/Q}^{w_0} \cap O_{K(Q)}$ is a power of the prime ideal of $O_{K(Q)}$ corresponding to w_0, we conclude that

$$\alpha := \prod_{v \in S} \alpha_v$$

satisfies the claim. □

Example 10.3.8. Let us consider the morphism $\varphi : \mathbb{G}_m \to \mathbb{G}_m$ over $\mathbb{Q}$ given by $x \mapsto x^2$. Then φ is unramified and finite. Now let d be a square-free integer, $|d| \geq 2$. If we view d as a point Q in $\mathbb{G}_m$ with coordinate $x = d$, then $P = \pm\sqrt{d}$ and $K(P) = \mathbb{Q}(\sqrt{d})$ is a quadratic extension of $K(Q) = \mathbb{Q}$. By Example B.1.16, we have

$$\mathfrak{d}_{\mathbb{Q}(\sqrt{d})/\mathbb{Q}} = \begin{cases} d & \text{if } d \equiv 1 \mod 4, \\ 4d & \text{if } d \equiv 2, 3 \mod 4. \end{cases}$$

We conclude that $\mathfrak{d}_{P/Q}$ is not bounded on integer valued points. However, this is not in contrast with the Chevalley–Weil theorem because integer valued points do not form a bounded set in $\mathbb{G}_m$. If we view φ as a map $\mathbb{G}_m \to \mathbb{P}^1_{\mathbb{Q}}$, then the above points form a bounded set in $\mathbb{P}^1_{\mathbb{Q}}$, but now φ is no longer a finite map.

Finally, we apply the Chevalley–Weil theorem to an abelian variety and the morphism multiplication by m. In 10.3.9 and in Proposition 10.3.10, we relate the notion of good reduction to the discriminants $\mathfrak{d}_{P/Q}$ introduced before. Here, the reader should be familiar with schemes. The main reference is [**44**]. We shall not need these results in the further course of this book, and the reader may move immediately to Theorem 10.3.11.

10.3.9. Let A be an abelian variety over the field K and let R_v be a discrete valuation ring in K. We say that A has **good reduction** in v if there is a proper smooth scheme $\overline{A}$ over R_v such that we may identify the generic fibre $\overline{A}_K$ with A.

Assume also that A has good reduction in v and let Y be any smooth scheme over R_v. A morphism $Y_K \to A$ over K induces a rational map $Y \dashrightarrow \overline{A}$, defined on the points of codimension 1 by the valuative criterion of properness (see [**148**], Theorem II.4.7). By the analogue of Theorem 8.2.21 for group schemes, it is a morphism. Then $\overline{A}$ is unique up to isomorphisms which extend the identity and is called the **Néron model** of A. For $m \in \mathbb{Z} \setminus \{0\}$, the morphism multiplication by m extends to $\overline{A}$, as well as the group structure. It can be shown that this extension $[m]_{\overline{A}}$ is flat ([**44**], 7.3.2). If m is not divisible by the characteristic of the residue field relative to v, then $[m]_{\overline{A}}$ is fibre-wise étale (by Proposition 8.7.2) and it is even an étale morphism ([**44**], 7.3.2).

The **local Chevalley–Weil theorem for abelian varieties** may be stated in the following precise form:

Proposition 10.3.10. *Let A be an abelian variety over K, let $m \in \mathbb{Z} \setminus \{0\}$ and let v be a discrete valuation on K. If A has good reduction in v and if the characteristic of the residue field does not divide m, then for any $P \in A(\overline{K})$, the extension $K(P)/K([m]P)$ is unramified at all places lying over v.*

Proof: By base change, we may assume that $Q := [m]P$ is K-rational. Let w be a place of $K(P)$ with $w|v$ and valuation ring R_w. By 10.3.9 and the valuative criterion of properness ([**148**], Theorem II.4.7), P extends to an R_w-valued point of $\overline{A}$. Then Proposition B.3.6 yields the claim. □

The following combination of the global Chevalley–Weil theorem for number fields and of Hermite's theorem on discriminants of number fields is most important for applications. This is usually called the **Chevalley–Weil theorem.**

Theorem 10.3.11. *Let K be a number field, $\overline{K}$ an algebraic closure of K, and let $\varphi : Y \longrightarrow X$ be a finite unramified morphism of K-varieties. If X is complete, then there is a number field L, $K \subset L \subset \overline{K}$ such that $P \in Y(L)$ for any $P \in Y(\overline{K})$ with $\varphi(P) \in X(K)$.*

Proof: We may assume Y irreducible. Let $P \in Y(\overline{K})$ with $Q := \varphi(P) \in X(K)$. By Theorem 10.3.7, there is $\alpha \in O_K$, independent of P, such that $\alpha \in \mathfrak{d}_{P/Q}$. Using B.2.13, we conclude that $K(P)/K$ is unramified outside the finite set of places dividing α. By Hermite's theorem (see Corollary B.2.15), there are only finitely many possibilities for $K(P)$. This proves the claim. □

10.3.12. Let A be an abelian variety over K. For a non-zero integer m, we denote by $\frac{1}{m}A(K)$ the subset $[m]^{-1}A(K)$ of $A(\overline{K})$. For $S \subset A(\overline{K})$, the field $K(S)$ is the smallest intermediate field $K \subset L \subset \overline{K}$ with $S \subset A(L)$.

The **Chevalley–Weil theorem for abelian varieties** states the following:

Corollary 10.3.13. *Let A be an abelian variety defined over a number field K. Then $[K(\frac{1}{m}A(K)) : K] < \infty$.*

Proof: By Proposition 8.7.2, the morphism $[m]$ is finite and étale. The claim is now obvious from Theorem 10.3.11. □

10.4. The weak Mordell–Weil theorem for abelian varieties

Let A be an abelian variety over the field K. The goal of this section is the proof of the following statement, known as the **weak Mordell–Weil theorem.**

Theorem 10.4.1. *Let A be an abelian variety over a number field K and let m be a positive integer. Then $A(K)/mA(K)$ is finite.*

10.4.2. To begin with we need to introduce some notation. As usual, $\mathrm{Gal}(L/K)$ denotes the Galois group of an intermediate field extension $K \subset L \subset \overline{K}$. Let $g \in \mathrm{Gal}(L/K)$ and let X be a variety over K. We view a point $x \in X(L)$ as belonging to some affine chart, with affine coordinates in L. Applying g^{-1} to the coordinates, we get a well-defined point $x^g \in X(L)$. Clearly, $x^{gh} = (x^g)^h$ and hence we have an action of the Galois group on $X(L)$. If $\varphi : X \to Y$ is a morphism over K, then $\varphi(x^g) = \varphi(x)^g$. If F denotes the fixed field of

$\mathrm{Gal}(L/K)$, then $x \in X(F)$ is equivalent to $x^g = x$ for every $g \in \mathrm{Gal}(L/K)$. In particular, if X is an abelian variety A, we have $(ma)^g = ma^g$ and $(a+b)^g = a^g + b^g$ for $a, b \in A(L)$ and $m \in \mathbb{Z}$.

Recall that $A[m]$ denotes the group of m-torsion points of A. The next statement is contained in 10.2.13. We give an alternative proof using methods of Kummer theory.

Lemma 10.4.3. *Let L be a finite Galois extension of K and let $m \in \mathbb{Z} \setminus \{0\}$. If $A(L)/mA(L)$ is finite, then $A(K)/mA(K)$ is finite.*

Proof: The inclusion $A(K) \subset A(L)$ induces a homomorphism

$$A(K)/mA(K) \longrightarrow A(L)/mA(L)$$

of abelian groups. Let N be its kernel. It is enough to show that N is finite. Choose a system of representatives in $A(K)$ for N. For each representative a, choose $b_a \in A(L)$ such that $a = mb_a$. Consider an element $g \in \mathrm{Gal}(L/K)$ and define

$$\lambda_a(g) := b_a^g - b_a.$$

By 10.4.2, we have

$$m\lambda_a(g) = (mb_a)^g - mb_a = a^g - a.$$

By K-rationality of a, this is zero. Using our system of representatives, the rule $a \mapsto \lambda_a$ defines a map from N to the set of maps

$$\mathrm{Gal}(L/K) \longrightarrow A[m].$$

N will be finite if the map is injective and the range is finite. The latter statement follows from Proposition 8.7.2. In order to prove the former, let us suppose that $\lambda_a = \lambda_{a'}$ for representatives a, a'. We have

$$b_{a'}^g - b_{a'} = b_a^g - b_a$$

and hence

$$(b_{a'} - b_a)^g = b_{a'} - b_a$$

for every g, or equivalently $b_{a'} - b_a \in A(K)$ (by 10.4.2). Therefore, by applying $[m]$ we get $a = a'$. □

10.4.4. An important step in the proof of the weak Mordell–Weil theorem is the generalization of some aspects of Kummer theory to abelian varieties.

Let $m \in \mathbb{Z} \setminus \{0\}$ be not divisible by $\mathrm{char}(K)$ and assume that $A[m] \subset A(K)$. We denote the separable algebraic closure of K in $\overline{K}$ by K^s. For $a \in A(K)$, there is $b \in A(K^s)$ such that $a = mb$ (using $[m]$ unramified from Proposition 8.7.2, every such $b \in A(\overline{K})$ is in $A(K^s)$). If $g \in \mathrm{Gal}(K^s/K)$, then we define

$$\langle a, g \rangle := b^g - b.$$

By 10.4.2, we have $\langle a, g \rangle \in A[m]$.

Let $a' \in A(K)$ and $b' \in A(K^s)$ with $a' = mb'$, then

$$(b+b')^g - (b+b') = (b^g - b) + (b'^g - b').$$

This shows that $\langle a, g \rangle$ is independent of the choice of b (choose $b' \in A[m]$ and use that $b' \in A(K)$ by assumption). Moreover, we see that $\langle \ , \ \rangle$ is linear in the first variable.

The map

$$\langle \ , \ \rangle : A(K) \times \mathrm{Gal}(K^s/K) \longrightarrow A[m]$$

is called the **Kummer pairing**. The right-kernel of $\langle \ , \ \rangle$ is defined by

$$\{g \in \mathrm{Gal}(K^s/K) \mid \langle a, g \rangle = 0 \quad \text{for every } a \in A(K)\}$$

and the left-kernel is defined in a similar fashion by

$$\{a \in A(K) \mid \langle a, g \rangle = 0 \quad \text{for every } g \in \mathrm{Gal}(K^s/K)\}.$$

As in Corollary 10.3.13, let $K\left(\frac{1}{m}A(K)\right)$ be the smallest intermediate field $K \subset L \subset \overline{K}$ such that any $b \in A(\overline{K})$ with $mb \in A(K)$ is rational over L.

Proposition 10.4.5. *The Kummer pairing is bilinear, with left-kernel* $mA(K)$ *and right-kernel the subgroup* $\mathrm{Gal}\left(K^s/K(\frac{1}{m}A(K))\right)$ *of* $\mathrm{Gal}(K^s/K)$.

Proof: Let $g, g' \in \mathrm{Gal}(K^s/K)$. Using the notion and arguments of 10.4.2 and 10.4.4, we have

$$\langle a, gg' \rangle = b^{gg'} - b = (b^g - b)^{g'} + b^{g'} - b.$$

Since $\langle a, g \rangle$ is K-rational by assumption, we get

$$\langle a, gg' \rangle = \langle a, g \rangle + \langle a, g' \rangle.$$

This proves linearity in the second variable and thus $\langle \ , \ \rangle$ is bilinear (see 10.4.4).

For $a \in mA(K)$, choose $b \in A(K)$ such that $a = mb$. By K-rationality of b, we have

$$\langle a, g \rangle = b^g - b = 0$$

for every $g \in \mathrm{Gal}(K^s/K)$. Conversely, let a be in the left-kernel. For any $b \in A(K^s)$ with $a = mb$, we have

$$0 = \langle a, g \rangle = b^g - b.$$

Since this is true for every $g \in \mathrm{Gal}(K^s/K)$ and since K is the fixed field of the Galois group, we conclude $b \in A(K)$ (see 10.4.2). So the left-kernel is equal to $mA(K)$.

Obviously, $\mathrm{Gal}\left(K^s/K(\frac{1}{m}A(K))\right)$ is contained in the right-kernel H.

On the other hand, let g be an element of the right-kernel. For $b \in A(K^s)$ with $mb \in A(K)$, we have $b^g = b$. It follows that the restriction of g to the residue field $K(b)$ is equal to the identity, hence the same is true for the restriction of g

to $K(\frac{1}{m}A(K))$. This proves $H \subset \mathrm{Gal}\left(K^s/K(\frac{1}{m}A(K))\right)$. We conclude that equality holds. □

Remark 10.4.6. It follows from Proposition 10.4.5 that the right-kernel is a closed normal subgroup of $\mathrm{Gal}(K^s/K)$. By Galois theory, $K(\frac{1}{m}A(K))$ is a Galois extension of K. By the same Proposition 10.4.5, we conclude that the Kummer pairing induces a non-degenerate pairing

$$\left(A(K)/mA(K)\right) \times \mathrm{Gal}\left(K(\tfrac{1}{m}A(K))/K\right) \longrightarrow A[m]$$

(i.e. left- and right-kernel are zero). Thus in order to prove the finiteness of the group $A(K)/mA(K)$, it is enough to show that $\mathrm{Gal}\left(K(\frac{1}{m}A(K))/K\right)$ is finite.

Proof of Theorem 10.4.1: By Lemma 10.4.3 and Proposition 8.7.2, we may assume that $A[m] \subset A(K)$. Since here K is a number field, we see that $K\left(\frac{1}{m}A(K)\right)/K$ is finite by Corollary 10.3.13. As we have seen in Remark 10.4.6, this is enough to prove the weak Mordell–Weil theorem. □

10.5. Kummer theory and Galois cohomology

In this section, we give additional information to the previous sections. The reader may skip it in a first read because the results are of minor importance in our book. First, we recall Kummer theory of algebraic field extensions. For completeness, we give the classical interpretation of the Kummer pairing in 10.4.4 in terms of Galois cohomology. This is essential for a deeper understanding of the group $A(K)/mA(K)$, going beyond its finiteness. Then we use Kummer theory to give a generalization of the proof of the weak Mordell–Weil theorem working also for a curve over an algebraically closed field.

10.5.1. Let us recall the basic facts from **Kummer theory**.

Let K be a field and $m \in \mathbb{N}$, $m > 0$. Assume that the group $\mu_m(K)$ of mth roots of unity in K has m elements (hence m is not divisible by $\mathrm{char}(K)$). A finite-dimensional extension L/K is called **abelian of exponent m** if L/K is a Galois extension and $\mathrm{Gal}(L/K)$ is abelian and if the least common multiple of the orders of the elements of $\mathrm{Gal}(L/K)$ is m. For $S \subset K^\times$, we denote by $K(S^{1/m})$ the smallest subfield of K^s containing $\{\alpha \in K^s \mid \alpha^m \in S\}$ and K. Moreover, we write $K^{\times m}$ for the subgroup of mth powers in $K^\times$ and, more generally, for any subgroup H of $K^\times$, we will write H^m for the group of mth powers in H.

Theorem 10.5.2. *There is an inclusion preserving bijection of the set of subgroups H of $K^\times$ containing $K^{\times m}$ onto the set of abelian extensions L/K of exponent dividing m with $L \subset \overline{K}$, given by $H \longmapsto L := K(H^{1/m})$. The inverse map is $L \longmapsto H := (L^\times)^m \cap K^\times$. There is a well-defined pairing*

$$\langle \ , \ \rangle : H \times \mathrm{Gal}(L/K) \longrightarrow \mu_m(K),$$

given by $\langle \alpha, g \rangle := g(\beta)/\beta$, where $\beta \in L$ with $\alpha = \beta^m \in H$ and $g \in \mathrm{Gal}(L/K)$. The map

$$\mathrm{Gal}(K(H^{1/m})/K) \longrightarrow \mathrm{Hom}(H/K^{\times m}, \mu_m(K)), \quad g \mapsto \langle \cdot, g \rangle$$

is an isomorphism of groups. The index of $K^{\times m}$ in H is equal to $[L : K]$ (possibly ∞).

For a proof the reader may consult **[49]**, Ch.5,§11.8.

10.5.3. We recall here some facts from **Galois cohomology** (see J.-P. Serre **[278]**,§2 for details). Let G be a profinite group, namely a projective limit of finite groups with the topology induced from the product topology of the finite discrete groups, hence G is compact. A discrete right-G-module M is an abelian group endowed with the discrete topology and with a continuous right-G-action. Let us view $\mathbb{Z}$ as a discrete right-G-module with trivial G-action (i.e. $n^g = n$). Then we denote the set of G-homomorphisms of $\mathbb{Z}$ into M by $H^0(G, M)$. This is a left-exact covariant functor on the category of discrete right-G-modules and we can form the higher cohomology groups $H^i(G, M)$. There is a down-to-earth description of these G-modules. For example, $H^0(G, M)$ is equal to the fixed points of M under the action of G and $H^1(G, M)$ may be viewed as the set of continuous derivations of G with values in M modulo inner derivations. Here, a **derivation** is a function $d : G \longrightarrow M$, satisfying **Leibniz's rule**

$$d(xy) = (dx)^y + dy$$

(the left action of G on M is viewed as trivial). In particular, for $a \in M$, $d_a(x) := a^x - a$ is a derivation. Obviously, it is a continuous derivation called an **inner derivation.**

10.5.4. We return to the case when A is an abelian variety over the field K. Here $m \in \mathbb{Z} \setminus \{0\}$ is not divisible by $\mathrm{char}(K)$.

We note that the Galois group $\mathrm{Gal}(K^s/K)$ is a profinite group. By 10.4.4, we have a short exact sequence of $\mathrm{Gal}(K^s/K)$-modules

$$0 \longrightarrow A[m] \longrightarrow A(K^s) \xrightarrow{[m]} A(K^s) \longrightarrow 0.$$

Here $A(K^s)$ has the discrete topology. By 10.4.2, the fixed part of $A(K^s)$ and $A[m]$ under the action of $\mathrm{Gal}(K^s/K)$ are $A(K)$ and $A(K)[m]$ respectively. The first part of the long exact cohomology sequence reads then

$$\begin{aligned} 0 \longrightarrow A(K)[m] \longrightarrow A(K) \longrightarrow A(K) \longrightarrow H^1(\mathrm{Gal}(K^s/K), A[m]) \\ \longrightarrow H^1(\mathrm{Gal}(K^s/K), A(K^s)) \xrightarrow{[m]} H^1(\mathrm{Gal}(K^s/K), A(K^s)), \end{aligned}$$

where the last map is multiplication by m. For $a \in A(K)$, a closer look at the definition of the map $\delta : a \mapsto \delta_a$ shows that δ_a is the following derivation: Choose $b_a \in A(K^s)$ such that $mb_a = a$, and set

$$\delta_a(g) = b_a^g - b_a, \qquad g \in \mathrm{Gal}(K^s/K).$$

Since a is K-rational and $[m]$ is defined over K, $\delta_a(g)$ is indeed an element of $A[m]$. It is easily seen that δ_a is a continuous derivation. We get a short exact sequence

$$\begin{aligned} 0 \longrightarrow A(K)/mA(K) \longrightarrow H^1(\mathrm{Gal}(K^s/K), A[m]) \\ \longrightarrow H^1(\mathrm{Gal}(K^s/K), A(K^s))[m] \longrightarrow 0 \end{aligned}$$

called the **Kummer sequence**. If $A[m] \subset A(K)$, then

$$H^1(\mathrm{Gal}(K^s/K), A[m]) = \mathrm{Hom}\,(\mathrm{Gal}(K^s/K), A[m])$$

(the action of the Galois group is trivial, use Leibniz's rule). Hence the Kummer sequence may be viewed as an analogue of the Kummer pairing when we do not assume $A[m] \subset A(K)$.

10.5.5. A basic step in the proof of the weak Mordell–Weil theorem is the finiteness of the group $K\left(\frac{1}{m}A(K)\right)/K$. This was shown in Corollary 10.3.13, using the theorems of Chevalley–Weil and Hermite. Next, we give an alternative proof of the finiteness, which is also valid when K is the function field of a curve. Here $m \in \mathbb{N} \setminus \{0\}$ is always an integer not divisible by $\mathrm{char}(K)$ and we also assume that the field K contains all mth roots of unity.

By Kummer theory (see Theorem 10.5.2), there is a unique maximal abelian extension K^{ab}/K of exponent dividing m, contained in $\overline{K}$.

Let v be a **valuation** on K, i.e. there is a non-archimedean absolute value $|\ |$ on K with $v = -\log|\ |$. We assume v to be a non-trivial valuation and we will usually identify v with the corresponding place of K.

Let K_v^{nr}/K be the maximal subextension of K^{ab}/K, which is unramified over v (see B.2.8). By Kummer theory, we have the corresponding subgroup $H_v := (K_v^{nr})^{\times m} \cap K^\times$ of $K^\times$.

Lemma 10.5.6. *Suppose that the characteristic of the residue field of v does not divide m. Then*

$$H_v = \{\alpha \in K \mid v(\alpha) \in mv(K^\times)\}.$$

Proof: Let $\alpha = \beta^m \in H_v$ with $\beta \in K_v^{nr}$. Since $K(\beta)/K$ is unramified over v, m divides $v(\alpha)$ in the value group of v (by Lemma B.2.6). On the other hand, let $\alpha \in K^\times$ be with $m|v(\alpha)$ in the value group of v. Again by Lemma B.2.6, $K(\alpha^{1/m})$ is unramified over K, whence $K(\alpha^{1/m}) \subset K_v^{nr}$. This proves $\alpha \in H_v$ and the claim. □

Remark 10.5.7. Let M be a set of non-trivial valuations on K. An extension L/K is said to be **M-unramified** if L/K is unramified over all $v \in M$. The unique maximal element K_M^{nr}/K in the set of M-unramified subextensions of K^{ab}/K is given by

$$K_M^{nr} := \bigcap_{v \in M} K_v^{nr}.$$

Suppose that none of the residue characteristics of $v \in M$ divides m. By Lemma 10.5.6, we get

$$H_M := (K_M^{nr})^{\times m} \cap K^\times = \{\alpha \in K^\times \mid v(\alpha) \in mv(K^\times) \text{ for every } v \in M\}.$$

If K does not contain all mth roots of unity, we proceed as follows. Let L/K be the extension generated by the mth roots of unity. We denote by K_M^{nr} the maximal subextension of L^{ab}/K, which is M-unramified. If none of the residue characteristics of $v \in M$ divides m, then Lemma B.2.6 shows that L/K is unramified over $v \in M$ and hence L is contained in $K_M^{nr} = L_{M_L}^{nr}$ (see Proposition B.2.3), where M_L is the set of valuations on L restricting to M in K.

10.5.8. Next, we need an analogue of the class group of a number field. We assume that M satisfies the following finiteness condition

(F) *For every $\alpha \in K^\times$, we have $v(\alpha) = 0$ up to finitely many $v \in M$.*

An **M-divisor** is a finite formal sum

$$\sum n_v\,[v],$$

where $n_v \in \mathbb{Z}$ and $v \in M$. In other words, it is an element of the free abelian group $\mathrm{Div}_M(K)$ on M. For $f \in K^\times$, we define

$$\mathrm{div}_M(f) := \sum_{v \in M} v(f)\ [v].$$

By assumption (F), this is a finite sum. The set $\{\mathrm{div}_M(f) \mid f \in K^\times\}$ forms a subgroup of $\mathrm{Div}_M(K)$. The quotient of $\mathrm{Div}(K)$ by this subgroup is called the **M-class group of** K and will be denoted by $\mathrm{Cl}_M(K)$.

For the next claim, we need the subgroup

$$U_M := \{\alpha \in K^\times \mid v(\alpha) = 0 \text{ for every } v \in M\}$$

of $K^\times$, which is the kernel of the homomorphism div_M.

Proposition 10.5.9. *Let $m \in \mathbb{N} \setminus \{0\}$, let K be a field containing all mth roots of unity and let M be a set of valuations on K, satisfying the finiteness condition (F) from 10.5.8 and with residue characteristics not dividing m. Then*

$$[K_M^{nr} : K] = [U_M : U_M^m] \cdot |\mathrm{Cl}_M(K)[m]|$$

(possibly ∞).

Proof: By Kummer theory (see Theorem 10.5.2) and Remark 10.5.7, we have

$$[K_M^{nr} : K] = [H_M : K^{\times m}].$$

For every $\alpha \in H_M$ there is an element $a \in \mathrm{Div}_M(K)$ with $\mathrm{div}_M(\alpha) = ma$ (see Remark 10.5.7). The map $\alpha \mapsto a$ yields a surjective homomorphism

$$H_M \longrightarrow \mathrm{Cl}_M(K)[m].$$

Obviously, $K^{\times m}$ is in the kernel. Let us consider the group homomorphism

$$\varphi : H_M / K^{\times m} \longrightarrow \mathrm{Cl}_M(K)[m]$$

and let α be in its kernel. There is $f \in K^\times$ such that

$$\mathrm{div}_M(\alpha) = m\, \mathrm{div}_M(f) = \mathrm{div}_M(f^m),$$

showing that $\alpha / f^m \in U_M$. It follows that we have a natural exact sequence

$$0 \longrightarrow U_M / U_M^m \longrightarrow H_M / K^{\times m} \longrightarrow \mathrm{Cl}_M(K)[m] \longrightarrow 0$$

and this proves the claim. □

Lemma 10.5.10. *Let K be a field with a set M of discrete valuations and let m be a positive integer not divisible by any of the residue characteristics of $v \in M$. We denote by M_L the set of valuations on an extension L/K restricting to M. If $[L_{M_L}^{nr} : L] < \infty$ for every finite extension L/K, then $[K_{M \setminus S}^{nr} : K] < \infty$ for every finite subset $S \subset M$.*

Proof: By Remark 10.5.7, we may assume that K contains all mth roots of unity. Let π_v be a local parameter of v and let L be the field extension of K generated by $\{\pi_v^{1/m} \mid v \in S\}$. Then L/K is a finite extension. It is obvious that, if $w \in M_L$, $w|v \in S$, and $\alpha \in K^\times$, then m belongs to the value group of w and that m divides $v(\alpha)$. By Lemma B.2.6, $L(\alpha^{1/m})/L$ is unramified over w. Therefore, $L(H_{M \setminus S}^{1/m})$ is unramified over w (see Definition B.2.8). By Kummer theory (see Theorem 10.5.2) and Proposition B.2.4, this is true also for $w \in M_L$, $w \mid v \notin S$. Since $L(H_{M\setminus S}^{1/m})$ has exponent dividing m, we conclude that $L(H_{M\setminus S}^{1/m}) \subset L_{M_L}^{nr}$. The claim follows from

$$K_{M\setminus S}^{nr} = K\,(H_{M\setminus S}^{1/m}) \subset L(H_{M\setminus S}^{1/m}) \subset L_{M_L}^{nr}. \qquad \square$$

Example 10.5.11. Let K be a number field, $M := M_K$ the canonical set of places, and S a finite subset of M containing all archimedean places. In such a situation, L/K is called **unramified outside** S, if the extension is $(M \setminus S)$-unramified. The group $\mathrm{Cl}_{M\setminus S}(K)$ is the class group of the S-integers, which is finite ([**162**], Theorem 2.7.1). $U_{M\setminus S}$ is the group of S-units, which is finitely generated by Dirichlet's unit theorem in 1.5.13. Therefore, $U_{M\setminus S}^m$ is of finite index in $U_{M\setminus S}$. This proves the finiteness of $[K_{M\setminus S}^{nr} : K]$. Indeed, we may enlarge S to include all places where the residue characteristic divides m. By Remark 10.5.7, we may assume that K contains all mth roots of unity. Then the result follows from Proposition 10.5.9.

Example 10.5.12. Let $k(X)$ be a function field of a projective geometrically irreducible smooth variety X over the field k and let $M = M_X$ be its set of discrete valuations, as in 1.4.6. If S is a finite subset of M (i.e. finitely many prime divisors), then A.9.18 shows that

$$\mathrm{Cl}_{M\setminus S}(K) \cong \mathrm{Pic}(X \setminus S)$$

and

$$U_{M\setminus S} = \{f \in k(X)^\times \mid \mathrm{supp}(\mathrm{div}(f)) \subset S\}.$$

Here, we identify S with the union of its prime divisors. Assume that $S = \emptyset$. Then $\mathrm{Cl}_M(K)$ is isomorphic to the Picard group of X. We claim that $U_M \cong k^\times$. By A.8.21, it is clear that $f \in U_M$ is regular on X. Using geometric irreducibility and A.6.15, we conclude that f is constant. By A.4.11, k has to be algebraically closed in $k(X)$ proving $f \in k^\times$.

Now let $X = C$ be an irreducible projective smooth curve of genus g over an algebraically closed field k. We assume that the positive integer m is not divisible by $\mathrm{char}(k)$. Then we claim that

$$[k(C)_M^{nr} : k(C)] \leq m^{2g}.$$

We note that $\mathrm{Pic}(C)[m]$ is a subset of the Jacobian variety J of C. By Proposition 8.7.2, we have

$$|J(k(C))[m]| = m^{2g}.$$

The claim now follows from Proposition 10.5.9 and $[k^\times : k^{\times m}] = 1$.

By the claim above and Lemma 10.5.10, $k(C)_{M\setminus S}^{nr} \,/\, k(C)$ is finite for every finite subset $S \subset M$.

Proposition 10.5.13. *Let A be an abelian variety over the field K with $A[m] \subset A(K)$ for a positive integer m not divisible by* $\mathrm{char}(K)$. *Then, given a set M of discrete valuations on K satisfying (F) from 10.5.8, there is a finite subset S of M such that $K(\frac{1}{m}A(K)) \subset K_{M\setminus S}^{nr}$.*

Proof: By assumption (F) and $m \neq 0$ in K, we may assume that none of the residue characteristics of $v \in M$ divides m. Thus we may assume that K contains all mth roots of unity (use Remark 10.5.7). By Remark 10.4.6, the Galois group $\mathrm{Gal}\,(K(\frac{1}{m}A(K))/K)$ is abelian of exponent dividing m and thus $K(\frac{1}{m}A(K)) \subset K^{ab}$. The morphism $[m]$

on A is finite and unramified (see Proposition 8.7.2). By Corollary 10.3.6, there is a finite $S \subset M$ such that $K(P)/K$ is unramified outside S for any $P \in A(\overline{K})$ with $Q = [m]P \in A(K)$. This proves the claim. □

Now we are ready to prove a more general version of the **weak Mordell–Weil theorem**.

Theorem 10.5.14. *Let K be a number field or a function field of an irreducible curve over an algebraically closed field. For any abelian variety A over K and any positive integer m not divisible by* $\mathrm{char}(K)$, *the quotient $A(K)/mA(K)$ is finite.*

Proof: In the function field case, we may assume that the curve is projective and smooth (see A.13.2, A.13.3). By a finite base change, we may also assume that K contains all mth roots of unity and that $A[m] \subset A(K)$ (see Lemma 10.4.3, Lemma 1.4.10). Now use Examples 10.5.11, 10.5.12 and Proposition 10.5.13 to show that $K(\frac{1}{m}A(K))$ is a finite extension of K. By Remark 10.4.6, this proves the claim. □

Remark 10.5.15. If K is a number field or a function field of a curve over an algebraically closed field, then we may choose S in Proposition 10.5.13 as follows. First, note that M denotes then the set of standard non-archimedean places of K. By Proposition 10.3.10, we may choose

$$S = \{v \in M \mid v(m) \geq 1 \text{ or } A \text{ has bad reduction}\}.$$

Obviously, the first condition is only necessary in the number field case. Note also that S is finite (see [44], 1.4.3).

10.6. The Mordell–Weil theorem

We have the **Mordell–Weil theorem**:

Theorem 10.6.1. *If A is an abelian variety over a number field K, then $A(K)$ is a finitely generated abelian group.*

In order to prove this fundamental theorem we need first the **Fermat descent**:

Lemma 10.6.2. *Let G be an abelian group and let $m \geq 2$ be a positive integer. Let also $\| \ \|$ be a real function on G satisfying*

$$\|x - y\| \leq \|x\| + \|y\|, \quad \|mx\| = m\,\|x\|$$

for any $x, y \in G$. Assume that S is a set of representatives for G/mG, bounded relative to $\| \ \|$ by a constant C. Then for any $x \in G$, there is a decomposition

$$x = \sum_{i=0}^{l} m^i y_i + m^{l+1} z,$$

where $y_i \in S$ and where $z \in G$ satisfies $\|z\| \leq C + 1$. In particular, G is generated by elements in the ball

$$\{x \in G \mid \|x\| \leq C + 1\}.$$

Proof: There are $y_0 \in S$, $x_0 \in G$ such that $x = y_0 + mx_0$. We have

$$\|x_0\| \leq \frac{1}{m}(C + \|x\|).$$

Proceeding by induction, there are $y_l \in S$, $x_l \in G$ such that $x_{l-1} = y_l + mx_l$ and

$$\|x_l\| \leq \left(\sum_{i=1}^{l+1} \frac{1}{m^i}\right) \cdot C + \frac{1}{m^{l+1}} \|x\|.$$

We choose l so large that $\|x\| \leq m^{l+1}$ and set $z := x_l$, getting

$$\|z\| \leq \frac{1}{m-1} C + 1 \leq C + 1.$$

Moreover, we have

$$x = y_0 + my_1 + \ldots + m^l y_l + m^{l+1} z,$$

which proves the first claim. The second claim is a trivial consequence of the first. □

Remark 10.6.3. Fermat's famous proof that $x^4 + y^4 = z^2$ has no non-trivial integer solution was done by constructing a non-trivial smaller solution starting from any non-trivial integer solution, thereby leading to a contradiction (see H. Edwards [**97**],§1.5). This was the original Fermat's ***descente infinie***. Its structure is indeed very similar to the proof of Lemma 10.6.2, where we start with a group element $x = x_0$ and produce new group elements $x_1, x_2, \ldots$ which get smaller and smaller, ending with $z = x_l$ of norm $\leq C + 1$.

10.6.4. *Proof of Theorem 10.6.1:* Choose an integer $m \geq 2$. The weak Mordell–Weil theorem in 10.4.1 gives the finiteness of $A(K)/mA(K)$. By Proposition 8.6.4, there is an even ample $\mathbf{c} \in \mathrm{Pic}(A)$. By Theorem 9.3.5, the assumptions of Lemma 10.6.2 for $\|\ \| := \widehat{h}_{\mathbf{c}}^{1/2}$ on $G := A(K)$ are satisfied. Therefore Lemma 10.6.2 shows that the group $A(K)$ is generated by a bounded set. Finally, Northcott's theorem in 2.4.9 shows that $A(K)$ is finitely generated. □

Remark 10.6.5. More generally, the Mordell–Weil theorem is true for fields finitely generated over the prime field. We give below the argument for the function field K of an irreducible curve over a finite field. For the higher-dimensional case and further generalizations, which include Néron's proof of Severi's theorem of the base for curves on an algebraic surface, see [**169**], Ch. 6.

We give the proof of the Mordell–Weil theorem for $K = k(C)$, where k is a finite field and C is an irreducible curve over k. By a finite base change using Lemma 1.4.10, we may assume that k contains all mth rooth of unity and that $A[m] \subset A(K)$. Here, we have used that a subgroup of a finitely generated abelian group is again finitely generated.

Note that $C_{\overline{k}}$ is birational to a disjoint finite union of irreducible smooth projective curves C_j (see A.13.2, A.13.3) and it is clear that $k(C)$ is a subfield of any $\overline{k}(C_j)$. Since C_j is defined over a finite field, we may reduce to the case where C is a geometrically irreducible smooth projective curve with a k-rational base point.

Then we note that the weak Mordell–Weil theorem holds for $K = k(C)$. Indeed the same proof as in Theorem 10.5.14 applies with the only exception that we have no longer $[k^\times : k^{\times m}] = 1$ but this index is trivially finite.

The same proof as in 10.6.4 then shows that the Mordell–Weil theorem holds for $K = k(C)$. It suffices to remark that Northcott's theorem holds by Proposition 9.4.16 because the Northcott property (N) is valid for K (see Example 9.4.20).

10.7. Bibliographical notes

Our presentation in Section 10.2 is an expansion of J.W.S. Cassels's elegant elementary account [**61**] of the proof of the weak Mordell–Weil theorem for elliptic curves.

C. Chevalley and A. Weil [**65**] considered in their original paper a finite unramified covering of a plane curve. Later, A. Weil [**326**] extended the Chevalley–Weil theorem to higher-dimensional varieties. The proofs are always based on Weil's decomposition theorem. Our presentation is a little bit more general than the usual presentations by working first with arbitrary valuations.

Vojta ([**307**], Th.5.1.6) gave a generalization of the Chevalley–Weil theorem to ramified coverings, where the ramification is measured by the counting function of the ramification divisor. Hence the ramification of $K(P)/K(\varphi(P))$ may be unbounded along the ramification of the covering φ. For a generalization of the Chevalley–Weil theorem to maps of one-dimensional orbifolds, we refer to H. Darmon [**76**]. The idea is that the boundedness of the original Chevalley–Weil theorem still holds if we allow modest ramification. This subject will be touched upon in the proof of the finiteness of solutions of the generalized Fermat equation in Section 12.6.

The deduction of the weak Mordell–Weil theorem from Kummer theory and from the Chevalley–Weil theorem is fairly standard, see for example Serre [**277**], §4.2. The exposition of Section 10.5 follows Lang [**169**], Ch.6, where the reader will also find generalizations to higher-dimensional function fields. The generalization of the Mordell–Weil theorem to finitely generated fields over the prime field is due to A. Néron [**216**].

As mentioned in the introduction, the Mordell–Weil theorem is not effective. For effective upper bounds of the order of the group $A(K)/mA(K)$ and its relation to the rank of $A(K)$, see M. Hindry and J.H. Silverman [**153**], Th.C.1.9.

11 FALTINGS'S THEOREM

11.1. Introduction

This chapter contains a detailed proof of **Faltings's theorem**:

Theorem 11.1.1. *Let C be a geometrically irreducible smooth projective curve of genus $g \geq 2$, defined over a number field K. Then the number of K-rational points of C is finite.*

The assumption $g \geq 2$ is crucial, the theorem fails for $C = \mathbb{P}^1_K$ and for elliptic curves of positive rank. We may easily dispense with the other assumptions on C. If C is any curve over the number field K, then A.13.2 and A.13.3 show the existence of a finite extension L/K such that C_L is birational to a finite disjoint union of geometrically irreducible smooth projective curves C_j over L and we may apply Faltings's theorem to $C_j(L)$ assuming that the genus of C_j is ≥ 2.

The above theorem was conjectured by Mordell (for the rational field $K = \mathbb{Q}$) at the end of his paper [**207**] proving the finite generation of the group of rational points of an elliptic curve defined over $\mathbb{Q}$. Its function field analogue was proved by H. Grauert [**129**] in 1965 (the important earlier paper by Yu.I. Manin [**189**], which claimed a proof of this result, was much later recognized to contain a serious gap, which was eventually corrected by R. Coleman [**69**] in 1990). The **Mordell conjecture** was at last proved by G. Faltings [**113**] in 1983, as a consequence of his proofs of the Tate conjecture and the Shafarevich conjecture. He used Arakelov theory on moduli spaces; we refer to the books of G. Faltings and G. Wüstholz [**116**] and L. Szpiro [**295**] for details.

A completely new proof was then given by P. Vojta, first in the function field case [**308**] and then in the arithmetic case [**310**]. In this chapter, we shall give **Vojta's proof** with the simplifications given in [**29**]. In view of the complicated proof, it is worthwhile to give an outline of the basic ideas behind it.

The precursor of Vojta's proof goes back to Mumford's paper [**211**] of 1965, with the results, which have been already described and proved in detail in Theorem 9.4.14. There we show that the height h_Δ on $C \times C$ can be expressed, up to bounded quantities, in terms of Néron–Tate heights on the Jacobian J of C. It then follows, by the quadraticity of heights on abelian varieties, that

$$h_\Delta(P,Q) = (\text{quadratic form}) + (\text{linear form}) + O(1).$$

Here we see directly that the quadratic form in the right-hand side of this equation is indefinite, if the genus g is at least 2.

On the other hand, since Δ is an effective curve on $C \times C$, the height $h_\Delta(P,Q)$ is bounded below, away from the diagonal. This puts strong restrictions on the pair (P,Q) because it means that P and Q, considered as lattice points in the euclidean space $J(K) \otimes \mathbb{R}$, can never be too close to each other with respect to the positive definite inner product determined by the canonical form. A simple geometric argument now shows that the values of the height of rational points on C, arranged in increasing order, grow at least exponentially. This is in sharp contrast with the quadratic growth of rational points on elliptic curves, and shows that rational points on curves of genus at least 2 are much harder to come by.

Mumford's argument works over any field and is not limited to zero characteristic. Since there are examples of curves of genus at least 2 over a function field of positive characteristic having infinitely many rational points with height increasing at an exponential rate, as shown here in 9.4.19 and the following examples, some new idea was needed to attack the Mordell conjecture along Mumford's line.

This new idea was provided by Vojta. We may look at other divisors than the diagonal, and more precisely the set of divisors D for which the quadratic part of the height $h_D(P,Q)$ is indefinite. If C is a general curve (in particular smooth), the Néron–Severi group of numerical equivalence classes of divisors on the surface $C \times C$ is generated by $\{P_0\} \times C$, $C \times \{P_0\}$ (here P_0 is any point on C) and the diagonal Δ, so we have to deal with a divisor D numerically equivalent to $l\,\{P_0\}\times C + m\,C\times\{P_0\} + n\,\Delta$ (see [**130**], pp.285–286). If $(l+n)(m+n) < g^2n^2$, the quadratic form associated to h_D is indefinite, and, moreover, for large k, the Riemann–Roch theorem shows that the linear system $|kD|$ is not empty and of positive dimension if $gn^2 < (l+n)(m+n)$. Thus, replacing D by a divisor in the rational equivalence class of kD, we may suppose that D is effective and therefore h_D is bounded below away from the support of D. Hence the idea is to choose the parameters l, m, n so as to force the quadratic form to be very negative at (P,Q), and then use the fact that h_D is bounded below to conclude that P and Q have bounded height.

The new problem we face here is the fact that the choice of l, m, n depends on the ratio of the heights of P and Q, so that we need to show that not only h_D is bounded below, but also that this lower bound has a sufficiently good uniformity with respect to the divisor D. This difficulty was overcome by Vojta using the arithmetic intersection theory [**126**] and the arithmetic Riemann–Roch theorem obtained by H. Gillet and C. Soulé in [**126**] and [**127**].

As Vojta's paper clearly shows, this idea is overly simple as such and there is one more big obstacle to overcome. The difficulty is that the lower bound fails if (P, Q) belongs to the support of the effective divisor D. In order to obtain a small lower bound, the divisor D must be defined locally by means of equations with small height, and there is little room to move it away from (P, Q). In characteristic zero, by an appropriate use of derivations, we see that this is not too serious a difficulty unless the divisor D goes through (P, Q) with very high multiplicity. Note that in positive characteristic the argument using derivations fails (as it must), and it is here that characteristic zero becomes part of the proof.

This situation is reminiscent of a familiar point, which occurs in many proofs in diophantine approximation and transcendence theory, namely the non-vanishing at specific points of functions arising from auxiliary constructions. In the classical case, there are various techniques at our disposal: Roth's lemma, which is arithmetic in nature, the algebro-geometric Dyson's lemma [**102**], and the powerful and flexible product theorem of Faltings [**114**] (see Theorem 7.6.4).

In our situation, application of any of these methods requires the important condition that the ratio of the heights of Q and P be sufficiently large in order to conclude that the divisor D cannot go through (P, Q) with high multiplicity.

More precisely, Vojta uses a suitable extension of Dyson's lemma for the product of two curves and shows that, if $(l + n)(m + n)$ is sufficiently close to gn^2 and $(l + n)/(m + n)$ is sufficiently small, then any effective D as above does not vanish too much at (P, Q), thereby completing the proof.

Faltings, in his solution of the Lang conjecture for subvarieties of abelian varieties, uses his product theorem. In our particular situation, it shows that either (P, Q) belongs to a finite union of proper product subvarieties of $C \times C$, and in particular P or Q belong to a finite set, or again D does not vanish too much at (P, Q).

The paper [**29**] simplified Vojta's proof by showing that a direct application of Roth's lemma also suffices for obtaining the required small vanishing of D at (P, Q). The proof uses only the elementary theory of heights developed in Chapters 1 and 2 and replaces the difficult arithmetic Riemann–Roch theorem in Vojta's proof by the algebro-geometric Riemann–Roch theorem on the surface $C \times C$ and by the classical Siegel lemma.

This chapter is organized as follows. In Section 11.2 we study Vojta divisors and the associated heights. The short Section 11.3 extends Mumford's method, already presented in Chapter 9, to Vojta divisors, getting the required upper bound for the height.

The short Section 11.4 gives a simple proof of a local version of Eisenstein's theorem on the coefficients of Taylor series of algebraic functions, which is used to control derivations.

Section 11.5 introduces norms on certain spaces of power series and proves a generalization of Gauss's lemma in 1.6.3. These results are used to give a less *ad hoc* proof of the Eisenstein theorem, which is now derived as an application of Banach's fixed point theorem. Our goal in including this material here has been to provide the reader with additional information, which may be useful in other contexts.

Section 11.6 obtains the crucial lower bound for the height, in terms of the height of a set of defining equations of the Vojta divisor and of its multiplicity or, more precisely, index at (P, Q). Sections 11.7 and 11.8 construct a Vojta divisor of small height, apply Roth's lemma and show that the Vojta divisor has low multiplicity at (P, Q). Section 11.9 compares the upper and lower bounds so obtained and completes the proof of Faltings's theorem outlined before.

We conclude this chapter by describing in Section 11.10, without proofs, two further important results. The first is **Faltings's big theorem** [**114**], [**115**] dealing with rational points of subvarieties of abelian varieties, and we give two applications. The second result, conjectured by Bogomolov in the case of curves, deals with small points on subvarieties of abelian varieties and is due to S.W. Zhang [**342**] and L. Szpiro, E. Ullmo, and S.W. Zhang [**296**]. The much easier corresponding results for small points on subvarieties of a linear torus were treated in some detail in Chapter 4. This section may be skipped in a first reading.

In the proof of Faltings's theorem, we have assumed for simplicity that C has a point P_0 defined over K, and made P_0 as part of our data; for example, we embed C into its Jacobian by means of the Albanese map $P \mapsto \mathrm{cl}([P] - [P_0])$. Thus certain constants appearing in the proof will depend *a priori* not only on the curve C, but also on the choice of P_0. This can be avoided by fixing instead a divisor D_0 of small degree and small height (for example a suitable canonical divisor), and working with the map $P \mapsto \mathrm{cl}((\deg D_0)[P] - D_0)$ instead of the Albanese map, at the cost of introducing additional complications in the application of Mumford's method, to get an upper bound for the height (for an application, see E. Bombieri, A. Granville, and J. Pintz [**33**]).

The finiteness result of Faltings's theorem is ineffective, in the sense that it provides no upper bound on the height of solutions. However, we can get bounds for the number of solutions. An examination of the proof shows that solutions can be viewed as the union of two sets, namely small solutions, for which an explicit bound for the height can be given, and large solutions. In order to study large solutions, we work in an euclidean space $\mathbb{R}^r$, where r is the rank of the **Mordell–Weil group** $J(C)(K)$ of the Jacobian of C. The group $J(C)(K)/\mathrm{tors}$ is a lattice in $\mathbb{R}^r$ and the euclidean metric induces the Néron–Tate height on this lattice. Now the result of the Vojta construction is that any two large solutions in $C(K)$ either have comparable height within a constant factor, or determine two vectors at an angle of at least say 40°, which suffices for proving finiteness in any cone with center O and opening 40°. However, just to start Vojta's method we need at least two solutions in a cone and if there is only one solution we cannot say anything on the height. Of course, in this case we obtain finiteness for free and a good bound for the number of solutions.

We have chosen not to give explicit bounds for the constants c, $C_1, C_2, \ldots$ and also other constants involved in the symbols $\ll$ and $O(\ )$, which appear in the course of the proof. However, all such constants are effectively computable and, at the end of Section 11.9, we mention some explicit bounds for the number of solutions.

This chapter is based on Chapters 2, 8, 9, and 10.

11.2. The Vojta divisor

Let C be an irreducible smooth projective curve of genus g over the field K with a K-rational point. Obviously, this is no restriction for the proof of Faltings's theorem. Thus we begin by fixing a point $P_0 \in C(K)$. In fact, this section is devoted to purely geometric properties of certain divisors on $C \times C$, and we never use the assumption that K is a number field.

By Δ we denote the diagonal of $C \times C$ and for simplicity of notation we shall also write

$$\Delta' := \Delta - \{P_0\} \times C - C \times \{P_0\}.$$

We study here properties of divisors on $C \times C$, which are expressed as linear combinations of the divisors $\{P_0\} \times C$, $C \times \{P_0\}$, and Δ. It is worth noting that, since we are in characteristic 0, for a general curve C, these three divisors generate the full group of divisors of $C \times C$ up to algebraic equivalence, hence the apparently special situation considered here is in fact typical for the general case.

Lemma 11.2.1. *The following table gives the intersection numbers:*

	$\{P_0\} \times C$	$C \times \{P_0\}$	Δ'
$\{P_0\} \times C$	0	1	0
$C \times \{P_0\}$	1	0	0
Δ'	0	0	$-2g$

Proof: In order to compute the intersection numbers, we may assume that K is algebraically closed, see A.9.25. The identities

$$(\{P_0\} \times C) \cdot (C \times \{P_0\}) = 1, \quad (\{P_0\} \times C) \cdot \Delta = 1$$

were shown in Example A.9.23. Next, we show $(\{P_0\} \times C) \cdot (\{P_0\} \times C) = 0$. In fact, the divisor $\{P_0\} \times C$ is algebraically equivalent to $\{P_1\} \times C$ for any point $P_1 \in C(\overline{K})$ (cf. A.9.40), therefore A.9.38 shows

$$(\{P_0\} \times C) \cdot (\{P_0\} \times C) = (\{P_0\} \times C) \cdot (\{P_1\} \times C).$$

Since $\{P_0\} \times C$ and $\{P_1\} \times C$ have disjoint supports if $P_1 \neq P_0$, we must have $(\{P_0\} \times C) \cdot (\{P_1\} \times C) = 0$, proving what we want.

We claim that $\Delta \cdot \Delta = 2 - 2g$. By symmetry and linearity, this will complete the intersection table. Let N_Δ be the normal bundle of Δ in $C \times C$. By Proposition A.9.19, we have

$$\Delta \cdot \Delta = \deg N_\Delta.$$

The pullback of N_Δ under the diagonal map is isomorphic to the tangent bundle on C. The degree of the latter equals

$$-\deg(\Omega_{C/K}) = 2 - 2g$$

by the Riemann–Roch theorem on C (cf. A.13.6). □

Definition 11.2.2. *For $d_1, d_2, d \in \mathbb{N}$, the divisor*

$$V := d_1 \{P_0\} \times C + d_2\, C \times \{P_0\} + d\, \Delta'$$

*is called a **Vojta divisor**.*

11.2.3. We are interested in expressing the divisor V as a difference of two well-chosen, very ample divisors on $C \times C$, in order to calculate an associated height. Fix $N \geq 2g + 1$ for the rest of this chapter. By A.13.7, the divisor $N[P_0]$ is very ample on C. Let

$$\varphi_{N[P_0]} : C \longrightarrow \mathbb{P}^n_K$$

be the corresponding closed embedding with $O(N[P_0]) \cong \varphi^* O_{\mathbb{P}^n}(1)$ (see for instance Remark A.6.11). It follows easily (see [134], Prop.4.3.1) that the product $\varphi_{N[P_0]} \times \varphi_{N[P_0]}$ gives a closed embedding

$$\psi : C \times C \longrightarrow \mathbb{P}^n_K \times \mathbb{P}^n_K$$

and we have

$$O_{C \times C}(\delta_1 N \{P_0\} \times C + \delta_2 N\, C \times \{P_0\}) \cong \psi^* O_{\mathbb{P}^n \times \mathbb{P}^n}(\delta_1, \delta_2). \tag{11.1}$$

For notational convenience, we shall abbreviate throughout $O(d) := O_{\mathbb{P}^m}(d)$ and $O(\delta_1, \delta_2) := O_{\mathbb{P}^n \times \mathbb{P}^n}(\delta_1, \delta_2)$.

Lemma 11.2.4. *For integers $\delta_1, \delta_2 \geq 1$, the divisor $\delta_1 N \{P_0\} \times C + \delta_2 N\, C \times \{P_0\}$ is very ample.*

Proof: Since $O(\delta_1, \delta_2)$ is very ample (see A.6.13), this follows from (11.1). □

Lemma 11.2.5. *If M is a sufficiently large integer, then*

$$B := M\left(\{P_0\} \times C + C \times \{P_0\}\right) - \Delta'$$

is a very ample divisor on $C \times C$.

Proof: This follows from Lemma 11.2.4 and A.6.10 (a). □

11.2.6. We fix such an M once for all. Let

$$\phi_B : C \times C \longrightarrow \mathbb{P}^m_K$$

be a corresponding closed embedding such that $\phi_B^* O_{\mathbb{P}^m}(1) \cong O(B)$ (cf. Remark A.6.11). The coordinates of $\mathbb{P}^m_K$ and $\mathbb{P}^n_K \times \mathbb{P}^n_K$ will be denoted by $\mathbf{y}$ and $(\mathbf{x}, \mathbf{x}')$. We consider $C \times C$ as a closed subvariety of $\mathbb{P}^m_K$ or $\mathbb{P}^n_K \times \mathbb{P}^n_K$, as the case may be, without mentioning the closed embeddings ϕ_B or ψ. Because they are defined with a basis of global sections, no coordinate x_j, x'_j, or y_i vanishes identically on $C \times C$. Let us consider the condition:

(V1) *$\delta_1 := (d_1 + Md)/N$ and $\delta_2 := (d_2 + Md)/N$ are positive integers.*

By adding to d_1 and d_2 integers bounded by N, (V1) will be satisfied. This is only a technical condition of minor import. Then we get a decomposition

$$V = \left(\delta_1 N \{P_0\} \times C + \delta_2 N\, C \times \{P_0\}\right) - dB$$

of the Vojta divisor V into the difference of two very ample divisors (Lemma 11.2.4 and Lemma 11.2.5). It follows from A.10.38 that for sufficiently large δ_1, δ_2, d the following condition is satisfied:

(V2) *The first cohomology groups of $\mathcal{J}_{\psi(C\times C)}(\delta_1, \delta_2)$ and $\mathcal{J}_{\phi_B(C\times C)}(d)$ vanish.*

Here, as usual, $\mathcal{J}_X$ denotes the ideal sheaf of a closed subvariety X and $\mathcal{F}(\delta_1, \delta_2)$, $\mathcal{F}(d)$ denote the tensor product of a sheaf $\mathcal{F}$ with the corresponding standard very ample sheaf of a multiprojective space.

Now the long exact cohomology sequence shows that the natural maps

$$\psi^* : \Gamma\left(\mathbb{P}^n_K \times \mathbb{P}^n_K, \mathcal{O}(\delta_1, \delta_2)\right) \longrightarrow \Gamma\left(C \times C, \psi^* \mathcal{O}(\delta_1, \delta_2)\right)$$

and

$$\phi_B^* : \Gamma\left(\mathbb{P}^m_K, \mathcal{O}(d)\right) \longrightarrow \Gamma\left(C \times C, \mathcal{O}(dB)\right)$$

are surjective (use Example A.10.20, A.10.22, and A.10.25). This is the property we shall use in the following lemma.

Lemma 11.2.7. *Suppose that V is a Vojta divisor satisfying* (V1) *and* (V2). *For any global section s of $O(V)$, there are polynomials $F_i(\mathbf{x}, \mathbf{x}')$, $i = 0, \ldots, m$, bihomogeneous of bidegree (δ_1, δ_2), such that*

$$s = F_i(\mathbf{x}, \mathbf{x}')/y_i^d\big|_{C\times C} \tag{11.2}$$

for $i = 0, \ldots, m$.

Conversely, assume that $F_i(\mathbf{x}, \mathbf{x}')$, $i = 0, \ldots, m$, *are bihomogeneous polynomials of bidegree* (δ_1, δ_2), *satisfying*

$$F_i(\mathbf{x}, \mathbf{x}')/y_i^d = F_j(\mathbf{x}, \mathbf{x}')/y_j^d \tag{11.3}$$

on $C \times C$ *for every* i, j. *Then there is a unique global section* s *on* $O(V)$ *such that* (11.2) *is valid for every* i.

Proof: Let s be a global section of $O(V)$. Then

$$s \otimes (y_i^d|_{C\times C})$$

is a global section of the line bundle associated to the divisor

$$V + dB = \delta_1 N \{P_0\} \times C + \delta_2 N\, C \times \{P_0\}.$$

By (11.1) on page 357, we get bihomogeneous polynomials $F_i(x, x')$ of bidegree (δ_1, δ_2) satisfying (11.2). Conversely, assume that $F_i(x, x')$, $i = 0, \ldots, m$, are bihomogeneous polynomials of bidegree (δ_1, δ_2) fulfilling (11.3) on $C \times C$ for every i, j. Then we define a meromorphic section of $O(V)$ by formula (11.2). This is independent of the choice of $i \in \{0, \ldots, m\}$ by (11.3). Note that the poles of s are contained in the subvariety $y_i = 0$. Since $y_0, \ldots, y_m$ have no common zero, s is a global section (see A.8.21). □

11.3. Mumford's method and an upper bound for the height

Let C be an irreducible projective smooth curve of genus $g \geq 1$ over a field K with product formula and let us fix $P_0 \in C(K)$ as in Section 11.2. The Jacobian variety of C, defined and studied in Section 8.10, is denoted by J. We have the closed embedding

$$j : C \longrightarrow J, \qquad P \longmapsto \mathrm{cl}([P] - [P_0]),$$

which is defined over K, since $P_0 \in C(K)$. Let $\langle\ ,\ \rangle$ be the canonical form of J as defined in 9.4.2; it is a symmetric positive semidefinite form on $J(\overline{K})$. The corresponding seminorm is denoted by $|\ \ |$; more explicitly, we have $|a|^2 = \langle a, a\rangle$ for $a \in J(\overline{K})$.

The next proposition generalizes Mumford's formula (Proposition 9.4.3) to the Vojta divisor V. It will be used as an upper bound for a height function $h_V(P, Q)$.

Proposition 11.3.1. *Let* $P, Q \in C(\overline{K})$ *and* $z = j(P)$, $w = j(Q)$. *Then*

$$\begin{aligned} h_V(P, Q) = {} & \frac{d_1}{2g}|z|^2 + \frac{d_2}{2g}|w|^2 - d\,\langle z, w\rangle \\ & + d_1 O(|z|) + d_2 O(|w|) + (d_1 + d_2 + d + 1) O(1). \end{aligned}$$

Proof: We have seen in the proof of Proposition 9.4.3 that

$$h_{\Delta'}(P,Q) = -\langle z,w\rangle + O(1)$$

and

$$h_{\{P_0\}\times C}(P,Q) = \frac{1}{2g}\,|z|^2 - \frac{1}{2g}\,\hat{h}_{\theta-\theta^-}(z) + O(1),$$

with a similar formula for $h_{C\times\{P_0\}}(P,Q)$. Therefore

$$\begin{aligned} h_V(P,Q) &= \frac{d_1}{2g}\,|z|^2 + \frac{d_2}{2g}\,|w|^2 - d\cdot\langle z,w\rangle \\ &\quad - \frac{d_1}{2g}\,\hat{h}_{\theta-\theta^-}(z) - \frac{d_2}{2g}\,\hat{h}_{\theta-\theta^-}(w) + (d_1+d_2+d+1)O(1). \end{aligned}$$

Since $\theta-\theta^- \in \mathrm{Pic}^0(J)$ (see Theorem 8.8.3) and since, by the proof of Proposition 9.4.1, $\theta+\theta^-$ is an even ample class with $\langle a,a\rangle = \hat{h}_{\theta+\theta^-}(a)$ for $a\in J$, we get the claim by Corollary 9.3.8. □

Remark 11.3.2. By the preceding Proposition 11.3.1, there is a natural quadratic form on $J(\overline{K})\times J(\overline{K})$ associated to $h_V(P,Q)$, namely

$$\frac{d_1}{2g}\,|z|^2 + \frac{d_2}{2g}\,|w|^2 - d\langle z,w\rangle.$$

Using the Cauchy–Schwarz inequality, this form is indefinite if and only if $d_1d_2 < g^2d^2$.

11.4. The local Eisenstein theorem

In this section we give a quick proof of a local version of a well-known theorem of Eisenstein on the coefficients of the Taylor series expansion of an algebraic function of one variable.

Theorem 11.4.1. *Let K be a field of characteristic 0 complete with respect to an absolute value $|\ |$ and let $p(x,t)\in K[x,t]$ be a polynomial in two variables with partial degrees at most d. Let ξ be an algebraic function of x such that $p(x,\xi(x))=0$. We suppose that $\xi(0)\in K$, $|\xi(0)|\le 1$ and*

$$\frac{\partial p}{\partial t}(0,\xi(0)) \neq 0.$$

Then the Taylor series expansion $\xi(x) = \sum_{k=0}^{\infty} a_k x^k$ has coefficients in K and the following bound holds for $l\ge 1$

$$|a_l| \le C^l \left(|p| \Big/ \left|\frac{\partial p}{\partial t}(0,\xi(0))\right|\right)^{2l-1},$$

where $|p| = \max|\text{coefficients of } p|$ *is the Gauss norm of* p, *and*

$$C = \begin{cases} 1 & \textit{in the non-archimedean case} \\ |8(d+1)^7| & \textit{in the archimedean case.} \end{cases}$$

Proof: Suppose first that $|\ |$ is not archimedean. We abbreviate $p_t = \partial p/\partial t$ and $\partial_l = (1/l!)(\frac{\mathrm{d}}{\mathrm{d}x})^l$. By Leibniz's formula

$$\partial_l(p(x,\xi(x))) = \sum_{ij}\sum p_{ij}\partial_{l_0}x^i\partial_{l_1}\xi\cdots\partial_{l_j}\xi\,,$$

where the inner sum is over all solutions of $l_0 + \cdots + l_j = l$; the p_{ij} s are the coefficients of the polynomial p. The sum of the terms with $l_\lambda = l$ and $\lambda \neq 0$ is simply $p_t \cdot \partial_l\xi$; since $\partial_l(p(x,\xi(x)))$ is identically 0 and the absolute value is ultrametric, we get

$$|p_t(0,\xi(0))| \cdot |\partial_l\xi(0)| \le |p| \max|\partial_{l_1}\xi(0)|\cdots|\partial_{l_j}\xi(0)|\,,$$

where max runs over $l_1 + \cdots + l_j \le l$ with each $l_\lambda < l$. On the other hand, by hypothesis $p_t(0,\xi(0)) \neq 0$ and we also have $|\xi(0)| \le 1$. This means that in the last displayed inequality we need only consider products in which each l_λ is at least 1; noting that $a_l = \partial_l\xi(0)$ and $|p_t(0,\xi(0))| \le |p|$, we get Theorem 11.4.1 by induction.

We treat the case in which $|\ |$ is archimedean in a different fashion. We assume that $|\ |$ is normalized as usual. Let us abbreviate $\xi^{(l)} = (\frac{\mathrm{d}}{\mathrm{d}x})^l\xi$. By induction on l we establish that there is a polynomial $q_l(x,t)$ such that

$$\xi^{(l)}(x) = -\frac{q_l(x,\xi(x))}{p_t(x,\xi(x))^{2l-1}}$$

for $l \ge 1$. We have $q_1 = p_x$ and

$$q_{l+1} = (q_l)_x p_t^2 - (q_l)_t p_t p_x + (2l-1)q_l(p_{tt}p_x - p_{xt}p_t)\,. \tag{11.4}$$

Note that the partial degree of q_l with respect to x (resp. t) is bounded by $l(2d-1)-d$ (resp. $l(2d-2)+2-d$). For two polynomials f, g in any number of variables, we have $|fg| \le N \cdot |f| \cdot |g|$, where N is the number of monomials in g. Thus, from the recurrence (11.4), we estimate the Gauss norm $|q_{l+1}|$ by

$$|q_{l+1}| \le A\,|p|^2\,|q_l|\,,$$

where

$$\begin{aligned} A &= d^4(d+1)^2\,(\deg_x(q_l) + \deg_t(q_l)) + 2(2l-1)d^6(d+1) \\ &\le 8(d+1)^7 l\,. \end{aligned}$$

Noting that $|q_1| \le d\,|p|$, this yields

$$|q_l| \le (l-1)!\,8^{l-1}(d+1)^{7l-6}|p|^{2l-1}\,.$$

The required estimate for $a_l = \partial_l \xi(0)$ follows from

$$\partial_l \xi(x) = -\frac{1}{l!}\frac{q_l(x,\xi(x))}{p_t(x,\xi(x))^{2l-1}}$$

and $|q_l(0,\xi(0))| \le |q_l|(\deg_t(q_l)+1)$. This proves the theorem. □

In the next section, which is not essential for the understanding of this chapter, we give a more conceptual proof of a slightly stronger result, valid in the case of several variables and also in any characteristic.

11.5. Power series, norms, and the local Eisenstein theorem

In this section K is a field complete with respect to an absolute value $|\ |$. We shall obtain here some simple but useful facts about norms of polynomials and power series, and deduce, by an application of Banach's fixed point theorem, a more general version of Theorem 11.4.1 covering the case of an algebraic function of several variables over a field of arbitrary characteristic.

11.5.1. Let us fix n and let $\mathbf{x} = (x_1,\dots,x_n)$. The ring of formal power series in $\mathbf{x}$ is denoted by $K[[\mathbf{x}]]$. For

$$f(\mathbf{x}) := \sum_{\alpha\in\mathbb{N}^n} a_\alpha\, \mathbf{x}^\alpha \in K[[\mathbf{x}]]$$

and $\mathbf{r} = (r_1,\dots,r_n)$ with $r_i > 0$, we define

$$\|f\|_{\mathbf{r}} := \sum_{\alpha\in\mathbb{N}^n} |a_\alpha|\mathbf{r}^\alpha .$$

Then

$$K\langle \mathbf{r}^{-1}\mathbf{x}\rangle := \{f \in K[[\mathbf{x}]] \mid \|f\|_{\mathbf{r}} < \infty\}$$

is a Banach algebra over K, complete with respect to the norm $\|\ \|_{\mathbf{r}}$, satisfying $\|fg\|_{\mathbf{r}} \le \|f\|_{\mathbf{r}} \cdot \|g\|_{\mathbf{r}}$. The spectral norm of f is defined by

$$\rho_{\mathbf{r}}(f) := \inf_{k\in\mathbb{N}\setminus\{0\}} \|f^k\|_{\mathbf{r}}^{1/k}.$$

It follows easily from $\|fg\|_{\mathbf{r}} \le \|f\|_{\mathbf{r}} \cdot \|g\|_{\mathbf{r}}$ that

$$\rho_{\mathbf{r}}(f) = \lim_{k\to\infty} \|f^k\|_{\mathbf{r}}^{1/k}.$$

If $\mathbf{r} = (1,\dots,1)$, we shall omit the suffix $\mathbf{r}$ in what follows. It is notationally convenient to study only the case in which $\mathbf{r} = (1,\dots,1)$. If each r_i belongs to the value group of K, then we can find elements z_i such that $|z_i| = r_i$ and replacing $(x_1,\dots,x_n)$ by $(x_1/z_1,\dots,x_n/z_n)$ renormalizes the situation to the case in which $\mathbf{r} = (1,\dots,1)$.

However, this need not be the case in general, even if we assume K to be algebraically closed. The following simple lemma shows that this difficulty disappears if we go to a suitable extension L/K of K and a suitable extension of $|\ |$:

Lemma 11.5.2. *Given $r > 0$, there is always a field extension L/K and an extension of $|\ |$ to L such that r is in the value group of this extension. In the non-archimedean case, the value group of $\overline{K}$ is given by*

$$|\overline{K}^\times| = \{|\alpha|^{1/m} \mid \alpha \in K^\times, m \in \mathbb{N} \setminus \{0\}\}.$$

Proof: We begin by proving the second claim. The inclusion $\supset$ is clear. Now let $\beta \in \overline{K}^\times$ be a zero of the polynomial

$$f(t) := a_n t^n + a_{n-1} t^{n-1} + \cdots + a_0$$

with coefficients in K, not all 0. Since $f(\beta) = 0$ and $|\ |$ is ultrametric, there must be two distinct indices i, j such that

$$|a_i \beta^i| = |a_j \beta^j| = \text{maximum}.$$

This yields

$$|\beta| = |a_i / a_j|^{1/(j-i)},$$

completing the proof of the second claim. The first claim is a consequence of the second one and of [47], Ch.VI, §10, no.1, Prop.1. In the archimedean case, we do not need an extension by Ostrowski's theorem (see Theorem 1.2.6). □

In view of this lemma, we see that going to a field extension we may always renormalize everything so that $\mathbf{r} = (1, \ldots, 1)$.

Lemma 11.5.3. *If the absolute value on K is not archimedean, then*

$$\rho_{\mathbf{r}}(f) = \sup_{\alpha \in \mathbb{N}^n} |a_\alpha| \mathbf{r}^\alpha.$$

Moreover, $\rho_{\mathbf{r}}(f)$ is multiplicative, i.e. $\rho_{\mathbf{r}}(fg) = \rho_{\mathbf{r}}(f)\rho_{\mathbf{r}}(g)$ for $f, g \in K\langle \mathbf{r}^{-1}\mathbf{x}\rangle$.

Proof: By Lemma 11.5.2, it suffices to deal with the case $\mathbf{r} = (1, \ldots, 1)$, hence we drop the suffix $\mathbf{r}$ in what follows.

Let us denote the right-hand side of the claim by $m(f)$. We shall see below that m is a multiplicative norm. In particular, m is power multiplicative

$$m(f^k) = m(f)^k$$

for all positive integers k. Since $m(f) \leq \|f\|$, we conclude that $m(f) \leq \rho(f)$. In order to prove $m(f) \geq \rho(f)$, we may assume that f is a polynomial. Now for a polynomial f in n variables of degree d, we have

$$\|f\| \leq (d+1)^n \, m(f).$$

Applying this to the polynomial f^k, we get

$$\|f^k\|^{1/k} \leq (kd+1)^{n/k} m(f)$$

and, letting $k \to \infty$, we deduce $\rho(f) \leq m(f)$, completing the proof of the first statement.

Finally, we prove that m is multiplicative. This is an extension of Gauss's lemma (see Lemma 1.6.3) from polynomials to formal power series. The proof is immediate, applying Gauss's lemma to arbitrarily large truncations of the power series.

Another argument is as follows. We may suppose $m(f) = m(g) = 1$ and we need to prove that $m(fg) = 1$. Let R be the valuation ring of $|\ |$, and $k(v)$ be its residue field. Reduction of the coefficients modulo the maximal ideal gives a homomorphism

$$\pi : R\langle \mathbf{x} \rangle \longrightarrow k(v)[\mathbf{x}].$$

Since $k(v)[\mathbf{x}]$ is an integral domain, we have $\pi(fg) = \pi(f)\pi(g) \neq \pi(0)$. This proves $m(fg) = 1$, hence m is multiplicative. □

Lemma 11.5.4. *Assume that* $\mathbf{r} \in |\overline{K}^{\times}|^n$ *and let* $P(\mathbf{r})$ *denote the closed polydisk*

$$P(\mathbf{r}) := \{\mathbf{x} \in \overline{K}^n \mid |x_i| \leq r_i, \quad i = 1, \dots, n\}.$$

Then for $f \in K\langle \mathbf{r}^{-1}\mathbf{x} \rangle$ *it holds*

$$\rho_{\mathbf{r}}(f) = \sup_{\mathbf{x} \in P(\mathbf{r})} |f(\mathbf{x})|$$

and there is $\mathbf{x} \in P(\mathbf{r})$ *with* $\rho_{\mathbf{r}}(f) = |f(\mathbf{x})|$.

Proof: Again, we may assume that $\mathbf{r} = (1, \dots, 1)$ and that K is algebraically closed. Let

$$s(f) := \sup_{\mathbf{x} \in P} |f(\mathbf{x})|.$$

The function s is a norm on $K\langle \mathbf{x} \rangle$ satisfying

$$s(f^k) = s(f)^k$$

for all positive integers k. Since $s \leq \|\ \|$, we conclude $s(f) \leq \rho(f)$.

First, we conclude the proof in the non-archimedean case. The argument is the same as at the end of proof of the preceding Lemma 11.5.3. We may assume that $\rho(f) = 1$ and need to prove $s(f) = 1$.

It is an easy consequence of Gauss's lemma in 1.6.3 that the residue field $k(v)$ is algebraically closed. By Lemma 11.5.3, the coefficients of f have absolute value at most 1 and $\pi(f)$ is not 0, thus there is $\pi(\mathbf{x}) \in k(v)^n$ with $\pi(f)(\pi(\mathbf{x})) \neq \pi(0)$. Now the claim is a consequence of $|f(\mathbf{x})| = 1$.

It remains to prove $\rho(f) \leq s(f)$ and the existence of $\mathbf{x} \in P$ with $|f(\mathbf{x})| = s(f)$ in the archimedean case. By Ostrowski's theorem in 1.2.6, we have $K = \mathbb{C}$. The existence of $\mathbf{x}$ follows from continuity of f and the compactness of P.

By the Cauchy inequalities for $f(\mathbf{x}) = \sum a_{\alpha}\mathbf{x}^{\alpha}$, we have $|a_{\alpha}| \leq s(f)$. This is well known for one variable and follows in the general case by induction on the number of variables. In order to prove the desired inequality, we may assume that f is a polynomial in n variables, of degree d. Then Cauchy's inequality yields *a fortiori*

$$\|f\| \leq (d+1)^n\, s(f)$$

and the claim follows as in the proof of Lemma 11.5.3. □

Definition 11.5.5. *The algebra* $K\{\mathbf{r}^{-1}\mathbf{x}\}$ *of* ***strictly convergent power series*** *is the completion of* $K\langle \mathbf{r}^{-1}\mathbf{x} \rangle$ *with respect to the spectral norm* $\rho_{\mathbf{r}}$.

Proposition 11.5.6. *The completion* $K\{\mathbf{r}^{-1}\mathbf{x}\}$ *of* $K\langle \mathbf{r}^{-1}\mathbf{x} \rangle$ *is:*

(a) *If* $|\ |$ *is not archimedean, then*

$$K\{\mathbf{r}^{-1}\mathbf{x}\} = \left\{\sum a_\alpha \mathbf{x}^\alpha \in K[[\mathbf{x}]] \mid \lim_{|\alpha|\to\infty} |a_\alpha|\mathbf{r}^\alpha = 0\right\}$$

and $\rho_\mathbf{r}$ *is the multiplicative Gauss norm, i.e.*

$$\rho_\mathbf{r}\left(\sum a_\alpha \mathbf{x}^\alpha\right) = \max_{\alpha\in\mathbb{N}^n} |a_\alpha|\mathbf{r}^\alpha .$$

(b) *If* $K = \mathbb{C}$, *then* $\mathbb{C}\{\mathbf{r}^{-1}\mathbf{x}\}$ *is equal to the set of continuous complex functions on* $P(\mathbf{r})$, *which are analytic in the interior and* $\rho_\mathbf{r}$ *is the supremum norm on* $P(\mathbf{r})$.

(c) *If* $K = \mathbb{R}$, *then* $\mathbb{R}\{\mathbf{r}^{-1}\mathbf{x}\}$ *is the Banach subalgebra of* $\mathbb{C}\{\mathbf{r}^{-1}\mathbf{x}\}$ *given by the functions that have Taylor series at* $\mathbf{0}$ *with real coefficients.*

Proof: The non-archimedean case follows easily from Lemma 11.5.3. If $K = \mathbb{C}$, then Lemma 11.5.4 and Weierstrass's theorem show that $\mathbb{C}\{\mathbf{r}^{-1}\mathbf{x}\}$ consists of continuous functions on $P(\mathbf{r})$, which are analytic in the interior, and that $\rho_\mathbf{r}$ is the supremum norm on $P(\mathbf{r})$.

Conversely, let f be a continuous function on $P(\mathbf{r})$, which is analytic in the interior. Clearly, we may assume $r_1 = \cdots = r_n = 1$. Let $\sum a_\alpha \mathbf{x}^\alpha$ be the Taylor series of f at $\mathbf{0}$. By the Cauchy formula and continuity of f, we have

$$a_\alpha = \left(\frac{1}{2\pi i}\right)^n \int_{|y_1|=1} \cdots \int_{|y_n|=1} \frac{f(\mathbf{y})}{y_1^{\alpha_1+1}\cdots y_n^{\alpha_n+1}}\, dy_1\cdots dy_n .$$

We conclude that $\sum a_\alpha e^{i\alpha\cdot t}$ is the Fourier series of the periodic function $f(e^{it_1},\ldots,e^{it_n})$, $t_i \in \mathbb{R}$. For $\beta \in \mathbb{N}^n$, let us consider the partial sum

$$s_\beta(\mathbf{x}) = \sum_{\alpha_1\le\beta_1} \cdots \sum_{\alpha_n\le\beta_n} a_\alpha \mathbf{x}^\alpha .$$

By the generalization of Féjer's theorem to several variables (cf. [**343**], Th.XVII.1.20), the arithmetic means

$$\sigma_\gamma(e^{it_1},\ldots,e^{it_n}) = \frac{1}{\gamma_1+1}\cdots\frac{1}{\gamma_n+1}\sum_{\beta_1\le\gamma_1}\cdots\sum_{\beta_n\le\gamma_n} s_\beta(e^{it_1},\ldots,e^{it_n})$$

converge uniformly to $f(e^{it_1},\ldots,e^{it_n})$ for $\gamma_1,\ldots,\gamma_n \to \infty$. The maximum modulus principle shows that the polynomials $\sigma_\gamma(z_1,\ldots,z_n)$ converge uniformly to $f(z_1,\ldots,z_n)$ on $\max|z_j| \le 1$. This proves the claim for $K = \mathbb{C}$.

If $K = \mathbb{R}$, then it is clear that uniform limits of real power series remain real. Conversely, if $f \in \mathbb{C}\{\mathbf{r}^{-1}\mathbf{x}\}$ has a real Taylor series at $\mathbf{0}$, then the above Féjer polynomials $\sigma_\gamma(z_1,\ldots,z_n)$ have real coefficients proving $f \in \mathbb{R}\{\mathbf{r}^{-1}\mathbf{x}\}$. □

11.5.7. Let $f(\mathbf{x}) = \sum a_\alpha \mathbf{x}^\alpha \in K\{\mathbf{r}^{-1}\mathbf{x}\}$ and let $\mathbf{x}_0 \in P(\mathbf{r})$ if $|\ |$ is not archimedean, and $\mathbf{x}_0$ be in the interior of the polydisk $P(\mathbf{r})$ if $|\ |$ is archimedean. We define

$$\mathbf{r}_0 := \begin{cases} \mathbf{r} & \text{if } |\ | \text{ is not archimedean} \\ \mathbf{r} - |\mathbf{x}_0| = (r_1 - |x_{01}|,\ldots,r_n - |x_{0n}|) & \text{if } |\ | \text{ is archimedean.} \end{cases}$$

With the **Taylor coefficients** of $f(\mathbf{x})$ at $\mathbf{x}_0$ defined by

$$\partial_{\boldsymbol{\alpha}} f(\mathbf{x}_0) := \sum_{\boldsymbol{\beta}} \binom{\boldsymbol{\alpha}+\boldsymbol{\beta}}{\boldsymbol{\alpha}} a_{\boldsymbol{\alpha}+\boldsymbol{\beta}} \mathbf{x}_0^{\boldsymbol{\beta}}$$

(here, as usual, $\binom{\boldsymbol{\alpha}+\boldsymbol{\beta}}{\boldsymbol{\alpha}} = \prod \binom{\alpha_i+\beta_i}{\alpha_i}$), we have the **Taylor series**

$$f(\mathbf{x}+\mathbf{x}_0) = \sum_{\boldsymbol{\alpha}} \partial_{\boldsymbol{\alpha}} f(\mathbf{x}_0)\mathbf{x}^{\boldsymbol{\alpha}}. \tag{11.5}$$

There is a unique element $g(\mathbf{x})$ of $K\{\mathbf{r}_0^{-1}\mathbf{x}\}$ such that

$$f(\mathbf{x}+\mathbf{x}_0) = g(\mathbf{x})$$

for every $\mathbf{x} \in P(\mathbf{r}_0)$. If K is a field of characteristic 0, we have

$$\partial_{\boldsymbol{\alpha}} f(\mathbf{x}_0) = \frac{1}{\alpha_1! \cdots \alpha_n!} \frac{\partial^{|\boldsymbol{\alpha}|} f}{\partial x_1^{\alpha_1} \cdots \partial x_n^{\alpha_n}}(\mathbf{x}_0),$$

which is the familiar formula for the coefficients of the Taylor expansion, and justifies the notation $\partial_{\boldsymbol{\alpha}}$ for them. The Taylor series in the non-archimedean case cannot be used for the purpose of analytic continuation, contrary to what happens if $K = \mathbb{C}$.

If $|\ |$ is not archimedean, we have $\rho_{\mathbf{r}}(f) = \rho_{\mathbf{r}}(g)$ (by Lemma 11.5.4), therefore

Proposition 11.5.8. *In the non-archimedean case the norm $\rho_{\mathbf{r}}$ on $K\{\mathbf{r}^{-1}\mathbf{x}\}$ is invariant by the translation $\mathbf{x} \mapsto \mathbf{x}+\mathbf{x}_0$, for any $\mathbf{x}_0 \in P(\mathbf{r})$.*

For convenience, we now state the **Cauchy inequalities** in terms of the norm $\rho_{\mathbf{r}}$:

Proposition 11.5.9. *If $f(\mathbf{x}) = \sum a_{\boldsymbol{\alpha}} \mathbf{x}^{\boldsymbol{\alpha}} \in K\{\mathbf{r}^{-1}\mathbf{x}\}$, then*

$$|a_{\boldsymbol{\alpha}}| \leq \rho_{\mathbf{r}}(f)\mathbf{r}^{-\boldsymbol{\alpha}}$$

for every $\boldsymbol{\alpha}$.

Proof: If $|\ |$ is not archimedean, this follows from Lemma 11.5.3. If instead $|\ |$ is archimedean then, as noted before, the result in one variable is classical and due to Cauchy, and in general follows by induction on the number of variables. □

Corollary 11.5.10. *In the notation of 11.5.7, it holds*

$$|\partial_{\boldsymbol{\alpha}} f(\mathbf{x}_0)| \leq \rho_{\mathbf{r}}(f)\mathbf{r}_0{}^{-\boldsymbol{\alpha}}.$$

Proof: Clear from Proposition 11.5.8, Proposition 11.5.9, and the Taylor series expansion (11.5) around $\mathbf{x}_0$. □

A related estimate is **Schwarz's lemma**:

Proposition 11.5.11. *Let $f \in K\{\mathbf{r}^{-1}\mathbf{x}\}$ vanish to order k at $\mathbf{0}$. Then for $0 < t \leq 1$ it holds*

$$\rho_{t\mathbf{r}}(f) \leq t^k \rho_{\mathbf{r}}(f).$$

Proof: In the non-archimedean case, the claim follows directly from Lemma 11.5.3. If instead $|\ |$ is archimedean, we may assume that $K = \mathbb{C}$. By Lemma 11.5.4, there is $\mathbf{x}^* \in P(t\mathbf{r})$ with $|f(\mathbf{x}^*)| = \rho_{t\mathbf{r}}(f)$. Since f vanishes to order k at $\mathbf{0}$, the function $g(\zeta) = \zeta^{-k} f(\zeta\,\mathbf{x}^*/t)$ is an element of $K\{\zeta\}$. By the maximum principle, we have

$$t^{-k}\rho_{t\mathbf{r}}(f) = |g(t)| \leq \max_{|\zeta|=1} |g(\zeta)| = \max_{|\zeta|=1} |f(\zeta\mathbf{x}^*/t)|,$$

and the result follows from Lemma 11.5.4. □

Proposition 11.5.12. *Let* $\mathbf{x} = (x_1, \ldots, x_n)$, $\mathbf{y} = (y_1, \ldots, y_m)$, *let* $f \in K\{\mathbf{r}^{-1}\mathbf{x}\}$. *For* $j = 1, \ldots, n$, *let* $g_j \in K\{\mathbf{s}^{-1}\mathbf{y}\}$ *with* $\rho_{\mathbf{s}}(g_j) \leq r_j$. *Then* $f(g_1, \ldots, g_n)$ *is an element of* $K\{\mathbf{s}^{-1}\mathbf{y}\}$ *with*

$$\rho_{\mathbf{s}}\left(f(g_1, \ldots, g_n)\right) \leq \rho_{\mathbf{r}}(f).$$

Proof: Again, by the rescaling described at the end of 11.5.2, we may assume $\mathbf{r} = (1, \ldots, 1)$ and $\mathbf{s} = (1, \ldots, 1)$. If f is a polynomial, then $f(g_1, \ldots, g_n) \in K\{\mathbf{y}\}$. In general, f is a uniform limit of polynomials on the closed polydisk P and the claim follows from the fact that ρ is the supremum norm on P. □

11.5.13. Given a polynomial $p(\mathbf{x}, t)$ in $n+1$ variables with coefficients in the complete field K and $p(\mathbf{0}, 0) = 0$, $\frac{\partial p}{\partial t}(\mathbf{0}, 0) \neq 0$, we know by the implicit function theorem that, at least in characteristic 0, the equation $p(\mathbf{x}, \xi) = 0$ has a unique solution $\xi = \xi(\mathbf{x})$, with $\xi(\mathbf{0}) = 0$, in a neighborhood of $\mathbf{0}$. In fact, $\xi(\mathbf{x})$ has a Taylor series expansion

$$\xi(\mathbf{x}) = \sum a_\alpha \mathbf{x}^\alpha$$

convergent in a neighborhood of $\mathbf{0}$.

We give here a proof of the **local Eisenstein theorem for a strictly convergent power series** $p(\mathbf{x}, t)$. Our argument is as for the implicit function theorem in calculus, but nonetheless we give here a detailed proof.

Theorem 11.5.14. *Let* $p(\mathbf{x}, t) \in K\{\mathbf{R}^{-1}\mathbf{x}, S^{-1}t\}$. *Suppose that*

$$p(\mathbf{0}, 0) = 0 \quad \text{and} \quad \frac{\partial p}{\partial t}(\mathbf{0}, 0) \neq 0.$$

There is a unique $\xi(\mathbf{x}) \in K[[\mathbf{x}]]$ *with the properties*

$$p(\mathbf{x}, \xi(\mathbf{x})) = 0, \qquad \xi(\mathbf{0}) = 0.$$

Let

$$B := S \left| \frac{\partial p}{\partial t}(\mathbf{0}, 0) \right| \rho_{\mathbf{R}, S}(p)^{-1}$$

and let $r < B^2$ *in the non-archimedean case, resp.* $r = \frac{1}{|32|} B^2$ *in the archimedean case. Then* $\xi(\mathbf{x}) \in K\{(r\mathbf{R})^{-1}\mathbf{x}\}$ *and*

$$\sup_{\mathbf{x} \in P(r\mathbf{R})} |\xi(\mathbf{x})| \leq \begin{cases} BS & \text{if } |\ | \text{ is not archimedean} \\ \frac{1}{|16|} BS & \text{if } |\ | \text{ is archimedean.} \end{cases}$$

Existence and uniqueness of ξ will follow from an application of Banach's fixed point theorem, and we will also get a polydisk P on which ξ is strictly convergent and an upper bound for the supremum norm of ξ on P. Then Cauchy's inequalities will give the desired estimate for the Taylor coefficients a_α.

We first recall **Banach's fixed point theorem:**

Theorem 11.5.15. *Let (X,d) be a complete metric space and let $\varphi : X \longrightarrow X$ be a contractive map, i.e. there is $\theta < 1$ such that*

$$d\left(\varphi(x),\varphi(y)\right) \leq \theta \cdot d(x,y) \quad \text{for all } x,\, y \in X.$$

Then φ has a unique fixed point.

Proof: We start with an arbitrary point x_0 of X. Then we apply φ iteratively to define $x_n := \varphi^n(x_0)$. The triangle inequality shows

$$d(x_{k+l},x_k) \leq \sum_{n=0}^{l-1} d(x_{k+n+1},x_{k+n}) \leq \sum_{n=0}^{l-1} \theta^{k+n} d(x_1,x_0) \leq \frac{\theta^k}{1-\theta} \cdot d(x_1,x_0).$$

Hence we have a Cauchy sequence converging to $x \in X$ and it follows from continuity of φ that x is a fixed point. Uniqueness is clear for a contraction. □

Proof of Theorem 11.5.14: We indicate by a dot the partial derivative $\frac{\partial}{\partial t}$, as in $\dot{f} := \frac{\partial f}{\partial t}$. By the Cauchy inequalities (see Proposition 11.5.9), we have

$$B := S\,|\dot{p}(\mathbf{0},0)|\,\rho_{\mathbf{R},S}(p)^{-1} \leq 1.$$

If the absolute value is archimedean, we may assume that $|\ \ |$ is the usual euclidean absolute value. Let

$$g(\mathbf{x},t) := t - \dot{p}(\mathbf{0},0)^{-1}p(\mathbf{x},t).$$

We shall apply Banach's fixed point theorem to the map

$$\varphi : f(\mathbf{x}) \longmapsto g(\mathbf{x}, f(\mathbf{x})).$$

We have to choose an appropriate set X of power series and a metric $d(x,y)$ in X which satisfy the hypothesis of the Banach fixed point theorem. The set X will depend on positive parameters $r_1,\ldots,r_n,s$ to be chosen later, which at this moment are required to satisfy $r_j/R_j \leq s/S \leq 1$ and, in the archimedean case, the stronger condition $r_j/R_j \leq s/S \leq 1/4$. As usual, we set $\mathbf{r} := (r_1,\ldots,r_n)$. We define

$$X = X(\mathbf{r},s) := \{f \in K\{\mathbf{r}^{-1}\mathbf{x}\} \mid \rho_{\mathbf{r}}(f) \leq s\}.$$

The set X is a closed subset of $K\{\mathbf{r}^{-1}\mathbf{x}\}$, hence X is complete with respect to the supremum norm $\rho_{\mathbf{r}}$ in the polydisk $P(\mathbf{r})$. We shall determine the parameters $\mathbf{r}$, s in such a way that φ is a contraction on X with respect to the distance function $d(f_1,f_0) = \rho_{\mathbf{r}}(f_1 - f_0)$. In order to achieve this, we may always assume, after a suitable field extension as in Lemma 11.5.2, that $r_j, R, s, S \in |K|$. This has the advantage that the spectral norm of the occuring strictly convergent power series equals the maximum norm on the corresponding polydisk (see Lemma 11.5.4). At the end of the argument for applying Banach's fixed point theorem, we will go back to the original K to be sure that $\xi(\mathbf{x})$ will have coefficients in K.

First, we claim that for $t, t_0 \in \overline{K}$ with $|t| \leq s$, $|t_0| \leq s$, we have

$$\rho_{\mathbf{r}}\left(g(\mathbf{x},t) - g(\mathbf{x},t_0)\right) \leq \theta\,|t - t_0|, \tag{11.6}$$

where

$$\theta := \begin{cases} \frac{s}{S} B^{-1} & \text{in the non-archimedean case} \\ \frac{8s}{S} B^{-1} & \text{in the archimedean case.} \end{cases}$$

Suppose first that the absolute value is archimedean. By Ostrowski's theorem in 1.2.6, we may assume $K = \mathbb{C}$. For $\mathbf{x} \in P(\mathbf{r})$, we have

$$g(\mathbf{x}, t) - g(\mathbf{x}, t_0) = \int_{t_0}^{t} \dot{g}(\mathbf{x}, u) du, \tag{11.7}$$

where the integral is on the line segment from t_0 to t; note that if $|t_0|, |t| \leq s$ then $|u| \leq s$ follows by convexity, and also $\dot{g}$ is well defined in any closed polydisk in the interior of $P(\mathbf{R}, S)$. By construction, we have $\dot{g}(\mathbf{0}, 0) = 0$. We apply Schwarz's lemma from 11.5.11 and Cauchy's inequality (see Corollary 11.5.10), getting (note that $r_j/R_j \leq s/S \leq 1/4$ and $B \leq 1$)

$$\rho_{\mathbf{r},s}(\dot{g}) \leq \frac{2s}{S}\, \rho_{\frac{S\mathbf{r}}{2s}, \frac{S}{2}}(\dot{g}) \leq \frac{4s}{S^2}\, \rho_{(\mathbf{R},S)}(g) \leq \frac{4s}{S}\,(1 + B^{-1}) \leq \frac{8s}{S}\, B^{-1}.$$

Using this bound in (11.7), we get (11.6).

In the non-archimedean case, let us introduce the operator D_t acting on power series in t by $f(t) \mapsto \frac{1}{t}(f(t) - f(0))$.

For $\mathbf{x} \in P(\mathbf{r})$ and $|t| \leq s$, $|t_0| \leq s$, an easy calculation shows

$$|g(\mathbf{x}, t) - g(\mathbf{x}, t_0)| \leq \rho_{(\mathbf{r},s)}(D_t g)\, |t - t_0|,$$

whence Schwarz's lemma in 11.5.11 implies

$$|g(\mathbf{x}, t) - g(\mathbf{x}, t_0)| \leq \frac{s}{S}\, \rho_{(\mathbf{R},S)}(D_t g)\, |t - t_0|. \tag{11.8}$$

We have

$$\rho_{(\mathbf{R},S)}(D_t g) \leq S^{-1} \rho_{(\mathbf{R},S)}(g) \leq B^{-1}.$$

Now (11.6) follows from this inequality and (11.8).

By Proposition 11.5.12, φ maps X to $K\{\mathbf{r}^{-1}\mathbf{x}\}$. The condition $\theta < 1$, as needed for the application of Banach's fixed point theorem, follows from (11.6) if $\frac{s}{S} < B$ or $\frac{s}{S} < B/8$ according as we are in the non-archimedean or archimedean case.

In order to complete the proof, we need to check that φ maps X into itself for $\mathbf{r}, s$ sufficiently small. This means checking that

$$|g(\mathbf{x}, t)| \leq s$$

for $\mathbf{x} \in P(\mathbf{r})$ and $|t| \leq s$.

We begin by noting that, since $g(\mathbf{0}, 0) = 0$, Schwarz's lemma from 11.5.11 yields

$$\sup_{\mathbf{x} \in P(\mathbf{r})} |g(\mathbf{x}, 0)| \leq \lambda \sup_{\mathbf{x} \in P(\mathbf{R})} |g(\mathbf{x}, 0)| \leq \lambda S B^{-1}, \tag{11.9}$$

where $\lambda = \max\{r_j/R_j\}$. Suppose first that $|\ |$ is archimedean. By (11.6) and (11.9) we have

$$|g(\mathbf{x}, t)| \leq \theta s + |g(\mathbf{x}, 0)| \leq \theta s + \lambda S B^{-1}. \tag{11.10}$$

In the non-archimedean case, (11.6) and the ultrametric inequality give

$$|g(\mathbf{x},t)| \le \max\left(\theta s, |g(\mathbf{x},0)|\right) \le \max\left(\theta s, \lambda S B^{-1}\right). \tag{11.11}$$

Now let $\mathbf{r}, s$ be given by

$$\frac{r_1}{R_1} = \cdots = \frac{r_n}{R_n} = a\,B^2, \qquad \frac{s}{S} = b\,B,$$

where we choose for example any $0 < a = b < 1$ in the non-archimedean case, and $a = 1/32$ and $b = 1/16$ in the archimedean case. Thus we see from (11.10) and (11.11) that

$$\rho_{(\mathbf{r},s)}(g(\mathbf{x},t)) \le s, \qquad \theta < 1$$

hold. We apply Banach's fixed point theorem, obtaining a unique $\xi \in K\{\mathbf{r}^{-1}\mathbf{x}\}$ with

$$p(\mathbf{x},\xi(\mathbf{x})) = 0, \qquad \rho_{\mathbf{r}}(\xi) \le s.$$

This proves the existence of a solution $\xi(\mathbf{x})$ satisfying the bound stated in the theorem.

It is also clear that $\xi(\mathbf{0}) = 0$. In fact, since φ maps the closed set

$$X_0 := \{f \in X \mid f(\mathbf{0}) = 0\}$$

into itself, we may apply Banach's fixed point theorem to X_0 as well. Uniqueness of the fixed point shows that $\xi \in X_0$, hence $\xi(\mathbf{0}) = 0$.

We leave the proof of the uniqueness of ξ as a formal power series as an exercise for the reader. For example, we can apply once more the Banach fixed point theorem, in the space $K[[\mathbf{x}]]$ of formal power series with the norm

$$|f| = \max_{a_\alpha \ne 0} e^{-|\alpha|},$$

to the set $X := \{f \in K[[\mathbf{x}]] \mid f(\mathbf{0}) = 0\}$, with the same map φ. □

The next statement is a sharper form of the **local Eisenstein theorem for a polynomial.**

Corollary 11.5.16. *Let $p \in K[\mathbf{x},t]$ with $p(\mathbf{0},0) = 0$ and $\frac{\partial p}{\partial t}(\mathbf{0},0) \ne 0$. Denote by $|p|$ the Gauss norm of p. Let $d_1,\dots,d_{n+1}$ be the partial degrees of p and let*

$$D := \prod_{j=1}^{n+1} (d_j + 1).$$

Then there is a unique

$$\xi = \sum_{\alpha \in \mathbb{N}^n} a_\alpha \mathbf{x}^\alpha \in K[[\mathbf{x}]]$$

with $\xi(\mathbf{0}) = 0$ and $p(\mathbf{x},\xi(\mathbf{x})) = 0$. For $\alpha \in \mathbb{N}^n$, we have

$$|a_\alpha| \le C^{|\alpha|} \cdot \left(|p| \Big/ \left|\frac{\partial p}{\partial t}(\mathbf{0},0)\right|\right)^{2|\alpha|-1},$$

where

$$C = \begin{cases} 1 & \text{in the non-archimedean case} \\ |32D^2| & \text{in the archimedean case.} \end{cases}$$

Proof: This follows from Theorem 11.5.14 and the Cauchy inequalities in 11.5.9, taking $(\mathbf{R}, S) = (1,\dots,1)$. In the non-archimedean case, we have $\rho(p) = |p|$. In the archimedean case, the claim is a consequence of $\rho(p) \le |Dp|$. □

11.5.17. The bound given by the local Eisenstein theorem is sharp in the non-archimedean case, and not far from the truth in the archimedean case.

Consider the polynomial $p(x,t) = x - 2at + t^2$, where $a \neq 0$ is a parameter satisfying $|a| \leq 1$. The formal power series solution with $\xi(0) = 0$ is given by

$$\xi(x) = a - a\sqrt{1 - x/a^2} = \sum_{j=1}^{\infty} (-1)^{j-1} \binom{\frac{1}{2}}{j} a^{1-2j} x^j.$$

It is easy to see that $(-1)^{j-1} 2^{2j-1} \binom{\frac{1}{2}}{j}$ is a positive integer for $j \geq 1$.

Suppose that $|\ |$ is not archimedean, with valuation ring R and residue field $k(v)$, and suppose that $|2| = 1$. Then

$$|\dot{p}(0,0)|/|p| = \frac{|2a|}{\max(1, |2a|)} = |a|,$$

and it follows that

$$|\partial_j \xi(0)| = \left(|p| \Big/ \left| \frac{\partial p}{\partial t}(0,0) \right| \right)^{2j-1}$$

whenever $|\binom{\frac{1}{2}}{j}| = 1$. This indeed happens for infinitely many values of j if $|2| = 1$. Suppose this is not the case. We choose $a \in K$ with $|a| = 1$ and look at the reduction $\pi(\xi(x))$ of $\xi(x) \in R[[x, a^{-1}]]$ modulo the maximal ideal of R. Then this reduction would be a polynomial in x and a^{-1}, contradicting the fact that the polynomial $x - 2\pi(a)t + t^2$ is irreducible over $k(v)(x)$.

The same example shows that, apart for the value of the numerical constant C, the bound given in the local Eisenstein theorem is also sharp in the archimedean case.

11.6. A lower bound for the height

Let C be an irreducible projective smooth curve over a number field K and let us fix $P_0 \in C(K)$. Let V be a Vojta divisor satisfying (V1) and (V2). We are interested in getting a lower bound for the height $h_V(P,Q)$, where $P, Q \in C(\overline{K})$. This is obtained by means of Lemma 11.6.5 and Lemma 11.6.7. The first lemma gives an explicit lower bound in term of the Taylor coefficients of local coordinates for C viewed as algebraic functions of a uniformizing parameter of C at P or Q. The second lemma applies the local Eisenstein theorem to bound these Taylor coefficients. The reader is assumed to be familiar with the concept of tangent spaces provided by A.7.

11.6.1. Let ∂, ∂' be non-zero vectors in the tangent space of C at P, Q. As in 11.5.7, we abbreviate

$$\partial_i := \frac{1}{i!}\partial^i, \quad \partial'_i := \frac{1}{i!}\partial'^i.$$

Any differential operator on $\mathcal{O}_{C\times C,(P,Q)}$ of degree k with values in $K(P,Q)$ is a homogeneous polynomial of degree k in the variables ∂, ∂' with coefficients in $K(P,Q)$. In fact, they act on $K(C \times C)$. The advantage of the normalizations above is that Leibniz's rule has an easier form. If $f_1, \dots, f_r$ are rational functions on C, then

$$\partial_i(f_1 \cdots f_r) = \sum \partial_{i_1} f_1 \cdots \partial_{i_r} f_r,$$

where the sum ranges over all $(i_1, \dots, i_r)$ with $i_1 + \cdots + i_r = i$. This is easily proved by induction on r, starting with $r = 2$.

11.6.2. Let $s \in \Gamma(C \times C, O(V)) \setminus \{0\}$. A pair $(i_1^*, i_2^*) \in \mathbb{N}^2$ is called **admissible** if and only if

$$\partial_{i_1^*} \partial'_{i_2^*} s(P, Q) \neq 0$$

and

$$\partial_{i_1} \partial'_{i_2} s(P, Q) = 0$$

whenever $i_1 \leq i_1^*$, $i_2 \leq i_2^*$ and $(i_1, i_2) \neq (i_1^*, i_2^*)$. In order to make sense of the above formulas, we should choose a trivialization of $O(V)$ in (P, Q). It is also clear that admissibility is independent of the choice of the trivialization and the choice of ∂, ∂'. By 11.2.6

$$\xi_{ij} := (x_i/x_j)|_C, \qquad \xi'_{ij} := (x'_i/x'_j)|_C$$

are well-defined non-zero rational functions on C for $i, j = 0, \dots, n$. We also write $\boldsymbol{\xi}_j$ for the vector with components ξ_{ij}, $i = 0, \dots, n$, and similarly for $\boldsymbol{\xi}'_{j'}$.

11.6.3. We are going to choose an explicit height function relative to $O(V)$. Choose a finite extension L/K such that $P, Q \in C(L)$. We use the decomposition

$$V = (\delta_1 N \{P_0\} \times C + \delta_2 N C \times \{P_0\}) - dB$$

of V as a difference of two very ample divisors (see 11.2.6). Now we have generating sections $\mathbf{x}^{\mathbf{h}} \mathbf{x}'^{\mathbf{h}'}$ ($|\mathbf{h}| = \delta_1, |\mathbf{h}'| = \delta_2$) of $O(\delta_1 N \{P_0\} \times C + \delta_2 N C \times \{P_0\})$ and $\mathbf{y}^{\mathbf{i}}$ ($|\mathbf{i}| = d$) of $O(dB)$. With respect to the presentation (see 2.2.1)

$$\left(s_V; O(\delta_1 N \{P_0\} \times C + \delta_2 N C \times \{P_0\}), \mathbf{x}^{\mathbf{h}} \mathbf{x}'^{\mathbf{h}'}; O(dB), \mathbf{y}^{\mathbf{i}}\right),$$

we have the global height function (see 2.4.6)

$$\begin{aligned} h_V(P, Q) :&= \sum_{v \in M_L} \max_{\mathbf{h}, \mathbf{h}'} \min_{\mathbf{i}} \log \left| \frac{\mathbf{x}^{\mathbf{h}} \mathbf{x}'^{\mathbf{h}'}}{\mathbf{y}^{\mathbf{i}}}(P, Q) \right|_v \\ &= \sum_{v \in M_L} \max_{j, j'} \min_{i} \log \left| \frac{x_j^{\delta_1} x'^{\,\delta_2}_{j'}}{y_i^d}(P, Q) \right|_v . \end{aligned}$$

Note that the vectors $\mathbf{x}(P)$, $\mathbf{x}'(Q)$ and $\mathbf{y}(P, Q)$ are only defined up to a multiple. By the product formula, $h_V(P, Q)$ is well-defined; it is the difference of two Weil heights, the first given by the closed embedding

$$(P, Q) \longmapsto \left(\mathbf{x}^{\mathbf{h}}(P) \mathbf{x}'^{\mathbf{h}'}(Q)\right)_{|\mathbf{h}| = \delta_1,\ |\mathbf{h}'| = \delta_2}$$

and the second given by the closed embedding

$$(P,Q) \longmapsto \left(\mathbf{y}^{\mathbf{i}}(P,Q)\right)_{|\mathbf{i}|=d}.$$

11.6.4. From Lemma 11.2.7, we get bihomogeneous polynomials $F_i(\mathbf{x},\mathbf{x}')$ of bidegree (δ_1,δ_2) with

$$s = \left(F_i(\mathbf{x},\mathbf{x}')/y_i^d\right)|_{C\times C}, \quad i=0,\dots,m.$$

Let $h(\mathbf{F})$ be the height of the point in appropriate projective space whose coordinates are given by all the coefficients of $F_0,\dots,F_m$. For $v \in M_K$, let j_v be the index j for which $|\xi_{j0}(P)|_v$ is largest and similarly j'_v for $|\xi'_{j0}(Q)|_v$.

Lemma 11.6.5. *Let s be a non-zero global section of $O(V)$ and let (i_1^*,i_2^*) be admissible for s at (P,Q). With the notation introduced above, we have*

$$\begin{aligned} h_V(P,Q) \geq &- h(\mathbf{F}) - n\log\left((\delta_1+n)(\delta_2+n)\right) \\ &- \sum_{v\in M_L} \max_{\{i_\lambda\}}\left(\sum_\lambda \max_\nu \log|\partial_{i_\lambda}\xi_{\nu j_v}(P)|_v\right) \\ &- \sum_{v\in M_L} \max_{\{i'_\lambda\}}\left(\sum_\lambda \max_\nu \log|\partial'_{i'_\lambda}\xi'_{\nu j'_v}(Q)|_v\right) \\ &- (\delta_1+\delta_2+i_1^*+i_2^*), \end{aligned}$$

where $\{i_\lambda\}$ and $\{i'_\lambda\}$ run over all partitions of i_1^* and i_2^*.

Proof: We fix trivializations of $O_{\mathbb{P}^n}(1)$ at P and Q and of $O_{\mathbb{P}^m}(1)$ at (P,Q). By tensor product, this gives trivializations of all the bundles in question, in particular of V by 11.2.3 and 11.2.6. In the following, we use the trivializations without mention. For example s is viewed as a regular function at (P,Q).

We have seen in 11.6.3 that

$$h_V(P,Q) = -\sum_v \max_i \min_{j,j'} \log\left|\frac{y_i^d}{x_j^{\delta_1}x'^{\delta_2}_{j'}}(P,Q)\right|_v.$$

Assume that $x_j(P)x'_{j'}(Q)y_i(P,Q) \neq 0$. Only such i,j,j' are of importance in the above formula. By admissibility and Leibniz's rule, we have

$$\left(\frac{y_i^d}{x_j^{\delta_1}x'^{\delta_2}_{j'}}\partial_{i_1^*}\partial'_{i_2^*}s\right)(P,Q) = \partial_{i_1^*}\partial'_{i_2^*}\left(\frac{y_i^d}{x_j^{\delta_1}x'^{\delta_2}_{j'}}s\right)(P,Q)$$

and the right-hand side equals

$$\partial_{i_1^*}\partial'_{i_2^*}F_i(\boldsymbol{\xi}_j,\boldsymbol{\xi}'_{j'})(P,Q).$$

Using $(\partial_{i_1^*}\partial'_{i_2^*}s)(P,Q) \neq 0$ and the product formula, we get

$$h_V(P,Q) = -\sum_v \max_i \min_{j,j'} \log\left|\partial_{i_1^*}\partial'_{i_2^*}F_i(\boldsymbol{\xi}_j,\boldsymbol{\xi}'_{j'})(P,Q)\right|_v.$$

The number of monomials of F_i is bounded by $\binom{\delta_1+n}{n}\binom{\delta_2+n}{n} \le (\delta_1+n)^n(\delta_2+n)^n$. We conclude that

$$\begin{aligned} h_V(P,Q) \ge & - h(\mathbf{F}) - n\,\log\left((\delta_1+n)(\delta_2+n)\right) \\ & - \sum_v \min_j \max_{|\mathbf{l}|=\delta_1} \log \left|\partial_{i_1^*} \boldsymbol{\xi}_j^{\mathbf{l}}(P)\right|_v \\ & - \sum_v \min_{j'} \max_{|\mathbf{l}'|=\delta_2} \log \left|\partial'_{i_2^*} \boldsymbol{\xi}'^{\mathbf{l}'}_{j'}(Q)\right|_v . \end{aligned}$$

Now, since for each v we take the minimum with respect to j and j', we may take instead $j = j_v$ and $j' = j'_v$. This remark is quite important in what follows.

Let us consider $\log|\partial_{i_1^*}\boldsymbol{\xi}_j^{\mathbf{l}}(P)|_v$ for $v \in M_L$. By Leibniz's rule, we have

$$\partial_{i_1^*}\boldsymbol{\xi}_j^{\mathbf{l}} = \sum \prod_{\nu=0}^{n} \prod_{\mu=1}^{l_\nu} \partial_{i_{\mu\nu}} \xi_{\nu j},$$

where $\sum_{\mu\nu} i_{\mu\nu} = i_1^*$. Since the total number of pairs $\mu\nu$ equals δ_1, the number of possibilities for $i_{\mu\nu}$ equals $\binom{\delta_1+i_1^*-1}{i_1^*} \le 2^{\delta_1+i_1^*}$. We are interested in the case in which $j = j_v$ and $j' = j'_v$. We note that, since $|\xi_{j_v0}(P)|_v$ is the largest $|\xi_{j0}(P)|_v$, we have $|\xi_{\nu j_v}(P)|_v = |\xi_{\nu 0}(P)/\xi_{j_v 0}(P)|_v \le 1$ for every ν, allowing us to get rid of the terms with $i_{\mu\nu} = 0$ in estimating derivatives by means of Leibniz's rule; the same remark of course applies to $|\boldsymbol{\xi}'_{j'_v}|_v$.

This gives the bound

$$\log\left|\partial_{i_1^*}\boldsymbol{\xi}_{j_v}{}^{\mathbf{l}}(P)\right|_v \le \max_{\{i_\lambda\}} \left(\sum_\lambda \max_\nu \log|\partial_{i_\lambda}\xi_{\nu j_v}(P)|_v \right) + (\delta_1 + i_1^*)\,\varepsilon_v \log 2,$$

where $\varepsilon_v = [L_v : \mathbb{R}]/[L : \mathbb{Q}]$ if $v|\infty$ and $\varepsilon_v = 0$ otherwise, and where $\{i_\lambda\}$ runs over all partitions of i_1^*. An entirely analogous estimate holds for the sum involving $\boldsymbol{\xi}'_{j'}$, and Lemma 11.6.5 follows. □

11.6.6. Our next task consists in majorizing the sums appearing in Lemma 11.6.5; this we do by an application of the local Eisenstein theorem. Let us fix a non-constant $f \in K(C)$. For any $P \in C(\overline{K})$ which is neither a pole of f nor a zero of df, the function $\zeta = f - f(P)$ is a local uniformizer at P (i.e. a local parameter in the local ring). Therefore the completion of $\mathcal{O}_{C,P}$ with respect to its maximal ideal is isomorphic to $K(P)[[\zeta]]$ (by the Cohen structure theorem, see [**148**], Th.I.5.5A). Moreover, we may differentiate with respect to ζ.

By assumption and A.4.11, $K(C)$ is a finite extension of $K(f)$. Choose $g_{ij}(x,t) \in K[x,t]$ such that $g_{ij}(f,t) \in K(f)[t]$ is a minimal polynomial of ξ_{ij} over $K(f)$. Since $\mathrm{char}(K) = 0$, we have $\frac{\partial}{\partial t} g_{ij}(f,\xi_{ij}) \neq 0$ in $K(C)$ (irreducible polynomials are separable). Let $\deg(g_{ij})$ be the total degree of $g_{ij}(x,t)$.

Let us denote by Z the finite subset of $C(\overline{K})$ consisting of:

(a) all zeros of x_j, for $j = 0, \dots, n$;

(b) all poles of f;

(c) the support of $\operatorname{div}(df)$;

(d) the zeros of $\frac{\partial}{\partial t} g_{ij}(f, \xi_{ij})$, for $i, j = 0 \dots, n$.

We are going to apply the local Eisenstein theorem to the polynomials

$$p_{ij}(x,t) := g_{ij}(x + f(P),\, t + \xi_{ij}(P))$$

for any $P \notin Z$. Note that $p_{ij}(0,0) = 0$ and $\frac{\partial}{\partial t} p_{ij}(0,0) \neq 0$.

Let L/K be a finite extension with $P \in C(L)$. For $v \in M_L$, we have

$$|p_{ij}|_v \leq |C_1|_v^{\eta_v} \max\left(1, |f(P)|_v, |\xi_{ij}(P)|_v\right)^{\deg(g_{ij})} |g_{ij}|_v, \tag{11.12}$$

where $C_1 \leq (\deg(g_{ij}) + 1)^2 2^{\deg(g_{ij})}$ and

$$\eta_v = \begin{cases} 1 & \text{if } v \text{ is archimedean} \\ 0 & \text{if } v \text{ is not archimedean.} \end{cases}$$

Since ξ_{ij} is regular at P, we get from the local Eisenstein theorem in 11.4.1

$$|\partial_k \xi_{ij}(P)|_v \leq |C_2|_v^{k\eta_v} \left(\frac{|p_{ij}|_v}{|\frac{\partial}{\partial t} p_{ij}(0,0)|_v} \right)^{2k-1} \tag{11.13}$$

for $k \geq 1$ and $\partial_k = (1/k!)(\partial/\partial\zeta)^k$, with for example $C_2 = \max 32(\deg(g_{ij}) + 1)^4$ (using the sharper estimate in Corollary 11.5.16). Now we relate the sum in Lemma 11.6.5 with the canonical form on the Jacobian.

Lemma 11.6.7. *If $P \notin Z$, then*

$$\sum_{v \in M_L} \max_{\{i_\lambda\}} \left(\sum_\lambda \max_\nu \log |\partial_{i_\lambda} \xi_{\nu j_\nu}(P)|_v \right) \ll i_1^* \left(|\jmath(P)|^2 + 1 \right),$$

where the maximum runs over all partitions $\{i_\lambda\}$ of i_1^*. The constant implied in the symbol $\ll$ is independent of P and i_1^*.

Proof: We denote by a dot differentiation with respect to t, as in $\dot{f} = \frac{\partial f}{\partial t}$.

It is clear that $(f,\, \xi_{ij})$ may be viewed as the first two affine coordinates of a morphism φ from C into some projective space. Since $N[P_0]$ is ample on C, there is $k \in \mathbb{N}$ such that $O(kN[P_0]) \otimes \varphi^* O(-1)$ is very ample (see A.6.10). By Theorem 2.3.8, this proves

$$h\left((1 : f(P) : \xi_{ij}(P))\right) \leq h_\varphi(P) \ll h_{N[P_0]}(P) + 1. \tag{11.14}$$

From (11.13), we have

$$\sum_v \max_{\{i_\lambda\}} \sum_\lambda \max_\nu \log |\partial_{i_\lambda} \xi_{\nu j_\nu}(P)|_v$$

$$\leq i_1^* \log(C_2) + \sum_v \max_{\{i_\lambda\}} \sum_\lambda \max_\nu (2i_\lambda - 1)\left(\log^+ |p_{\nu j_\nu}| + \log^+ \left|\frac{1}{\dot{p}_{\nu j_\nu}(0,0)}\right|_v\right).$$

By (11.12) and (11.14), the right-hand side is bounded by

$$2i_1^* h(\mathbf{g}) + 2i_1^* \sum_{ij} \sum_v \log^+ \left(\frac{1}{|\dot{g}_{ij}(f(P), \xi_{ij}(P))|_v}\right)$$
$$+ 2i_1^* \log(C_2 C_1) + 2i_1^* C_3(h_{N[P_0]}(P) + 1),$$

where $h(\mathbf{g})$ is the height of the vector formed with the coefficients of all polynomials g_{ij} and 1, and where C_3 is the largest constant involved in (11.14) times the maximum of all $\deg(g_{ij})$. Now the claim follows from

$$h_{N[P_0]}(P) = \frac{N}{2g} |\jmath(P)|^2 + O(|\jmath(P)|) + O(1),$$

which is a consequence of Proposition 11.3.1 applied to the Vojta divisor $N\{P_0\} \times C$, and from the fact that, by (11.14) and $h(a) = h(1/a)$, we have

$$\begin{aligned}\sum_{ij} \sum_{v \in M_L} \log^+ \frac{1}{|\dot{g}_{ij}(f(P), \xi_{ij}(P))|_v} &= \sum_{ij} h\left(\frac{1}{\dot{g}_{ij}(f(P), \xi_{ij}(P))}\right) \\ &= \sum_{ij} h(\dot{g}_{ij}(f(P), \xi_{ij}(P))) \\ &\ll h((1 : f(P) : \xi_{ij}(P))) \ll h_{N[P_0]}(P) + 1.\end{aligned}$$

This completes the proof. □

11.7. Construction of a Vojta divisor of small height

In this section, we prove the crucial Lemma 11.7.3, which gives the existence of a section of $O(V)$, with V a Vojta divisor, of small height. The argument is fairly standard. The space of sections of $O(V)$ is presented as a subspace, given by linear relations with small height, of a vector space with a standard basis. The Riemann–Roch theorem shows that this subspace has large dimension, and the existence of a small section follows by Siegel's lemma or, equivalently, by geometry of numbers.

Let C be an irreducible projective smooth curve of genus g over a number field K and let us fix $P_0 \in C(K)$. We shall use the notation of Section 11.2, in particular V will be a Vojta divisor satisfying (V1) and (V2).

Lemma 11.7.1. *The following holds*

$$\dim \Gamma(C \times C, \psi^* O(\delta_1, \delta_2)) = (N\delta_1 + 1 - g)(N\delta_2 + 1 - g)$$

and, for $d_1 + d_2 > 4g - 4$

$$\dim \Gamma(C \times C, O(V)) \geq d_1 d_2 - g d^2 + O(d_1 + d_2).$$

Proof: By our assumptions in 11.2.3, the divisors $\delta_i N[P_0]$ are very ample on C. By the Riemann–Roch theorem (see Theorem A.13.5) on the curve C and $N \geq 2g + 1$, we obtain

$$\dim \Gamma(C, O(\delta_i N[P_0])) = N\delta_i + 1 - g,$$

for $i = 1, 2$. Since $O(\delta_1 N \{P_0\} \times C + \delta_2 N C \times \{P_0\})$ is the pull-back of $O(\delta_1, \delta_2)$ (see (11.1) on page 357), we get

$$\dim \Gamma(C \times C, \psi^* O(\delta_1, \delta_2)) \geq (N\delta_1 + 1 - g)(N\delta_2 + 1 - g).$$

The right-hand side is the dimension of the subspace of $\Gamma(C \times C, \psi^* O(\delta_1, \delta_2))$ generated by tensor products $s_1 \otimes s_2$, where s_i is a global section of $O_{\mathbb{P}^n}(\delta_i N[P_0])$. By (V2), it follows that any global section of $O(\delta_1 N \{P_0\} \times C + \delta_2 N C \times \{P_0\})$ is the restriction of a section of $O(\delta_1, \delta_2)$ (see 11.2.6, end). Because of

$$\Gamma(\mathbb{P}^n_K \times \mathbb{P}^n_K, O(\delta_1, \delta_2)) \cong \Gamma(\mathbb{P}^n_K, O_{\mathbb{P}^n}(\delta_1)) \otimes \Gamma(\mathbb{P}^n_K, O_{\mathbb{P}^n}(\delta_2))$$

(see Example A.6.13), the above-mentioned subspace equals $\Gamma(C \times C, \psi^* O(\delta_1, \delta_2))$ and we get the first claim.

For the proof of the second claim, we need the Riemann–Roch theorem on the surface $C \times C$. First note that $H = \{P_0\} \times C + C \times \{P_0\}$ is ample (from Lemma 11.2.4) and that

$$V \cdot H = d_1 + d_2$$

(from Lemma 11.2.1). Let K_C be a canonical divisor on C; then $K_{C \times C} := K_C \times C + C \times K_C$ is a canonical divisor on $C \times C$, numerically equivalent to $(2g - 2)H$. The former follows immediately from $T_{C \times C} = T_C \oplus T_C$. In order to verify the latter, we may assume that we are working with varieties over an algebraically closed field (see A.9.25). A canonical divisor K_C of the curve C, of genus g, has degree $2g - 2$ (see A.13.6), hence it is algebraically equivalent to $(2g - 2)[P_0]$ because the degree characterizes algebraic equivalence of divisors on a curve (see A.9.40). Therefore, $(2g - 2)H$ is algebraically equivalent to $K_{C \times C}$ (use A.9.36) and *a fortiori* numerically equivalent to it (see A.9.41). If

$$d_1 + d_2 > K_{C \times C} \cdot H = 4g - 4,$$

we have $(K_{C \times C} - V) \cdot H < 0$. Therefore

$$\dim \Gamma(C \times C, O(K_{C \times C} - V)) = 0,$$

because an effective divisor cannot have a negative intersection number with the ample divisor H. By the Riemann–Roch theorem in A.13.9, we get

$$\dim \Gamma(C \times C, O(V)) \geq \frac{1}{2} V \cdot (V - K_{C\times C}) + 1 + p_a(C \times C),$$

where $p_a(C \times C)$ is the arithmetic genus. Again by Lemma 11.2.1, we get

$$V \cdot V = 2d_1 d_2 - 2gd^2.$$

Together with

$$V \cdot K_{C\times C} = (2g-2) V \cdot H = (2g-2)(d_1 + d_2),$$

we get the second claim. □

11.7.2. We need now another assumption for the parameters of the Vojta divisor.

(V3) $d_1 + d_2 > 4g - 4$ *and* $d_1 d_2 - gd^2 > \gamma\, d_1 d_2$ *for some* $\gamma > 0$.

Here, γ is independent of d_1, d_2, d. As we have seen at the beginning of A.11.6, we may map $C \times C$ by a birational morphism onto a hypersurface of degree D in $\mathbb{P}^3_K$. We denote the projective coordinates in $\mathbb{P}^3_K$ by $\mathbf{z}$ and we may assume that the polynomial giving the hypersurface has the same form as in A.11.7 or in 2.5.1. All presentations (see 2.5.4) will refer to this set up.

Now we are ready to construct a **Vojta divisor with small height**.

Lemma 11.7.3. *There are two positive constants C_4, C_5 independent of d_1, d_2, d and γ with the following property. Let V be a Vojta divisor satisfying (V1), (V2), (V3), and $d_1, d_2 \geq C_4/\gamma$. Then there is a non-zero global section s of $O(V)$ such that the polynomials $F_0, \ldots, F_m$ in Lemma 11.2.7 may be chosen with*

$$h(\mathbf{F}) \leq C_5 (d_1 + d_2)/\gamma.$$

Proof: The idea is to apply Siegel's lemma to get a section of small height. Thus we have to transfer the equations in Lemma 11.2.7 into a linear system of equations with coefficients in K. Thereby, we must be careful not to increase the height of the linear system too much.

We consider C as a curve in $\mathbb{P}^n_K$ of degree N (via the closed embedding $\varphi_{N[P_0]}$) and we may also assume, by a linear change of coordinates, that the projection $p(\mathbf{x}) = (x_0 : x_1 : x_2)$ maps C by a birational morphism into $\mathbb{P}^2_K$. The purpose of this projection is to reduce the number of linear equations to be considered in the application of Siegel's lemma; this is a rather important step in our proof. Moreover, we may also assume that $p(C)$ is explicitly given by a homogeneous polynomial

$$f(x_0, x_1, x_2) = a_0 + a_1 x_2 + \ldots + a_{N-1} x_2^{N-1} + x_2^N$$

with $a_i \in K[x_0, x_1]$ homogeneous polynomials of degree $N - i$ (see A.11.5–A.11.7 for details).

Now let U be the subspace of $\Gamma(C \times C, O(V))$ consisting of sections of the form

$$s = y_i^{-d} F_i(x_0, x_1, x_2; x_0', x_1', x_2')\Big|_{C\times C}$$

for $i = 0, \ldots, m$ and bihomogeneous polynomials F_i of bidegree (δ_1, δ_2) with the additional restriction that

$$\deg_{x_2} F_i < N, \qquad \deg_{x_2'} F_i < N.$$

We may assume $\delta_i \geq N$, then the number of possible monomials so involved in F_i is

$$N^2 \left(\delta_1 - \frac{N-3}{2}\right)\left(\delta_2 - \frac{N-3}{2}\right).$$

Since f is irreducible over $K(x_0, x_1)$, we see that the restrictions to $C \times C$ of the above monomials are linearly independent over K. By Lemma 11.2.7, the vector space $\Gamma(C \times C, O(V))$ may be identified with

$$W := \left\{(F_i) \in \Gamma(C \times C, O(\delta_1, \delta_2))^{m+1} \mid (F_i) \text{ satisfies (11.2) on page 358}\right\}.$$

The above considerations and Lemma 11.7.1 show that

$$\operatorname{codim}(U^{m+1}, \Gamma(C \times C, O(\delta_1, \delta_2))^{m+1}) = O(\delta_1 + \delta_2);$$

therefore, we do not lose too many sections if we restrict our attention to $U^{m+1} \cap W$. More precisely, Lemma 11.7.1 shows that

$$\begin{aligned}\dim(U^{m+1} \cap W) &\geq \dim(W) - \operatorname{codim}(U^{m+1}, \Gamma(C \times C, O(\delta_1, \delta_2))^{m+1}) \\ &\geq d_1 d_2 - g d^2 - O(d_1 + d_2).\end{aligned} \tag{11.15}$$

Note also that $d = O(d_1 + d_2)$ by (V3), whence $\delta_1 + \delta_2 = O(d_1 + d_2)$.

Let $(F_i) \in U^{m+1} \cap W$, that is

$$y_0^d F_i(x_0, x_1, x_2; x_0', x_1', x_2') = y_i^d F_0(x_0, x_1, x_2; x_0', x_1', x_2'). \tag{11.16}$$

Let $\mathbf{p}$ be a presentation of the morphism ϕ_B, hence

$$\mathbf{p} \in K[x_0, x_1, x_2; x_0', x_1', x_2']^{m+1},$$

where all entries are bihomogeneous of the same bidegree $(d_1(\mathbf{p}), d_2(\mathbf{p}))$ with

$$\deg_{x_2} p_i < N, \quad \deg_{x_2'} p_i < N$$

and

$$(y_0 : \cdots : y_m) = (p_0(\mathbf{x}; \mathbf{x}') : \cdots : p_m(\mathbf{x}; \mathbf{x}'))$$

on $C \times C$. Then $(d_1(\mathbf{p}), d_2(\mathbf{p}))$ is by definition the bidegree of this presentation. The height $h(\mathbf{p})$ is the height of the vector of coefficients of all polynomials p_i.

We consider equation (11.16) as a linear system, where the unknowns are the coefficients $\mathbf{a}_i = (a_{i\beta\beta'})$ of the polynomials

$$F_i(x_0, x_1, x_2; x'_0, x'_1, x'_2) = \sum \cdots \sum a_{i\beta\beta'} \mathbf{x}^{\beta} \mathbf{x}'^{\beta'}.$$

Our goal is to transform (11.16) into a linear system with coefficients in K. We may replace y_0 and y_i by p_0 and p_i. Then, using the relation $f(x_0, x_1, x_2) = 0$ to reduce the exponents in x_2, x'_2 to degree strictly less than N, we transform (11.16) into a linear system

$$\sum_{\alpha,\alpha'} L_{0\alpha\alpha'}(\mathbf{a}_i) \mathbf{x}^{\alpha} \mathbf{x}'^{\alpha'} = \sum_{\alpha,\alpha'} L_{i\alpha\alpha'}(\mathbf{a}_0) \mathbf{x}^{\alpha} \mathbf{x}'^{\alpha'} \tag{11.17}$$

for $i = 0, \ldots, m$, where the coefficients $L_{i\alpha\alpha'}(\mathbf{a}_j)$ are linear forms with unknowns the vector $\mathbf{a}_j$ of coefficients of the polynomial F_j. Here the indices $\boldsymbol{\alpha}$, $\boldsymbol{\alpha}'$ range over

$$\begin{aligned} &\boldsymbol{\alpha} = (\alpha_0, \alpha_1, \alpha_2), && \boldsymbol{\alpha}' = (\alpha'_0, \alpha'_1, \alpha'_2), \\ &|\boldsymbol{\alpha}| = \delta_1 + d \cdot d_1(\mathbf{p}), && |\boldsymbol{\alpha}'| = \delta_2 + d \cdot d_2(\mathbf{p}), \\ &\alpha_2 < N, \quad \alpha'_2 < N. \end{aligned}$$

It remains to compute a bound for the height of the linear forms $L_{i\alpha\alpha'}$. This is a routine procedure, which we have already used in Chapter 2 in analyzing the height of presentations, specifically in Lemma 2.5.6. More precisely, let $\mathbf{q}$ be the presentation

$$\mathbf{q} = (x_0^{\beta_0} x_1^{\beta_1} x_2^{\beta_2} x_0'^{\beta'_0} x_1'^{\beta'_1} x_2'^{\beta'_2}), \ |\boldsymbol{\beta}| = \delta_1, \ |\boldsymbol{\beta}'| = \delta_2, \ \beta_2 < N, \ \beta'_2 < N.$$

We consider the new presentation

$$(p_i^d q_{\beta\beta'}), \ i = 0, \ldots, m$$

with $\boldsymbol{\beta}$, $\boldsymbol{\beta}'$ as above. Using the equations $f(x_0, x_1, x_2) = 0$ and $f(x'_0, x'_1, x'_2) = 0$, we express the restriction to $C \times C$ of monomials appearing with exponents greater or equal to N as linear combinations of monomials in $x_0, \ldots, x'_2$ involving only x_2 and x'_2 with exponents strictly less than N. In this way we obtain the new presentation

$$\left(\sum_{\alpha,\alpha'} L_{i\alpha\alpha',\beta\beta'} \mathbf{x}^{\alpha} \mathbf{x}'^{\alpha'} \right),$$

where the $L_{i\alpha\alpha',\beta\beta'}$ are the coefficients of the linear form $L_{i\alpha\alpha'}(\mathbf{a})$.

Now the same proof as in Lemma 2.5.6 for bigraded presentations of morphisms shows that the height of the linear forms $L_{i\alpha\alpha'}(\mathbf{a})$ is bounded by

$$d\,h(\mathbf{p}) + h(\mathbf{q}) + o(d) = d\,h(\mathbf{p}) + o(d).$$

Here we have used our special model and $F \in U$ to get $h(\mathbf{q}) = 0$, otherwise the upper bound would have been only $\ll d_1 + d_2 + d$, not sufficient for the applications we have in mind.

From (11.17) we have to solve the linear system

$$L_{0\boldsymbol{\alpha}\boldsymbol{\alpha}'}(\mathbf{a}_i) = L_{i\boldsymbol{\alpha}\boldsymbol{\alpha}'}(\mathbf{a}_0)$$

with $i = 0, \ldots, m$ and $\boldsymbol{\alpha}$, $\boldsymbol{\alpha}'$ as above. The number of unknowns is

$$(m+1)N^2\left(\delta_1 - \frac{N-3}{2}\right)\left(\delta_2 - \frac{N-3}{2}\right) \leq (m+1)N^2\delta_1\delta_2.$$

By (11.15) on page 379 and assumption (V3), the dimension of the space of solutions is bounded below by

$$d_1d_2 - gd^2 - O(d_1 + d_2) \geq \gamma d_1 d_2 - O(d_1 + d_2).$$

There is a constant $C_4 > 0$ such that this is bounded below by $\frac{\gamma}{2}d_1d_2$ for $d_1, d_2 \geq C_4/\gamma$. Therefore, by Siegel's lemma in Corollary 2.9.9 (even the simple version in Corollary 2.9.2 is enough if we replace the height $h(\mathbf{p})$ by $\log \max_{\sigma,i} |\sigma(p_i)|$), there is a solution yielding an $\mathbf{F} = (F_i) \in U^{m+1} \cap W$ with

$$h(\mathbf{F}) \leq 2(m+1)N^2 \frac{\delta_1\delta_2}{\gamma d_1 d_2}(h(\mathbf{p})d + \log(\delta_1) + \log(\delta_2) + o(d)).$$

Since $d_1d_2 > gd^2$, we easily get

$$\frac{\delta_1\delta_2 d}{d_1d_2} = O(d_1 + d_2) \quad \text{and} \quad \frac{\delta_1\delta_2\log(\delta_i)}{d_1d_2} = O(\sqrt{d_1+d_2}\log(d_1+d_2))$$

proving our claim. □

11.8. Application of Roth's lemma

Let C be an irreducible projective smooth curve over the number field K and let $P_0 \in C(K)$. With the notation introduced in Section 11.2, let V be a Vojta divisor satisfying (V1) and (V2). Let $F_i(\mathbf{x}, \mathbf{x}')$, $i = 0, \ldots, m$ denote bihomogeneous polynomials of bidegree (δ_1, δ_2), describing a non-trivial global section s of $O(V)$ as in Lemma 11.2.7, hence

$$s = F_i(\mathbf{x}, \mathbf{x}')/y_i^d|_{C\times C}$$

for $i = 0, \ldots, m$. We are looking for an upper bound of the admissible pair (i_1^*, i_2^*) in the point $(P, Q) \in (C \times C)(\overline{K})$, defined in 11.6.2. The idea is to project down to $\mathbb{P}^1_K \times \mathbb{P}^1_K$ to get a bihomogeneous polynomial instead of s and then apply Roth's lemma to that polynomial. In 11.8.2–11.8.5, we describe the push-down of sy_i^d and show that it has similar properties as s. Lemma 11.8.6, the goal of this section, is the application of Roth's lemma.

11.8.1. For the reader's convenience, we make the following simplifications: By a change of coordinates, we may assume that the projection

$$p : \mathbb{P}^n_K \longrightarrow \mathbb{P}^2_K, \qquad \mathbf{x} \longmapsto (x_0 : x_1 : x_2)$$

is well defined on C and maps C birationally onto its image. Since C has degree N in $\mathbb{P}^n_K$, we may assume that $p(C)$ is given by an irreducible polynomial

$$f(x_0, x_1, x_2) = a_0 + a_1 x_2 + \cdots + a_{N-1} x_2^{N-1} + x_2^N,$$

where $a_i \in K[x_0, x_1]$ is homogeneous of degree $N - i$. Additionally, we may assume that none of the coordinates $x_0, \ldots, x_n$ vanishes identically on C. We also consider the finite morphism $\pi : C \to \mathbb{P}^1_K$ given by $\pi(\mathbf{x}) = (x_0 : x_1)$, of degree N. For details, we refer to A.11.5–A.11.7.

Since p is well defined on C and f vanishes at $(x_0(P) : x_1(P) : x_2(P))$, either $x_0(P) \neq 0$ or $x_1(P) \neq 0$. Without loss of generality, we may assume $x_0(P) \neq 0$. On $\{\mathbf{x} \in C \mid x_0 \neq 0\}$, we use the affine coordinates $\xi_1, \ldots, \xi_n$, where

$$\xi_j := x_j / x_0|_C, \quad j = 1, \ldots, n.$$

Since $f(x_0, x_1, x_2)$ is a monic polynomial in x_2, we see that ξ_2 is integral over $K[\xi_1]$. Similarly, we may assume that ξ_j is integral over $K[\xi_1]$ for $j \geq 3$. For simplicity, we may also assume $x_0(Q) \neq 0$. The affine coordinates on $\mathbb{P}^n_K \times \mathbb{P}^n_K$ are denoted by $(\boldsymbol{\xi}, \boldsymbol{\xi}')$.

In order to understand the following, it is convenient for the reader to be familiar with the intersection theory of divisors, as in Section A.9.

Lemma 11.8.2. *There is a bihomogeneous polynomial* $G_i \in K[x_0, x_1, x_0', x_1']$ *of bidegree* $(N^2\delta_1, N^2\delta_2)$ *such that*

$$(\pi \times \pi)_* \operatorname{div}(F_i|_{C \times C}) = \operatorname{div}(G_i).$$

Here $\pi \times \pi$ *denotes the natural morphism* $C \times C \longrightarrow \mathbb{P}^1_K \times \mathbb{P}^1_K$ *induced by* π.

Proof: Since $\pi \times \pi$ is finite, $(\pi \times \pi)_* \operatorname{div}(F_i|_{C \times C})$ is an effective divisor on $\mathbb{P}^1_K \times \mathbb{P}^1_K$, and is not 0. Recall that $\operatorname{Pic}(\mathbb{P}^1_K \times \mathbb{P}^1_K) \cong \mathbb{Z} \times \mathbb{Z}$, with elements represented by $O(m, n)$ (cf. A.9.28). Therefore, we have a bihomogeneous polynomial $G_i \in K[x_0, x_1, x_0', x_1']$ such that $\operatorname{div}(G_i)$ is equal to that divisor (see Example A.6.13). Denoting the bidegree of G_i by (k_1, k_2) and by P_1 a point on $\mathbb{P}^1_K(K)$, we have a rational equivalence

$$\operatorname{div}(G_i) \sim k_1 \{P_1\} \times \mathbb{P}^1_K + k_2 \mathbb{P}^1_K \times \{P_1\}.$$

Using Lemma 11.2.1 for $\mathbb{P}^1_K \times \mathbb{P}^1$, we have

$$k_2 = \operatorname{div}(G_i) \cdot (\{P_1\} \times \mathbb{P}^1_K).$$

By the projection formula, the right-hand side equals

$$\operatorname{div}(F_i|_{C \times C}) \cdot (H \times \mathbb{P}^n_K)|_{C \times C},$$

where H denotes a hyperplane in $\mathbb{P}^n_K$. Now from (11.1) on page 357 it follows that the divisor $\operatorname{div}(F_i|_{C\times C})$ is rationally equivalent to $\delta_1 N\{P_0\}\times C+\delta_2 N\, C\times\{P_0\}$, while $N\{P_0\}\times C$ is in the class of $H\times\mathbb{P}^n_K|_{C\times C}$. Using Lemma 11.2.1 for $C\times C$, we get $k_2=N^2\delta_2$. Similarly, we have $k_1=N^2\delta_1$. □

Remark 11.8.3. Let Norm denote the norm with respect to the field extension $K(C\times C)/K(\mathbb{P}^1_K\times\mathbb{P}^1_K)$. Then we may choose

$$G_i(\xi_1,\xi_1')=\operatorname{Norm}(F_i(\boldsymbol{\xi},\boldsymbol{\xi}')). \tag{11.18}$$

In order to see this, note that

$$\begin{aligned}\operatorname{div}(\operatorname{Norm}(F_i(\boldsymbol{\xi},\boldsymbol{\xi}')) &= (\pi\times\pi)_*\operatorname{div}(F_i(\boldsymbol{\xi},\boldsymbol{\xi}')))\\ &= \operatorname{div}(G_i(\mathbf{x},\mathbf{x}'))-(\pi\times\pi)_*\operatorname{div}(x_0^{\delta_1}{x_0'}^{\delta_2}|_{C\times C})\end{aligned}$$

by A.9.11. Since π has degree N and $\operatorname{div}(x_0|_C)=\pi^*([\infty])$ on C, the projection formula for proper intersections ([**125**], Prop.2.3) yields $\pi_*(\operatorname{div}(x_0|_C))=N[\infty]$. This is equal to $\operatorname{div}(x_0^N)$ on $\mathbb{P}^1_K$ and a similar identity holds for x_0'. This easily proves (11.18).

Lemma 11.8.4.

$$h(G_i)\le N^2h(F_i)+O(\delta_1+\delta_2).$$

Proof: Using 11.8.1, it is clear that $1,\xi_2,\dots,\xi_2^{N-1}$ form a basis of $K(C)$ over $K(\xi_1)$. There are $a_{jk}\in K(\xi_1)$, $k=0,\dots,N-1$, such that

$$\xi_j=\sum_{k=0}^{N-1}a_{jk}\,\xi_2^k,\qquad j=3,\dots,n$$

and a similar relation holds for ξ_j'.

There are polynomials $b(x_0,x_1),b_{jk}(x_0,x_1)\in K[x_0,x_1]$, all homogeneous of the same degree N', such that $a_{jk}=b_{jk}/b$; then we have a presentation $\mathbf{p}$ of the closed embedding $C\longrightarrow\mathbb{P}^n_K$ given by

$$p_j(x_0,x_1,x_2)=\begin{cases}bx_0^{N-1} & \text{if } j=0\\ bx_0^{N-2}x_j & \text{if } j=1,2\\ \sum_{k=0}^{N-1}b_{jk}x_0^{N-k-1}x_2^k & \text{if } j\ge 3.\end{cases}$$

The relations

$$\left.\frac{p_j}{p_0}=\frac{x_j}{x_0}\right|_{C\times C}$$

are obvious, hence $\mathbf{p}$ is indeed a presentation of degree $N''=N'+N-1$. By Lemma 2.5.6, any monomial in $\xi_1,\dots,\xi_n$ of degree $\le\delta_1$ has the form

$$\frac{1}{b(\xi_1)^{\delta_1}}\sum_{\substack{k_1+k_2\le N''\delta_1\\ k_2<N}}c_{k_1k_2}\,\xi_1^{k_1}\xi_2^{k_2}$$

with $c_{k_1k_2} \in K$ and $h(\mathbf{c}) \le \delta_1 h(\mathbf{p}) + O(\delta_1) = O(\delta_1)$. We get a similar expression for a monomial in $\xi'_1, \dots, \xi'_n$ of degree $\le \delta_2$. By Proposition 1.6.2, any monomial in $\boldsymbol{\xi}, \boldsymbol{\xi}'$ of bidegree $\le (\delta_1, \delta_2)$ has the form

$$\frac{1}{b(\xi_1)^{\delta_1} b'(\xi'_1)^{\delta_2}} \sum_{\substack{k_1+k_2 \le N''\delta_1 \\ k_2 < N}} \sum_{\substack{k'_1+k'_2 \le N''\delta_2 \\ k'_2 < N}} c_{k_1k_2k'_1k'_2} \xi_1^{k_1} \xi_1'^{k'_1} \xi_2^{k_2} \xi_2'^{k'_2} \tag{11.19}$$

with $h(\mathbf{c}) \le O(\delta_1 + \delta_2)$. Using the very definition of the height of a polynomial, it becomes clear that $F_i(\boldsymbol{\xi}, \boldsymbol{\xi}')$ may be written in the form (11.19) with

$$h(\mathbf{c}) \le h(F_i) + O(\delta_1 + \delta_2). \tag{11.20}$$

Similar considerations also apply to the polynomials $\xi_2^k \xi_2'^{k'} F_i(\boldsymbol{\xi}, \boldsymbol{\xi}')$ instead of $F_i(\boldsymbol{\xi}, \boldsymbol{\xi}')$, we get expressions as in (11.19) with bounds as in (11.20).

The computation of the norm is done using the basis $\xi_2^k \xi_2'^{k'}$, $0 \le k, k' < N$, of $K(C \times C)$ over $K(\xi_1, \xi'_1)$. Since we may assume $G_i(\xi_1, \xi'_1) = \mathrm{Norm}(F_i(\boldsymbol{\xi}, \boldsymbol{\xi}'))$ (cf. Remark 11.8.3), it is the determinant of a $N^2 \times N^2$ matrix A with entries

$$A_{\mu\nu}(\xi_1, \xi'_1) \in K(\xi_1, \xi'_1)$$

whose numerators $B_{\mu\nu} \in K[\xi_1, \xi'_1]$ have degree of order $O(\delta_1 + \delta_2)$ and height bounded by $h(F_i) + O(\delta_1 + \delta_2)$. So far, we have given the argument for the height, but if we follow the arguments carefully then we see that

$$|B_{\mu\nu}|_v \le C_v^{\delta_1+\delta_2} |F_i|_v$$

and $C_v = 1$ for all but finitely many $v \in M_K$. Therefore, there are $B_1, B_2 \in K[\xi_1, \xi'_1]$ of degree $O(\delta_1 + \delta_2)$ such that

$$G_i = B_1 / B_2$$

and, with a new C_v still such that $C_v = 1$ for all but finitely many $v \in M_K$, we have

$$|B_1|_v \le C_v^{\delta_1+\delta_2} |F_i|_v^{N^2}$$

for all $v \in M_K$. This shows

$$h(B_1) \le N^2 h(F_i) + O(\delta_1 + \delta_2).$$

Then appealing to Theorem 1.6.13 we get

$$h(G_i) \le N^2 h(F_i) + O(\delta_1 + \delta_2).$$

This proves the claim. □

Lemma 11.8.5. *There is a bihomogeneous polynomial $E(x_0, x_1, x'_0, x'_1)$, of bidegree (Nd_1, Nd_2), with the following properties:*

(a) $(\pi \times \pi)_*(\mathrm{div}(s)) = \mathrm{div}(E)$.

(b) *If (j_1^*, j_2^*) is an admissible pair of E in $(\pi(P), \pi(Q))$, then there is an admissible pair (i_1^*, i_2^*) of s in (P, Q) such that $i_1^* \le j_1^*$, $i_2^* \le j_2^*$.*

(c) $h(E) \leq N^2 h(\mathbf{F}) + O(d_1 + d_2 + d)$.

Proof: Similarly as in the proof of Lemma 11.8.2, there are k_1, k_2 with

$$(\pi \times \pi)_*(\mathrm{div}(y_i|_{C\times C}) \sim k_1(\{P_1\} \times \mathbb{P}^1_K) + k_2(\mathbb{P}^1_K \times \{P_1\}).$$

Note that

$$\mathrm{div}(y_i|_{C\times C}) \sim B = M(\{P_0\} \times C + C \times \{P_0\}) - \Delta'$$

by definition of ϕ_B. Similarly as in the proof of Lemma 11.8.2, we get $k_1 = k_2 = NM$. Using

$$\mathrm{div}(s) = \mathrm{div}(F_i|_{C\times C}) - d\mathrm{div}(y_i|_{C\times C})$$

and Lemma 11.8.2, we see that

$$\begin{aligned}(\pi \times \pi)_*\mathrm{div}(s) &\sim N(N\delta_1 - Md)(\{P_1\} \times \mathbb{P}^1_K) + N(N\delta_2 - Md)(\mathbb{P}^1_K \times \{P_1\}) \\ &\sim Nd_1(\{P_1\} \times \mathbb{P}^1_K) + Nd_2(\mathbb{P}^1_K \times \{P_1\}).\end{aligned}$$

Hence we have a global section of $O_{\mathbb{P}^1\times\mathbb{P}^1}(Nd_1, Nd_2)$ with divisor equal to the left-hand side. This global section may be viewed as a bihomogeneous polynomial $E(x_0, x_1, x_0', x_1')$ of bidegree (Nd_1, Nd_2) proving (a) (see Example A.6.13).

It is clear that there is a hyperplane $\ell(\mathbf{y}) = 0$ of $\mathbb{P}^m_K$ disjoint from the fibre $(\pi \times \pi)^{-1}(\pi(P), \pi(Q))$. As in Lemma 11.2.7, there is a bihomogeneous polynomial $F(\mathbf{x}, \mathbf{x}')$ of bidegree (δ_1, δ_2) with

$$s = F(\mathbf{x}, \mathbf{x}')/\ell(\mathbf{y})^d|_{C\times C}.$$

Again as in Lemma 11.8.2 and Remark 11.8.3, there is a bihomogeneous polynomial $G(x_0, x_1, x_0', x_1')$ of bidegree $(N^2\delta_1, N^2\delta_2)$ with

$$(\pi \times \pi)_*\mathrm{div}(F|_{C\times C}) = \mathrm{div}(G).$$

From the proof of (a), we get a bihomogeneous polynomial $H(x_0, x_1, x_0', x_1')$ of bidegree (NM, NM) such that

$$(\pi \times \pi)_*\mathrm{div}(\ell(\mathbf{y})|_{C\times C}) = \mathrm{div}(H).$$

We may assume that

$$EH^d = G.$$

Since the hyperplane $\{\ell(\mathbf{y}) = 0\}$ does not meet the fibre over $(\pi(P), \pi(Q))$, we conclude that H doesn't vanish at $(\pi(P), \pi(Q))$. By Leibniz's rule, (j_1^*, j_2^*) is an admissible pair for G in $(\pi(P), \pi(Q))$. As in Remark 11.8.3, we may assume

$$G(\xi_1, \xi_1') = \mathrm{Norm}(F(\boldsymbol{\xi}, \boldsymbol{\xi}')).$$

Integral elements form a ring and so $F(\boldsymbol{\xi}, \boldsymbol{\xi}')$ is integral over $K[\xi_1, \xi_1']$ using 11.8.1. By passing to the minimal polynomial of $F(\boldsymbol{\xi}, \boldsymbol{\xi}')$, we find a polynomial $a \in K[\boldsymbol{\xi}, \boldsymbol{\xi}']$ such that

$$a(\boldsymbol{\xi}, \boldsymbol{\xi}')F(\boldsymbol{\xi}, \boldsymbol{\xi}') = G(\boldsymbol{\xi}, \boldsymbol{\xi}').$$

This follows from the fact that the last coefficient of the minimal polynomial equals the norm up to a sign. By Leibniz's rule, we see that

$$\partial_{i_1}\partial'_{i_2}F(P,Q) \neq 0$$

for some $i_1 \leq j_1^*, i_2 \leq j_2^*$. Since $\ell(\mathbf{y})$ does not vanish at (P,Q), an admissible pair for F in (P,Q) is the same as an admissible pair for s in (P,Q). This proves (b).

Let H_i be the bihomogeneous polynomial of bidegree (NM, NM) such that

$$(\pi \times \pi)_* \mathrm{div}(y_i^d|_{C\times C}) = \mathrm{div}(H_i).$$

Then we may assume

$$EH_i^d = G_i$$

as above. By Theorem 1.6.13, we have

$$h(H_i^d) = O(d)$$

and

$$h(E) + h(H_i^d) = h(G_i) + O(d_1 + d_2 + d).$$

Together with Lemma 11.8.4, we conclude

$$h(E) \leq N^2 h(F_i) + O(d_1 + d_2 + d)$$

proving (c). □

Lemma 11.8.6. *There is a constant $C_6 > 0$, independent of d_1, d_2, d, and γ, such that for $0 < \varepsilon \leq 1/\sqrt{2}$, for any Vojta divisor satisfying (V1), (V2), (V3), with*

$$d_2 \geq C_4/\gamma, \quad d_2/d_1 \leq \varepsilon^2$$

and for any $P, Q \in C(\overline{K})$ with

$$\min\left(d_1\, h_{N[P_0]}(P),\, d_2\, h_{N[P_0]}(Q)\right) \geq C_6\, \frac{d_1}{\gamma\varepsilon^2} \tag{11.21}$$

there exists a global section s of $O(V)$ with an admissible pair (i_1^*, i_2^*) in (P,Q) such that

$$h(\mathbf{F}) \leq C_5 \frac{d_1 + d_2}{\gamma}, \quad \frac{i_1^*}{d_1} + \frac{i_2^*}{d_2} \leq 4N\varepsilon.$$

Proof: By Lemma 11.7.3, there is a non-zero global section s of $O(V)$ with

$$h(\mathbf{F}) \leq C_5 \frac{d_1 + d_2}{\gamma}.$$

We apply 11.8.1–11.8.5 to this s. The goal is to use Roth's lemma in 6.3.7 for the polynomial $E(\xi_1, \xi_1')$. By Lemma 11.8.5, the partial degrees of E are bounded by $r_1 := Nd_1$ and $r_2 := Nd_2$, respectively, and

$$h(E) + 8r_1 \ll \frac{d_1}{\gamma} \tag{11.22}$$

(using $d_1 \geq d_2$). On the other hand, we have

$$\min\{r_1 h(\pi(P)), r_2 h(\pi(Q))\} = N \min\{d_1 h_{N[P_0]}(P), d_2 h_{N[P_0]}(Q)\} + O(d_1) \tag{11.23}$$

using the elementary fact

$$h_{N[P_0]} = h \circ \pi + O(1)$$

following from 11.2.3 and Theorem 2.3.8. Using (11.22) and (11.23), we find easily a constant $C_6 > 0$ such that (11.21) implies

$$\epsilon^{-2}\left(h(E) + 8r_1\right) \leq \min\{r_1 h(\pi(P)), r_2 h(\pi(Q))\}.$$

Hence the assumptions in Roth's lemma are satisfied and we get an admissible pair (j_1^*, j_2^*) for E in $(\pi(P), \pi(Q))$ with

$$\frac{j_1^*}{r_1} + \frac{j_2^*}{r_2} \leq 4\varepsilon.$$

Now the claim follows from Lemma 11.8.5 (b). □

11.9. Proof of Faltings's theorem

Let C be an irreducible projective smooth curve of genus $g \geq 2$, defined over a number field K, with a point P_0 defined over K. The next result is **Vojta's theorem.**

Theorem 11.9.1. *There are constants C_7, C_8, depending only on C and P_0, with the following property: Let $P, Q \in C(\overline{K})$ and $z = j(P)$, $w = j(Q)$. Then one of $|z| \leq C_7$, $|w| \leq C_8|z|$, $\langle z, w\rangle \leq \frac{3}{4}|z|\,|w|$ holds.*

Remark 11.9.2. The constant $\frac{3}{4}$ has no special significance and can be replaced by any constant in $(\frac{1}{\sqrt{g}}, 1]$; what matters here is that it is strictly less than 1. This follows from the proof.

Proof: We consider $P, Q \in C(\overline{K})$ with $|z| \geq C_7$, $|w| \geq C_8|z|$ for large constants C_7, C_8 to be determined later. Since the set Z defined in 11.6.6 is finite and effectively determinable, we may assume that the constants C_7 and C_8 are so large that $P, Q \notin Z$. Suppose that V is a Vojta divisor satisfying (V1), (V2), (V3), and $d_1, d_2 \geq C_4/\gamma$ (cf. 11.2.6 and 11.7.2). Then Proposition 11.3.1, Lemma 11.6.5, Lemma 11.6.7 and Lemma 11.7.3 show that, for a positive constant C_9 depending only on C and P_0, we have

$$-C_9 \cdot \left(\frac{d_1 + d_2}{\gamma} + i_1^*\,|z|^2 + i_2^*\,|w|^2 + i_1^* + i_2^*\right)$$
$$\leq \frac{d_1}{2g}\,|z|^2 + \frac{d_2}{2g}\,|w|^2 - d\,\langle z, w\rangle + O(d_1|z| + d_2|w| + d_1 + d_2).$$

Now we also assume that there is an ε, $0 < \varepsilon \leq 1/\sqrt{2}$, with

$$d_2/d_1 \leq \varepsilon^2 \tag{11.24}$$

and

$$\min\left(d_1\, h_{N[P_0]}(P),\, d_2\, h_{N[P_0]}(Q)\right) \geq C_6\, \frac{d_1}{\gamma\varepsilon^2} \tag{11.25}$$

as in Lemma 11.8.5. Applying this lemma, we get

$$\begin{aligned}&-C_9 \cdot \left(2\frac{d_1}{\gamma} + 4N\varepsilon d_1|z|^2 + 4N\varepsilon d_2|w|^2\right)\\ &\quad\leq \frac{d_1}{2g}\,|z|^2 + \frac{d_2}{2g}\,|w|^2 - d\,\langle z, w\rangle + O(d_1|z| + d_2|w| + d_1).\end{aligned} \tag{11.26}$$

For a small positive number $\gamma_0 < 1$ and $D \in \mathbb{N}$, we choose

$$d_1 = \sqrt{g+\gamma_0}\;\frac{D}{|z|^2} + O(1),$$

$$d_2 = \sqrt{g+\gamma_0}\;\frac{D}{|w|^2} + O(1)$$

and

$$d = \frac{D}{|z|\;|w|} + O(1).$$

The $O(1)$ terms are for small adjustments so that $d_1, d_2, d, \delta_1, \delta_2$ are all non-zero natural numbers. This is a choice which makes (11.26) relatively sharp and fulfills (V1), (V2), (V3), for D sufficiently large as a function of $|z|$, $|w|, \gamma_0$. It is immaterial here how this notion of D being large depends on $|z|$, $|w|$, or γ_0, since in the end we shall let $D \to \infty$. Note that we have

$$d_1 d_2 - g d^2 \geq \gamma\, d_1 d_2$$

for

$$\gamma = \frac{\gamma_0}{g+\gamma_0} + o(1),$$

where the term implicit in $o(1)$ tends to 0 as $D \to \infty$. Using

$$\frac{d_2}{d_1} = \frac{|z|^2}{|w|^2} + o(1),$$

condition (11.24) becomes

$$\frac{|z|}{|w|} \leq \varepsilon + o(1). \tag{11.27}$$

As remarked in the proof of Lemma 11.6.7, we have

$$h_{N[P_0]}(P) = \frac{N}{2g}|z|^2 + O(|z|) + O(1),$$

with a similar equation for Q and $|w|$. Thus the condition

$$|z| \geq C_{10}/(\varepsilon\sqrt{\gamma}), \tag{11.28}$$

with $C_{10} \geq 1$ a positive constant depending on C and P_0, implies (11.25) for sufficiently large D.

We substitute the values for d_1, d_2, d, γ in (11.26) to derive

$$-C_{11} \cdot \left(\frac{1}{\gamma_0 |z|^2} + \varepsilon \right) D \leq \frac{\sqrt{g + \gamma_0}}{g} D - \frac{\langle z, w \rangle}{|z|\,|w|} D + O\left(\left(\frac{1}{|z|} + \frac{1}{|w|} \right) D \right) + o(D),$$

for a certain constant C_{11} depending only on C and P_0. Assuming (11.28), we divide by D, let D tend to ∞, simplify, and find after rearranging terms

$$\frac{\langle z, w \rangle}{|z|\,|w|} - \frac{\sqrt{g + \gamma_0}}{g} \leq C_{12}\varepsilon \tag{11.29}$$

with C_{12} depending only on C, P_0. We still need conditions (11.28) and $|z| < \varepsilon\,|w|$, the limit of (11.27) for $D \to \infty$. To this end, we choose first γ_0 so small that

$$\frac{3}{4} - \frac{\sqrt{g + \gamma_0}}{g} > 0,$$

and then ε so small that

$$\frac{3}{4} - \frac{\sqrt{g + \gamma_0}}{g} > C_{12}\varepsilon. \tag{11.30}$$

Here we have used $\frac{3}{4} > \frac{\sqrt{2}}{2}$ and $g \geq 2$. Let

$$C_7 > C_{10} \Big/ \left(\varepsilon \sqrt{\frac{\gamma_0}{g + \gamma_0}} \right)$$

and

$$C_8 > \frac{1}{\varepsilon}.$$

For $P, Q \in C(\overline{K})$ satisfying

$$|z| \geq C_7, \qquad |w| \geq C_8\,|z|,$$

(11.27) and (11.28) are both satisfied and inequality (11.29) holds. Finally, because of (11.30) and $C_{10} \geq 1$, we see that (11.29) implies

$$\frac{\langle z, w \rangle}{|z|\,|w|} \leq \frac{3}{4},$$

proving the theorem. □

Now we are ready to prove Faltings's theorem (see Theorem 11.1.1).

Proof: We may assume that C has a base point $P_0 \in C(K)$. From Theorem 9.3.5 and the Mordell–Weil theorem in 10.6.1, we know that $J(K) \otimes_{\mathbb{Z}} \mathbb{R}$ is a finite-dimensional euclidean space. By Lemma 5.2.19, we may cover it by finitely many cones T centered at 0 with angle $\alpha/2$ from the axis to the ending, where $\cos\alpha > \frac{3}{4}$. Let C_7, C_8 be the constants in Theorem 11.9.1.

By Proposition 9.4.5 (Mumford's gap principle with $\varepsilon \leq 2g\cos\alpha - 3$), there is a constant C_{13}, depending only on C and P_0, such that for any pair of distinct points P, Q in a same cone T, with $C_{13} \leq |\jmath(P)| \leq |\jmath(Q)|$, we have

$$|\jmath(Q)| \geq 2\,|\jmath(P)|. \tag{11.31}$$

Let $C_{14} = \max(C_7, C_{13})$. The set of K-rational points in the ball with center 0 and radius C_{14} is finite by Northcott's theorem in 2.4.9, so it remains to see that

$$S := T \cap \{P \in C(K) \mid |\jmath(P)| > C_{14}\}$$

is finite. Suppose $P_0, P_1, \ldots, P_k$ are different points of S such that

$$|\jmath(P_i)| \leq |\jmath(P_{i+1})|, \qquad i = 0, \ldots, k-1.$$

By (11.31), we have

$$|\jmath(P_{i+1})| \geq 2\,|\jmath(P_i)|,$$

yielding

$$|\jmath(P_k)| \geq 2^k\,|\jmath(P_0)|.$$

On the other hand, Theorem 11.9.1 shows that

$$|\jmath(P_k)| \leq C_8\,|\jmath(P_0)|.$$

This proves

$$k \leq \log(C_8)/\log 2$$

and the theorem. □

11.9.3. In much the same way as in Section 9.4, Lemma 5.2.19 shows easily that the number of cones may be bounded by 7^r, where r is the rank of the Mordell–Weil group over K of the Jacobian of C. Hence $C(K)$ is the union of two finite sets, namely the set of small points P with $h(P) \leq C_{14}$, and the set of large points P with $h(P) > C_{14}$. The former set is finite by Northcott's theorem, and the latter contains not more than $(\lfloor\log(C_8)/\log 2\rfloor + 1)\,7^r$ elements. The constants C_{14} and C_8 are effectively computable. As mentioned in the introduction to this chapter, we can dispense with the choice of a K-rational point P_0. It turns out that we can take $C_8 = C_{15}g^{C_{16}}$ and $C_{14} = C_{17}g^{C_{18}}(h(C)+1)$ for suitable absolute effectively computable constants $C_{15}, \ldots, C_{18}$; here $h(C)$ is the height of a presentation of C by means of (for example) a bicanonical closed embedding. A sketch of the additional arguments needed can be found in [**33**]. In any case, it is noteworthy that the bound for the height of small solutions is independent of the field K and is linear in $h(C)+1$, and that the dependence on the field K shows up only through the rank of the Mordell–Weil group.

11.10. Some further developments

Inspired by Vojta's proof, G. Faltings [114], [115], proved the following generalization of the Mordell conjecture called now **Faltings's big theorem:**

Theorem 11.10.1. *Let A be an abelian variety over a number field K, let $\Gamma = A(K)$ and let X be a geometrically irreducible closed subvariety of A, which is not a translate of an abelian subvariety over $\overline{K}$. Then $X \cap \Gamma$ is not Zariski dense in X.*

We will not prove this theorem here, since the proof is better understood in the language of arithmetic geometry and we refer instead to the original papers by Faltings [114], citeFa2, as well as to B. Edixhoven and J.-H. Evertse [96] or P. Vojta [313]. We state two applications of this theorem, with only a sketch of proofs. For the properties of curves needed to understand the arguments, we refer to [13]. The first result is due to J. Harris and J.H. Silverman [147], Th.2, Cor.3.

11.10.2. Recall that a smooth geometrically irreducible curve C of genus $g \geq 2$, defined over a number field K, is **hyperelliptic** if it is a double cover of $\mathbb{P}^1_L$ and **bi-elliptic** if it is a double cover of an elliptic curve over some finite extension L/K. The hyperelliptic cover is indeed defined over K and unique up to $GL(2,K)$ (see [148], Prop.IV.5.3). If $g \geq 6$, the bi-elliptic cover is unique up to translation ([13] Ch.VIII, C-2, p.366) and also defined over K ([153], Lemma 5).

Theorem 11.10.3. *Let C be a smooth geometrically irreducible curve over a number field K, of genus $g \geq 3$. Let $\mathcal{D}(K,2)$ be the set of effective divisors of degree 2 on C, defined over K. Then:*

(a) *if C as above is not hyperelliptic or bi-elliptic, then $\mathcal{D}(K,2)$ is a finite set;*

(b) *if C is hyperelliptic and the genus is at least 9, then $\mathcal{D}(K,2)$ consists of a finite set together with all divisors $[Q]+[Q'] = f^*([P])$ with $P \in \mathbb{P}^1(K)$ and $f : C \to \mathbb{P}^1$ the hyperelliptic morphism of degree 2;*

(c) *if C is bi-elliptic and the genus is at least 9, then $\mathcal{D}(K,2)$ consists of a finite set together with all divisors $[Q] + [Q'] = f^*([P])$ with $P \in E(K)$ and $f : C \to E$ the bi-elliptic morphism of degree 2.*

Outline of proof: The proof is based on the following observations.

We may assume that C has a K-rational point Q_0, otherwise we perform a quadratic base change, which is harmless for our statements.

Effective divisors of degree 2 are parametrized by a smooth surface $C^{(2)}$, the symmetric square of C, and points of $C^{(2)}(K)$ are in one-to-one correspondence with effective divisors of degree 2, rational over K. Now the map $Q \mapsto \mathrm{cl}([Q] - [Q_0])$, extended by linearity to divisors, yields a morphism $j_{(2)} : C^{(2)} \to A$ of $C^{(2)}$ into an abelian variety A of dimension g, namely the Jacobian of C (see 8.10.3 and Remark 8.10.12). Let X be the image of $C^{(2)}$ in A and denote by $\phi : C^{(2)} \to X$ the restriction of $j_{(2)}$ to X. Since Q_0 is defined over K, the image of a rational point of $C^{(2)}$ is a rational point of X. Therefore, applying Faltings's big theorem, we infer that the image of such a point lies in the union of a finite set and finitely many translates of elliptic curves. Indeed, Remark 8.10.12

implies that X cannot be a translate of an abelian surface. Thus it remains to describe the set $\phi^{-1}(X(K))$.

If we have two distinct divisors $[Q_1]+[Q_2]$ and $[Q_1']+[Q_2']$ with the same image x_0 under ϕ, then $H^0(C, O([Q_1]+[Q_2]))$ is two-dimensional and the quotient of independent global sections leads to a hyperelliptic structure on C, which is unique ([**148**], Prop.IV.5.3). We conclude that $|\phi^{-1}(x_0)| \geq 2$ implies that $\phi^{-1}(x_0)$ is a curve of genus 0. Conversely, if there is an hyperelliptic morphism $f : C \to \mathbb{P}^1$, then $x_0 := \phi(f^*([P]))$ remains constant as P varies on $\mathbb{P}^1$, hence $\phi^{-1}(x_0)$ is a curve of genus 0 in $C^{(2)}$. Thus ϕ is one-to-one, except at the point x_0 if C is hyperelliptic.

Note that, if $g \geq 4$, the curve C cannot be both hyperelliptic and bi-elliptic ([**13**],* Ch.VIII, C-2, p.366). Hence, if the curve C is bi-elliptic, the above shows that the image Y of E under the map $P \mapsto \phi(f^*([P]))$ is a curve. By Corollary 8.2.9, it is a translate of an elliptic curve in A. Conversely, we can show that, if C has genus $g \geq 9$, any translate $Y = E + b \subset X$ of an elliptic curve determines a bi-elliptic structure $C \to Y$ ([**147**], proof of Th.2 (b)). Moreover, since $g \geq 6$, this bi-elliptic structure is unique and defined over K. This yields (b) and (c). For statement (a), we prove that, if X contains a curve of genus 1, then the curve C is either hyperelliptic or bi-elliptic ([**147**], Th.2 (a)). □

11.10.4. In our second application of Faltings's big theorem, we consider a variety with many rational points. The study of such varieties is an important branch of diophantine geometry, which could not be treated in this book. For a survey of the theory, the reader is referred to E. Peyre [**234**].

For a projective variety X over a number field K and a fixed height function h_L with respect to an ample line bundle L, we consider the multiplicative height $H_L := e^{h_L}$ and

$$N_L(X,T) := \{P \in X(K) \mid H_L(P) \leq T\}$$

for $T \geq 0$. Sometimes X is replaced by a subset and we are interested in the asymptotics of this counting function for $T \to \infty$.

11.10.5. The simplest example of a variety with many rational points is $\mathbb{P}_K^{n-1}$ for $n \geq 2$. Let $N(\mathbb{P}_K^{n-1}, T)$ be the counting function of points of bounded height with respect to the standard height. The case $K = \mathbb{Q}$ was considered by Dedekind and Weber, and extended to the counting of rational points on Grassmannians by W.M. Schmidt [**265**]. For general K, let d be the degree, r (resp. s) the number of real (resp. complex) places, R_K the regulator, $D_{K/\mathbb{Q}} > 0$ the discriminant, w_K the number of roots of unity, h_K the class number, and ζ_K the zeta function of our number field K (see [**172**] for definitions). If the reader is not willing to enter into the terminology of algebraic number theory, he may just consider the special case $K = \mathbb{Q}$ and hence $d = r = R_\mathbb{Q} = D_{K/\mathbb{Q}} = h_K = 1$, $w_K = 2$, $s = 0$ and ζ_K is the Riemann zeta function.

* The result is stated to hold for $g \geq 3$, but the proof there gives the condition $g \geq 4$. The example $y^2 = x^8 + 1$ shows that $g \geq 4$ is indeed necessary.

Schanuel's theorem (see [**257**]) states

$$N(\mathbb{P}_K^{n-1}, T) = \frac{h_K R_K}{w_K \zeta_K(n)} \left(\frac{2^r (2\pi)^s}{D_{K/\mathbb{Q}}^{1/2}} \right)^n n^{r+s-1} T^{nd} + \begin{cases} O(T(1+\log T)) & \text{if } d=1, n=2, \\ O(T^{nd-1}) & \text{otherwise,} \end{cases}$$

for $T \geq 1$, where the implied constant may depend on n and K.

11.10.6. Schanuel's theorem was generalized by J. Franke, Yu.I. Manin, and Y. Tschinkel ([**122**]) to flag manifolds, see also J.L. Thunder [**299**] for a more elementary approach with a concrete error term. We expect also many rational points on Fano varieties, at least for K sufficiently large. Recall that an irreducible projective smooth variety X is called a **Fano variety**, if the anticanonical divisor $-K_X$ is ample. Then Manin has given a conjecture about the asymptotics of $N_L(U,T)$ for a sufficiently small open dense subset U of X (see [**122**] and [**234**]).

11.10.7. Let X be a smooth cubic threefold in $\mathbb{P}_K^4$, defined over a number field K. The canonical divisor of X is the restriction of $O_{\mathbb{P}^4}(-2)$ to X ([**148**], II.8.20) and hence X is a Fano variety. Manin's conjecture predicts an asymptotic $\sim c_K T^{2d}$ for the number of points of height at most T in a suitable dense open set of X, and it is expected that for X itself we still have a bound $O(T^{2d})$, where we use the standard height on $\mathbb{P}_K^4$. Since no precise asymptotics are used here, it is easy to see that we may choose a different height function with respect to $O_{\mathbb{P}^4}(1)$. It will be more convenient to use the Arakelov height h_{Ar} on $\mathbb{P}_K^4$ (replacing the max-norm by the L^2-norm at archimedean places, see Section 2.8) and we denote by $N_{\mathrm{Ar}}(X,T)$ the corresponding counting function of points of bounded height.

The solution of this problem is of particular importance in additive number theory with Waring's problem with cubes, and analytic number theory with cubic Weyl sums. General Weyl sums

$$S(\alpha) = \sum_{n=1}^{N} e^{2\pi i \alpha f(n)}$$

with $f(x) \in \mathbb{Z}[x]$ a polynomial of degree $k \geq 1$ occur quite often in analytic number theory and harmonic analysis, and the problem of estimating such sums, both pointwise or *via* high moments, is of central importance. For polynomials of high degree, Vinogradov's method and its variants lead to the best known results. However, for small degrees, in particular for cubic polynomials, no improvements over Weyl's original bounds (dating back to 1910) in the case of pointwise estimates, or L.K. Hua's bounds for moments [**154**] (dating back to 1938) have been obtained.

The simplest non-trivial case for moments is $f(n) = n^3$ and the sixth moment, with the conjectural bound

$$\int_0^1 |S(\alpha)|^6 \mathrm{d}\alpha \ll_\varepsilon N^{3+\varepsilon},$$

which if true would be the best possible (an asymptotic $\sim cN^3$ is actually expected to hold). As yet, no improvement in the exponent over the old Hua's bound $O(N^{\frac{7}{2}+\varepsilon})$ of 1938 has been obtained. (Hua's exponent is the critical one; any improvement, no matter how small, would have interesting consequences.)

The above integral is the number of integer solutions of

$$x_1^3 + x_2^3 + x_3^3 = x_4^3 + x_5^3 + x_6^3$$

for $1 \leq x_i \leq N$, $i = 1, \ldots, 6$. This defines a non-singular cubic projective fourfold and the problem amounts to counting the number of rational points of height at most T on the subset of points with positive coordinates. By slicing, we are led to the problem of counting rational points in a cubic threefold.

One of the difficulties in studying the distribution of points on a cubic hypersurface is that the obvious composition law for pairs of points, namely the residual intersection with the line through two points, is not defined if the line lies on the cubic hypersurface. A thorough study of this composition defined in the complement of rational lines is in the interesting monograph **[190]** by Yu.I. Manin. Thus the set of rational lines in a cubic hypersurface is an exceptional set, which requires separate study.

11.10.8. First, we need some facts about the geometry of X, which are due to G. Fano **[118]**. We consider the set Σ of lines of $\mathbb{P}_K^4$ contained in X. Then Σ is a subset of the Grassmannian $G(2,5)$ of lines in four-dimensional projective space $\mathbb{P}^4$. If $\mathbf{x} = (x_0 : \cdots : x_4)$ and $\mathbf{x}' = (x_0' : \cdots : x_4')$ are two points determining a line L, the Grassmann coordinate of L is the vector

$$\Lambda := \left(\det \begin{pmatrix} x_i & x_j \\ x_i' & x_j' \end{pmatrix} \right)_{0 \leq i < j \leq 4}$$

(with for example (i,j) in lexicographic order) determining a closed embedding of $G(2,5)$ in $\mathbb{P}^9$. It is easy to see that Σ becomes a closed projective subvariety of $\mathbb{P}_K^9$ (see J. Harris **[146]**, Example 6.19). This variety, of paramount importance in the study of the cubic threefold X, is called in honor of Fano the **Fano surface**† **associated to** X. The surface Σ is a smooth geometrically irreducible surface (see **[34]**, Lemma 3).

11.10.9. We assume that $\Sigma(K) \neq \emptyset$ and we fix a base point $Q_0 \in \Sigma(K)$. Then there is a canonical abelian variety A, called the **Albanese variety** of Σ, and a canonical map $a : \Sigma \to A$ mapping Q_0 to 0 factoring through every such morphism from Σ to arbitrary abelian varieties (see **[165]**, II.3). It is obvious that $a(\Sigma)$ generates A. This construction works for any irreducible smooth projective variety with a base point and it may be shown that the Albanese variety is the dual of the Picard variety (**[165]**, VI, §1, Th.1). It was also used in the deduction of Corollary 9.3.10.

Complex analytically, the Albanese variety is given as a complex torus $H^0(\Sigma, \Omega_\Sigma^1)^* / H^1(\Sigma, \mathbb{Z})$ and the canonical map a is given by $a(Q)(\omega) = \int_{\gamma_Q} \omega$, where $\omega \in H^0(\Sigma, \Omega_\Sigma^1)$ and γ_P is a path from Q_0 to Q (see **[130]**, II.6). The reader will note that the Albanese variety is a generalization of the Jacobian variety to higher dimensions.

Now we come back to the case of the cubic threefold. Then the Albanese variety A of the Fano surface Σ has dimension 5 (**[34]**, Lemma 5). Using the complex analytic description of the Albanese map a above and that K_Σ is very ample, it is easy to see that a has finite fibres and hence it is a finite morphism (see A.6.15, A.12.4). Moreover, C.H. Clemens and

† We should not be confused with the notion of Fano variety above, the surface Σ has a very ample canonical bundle and therefore is of general type, see E. Bombieri and H.P.F. Swinnerton-Dyer **[34]**, proof of Lemma 5, **[68]**, Lemma 10.13.

P. Griffiths have shown that φ is an immersion (i.e. a local embedding, see [68], §12), but we do not need this fact for our purposes.

11.10.10. On Σ, we use the Arakelov height induced by restricting the Arakelov height h_{Ar} from $\mathbb{P}^9_K$. If $Q \in \Sigma$ with corresponding line L_Q in X, then we have

$$h_{\mathrm{Ar}}(Q) = h_{\mathrm{Ar}}(L_Q),$$

where on the right-hand side we have the Arakelov height of the line in $\mathbb{P}^4_K$. Here and in the following, the Arakelov height of a projective linear subspace is defined as the Arakelov height of the corresponding linear subspace in the sense of Section 2.8. Let $N_{\mathrm{lines}}(X,T)$ be the number of points $P \in X(K)$ of height $H_{\mathrm{Ar}}(P) \leq T$, which are on K-rational lines in X.

Theorem 11.10.11. *If $K \neq \mathbb{Q}$, then for $T \geq 1$ the estimate*

$$N_{\mathrm{lines}}(X,T) = c_2\, \gamma\, T^{2d} + O\Big(T^{2d-1}(1+\log T)^{r/2}\Big)$$

holds, where

$$\gamma = \sum_{Q \in \Sigma(K)} H_{\mathrm{Ar}}(Q)^{-d} < \infty \quad \text{and} \quad c_2 = \frac{h_K R_K}{w_K \zeta_K(2)} \frac{2^{d-1}\pi^d}{D_{K/\mathbb{Q}}}$$

and where r is the maximum rank of $E(K)$ for all elliptic curves E in Σ. If $K = \mathbb{Q}$, then the error term has to be replaced by $O\Big(T(1+\log T)^{1+r/2}\Big)$.

Before we come to the proof, we need some useful results. First, we interpretate the Arakelov height of a subspace in terms of the associated lattice.

11.10.12. Let W be an n-dimensional subspace of K^N. The diagonal embedding $K^N \to \prod_{v|\infty} K_v^N \cong \mathbb{R}^{Nd}$ maps the K-lattice $\Lambda := W \cap O_K^N$ to an nd-dimensional $\mathbb{R}$-lattice Λ_∞ in the closure of $\overline{W}$ in $\mathbb{R}^{Nd}$ (cf. Corollary C.2.7). Let $\mathrm{vol}(\Lambda_\infty)$ be the volume of a fundamental domain of $\overline{W}/\Lambda_\infty$ with respect to the Lebesgue measure. We denote by λ_1 the length of the shortest non-zero vector in Λ_∞ with respect to the euclidean norm on $\mathbb{R}^{Nd}$.

We have the following result of Schmidt [263], Th.1, which we quote without proof.

Theorem 11.10.13. $\quad 2^{ns}\mathrm{vol}(\Lambda_\infty) = D_{K/\mathbb{Q}}^{n/2} H_{\mathrm{Ar}}(W)^d$.

Next, we need a generalization of Schanuel's theorem to an $n-1$-dimensional linear subspace L of $\mathbb{P}^{N-1}$, defined over K, which is uniform with respect to L. We denote by

$$\alpha(n) = \frac{\pi^{n/2}}{\Gamma(\frac{n}{2}+1)}$$

the volume of the unit ball in $\mathbb{R}^n$.

Theorem 11.10.14. *Under the assumptions of 11.10.12 and with $n \geq 2$, let L be the projective linear subspace of $\mathbb{P}^{N-1}_K$ induced by W. Then for $T \geq 1$ the following estimate* *holds*

$$N_{\mathrm{Ar}}(L,T) = c_n \frac{T^{nd}}{H_{\mathrm{Ar}}(L)^d} + \begin{cases} O\left(\frac{T}{\lambda_1}(1+\log^+(\frac{T}{\lambda_1}0))\right) & \text{if } d=1, n=2, \\ O\left((\frac{T}{\lambda_1})^{nd-1}\right) & \text{otherwise,} \end{cases}$$

where the implicit constant in the bound may depend on n *and* K *and where*

$$c_n = \frac{h_K R_K}{w_K \zeta_K(n)} D_{K/\mathbb{Q}}^{-n/2} n^{r+s-1} \alpha(n)^r \{2^n \alpha(2n)\}^s.$$

For the proof, we need the following lemma from geometry of numbers. Let Ω be a bounded measurable set in $\mathbb{R}^n$ with Lipschitz parametrizable boundary meaning that there are finitely many Lipschitz continuous maps, defined on bounded subsets of $\mathbb{R}^{n-1}$ and with images covering $\partial\Omega$.

Lemma 11.10.15. *Let* Λ *be a lattice in* $\mathbb{R}^n$, *let* λ_1 *be the length of the shortest non-zero lattice vector with respect to the euclidean norm* $\| \ \|$. *For* $T > 0$, *the number* $N(\Lambda, T)$ *of lattice points in* $T\Omega$ *satisfies*

$$N(\Lambda, T) = \frac{\mathrm{vol}(\Omega)}{\mathrm{vol}(\Lambda)} T^n + O\left(\left(\frac{T}{\lambda_1}\right)^{n-1}\right) + 1,$$

where the implied constant depends on n *and* Ω *but not on* Λ.

Proof: We refine the standard counting argument from **[172]**, Ch.6, §2, Th.2 to make the error term uniform with respect to the lattice. First, we use a result of geometry of numbers (see C.G.Lekkerkerker **[181]**, Ch.2, §10, Th.4) confirming that Λ has a basis $v_1, \dots, v_n$ such that the fundamental domain $F_\Lambda := \{\sum_{i=1}^n m_i v_i \mid 0 \le m_i < 1\}$ is not too skew, namely

$$\frac{\|v_1\| \cdots \|v_n\|}{\mathrm{vol}(F_\Lambda)} = O(1).$$

Hence there is a change of coordinates of norm bounded by a constant C_1 independent of Λ, which transforms Λ into the orthogonal lattice $\sum_{i=1}^n \mathbb{Z}\|v_i\| e_i$, where (e_i) is the standard basis of $\mathbb{R}^n$. We conclude that a ball of diameter $< \lambda_1/C_1$ intersects at most 2^n translates $\boldsymbol{\lambda} + F_\Lambda, \boldsymbol{\lambda} \in \Lambda$.

Now let $N_{\mathrm{int}}(\Lambda, T)$ (resp. $N_{\mathrm{bd}}(\Lambda, T)$) be the number of lattice points $\boldsymbol{\lambda} \in \Lambda$ with $\boldsymbol{\lambda} + F_\Lambda$ contained in the interior or Ω (resp. with $(\lambda + F_\Lambda) \cap \partial(T\Omega) \neq \emptyset$). Then

$$N_{\mathrm{int}}(\Lambda, T) \le N(\Lambda, T) \le N_{\mathrm{int}}(\Lambda, T) + N_{\mathrm{bd}}(\Lambda, T).$$

Obviously, we have

$$N_{\mathrm{int}}(\Lambda, T)\mathrm{vol}(\Lambda) \le \mathrm{vol}(T\Omega) = T^n \mathrm{vol}(\Omega),$$

hence it is enough to show that $N_{\mathrm{bd}}(\Lambda, T)$ may be estimated by the error term in the claim. The Lipschitz parametrizations of the boundary yield easily that $\partial\Omega$ may be covered by at most C_2/ν^{n-1} balls of diameter $< \nu \le 2$. If we set $\nu = \lambda_1/(C_1 T)$, then the above yields

$$N_{\mathrm{bd}}(\Lambda, T) \le 2^n C_1^{n-1} C_2 \left(\frac{T}{\lambda_1}\right)^{n-1}$$

at least for $\nu \le 2$. Since $N(\Lambda, T) = 1$ for $T < \lambda_1$, this proves the claim. □

Proof of Theorem 11.10.14: For simplicity, we restrict the proof to the case $K = \mathbb{Q}$. The extension to arbitrary number fields may be done by S.H. Schanuel's method in [257].

Let $N(W,T)$ be the number of $\mathbf{x} \in \mathbb{Z}^N \cap W$ with $|x_1|^2 + \cdots + |x_N|^2 \leq T^2$. Let Ω be the intersection of $\overline{W}$ with the unit ball in $\mathbb{R}^N$. Then Lemma 11.10.15 and Theorem 11.10.13 yield

$$N(W,T) = \frac{\alpha(n)}{H_{\mathrm{Ar}}(L)} T^n + O((\tfrac{T}{\lambda_1})^{n-1}) + 1.$$

Note that the implied constant is independent of W because all Ωs are isometric. Let $N^*(W,T)$ be the number of primitive solutions, i.e. $\mathbf{x} \in \mathbb{Z}^N \cap W$ with $|x_1|^2 + \cdots + |x_N|^2 \leq T^2$ and $\mathrm{GCD}(x_1, \ldots, x_N) = 1$. It is clear that

$$N(W,T) - 1 = \sum_{k=1}^{\infty} N^*\left(W, \frac{T}{k}\right)$$

and hence the Möbius inversion formula gives

$$N^*(W,T) = \sum_{k=1}^{\infty} \mu(k) \left(N\left(W, \frac{T}{k}\right) - 1 \right),$$

where μ is the Möbius function. We have $N(W, T/k) = 1$ if $k > T/\lambda_1$. We must count only one half of the primitive solutions for $N(L,T)$, hence we get

$$N_{\mathrm{Ar}}(L,T) = \frac{1}{2} N^*(W,T) = \frac{\alpha(n)}{2} \frac{T^n}{H_{\mathrm{Ar}}(L)} \sum_{k=1}^{\lfloor T/\lambda_1 \rfloor} \frac{\mu(k)}{k^n} + O((\tfrac{T}{\lambda_1})^{n-1}) \sum_{k=1}^{\lfloor T/\lambda_1 \rfloor} \frac{|\mu(k)|}{k^{n-1}}.$$

By the product development

$$\frac{1}{\zeta(s)} = \prod_p (1 - p^{-s}) = \sum_{k=1}^{\infty} \frac{\mu(k)}{k^s}$$

for the Riemann zeta function ([8], Ch.5, Sec.4.1, Th.9) and using Minkowski's first theorem in C.2.19 together with Theorem 11.10.13, we easily deduce the claim. $\square$

Proof of Theorem 11.10.11: We may assume that Σ has a K-rational base point Q_0, otherwise the theorem is trivial. K-rational lines on X correspond to K-rational points on Σ and we can apply Faltings's big theorem to the image $a(\Sigma)$ of Σ in the Albanese variety A. Since $a(\Sigma)$ generates A, no translate of an abelian subvariety is equal to $a(\Sigma)$. Hence $a(\Sigma(K))$ is contained in the union of finitely many elliptic curves Y_i and a finite set S. Here, the elliptic curves are allowed to have origin different from the origin of A and so we may assume that they are defined over K. Let $\widehat{h}$ be the Néron–Tate height on an elliptic curve Y_i with respect to an even ample line bundle H_i; then we have

$$\widehat{h}(a(Q)) \ll h_{\mathrm{Ar}}(Q) \ll \widehat{h}(a(Q))$$

for all $Q \in a^{-1}(Y_i)$ (since $a^*(H_i)$ is ample on $a^{-1}(Y_i)$, see A.12.7).

Now recalling that the Néron–Tate height yields a norm on $Y_i(K)/\mathrm{tors}$, we deduce that for any fixed $\kappa > 0$ the number of points $Q \in \Sigma(K)$ with $H_{\mathrm{Ar}}(Q) \leq T^\kappa$ is $\ll (1 + \kappa \log T)^{r/2}$. Moreover, $H_{\mathrm{Ar}}(Q)$ grows faster than polynomial and hence γ is finite. By

Theorem 11.10.14 and noticing that the first successive minimum is uniformly bounded from below (by the first successive minimum of O_K^{10}), we have

$$\left|\{P \in L_Q(K) \mid H_{\mathrm{Ar}}(P) \le T\}\right| = c_2 \frac{T^{2d}}{H_{\mathrm{Ar}}(Q)^d} + O\left(T^{2d-1}\right) \tag{11.32}$$

with T^{2d-1} replaced by $T(1+\log T)$ in case of $K = \mathbb{Q}$.

In order to finish the proof, we shall prove that there are constants $\kappa, \rho > 0$, and a (possibly empty) finite set $\mathcal{P}$ of points $P \in X(\overline{K})$, such that for every $T \ge \rho$ and every $Q \in \Sigma(\overline{K})$ with $H_{\mathrm{Ar}}(Q) > T^\kappa$, the set

$$\{P \in L_Q(\overline{K}) \mid H_{\mathrm{Ar}}(P) \le T\}$$

either is empty or consists of exactly one point $P \in \mathcal{P}$. Thus the contribution to the counting of $N_{\mathrm{lines}}(X,T)$ due to lines L_Q with $H_{\mathrm{Ar}}(Q) > T^\kappa$ is at most $|\mathcal{P}|$.

We have already remarked that there are $\ll (1+\kappa \log T)^{r/2}$ points $Q \in \Sigma(K)$ with $H_{\mathrm{Ar}}(Q) \le T^\kappa$, hence summing (11.32) over Q we get the theorem, because the effect of intersection of lines is $\ll (1+\kappa\log T)^r$ and may be neglected in the counting.

Consider lines L_Q with

$$H_{\mathrm{Ar}}(L_Q) = H_{\mathrm{Ar}}(Q) > T^\kappa.$$

If $\mathbf{x}$ and $\mathbf{x}'$ are two distinct points of L_Q, 2.9.8 yields

$$H_{\mathrm{Ar}}(L_Q) \le H_{\mathrm{Ar}}(\mathbf{x}) H_{\mathrm{Ar}}(\mathbf{x}'),$$

hence if $\kappa > 2$ we cannot have two points on L_Q with Arakelov height at most T. Suppose now that $\mathbf{x}_0 \in L_Q(\overline{K})$ is a (necessarily unique) point such that $H_{\mathrm{Ar}}(\mathbf{x}_0) \le T$ and let $\mathbf{y}$ be any other point on $L_Q(\overline{K})$. The line L_Q has a parametrization $\mathbf{x}_0 u + \mathbf{y} v$ and lies in X, a cubic threefold given by an equation

$$\sum_{ijk} a_{ijk} x_i x_j x_k = 0.$$

Substituting the parametrization of L_Q into this equation and looking at the coefficient of $u^2 v$, we infer that

$$\sum_{k=0}^{4} \left(\sum_{ij\pi} a_{\pi(ijk)} x_{0i} x_{0j} \right) y_k = 0, \tag{11.33}$$

where π ranges over all permutations. In other words, the line L_Q is contained in the cubic surface S in $\mathbb{P}^3$, which is the intersection of X with the hyperplane Π_Q defined by (11.33). Now we distinguish cases.

If S contains only finitely many lines, these lines are determined algebraically by the coefficients of the defining equations of X and Π_Q, hence they have height of polynomial size in the heights of these equations, hence $\ll H_{\mathrm{Ar}}(\mathbf{x}_0)^\rho \le T^\rho$ for some absolute constant ρ (for a precise result, we may use some sort of arithmetic Bézout theorem, see [**45**], Th.5.5.1, for an advanced version). Since L_Q is one of these lines, this contradicts the lower bound $H_{\mathrm{Ar}}(L_Q) > T^\kappa$ if $\kappa > \rho$ and T is larger than some constant depending only on X, which we can assume by taking ρ large enough. Thus we need only deal with the case in which S contains infinitely many lines.

Note that the smooth cubic surface contains exactly 27 lines ([**148**], Th.V.4.7). Therefore, if S contains infinitely many lines it must be singular and, intersecting S with a generic place in Π_Q, we verify that S can be only one of the following possibilities:

(a) a cone over an elliptic curve;

(b) a cone over a rational cubic curve;

(c) a rational cubic ruled surface;

(d) reducible, with a projective plane as a component.

Cases (b), (c), (d) are immediately excluded because they would give rise to a family of lines of X parametrized by a rational curve. This would lead to a rational curve in the Fano surface Σ. Since the Albanese map is finite, the image under the Albanese map a would give a rational curve inside an abelian variety contradicting 8.2.18 or Proposition 8.2.19.

Case (a) can actually occur. The maximum number of such cones is 30, attained for example with the Fermat cubic threefold $x_0^3 + x_1^3 + x_2^3 + x_3^3 + x_4^3 = 0$ for the complete intersections with the hyperplanes $x_i + \eta x_j = 0, \eta^3 = 1$ and $i \neq j$, see [**68**], Lemma 8.1 and p.315.

We conclude that the vertices $\mathbf{x}_0$ of these cones belong to a finite set $\mathcal{P}$ and by choosing T sufficiently large only the vertices contribute to the counting, proving the claim. □

11.10.16. We have studied in Chapter 4 the theory of small points on subvarieties of a linear torus. There is a similar theory on an abelian variety A over a number field K, which we mention briefly. For details, we refer to the overview article of Abbes [**1**].

Let X be a closed subvariety of A. As in 3.2.2, a **torsion coset** of X is a translate of an abelian subvariety by a torsion point and we define X^* to be the complement of the union of all torsion cosets in X. We have the following analogue of Theorem 4.2.2, also due to Zhang [**342**]:

Theorem 11.10.17. *Let $\widehat{h}_L$ be the Néron–Tate height on A with respect to an ample symmetric line bundle L. Then*

(a) *There are only finitely many maximal torsion cosets in X and its union is $X \setminus X^*$.*
(b) *There is a positive lower bound for the restriction of $\widehat{h}_L$ to X^*.*

11.10.18. This theorem is called the **Bogomolov conjecture**. In fact, Bogomolov conjectured it only for a subcurve of A. We omit the proof of Theorem 11.10.17, which is best done in the framework of Arakelov theory. It relies on an equidistribution theorem due to Szpiro, Ullmo, and Zhang [**296**]. This inspired Bilu's approach in Section 4.3.

Since torsion points are characterized by Néron–Tate height 0, Theorem 11.10.17 yields immediately:

Corollary 11.10.19. *If X is not a torsion coset, then the torsion points are not Zariski-dense in X.*

11.10.20. A. Moriwaki [**208**] generalized Theorem 11.10.17 to finitely generated fields F over $\mathbb{Q}$ using a new type of heights given in terms of Arakelov geometry. This proves

Corollary 11.10.20 over F, which is the **Manin–Mumford conjecture**. The latter was first proved by M. Raynaud [**237**], [**238**], using different methods.

11.11. Bibliographical notes

For a historical account, we refer to the introduction. Our presentation follows quite closely [**29**].

There is a version of the local Eisenstein theorem (without the existence statement) which does not require the condition $\frac{\partial p}{\partial t}(\mathbf{0},0) \neq 0$, as shown by W.M. Schmidt [**270**]; his proof is more difficult and uses techniques from p-adic linear differential equations. The sharpest result of this type is due to B.M. Dwork and A.J. van der Poorten [**95**], also obtained with similar methods. These deeper results are not needed here.

P. Vojta [**314**] proved Faltings's big theorem for a semiabelian variety A over any field K of characteristic 0 and with Γ any finitely generated subgroup Γ of $A(K)$. M. McQuillan [**199**] extended Vojta's result to the division group of Γ. B. Poonen [**235**] proved a generalization of Faltings's big theorem which includes also Bogomolov's conjecture in the case of A isogeneous to a product of an abelian variety with a torus, and G. Rémond [**242**] showed it for all semiabelian varieties.

The problem of obtaining an effective bound for the number of points in Faltings's big theorem remained open for quite a while. Eventually, G. Rémond [**240**] proved that the number of translates of maximal abelian subvarieties contained in X is effectively bounded in terms of A, $\mathrm{rank}(\Gamma)$ and $\deg(X)$ in the number field case. The main difficulty in Rémond's proof arises from the fact that the natural generalization of Mumford's gap principle does not hold for varieties of higher dimension and he had to introduce new ideas to overcame this problem. In the end, his bound turns out to be simply exponential in $\mathrm{rank}(\Gamma)$.

The conjecture of Manin on the number of rational points on a Fano manifold has now been proved in several cases, see Peyre [**234**], although counterexamples were found by V.V. Batyrev and Y. Tschinkel [**17**].

Theorem 11.10.11 is previously unpublished.

A proof of Bogomolov's conjecture for subvarieties of an abelian variety of CM type, following the method used in Chapter 4 for the torus case, was given in E. Bombieri and U. Zannier [**39**]; however, the proof does not extend in an obvious way to the general case. The case of curves in an arbitrary abelian variety was then proved by E. Ullmo [**301**] introducing new ideas, and the general case was finally settled in [**342**] and [**296**].

S. David and P. Philippon have given good explicit quantitative versions of the Bogomolov conjecture (see [**82**], [**84**]).

12 THE *abc*-CONJECTURE

12.1. Introduction

The abc-conjecture of Masser and Oesterlé is a typical example of a simple statement that can be used to unify and motivate many results in number theory, which otherwise would be scattered statements without a common link. As such, it deserves to be discussed, first by showing its power and then by generalizing it and showing how it fits into the much more general and coherent set of conjectures provided by Vojta in his thesis.

Although a pessimist may conclude that the ease with which the abc-conjecture may be applied to solve notoriously difficult problems is only a reflection of how difficult its proof is likely to be, we should keep in mind that its function field analogue is quite easy to prove and provides a unified method of attack for many problems in the arithmetic of function fields. Moreover, whatever its status in the classical case, it is likely that exceptions, if any, will be extremely rare and most of the conclusions obtained by its application are also likely to be valid and provable in some instances by different methods. The abc-conjecture is also a useful tool for guessing the right answer when analysing specific problems, hence its significance should not be too easily discounted.

The content of this chapter is as follows. In Section 12.2 we recall the formulation of the abc-conjecture over $\mathbb{Q}$ and prove some consequences of it, including Elkies's proof that the strong abc-conjecture implies Roth's celebrated theorem on approximation of algebraic irrationals by rational numbers, in fact effectively if we assume an effective abc-conjecture.

Elkies's proof is based on an interesting result, Belyĭ's lemma, which proves that any non-trivial morphism $C \to C'$ of irreducible smooth curves over a number field K can be extended to $C \to C' \to \mathbb{P}^1$, where now the composition $C \to \mathbb{P}^1$ is ramified only at $\{0, 1, \infty\}$. A very interesting feature of this lemma is that it holds only for curves defined over a number field. Belyĭ's lemma will also be the main tool in Section 12.3 to prove Belyĭ's theorem, which is a necessary and sufficient criterion for a complex curve to be defined over $\overline{\mathbb{Q}}$. In Section 12.4, we prove first the analogue of the abc-conjecture for polynomials. Then we give examples for the strong abc-conjecture over $\mathbb{Q}$ as well as counterexamples for stronger formulations.

In Section 12.5, we deal with the equivalence of the abc-conjecture with other conjectures, including Szpiro's conjecture about conductors and discriminants of elliptic curves over $\mathbb{Q}$. This chapter concludes with Section 12.6 dealing with a result of Darmon and Granville that the generalized Fermat equation has only finitely many integer solutions. This comes from a non-trivial application of Faltings's theorem based on the theory of ramified coverings presented in Section 12.3.

We use several previous results, but otherwise this chapter may be read to a large extend independently of the other parts of the book.

12.2. The abc-conjecture

This section begins with the formulation of the abc-conjecture over the rational numbers. Then we give a weak explicit form and indicate how to deduce Fermat's and Catalan's conjectures from it. The standard formulation of the abc-conjecture involves a positive small parameter ε and an unspecified positive constant $C(\varepsilon)$, which depends only on ε, to rule out the possibility of disproving the conjecture by finding numerical counterexamples. The main drawback of such an approach is that algebraic geometry has not been of any help in producing plausible heuristics about the behavior of $C(\varepsilon)$. However, considerations of diophantine approximation have led A. Baker to suggest a specific dependence of $C(\varepsilon)$ on ε, and we mention his conjecture in 12.2.6.

The remaining part is dedicated to an argument, due independently to Langevin and Elkies, that the abc-conjecture implies Roth's original theorem over $\mathbb{Q}$, in fact in a stronger form. Its importance also lies in the fact that an effective abc-theorem would make Roth's theorem effective. The proof is based on Belyĭ's lemma from 12.2.7. In Section 14.4, Elkies's argument will be extended to number fields and the proof becomes more conceptual. We conclude this section with an amusing application of the abc-conjecture to a classical question of analytic number theory.

Definition 12.2.1. *The* ***radical*** $\mathrm{rad}(N)$ *of an integer* N *is the product of all distinct primes dividing* N

$$\mathrm{rad}(N) = \prod_{p|N} p.$$

The following conjecture is called the **abc-conjecture in the strong form**:

Conjecture 12.2.2. *Let* $\varepsilon > 0$ *be a positive real number. Then there is a constant* $C(\varepsilon)$ *such that, for any triple* a, b, c *of coprime positive integers with* $a + b = c$, *the inequality*

$$c \leq C(\varepsilon)\,\mathrm{rad}(abc)^{1+\varepsilon}$$

holds.

12.2.3. The adjective "strong" here refers to the fact that this statement is supposed to hold for every positive ε. If we assume that it only holds for some fixed $\varepsilon > 0$, for example $\varepsilon = 1$, then we refer to it as the **weak abc-conjecture.**

In this respect, we note that for applications the weak abc-conjecture often is as useful as in its strong formulation. The statement

$$a + b \leq \operatorname{rad}(ab(a+b))^2 \tag{12.1}$$

for every pair a, b of positive coprime integers has in fact been conjectured by several authors as a likely explicit form of the weak abc-conjecture.

Example 12.2.4. Suppose $x^n + y^n = z^n$ is a non-trivial solution in coprime positive integers of the famous **Fermat equation**. Let us take $a = x^n$, $b = y^n$, $c = z^n$ in (12.1). Since $abc = (xyz)^n$ we deduce

$$z^n \leq \operatorname{rad}((xyz)^n)^2 = \operatorname{rad}(xyz)^2 \leq (xyz)^2 < z^6.$$

Since $z > 1$, this implies $n \leq 5$. It is well known that the Fermat equation has no non-trivial solutions for $n = 3$ (Euler), $n = 4$ (Fermat), $n = 5$ (Dirichlet, Legendre). For a proof of these classical cases, we refer to [**97**]. We conclude that (12.1) implies **Fermat's last theorem**. Any weak abc-conjecture would lead to a proof of the asymptotic Fermat conjecture, namely a proof for all sufficiently large exponents n.

Fermat's last theorem is now proved, at last, by A. Wiles and R. Taylor [**331**], [**297**]. For an account of the proof, we refer to the book of H. Darmon, F. Diamond, and R. Taylor [**77**].

Example 12.2.5. The same argument applies to **Catalan's conjecture** that 8 and 9 are the only two consecutive perfect powers in the sequence of positive natural integers. If we apply (12.1) to the Catalan equation $x^m + 1 = y^n$, we find

$$y^n \leq (xy)^2 < y^{2\frac{n}{m}+2}$$

and $n(m-2) < 2m$. As we may restrict to m, n prime, this leaves us with the well-known possibilities $m = 2$ (V.A. Lebesgue), or $n = 2$ (Chao Ko), or (m, n) one of the pairs $(3,3)$ (trivial), $(3,5)$, $(5,3)$ (Nagell). For an account of these cases, we refer to P. Ribenboim [**243**]. Hence the Catalan conjecture follows from (12.1).

The Catalan conjecture has recently been established unconditionally by P. Mihăilescu [**203**] (see also Yu.F. Bilu [**25**] for an exposition of his proof). It had been shown earlier unconditionally by R. Tijdeman [**300**] that the Catalan equation has only a finite number of solutions, and effective bounds for the size of the solutions x, y and of the exponents m, n could also be given. It suffices to consider the equation $x^p - y^q = \pm 1$ with $p > q$ odd primes and what really matters is to give a upper bound for the exponent p. Tijdeman's proof achieves this by appealing to Baker's theory of linear forms in logarithms. A noteworthy aspect of his method is that it extends to study the equation $x^p - y^q = \pm 1$ over an arbitrary number field.

The full solution required additional considerations from the theory of cyclotomic fields so to impose severe restrictions on the possible pairs of exponents (q, p). In particular, M. Mignotte and Y. Roy [**201**] proved its validity for $q < 10^5$, while P. Mihăilescu [**202**] proved that if odd primes $q < p$ allow a solution of Catalan's equation, then (q, p) satisfies the two congruences $p^{q-1} \equiv 1 \pmod{q^2}$ and $q^{p-1} \equiv 1 \pmod{p^2}$, forming a so-called **Wieferich pair**. A few examples of such pairs are known, namely

$$(2, 1093), (3, 1006003), (5, 1645333507), (83, 4871), (911, 318917), (2903, 18787),$$

but they appear to be rather uncommon.

Finally, Mihăilescu [**25**] was able to prove the further congruence $p \equiv 1 \pmod{q}$, hence $p \equiv 1 \pmod{q^2}$ followed because of the previously established congruence $p^{q-1} \equiv 1 \pmod{q^2}$. Therefore, $p \geq q^2 + 1$. A slightly more accurate argument also proved $p \geq 4q^2 + 1$. This clean lower bound for p could be combined with the upper bound for p in terms of q, again obtained by means of Baker's theory, showing that q had to be within the range covered by the Mignotte and Roy result. This completed the proof of Catalan's conjecture.

Variants of this proof, avoiding the use of Baker's theory and using instead the theory of cyclotomic fields to obtain the required upper bound for p, have been described by several authors and we refer to Mihăilescu [**203**] for further details.

The strong abc-conjecture as formulated before is unsatisfactory because it does not make precise the constant $C(\varepsilon)$. A. Baker [**15**] proposed the following more explicit statement:

Conjecture 12.2.6. *There is an absolute constant $\mathcal{K}$, such that if a, b, c are three coprime integers with $a + b + c = 0$ the inequality*

$$\max(|a|, |b|, |c|) \leq \mathcal{K} \cdot \Bigl(\prod_{p|abc} (p/\varepsilon) \Bigr)^{1+\varepsilon}$$

holds for every ε with $0 < \varepsilon \leq 1$.

Numerical experiments are consistent with a small value for the constant $\mathcal{K}$.

We continue with the application of the abc-conjecture to Roth's theorem, beginning with **Belyĭ's lemma**:

Lemma 12.2.7. *Let $g : C \to C'$ be a non-constant morphism between two irreducible smooth curves C, C', defined over a number field K. Let S be any finite set of points on $C(\overline{K})$. Then there is a non-constant rational function $h : C' \to \mathbb{P}^1_K$ such that the composite morphism $f = h \circ g : C \to \mathbb{P}^1_K$ is unramified outside of $f^{-1}(\{0, 1, \infty\})$ and moreover $f(S) \subset \{0, 1, \infty\}$.*

Unramified morphsims are studied in A.12, B.4, and B.3. Here, we deal with unramified morphisms $\varphi : C \to C'$ of smooth curves over a number field. Then it suffices to know that φ is unramified in $x \in C$ if and only if $\mathrm{d}(w \circ \varphi)/\mathrm{d}z(x) \neq 0$ with respect to local analytic coordinates z at x and w at $\varphi(x)$ (see Proposition A.12.18 and Example B.4.8).

Proof: **Reduction to $C = \mathbb{P}^1_{\mathbb{Q}}$ and to the identity morphism.** We choose any non-constant rational function $g_1 : C' \to \mathbb{P}^1_K$ and replace g by the composition $g_1 \circ g$. This reduces the problem to the case $C' = \mathbb{P}^1_K$. By increasing S so as to include the ramification set of g and setting $S' = g(S)$, it is enough to find a non-constant $h \in \mathbb{Q}(\mathbb{P}^1)$ which is unramified outside of $h^{-1}(\{0,1,\infty\})$ and maps S' into $\{0,1,\infty\}$.

The proof is completed by the following descending double induction on the degree of components of S and on the cardinality of S.

Lowering the degree of a point in S. Let $\alpha \in S$ be algebraic of highest degree $d \geq 2$. The minimal polynomial $p(x)$ of α defines a morphism $p : \mathbb{P}^1_{\mathbb{Q}} \to \mathbb{P}^1_{\mathbb{Q}}$ of degree d. The ramification set S_1 of p consists of the roots of $p'(x) = 0$ and ∞, hence its elements have degree at most $d-1$. After composition with the morphism p, by the chain rule we may replace S by the set $S' = p(S \cup S_1)$.

Note that $p(\alpha) = 0$ and that $p(\beta)$, $\beta \in \overline{\mathbb{Q}}$, has degree not exceeding the degree of β. Therefore, the number of ramification points of highest degree $d \geq 2$ has gone down at least by 1, replacing S by S'. By composition of such morphisms we reach a situation in which S consists only of rational points and ∞.

Lowering the cardinality of S. Suppose now that S consists only of rational points and ∞ and has cardinality $|S| \geq 4$. By applying a projective automorphism of $\mathbb{P}^1_{\mathbb{Q}}$ we may assume that S contains $\{0,1,\infty\}$. Let $\lambda = A/(A+B)$ be a fourth point in S, where A, B are integers with A, B, $A+B \neq 0$. Consider the rational function $h(x) = cx^A(1-x)^B$, where c is a non-zero constant to be determined. Since

$$\frac{h'}{h} = \frac{A}{x} - \frac{B}{1-x}$$

vanishes only at $x = \infty$ or $x = \lambda$, the morphism $h : \mathbb{P}^1_{\mathbb{Q}} \to \mathbb{P}^1_{\mathbb{Q}}$ is ramified at most over $\{0,1,\infty,\lambda\}$. We have $h(\{0,1,\infty\}) \subset \{0,\infty\}$ because A, B, $A+B \neq 0$. Note that

$$h(\lambda) = c\lambda^A(1-\lambda)^B,$$

hence choosing $c = \lambda^{-A}(1-\lambda)^{-B}$ we get $h(\lambda) = 1$. Moreover, h is unramified outside of $h^{-1}(\{0,1,\infty\})$. Therefore, composition by h replaces S by $\{0,1,\infty\} \cup h(S)$, decreasing the cardinality of S at least by 1 because $h(\lambda) = 1$. This completes the second induction step and the proof of the lemma. □

Example 12.2.8. Consider the morphism $g : \mathbb{P}^1_{\mathbb{Q}} \to \mathbb{P}^1_{\mathbb{Q}}$, where $g(x) = 2x^3 - 3ax^2 + 1$ and $a \in \mathbb{Z}$, $a \neq 0, 1$. Let S be the set of roots of $g(x)$. Since $a \neq 1$, the elements of S are algebraic numbers of degree 3. In order to determine the morphism h, we have to replace S by $g(S \cup S_1)$, where S_1 is the ramification set of g. In this case, since $g'(x) = 6x^2 - 6ax$, the ramification set of g is $S_1 = \{\infty, 0, a\}$. We have

$$S' = g(S \cup S_1) = \{0, 1, \infty, 1 - a^3\}.$$

By our choice of S and g, this makes the second step redundant. To lower the cardinality, we may take $A = 1 - a^3$, $B = a^3$ whence

$$h(x) = (1-a^3)^{-1+a^3}(a^3)^{-a^3}x^{1-a^3}(1-x)^{a^3}.$$

The composite map $f = h \circ g$ yields the desired morphism $f : \mathbb{P}^1_{\mathbb{Q}} \to \mathbb{P}^1_{\mathbb{Q}}$ with the property that f is unramified outside of $f^{-1}(\{0,1,\infty\})$ and $f(S) \subset \{0,1,\infty\}$.

If we further specialize $a = -1$, we find

$$f(x) = \frac{(1+3x^2+2x^3)^2}{4x^2(3+2x)},$$

$$1 - f(x) = -\frac{(1+x)^4(1-2x)^2}{4x^2(3+2x)},$$

yielding the identity

$$(1+3x^2+2x^3)^2 - (1+x)^4(1-2x)^2 = 4x^2(3+2x).$$

In general, this procedure based on composition of maps quickly leads to rational functions with gigantic degree and height. Note that the procedure followed here is not necessarily the best, for example if $a = -1$ the polynomial $h(x) = (x-1)^2$ does the job as well, with a corresponding $f(x) = (3x^2+2x^3)^2$ and polynomial identity

$$x^4(3+2x)^2 + (1+x)^2(1-2x)(1+3x^2+2x^3) = 1.$$

Theorem 12.2.9. *The strong abc-conjecture over* $\mathbb{Q}$ *implies Roth's theorem (see Theorem 6.2.3), in the special case* $K = \mathbb{Q}$ *and* $S = \{\infty\}$.

Remark 12.2.10. The proof is constructive; hence an effective version of the strong abc-conjecture implies the effective Roth theorem, indeed a stronger version of it which takes into account arithmetic ramification (see Section 14.4).

Proof: Let α be an algebraic number of degree $n \geq 2$ (the case $n = 1$ is trivial) and let $g(x)$ be its minimal polynomial over $\mathbb{Z}$. We apply Belyĭ's lemma from 12.2.7 to the morphism $g : \mathbb{P}^1_{\mathbb{Q}} \to \mathbb{P}^1_{\mathbb{Q}}$ and the set S consisting of the roots of $g(x)$, obtaining a rational function $h(x)$, defined over $\mathbb{Q}$, such that the composition $f(x) = h(g(x))$ has the following properties:

(a) the morphism $f : \mathbb{P}^1_{\mathbb{Q}} \to \mathbb{P}^1_{\mathbb{Q}}$ is unramified outside of $\{0,1,\infty\}$;

(b) $f(S) = h(0) \subset \{0,1,\infty\}$.

Without loss of generality, we may suppose that $h(0) = 0$. Let $f(x) = u(x)/w(x)$, where u, w are polynomials with integral coefficients without common factors, and let $v(x) := w(x) - u(x)$. Let d be be the degree of the morphism f, hence $d = \max(\deg(u), \deg(w))$, and let

$$U(X,Y) = Y^d u(X/Y), \quad V(X,Y) = Y^d v(X/Y), \quad W(X,Y) = Y^d w(X/Y)$$

be the associated homogeneous forms of degree d. Note that $U + V = W$ and they have no common factors as well.

Consider the factorizations of U, V, W into irreducible factors U_i of degree $n_i \geq 1$, namely

$$\begin{aligned} U(X,Y) &= u_0\, U_1(X,Y)^{m_1} \cdots U_r(X,Y)^{m_r} \\ V(X,Y) &= v_0\, U_{r+1}(X,Y)^{m_{r+1}} \cdots U_s(X,Y)^{m_s} \\ W(X,Y) &= w_0\, U_{s+1}(X,Y)^{m_{s+1}} \cdots U_t(X,Y)^{m_t} \end{aligned}$$

in the ring $\mathbb{Z}[X,Y]$ and with $u_0, v_0, w_0 \in \mathbb{Z}$.

Since we assume $h(0) = 0$, we see that we may take $n_1 = n$ and $U_1(X,Y) = G(X,Y) = Y^n g(X/Y)$, the irreducible homogeneous binary form associated to the algebraic number α.

We need a simple lemma based on the theory of Weil heights from Section 2.4. The height of f is by definition the height of $u(x) + w(y)$.

Lemma 12.2.11. *There is a positive integer D, bounded in terms of the height and degree of the rational function $f(x)$, such that, if k, l are coprime integers, then*

$$\mathrm{GCD}(U(k,l),\, V(k,l),\, W(k,l)) \mid D.$$

Proof: Let us consider the morphism $\varphi : \mathbb{P}^1_{\mathbb{Q}} \to \mathbb{P}^2_{\mathbb{Q}}$ given by $\varphi(\mathbf{x}) = (U(\mathbf{x}) : V(\mathbf{x}) : W(\mathbf{x}))$. Then we have $\varphi^* O_{\mathbb{P}^2}(1) \cong O_{\mathbb{P}^1}(d)$. In terms of the multiplicative height H, Theorem 2.3.8 gives a constant $C < \infty$ with

$$H(\mathbf{x})^d \leq C H_\varphi(\mathbf{x}).$$

Thus setting $\mathbf{x} = (k : l)$, the above inequality yields

$$\begin{aligned} \max(|k|,|l|)^d &\leq C \frac{\max\{|U(k,l)|, |V(k,l)|, |W(k,l)|\}}{\mathrm{GCD}(U(k,l), V(k,l), W(k,l))} \\ &\leq C' \frac{\max(|k|,|l|)^d}{\mathrm{GCD}(U(k,l), V(k,l), W(k,l))} \end{aligned}$$

with a constant C' depending only on the height and the degree of the rational function f. This proves the claim. □

Now we complete the proof of Theorem 12.2.9 as follows.

We may suppose that α is real. Let k/l be a rational approximation to α. We may assume k, l coprime, $l > 0$ and that $U(k,l)V(k,l)W(k,l) \neq 0$. We abbreviate

$$D_0 := \mathrm{GCD}(U(k,l),\, V(k,l),\, W(k,l)),$$

set

$$a = U(k,l)/D_0, \quad b = V(k,l)/D_0, \quad c = W(k,l)/D_0$$

and apply the abc-inequality to the relation $a + b = c$. The radical $\mathrm{rad}(abc)$ is a divisor of

$$u_0 v_0 w_0\, U_1(k,l) U_2(k,l) \cdots U_t(k,l);$$

therefore, recalling that $U_1(k,l) = G(k,l)$ and $|U_i(k,l)| \ll_f l^{\deg(U_i)}$, we get

$$\operatorname{rad}(abc) \ll_f |G(k,l)|\, l^K,$$

where

$$K = -\deg(U_1) + \sum_{i=1}^{t} \deg(U_i) = -n + \sum_{i=1}^{t} \deg(U_i).$$

Next, we note that, since $f(\alpha) = 0$ and $|\alpha - k/l|$ is assumed to be sufficiently small, we must have that k/l is bounded away from the zeros of v. It then follows

$$|V(k,l)| = l^d \left| v\left(\frac{k}{l}\right)\right| \gg_f l^d.$$

By Lemma 12.2.11, we conclude $|b| \gg_f l^d$, with an implied constant depending only on the height and degree of $f(x)$.

In view of these considerations, the abc-inequality yields

$$l^d \ll_{\varepsilon,f} (|G(k,l)|\, l^K)^{1+\varepsilon}$$

whence

$$|G(k,l)| \gg_{\varepsilon,f} l^{d-K-d\varepsilon/(1+\varepsilon)}.$$

It remains to evaluate $d - K$. To this end, we apply Hurwitz's theorem from B.4.6 to the ramified covering $f : \mathbb{P}^1_{\mathbb{Q}} \to \mathbb{P}^1_{\mathbb{Q}}$. The ramification occurs only over $\{0, 1, \infty\}$ and we get

$$-2 = d \cdot (-2) + \sum_{i=1}^{t} (m_i - 1) \deg(U_i).$$

Moreover, we have

$$3d = \sum_{i=1}^{t} m_i \deg(U_i)$$

leading to

$$\sum_{i=1}^{t} \deg(U_i) = d + 2$$

and finally $d - K = d + n - \sum \deg(U_i) = n - 2$. We conclude that

$$|G(k,l)| \gg_{\varepsilon,f} l^{n-2-\varepsilon d/(1+\varepsilon)}. \tag{12.2}$$

Since $G(k,l) = l^n g(k/l)$ and $g(\alpha) = 0$, the mean-value theorem shows that

$$G(k,l) = l^n g(k/l) = l^n (g(k/l) - g(\alpha)) = l^n \left(\frac{k}{l} - \alpha\right) g'(\xi)$$

for some point ξ between k/l and α. Since k/l is close to α and $g'(\alpha) \neq 0$, we see that $|g'(\xi)|$ is bounded away from 0, giving

$$\left|\frac{k}{l} - \alpha\right| \gg_g l^{-n}\, |G(k,l)|.$$

Comparison with (12.2) yields the theorem. □

As already noted by M. Langevin [**176**], [**177**] and as we will further explain in Remark 14.4.17, this argument proves much more than Roth's theorem, namely

Theorem 12.2.12. *Let $\varepsilon > 0$ and $F(x,y) \in \mathbb{Z}[x,y]$ be a homogeneous polynomial of degree d with distinct linear factors over $\mathbb{C}$. Then for all coprime integers m, n with $F(m,n) \neq 0$, the strong abc-conjecture over $\mathbb{Q}$ implies*

$$\mathrm{rad}(F(m,n)) \gg_{\varepsilon,F} \max(|m|,|n|)^{d-2-\varepsilon}.$$

In particular, if we take $F(x,y) = xy(x+y)$, we recover the strong *abc*-conjecture over $\mathbb{Q}$. If we apply Theorem 12.2.12 to the homogenization of the minimal polynomial of an algebraic number, then we easily deduce Roth's theorem over $\mathbb{Q}$ and $S = \{\infty\}$. An immediate consequence of Theorem 12.2.12 is

Corollary 12.2.13. *Let $\varepsilon > 0$ and $f(x) \in \mathbb{Z}[x]$ be a polynomial of degree d, with distinct roots over $\mathbb{C}$. Then the strong abc-conjecture over $\mathbb{Q}$ implies*

$$\mathrm{rad}(f(n)) \gg_{\varepsilon,f} |n|^{d-1-\varepsilon}$$

for non-zero $n \in \mathbb{Z}$ with $f(n) \neq 0$.

Proof: The polynomial $F(x,y) = y^{d+1} f(x/y)$ is a homogeneous polynomial of degree $d+1$ with distinct linear factors over $\mathbb{C}$. Then apply the preceding theorem to $F(n,1) = f(n)$. □

Example 12.2.14. The following conditional result of analytic number theory is a unusual application of the *abc*-conjecture, due to A. Granville [**128**].

Let $f(x) \in \mathbb{Z}[x]$ be a polynomial without multiple roots. Assume also that $f(n)$ has no fixed square divisor for $n \in \mathbb{Z}$. Then it is conjectured that $f(n)$ takes squarefree values infinitely often, in fact for a set of integers of positive density. This is easy to prove for degree 1, not difficult for degree 2 using sieve methods, and more delicate arguments can be used for dealing with degree 3, but not much more is known for larger degrees. We have

Theorem 12.2.15. *Assume the strong abc-conjecture and let $f(x)$ be as above. Then the sequence $(f(n))_{n\in\mathbb{N}}$ contains infinitely many squarefree integers. More precisely, the sequence of positive integers n such that $f(n)$ is squarefree has positive density*

$$|\{n \leq x \mid f(n) \text{ is squarefree}\}| \sim c(f)\, x$$

for some constant $c(f) > 0$.

Proof: We may assume that f is not a constant. Let $\omega(p)$ be the number of solutions of the congruence $f(a) \equiv 0 \pmod{p^2}$, hence the integers n with $f(n)$ not divisible by p^2 form a sequence of density

$$c_p(f) = 1 - \frac{\omega(p)}{p^2}.$$

If $c_p(f) = 0$, then p^2 divides $f(n)$ for every n, which was excluded by hypothesis.

If p does not divide the discriminant of f, any solution of the congruence $f(a) \equiv 0 \pmod{p}$ lifts uniquely to a solution $\pmod{p^2}$, as we see using Hensel's lemma in 1.2.10.

This proves $\omega(p) \leq \deg(f)$ and hence the infinite product $\prod_p c_p(f)$ is absolutely convergent.

Let x be a large integer and let $M := \prod_{p \leq \sqrt{\log x}} p^2$. By the Chinese remainder theorem, in any interval $\{k, \ldots, k + M - 1\}$ of M consecutive integers there are exactly $\prod_{p \leq \sqrt{\log x}} (p^2 - \omega(p))$ integers n such that $f(n)$ is not divisible by p^2 for every prime $p \leq \sqrt{\log x}$. As usual, we denote the number of primes up to z by $\pi(z)$. By the elementary estimate $\pi(z) \ll z/\log z$, we have $\sum_{p \leq z} \log p \leq \pi(z) \log z \ll z$; hence

$$M = e^{O(\sqrt{\log x})} = o(x).$$

Therefore, the number of integers $n \leq x$ for which $f(n)$ has no prime factors $p^2 | f(n)$ with $p \leq \sqrt{\log x}$ is equal to

$$\left\lfloor \frac{x}{M} \right\rfloor M \prod_{p \leq \sqrt{\log x}} \left(1 - \frac{\omega(p)}{p^2}\right) + O(M) \sim c(f) \cdot x,$$

where $c(f) := \prod_p c_p(f)$ is non-zero, as we have seen above.

The number of integers $n \leq x$ for which $p^2 | f(n)$ and $\sqrt{\log x} < p < x$ is majorized by

$$\left(\sum_{p > \sqrt{\log x}} \frac{\omega(p)}{p^2} \right) \cdot x + \sum_{p < x} \omega(p) \ll \frac{x}{\sqrt{\log x}}$$

again because of the elementary estimate $\pi(x) \ll x/\log x$, is also negligible in our counting.

It remains to show that the sequence of integers n, for which $p^2 | f(n)$, for some prime $p \geq n$, has zero density. This is the difficult step and sieve methods, combined with additional ingenious ideas, have been used to prove this for polynomials $f(x)$ of degree at most 3.

Unfortunately, this approach fails if the degree of $f(x)$ is 4 or more. However, on the assumption of the *abc*-conjecture we can use a clever trick and Corollary 12.2.13 to conclude the proof in a single stroke. We choose an integer m larger than the distance of any two roots of $f(x)$ and let l be another positive integer, which is at our disposal. Consider the polynomial

$$g(x) = f(x) f(x + m) \cdots f(x + lm)$$

and note that the assumption on m ensures that $g(x)$ too has no multiple roots.

By Corollary 12.2.13, the strong *abc*-conjecture implies that

$$\prod_{p | g(n)} p \gg_{g,\varepsilon} |n|^{\deg(g) - 1 - \varepsilon}$$

for $n \in \mathbb{Z}$ with $g(n) \neq 0$. Therefore, if $g(n) = uv^2$ we must have

$$|uv| \gg_{g,\varepsilon} |n|^{\deg(g) - 1 - \varepsilon}.$$

Noting that $g(n) = uv^2$ has precise order $|n|^{\deg(g)}$, we get $|v| \ll_{g,\varepsilon} |n|^{1+\varepsilon}$.

For n sufficiently large, this shows that of the integers $f(n), f(n + m), f(n + 2m), \ldots, f(n+lm)$ only one can be divisible by p^2 for some prime $p \geq n$ and, splitting the integers into m progressions modulo m, only m of the integers $f(n), f(n + 1), \ldots, f(n + lm)$

may admit such a square factor. In particular, the density of integers n for which $f(n)$ admits a square factor p^2 with $p \geq n$ is at most $m/(lm) = 1/l$. Since l can be chosen arbitrarily large, the set of such integers n has density 0, concluding the proof. □

12.3. Belyĭ's theorem

Our next goal is Belyĭ's striking theorem that a complex projective curve is defined over $\overline{\mathbb{Q}}$ if and only if it is a covering over $\mathbb{P}^1_{\mathbb{C}}$, unramified outside of $\{0, 1, \infty\}$. It will be of minor importance in our book and the reader may skip it in a first reading.

We begin by gathering the necessary facts about coverings, referring for details to W.S. Massey **[196]**, Ch.5, and J.B. Conway **[70]**, Ch.16. The reader is assumed to be familiar with the connexion between algebraic and analytic structures on a smooth complex variety as provided by A.14. Then we will prove Belyĭ's theorem using the language of schemes. An extension of the last part of the proof will yield a result of Grothendieck.

12.3.1. A **topological covering** is a continuous map $\pi : Y \to X$ of non-empty topological spaces which is locally trivial, i.e. every $x \in X$ has an open neighbourhood U and a discrete non-empty topological space F such that $\pi^{-1}(U)$ is homeomorphic to $U \times F$ with π corresponding to the first projection.

12.3.2. A **morphism of topological coverings** $\pi_1 : Y_1 \to X$, $\pi_2 : Y_2 \to X$ is a continuous map $\varphi : Y_1 \to Y_2$ with $\pi_1 = \pi_2 \circ \varphi$. If $Y_1 = Y_2$ and φ is a homeomorphism, then φ is called an automorphism of the covering. The group of automorphisms of a topological covering is called the **covering group**.

12.3.3. We always assume in this section that X is a connected locally contractible topological space with a base point $x_0 \in X$. The **fundamental group** $\pi_1(X, x_0)$ is the group of homotopy classes of loops with origin x_0. For a topological covering $\pi : Y \to X$, the fundamental group $\pi_1(X, x_0)$ acts from the right on the fibre $\pi^{-1}(x_0)$. Explicitly, for $y \in \pi^{-1}(x_0)$ and $\gamma \in \pi_1(X, x_0)$, we define y^γ as the end point of the unique lift of γ to Y with starting point y. Every morphism of topological coverings of X restricts to a map of fibres over x_0 compatible with the action of $\pi_1(X, x_0)$.

12.3.4. There is a bijective correspondence between isomorphism classes of connected topological coverings with base point over x_0 and subgroups of $\pi_1(X, x_0)$, similar to the Galois correspondence in the theory of algebraic field extensions. A connected topological covering $\pi : Y \to X$ with base point y_0 over x_0 corresponds to the subgroup $\pi_1(Y, y_0)$ of $\pi_1(X, x_0)$. Note that the identification of the elements of $\pi_1(Y, y_0)$ with its images in X is allowed because this homomorphism is injective. If $\varphi : Y_1 \to Y_2$ is a morphism of connected topological coverings mapping the base point of Y_1 to the base point of Y_2, then it has to be surjective and induces an inclusion $H_1 \subset H_2$ of corresponding fundamental groups. Moreover, every inclusion of subgroups arises this way and the corresponding φ is unique up to isomorphisms.

12.3.5. Let Y be a connected topological covering of X with base point y_0 over x_0. Then the (right) action of $G := \pi_1(X, x_0)$ on the fibre $\pi^{-1}(x_0)$ is transitive. The stabilizer H of y_0 is equal to $\pi_1(Y, y_0)$. Let $N(H) := \{g \in G \mid gHg^{-1} = H\}$ be the normalizer of H in G. Then $N(H)/H$ is isomorphic to the covering group Γ. In fact, the choice of

y_0 leads to an identification of $\pi^{-1}(x_0)$ with the right coset space $H\backslash G$ and $N(H)/H$ operates from the left giving rise to the isomorphism with Γ.

If H is a normal subgroup of G, then $\Gamma \cong G/H$ operates transitively and freely on $\pi^{-1}(x_0)$ and we may write $X = \Gamma\backslash Y$, as a quotient of Y by the left Γ-action.

12.3.6. If we change the base point from x_0 to x_1, then we choose a path ρ from x_0 to x_1 getting an isomorphism

$$\pi_1(X, x_0) \longrightarrow \pi_1(X, x_1), \quad \gamma \mapsto \rho^{-1}\gamma\rho,$$

where ρ^{-1} is obtained from ρ by following the reverse direction. Now fixing x_0 but varying the base point y_0 of Y in 12.3.4, we get a conjugated subgroup $\pi_1(Y, y_1)$ of $\pi_1(Y, y_0)$ in $\pi_1(X, x_0)$. Hence there is a bijective correspondence between isomorphism classes of connected topological coverings of X and conjugation classes of subgroups of $\pi_1(X, x_0)$.

The connected topological covering corresponding to the trivial subgroup $\{1\}$ is the **universal covering** $\widetilde{X}$ of X. By 12.3.4 and 12.3.5, the covering group $\widetilde{\Gamma}$ of $\widetilde{X}$ is isomorphic to $\pi_1(X, x_0)$ and there is a bijective correspondence between conjugation classes of subgroups Γ of $\widetilde{\Gamma}$ and isomorphism classes of connected topological coverings Y of X given by $Y = \Gamma\backslash\widetilde{X}$.

The covering Y is called **finite** if the number of fibre points is finite. The number of fibre points in Y of X is equal to the index $[\widetilde{\Gamma} : \Gamma]$, hence the covering Y is finite over X if and only if Γ has finite index in $\widetilde{\Gamma}$.

12.3.7. Now let X be a connected Riemann surface, in other words a connected complex manifold of dimension 1. Then every topological covering $\pi : Y \to X$ has a canonical analytical structure: By 12.3.1, there is an analytic atlas $(U_\iota)_{\iota \in I}$ of X such that Y is locally trivial over U_ι. Then we use the atlas $(\pi^{-1}(U_\iota))_{\iota \in I}$ of Y to define Y as a complex manifold. Moreover, it is clear that every morphism of topological coverings of X will be analytic.

12.3.8. Let us also assume that X is algebraic and $\pi : Y \to X$ is a finite connected topological covering. We have seen in 12.3.7 that the covering is analytic. Let $\overline{X}$ be the irreducible smooth projective curve containing X as an open subvariety (see A.13.3). We use the complex manifold structure on X and $\overline{X}$. A local application of 12.3.6 to $\{z \in \mathbb{C} \mid 0 < |z| < 1\}$ proves that π extends uniquely to a finite ramified covering $\overline{\pi} : \overline{Y} \to \overline{X}$, namely to a holomorphic map of connected Riemann surfaces locally of the form $z \mapsto z^n$ for some $n \in \mathbb{N}$, $n \geq 1$.

The **Riemann existence theorem** says that every compact Riemann surface is projective algebraic (see **[148]**, Th.B.3.1). This applies to $\overline{Y}$ and the GAGA-principle in A.14.7 shows that $\overline{\pi}$ is algebraic. We conclude that every finite topological covering of X is algebraic and that every morphism of topological coverings of X extends uniquely to a ramified algebraic morphism of the smooth projective compactifications. As an algebraic morphism, π is finite (because $\overline{\pi}$ is finite) and also étale. The latter follows from the fact that π is analytically a local isomorphism and from A.12.18.

12.3.9. Conversely, we claim that every finite étale algebraic morphism π of a variety X' on to a smooth complex variety X gives rise to a finite topological covering $\pi_{\mathrm{an}} : X'_{\mathrm{an}} \to$

X_{an}. To see this, note first that π_{an} is a local isomorphism (use A.12.18). For $x \in X$ with $\pi_{\mathrm{an}}^{-1}(x) = \{x'_1, \dots, x'_n\}$, we get disjoint open neighbourhoods U'_j of x'_j in the complex topology such that π_{an} maps U'_j biholomorphically onto an open neighbourhood U of x. We need to prove, for U sufficiently small, that $\pi_{\mathrm{an}}^{-1}(U) = U'_1 \cup \dots \cup U'_n$, yielding local triviality of π_{an}, but this is an easy consequence of properness of π_{an} (cf. A.14.6).

12.3.10. The universal covering space of $\mathbb{P}^1_{\mathrm{an}} \setminus \{0, 1, \infty\}$ is the upper half plane $\mathbb{H} := \{z \in \mathbb{C} \mid \Im z > 0\}$. The covering map is the modular function λ known from the proof of Picard's little theorem (see L. Ahlfors **[8]**; in 13.2.35 we give another proof using Nevanlinna theory). Let $\Gamma(2)$ be the kernel of the reduction modulo 2 on $SL(2, \mathbb{Z})$ and let $\overline{\Gamma}(2)$ be the image of $\Gamma(2)$ in $PSL(2, \mathbb{Z}) = SL(2, \mathbb{Z})/\{\pm 1\}$. Then the covering group of $\lambda : \mathbb{H} \to \mathbb{P}^1_{\mathrm{an}} \setminus \{0, 1, \infty\}$ is $\overline{\Gamma}(2)$. The next result is **Belyĭ's theorem**:

Theorem 12.3.11. *Let C be an irreducible smooth projective curve defined over $\mathbb{C}$. The following conditions are equivalent:*

(a) *There is a curve C' defined over $\overline{\mathbb{Q}}$ with C isomorphic to the base change $C'_{\mathbb{C}}$.*

(b) *There exists a non-constant rational function $f : C \to \mathbb{P}^1_{\mathbb{C}}$ ramified at most over three points.*

(c) *There is a subgroup Γ of finite index in $\overline{\Gamma}(2)$ such that $\Gamma \backslash \mathbb{H}$ is isomorphic to U_{an} for a Zariski open subset U of C.*

Moreover, if (a) *holds, if we identify $C'_{\mathbb{C}}$ with C, and if S is a finite subset of $C'(\overline{\mathbb{Q}})$, then Γ in* (c) *can be chosen such that $S \subset C \setminus U$.*

Proof: (a) $\Rightarrow$ (b). This follows from Belyĭ's lemma in 12.2.7.

(b) $\Rightarrow$ (c). Let $f : C \to \mathbb{P}^1_{\mathbb{C}}$ be the rational function in (b); we may assume, after a projective linear transformation of $\mathbb{P}^1_{\mathbb{C}}$, that f is unramified outside of $\{0, 1, \infty\}$. For $U := f^{-1}(\mathbb{P}^1_{\mathbb{C}} \setminus \{0, 1, \infty\})$, the map $f : U_{\mathrm{an}} \longrightarrow \mathbb{P}^1_{\mathrm{an}} \setminus \{0, 1, \infty\}$ is a finite unramified connected covering. Since $\mathbb{H}$ is the universal covering space of $\mathbb{P}^1_{\mathrm{an}} \setminus \{0, 1, \infty\}$ with covering group $\overline{\Gamma}(2)$ (cf. 12.3.10), there is a subgroup Γ of $\overline{\Gamma}(2)$ of finite index such that $U_{\mathrm{an}} \cong \Gamma \backslash \mathbb{H}$ (cf. 12.3.6). This proves (c).

Moreover, if (a) holds and S is a finite subset of $C'(\overline{\mathbb{Q}})$, then Belyĭ's lemma proves that there is a non-constant rational function f on C unramified outside of $\{0, 1, \infty\}$ with $f(S) \subset \{0, 1, \infty\}$. With this f in the proof of the implication (b) $\Rightarrow$ (c), we get $S \subset C \setminus U$.

(c) $\Rightarrow$ (a). Let Γ be a subgroup of $\overline{\Gamma}(2)$ of finite index. By 12.3.6 and 12.3.10, we have a finite topological covering $f : U_{\mathrm{an}} = \Gamma \backslash \mathbb{H} \to \mathbb{P}^1_{\mathrm{an}} \setminus \{0, 1, \infty\}$.

We first give a brief sketch of the proof. Evidently, f is defined over a finitely generated $\overline{\mathbb{Q}}$-algebra $R \subset \mathbb{C}$. Then the affine $\overline{\mathbb{Q}}$-variety S with coordinate ring R parametrizes a family $(f_s)_{s \in S}$ of coverings of $\mathbb{P}^1 \setminus \{0, 1, \infty\}$ with generic fibre f. By shrinking S, we will show that this family is étale leading to unramified coverings. Proving that the fibres have the same monodromy, we will conclude that the coverings are isomorphic. Comparing the generic fibre with the fibre over $s \in S(\overline{\mathbb{Q}})$, we will get the claim.

The details are as follows. By 12.3.8, f is a finite étale algebraic morphism and therefore U is affine. The coordinate ring of $X := \mathbb{P}^1 \setminus \{0, 1, \infty\}$ over $\mathbb{C}$ is

$$\mathbb{C}[X] = \mathbb{C}\left[x, \frac{1}{x}, \frac{1}{x-1}\right]$$

and $\mathbb{C}[U]$ is a finitely generated $\mathbb{C}[X]$-module. There are variables $\mathbf{y} = (y_1, \ldots, y_N)$ and $\mathbf{x} = (x, \frac{1}{x}, \frac{1}{1-x})$ such that $\mathbb{C}[U] = \mathbb{C}[\mathbf{x}, \mathbf{y}]/I$ for an ideal I. By finiteness, we may assume that $y_1, \ldots, y_N$ generate $\mathbb{C}[U]$ as a $\mathbb{C}[X]$-module, hence

$$y_i y_j - \sum_{k=1}^{N} \lambda_k(\mathbf{x}) y_k \in I \tag{12.3}$$

for suitable $\lambda_k(\mathbf{x}) \in \mathbb{C}[X]$. Hence I is generated by polynomials $p_1(\mathbf{x}, \mathbf{y}), \ldots, p_r(\mathbf{x}, \mathbf{y})$ consisting of all polynomials in (12.3) and some other polynomials of degree 1 in $\mathbf{y}$. The coefficients of these polynomials are contained in a subring R of $\mathbb{C}$ containing $\overline{\mathbb{Q}}$ such that R is a finitely generated $\overline{\mathbb{Q}}$-algebra. We note that R is an integral domain.

Now it is convenient to use the language of schemes. We consider the affine scheme

$$U_0 := \operatorname{Spec}\left(R[\mathbf{x}, \mathbf{y}]/(p_1(\mathbf{x}, \mathbf{y}), \ldots, p_r(\mathbf{x}, \mathbf{y}))\right)$$

of finite type over the integral affine scheme $S := \operatorname{Spec}(R)$ and let us denote by $\pi : U_0 \to S$ the morphism of structure. By construction, we have a canonical finite morphism $f_0 : U_0 \to X_R$ defined over R whose base change from R to $\mathbb{C}$ is f. Note that $R \subset \mathbb{C}$ induces a geometric point $z \in S(\mathbb{C})$, namely a morphism $z : \operatorname{Spec}(\mathbb{C}) \to S$. By construction, the fibre $(f_0)_z$ of f_0 over z is f and hence étale. The image of z is the generic point ζ of S and $(f_0)_z$ is the base change of the fibre $(f_0)_\zeta$ to $\mathbb{C}$. By flat descent, we conclude that $(f_0)_\zeta$ is an étale morphism (cf. A. Grothendieck and J.A. Dieudonné [**137**], Prop.17.7.1). Obviously, $(f_0)_\zeta$ is also surjective.

Note that f_0 is flat in every point u over ζ. This follows from the fact that $\mathcal{O}_{U_0,u}$ is canonically isomorphic to the local ring of u in the fibre $(U_0)_\zeta$, from the similar fact for $f_0(u)$ and from flatness of $(f_0)_\zeta$ (for details about fibres, cf. [**139**], 3.4). Clearly, f_0 is also unramified in u because this is a property of the fibre over $\pi(u) = \zeta$. This implies that f_0 is étale in all the points of $(U_0)_\zeta$. The étale points of f_0 form an open subset V_0 of U_0 (cf. B.3.2).

Note that there is an open dense subset S_0 of S such that $\pi^{-1}(S_0) \subset V_0$. This follows from a theorem of Chevalley stating that the image of a morphism is a constructible subset, i.e. a finite disjoint union of intersections of a closed and an open subset (cf. [**148**], Ex.s II.3.18, II.3.19). Since $(U_0)_\zeta \subset V_0$, we get $\zeta \notin T_0 := \pi(U_0 \setminus V_0)$. Then the constructibility of T_0 implies that ζ is not contained in the closure of T_0 in S and, by setting $S_0 := S \setminus \overline{T_0}$, we get our claim.

Now we consider S_0 as an irreducible variety over $\overline{\mathbb{Q}}$ and, by passing to a dense open subset, we may assume that S_0 is smooth (cf. A.7.16). Thus $X \times S_0$ is a smooth irreducible variety over $\overline{\mathbb{Q}}$ (cf. A.4.11 and A.7.17). Since the restriction of f_0 to $\pi^{-1}(S_0)$ is étale, we conclude that $\pi^{-1}(S_0)$ is a smooth variety over $\overline{\mathbb{Q}}$ (cf. [**137**], Prop.17.3.3). By Hilbert's Nullstellensatz in A.2.2, there is an algebraic point $s \in S_0(\overline{\mathbb{Q}})$.

By base change of f_0 from $\overline{\mathbb{Q}}$ to $\mathbb{C}$, we obtain a finite étale morphism

$$F : \pi^{-1}(S_0)_{\mathbb{C}} \longrightarrow X \times (S_0)_{\mathbb{C}}$$

over $(S_0)_{\mathbb{C}}$. By construction, the fibre F_z is f and the fibre F_s is equal to the base change of $(f_0)_s$ from $\overline{\mathbb{Q}}$ to $\mathbb{C}$. By 12.3.9 and connectedness of U_{an}, we conclude that F_{an} is a connected topological covering between associated complex manifolds.

We choose $x \in X(\mathbb{C})$. Let ρ be a path in $(S_0)_{\mathrm{an}}$ from z to s and let $\rho_x = \{x\} \times \rho$. For $y \in F^{-1}(x, z)$, let y^ρ be the endpoint of the lift of ρ_x to a path with origin y.

The fibres of F over S are finite étale morphisms, therefore $(f_z)_{\mathrm{an}}$ and $(F_s)_{\mathrm{an}}$ are also topological coverings of $X_{\mathrm{an}} = \mathbb{P}^1_{\mathrm{an}} \setminus \{0, 1, \infty\}$. For $\gamma \in G := \pi_1(X_{\mathrm{an}}, x)$, let $\gamma_z := \gamma \times \{z\}$ and $\gamma_s := \gamma \times \{s\}$. Obviously, we have

$$\rho_x^{-1} \gamma_z \rho_x = \gamma_s \in \pi_1(X_{\mathrm{an}} \times (S_0)_{\mathrm{an}}, (x, s)).$$

For $y \in F_z^{-1}(x) = F^{-1}(x, z)$, we get

$$(y^\gamma)^{\rho_x} = (y^{\gamma_z})^{\rho_x} = (y^{\rho_x})^{\gamma_s} = (y^{\rho_x})^\gamma$$

meaning that the map $y \mapsto y^{\rho_x}$ is a G-equivariant bijection of $F^{-1}(x, z)$ onto $F^{-1}(x, s)$. Because $(F_z)_{\mathrm{an}} = f_{\mathrm{an}}$ is a connected topological covering, G operates transitively on $F^{-1}(x, z)$ and hence the same is true for the action on $F^{-1}(x, s)$. This proves that $(F_s)_{\mathrm{an}}$ is also a connected covering. Moreover, it follows that both coverings correspond to conjugated subgroups of G (given as stabilizers of the actions). By the Galois correspondence in 12.3.6, the topological coverings $(F_z)_{\mathrm{an}}$ and $(F_s)_{\mathrm{an}}$ are isomorphic. By 12.3.8, the isomorphism is algebraic. This proves (a) with $C' = (U_0)_s$. ☐

Theorem 12.3.12. *Let X be a variety over $\overline{\mathbb{Q}}$ and let φ be a finite topological covering of $(X_{\mathbb{C}})_{\mathrm{an}}$. Then there is a finite étale morphism $\psi : Y \to X$ of varieties over $\overline{\mathbb{Q}}$ such that $(\psi_{\mathbb{C}})_{\mathrm{an}} = \varphi$. Moreover, Y and ψ are unique up to isomorphism.*

Proof: The arguments are similar to the proof of (c) $\Rightarrow$ (a) for Belyĭ's theorem in 12.3.11. The details are left to the reader. We give the following hints:

(a) By passing to the connected components, we may assume that X and Y are both connected.

(b) There is a unique way to endow Y with the structure of a complex space such that φ is a finite analytic morphism which is locally biholomorphic.

(c) For algebraicity, we have to use **Grothendieck's generalization of the Riemann existence theorem** (cf. A. Grothendieck [**140**], Exposé XII, Th.5.1; see also [**148**], Th.B.3.2 for the original version of Grauert–Remmert):

Theorem 12.3.13. *Let us assign to every finite étale covering of a complex algebraic variety Z the associated analytic morphism. Then this gives an equivalence from the category of finite étale coverings of Z to the category of finite analytic coverings of Z_{an} (meaning finite morphisms which are locally biholomorphic).*

12.4. Examples

We start with the abc-theorem for polynomials which is a direct consequence of Hurwitz's theorem and holds for $\varepsilon = 0$ and $C(\varepsilon) = 1$. The proof easily extends to function fields and will imply the abc-theorem of W.W. Stothers and R.C. Mason in Section 14.5. The analogy with number fields gives some evidence for the abc-conjecture.

On the other hand, the theorem of C.L. Stewart and R. Tijdeman [**291**], which will be our next goal, shows that the abc-conjecture over $\mathbb{Q}$ does not hold for $\varepsilon = 0$ whatever constant $C(\varepsilon)$ we choose. This is a clear instance that the analogy between function fields and number fields can have some very subtle points to it and should not be followed blindly as a wholesale tool for making conjectures. The proof of this theorem relies only on Dirichlet's pigeon-hole principle and the prime number theorem. There is a connexion from the proof to the birthday paradox and we sketch some heuristics of Granville on this theme.

The **abc-theorem for polynomials** has the following form:

Theorem 12.4.1. *Let K be a field of characteristic 0 and let $a(x), b(x), c(x) \in K[x]$ be not all constant, coprime in pairs and such that $a(x) + b(x) + c(x) = 0$. Let $\mathrm{rad}(abc)$ be the monic polynomial with simple zeros at the zeros of abc. Then*

$$\max\{\deg(a), \deg(b), \deg(c)\} \leq \deg(\mathrm{rad}(abc)) - 1.$$

We give here two simple proofs, which will be reinterpreted and extended to more general cases in Section 14.5.

First proof: Let us first observe that on the right-hand side, we count the roots z without their multiplicity ord_z. Clearly, we may assume that K algebraically closed and, by a permutation, $\deg(a) = \deg(b) \geq \deg(c)$.

The hypotheses of the theorem imply that none of a, b, c is identically 0 and the rational functions $u := -a/c$, $v := -b/c$ are not constant, with

$$u + v = 1.$$

First, note that the degree of u as a morphism $\mathbb{P}^1 \to \mathbb{P}^1$ satisfies

$$\deg(u) = \max\{\deg(a), \deg(b), \deg(c)\}.$$

For the ramification divisor R_u, we have

$$\mathrm{ord}_z(R_u) \begin{cases} \mathrm{ord}_z(u) - 1 & \text{if } z \in u^{-1}(0), \\ \mathrm{ord}_z(v) - 1 & \text{if } z \in u^{-1}(1), \\ -\mathrm{ord}_z(u) + 1 & \text{if } z \in u^{-1}(\infty). \end{cases}$$

By Hurwitz's theorem in B.4.6, we have

$$-2 \geq \deg(u) \cdot (-2) + \sum_{z \in u^{-1}(\{0,1,\infty\})} \operatorname{ord}_z(R_u).$$

Since $\deg(u) = \deg(v)$ and

$$\sum_{z \in u^{-1}(\{\lambda\})} \operatorname{ord}_z(u) \begin{cases} \deg(u) & \text{if } \lambda \neq \infty, \\ -\deg(u) & \text{if } \lambda = \infty, \end{cases}$$

we conclude that

$$-2 \geq \deg(u) - \left| u^{-1}(\{0,1,\infty\}) \right|. \tag{12.4}$$

It remains to give a upper bound for $\left| u^{-1}(\{0,1,\infty\}) \right|$.

We distinguish two cases. Suppose first that $\deg(u) = \deg(a) = \deg(b) = \deg(c)$. Then

$$\begin{aligned} \operatorname{supp}(u^{-1}(0)) &= \{z \in \overline{K} \mid a(z) = 0\}, \\ \operatorname{supp}(u^{-1}(1)) &= \{z \in \overline{K} \mid b(z) = 0\}, \\ \operatorname{supp}(u^{-1}(\infty)) &= \{z \in \overline{K} \mid c(z) = 0\}, \end{aligned}$$

and, since a, b, c are coprime, we get

$$-2 \geq \deg(u) - \deg(\operatorname{rad}(abc)),$$

which is even better than the conclusion of the theorem.

Otherwise, if $\deg(u) = \deg(a) = \deg(b) > \deg(c)$, then the support of $u^{-1}(\infty)$ must be increased by adding the point ∞. This increases the final counting by 1, completing the proof. □

Second proof: Consider the equation $a(x) + b(x) = c(x)$ and differentiate with respect to x. We denote differentiation by $'$. Then

$$\begin{aligned} 1 \cdot a \quad + \quad 1 \cdot b \quad + \quad 1 \cdot (-c) \quad &= 0 \\ \tfrac{a'}{a} \cdot a \quad + \quad \tfrac{b'}{b} \cdot b \quad + \quad \tfrac{c'}{c} \cdot (-c) \quad &= 0, \end{aligned}$$

which we view as a homogeneous linear system of two equations with a solution $(a, b, -c)$. The associated matrix is

$$\begin{pmatrix} 1 & 1 & 1 \\ \frac{a'}{a} & \frac{b'}{b} & \frac{c'}{c} \end{pmatrix}.$$

Clearly, its rank is 2 otherwise $\frac{a'}{a} = \frac{b'}{b} = \frac{c'}{c}$, contradicting the assumption that a, b, c are coprime and not all constant. By Cramer's rule, it follows that the solution

$(a, b, -c)$ of the above linear system is proportional to the vector of cofactors of the matrix. Let $\lambda \in K(x)$ be the the associated proportionality factor, hence

$$\frac{c'}{c} - \frac{b'}{b} = \lambda \cdot a,$$
$$\frac{a'}{a} - \frac{c'}{c} = \lambda \cdot b,$$
$$\frac{b'}{b} - \frac{a'}{a} = \lambda \cdot (-c).$$

Recall that $\mathrm{rad}(abc)$ is the monic polynomial with simple zeros at the zeros of abc. By the basic property $\mathrm{d}\log(fg) = \mathrm{d}\log(f) + \mathrm{d}\log(g)$, it is clear that the product of $\mathrm{rad}(abc)$ with each cofactor on the left is a polynomial. Since a, b, c are coprime, we conclude that the denominator of λ must be a divisor of $\mathrm{rad}(abc)$. Since each cofactor vanishes at ∞, taking degrees we infer that $\max(\deg(a), \deg(b), \deg(c)) \leq \deg(\mathrm{rad}(abc)) - 1$. □

Remark 12.4.2. Theorem 12.4.1 is sharp precisely whenever $u := -a/c$ defines a Belyĭ morphism $u : \mathbb{P}^1_{\mathrm{an}} \to \mathbb{P}^1_{\mathrm{an}}$ ramified only over $\{0, 1, \infty\}$ and a, b, c are not all of the same degree. This is clear from the first proof of the theorem, because additional ramification will contribute an additional positive term to the right-hand side of (12.4).

Remark 12.4.3. A more natural geometric formulation of Theorem 12.4.1 is obtained by replacing a, b, c by non-zero elements of the function field $K(x)$ with $a + b + c = 0$ and introducing the obvious projective height function

$$h((a : b : c)) = \sum_{z \in \mathbb{P}^1(\overline{K})} -\min(\mathrm{ord}_z(a), \mathrm{ord}_z(b), \mathrm{ord}_z(c)),$$

with $\mathrm{ord}_z = -\deg$ if z is the point at ∞ of $\mathbb{P}^1(\overline{K})$, thus with a contribution at $z = \infty$ of $\max(\deg(a), \deg(b), \deg(c))$. Then, if S is a finite subset of $\mathbb{P}^1(\overline{K})$ such that a, b, c are S-units, Theorem 12.4.1 is equivalent to the inequality

$$h((a : b : c)) \leq |S| - 2.$$

Equality holds if and only if $u := -a/c$ defines a Belyĭ morphism and S is the set of valuations for which either u or $1 - u$ are not units.

This formulation admits a natural extension in which we replace the field of rational functions $K(x)$ by a function field in one variable, of characteristic 0 and genus g. Under the same hypotheses on a, b, c, we have the **theorem of Stothers and Mason**

$$h((a : b : c)) \leq |S| + 2g - 2,$$

which will be proved in Theorem 14.5.13.

The second proof of the abc-theorem for polynomials easily extends to prove the **general *abc*-theorem for polynomials.**

Theorem 12.4.4. *Let K be a field of characteristic 0 and let $a_i(x) \in K[x]$, $i = 1, \dots, n$, where $n \geq 3$, be polynomials such that $a_1(x) + \dots + a_n(x) = 0$. Suppose that the polynomials $a_i(x)$ have no common zero for all $i = 1, \dots, n$ and also that no proper subsum of $a_1(x) + \dots + a_n(x)$ vanishes identically. Then*

$$\max_i \deg(a_i) \leq \frac{(n-1)(n-2)}{2} \max\left\{\deg \operatorname{rad}\left(\prod_{i=1}^{n} a_i\right) - 1, 0\right\}.$$

All these generalizations will be proved, in more detail and in a more general setting, in Section 14.5.

Remark 12.4.5. The coefficient is sharp if $n = 3$ or 4. If $n \geq 5$, it is not clear what is the best coefficient in Theorem 12.4.4. The simple example

$$\sum_{j=0}^{n-2} \binom{n-2}{j} x^{ej} - (x^e + 1)^{n-2} = 0$$

shows that $\frac{1}{2}(n-1)(n-2)$ cannot be replaced by any constant smaller than $n-2$.

Better lower bounds can be provided for $n \geq 4$. The following clever example is in J. Browkin and J. Brzeziński [**51**]. Let $P_k(x)$ be the polynomial of degree k such that

$$\frac{x^{2k+1} - 1}{x - 1} = x^k P_k\left(\frac{(x-1)^2}{x}\right).$$

It is easy to see that all roots of P_k are negative, hence P_k has positive coefficients. If we take $k = n - 3$ and set x^e in place of x, we obtain an identity

$$x^{(2n-5)e} - 1 - \sum_{j=0}^{n-3} s_j (x^e - 1)^{2j+1} x^{(n-3-j)e} = 0$$

with coefficients $s_j > 0$. This is a vanishing sum of n polynomials and no proper subsum vanishes, as we verify for example by specializing $x = 2$, since then only the first term in the identity is positive and all others are negative. Now the maximum degree is $(2n-5)e$, while the radical $(x^e - 1)x$ has degree $e + 1$. Therefore, the best coefficient in Theorem 12.4.4 is at least $2n - 5$. In particular, the precise coefficient is 3 if $n = 4$.

The exponent $1 + \varepsilon$ is necessary in the abc-conjecture over $\mathbb{Q}$, as we see by the following interesting result, due to Stewart and Tijdeman [**291**].

Theorem 12.4.6. *For every $\delta > 0$, there are infinitely many triples a, b, c of coprime positive numbers with $a + b = c$ and*

$$c > e^{(4-\delta)\frac{\sqrt{\log c}}{\log \log c}} \prod_{p \mid abc} p.$$

Note that the factor is of order $O(c^{\varepsilon})$ for any $\varepsilon > 0$ and hence this lower bound is compatible with the strong abc-conjecture in 12.2.2.

12.4.7. The strategy of proof is quite simple. If we restrict a and c to integers whose prime factors belong to a rather small set, then the radical of ac is small. On the other hand, the number of such integers up to x is large, and an application of Dirichlet's pigeonhole principle shows that we can make $b = c - a$ fairly small. It does not matter much which absolute value we use for this purpose, and we shall use the 2-adic topology in our argument. If b is small in this sense, then its radical is smaller than expected, thus giving a non-trivial example of an (a, b, c)-triple.

We begin with counting the number of integers up to N all of whose prime factors belong to a fixed set $\mathcal{P}$, provided the set $\mathcal{P}$ is not too large.

Proposition 12.4.8. *Let $\mathcal{P}$ be a finite set of primes and let $\vartheta_{\mathcal{P}} = \sum_{p\in\mathcal{P}} \log p$. Then the number $\Psi(N, \mathcal{P})$ of integers $n \leq N$ with prime factors only from $\mathcal{P}$ is bounded by*

$$\frac{(\log N)^{|\mathcal{P}|}}{|\mathcal{P}|!}\Big(\prod_{p\in\mathcal{P}} \log p\Big)^{-1} \leq \Psi(N, \mathcal{P}) \leq \frac{(\log N + \vartheta_{\mathcal{P}})^{|\mathcal{P}|}}{|\mathcal{P}|!}\Big(\prod_{p\in\mathcal{P}} \log p\Big)^{-1}.$$

Proof: Let $n = \prod_{p\in\mathcal{P}} p^{a_p}$ and let us associate to n the integer vector $\mathbf{a} = (a_p)_{p\in\mathcal{P}}$. This makes it clear that $\Psi(N, \mathcal{P})$ is the number of lattice points in

$$\mathcal{T}(\log N) := \Big\{\mathbf{x} \in \mathbb{R}^{\mathcal{P}} \mid \sum_{p\in\mathcal{P}} (\log p) x_p \leq \log N,\ x_p \geq 0 \text{ for } p \in \mathcal{P}\Big\}.$$

Let $\mathcal{A}$ be the set of lattice points in $\mathcal{T}(\log N)$. We associate to a lattice point $\mathbf{a} \in \mathcal{A}$ the unit cube $\mathcal{C}(\mathbf{a}) = \{\mathbf{z} \mid a_p \leq z_p < a_p + 1,\ p \in \mathcal{P}\}$. Then it is clear that

$$\mathcal{T}(\log N) \subset \bigcup_{\mathbf{a}\in\mathcal{A}} \mathcal{C}(\mathbf{a}) \subset \mathcal{T}(\log N + \vartheta_{\mathcal{P}}).$$

Taking volumes, we obtain what we want. □

Let $\Psi_{\mathrm{odd}}(x, y)$ be the number of y-smooth odd integers up to x, in other words the number of odd integers $n \leq x$ which admit only prime divisors $p \leq y$. Let $\pi(y)$ be the number of primes up to y. We have

Lemma 12.4.9. *If $y = o(\log x)$, then*

$$\log \Psi_{\mathrm{odd}}(x, y) = (\pi(y) - 1) \log\log x - y + o\Big(\frac{y}{\log y}\Big).$$

Proof: We apply Proposition 12.4.8 with $N = x$, taking for $\mathcal{P}$ the set of all odd primes not exceeding y, which has cardinality $|\mathcal{P}| = \pi(y) - 1$. We get

$$\frac{(\log x)^{|\mathcal{P}|}}{|\mathcal{P}|!}\Big(\prod_{p\in\mathcal{P}} \log p\Big)^{-1} \leq \Psi_{\mathrm{odd}}(x, y) \leq \frac{(\log x + \vartheta_{\mathcal{P}})^{|\mathcal{P}|}}{|\mathcal{P}|!}\Big(\prod_{p\in\mathcal{P}} \log p\Big)^{-1}. \quad (12.5)$$

Partial summation of

$$\sum_{2<p\leq y} \log\log p = \sum_{2<n\leq y} (\pi(n) - \pi(n-1)) \log\log n$$

yields

$$\begin{aligned}\log\Big(\prod_{p\in\mathcal{P}}\log p\Big) &= \sum_{2<n\le y}(\pi(n)-\pi(n-1))\log\log y\\ &\quad+\sum_{2<m\le y-1}\int_{m+1}^{m}\sum_{2<n\le u}(\pi(n)-\pi(n-1))\,\frac{du}{u\log u}\\ &= (\pi(y)-1)\log\log y-\int_3^y\frac{\pi(u)-1}{u\log u}\,du.\end{aligned}$$

By the prime number theorem and integration by parts, we deduce

$$\log\Big(\prod_{p\in\mathcal{P}}\log p\Big)\pi(y)\log\log y+O\Big(\frac{y}{(\log y)^2}\Big). \tag{12.6}$$

Similarly, we find

$$\vartheta_{\mathcal{P}}=\sum_{2<p\le y}\log p=y+o\Big(\frac{y}{\log y}\Big). \tag{12.7}$$

Then the hypothesis $y=o(\log x)$ implies

$$|\mathcal{P}|\log(\log x+\vartheta_{\mathcal{P}})(\pi(y)-1)\log\log x+o\Big(\frac{y}{\log y}\Big). \tag{12.8}$$

Now (12.5), (12.6), (12.8) and Stirling's formula show that

$$\log\Psi_{\mathrm{odd}}(x,y)=(\pi(y)-1)\log\log x-\pi(y)\log\pi(y)+\pi(y)-\pi(y)\log\log y+o\Big(\frac{y}{\log y}\Big).$$

The lemma follows from the prime number theorem in the stronger form (see G.J.O. Jameson **[158]**, Exercise 1.5.4, Th. 5.1.8)

$$\pi(y)=\frac{y}{\log y}+\frac{y}{(\log y)^2}+o\Big(\frac{y}{(\log y)^2}\Big). \qquad \square$$

12.4.10. *Proof of Theorem 12.4.6:* We choose $y=\sqrt{\log x}$ and for simplicity we abbreviate $M=\Psi_{\mathrm{odd}}(x,y)$. We also define k by $2^k<M\le 2^{k+1}$. Now consider the sequence $1=n_1<n_2<\cdots<n_M\le x$ of odd integers up to x without prime factors larger than y. Since $2^k<M$, by Dirichlet's pigeon-hole principle there are two distinct elements $n_i<n_j$ in this sequence, having the same residue class modulo 2^k. Let $d=\mathrm{GCD}(n_i,n_j)$ and set

$$a=n_i/d,\qquad b=(n_j-n_i)/d,\qquad c=n_j/d.$$

The triple (a,b,c) so obtained is the desired good example for the abc-conjecture.

By construction, 2^k divides $n_j-n_i=d\cdot b$. On the other hand, n_i and n_j are both odd, thus d is odd. Hence 2^k divides b, giving

$$\mathrm{rad}(b)\le 2^{-k+1}b\le 4M^{-1}b.$$

Also, every prime factor of n_in_j does not exceed y, whence

$$\mathrm{rad}(ac)\le\prod_{p\in\mathcal{P}}p=e^{\vartheta_{\mathcal{P}}}.$$

It now follows from the last two displayed equations that

$$\mathrm{rad}(abc) \leq 4M^{-1}e^{\vartheta_{\mathcal{P}}}b. \tag{12.9}$$

By Lemma 12.4.9, we have

$$\log M = y + \frac{2y}{\log y} + o\left(\frac{y}{\log y}\right),$$

therefore from (12.7) and (12.9) we infer that

$$b > K \prod_{p|abc} p$$

with (note that $c \leq x$)

$$K = e^{(4+o(1))\frac{\sqrt{\log c}}{\log \log c}}.$$

The result follows. □

Remark 12.4.11. The comparison of this example with Baker's Conjecture 12.2.6 is of some interest. In a triple (a, b, c) such as in the example the number of distinct prime factors of ac is $O(y/\log y)$, with $y = \sqrt{\log x}$. The most unfavorable case for the comparison occurs when $d = \mathrm{GCD}(a, c)$ is negligible, c is of order x, and the number of prime factors of b is also $O(y/\log y)$. Then, choosing the parameter ε in Baker's conjecture optimally, we would get the upper bound

$$c \leq e^{\kappa\sqrt{\log c}} \prod_{p|abc} p$$

for some positive constant κ. This hypothetical bound is not too far away from the actual lower bound. Thus Baker's conjecture, if true, would probably be close to optimal.

The same method proves:

Proposition 12.4.12. *Let $\mathcal{P}$ be a finite set of odd primes and let*

$$\vartheta_{\mathcal{P}} = \sum_{p \in \mathcal{P}} \log p.$$

Let N be a positive integer. Then we can find coprime positive integers a, b, $c = a+b$ not exceeding N, such that the prime factors of a and c are only from the set $\mathcal{P}$ and moreover

$$c > \frac{1}{4}\frac{(\log N)^{|\mathcal{P}|}}{|\mathcal{P}|!} e^{-\vartheta_{\mathcal{P}}} \left(\prod_{p\in\mathcal{P}} \log p\right)^{-1} \mathrm{rad}(abc).$$

Proof: Let $M = \Psi(N, \mathcal{P})$ and let $n_1 < n_2 < \cdots < n_M$ be the set of integers up to N all of whose prime factors are from the set $\mathcal{P}$. Then Proposition 12.4.8 yields

$$M \geq \frac{(\log N)^{|\mathcal{P}|}}{|\mathcal{P}|!} \left(\prod_{p\in\mathcal{P}} \log p\right)^{-1}.$$

If $2^k < M \leq 2^{k+1}$, then by Dirichlet's pigeon-hole principle we can find two distinct elements n_i, n_j in the same class $\pmod{2^k}$; the rest of the proof is as in 12.4.10. □

Example 12.4.13. Consider the special case $a = 1$, $b = 3^{2^n} - 1$, $c = 3^{2^n}$. Then Euler's theorem shows that 2^{n+1} divides b, giving the explicit example

$$c > \frac{\log c}{3\log 3}\,\mathrm{rad}(abc).$$

This is comparable in strength with what we can obtain from Proposition 12.4.12 if $\mathcal{P}$ consists of only one prime.

Example 12.4.14. We note here the example $2 + 109 \cdot 3^{10} = 23^5$, which has high ***abc*-ratio** $\log c/\log \mathrm{rad}(abc)$, namely 1.62991. This was found by E. Reyssat, by searching for very good rational approximations to numbers of type $a^{1/n}$. Since

$$109^{\frac{1}{5}} = 2 + \cfrac{1}{1 + \cfrac{1}{1 + \cfrac{1}{4 + \cfrac{1}{77733 + \dots}}}}$$

we see that $\frac{23}{9}$ is a very good convergent to $109^{1/5}$, corresponding to the example (note that since 9 is a square this example is rather more effective than others of similar type).

Remark 12.4.15. For any coprime integers a, b, c with $a+b = c$, the ***abc*-ratio** is defined by

$$\frac{\log\max(|a|,|b|,|c|)}{\log\mathrm{rad}(abc)}.$$

The strong abc-conjecture implies that the set of accumulation points of all abc-ratios is the interval $[\frac{1}{3}, 1]$. For a proof, note that the lower bound $\frac{1}{3}$ is obvious and the upper bound 1 is the strong abc-conjecture. Now take $a = n^k g(n)$, $b = h(n)$, $c = n^k g(n) + h(n)$ with suitable polynomials g, h, and apply Theorem 12.2.15 to $f(x) := xg(x)h(x)(x^k g(x) + h(x))$ to ensure that $f(n)$ is square free infinitely often. For such values of n, we have $\mathrm{rad}(abc) = f(n)$. As $n \to \infty$, the abc-ratio tends to

$$\frac{\max\{k + \deg(g), \deg(h), \deg(x^k g + h)\}}{1 + \deg(g) + \deg(h) + \deg(x^k g + h)}.$$

The set of these rational numbers is contained in $[\frac{1}{3}, 1]$ and is dense there. □

12.4.16. A simple heuristic argument has been proposed, suggesting that statements stronger than Proposition 12.4.12 may be true. This is based on the so-called **birthday paradox**.

There are precisely m^s integral vectors $\{n_1, \dots, n_s\}$ with $1 \le n_i \le m$. On the other hand, the number of such vectors where all components n_i are distinct is $m(m - 1)\cdots(m - s + 1)$. By Stirling's formula, we have

$$m(m-1)\cdots(m-s+1)/m^s \sim e^{-s^2/2m}$$

uniformly for $s = o(m^{2/3})$. This means that, if we look at vectors of length $s \sim \alpha\sqrt{m}$ with $\alpha = o(m^{1/6})$, then the probability of such a vector of having two equal components is asymptotic to $1 - e^{-\alpha^2/2}$. In other words, a relatively short vector has a positive probability of having two equal components. For example, in a group of 40 people there is an $89\,\%$ chance that two persons will be born on the same day of the year (whence the name "birthday paradox," since such a conclusion appears to be implausible at first sight).

In view of the above, if the sequence n_i, $i = 1, \ldots, M$ appearing in 12.4.10 were sufficiently random, we should find two elements in the same class $\pmod{2^k}$ as soon as $M \gg (\sqrt{2})^k$, rather than the condition $M > 2^k$ required by Dirichlet's pigeon-hole principle. Then the construction in 12.4.10 should lead to the bound

$$\mathrm{rad}(abc) \ll M^{-2} e^{\vartheta_{\mathcal{P}}} b$$

rather than (12.9) on page 422. If this were the case, then there would be a corresponding improvement in Theorem 12.4.6 with $y = (\log x)^{\frac{2}{3}}$, yielding hypothetically

$$c > e^{(\frac{9}{4} - \delta)\frac{(\log c)^{\frac{2}{3}}}{\log \log c}} \prod_{p \mid abc} p.$$

12.4.17. On the other hand, it has been pointed out by Granville that such a sequence n_i is unlikely to have the randomness property needed to apply the birthday paradox with confidence. In fact, the construction proposed before can be rephrased as follows:

Let $p_1, \ldots, p_t$ be the elements of $\mathcal{P}$ (we assume $2 \notin \mathcal{P}$). We want to solve the congruence

$$p_1^{a_1} \cdots p_t^{a_t} \equiv p_1^{b_1} \cdots p_t^{b_t} \pmod{2^k}.$$

Let G be the group $(\mathbb{Z}/2^k\mathbb{Z})^\times$ of units of the ring $\mathbb{Z}/2^k\mathbb{Z}$. Then, if $g_1, \ldots, g_t$ are the images of $\mathcal{P}$ in G, the above congruence becomes a relation

$$g_1^{a_1 - b_1} \cdots g_t^{a_t - b_t} = 1$$

in the group G. Now there are too many choices of pairs $(a_1, \ldots, a_t)$, $(b_1, \ldots, b_t)$ leading to the same difference $(a_1 - b_1, \ldots, a_t - b_t)$ and the argument on which the birthday paradox was built collapses. A closer analysis of this argument only shows that the constant $4 - \delta$ in the theorem of Stewart and Tijdeman should be replaced by a larger one.

Remark 12.4.18. Indeed, M. van Frankenhuysen [**304**], using a packing of spheres argument, has improved the constant $4 - \delta$ in Theorem 12.4.10 unconditionally to 6.068.

12.5. Equivalent conjectures

We start with Hall's conjecture and its strong version. Then we recall some basic facts about elliptic curves over Dedekind domains to introduce notions such as reduction, minimal discriminant, and conductor. This enables us to formulate Szpiro's conjecture bounding the discriminant in terms of the conductor of an elliptic curve over $\mathbb{Q}$. Finally, we prove the equivalence of the *abc*-conjecture, the strong Hall conjecture, and the generalized Szpiro conjecture, *via* Frey's famous elliptic curve $y^2 = x(x+a)(x-b)$. At the end, we give the generalization of Hall's conjecture due to Hall–Lang–Waldschmidt–Szpiro.

12.5.1. In 1969, on the basis of what at that time was considered extensive numerical evidence, Marshall Hall [**144**] conjectured that there is a positive constant C such that $|x^3 - y^2| \geq C|x|^{\frac{1}{2}}$ for $x, y \in \mathbb{Z}$ with $x^3 - y^2 \neq 0$. The exponent $1/2$ in this statement cannot be improved upon, as shown a little later by L.V. Danilov, who proved in [75] that $0 < |x^3 - y^2| < 0.97|x|^{1/2}$ has infinitely many solutions in integers x, y.

The Hall conjecture is unlikely to be true as originally formulated and nowadays we refer to **Hall's conjecture** as the slightly weaker statement in which the exponent $\frac{1}{2}$ is replaced by $\frac{1}{2} - \varepsilon$ and C by some $C(\varepsilon) > 0$, for every fixed $\varepsilon > 0$. An equivalent formulation is that, for any solution of $y^2 = x^3 - z$ with $x, y, z \in \mathbb{Z}$ and $z \neq 0$ viewed as a parameter, we have $|x| \ll_\varepsilon |z|^{2+\varepsilon}$, $|y| \ll_\varepsilon |z|^{3+\varepsilon}$.

12.5.2. What is of interest to us here is a stronger form of the conjecture, the **strong Hall conjecture** below, claiming a bound in terms of $\operatorname{rad}(z)$ rather than $|z|$. However, we have to avoid the counterexample

$$(2p^4)^3 - (3p^6)^2 = -p^{12}.$$

Note that every solution of $z = x^3 - y^2 \neq 0$ with $\operatorname{GCD}(x^3, y^2)$ divisible by a sixth power g^6 may be reduced to a solution $x' := x/g^2$, $y' := y/g^3$, $z' := z/g^6$ such that $\operatorname{GCD}(x'^3, y'^2)$ is free from sixth powers. We call (x', y', z') a **primitive solution.**

Conjecture 12.5.3. *Given $\varepsilon > 0$, every primitive solution of $x^3 - y^2 = z \neq 0$ satisfies*

$$|x| \ll_\varepsilon \operatorname{rad}(z)^{2+\varepsilon}, \qquad |y| \ll_\varepsilon \operatorname{rad}(z)^{3+\varepsilon}.$$

12.5.4. Another conjecture concerns the discriminant and the conductor of an elliptic curve. We recall first some basic facts (cf. [284] for more details).

Let K be any field. By (8.2) on page 241, an elliptic curve over K may be given by an affine Weierstrass equation

$$y^2 + a_1xy + a_3y = x^3 + a_2x^2 + a_4x + a_6. \tag{12.10}$$

Using the quantities

$$b_2 = a_1^2 + 4a_2, \quad b_4 = a_1a_3 + 2a_4, \quad b_6 = a_3^2 + 4a_6$$

and

$$b_8 = a_1^2a_6 + 4a_2a_6 - a_1a_3a_4 + a_2a_3^2 - a_4^2,$$

we define

$$c_4 = b_2^2 - 24b_4, \quad c_6 = -b_2^3 + 36b_2b_4 - 216b_6$$

and the **discriminant**

$$\Delta = -b_2^2b_8 - 8b_4^3 - 27b_6^2 + 9b_2b_4b_6.$$

If $\operatorname{char}(K) \neq 2$, then replacing y by $\frac{1}{2}(y - a_1x - a_3)$ leads to the Weierstrass normal form

$$y^2 = 4x^3 + b_2x^2 + 2b_4x + b_6.$$

If $\mathrm{char}(K) \neq 3$ as well, then replacing (x, y) by $(\frac{1}{36}(x - 3b_2), \frac{1}{108}y)$ leads to the Weierstrass equation

$$y^2 = x^3 - 27c_4x - 54c_6.$$

In any case, the quantities are defined in such a way that they are suitable in every characteristic. We have the relation

$$1728\Delta = c_4^3 - c_6^2. \tag{12.11}$$

If $a_1 = a_3 = 0$, then Δ is 16 times the discriminant of $x^3 + a_2x^2 + a_1x + a_0$. This normalization leads to the extension of Proposition 10.2.3 that an affine Weierstrass equation (12.10) describes an elliptic curve if and only if $\Delta \neq 0$ (see [**284**], Prop.III.1.4). It is easy to see that a Weierstrass equation (12.10) is unique up to a coordinate transformation of the form

$$x = u^2x' + r, \quad y = u^3y' + su^2x' + t \tag{12.12}$$

for $r, s, t, u \in K, u \neq 0$. Then a direct calculation shows

$$u^4c_4' = c_4, \quad u^6c_6' = c_6, \quad u^{12}\Delta' = \Delta. \tag{12.13}$$

If $\Delta = 0$, then there is exactly one singular point on the projective curve described by (12.10). It is either a cusp or a node (see [**284**], Prop.III.1.4). The same formulas as in Proposition 8.3.8 show that the smooth points of the projective curve form a one-dimensional commutative group variety, hence it is isomorphic to the torus $\mathbb{G}_m$ (case of a node) or to $\mathbb{A}^1$ (case of a cusp) over $\overline{K}$ (see [**284**], Prop.III.2.5).

12.5.5. Now let v be a discrete valuation on K with valuation ring R and local parameter π. As usual, we assume that the valuation is normalized by $v(\pi) = 1$. Then $v(x)$ is the order of an element $x \in K$, namely the largest integer for which $x \in R\pi^{v(x)}$.

It is clear that the elliptic curve E over K may be given by a Weierstrass equation (12.10) with every $a_i \in R$. Such a Weierstrass equation is called minimal if $v(\Delta)$ is minimal. Obviously, a **minimal Weierstrass equation** exists and (12.13) shows that the corresponding **minimal discriminant** is unique up to units in R. Using the transformation formulas for the quantities b_i, c_4, Δ, we prove that the minimal Weierstrass equation is unique up to transformations of the form (12.12) with $r, s, t \in R$ and $u \in R^\times$ ([**284**], Prop.VII.1.3). Let $\overline{a} \in k(v)$ be the reduction of $a \in R$. Then the reduction

$$y^2 + \overline{a}_1xy + \overline{a}_3y = x^3 + \overline{a}_2x^2 + \overline{a}_4x + \overline{a}_6$$

of a minimal Weierstrass equation is an affine Weierstrass equation over the residue field $k(v)$ and the corresponding projective curve $\overline{E}$ over $k(v)$ is unique up to isomorphism.

We say that E has **good reduction** if $\overline{E}$ is an elliptic curve and otherwise we speak of **bad reduction**. In the latter case, if the singular point is a node, the reduction is called **multiplicative** (or semistable) and if the singular point is a cusp, the reduction is called **additive**.

Multiplicative reduction is characterized by $v(\Delta) > 0$ and $v(c_4) = 0$, while additive reduction is characterized by $v(\Delta) > 0$ and $v(c_4) > 0$ (see **[284]**, Prop.VII.5.1). If $\operatorname{char}(k(v)) \neq 2, 3$ the last condition is equivalent to $v(c_4) > 0$ and $v(c_6) > 0$.

We also have the following useful necessary condition for minimality.

Proposition 12.5.6. *Let K, v, R, π be as before and suppose that* $\operatorname{char}(K) \neq 2, 3$. *Let $y^2 + a_1xy + a_3y = x^3 + a_2x^2 + a_4x + a_6$ be an R-Weierstrass equation for E. Then, if $48\pi^4 | c_4$ and $864\pi^6 | c_6$, the given R-Weierstrass equation for E is not minimal.*

As an immediate consequence, we note:

Corollary 12.5.7. *For a minimal Weierstrass equation as before, the following estimate holds*

$$\min(3v(c_4), 2v(c_6)) < \begin{cases} 12 & \textit{if } \operatorname{char}(k(v)) \neq 2, 3, \\ 12 + 6v(3) & \textit{if } \operatorname{char}(k(v)) = 3, \\ 12 + 12v(2) & \textit{if } \operatorname{char}(k(v)) = 2. \end{cases}$$

Proof of Proposition 12.5.6: The change of variables

$$(x, y) = \left(x' - \frac{1}{12}\, b_2, y' - \frac{1}{2}\, a_1 x' + \frac{1}{24}\, a_1 b_2 - \frac{1}{2}\, a_3\right) \tag{12.14}$$

transforms the Weierstrass equation

$$y^2 + a_1xy + a_3y = x^3 + a_2x^2 + a_4x + a_6$$

into

$$(y')^2 = (x')^3 - \frac{1}{48}\, c_4\, x' - \frac{1}{864}\, c_6. \tag{12.15}$$

The discriminant Δ' of (12.15) is $\Delta' = \Delta$, because the change of variables (12.14) has determinant 1.

The coefficients of (12.14) are *a priori* only in the ring $R\left[\frac{1}{2}, \frac{1}{3}\right]$, but if

$$\frac{1}{12}\, b_2 \in R, \quad \frac{1}{2}\, a_1 \in R, \quad \frac{1}{2}\, a_3 \in R, \tag{12.16}$$

then (12.14) has coefficients in R and (12.15) is again an R-Weierstrass equation equivalent to the original one.

Now we claim that the conditions (12.16) already follow from $\frac{1}{48}\, c_4 \in R$ and $\frac{1}{864}\, c_6 \in R$. Assuming this claim the proof of the proposition is immediate, because if $\frac{1}{48}\, c_4 \in \pi^4 R$ and $\frac{1}{864}\, c_6 \in \pi^6 R$ then we can make the further change of variables $(x', y') = (\pi^2 x'', \pi^3 y'')$, obtaining a new R-Weierstrass equation with discriminant $\Delta'' = \pi^{-12}\Delta$, hence with $v(\Delta'') = v(\Delta) - 12$ and thereby proving that our original R-Weierstrass equation is not minimal. Thus it only remains to prove our claim.

If $\mathrm{char}(k(v)) \notin \{2, 3\}$, this is obvious because then 2 and 3 are invertible in R.

If $\mathrm{char}(k(v)) = 3$, then 2 is invertible in R and we need to prove that $\frac{1}{3}\, c_4 \in R$ and $\frac{1}{27}\, c_6 \in R$ imply $\frac{1}{3}\, b_2 \in R$. This is easy. Since $c_6 = -b_2^3 + 36b_2b_4 - 216b_6$ and $27|216$, we see that $\frac{1}{27}\, c_6 \in R$ implies

$$-\left(\frac{1}{3}\, b_2\right)^3 + 4b_4\left(\frac{1}{3}\, b_2\right) \in R,$$

yielding $\frac{1}{3}\, b_2 \in R$, as wanted.

If $\mathrm{char}(k(v)) = 2$, then 3 is invertible in R and the argument becomes rather intricate. We do this in several steps. The assumption now is $\frac{1}{16}\, c_4 \in R$ and $\frac{1}{32}\, c_6 \in R$, and we want to show that

$$\frac{1}{2}\, a_1 \in R, \quad \frac{1}{2}\, a_3 \in R, \quad \frac{1}{4}\, b_2 \in R.$$

Since $c_4 = b_2^2 - 24b_4$ and $v(c_4) \geq 4v(2)$, we get $v(b_2^2) \geq 3v(2)$ and $v(b_2) \geq \frac{3}{2}v(2)$. Now $b_2 = a_1^2 + 4a_2$ and we conclude that $v(a_1) \geq \frac{3}{4}v(2)$.

Next, $c_6 = -b_2^3 + 36b_2b_4 - 216b_6$ and $v(c_6) \geq 5v(2)$, whence

$$\begin{aligned} v(216b_6) = v(b_6) + 3v(2) &\geq \min(v(c_6), v(b_2^3), v(36b_2b_4)) \\ &\geq \min\left(5v(2), \frac{9}{2}\, v(2), 2v(2) + \frac{3}{2}\, v(2)\right) = \frac{7}{2}\, v(2) \end{aligned}$$

and $v(b_6) \geq \frac{1}{2}v(2)$ follows. Since $b_6 = a_3^2 + 4a_6$, we see that $v(a_3) \geq \frac{1}{4}v(2)$. Therefore, $v(a_1a_3) \geq \frac{3}{4}v(2) + \frac{1}{4}v(2) = v(2)$ and, recalling that $b_4 = a_1a_3 + 2a_4$, we infer that $v(b_4) \geq v(2)$. Now $c_4 = b_2^2 - 24b_4$ gives

$$v(b_2^2) \geq \min(v(c_4), v(24b_4)) \geq 4v(2)$$

and $v(b_2) \geq 2v(2)$. Hence $\frac{1}{4}\, b_2 \in R$, as wanted.

Now $b_2 = a_1^2 + 4a_2$ shows that $v(a_1) \geq v(2)$ and $\frac{1}{2}\, a_1 \in R$, as wanted.

Finally

$$\begin{aligned} v(216b_6) = v(b_6) + 3v(2) &\geq \min(v(c_6), v(b_2^3), v(36b_2b_4)) \\ &\geq \min(5v(2), 6v(2), 2v(2) + 2v(2) + v(2)) = 5v(2) \end{aligned}$$

and we get $v(b_6) \geq 2v(2)$. Since $b_6 = a_3^2 + 4a_6$, we conclude that $v(a_3) \geq v(2)$ and $\frac{1}{2}\, a_3 \in R$, as wanted. □

12.5.8. Now let R be a Dedekind domain with quotient field K. Then every maximal ideal $\mathfrak{p}$ induces a discrete valuation $v_\mathfrak{p}$ and hence a minimal discriminant $\Delta_\mathfrak{p}$ of the elliptic curve E over K. The **minimal discriminant** of E is the ideal

$$\Delta := \prod_{\mathfrak{p}} \mathfrak{p}^{v_\mathfrak{p}(\Delta_\mathfrak{p})},$$

where $\mathfrak{p}$ ranges over all maximal ideals of R. Clearly, only finitely many prime ideals give a non-trivial contribution and Δ is well defined. In general, there is no minimal Weierstrass equation with coefficients in R working for all primes. However for $R = \mathbb{Z}$, there is such a minimal Weierstrass equation working for all primes (cf. [**284**], VIII.Cor.8.3) and we call it the **global minimal Weierstrass equation.**

12.5.9. We do not give the precise definition of the conductor of an elliptic curve over a number field K (the reader can find it in A. Ogg [**230**], T. Saito [**256**] or J.H. Silverman [**286**], IV, §10). All we need to know here is that it is an ideal

$$\operatorname{cond}(E) = \prod_{\mathfrak{p}} \mathfrak{p}^{f_\mathfrak{p}},$$

where $\mathfrak{p}$ ranges over the maximal ideals of O_K and the exponent $f_\mathfrak{p} \in \mathbb{N}$ has the following properties (cf. [**286**], Th.IV.10.2, Cor.IV.11.2):

(a) E has good reduction at $\mathfrak{p}$ if and only if $f_\mathfrak{p} = 0$.

(b) E has multiplicative reduction if and only if $f_\mathfrak{p} = 1$.

(c) E has additive reduction if and only if $f_\mathfrak{p} \geq 2$.

(d) The minimal discriminant Δ is a multiple of $\operatorname{cond}(E)$ and is supported in the same set of prime ideals.

Example 12.5.10. G. Frey's idea [**123**] for the proof of Fermat's last theorem was to associate to coprime $a, b, c \in \mathbb{Z} \setminus \{0\}$ with $a + b = c$ the curve

$$y^2 = x(x + a)(x - b).$$

By Proposition 10.2.3, it is the affine part of an elliptic curve E defined over $\mathbb{Q}$ called the **Frey curve**. The Weierstrass form is given by

$$y^2 = x^3 + (a - b)x^2 - abx. \tag{12.17}$$

We have

$$b_2 = 4(a - b), \quad b_4 = -2ab, \quad b_6 = 0, \quad b_8 = -(ab)^2$$

and

$$c_4 = 16(a^2 + ab + b^2), \quad \Delta = 16(abc)^2.$$

A prime number $p \neq 2$ divides at most one of c_4, Δ. Hence (12.13) on page 426 shows that (12.17) is a minimal Weierstrass equation for the prime p.

If $p = 2$ and $16 \nmid abc$, then $2^{11} \nmid \Delta$ and the transformation formula for Δ in (12.13) on page 426 shows that (12.17) is a minimal Weierstrass equation for $p = 2$ and hence a global minimal Weierstrass equation.

This is not necessarily true if one of a, b, c is divisible by 16. Up to evident changes of coordinates, there are two cases:

(a) $a \equiv 1 \pmod 4$ and $b \equiv 0 \pmod{16}$.

In this case, the transformation

$$x = 4x', \quad y = 8y' + 4x'$$

leads to the Weierstrass form

$$(y')^2 + x'y' = (x')^3 + \frac{a-b-1}{4}(x')^2 - \frac{ab}{16}x'. \tag{12.18}$$

By assumption, the coefficients are integers and (12.13) on page 426 shows

$$c_4' = a^2 + ab + b^2, \quad \Delta' = 2^{-8}(abc)^2.$$

Since c_4' is odd and using (12.13) on page 426 again, we conclude that (12.18) is a minimal Weierstrass equation for $p = 2$. Clearly, it is a global minimal Weierstrass equation.

(b) $a \equiv -1 \pmod 4$ and $b \equiv 0 \pmod{16}$.

Again, we prove that (12.17) is a minimal Weierstrass equation for $p = 2$, hence a global minimal Weierstrass equation. We argue by contradiction and assume that there is a transformation (12.12) on page 426 with $v_2(\Delta') < v_2(\Delta)$ leading to an R-Weierstrass equation. Using $v_2(c_4) = 4$ and (12.13) on page 426, we conclude $v_2(u) = 1$. The transformation formula

$$u^8 b_8' = b_8 + 3rb_6 + 3r^2 b_4 + r^3 b_2 + 3r^4 = -(ab)^2 - 6abr^2 + 4(a-b)r^3 + 3r^4$$

(cf. [**284**], Table III.1.2) implies now $v_2(r) \geq 2$. Finally, we consider the transformation formula

$$u^2 a_2' = a_2 - sa_1 + 3r - s^2 = a - b + 3r - s^2$$

showing that $v_2(s) \geq 0$. Moreover, reduction modulo 4 proves $0 \equiv -1 - s^2 \pmod 4$, which is impossible, completing the proof.

Note that we do not need this case for applications to the abc-conjecture, because we are free to replace a by $-c$ and c by $-a$ to go back to case (a).

Next, we consider the reduction $\overline{E}$ of E at a prime p. If p does not divide the minimal discriminant Δ, then E has good reduction. So let us assume $p|\Delta$. For $p \neq 2$, we have $p|abc$ and hence p does not divide $a - b$. It follows that $\overline{E}$ is given by the affine Weierstrass equation

$$\overline{y}^2 = \overline{x}^3 + (\overline{a} - \overline{b})\overline{x}^2.$$

Since the residue characteristic is not 2, we see that the singular point $(\overline{0}, \overline{0})$ is a node and there is multiplicative reduction.

For $p = 2$, we assume first that the minimal Weierstrass equation has the form (12.17) on page 429. Then $\Delta \equiv 0 \pmod 2$ and Frey's curve E has always bad reduction at 2. The singular point $(\overline{x}_0, \overline{y}_0)$ of $\overline{E}$ is determined by the unique solution of $\overline{x}_0^2 = \overline{a}\overline{b}$ (we are in characteristic 2). The transformation $\overline{x} = \overline{x}' + \overline{x}_0$, $\overline{y}\overline{y}' + \overline{y}_0$ leads to the Weierstrass equation

$$(\overline{y}')^2 = (\overline{x}')^3 + (\overline{x}_0 + \overline{a} - \overline{b})(\overline{x}')^2$$

for $\overline{E}$. As the residue characteristic is 2, the tangents in the singularity have the same direction and hence it is a cusp, and there is additive reduction at 2.

Finally, assume that the minimal Weierstrass equation of E in $p = 2$ has the form (12.18), hence $a \equiv 1 \pmod 4$ and $b \equiv 0 \pmod{16}$. Assuming also that E has bad reduction at 2, $\Delta \equiv 0 \pmod 2$ leads to $b \equiv 0 \pmod{32}$. Then $\overline{E}$ is given by the affine Weierstrass equation

$$\overline{y}^2 + \overline{x}\overline{y} = \overline{x}^3 + \overline{d}\overline{x}^2, \quad d = \frac{a - b - 1}{4}$$

and it is clear that the singular point $(\overline{0}, \overline{0})$ is a node. Hence we have multiplicative reduction.

We conclude that the conductor of E has the form

$$\operatorname{cond}(E) = 2^{f_2} \cdot \prod_{\substack{p \mid abc \\ p \neq 2}} p,$$

where $f_2 = 0$ if E has good reduction at $p = 2$ and $f_2 = 1$ if E has multiplicative reduction at $p = 2$. In the case of additive reduction at $p = 2$, we can only say that $2 \leq f_2 \leq v_2(\Delta)$.

Now we are ready to state the **generalized Szpiro conjecture:**

Conjecture 12.5.11. *Let E be an elliptic curve over $\mathbb{Q}$ with minimal Weierstrass equation* (12.10) *on page 425 over $\mathbb{Z}$ and let $\varepsilon > 0$. Then*

$$\max(|\Delta|, |c_4|^3) \ll_\varepsilon \operatorname{cond}(E)^{6+\varepsilon},$$

where the constant involved in the inequality is independent of E.

Theorem 12.5.12. *The following conjectures are equivalent:*

(a) *strong abc-conjecture in 12.2.2 over $\mathbb{Q}$;*

(b) *strong Hall conjecture in 12.5.3;*

(c) *generalized Szpiro conjecture in 12.5.11.*

Proof: (a) ⇒ (b). Let $\Gamma := \mathrm{GCD}(x^3, y^2)$. Applying the strong abc-conjecture to the relation $(x^3/\Gamma) + (-y^2/\Gamma) = z/\Gamma$, we get

$$\max(|x|^3, |y|^2) \ll_\varepsilon \Gamma \,\mathrm{rad}\left(\frac{x^3y^2z}{\Gamma^3}\right)^{1+\varepsilon}. \tag{12.19}$$

We claim that

$$\Gamma \,\mathrm{rad}\left(\frac{x^3y^2z}{\Gamma^3}\right) \le |xy| \,\mathrm{rad}(z). \tag{12.20}$$

For the proof, it is enough to show that

$$v_p(\Gamma) + v_p\left(\mathrm{rad}\left(\frac{x^3y^2z}{\Gamma^3}\right)\right) \le v_p(x) + v_p(y) + v_p(\mathrm{rad}(z)) \tag{12.21}$$

holds for every prime p. This will be done by checking the following three cases:

First case: $3v_p(x) < 2v_p(y)$

Then $x^3 - y^2 = z$ gives $3v_p(x) = v_p(\Gamma) = v_p(z)$ and (12.21) reads as

$$3v_p(x) + 1 \le v_p(x) + v_p(y) + \chi(v_p(x)), \tag{12.22}$$

where $\chi(t) = 0$ if $t \le 0$ and $\chi(t) = 1$ if $t > 0$. Since Γ is free from sixth powers, only the cases $v_p(x) = 0, 1$ occur and (12.22) obviously holds.

Second case: $2v_p(y) < 3v_p(x)$

Similarly as in the first case, we have $2v_p(y) = v_p(\Gamma) = v_p(z)$ and (12.21) is equivalent to

$$2v_p(y) + 1 \le v_p(x) + v_p(y) + \chi(v_p(y)).$$

Checking $v_p(y) = 0, 1, 2$, this is true.

Third case: $3v_p(x) = 2v_p(y)$

Since Γ is free from sixth powers, we have $v_p(x) = v_p(y) = 0$ and (12.21) holds.

Now we deduce the strong Hall conjecture from (12.19) and (12.20). We conclude that

$$|x|^3 \ll_\varepsilon |xy|^{1+\varepsilon}\mathrm{rad}(z)^{1+\varepsilon}, \tag{12.23}$$

and

$$|y|^2 \ll_\varepsilon |xy|^{1+\varepsilon}\mathrm{rad}(z)^{1+\varepsilon}. \tag{12.24}$$

By combining (12.23) and (12.24), we get

$$|xy|^6 \ll_\varepsilon |xy|^{5(1+\varepsilon)}\mathrm{rad}(z)^{5(1+\varepsilon)}.$$

By passing to a sufficiently small ε, as we are allowed to do, we get

$$|xy| \ll_\varepsilon \mathrm{rad}(z)^{5+\varepsilon}.$$

Inserting this in (12.23) and (12.24), we get (ii).

(b) $\Rightarrow$ (c). We choose a global minimal Weierstrass equation for E. Let

$$\Gamma := \mathrm{GCD}(c_4^3, c_6^2) = \Gamma' g^6$$

with Γ' sixth-power free. By (12.11) on page 426, we have the relation

$$1728\,\frac{\Delta}{g^6} = \left(\frac{c_4}{g^2}\right)^3 - \left(\frac{c_6}{g^3}\right)^2.$$

Applying the strong Hall conjecture, we obtain

$$\max(|c_4|^3, |c_6|^2) \ll_\varepsilon g^6 \mathrm{rad}(\Delta/g^6)^{6+\varepsilon}.$$

By Corollary 12.5.7, we deduce $g \ll \mathrm{rad}(g) \leq \mathrm{rad}(\Gamma)$ and hence

$$\max(|c_4|^3, |c_6|^2) \ll_\varepsilon (\mathrm{rad}(\Delta)\mathrm{rad}(\Gamma))^{6+\varepsilon}. \tag{12.25}$$

We have already remarked in 12.5.5 and 12.5.9 that if $p \neq 2$, 3 and p divides both c_4 and c_6 then E has additive reduction at p and $f_p \geq 2$. Hence 12.5.9 leads to

$$\mathrm{rad}(\Delta)\,\mathrm{rad}(\Gamma) \leq 6\,\mathrm{cond}(E),$$

proving what we want.

(c) $\Rightarrow$ (a): Let $a, b, c \in \mathbb{Z}$ be coprime with $a + b = c$ and let us consider the Frey curve E from Example 12.5.10 with affine equation

$$y^2 = x(x+a)(x-b).$$

First, we assume that $16 \nmid abc$. Then (12.17) on page 429 is the global minimal Weierstrass equation for E and the generalized Szpiro conjecture leads to

$$\max\left(16^3|a^2+ab+b^2|^3, 16|abc|^2\right) \ll_\varepsilon \mathrm{cond}(E)^{6+\varepsilon}. \tag{12.26}$$

By $16^3 \nmid \Delta$, our considerations in Example 12.5.10 show that

$$\mathrm{cond}(E) < 16^3 \prod_{\substack{p \mid abc \\ p \neq 2}} p.$$

Taking this into account in (12.26), we easily deduce that

$$\max(|a|, |b|, |c|) \ll_\varepsilon \mathrm{rad}(abc)^{1+\varepsilon}$$

as claimed by the strong abc-conjecture.

Finally, we assume $16|abc$. By permuting a, b, c, we may assume $a \equiv 1 \pmod 4$ and $b \equiv 0 \pmod{16}$. Then (12.18) on page 430 is the global minimal Weierstrass equation for E and the generalized Szpiro conjecture gives

$$\max\left(|a^2+ab+b^2|^3, \left|\frac{abc}{16}\right|^2\right) \ll_\varepsilon \mathrm{cond}(E)^{6+\varepsilon}.$$

Since E has multiplicative reduction at all primes $p|\Delta$, we have

$$\operatorname{cond}(E) = \operatorname{rad}(\Delta) \leq \prod_{p|abc} p$$

and, as in the first case, we get (a). □

Remark 12.5.13. In order to understand the following remark, the reader should be familiar with the proper minimal model $\mathcal{C}$ of the elliptic curve E over $\mathbb{Q}$ and the Kodaira classification of special fibres (see **[286]**, Ch.IV). The proof of the implication (a) ⇒ (b) yields a little more than the strong Hall conjecture and suggests that the generalized Szpiro conjecture may have a slightly stronger formulation in which the conductor is replaced by $\prod_p p^{f'_p}$, where p ranges over all primes ≥ 5 and where $f'_p = 0$ if E has good reduction at p, where $f'_p = 1$ if E has multiplicative reduction or the special fibre of $\mathcal{C}$ has Kodaira type II, III, IV at p, and where $f'_p = 2$ if the special fibre of $\mathcal{C}$ has Kodaira type I_ν^*, II*, III* or IV*.

In order to see this, we note first that f_p is bounded by $2+3v_p(3)+6v_p(2)$ for any prime p (see **[286]**, Th.IV.10.4), hence we may neglect the primes 2 and 3. Then replacing $\operatorname{rad}(\Gamma)$ in (12.25) by the stronger $\operatorname{rad}(g)$, we conclude that for primes $p \geq 5$ the exponent 2 in the conductor is only needed when $p^6|\Gamma$. This is equivalent to the statement that the special fibre of $\mathcal{C}$ has Kodaira type I_ν^*, II*, III*, or IV* at p (use Tate's algorithm in **[286]**, IV, §9).

In a similar vein to the Hall conjecture, the **Hall–Lang–Waldschmidt–Szpiro conjecture** claims the following:

Conjecture 12.5.14. *Let $A, B \in \mathbb{Z}\setminus\{0\}$, $m, n \in \mathbb{N}\setminus\{0,1\}$. Then for all $x, y \in \mathbb{Z}$ with x and y coprime and*

$$Ax^m + By^n = z \neq 0$$

and, for every $\varepsilon > 0$, it holds

$$|x|^{mn-m-n} \ll_\varepsilon \operatorname{rad}(z)^{n+\varepsilon}$$

and

$$|y|^{mn-m-n} \ll_\varepsilon \operatorname{rad}(z)^{m+\varepsilon}.$$

12.5.15. We leave to the reader the easy task of deducing Conjecture 12.5.14 from the abc-conjecture. We may just follow the arguments proving Theorem 12.5.12(b) from (12.19) and (12.20) on page 432.

Conversely, Conjecture 12.5.14 for the equation $x^3 - y^2 = z$ implies the strong abc-conjecture. For the proof we use the identity

$$(a^2 + ab + b^2)^3 - ((b-a)(a+2b)(2a+b)/2)^2 = 3^3\,(ab(a+b)/2)^2.$$

Since a, b are coprime, we see that $x := a^2 + ab + b^2$ and $y := (b-a)(a+2b)(2a+b)/2$ have no common prime divisors $\neq 3$. If $3|b-a$, then we may divide the identity by 3^3 to get coprime x and y. Then Conjecture 12.5.14 gives

$$a^2 + ab + b^2 \ll_\varepsilon \mathrm{rad}(abc)^{2+\varepsilon}$$

proving immediately the strong abc-conjecture for $c = a + b$.

Remark 12.5.16. The conjecture implies that the generalized Fermat equation $Ax^m + By^n = Cz^p$ has only finitely many solutions in coprime integers x, y, z provided $ABC \neq 0$ and $\frac{1}{m} + \frac{1}{n} + \frac{1}{p} < 1$. In the next section, we will show that this is actually a consequence of Faltings's theorem. It has been also conjectured that for fixed A, B, C there will be no non-trivial solutions (namely, satisfying a certain necessary coprimality condition) for sufficiently large m, n, and p, although this remains open.

If $A = B = C = 1$, we quickly find the solutions

$$1 + 2^3 = 3^2, \quad 2^5 + 7^2 = 3^4, \quad 7^3 + 13^2 = 2^9,$$

$$2^7 + 17^3 = 71^2, \quad 3^5 + 11^4 = 122^2.$$

However, Beukers did a more thorough computer search and found the amazing examples

$$17^7 + 76271^3 = 21063928^2, \quad 1414^3 + 2213459^2 = 65^7,$$

$$43^8 + 96222^3 = 30042907^2, \quad 33^8 + 1549034^2 = 15613^3$$

to which Zagier added

$$9262^3 + 15312283^2 = 113^7.$$

These five large solutions look unusual from the point of view of diophantine equations. Their respective abc-ratios $\log c/\log \mathrm{rad}(abc)$ are approximately

$$1.14125, \quad 1.12221, \quad 1.06109, \quad 1.05700, \quad 1.08836.$$

These values, fairly close to 1, are consistent with the strong abc-conjecture predicting that 1 is the largest accumulation point of abc-ratios.

12.6. The generalized Fermat equation

By Fermat's last theorem, the Fermat equation $x^n + y^n = z^n$ has no solutions in $\mathbb{Z} \setminus \{0\}$ for $n \geq 3$. More generally, we may consider the **generalized Fermat equation**

$$Ax^p + By^q = Cz^r$$

for given parameters $A, B, C \in \mathbb{Z} \setminus \{0\}$. In this section, we prove a result of H. Darmon and A. Granville ([**78**]) that if $\frac{1}{p} + \frac{1}{q} + \frac{1}{r} < 1$, then the generalized Fermat equation has only finitely many coprime solutions (meaning that $\mathrm{GCD}(x, y, z) = 1$).

For $p = q = r \geq 4$, the generalized Fermat equation describes a smooth projective curve of genus $\frac{1}{2}(p-1)(p-2) \geq 3$ (see A.13.4), hence Faltings's theorem implies immediately the claim. For $p = q = r = 3$, we get an elliptic curve and we may easily construct examples with infinitely many coprime solutions by essentially doubling the points (see **[78]**, §6).

In the general case, the argument is more involved because the equation no longer describes a projective curve. The idea of Darmon and Granville is to consider a finite Galois covering $\pi : C \to \mathbb{P}^1$, which is unramified outside $\pi^{-1}\{0,1,\infty\}$ and with ramification indices p, q, r over $0, 1, \infty$. The construction of π will use the topological and analytical means from Section 12.3 and Belyĭ's theorem will show that π is defined over a number field K.

By analysing the ramification of $K(P)/K(\beta)$ for $\beta = Ax^p/(Cz^r)$ and $P \in \pi^{-1}(\beta)$, we can deduce from Hermite's theorem that the fibre points over β are rational over a fixed number field. This may be seen as an extension of the Chevalley–Weil theorem to modestly ramified coverings. By Hurwitz's theorem, the genus of C will turn out to be ≥ 2 and hence Faltings's theorem will lead to the finiteness of the number of coprime solutions. This completes the discussion in Remark 12.5.16.

In this section, we assume that the reader is familiar with Section 12.3.

12.6.1. We begin by describing the fundamental group of $\mathbb{P}^1_{\text{an}} \setminus \{0,1,\infty\}$.

Let $x_0 \in \mathbb{P}^1_{\text{an}} \setminus \{0,1,\infty\}$ be the base point and let $\sigma_0 \in \pi_1(\mathbb{P}^1_{\text{an}} \setminus \{0,1,\infty\}, x_0)$ be represented by a path connecting x_0 with a point in a small neighbourhood of 0, then turning on a positive circle around 0 and then going backwards along the same way to x_0. In the same way, we define $\sigma_1, \sigma_\infty \in \pi_1(\mathbb{P}^1_{\text{an}} \setminus \{0,1,\infty\}, x_0)$ turning around 1 and ∞, respectively. Then we claim that $\pi_1(\mathbb{P}^1_{\text{an}} \setminus \{0,1,\infty\}, x_0)$ is generated by $\sigma_0, \sigma_1, \sigma_\infty$ with the single relation $\sigma_0\sigma_1\sigma_\infty = 1$.

In order to prove this statement, we may identify $\mathbb{P}^1_{\text{an}}$ with the Riemann sphere. We choose a closed disc D in $\mathbb{C}$ containing $0, 1$. Topologically, the closure E of $\mathbb{P}^1_{\text{an}} \setminus D$ is also a disc and $D \cap E$ is a circle. We choose the base point $x_0 \in D \cap E$ getting natural homomorphisms φ_D, φ_E from $\pi_1(D \cap E, x_0)$ to $\pi_1(D \setminus \{0,1\}, x_0)$ and $\pi_1(E \setminus \{\infty\}, x_0)$, respectively. Then van Kampen's theorem (see H. Seifert and W. Threlfall **[274]**, §52, Th.1) says that $\pi_1(\mathbb{P}^1_{\text{an}} \setminus \{0,1,\infty\}, x_0)$ is the free product of $\pi_1(D \setminus \{0,1\}, x_0)$ and $\pi_1(E \setminus \{\infty\}, x_0)$ modulo the relations $\varphi_D(\gamma) = \varphi_E(\gamma)$ with γ varying over generators of $\pi_1(D \cap E, x_0)$.

Clearly, we have $\pi_1(E \setminus \{\infty\}, x_0) = \sigma_\infty^{\mathbb{Z}}$ and another application of van Kampen's theorem shows that $\pi_1(D \setminus \{0,1\}, x_0)$ is the free group on the set $\{\sigma_0, \sigma_1\}$. Then, if γ denotes the boundary of D in positive direction, we have $\pi_1(D \cap E, x_0) = \gamma^{\mathbb{Z}}$ and the relations in van Kampen's theorem are $\gamma = \sigma_0\sigma_1 = \sigma_\infty^{-1}$, proving the claim. □

12.6.2. Our goal is to construct a ramified covering $\pi : C_{\text{an}} \to \mathbb{P}^1_{\text{an}}$ for a connected Riemann surface C_{an} which is unramified outside $\pi^{-1}\{0,1,\infty\}$. By 12.3.4 and 12.3.8, such a covering corresponds to the subgroup $H = \pi_1(C_{\text{an}} \setminus \pi^{-1}\{0,1,\infty\}, y_0)$ of $G = \pi_1(\mathbb{P}^1_{\text{an}} \setminus \{0,1,\infty\}, x_0)$, where the base point y_0 lies over x_0. We would like to have a Galois covering, which means that H is a normal subgroup. It follows easily from 12.3.5 and 12.3.8 that the covering group G/H acts also transitively on the ramified fibres of π.

For $x \in C_{\text{an}}$, we denote by $e_{x/\pi(x)}$ the ramification index of the corresponding valuations which is equal to the multiplicity of the fibre divisor $\pi^*([\pi(x)])$ in x (see Example

1.4.12 for an algebraic discussion). The ramification index should not be confused with the multiplicity of the ramification divisor at x which is $e_{x/\pi(x)} - 1$.

Lemma 12.6.3. *Let π be a covering as in 12.6.2. If π is finite, then $e_{x/\pi(x)}$ is equal to the cardinality of the stabilizer $(G/H)_x$ of x.*

Proof: The argument may be done complex analytically or complex algebraically by A.14.7. Since the covering group G/H acts transitively on the fibre $\pi^{-1}(y)$ for any $y \in \mathbb{P}^1_{\mathrm{an}}$, the ramification index $e_{x/\pi(x)}$ depends only on $\pi(x)$. By 12.3.6, the degree of π is equal to $[G:H]$. By Example 1.4.12, we have $\deg(\pi) = m e_{x/\pi(x)}$, where m is the cardinality of the set $\pi^{-1}(y)$. Since $m = [G:H]/|(G/H)_x|$, we get the claim. □

12.6.4. In hyperbolic geometry, we easily show that the disc $\mathbb{D} = \{|z| < 1\}$ contains a geodesic triangle with angles $\frac{\pi}{p}, \frac{\pi}{q}, \frac{\pi}{r}$. This means that the sides are contained in circles perpendicular to $\{|z| = 1\}$ and the sum of angles is $\pi(\frac{1}{p} + \frac{1}{q} + \frac{1}{r}) < \pi$ leading to hyperbolicity. Moreover, reflecting the triangles successively at their sides, we get a tessellation of $\mathbb{D}$. By the Riemann mapping theorem ([**8**], Ch.6, Th.s 1-4), the interior of the original geodesic triangle may be mapped biholomorphically on to the upper half plane $\{\Im(w) > 0\}$ and the boundary is mapped to the boundary. Moreover, by Schwarz's reflection principle, we get an extension to a holomorphic map $\Omega : \mathbb{D} \to \mathbb{C}$. The multivalued inverse may be described explicitly by a quotient of hypergeometric functions. For details, we refer to C. Carathéodory [**57**], Vol.II, Part 7, Ch.III.

By the description above, it is clear that Ω is ramified only over points over $0, 1, \infty$ with ramification indices p, q, and r, respectively. Let P_0, P_1, P_∞ be the vertices of the original geodesic triangle Δ lying over $0, 1, \infty$. Let τ_∞ be the hyperbolic reflection at the side through P_0 and P_1. Similarly, we denote the reflections at the other sides of Δ by τ_0, τ_1 and we set $\gamma_0' := \tau_1 \circ \tau_\infty$, $\gamma_1' := \tau_\infty \circ \tau_0$, $\gamma_\infty' := \tau_0 \circ \tau_1$. These are hyperbolic rotations around the vertices P_0, P_1, P_∞ with angles $\frac{2\pi}{p}, \frac{2\pi}{q}, \frac{2\pi}{r}$. So we have the relations

$$(\gamma_0')^p = (\gamma_1')^q = (\gamma_\infty')^r = \gamma_0'\gamma_1'\gamma_\infty' = 1 \tag{12.27}$$

in the subgroup Γ of the automorphism group of $\mathbb{D}$ generated by $\gamma_0', \gamma_1', \gamma_\infty'$. By construction, it is easy to see that Γ is the covering group of Ω. We call Γ a **triangle group.**

12.6.5. We can show that Γ is the free group on the set $\{\gamma_0', \gamma_1', \gamma_\infty'\}$ modulo the relations (12.27) (see S. Katok [**159**], Sec.4.3). It follows from 12.3.4 and 12.3.5 that a covering $\pi : C_{\mathrm{an}} \to \mathbb{P}^1_{\mathrm{an}}$ as in 12.6.2 has ramification indices over $0, 1, \infty$ dividing p, q, and r, respectively, if and only if π factors through Ω. We will use this only as a motivation to consider suitable normal subgroups of finite index in Γ to get our desired finite covering π of ramification indices p, q, and r. This will be provided by **Fox's theorem** solving a conjecture of Fenchel.

Theorem 12.6.6. *The triangle group Γ has a torsion-free normal subgroup H of finite index.*

For the quite elementary algebraic proof, we refer to the original article R.H. Fox [**121**].

Corollary 12.6.7. *There is a compact connected Riemann surface C_{an} and a finite Galois covering $\pi : C_{\mathrm{an}} \to \mathbb{P}^1_{\mathrm{an}}$ which is unramified outside of $\pi^{-1}\{0, 1, \infty\}$ and which has ramification indices p, q, r at the points over 0, 1, and ∞, respectively.*

Proof: By 12.3.4 and 12.3.5, the normal subgroup H from Fox's theorem lifts to a normal subgroup of $\pi_1(\mathbb{P}^1_{\rm an}\backslash\{0,1,\infty\},x_0)$ leading to a ramified Galois covering $\pi : C_{\rm an} \to \mathbb{P}^1_{\rm an}$. From 12.3.6, we deduce that π is finite and $\deg(\pi) = [\Gamma : H]$. By construction, π is unramified outside of $\pi^{-1}\{0,1,\infty\}$. The geometric description of Γ given in 12.6.4 shows that Γ contains an element γ_0' of order p. Since H is torsion free, we conclude that Γ/H contains also an element γ_0'' of order p. By 12.3.4 and 12.3.8, there is a ramified covering $\varphi : \mathbb{D} \to C_{\rm an}$ such that $\Omega = \pi\circ\varphi$. By 12.3.5, it is clear that Γ/H is the covering group of π. By construction, γ_0'' is in the stabilizer of $\varphi(P_0)$.

Thus Lemma 12.6.3 yields that p divides the ramification index $e_{\varphi(P_0)/0}$. On the other hand, we have $e_{\varphi(P_0)/0}|e_{P_0/0} = p$, hence $e_{\varphi(P_0)/0} = p$. We argue at points over 1 and ∞ much in the same way, proving the claim. □

12.6.8. By Belyĭ's theorem and its proof (see Theorem 12.3.11), there is a number field K and a morphism $\pi : C \to \mathbb{P}^1_K$ from an irreducible smooth projective curve C over K such that the ramified covering from Corollary 12.6.7 is induced by base change. We are free to enlarge K, hence we may assume that all fibre points over $0,1,\infty$ are K-rational. Since we are in characteristic 0, local parameters do not change under field extension (use A.4.6) and hence the ramification points of C lie over $0,1,\infty$ with ramification indices p, q, and r, respectively.

Proposition 12.6.9. *The genus of* C *satisfies* $g(C) \geq 2$.

Proof: The genus is invariant under base extension (use A.10.28), so we may work over the complex numbers. Since the covering is Galois, the covering group operates transitively on the fibres (see 12.6.2) and the ramification indices are the same in a given fibre. By Example 1.4.12, the number of fibre points over $0,1,\infty$ is equal to $\frac{d}{p}$, $\frac{d}{q}$, and $\frac{d}{r}$, respectively, where d is the degree of the covering morphism π. Then Hurwitz's theorem (see Theorem B.4.6) yields

$$2-2g(C) = 2d - \frac{d}{p}(p-1) - \frac{d}{q}(q-1) - \frac{d}{r}(r-1) = d(\frac{1}{p}+\frac{1}{q}+\frac{1}{r}-1) < 0$$

proving the claim. □

12.6.10. Now we use the language of schemes to construct an irreducible projective scheme $\mathcal{C}$ over O_K with generic fibre $C = \mathcal{C}_K$. This can be done by using a set of homogeneous equations with coefficients in O_K describing C in projective space $\mathbb{P}^n_K$. This leads to a closed subscheme of $\mathbb{P}^n_{O_K}$ and we choose $\mathcal{C}$ to be the irreducible component containing C.

The morphism π does not necessarily extend to a morphism $\mathcal{C} \to \mathbb{P}^1_{O_K}$, its domain is an open subset containing the generic fibre. Hence there is a closed finite subset S of $\mathrm{Spec}(O_K)$ such that π extends to a morphism $\overline{\pi} : \mathcal{C}_S \to \mathbb{P}^1_{O_{S,K}}$ over $O_{S,K}$, where $O_{S,K}$ is the ring of S-integers in O_K and $\mathcal{C}_S$ is the part of $\mathcal{C}$ lying over $\mathrm{Spec}(O_{S,K})$. Since the set of smooth points of $\mathcal{C}_S$ is open and contains C (use ([**137**], Th.17.5.1), the same argument shows that we may assume, after enlarging S, that $\mathcal{C}_S$ is a smooth scheme over $O_{S,K}$. This means that $\mathcal{C}$ has **good reduction** over the non-archimedean places outside of S. We may also assume that no place outside of S divides pqr.

The set of unramified points of $\mathcal{C}_S$ with respect to $\overline{\pi}$ is open (see B.3.2) and does not contain the points over $0,1,\infty$ of the generic fibre C. Hence the ramified points are contained in the closure of $\pi^{-1}\{0,1,\infty\}$ in $\mathcal{C}_S$ and in finitely many fibres of $\mathcal{C}$ over closed points of $\mathrm{Spec}(O_{S,K})$. By enlarging S again, we may exclude the latter.

For two points in C, their closures in $\mathcal{C}$ intersect at most in finitely many points lying over the closed points of $\mathrm{Spec}(O_K)$. Again, we may assume that the closures in $\mathcal{C}_S$ of different fibre points over 0 are disjoint and that the same holds for the fibres over 1 and ∞.

12.6.11. A point $P \in C$ may be viewed as an $O_{K(P)}$-integral point of $\mathcal{C}$ by the valuative criterion of properness ([**148**], Th.II.4.7). This means that there is a unique morphism $\overline{P}$: $\mathrm{Spec}(O_{K(P)}) \to \mathcal{C}$ mapping the generic point $\{0\}$ to P. For a non-archimedean place w of $K(P)$ with maximal ideal $\mathfrak{m}_w \in \mathrm{Spec}(O_{K(P)})$, let $P(w) := \overline{P}(\mathfrak{m}_w)$. We suppose that the restriction v of w to K is not contained in S and that $P(w)$ is a $k(v)$-rational point of the reduction $\mathcal{C}_{k(v)}$. Since $\mathcal{C}$ has good reduction in v, the reduction is a smooth curve over $k(v)$ and hence we have a local parameter $\widetilde{\zeta}$ in the discrete valuation ring $\mathcal{O}_{\mathcal{C}_{k(v)},P(w)}$. Let ζ be a lift to $\mathcal{O}_{\mathcal{C},P(w)}$.

Lemma 12.6.12. *Under the assumptions above and if we identify the completion K_v with a subfield of the completion $K(P)_w$, then $K(P)_w = K_v(\zeta(P))$.*

Proof: We need to show that $K(\zeta(P))$ is dense in $K(P)$ with respect to w. Let π_v be a local parameter in the valuation ring R_v of K, let π_w be similarly defined for w, and let $N \in \mathbb{N}$.

First step: *For every $a \in \mathcal{O}_{\mathcal{C},P(w)}$, there is $p(x) \in R_v[x]$ with*

$$a - p(\zeta) \in [\pi_v, \zeta]^N \lhd \mathcal{O}_{\mathcal{C},P(w)}.$$

To prove the first step, we proceed by induction on N. For $N = 0$, there is nothing to prove. Now let $N \geq 1$. Since $\widetilde{\zeta}$ is a local parameter in the smooth $k(v)$-rational point $P(w)$, there is a polynomial $q(x) \in R_v[x]$ with reduction $\widetilde{q}(x) \in k(v)[x]$ such that $\widetilde{a} \equiv \widetilde{q}(\widetilde{\zeta}) \pmod{\widetilde{\zeta}^N}$. By lifting, there is $b \in \mathcal{O}_{\mathcal{C},P(w)}$ with

$$a \equiv q(\zeta) + b\pi_v \pmod{\zeta^N}.$$

By induction applied to b, we get the first step.

To prove the lemma, we note that the image of $\mathcal{O}_{\mathcal{C},P(w)}$ under the evaluation map at P generates $K(P)$ as a field. Using $\zeta(P) \equiv 0 \pmod{\pi_w}$ and the first step, we conclude that every element in $K(P)$ may be approximated by rational functions in $\zeta(P)$ proving the claim. □

Lemma 12.6.13. *Let $|\ |_v$ be a complete non-archimedean absolute value on the field F and let $k \in \mathbb{N}$ be coprime to the characteristic of the residue field of v. For $\alpha \in F$ with $|\alpha|_v < 1$, let $u := 1 + \alpha$. Then there is $\lambda \in F$ with $u = \lambda^k$.*

Proof: Formally, we have the identity

$$1 + x = \left(\sum_{n=0}^{\infty} \binom{1/k}{n} x^n \right)^k$$

of power series with coefficients in $\mathbb{Q}$. It is easy to see that the coefficients $\binom{1/k}{n}$ are indeed in $\mathbb{Z}[\frac{1}{k}]$. Since k is a unit in R_v, we conclude that the identity also holds for power series with coefficients in R_v. Replacing x by α, the power series is convergent and we get a kth root of u in F. □

The following result of S. Beckmann [**19**] is the key to the proof of finiteness of the number of solutions of the generalized Fermat equation.

Lemma 12.6.14. *Let $\pi : C \to \mathbb{P}^1_K$ be the covering from 12.6.8 and let S be as in 12.6.10. We identify the place $v \notin S$ with the corresponding discrete valuation normalized by $v(O_K) = \mathbb{Z}$ and let $v^+ := \max(0, v)$. Then for $Q \in \mathbb{P}^1(K) \setminus \{0, 1, \infty\}$ with affine coordinate β satisfying*

$$v^+(\beta) \equiv 0 \pmod p, \quad v^+(\beta - 1) \equiv 0 \pmod q, \quad v^+(\tfrac{1}{\beta}) \equiv 0 \pmod r,$$

the extension $K(P)/K$ is unramified over v for every $P \in \pi^{-1}(Q)$.

Proof: First, we handle the case $v^+(\beta) = v^+(\beta - 1) = v^+(\frac{1}{\beta}) = 0$. Clearly, this implies $v(\beta) = v(\beta - 1) = 0$ and hence we have $Q(v) \notin \{0(v), 1(v), \infty(v)\}$ for the reductions. Let w be a place of $K(P)$ over v. Since $\pi(P) = Q$, we get $\overline{\pi}(P(w)) = Q(v) \notin \{0(v), 1(v), \infty(v)\}$. By our considerations in 12.6.10, we conclude that $\overline{\pi}$ is unramified in $P(w)$ and hence $K(P)/K$ is unramified over w (see Proposition B.3.6).

Now we assume that at least one congruence is not an equality. By a change of coordinates, we may assume $v(\beta) > 0$. This means $Q(v) = 0(v)$. Every point $P \in C$ may be viewed as an $O_{K(P)}$-integral point $\overline{P}$ of $\mathcal{C}$ (see 12.6.11) and hence as a prime divisor on the smooth model $\mathcal{C}_{R_v}$ over the discrete valuation ring R_v. By standard facts for divisors (see A.8 and [**125**], Ch.20 for the generalization over Dedekind rings), we have the identity

$$\overline{\pi}^*(\overline{0}) = \sum_{R \in \pi^{-1}(0)} p \cdot \overline{R} \tag{12.28}$$

of divisors on $\mathcal{C}_{R_v}$. This follows from surjectivity of $\mathcal{C}_{k(v)}$ onto $\mathbb{P}^1_{k(v)}$, so only horizontal components occur in $\pi^*(\overline{0})$ and hence the multiplicities may be computed in the generic fibre C. For the reduction $\widetilde{\pi}$ of $\overline{\pi}$ modulo v, we have the identity

$$\widetilde{\pi}^*[0(v)] = \sum_{\widetilde{R} \in \widetilde{\pi}^{-1}(0(v))} e_{\widetilde{R}/0(v)}[\widetilde{R}] \tag{12.29}$$

of divisors on $\mathcal{C}_{k(v)}$. Pulling (12.28) back to $\mathcal{C}_{k(v)}$ and comparing with (12.29), we conclude that reduction modulo v maps $\pi^{-1}(0)$ onto $\pi^{-1}(0(v))$. By our assumptions in 12.6.10, this reduction is also one-to-one and hence bijective. Moreover, the above comparison shows that the ramification index $e_{\widetilde{R}/0(v)}$ is equal to p. By our assumptions in 12.6.8, all fibre points over 0 are K-rational and hence all fibre points over $0(v)$ are $k(v)$-rational.

Now we consider $P \in \pi^{-1}(Q)$ and a place w of $K(P)$ over v. Since $Q(v) = 0(v)$, the above shows the $k(v)$-rationality of $P(w) \in \widetilde{\pi}^{-1}(0(v))$. There is a unique $R \in \pi^{-1}(0)$ with $R(v) = P(w)$. Let ζ be a local equation of the prime divisor $\overline{R}$ in $R(v)$. Hence $\zeta \in \mathcal{O}_{\mathcal{C},P(w)}$ with reduction $\widetilde{\zeta}$ equal to a local parameter of $P(w)$ inside the smooth curve $\mathcal{C}_{k(v)}$. Let x be the affine coordinate on $\mathbb{P}^1_K$, then x is a local equation for $\overline{0}$ and therefore (12.28) shows

$$x = u\zeta^p \tag{12.30}$$

for a unit $u \in \mathcal{O}_{\mathcal{C},P(w)}$.

Let L be the number field obtained from K by adjoining all pth roots of $u(R)$. Since by assumption the characteristic of $k(v)$ does not divide p, the extension L/K is unramified over v (Lemma B.2.6). By Proposition B.2.3, it is enough to show that $L(P)/L$ is unramified over v. Replacing K by L, we may assume that K contains all pth roots of $u(R)$.

We claim that $K_v(\beta^{1/p}) = K(P)_w$. Replacing u by $u/u(R)$ and ζ by $u(R)^{1/p}\zeta$, we may assume $u(R) = 1$. For the reduction $\widetilde{u}$ of u modulo v, we have $\widetilde{u}(P(w)) = \widetilde{u}(R(v)) = \overline{1} \in k(v)$ and hence $|u(P) - 1|_w < 1$. If u has a pth root, then our claim follows directly from (12.30) and Lemma 12.6.12. However, in general the pth root of the unit u exists only as a v-adic analytic function and we need to modify the proof slightly. First we note the formal identity

$$u^{\frac{1}{p}} = \sum_{n=0}^{\infty} \binom{1/p}{n} (u-1)^n, \tag{12.31}$$

which becomes an equation in the w-adic topology by evaluating it at $u = u(P)$ (see proof of Lemma 12.6.13). The first step in the proof of Lemma 12.6.12 shows that $u - 1$ may be expressed as a formal power series in ζ without the constant term and with coefficients in the valuation ring $\hat{R}_v$ of the completion K_v. Inserting this in the right-hand side of (12.31), and using (12.30), we get a formal identity

$$x^{\frac{1}{p}} = \zeta + \sum_{n=2}^{\infty} a_n \zeta^n \tag{12.32}$$

with coefficients $a_n \in \hat{R}_v$. Since $\widetilde{\zeta}(P(v)) = \overline{0} \in k(v)$, we have $|\zeta(P)|_w < 1$ and hence (12.32) gives an identity in $K(P)_w$ by evaluating it at $\zeta = \zeta(P)$. Now (12.32) implies that

$$\zeta = x^{\frac{1}{p}} + \sum_{n=2}^{\infty} b_n x^{\frac{n}{p}} \tag{12.33}$$

for unique coefficients $b_n \in \hat{R}_v$. This follows from putting (12.33) into (12.32) and comparison of coefficients. Inserting P in (12.32) and (12.33), completeness yields $K_v(\beta^{1/p}) = K_v(\zeta(P))$ and hence Lemma 12.6.12 implies $K_v(\beta^{1/p}) = K(P)_w$.

By assumption the residue characteristic of v does not divide p and $p|v(\beta)$, hence Lemma B.2.6 yields that $K(P)_w/K_v$ is unramified over v. Since residue fields do not change by passing to completions (see Proposition 1.2.11), we have proved that $K(P)/K$ is unramified over w. □

Now we are ready to prove the **theorem of Darmon and Granville.**

Theorem 12.6.15. *Let $p, q, r \in \mathbb{N} \setminus \{0\}$ with $\frac{1}{p} + \frac{1}{q} + \frac{1}{r} < 1$ and let $A, B, C \in \mathbb{Z} \setminus \{0\}$. Then there are only finitely many solutions $(x, y, z) \in \mathbb{Z}^3$ of the generalized Fermat equation*

$$Ax^p + By^q = Cz^r$$

with $\mathrm{GCD}(x, y, z) = 1$.

Proof: Let $\pi : C \to \mathbb{P}^1_K$ be the covering from 12.6.8 and let S be the finite set of non-archimedean places on K constructed in 12.6.10. By enlarging S, we may assume that it contains all places dividing ABC. Let (x,y,z) be a coprime solution of the generalized Fermat equation. We may assume $xyz \neq 0$ excluding at most finitely many solutions. For

$$\beta := \frac{Ax^p}{Cz^r},$$

our assumptions on S and $\mathrm{GCD}(x,y,z) = 1$ yield the congruences in Lemma 12.6.14. Note that $Q := \beta \in \mathbb{P}^1(\mathbb{Q}) \setminus \{0,1,\infty\}$ and Lemma 12.6.14 shows that $K(P)/K$ is unramified over v for all non-archimedean places $v \notin S$ of K and for all $P \in \pi^{-1}(Q)$. By Example 1.4.12, we have $[K(P) : K] \leq \deg(\pi)$. Hermite's theorem implies that there are only finitely many number fields in $\overline{K}/K$ of bounded degree and unramified outside S (see Corollary B.2.15). Hence there is a number field L extending K containing $K(P)$ for all possible coprime solutions. Since $g(C) \geq 2$ (see Proposition 12.6.9), Faltings's theorem (see Theorem 11.1.1) shows that $C(L)$ is finite, thus proving that there are only finitely many possibilities for β.

Let (x,y,z) and (x',y',z') be coprime solutions in $(\mathbb{Z} \setminus \{0\})^3$ of the generalized Fermat equation with $\beta = \beta'$. Let ℓ be a prime with ℓ-adic valuation v_ℓ. Using coprimeness, we get $v_\ell(x) = v_\ell(x')$ and $v_\ell(z) = v_\ell(z')$ for every ℓ not dividing ABC. Moreover, for $\ell | ABC$ we have $\min\{v_\ell(x), v_\ell(z)\} \leq v_\ell(B)$ and hence only finitely many coprime solutions give rise to the same β. This proves the claim. □

12.7. Bibliographical notes

Szpiro's conjecture, originally formulated only for the discriminant and with a undetermined exponent, was formulated at an exposition in Hannover in 1983. Influenced by this conjecture and the theorem of Stothers–Mason, the abc-conjecture came out from a discussion between Masser and Oesterlé in 1985. The original purpose of both conjectures was to give new insights into Fermat's last theorem.

N. Elkies [**98**] proved that the abc-conjecture implies Mordell's conjecture, effectively if an effective version of the abc-conjecture is available. He also mentioned that his argument using Belyĭ's lemma proves Vojta's height inequality and hence Roth's theorem as a special case. This aspect of the abc-conjecture over arbitrary number fields will be touched upon in Section 14.4. Proofs of Theorem 12.2.9 have also been given by Langevin in [**177**] and Oesterlé (in a seminar).

The implication (b) $\Rightarrow$ (a) in Belyĭ's theorem was well known to the experts, the argument uses a specialization argument due to Weil. The input of Grothendieck to Theorem 12.3.12 was the generalization of Riemann's existence theorem. In 1979, G.V. Belyĭ [**21**] proved the converse implication in Theorem 12.3.11 and caused Grothendieck to remark that such a deep result was never proved in such a simple and short way (see A. Grothendieck [**138**], where the study of *dessins d'enfant* was initiated).

In 1981 W.W. Stothers [**293**] obtained the abc-theorem for polynomials; our first proof follows essentially his arguments with Hurwitz's genus formula. Independently, Mason was inspired by an analogue of Baker's theory of linear forms of logarithms for function fields and found, in 1983, the generalization given in 12.4.3, which is a major tool in his studies of diophantine equations over function fields (R.C. [**192**], [**193**]).

The idea of the second proof appears, in a special case, in a Comptes Rendus note of A. Korkine in 1880 (see [**89**], Vol.II, p.750). The general abc-theorem for polynomials is a special case of a result of D. Brownawell and D. Masser [**52**], where also the first example in Remark 12.4.5 appears and, independently, of J.F. Voloch [**319**]. Granville's remark at the end of Section 12.4 is unpublished, so it remains to the reader to fill in the necessary details.

The equivalence of the various conjectures in Section 12.5 is given by Vojta [**307**], Ch.5, App.ABC. Note however that his version of the Hall–Lang–Waldschmidt–Szpiro conjecture is not true as stated, with the counterexample given in 12.5.2. If we assume x, y coprime as in Conjecture 12.5.14, then his proof of "Hall–Lang–Waldschmidt–Szpiro conj. $\Rightarrow$ generalized Szpiro conj." is incomplete because c_4 and c_6 may have common divisors for a minimal Weierstrass equation.

In J. Oesterlé [**229**], a sketch of proof for "abc $\Rightarrow$ generalized Szpiro conj." is given after ideas of Hindry, analysing the reduction of the elliptic curve at every place. This is similar to our proof of Theorem 12.5.12 (a) $\Rightarrow$ (b) based on the minimality criterion in Corollary 12.5.7, for which we could not find a complete reference in the literature. The implication "Hall–Lang–Waldschmidt–Szpiro conj. $\Rightarrow$ abc" is from A. Nitaj [**222**], where the reader will also find further applications of the abc-conjecture.

In Section 12.6, we follow the presentation of Darmon and Granville [**78**]. Beckmann's proof of the crucial Lemma 12.6.14 uses Galois representations. We expand here the sketch of proof given in [**78**]. Darmon [**76**] has also an interpretation of the lemma as a Chevalley–Weil theorem for orbifolds. For explicit examples and a similar treatment for the superelliptic equation, we refer to [**78**].

13 NEVANLINNA THEORY

13.1. Introduction

In 1987 Vojta formulated a sweeping set of precise conjectures about the structure of the set of rational points on algebraic varieties. The rationale about these conjectures was a rather precise analogy between the Nevanlinna theory of the distribution of values of meromorphic functions and diophantine approximation. In this way, Vojta motivated, clarified, and unified results and conjectures in diophantine approximation and diophantine equations. The analogy between Nevanlinna theory and diophantine approximation had also been noticed earlier by Ch. Osgood, in a somewhat different setting.

Here we discuss the Nevanlinna theory, while Vojta's conjectures, their connexion with the abc-conjecture studied in the preceding chapter, and their parallelism with Nevanlinna theory, will be dealt with in the next final chapter.

Section 13.2 is a brief introduction to the classical Nevanlinna theory in one variable, presenting the main results together with some examples. Next, Section 13.3 presents the Ahlfors–Shimizu elegant formulation of the theory, including Ahlfors's proof of the second main theorem. Holomorphic curves are the content of Section 13.4 dealing with geometric aspects of Nevanlinna theory, culminating with the conjectural second main theorem of Griffiths.

This chapter may be read independently of the other parts of the book.

13.2. Nevanlinna theory in one variable

Nevanlinna theory in one variable in its simplest form describes the value distribution theory of a non-constant meromorphic function $f : \mathbb{C} \to \mathbb{P}^1_{\mathrm{an}}$. First, we prove Jensen's formula, which is the basic tool. As a consequence, we will obtain Nevanlinna's first main theorem. Then we state without proof the lemma on the logarithmic derivative and the second main theorem of Nevanlinna. Next we present the notion of defect, deduce the basic defect inequality and Picard's little theorem as a corollary. The section ends by showing that the analogue of the abc-conjecture holds for meromorphic functions on $\mathbb{C}$.

Throughout this chapter, we shall suppose that the meromorphic function f is not a constant, unless specified otherwise. Proofs, when given, will be written in a concise form. Working them out in full detail is left as a useful exercise for the reader. We follow here the standard classical notation except that the ramification function and the counting function truncated at 1 are denoted by N_{ram} and $N^{(1)}$ instead of N_1 and $\overline{N}$, which we find in the classical literature.

The usual way to study the distribution of values of a meromorphic function $f(z)$ is to consider the number of solutions, counted with multiplicity, of the equation $f(z) = a$ in a disk $\{|z| < r\}$, as r varies. Recall that $\mathrm{ord}_z(f)$ denotes the order of f at $z \in \mathbb{C}$. The main tool at our disposal is the **Poisson–Jensen formula**:

Proposition 13.2.1. *Let f be meromorphic in the closed disk $|z| \leq R$ and assume that $f(z) \neq 0, \infty$. Then for $|z| < R$ it holds*

$$\begin{aligned} \log|f(z)| = &- \sum_{|a|<R, a\neq z} \mathrm{ord}_a(f) \log\left|\frac{R^2 - \overline{a}z}{R(z-a)}\right| \\ &+ \frac{1}{2\pi}\int_0^{2\pi} \log\left|f(Re^{i\theta})\right| \cdot \Re\left(\frac{Re^{i\theta}+z}{Re^{i\theta}-z}\right) d\theta. \end{aligned} \tag{13.1}$$

The case in which there are no zeros or poles is called **Poisson's formula** and the case $z = 0$ is called **Jensen's formula.**

Proof: We give only a quick sketch of proof. Consider first the special case in which $f(z)$ has no zeros or poles. Then $u(z) := \log|f(z)|$ is harmonic in an open neighborhood of the disk $\{|z| \leq R\}$, hence $u(0)$ is the average of $u(z)$ on the boundary $\{|z| = R\}$. The Poisson formula expressing $u(z)$ in terms of its boundary values reduces to this special case by composition with a Möbius transformation mapping the disk conformally onto itself and sending the origin to the point z.

In the general case, the Lebesgue dominated convergence theorem applied to the functions $f(rz)$ with $r \to 1$ implies that we may assume f without zeros and poles on the boundary. Finally, the claim is easily obtained by multiplying $f(z)$ by the finite Blaschke product $\prod\{R(z-a)/(R^2 - \overline{a}z)\}^{-\mathrm{ord}_a(f)}$ for $|a| < R$ and applying the formula to the new function (note that the product has absolute value 1 on $|z| = R$). For more details, we refer to W. K. Hayman [**149**], p.1 and [**8**], p.208. $\square$

The special case of the Poisson–Jensen formula in which $z = 0$ is particularly important to us and we proceed to rewrite it as follows. First, we introduce some notation. For a real-valued function $F(r)$, $r > 0$, quantitative estimates such as $F(r) = O(\log r)$ are always meant with respect to $r \to \infty$. Also we define

$$F^+(r) := \max\{F(r), 0\} \quad , \quad F^-(r) := -\min\{F(r), 0\},$$

so that $F(r) = F^+(r) - F^-(r)$.

Definition 13.2.2. *For $a \in \mathbb{C}$ and $r > 0$, the* ***enumerating function*** *is defined by*

$$n(r, a, f) := \sum_{|z|<r} \operatorname{ord}_z^+(f - a),$$

the number of solutions of $f(z) = a$ in the disk $|z| < r$ counted with their multiplicity. For $a = \infty$, we replace f by $1/f$ to get

$$n(r, \infty, f) := \sum_{|z|<r} \operatorname{ord}_z^-(f),$$

the number of poles of $f(z)$ in the disk $|z| < r$ counted with their multiplicity.

13.2.3. This function is too irregularly behaved and a logarithmic average

$$\int^r \frac{n(t, a, f)}{t} \,\mathrm{dt}$$

is a much better function which also arises in other contexts. However, care must be taken if $f(z) - a$ vanishes at the origin, because then the integral diverges at $r = 0$.

Definition 13.2.4. *For $a \in \mathbb{C}$ and $r > 0$, the* ***counting function*** *is defined by*

$$\begin{aligned} N(r, a, f) &:= \int_0^r \frac{n(t, a, f) - \operatorname{ord}_0^+(f - a)}{t} \,\mathrm{dt} + \operatorname{ord}_0^+(f - a) \log r \\ &= \operatorname{ord}_0^+(f - a) \log r + \sum_{0<|z|<r} \operatorname{ord}_z^+(f - a) \log \left|\frac{r}{z}\right|. \end{aligned}$$

For $a = \infty$, we replace f by $1/f$ and ∞ by 0 to get

$$\begin{aligned} N(r, \infty, f) &:= \int_0^r \frac{n(t, \infty, f) - \operatorname{ord}_0^-(f)}{t} \,\mathrm{dt} + \operatorname{ord}_0^-(f) \log r \\ &= \operatorname{ord}_0^-(f) \log r + \sum_{0<|z|<r} \operatorname{ord}_z^-(f) \log \left|\frac{r}{z}\right|. \end{aligned}$$

Remark 13.2.5. The modification in the definition of $N(r, a, f)$ if $\operatorname{ord}_0(f - a) \neq 0$ is, in Hayman's words, a tiresome but minor irritation of the theory. Its effect is to replace quantities such as $f(0)$ by $c(f, 0)$, the leading coefficient of the Laurent series of $f(z)$ at $z = 0$.

On the other hand, the function $N(r, a, f)$ so defined is perfectly suited for a compact reformulation of the Poisson–Jensen formula at $z = 0$, which is **Jensen's formula**:

Proposition 13.2.6. *Let*

$$c(f, 0) := \lim_{z \to 0} f(z) z^{-\operatorname{ord}_0(f)}$$

be the leading coefficient in the Laurent series of f at 0. Then

$$\frac{1}{2\pi}\int_0^{2\pi} \log\left|f(re^{i\theta})\right| d\theta + N(r,\infty,f) - N(r,0,f) = \log|c(f,0)|.$$

Proof: This is the special case $z = 0$ of the Poisson–Jensen formula (13.1) on page 445 applied to $z^{-\mathrm{ord}_0(f)}f(z)$, as we verify using the definition of $N(r,a,f)$ and the general equation $F(r) = F^+(r) - F^-(r)$. □

Definition 13.2.7. *The **proximity function** is*

$$m(r,a,f) := \frac{1}{2\pi}\int_0^{2\pi} \log^+ \frac{1}{\left|f(re^{i\theta}) - a\right|}\, d\theta.$$

For $a = \infty$, we replace f by $1/f$ to get

$$m(r,\infty,f) := \frac{1}{2\pi}\int_0^{2\pi} \log^+\left|f(re^{i\theta})\right| d\theta.$$

Remark 13.2.8. The counting function is a logarithmically weighted degree of the zero divisor of $f - a$ on the open disk $D(r) := \{z \in \mathbb{C} \mid |z| < r\}$. The proximity function is a logarithmic average on the boundary $\partial D(r)$ measuring how close $f(z)$ is to a. Note that $m(r,a,f) < \infty$ because values $f(re^{i\theta}) = a$ lead only to integrable logarithmic singularities on $\partial D(r)$.

Remark 13.2.9. Suppose that f is an entire function. Then the proximity function and $\log\|f\|_r$, where $\|f\|_r := \max_{|z|\le r}|f(z)|$, are comparable. In one direction, we have

$$m(r,\infty,f) \le \log^+\|f\|_r.$$

In the other direction, we apply (13.1) on page 445, noting that for $|z| = r < R$ we have

$$\left|\frac{R^2 - \bar{a}z}{R(z-a)}\right| \ge 1$$

and

$$0 \le \Re\left(\frac{Re^{i\theta} + z}{Re^{i\theta} - z}\right) \le \frac{R+r}{R-r}.$$

Since $f(z)$ is entire, we have $\mathrm{ord}_a(f) \ge 0$ and we conclude that

$$\log|f(z)| \le \frac{R+r}{R-r}\, m(R,\infty,f).$$

Setting $R = 2r$, we get

$$\log^+\|f\|_r \le 3m(2r,\infty,f). \tag{13.2}$$

Thus in the case of entire functions the proximity function at $a = \infty$ plays the same role as the logarithm of the maximum modulus. This is not so for meromorphic functions and we owe to R. Nevanlinna the discovery of a quantity that can take the place of the logarithm of the maximum modulus in the general case. This is expressed in his **first main theorem**.

Theorem 13.2.10. *For $a \in \mathbb{C}$ the following formula holds:*

$$m(r,a,f)+N(r,a,f) = m(r,\infty,f)+N(r,\infty,f) - \log|c(f-a,0)| + \epsilon(r,a,f),$$

with $|\epsilon(r,a,f)| \le \log^+|a| + \log 2$.

Proof: Immediate from Jensen's formula in 13.2.6 applied to $f-a$, noting first that $\log|f| = \log^+|f| - \log^-|f|$, $\mathrm{ord}_a(f) = \mathrm{ord}_a^+(f) - \mathrm{ord}_a^-(f)$, and then that $\left|\log^+|f-a| - \log^+|f|\right| \le \log^+|a| + \log 2$. □

In 13.4.9, we will give the argument in a more general setting in the higher-dimensional case.

Since $m(r,a,f) + N(r,a,f)$ turns out to be independent of a up to a bounded function, this suggests the following definition:

Definition 13.2.11. *The* ***characteristic function*** *of f is*

$$T(r,f) := m(r,\infty,f) + N(r,\infty,f).$$

This function turns out to be well behaved as a function of r.

Example 13.2.12. Let $f(z)$ be a polynomial of degree d. Then the fundamental theorem of algebra shows

$$N(r,a,f) = d\log r + O(1), \quad m(r,a,f) = O(1)$$

for $a \ne \infty$ and $r \to \infty$, the implied constant in the $O(1)$ symbol depends on f and a. For $a = \infty$, we have $N(r,\infty,f) = 0$, $m(r,\infty,f) = d\,\log r + O(1)$, hence $T(r,f) = d\,\log r + O(1)$.

The following interesting result, **Cartan's formula**, is an easy consequence of Jensen's formula.

Proposition 13.2.13. *Let $C := \log^+|f(0)|$ if $f(0) \ne \infty$ and $C := \log|c(f,0)|$ if $f(0) = \infty$. Then*

$$T(r,f) = \frac{1}{2\pi}\int_0^{2\pi} N(r,e^{i\theta},f)\,\mathrm{d}\theta + C.$$

Proof: We assume first that $f(0) \ne \infty$. By Jensen's formula in 13.2.6, if $f(0) \ne e^{i\theta}$ we have

$$\frac{1}{2\pi}\int_0^{2\pi} \log\left|f(re^{i\phi}) - e^{i\theta}\right|\mathrm{d}\phi + N(r,\infty,f) = N(r,e^{i\theta},f) + \log\left|f(0) - e^{i\theta}\right|.$$

Integrating with respect to θ, using Fubini's theorem and

$$\frac{1}{2\pi}\int_0^{2\pi} \log\left|c - e^{i\theta}\right|\mathrm{d}\theta = \log^+|c| \tag{13.3}$$

(apply Jensen's formula to $c - z$), we get Cartan's formula.

If $f(0) = \infty$, we have to replace $\log |f(0) - e^{i\theta}|$ above by $\log |c(f,0)|$ and the proof follows along the same lines. □

Corollary 13.2.14. *The Nevanlinna characteristic $T(r, f)$ is an increasing convex function of $\log r$.*

Proof: Note that $N(r, e^{i\theta}, f)$ is an increasing convex function of $\log r$, hence so is its mean value in θ. □

Remark 13.2.15. Hence $T(r, f)$ is increasing but it has not to be strictly increasing. For example, if $f\{|z| \le r\} \subset \{|w| < 1\}$, then $T(r, f) = 0$.

Corollary 13.2.16. *The proximity function is bounded on average on circles*

$$\frac{1}{2\pi}\int_0^{2\pi} m(r, a + Re^{i\theta}, f)\,\mathrm{d}\theta \le \log 2 + \log^+(1/R).$$

Proof: Replacing f by $f - a$ we may assume that $a = 0$. Next, note that we have a general inequality $m(r, Rb, f) \le m(r, b, f/R) + \log^+(1/R)$, as we readily see using $\frac{1}{f-Rb} = \frac{1}{R}\frac{1}{(f/R)-b}$ and the definition of proximity function. Hence we may assume that $R = 1$. Now we set $a = e^{i\theta}$ in the first main theorem, take its mean value with respect to θ, and conclude applying Cartan's formula and equation (13.3). □

Proposition 13.2.17. *If f is not a constant, then $T(r, f)$ is unbounded. If*

$$\liminf_{r\to\infty} \frac{T(r, f)}{\log r} < +\infty,$$

then f is a rational function.

Proof: For the second statement, the hypothesis entails that $T(r_j, f) = O(\log r_j)$ along an unbounded sequence $(r_j)_{j\in\mathbb{N}}$, hence $N(r_j, \infty, f) = O(\log r_j)$ and we conclude that f has only finitely many poles. Hence there is a polynomial $Q(z)$ such that $Q(z)f(z)$ is an entire function. By Example 13.2.12, we also have

$$m(r_j, \infty, Qf) \le m(r_j, \infty, Q) + m(r_j, \infty, f) = O(\log r_j).$$

Using (13.2) on page 447, we conclude that Qf is an entire function with

$$\sup_{|z|\le r_j} |Qf|(z) \ll r_j^d$$

for some $d \in \mathbb{N}$. The Cauchy inequalities applied to $\{|z| \le r_j\}$ show that Qf is a polynomial of degree $\le d$. This proves the second statement.

If $T(r, f)$ is bounded, then the above argument shows that f has no poles, hence we may choose $Q = 1$ and $d = 0$ proving that f is constant. □

The next example gives the converse of Proposition 13.2.17.

Example 13.2.18. Let f be a non-constant rational function. Then there are co-prime polynomials P, Q with $f(z) = P(z)/Q(z)$. Recall that the degree of f, considered as a finite morphism, is given by $\deg(f) = \max\{\deg(P), \deg(Q)\}$. Similarly as in Example 13.2.12, we may compute directly

$$N(r, a, f) = \deg(P - aQ) \log r + O(1),$$
$$m(r, a, f) = (\deg(f) - \deg(P - aQ)) \log r + O(1)$$

for $a \neq \infty$ and

$$N(r, \infty, f) = \deg(Q) \log r + O(1),$$
$$m(r, \infty, f) = (\deg(f) - \deg(Q)) \log r + O(1).$$

This illustrates the first main theorem and shows that, if f is a non-constant rational function, then $T(r, f) = \deg(f) \log r + O(1)$.

Definition 13.2.19. *The **order** $\rho(f)$ of a meromorphic function f is*

$$\rho(f) := \limsup_{r \to \infty} \frac{\log T(r, f)}{\log r}.$$

13.2.20. For meromorphic functions f, g not identically zero, we have

$$T(r, fg) \leq T(r, f) + T(r, g), \quad T(r, g) = T(r, 1/g) + \log |c(g, 0)|.$$

The first identity is obtained from the analogous relations for the proximity and counting functions. The second identity follows immediately from Jensen's formula. We conclude that

$$\rho(fg) \leq \max\{\rho(f), \rho(g)\}, \quad \rho(f/g) \leq \max\{\rho(f), \rho(g)\}.$$

13.2.21. The function $N(r, a, f)$ is a measure of the vanishing of $f - a$ in the circle $\{|z| < r\}$, counting multiplicity. It is very important to measure this multiplicity, in other words the ramification of f, and this is done as follows.

By the Weierstrass factorization theorem ([**8**], Ch.5, Th.7), there are entire functions f_0, f_1 without common zeros such that $f = f_1/f_0$.

The **Wronskian** $W(f_0, f_1)$ of $(f_0 : f_1)$ is

$$W(f_0, f_1) := \det \begin{pmatrix} f_0 & f_1 \\ f_0' & f_1' \end{pmatrix}$$

and we define

$$N_{\mathrm{ram}}(r, f) := N(W(f_0, f_1), 0, r).$$

This function is an average measure of the ramification of f in the disk $\{|z| < r\}$ and is independent of the choice of f_0, f_1 with $f = f_1/f_0$, as long as f_0 and f_1 have no common zeros. For $z \in \mathbb{C}$ and $a := f(z)$, let $m(z)$ be the multiplicity of z in the equation $f(z) = a$. Then it is easy to show that

$$\operatorname{ord}_z(W(f_0, f_1)) = m(z) - 1.$$

Therefore, we have

$$N_{\mathrm{ram}}(r,f) = N(r,0,f') + 2\,N(r,\infty,f) - N(r,\infty,f'). \qquad (13.4)$$

13.2.22. It is also clear how to define functions $n_{\mathrm{ram}}(r,a,f)$ and $N_{\mathrm{ram}}(r,a,f)$. The only difference consists in taking $\max\{\mathrm{ord}_z^+(f-a)-1,0\}$ instead of ord_z^+ $(f-a)$ in the definitions of $n(r,a,f)$ and $N(r,a,f)$ (if $a=\infty$ use $\max\{\mathrm{ord}_z^-$ $(f)-1,0\}$ instead of $\mathrm{ord}_z^-(f)$). Then we have

$$N_{\mathrm{ram}}(r,f) = \sum_{a\in\mathbb{P}^1_{\mathrm{an}}} N_{\mathrm{ram}}(r,a,f),$$

where the sum on the right is actually a finite sum for each r.

Fundamental in Nevanlinna theory is the **lemma on the logarithmic derivative.**

Lemma 13.2.23. *The estimate*

$$m(r,\infty,f'/f) = O(\log T(r,f)) + O(\log r)$$

holds for r outside a set E of finite Lebesgue measure.

The proof is elementary, but rather lengthy and certainly not obvious. We refer to the literature, see for example [**149**], Th.2.2 or W. Cherry and Z. Ye [**64**], Ch.3.

We have seen in Corollary 13.2.16 that the average of the proximity function $m(r,a,f)$ on a circle of radius R is at most $\log 2 + \log^+(1/R)$. This means that values of a for which $m(r,a,f)$ is large and comparable with $T(r,f)$ must be quite exceptional and we expect $N(r,a,f)/T(r,f)$ to be usually near to 1. The **second main theorem** of Nevanlinna theory makes this observation quantitative.

Theorem 13.2.24. *Let $a_1,\dots,a_q$ be different elements of $\mathbb{P}^1_{\mathrm{an}} = \mathbb{C}\cup\{\infty\}$. Then*

$$\sum_{j=1}^{q} m(r,a_j,f) + N_{\mathrm{ram}}(r,f) \le 2T(r,f) + O(\log T(r,f)) + O(\log r)$$

outside of a set E of finite Lebesgue measure. Moreover, if f has finite order the result holds for all large r without exception.

Proof: This is a consequence of the lemma on the logarithmic derivative as we will show more generally for the higher-dimensional case, in Theorem 13.4.16. A complete proof will be given in Section 13.3. □

Remark 13.2.25. Lang drew attention to the significance of the error term, pointing out that it has a definite counterpart in metric diophantine approximation (see S. Lang [**171**], p.199). The interested reader will find in [**64**] a thorough analysis of the error term, including numerically explicit estimates. So far, no significant application to the theory of meromorphic functions in the plane has come from these refinements.

Example 13.2.26. Let $f(z) = \exp(z)$. Clearly, we have $N(r,0,f) = N(r,\infty,f) = 0$. Further, we compute

$$m(r,\infty,f) = \frac{1}{2\pi}\int_0^{2\pi} \log^+ |e^{re^{i\theta}}|\,\mathrm{d}\theta = \frac{1}{2\pi}\int_{-\pi/2}^{\pi/2} r\cos\theta\,\mathrm{d}\theta = \frac{r}{\pi}$$

and hence $T(r,f) = r/\pi$. Similarly, we get $m(r,0,f) = r/\pi$. For $a \notin \{0,\infty\}$, the set of solutions of $f(z) = a$ is $\log a + 2\pi i\mathbb{Z}$, with multiplicity 1. From this it is easy to verify that

$$N(r,a,f) = \frac{r}{\pi} + O(\log r)$$

and a more accurate analysis shows that $N(r,a,f) = \pi^{-1}r + O(1)$ (see [**149**], p.7 or [**64**], Prop.1.6.1). Hence from the first main theorem we get $m(r,a,f) = O(1)$.

Definition 13.2.27. *The* ***defect*** *of f at a is*

$$\delta(a,f) := \liminf_{r\to\infty} \frac{m(r,a,f)}{T(r,f)}\,.$$

The ***ramification defect*** *of f at a is*

$$\theta(a,f) := \liminf_{r\to\infty} \frac{N_{\mathrm{ram}}(r,a,f)}{T(r,f)}\,.$$

Also we define

$$\Theta(a,f) := \liminf_{r\to\infty} \frac{m(r,a,f) + N_{\mathrm{ram}}(r,a,f)}{T(r,f)} \geq \delta(a,f) + \theta(a,f).$$

As an aside, note that $\Theta(a,f) > \delta(a,f) + \theta(a,f)$ may occur even for functions of finite order (contrary to what has been written in [**64**], p.125).

Since $N_{\mathrm{ram}}(r,f) = \sum_a N_{\mathrm{ram}}(r,a,f)$, as an almost immediate consequence of the second main theorem we have the celebrated **defect inequality** of Nevanlinna:

Theorem 13.2.28. *Let f in $\mathbb{C}$ be a meromorphic function. Then*

$$\sum_{a\in\mathbb{P}^1_{\mathrm{an}}} \{\delta(a,f) + \theta(a,f)\} \leq \sum_{a\in\mathbb{P}^1_{\mathrm{an}}} \Theta(a,f) \leq 2.$$

Moreover, $\delta(a,f) + \theta(a,f) \leq 1$ for every $a \in \mathbb{P}^1_{\mathrm{an}}$.

Proof: If f is not a rational function, we know that $T(r,f)/\log r \to \infty$ (see Proposition 13.2.17) and the result is immediate from the second main theorem. If instead f is a non-constant rational function, then Example 13.2.18 shows that the only deficient value is $a = f(\infty)$ with

$$\delta(a,f) = \frac{m(\infty)}{\deg(f)} \leq 1,$$

where $m(\lambda)$ is the multiplicity of $f(z) = a$ at $z = \lambda$. Hurwitz's theorem in B.4.6 applied to the morphism $f : \mathbb{P}^1_{\mathbb{C}} \to \mathbb{P}^1_{\mathbb{C}}$ yields

$$-2 = -2\deg(f) + \sum_{\lambda \in \mathbb{P}^1_{\mathbb{C}}} \{m(\lambda) - 1\},$$

where $m(\lambda) - 1$ is the multiplicity of the ramification divisor of f at the point $z = \lambda$. It is a simple exercise to rewrite this as

$$\sum_{a \in \mathbb{P}^1_{\mathbb{C}}} \{\delta(a, f) + \theta(a, f)\} = 2 - \frac{1}{\deg(f)}, \qquad (13.5)$$

hence the defect inequality continues to hold *a fortiori* for all non-constant rational functions. The last claim follows from $N_{\text{ram}}(r, a, f) \leq N(r, a, f)$ (use 13.2.22) and the first main theorem. □

Remark 13.2.29. The term $-1/\deg(f)$ in the right-hand side of (13.5) arises because we are dealing with a covering $f : \mathbb{C} \to \mathbb{P}^1_{\text{an}}$, while Hurwitz's theorem deals with the covering $f : \mathbb{P}^1_{\text{an}} \to \mathbb{P}^1_{\text{an}}$.

Remark 13.2.30. From the definition and the first main theorem, we have

$$1 - \delta(a, f) = \limsup_{r \to \infty} \frac{N(r, a, f)}{T(r, f)}.$$

The defect inequality implies that $\delta(a, f) = 0$ except for a countable set of values a, hence $\delta(a, f)$ measures the failure for $N(r, a, f)$ to be close to $T(r, f)$. This explains the terminology "defect" for $\delta(a, f)$ and **deficient value** for a whenever $\delta(a, f) > 0$.

Remark 13.2.31. The notion of deficient value is not invariant by translation. There are examples of meromorphic functions $f(z)$ such that $\delta(f(z), a) \neq \delta(f(z - z_0), a)$ may occur for a suitable translation z_0. This quite pathological phenomenon is associated to either infinite order or extremely irregular behaviour of $T(r, f)$ and does not occur if $T(r + 1, f)/T(r, f) \to 1$ as $r \to \infty$ (see R. Nevanlinna **[221]**, Ch.X, §2, 230).

Remark 13.2.32. In Example 13.2.26, the deficient values are $a = 0, \infty$, both with $\delta(a, f) = 1$. Thus Theorem 13.2.28 is sharp.

Example 13.2.33. Let $\wp(z)$ be the Weierstrass elliptic function associated to the elliptic curve $y^2 = 4x^3 - g_2 x - g_3 = 4(x - e_1)(x - e_2)(x - e_3)$ (see 8.3.10). Define $e_0 = \infty$. Then $\wp(z)$ is a doubly periodic meromorphic function of order 2 with no deficient values.

A doubly periodic meromorphic function on $\mathbb{C}$ is called **elliptic**. Let f be a non-constant elliptic function with fundamental domain Ω of the period lattice Λ. The order d of f is the number of solutions $z \in \Omega$, counted with multiplicity, of the equation $f(z) = a$ (the reader should be careful not to confuse this definition of order with the order of f as a meromorphic function, which is always $\rho(f) = 2$ as we will show below).

An application of the argument principle shows that the order does not depend on the choice of a. An easy covering argument shows

$$n(r,a,f) = \frac{\pi d}{\mathrm{vol}(\Omega)} \cdot r^2 + O(r)$$

and hence

$$N(r,a,f) = \frac{\pi d}{2\mathrm{vol}(\Omega)} \cdot r^2 + O(r).$$

Note that the bounds may be chosen independently of a, therefore Cartan's formula (Proposition 13.2.13) gives

$$T(r,f) = \frac{\pi d}{2\mathrm{vol}(\Omega)} \cdot r^2 + O(r).$$

By Remark 13.2.30, no deficient values occur for an elliptic function f.

Now we specialize $f = \wp$. Since $\wp$ has only the pole $0 \in \Omega$ and the pole is of order two, we infer that the order of $\wp$ is 2 as claimed at the beginning and which is, by accident, equal to $\rho(\wp)$. Moreover, $\wp'$ is an odd elliptic function of order 3 with the triple pole in 0, hence the set of zeros of $\wp'$ is $(\frac{1}{2}\Lambda)\setminus\Lambda$ and they are all simple. We conclude that $\frac{1}{2}\Lambda$ is the set of ramification points of $\wp$. They are the solutions of the equations $\wp(z) = e_j$ $(j = 0, \ldots, 3)$ which have only double roots. A geometric argument as above gives

$$N_{\mathrm{ram}}(r, e_j, \wp) = \frac{\pi}{2\mathrm{vol}(\Omega)} \cdot r^2 + O(r).$$

We conclude that $\theta(e_j, \wp) = \frac{1}{2}$ and hence $\sum \theta(e_j, \wp) = 2$ giving again an extremal example of Theorem 13.2.28.

Consider the function $z = z(w)$ defined by

$$z = \int_0^w (t-a)^{\frac{1}{m}-1}(t-b)^{\frac{1}{n}-1}(t-c)^{\frac{1}{p}-1}\,dt,$$

where a, b, c are distinct real numbers and m, n, p are integers such that $\frac{1}{m}+\frac{1}{n}+\frac{1}{p} = 1$. Then $z(w)$ maps the upper half plane biholomorphically onto an open triangle in the z-plane with angles π/m, π/n, π/p at the vertices $z(a), z(b), z(c)$. (Look first at the boundary. This is a special case of the Schwarz–Christoffel formula, see [**152**], Th.17.6.1, p.374.) The Schwarz reflection principle ([**8**], Th.6.5.26) proves that the image of the multi-valued function $z(w)$ with ramification points a, b, c is the whole plane. In fact, the possible values for (m, n, p) are $(2,3,6)$, $(2,4,4)$, $(3,3,3)$ up to ordering and the image triangles of the closed upper/lower plane cover $\mathbb{C}$. The inverse function $f(z) = w$ is a meromorphic elliptic function (Schwarz reflection principle again) with no deficient values and similar arguments as above show that

$$\theta(a,f) = 1 - \frac{1}{m}, \quad \theta(b,f) = 1 - \frac{1}{n}, \quad \theta(c,f) = 1 - \frac{1}{p},$$

providing other extremal examples of Theorem 13.2.28 (see [**221**], Ch.X, §3, 236).

Remark 13.2.34. The following **Nevanlinna inverse problem** was completely solved by D. Drasin, in a major paper [**92**]. Let $\{(a_j, \delta_j, \theta_j)\}$, $j = 1, 2, \ldots$ be any finite or countable sequence with distinct $a_j \in \mathbb{C} \cup \{\infty\}$, non-negative real numbers δ_j, θ_j satisfying $0 < \delta_j + \theta_j \le 1$ and $\sum_j (\delta_j + \theta_j) \le 2$. Then there is a meromorphic function f with $\{a_j\}$ as its set of deficient values and $\delta(a_j, f) = \delta_j$, $\theta(a_j, f) = \theta_j$. In general, such a function will have infinite order but the growth can always be bounded by $T(r, f) < r^{\omega(r)}$, with $\omega(r) \to \infty$ arbitrarily slowly. For finite order, there are additional restrictions. For example, a meromorphic function of order 0 can have at most one deficient value (this is an old result of Valiron from 1925, see **[149]**, Th.4.10, p.110). This extends the result, pointed out in the proof of Theorem 13.2.28, that $f(\infty)$ is the only possible deficient value for a rational function. According to Drasin (*loc.cit.*), his method also shows that the restricted inverse problem with the additional condition $\delta_j = 0$ for every j can always be solved with f of order 0.

Another deep result of D. Drasin **[93]**, solving a long-standing conjecture of F. Nevanlinna, is that if the sum of the deficiencies of a meromorphic function f of finite order ρ is $\sum \delta(a, f) = 2$, then 2ρ is an integer ≥ 2, $\rho\,\delta(a, f)$ is a positive integer, and every possibility can actually occur (see R. Nevanlinna **[220]**, p. 357, or L. Ahlfors **[4]**, p.406).

Meromorphic functions of finite order must also satisfy additional conditions. For example, a result of A. Weitsman **[330]** shows that $\sum \delta(a, f)^{1/3} < \infty$ whenever f has finite order.

13.2.35. Picard's little theorem, stating that a meromorphic function $f : \mathbb{C} \to \mathbb{P}^1_{\mathrm{an}}$ omitting three values is a constant, follows immediately from Theorem 13.2.28. Indeed, suppose that a is an omitted value. Then $N(r, a, f) = 0$ and $\delta(a, f) = 1$ by Remark 13.2.30. The defect inequality in Theorem 13.2.28 now shows that a non-constant f has at most two omitted values in $\mathbb{P}^1_{\mathrm{an}}$.

Example 13.2.36. Finally, we show that the analogue of the abc-conjecture in Nevanlinna theory is well known (as in the function field case). We begin by defining the analogue of the radical for a meromorphic function. We start with the notion of the truncated counting function:

Definition 13.2.37. *Let f be a non-constant meromorphic function and let $a \in \mathbb{C}$. The **truncated counting function** $N^{(1)}(r, a, f)$ is*

$$N^{(1)}(r, a, f) := \min\{1, \mathrm{ord}_0^+(f - a)\} \log r + \sum_{0<|z|<r} \min\{1, \mathrm{ord}_z^+(f - a)\} \log \left|\frac{r}{z}\right|.$$

For $a = \infty$ we define

$$N^{(1)}(r, \infty, f) := \min\{1, \mathrm{ord}_0^-(f)\} \log r + \sum_{0<|z|<r} \min\{1, \mathrm{ord}_z^-(f)\} \log \left|\frac{r}{z}\right|.$$

We have defined in Chapter 12 the radical of a non-zero integer m to be $\mathrm{rad}(m) = \prod_{p|m} p$, the product of all distinct primes dividing m. More generally, we define the radical of a rational number m/n (with m, n coprime) to be $\mathrm{rad}(m/n) = \mathrm{rad}(mn)$, the product of the distinct primes appearing in the factorization of the rational number. We define the following logarithmic analogue:

Definition 13.2.38. *Let f be a non-constant meromorphic function. The* ***conductor*** *of f in $|z| \le r$ is by definition*

$$\mathrm{cond}(r, f) := N^{(1)}(r, 0, f) + N^{(1)}(r, \infty, f) = N(r, \infty, f'/f).$$

The following result can be viewed as the analogue of an ***abc*-theorem for meromorphic functions.**

Theorem 13.2.39. *Let f, g be non-constant meromorphic functions such that*

$$f + g = 1.$$

Then

$$T(r, f) \le \mathrm{cond}(r, fg) + O(\log T(r, f)) + O(\log r)$$

for all r outside of a set E of finite Lebesgue measure. Moreover, if f has finite order, then we may choose E bounded.

Proof: We apply the second main theorem to f, with the three points $\{a_1, a_2, a_3\} = \{0, 1, \infty\}$. We infer that

$$\sum_{j=1}^{3} m(r, a_j, f) + N_{\mathrm{ram}}(r, f) \le 2T(r, f) + O(\log T(r, f)) + O(\log r) \quad (13.6)$$

for r outside of an exceptional set E of finite Lebesgue measure. Moreover, the set E is bounded if f has finite order.

By the first main theorem, we have

$$\sum_{j=1}^{3} \{m(r, a_j, f) + N(r, a_j, f)\} = 3T(r, f) + O(1)$$

and combining this with (13.6) we get

$$T(r, f) + N_{\mathrm{ram}}(r, f) \le \sum_{j=1}^{3} N(r, a_j, f) + O(\log T(r, f)) + O(\log r)$$

for $r \notin E$. In view of 13.2.22, we verify that

$$\begin{aligned} \sum_{j=1}^{3} N(r, a_j, f) &\le \sum_{j=1}^{3} N^{(1)}(r, a_j, f) + N_{\mathrm{ram}}(r, f) \\ &= \mathrm{cond}(r, fg) + N_{\mathrm{ram}}(r, f), \end{aligned}$$

the last step because $N^{(1)}(r, 1, f) = N^{(1)}(r, 0, g)$ and f, g have the same poles. The theorem follows by combining the last two displayed inequalities. ☐

13.3. Variations on a theme: the Ahlfors–Shimizu characteristic

There is a very nice geometric definition for a slightly different characteristic function, due independently to L. Ahlfors [**3**] and T. Shimizu [**281**]. It leads to a theory equivalent to Nevanlinna's, but also to simpler and more elegant proofs, well motivated by underlying geometric concepts. We still denote by $f : \mathbb{C} \to \mathbb{P}^1_{\text{an}}$ a non-constant meromorphic function.

13.3.1. We begin by recalling the **stereographic projection** and its main properties. Let S be the Riemann sphere of diameter 1 in $\mathbb{R}^3$, lying on the Gauss plane $\mathbb{C}$ with coordinates $z = x + iy$ and touching the plane at the origin 0. Let N denote the North Pole on S.

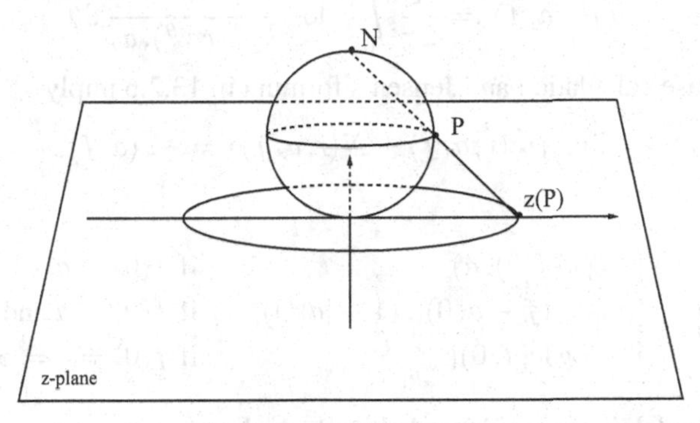

the stereographic projection

The stereographic projection maps a point $P \neq N$ on S to the point $z(P)$ in the Gauss plane $\mathbb{C}$ intersection with the line joining N and P. This map is conformal, i.e. preserving angles, and transforms circles into circles or lines. If $P \to N$, then $z(P) \to \infty$ and S becomes a model of the projective complex line $\mathbb{C} \cup \{\infty\} = \mathbb{P}^1_{\mathbb{C}}$. We define

$$k(z_1, z_2) := \begin{cases} \frac{|z_1 - z_2|}{\sqrt{1+|z_1|^2}\,\sqrt{1+|z_2|^2}} & \text{if } z_1, z_2 \in \mathbb{C}, \\ \frac{1}{\sqrt{1+|z|^2}} & \text{if } \{z_1, z_2\} = \{z, \infty\}. \end{cases}$$

Its interpretation is that, if P, Q, are two points on S and $z(P)$ and $z(Q)$ are their images in $\mathbb{C} \cup \{\infty\}$ by the stereographic projection, then the euclidean **chordal distance** $\|P - Q\|$ between P and Q is given by

$$\|P - Q\| = k(z(P), z(Q)).$$

This induces a distance function on $\mathbb{P}^1_{\text{an}}$. For the elements of area, we easily compute that the pull-back of the **Fubini–Study form** ω, given in affine coordinates w by

$$\omega(z) = \frac{i}{2\pi} \frac{dz \wedge d\overline{z}}{(1 + |z|^2)^2},$$

with respect to the stereographic projection, is equal to $\pi^{-1}\,d\sigma$, where $d\sigma$ is the euclidean spherical area form on S. The normalization is such that ω induces a probability measure on $\mathbb{P}^1_{\mathrm{an}}$. Let us consider the euclidean form

$$\Phi = \frac{i}{2\pi}\,dz \wedge d\bar{z} = \frac{1}{\pi}\,dx \wedge dy$$

on $\mathbb{C}$. Using the notation of 13.2.21, an easy computation shows that

$$f^*\omega = \frac{|f'|^2}{(1+|f|^2)^2}\,\Phi = \frac{|W(f_0,f_1)|^2}{(|f_0|^2+|f_1|^2)^2}\,\Phi. \tag{13.7}$$

Now we define the **spherical proximity function** $\mathring{m}(r,a,f)$ by

$$\mathring{m}(r,a,f) := \frac{1}{2\pi}\int_0^{2\pi} \log\frac{1}{k(f(re^{i\theta}),a)}\,d\theta. \tag{13.8}$$

A case-by-case calculation and Jensen's formula in 13.2.6 imply

$$\lim_{r\to 0+}\{\mathring{m}(r,a,f) + N(r,a,f)\} = -\nu(a,f), \tag{13.9}$$

where

$$\nu(a,f) = \begin{cases} \log k(f(0),a) & \text{if } f(0)\neq a,\\ \log(|c(f-a,0)|/(1+|a|^2)) & \text{if } f(0)=a \text{ and } a\neq\infty,\\ -\log|c(f,0)| & \text{if } f(0)=a=\infty. \end{cases}$$

Since the chordal distance is bounded by 1, we have

$$\mathring{m}(r,a,f) \geq 0.$$

With the new proximity function, the first main theorem takes the following very elegant form:

Theorem 13.3.2. *For every $a,\,b \in \mathbb{P}^1_{\mathrm{an}}$ it holds*

$$\mathring{m}(r,a,f) + N(r,a,f) + \nu(a,f) = \mathring{m}(r,b,f) + N(r,b,f) + \nu(b,f).$$

Definition 13.3.3. *The quantity $\mathring{T}(r,f) := \mathring{m}(r,a,f) + N(r,a,f) + \nu(a,f)$, which is independent of a, is the **Ahlfors–Shimizu characteristic**. It has been normalized so that $\lim_{r\to 0+}\mathring{T}(r,f) = 0$.*

Proof of theorem: It is enough to verify the claim for $a \in \mathbb{C}$ and $b=\infty$. By (13.8) and Jensen's formula in 13.2.6, we have

$$\begin{aligned}\mathring{m}(r,a,f) - \mathring{m}(r,\infty,f) &= \frac{1}{2\pi}\int_0^{2\pi}\log\frac{\sqrt{1+|a|^2}}{|f(re^{i\theta})-a|}\,d\theta\\ &= N(r,\infty,f) - N(r,a,f) - \log|c(f-a,0)| + \log\sqrt{1+|a|^2}.\end{aligned}$$

Let $\mathring{T}(r, a, f)$ denote the left-hand side in the theorem. The above shows that $\mathring{T}(r, a, f) - \mathring{T}(r, \infty, f)$ is constant in r. On the other hand, (13.9) shows that its limit for $r \to 0+$ is equal to 0, hence $\mathring{T}(r, a, f) = \mathring{T}(r, \infty, f)$. □

Lemma 13.3.4. *Let η be an integrable 2-form on $\mathbb{P}^1_{\mathrm{an}}$. Then*

$$\int_{|z| \leq r} f^* \eta = \int_{\mathbb{P}^1_{\mathrm{an}}} n(r, a, f)\, \eta(a).$$

Proof: Let γ_r be the curve $f(\{|z| = r\})$. Then the local mapping principle shows that the restriction of f to $\{|z| < r\} \setminus f^{-1}(\gamma_r)$ is a ramified covering. For a non-ramified $a \in \mathbb{P}^1_{\mathrm{an}} \setminus \gamma_r$, the number of sheets over a is equal to $n(r, a, f)$ and hence the transformation formula of integrals implies the claim. □

Remark 13.3.5. As a corollary of proof, we note that $n(r, a, f)$ is a locally constant function in $a \in \mathbb{P}^1_{\mathrm{an}} \setminus \gamma_r$. The same argument shows that $n(r, a, f)$ is a locally constant function in (r, a) outside of the closed set $\{(r, a) \mid r > 0, a \in \gamma_r\}$ of measure zero.

Theorem 13.3.6.

$$\mathring{T}(r, f) = \int_0^r \int_{|z| \leq t} f^* \omega \,\frac{\mathrm{d}t}{t}.$$

Proof: Since the chordal distance is continuous and bounded by 1, the spherical proximity function $\mathring{m}(r, a, f)$ is a non-negative measurable function on $\mathbb{R}_+ \times \mathbb{P}^1_{\mathrm{an}}$. Interchanging the order of integration using Fubini's theorem and recalling that the chordal distance is rotation invariant, we get

$$\int_{\mathbb{P}^1_{\mathrm{an}}} \mathring{m}(r, a, f)\, \omega(a) = \int_{\mathbb{P}^1_{\mathrm{an}}} \log\left(\frac{1}{k(f(0), a)}\right) \omega(a) = -\int_0^{2\pi} \nu(a, f)\, \omega(a).$$

This is clearly finite and so we may integrate the equation

$$\mathring{T}(r, f) = \mathring{m}(r, a, f) + N(r, a, f) + \nu(a, f)$$

on $\mathbb{P}^1_{\mathrm{an}}$ with respect to ω to get

$$\mathring{T}(r, f) = \int_{\mathbb{P}^1_{\mathrm{an}}} N(r, a, f)\, \omega(a).$$

By Remark 13.3.5, the integrand in the definition of $N(r, a, f)$ is a non-negative measurable function on $\mathbb{R}_+ \times \mathbb{P}^1_{\mathrm{an}}$ and hence Fubini's theorem implies

$$\mathring{T}(r, f) = \int_0^r \int_{\mathbb{P}^1_{\mathrm{an}}} n(t, a, f)\, \omega(a) \,\frac{\mathrm{d}t}{t}.$$

By Lemma 13.3.4, we get the claim including the convergence of the integral. □

Remark 13.3.7. By Lemma 13.3.4, $\int_{|z|\le r} f^*\omega$ is equal to the area of $f(\{|z| \le r\})$ on the sphere with respect to the spherical area form ω, counted with the sheet multiplicity. Hence $\mathring{T}(r,f)$ is a logarithmic average of the growth of the spherical area covered by f on disks of increasing radius.

Remark 13.3.8. Note that $\int_{|z|\le r} f^*\omega$ is a continuous positive strictly increasing function equal to $r\frac{d}{dr}\mathring{T}(r,f)$. We conclude that the Ahlfors–Shimizu characteristic $\mathring{T}(r,f)$ is a positive convex strictly increasing function of $\log(r)$ and hence $\lim_{r\to\infty} \mathring{T}(r,f) = \infty$. Moreover, using (13.7) on page 458 and polar coordinates, we easily deduce that $\mathring{T}(r,f)$ is a C^∞-function of r.

13.3.9. We have

$$\mathring{m}(r,\infty,f) = \frac{1}{2\pi}\int_0^{2\pi} \log\sqrt{1+\left|f(re^{i\theta})\right|^2}\,\mathrm{d}\theta,$$

hence

$$0 \le \mathring{m}(r,\infty,f) - m(r,\infty,f) \le \frac{1}{2}\log 2.$$

This proves

$$\nu(\infty,f) \le \mathring{T}(r,f) - T(r,f) \le \nu(\infty,f) + \frac{1}{2}\log 2$$

and hence $\mathring{T}(r,f)$ and $T(r,f)$ differ by a bounded function. We will prove this again in 13.4.8, in a general setting valid in higher dimensions.

Now we give L. Ahlfors's proof [5] of the second main theorem, in the following form.

Theorem 13.3.10. *Let f be a non-constant meromorphic function on $\mathbb{C}$ and $a_1,\dots,a_q$ be distinct points in $\mathbb{P}^1_{\mathrm{an}}$. Let $k \ge -1$ be a given real number and let $\varepsilon > 0$. Then*

$$\sum_{j=1}^{q} \mathring{m}(r,a_j,f) + N_{\mathrm{ram}}(r,f) < 2\mathring{T}(r,f) + (1+\varepsilon)\log\mathring{T}(r,f) + (k+\varepsilon)\log r,$$

for $r \to \infty$ outside of an open set E such that $\int_E t^k\,\mathrm{d}t < \infty$. Moreover, if f has finite order ρ, the result holds with the right-hand side $2\mathring{T}(r,f) + (2\rho - 1 + \varepsilon)\log r$, for all large r without exception.

Remark 13.3.11. The error term is sharp for $\rho = 0$. If we apply it to a non-constant rational function, hence $\rho = 0$, we recover Hurwitz's theorem in the form (13.5) on page 453. For $\rho > 0$, the error term was improved by Z. Ye to the sharp $(\rho - 1 + \varepsilon)\log r$, see [64], Th.4.3.1 and Z. Ye [334].

Proof: We start with a natural generalization of the spherical proximity function. Let us consider an absolutely continuous probability measure μ on $\mathbb{P}^1_{\text{an}}$, with associated density ρ given by $\mathrm{d}\mu = \rho\omega$.

The local mapping principle implies that $n(r, a, f)$ is a non-negative lower semi-continuous function in (r, a) (see Remark 13.3.5), which is bounded for bounded r. Therefore $n(r, a, f)$ is μ-integrable and we may define the enumerating function

$$n_\mu(r, f) := \int_{\mathbb{P}^1_{\text{an}}} n(r, a, f) \, \mathrm{d}\mu(a) < \infty.$$

We also define a non-negative proximity function

$$\mathring{m}_\mu(r, f) := \int_{\mathbb{P}^1_{\text{an}}} \mathring{m}(r, a, f) \, \mathrm{d}\mu(a).$$

Let $r > r_0 > 0$, where r_0 is fixed and will be chosen later. We assume that $\mathring{m}_\mu(r_0, f) < \infty$. Then the first main theorem in 13.3.2 yields

$$\mathring{m}_\mu(r, f) - \mathring{m}_\mu(r_0, f) + \int_{r_0}^{r} n_\mu(s, f) \frac{\mathrm{d}s}{s} = \mathring{T}(r, f) - T(r_0, f) \tag{13.10}$$

proving that $\mathring{m}_\mu(r, f) < \infty$. We have

$$\begin{aligned} n_\mu(r, f) &= \int_{\mathbb{P}^1_{\text{an}}} n(r, a, f) \, \mathrm{d}\mu(a) = \int_{|z| \le r} f^*(\rho\omega) \\ &= \frac{1}{\pi} \int_0^r \int_0^{2\pi} \frac{|f'(se^{i\theta})|^2}{\left(1 + |f(se^{i\theta})|^2\right)^2} \rho(f(se^{i\theta})) \, s \, \mathrm{d}s \, \mathrm{d}\theta, \end{aligned} \tag{13.11}$$

where we have used Lemma 13.3.4 and the last step was obtained by (13.7) on page 458 writing $f^*(\rho\omega)$ in polar coordinates.

By (13.7) on page 458, it is clear that $|f'|^2/(1 + |f|^2)^2$ is a C^∞-function on $\mathbb{C}$ and we define

$$\lambda(r) := \frac{1}{2\pi} \int_0^{2\pi} \frac{|f'(re^{i\theta})|^2}{(1 + |f(re^{i\theta})|^2)^2} \rho(f(re^{i\theta})) \, \mathrm{d}\theta.$$

Using (13.11) and the definition of λ in (13.10), we infer that

$$2 \int_{r_0}^{r} \int_0^s \lambda(t) \, t \, \mathrm{d}t \frac{\mathrm{d}s}{s} \le \mathring{T}(r, f) + \mathring{m}_\mu(r_0; f). \tag{13.12}$$

Now we proceed to the key step, namely obtaining a lower bound for $\lambda(r)$. To this end, we use **Jensen's inequality** (see [**252**], Th.3.3) applied to log which generalizes the inequality between arithmetic and geometric means:

Lemma 13.3.12. *If $F(x)$ is positive and integrable in the interval $[a,b]$, then*

$$\log\left(\frac{1}{b-a}\int_a^b F(x)\,\mathrm{d}x\right) \geq \frac{1}{b-a}\int_a^b \log F(x)\,\mathrm{d}x.$$

Hence from the definition of $\lambda(r)$ and Jensen's inequality we get

$$\frac{1}{2\pi}\int_0^{2\pi} \log\frac{|f'(re^{i\theta})|^2}{(1+|f(re^{i\theta})|^2)^2}\,\mathrm{d}\theta + \frac{1}{2\pi}\int_0^{2\pi}\log\rho(f(re^{i\theta}))\,\mathrm{d}\theta \leq \log\lambda(r). \tag{13.13}$$

The usefulness of this inequality lies in the decoupling of f and ρ resulting from it. The idea of using it in this context goes back to F. Nevanlinna.

We evaluate the first integral in (13.13) by splitting it as

$$\frac{1}{2\pi}\int_0^{2\pi} 2\,\log|f'(re^{i\theta})|\,\mathrm{d}\theta - \frac{1}{2\pi}\int_0^{2\pi} 4\,\log\sqrt{1+|f(re^{i\theta})|^2}\,\mathrm{d}\theta.$$

The result is

$$2N(r,0,f') - 2N(r,\infty,f') + 2\log|c(f',0)| - 4\mathring{m}(r,\infty,f),$$

as we verify by applying Jensen's formula in 13.2.6 to $\frac{1}{2\pi}\int\log|f'(re^{i\theta})|\,\mathrm{d}\theta$. By the first main theorem in the Ahlfors–Shimizu form, we also have $\mathring{m}(r,\infty,f) = \mathring{T}(r,f) - N(r,\infty,f) - \nu(\infty,f)$, hence the first term in the left-hand side of (13.13) is

$$\begin{aligned}2N(r,0,f') + 4N(r,\infty,f) - 2N(r,\infty,f') - 4\mathring{T}(r,f)\\ + 2\,\log|c(f',0)| + 4\nu(\infty,f).\end{aligned}$$

We have already noted in (13.4) on page 451 that

$$N_{\mathrm{ram}}(r,f) = N(r,0,f') + 2N(r,\infty,f) - N(r,\infty,f').$$

Therefore, we have proved the important inequality

$$\frac{1}{2\pi}\int_0^{2\pi}\log\rho(f(re^{i\theta}))\,\mathrm{d}\theta - 4\mathring{T}(r,f) + 2N_{\mathrm{ram}}(r,f) \leq \log\lambda(r) + O(1). \tag{13.14}$$

To complete the proof of the second main theorem, we combine the lower bound for $\lambda(r)$ given by (13.14) with the upper bound (13.12), together with a careful choice for μ.

The upper bound (13.12) controls only a certain double average of $\lambda(r)$, while we want a pointwise bound. We can achieve this by excluding an exceptional set of values of r. A simple way is as follows. We write for brevity

$$A(r) := \int_0^r \lambda(t)t\,\mathrm{d}t, \qquad B(r) := \int_{r_0}^r A(s)\,\frac{\mathrm{d}s}{s}.$$

Note that A is a non-negative increasing continuous function and $B(r)$ is a non-negative increasing C^1-function. We may choose $r_0 > 0$ such that $A(r_0) > 0$. Note that $rB'(r) = A(r)$ and hence $B(r)$ is a strictly increasing function of $\log r$ for $r \geq r_0$.

We fix a parameter $k \geq -1$ and $\varepsilon > 0$ and define

$$\begin{aligned} E_1 &:= \{r > r_0 \mid \lambda(r) > r^{k-1}A(r)^{1+\varepsilon}\} \\ E_2 &:= \{r > r_0 \mid A(r) > r^{k+1}B(r)^{1+\varepsilon}\} \\ E &:= E_1 \cup E_2. \end{aligned}$$

Let S be the set of $r > r_0$ such that A is strictly increasing in a neighbourhood of r. Clearly, S is an open subset of (r_0, ∞) and $E_1 \setminus S$ is a set of measure zero. Then

$$\int_{E_1} r^k \, \mathrm{d}r \leq \int_{E_1 \setminus S} \frac{r\lambda(r)}{A(r)^{1+\varepsilon}} \, \mathrm{d}r = \int_{E_1 \setminus S} \frac{\mathrm{d}A(r)}{A(r)^{1+\varepsilon}} \leq \frac{A(r_0)^{-\varepsilon}}{\varepsilon}$$

and in a similar way

$$\int_{E_2} r^k \, \mathrm{d}r = \int_{E_2} r^{k+1} \frac{\mathrm{d}B(r)}{A(r)} \leq \int_{E_2} \frac{\mathrm{d}B(r)}{B(r)^{1+\varepsilon}} \leq \frac{B(r_0)^{-\varepsilon}}{\varepsilon}.$$

Thus for $r > r_0$ and outside of the set $E := E_1 \cup E_2$ we have by (13.12) on page 461

$$\begin{aligned} \lambda(r) \leq r^{k-1}A(r)^{1+\varepsilon} &\leq r^{k-1+(k+1)(1+\varepsilon)}B(r)^{(1+\varepsilon)^2} \\ &\leq r^{2k+(k+1)\varepsilon}\left(\frac{1}{2}\mathring{T}(r,f) + O(1)\right)^{(1+\varepsilon)^2} \end{aligned}$$

and moreover

$$\int_E r^k \, \mathrm{d}r < \frac{1}{\varepsilon}\left(A(r_0)^{-\varepsilon} + B(r_0)^{-\varepsilon}\right).$$

We plug this upper bound for $\lambda(r)$ into (13.14) and easily obtain

$$\begin{aligned} \frac{1}{2\pi}\int_0^{2\pi} \log \rho(f(re^{i\theta}))\, \mathrm{d}\theta \leq 4\mathring{T}(r,f) - 2N_{\mathrm{ram}}(r,f) + (1+\varepsilon)^2 \log \mathring{T}(r,f) \\ + \{2k + (k+1)\varepsilon\}\log r + O(1) \end{aligned} \tag{13.15}$$

for $r \notin E \cup [0, r_0]$. The $O(1)$ term depends *a priori* on f, μ, and r_0 but is independent of r and ε.

Let $a_1, \ldots, a_q$ be distinct points in $\mathbb{P}^1_{\mathrm{an}}$. We define a non-negative function

$$R(a) := \sum_{j=1}^{q} \log \frac{1}{k(a, a_j)}$$

and set

$$\log \rho(a) := 2R(a) - \alpha \log(1 + R(a)) + K, \tag{13.16}$$

where $\alpha > 1$ and K is a constant such that $d\mu = \rho\omega$ is a probability measure on $\mathbb{P}^1_{\mathrm{an}}$. This choice is indeed admissible, because the singularities of ρ occur only at $a = a_j$. If we write $t = a - a_j$, $t = |t|e^{i\theta}$ for $a_j \neq \infty$, then in a neighborhood of $t = 0$ we have

$$\rho(a_j + t) = e^{O(1)}|t|^{-2}\left(\log \frac{1}{|t|}\right)^{-\alpha},$$

while $\omega(a_j + t)$ is comparable with $|t|\, d|t|\, d\theta$. Since $\alpha > 1$, then this and a similar calculation for $a_j = \infty$ show that ρ is integrable.

We claim that for this μ we can choose r_0 such that $\mathring{m}_\mu(r_0, f) < \infty$. In order to verify this, it suffices to have $a_j \notin f(\{|z| = r_0\})$ for $j = 1, \ldots, q$, which is obviously the case for almost every r_0. Indeed, the potential function

$$P(z) = \int_{\mathbb{P}^1_{\mathrm{an}}} \log \frac{1}{k(z, a)}\, d\mu(a)$$

is bounded for z away from $\{a_j \mid j = 1, \ldots, q\}$ and our claim follows noting that $\mathring{m}_\mu(r_0, f) = \frac{1}{2\pi}\int_0^{2\pi} P(f(r_0 e^{i\theta}))\, d\theta$.

By definition

$$\frac{1}{2\pi}\int_0^{2\pi} R(f(re^{i\theta}))\, d\theta = \sum_{j=1}^{q} \mathring{m}(r, a_j, f).$$

Moreover, Jensen's inequality yields

$$\begin{aligned}
\frac{1}{2\pi}\int_0^{2\pi} \log\left(1 + R(f(re^{i\theta}))\right) d\theta &\leq \log\left(\frac{1}{2\pi}\int_0^{2\pi} (1 + R(f(re^{i\theta})))\, d\theta\right) \\
&\leq \log\left(1 + \sum_{j=1}^{q} \mathring{m}(r, a_j, f)\right) \\
&\leq \log^+ \mathring{T}(r, f) + \log q + O(1).
\end{aligned}$$

Putting everything together, we conclude from our choice (13.16) that

$$\frac{1}{2\pi}\int_0^{2\pi} \log \rho(f(re^{i\theta}))\, d\theta \geq 2\sum_{j=1}^{q} \mathring{m}(r, a_j, f) - \alpha \log^+ \mathring{T}(r, f) - O(1). \tag{13.17}$$

By Remark 13.3.8, we know that $\mathring{T}(r, f)$ is a strictly increasing unbounded function and so we may assume $\mathring{T}(r, f) \geq 1$ for $r \geq r_0$.

If we combine (13.17) and (13.15) on page 463, we find

$$2\sum_{j=1}^{q}\mathring{m}(r,a_j,f) \le 4\mathring{T}(r,f) - 2N_{\mathrm{ram}}(r,f)$$
$$+\{\alpha+(1+\varepsilon)^2\}\log\mathring{T}(r,f)+\{2k+(k+1)\varepsilon\}\log r+O(1)$$

for r outside of a set E with $\int_E r^k\,\mathrm{d}r < +\infty$. Note that we may choose E open because outside of a countable set of $r \ge r_0$ without accumulation point, our function $\lambda(r)$ is continuous, hence E_1 is open and E_2 is open anyway. Since ε can be taken arbitrarily small and $\alpha > 1$ can be taken arbitrarily close to 1, we get the first statement of the second main theorem.

Finally, suppose that f has finite order ρ, hence $\mathring{T}(r,f) = O(r^{\rho+\varepsilon})$. We fix $k > \rho - 1$ and we may choose $\varepsilon > 0$ with $k > \rho - 1 + \varepsilon$. Then by the first main theorem it suffices to show that

$$(q-2)\mathring{T}(r,f) + N_{\mathrm{ram}}(r,f) \le \sum_{j=1}^{q} N(r,a_j,f) + (\rho+k)\log r$$

holds for sufficiently large r. This certainly holds outside of the set E. Now the left-hand side of this inequality is an increasing function of r, hence if $r \in E$ we have

$$(q-2)\mathring{T}(r,f) + N_{\mathrm{ram}}(r,f) \le \sum_{j=1}^{q} N(r_1,a_j,f) + (\rho+k)\log r_1 \qquad (13.18)$$

for any $r_1 > r$ with $r_1 \notin E$. On the other hand, since $\int_E r^k\,\mathrm{d}r < \infty$, we can always find such an r_1 with $r_1 < r + r^{-k}$ if r is large enough. This gives $\log(r_1/r) < r^{-k-1}$. Further, for any a and $r < R$, we have

$$n(r,a,f)\log\frac{R}{r} \le N(R,a,f) - N(r,a,f) \le n(R,a,f)\log\frac{R}{r}.$$

If we take $R = er_1$, we get *a fortiori* $n(r_1,a,f) \le \mathring{T}(er_1,f)+O(1) = O(r^{\rho+\varepsilon})$. If we take $R = r_1$, we get $N(r_1,a,f) - N(r,a,f) = O(r^{\rho+\varepsilon-k-1}) = o(1)$, because $k > \rho - 1 + \varepsilon$. This shows that having r_1 in the right-hand side of (13.18) has no effect on the final result and completes the proof. □

13.4. Holomorphic curves in Nevanlinna theory

In this section, we consider a holomorphic map $f : \mathbb{C} \to X$ into a complex variety X. We generalize the concepts from the case $X = \mathbb{P}^1_{\mathrm{an}}$ considered in 13.2. After introducing characteristic functions for metrized line bundles completely analogous to Section 2.7, the generalization of Nevanlinna's first main theorem corresponds to Weil's theorem in 2.7.14. The extension of Nevanlinna's second main theorem to holomorphic curves f remains conjectural, but the case of $\mathbb{P}^n_{\mathrm{an}}$ with D the sum of hyperplanes in general position was proved in Cartan's thesis. We deduce this special case from the lemma on the logarithmic derivative using Wronskian techniques. The case of a curve X is also well known and we finish by describing the generalization replacing $\mathbb{C}$ by finite ramified analytical coverings, which find their arithmetical counterpart in finite extensions of number fields. At the end, we give a reformulation of the abc-theorem for meromorphic functions, which makes the analogy to the abc-conjecture clearer.

For this section, the reader is assumed to be familiar with the analytic theory of complex varieties (see Section A.14 for an introduction and [**130**] for more details).

13.4.1. Let s be an invertible meromorphic section of a line bundle L giving rise to a Cartier divisor $D = \mathrm{div}(s)$. We suppose that $f(\mathbb{C})$ is not contained in the support of D. We use the notation $\mathrm{ord}_Y(D) := \mathrm{ord}_Y(s)$ to denote the multiplicity of D in the prime divisor Y.

Definition 13.4.2. *For $r > 0$, the **counting function** is defined by*

$$N_{f,D}(r) := \mathrm{ord}_0(f^*D)\log r + \sum_{0<|z|<r} \mathrm{ord}_z(f^*D)\log\left|\frac{r}{z}\right|$$

13.4.3. Let $\| \ \|$ be a continuous metric on L. Then

$$\lambda_D(x) := -\log \|s(x)\|$$

is a local height with respect to D (see Section 2.7). It is continuous outside of the support of D and has logarithmic singularities along D. For notational simplicity, we omit here the reference to the metric.

Definition 13.4.4. *The **proximity function** is defined by*

$$m_{f,D}(r) := \frac{1}{2\pi}\int_0^{2\pi} \lambda_D \circ f(re^{i\theta})\, d\theta$$

*and the **characteristic function** is*

$$T_{f,D}(r) := N_{f,D}(r) + m_{f,D}(r).$$

Example 13.4.5. Let $X = \mathbb{P}^n_{\mathrm{an}}$ and let $L = O_{\mathbb{P}^n}(1)$. We denote the coordinates on $\mathbb{P}^n_{\mathrm{an}}$ by $x_0, \dots, x_n$. We consider a global section $\ell(\mathbf{x}) = \ell_0 x_0 + \cdots + \ell_n x_n$ of $O_{\mathbb{P}^n}(1)$, whose divisor is a hyperplane H. On L, we choose the standard metric given by

$$\|\ell(\mathbf{x})\| = \frac{|\ell(\mathbf{x})|}{\max\{|x_0|, \dots, |x_n|\}}.$$

A holomorphic map $f : \mathbb{C} \to \mathbb{P}^n_{\mathrm{an}}$ is always given by entire functions $f_0, \dots, f_n$ without common zeros such that $f_0(z), \dots, f_n(z)$ are the homogeneous coordinates of $f(z)$. This follows from the Weierstrass factorization theorem ([**8**], Ch.5, Th.7). Assuming $f(\mathbb{C}) \not\subset H$, we verify that

$$N_{f,H}(r) = N(r, 0, \ell \circ \mathbf{f})$$

and

$$m_{f,H}(r) = \frac{1}{2\pi}\int_0^{2\pi} \log\max\{|f_0(re^{i\theta})|,\ldots,|f_n(re^{i\theta})|\}\,\mathrm{d}\theta - \frac{1}{2\pi}\int_0^{2\pi} \log|\ell\circ\mathbf{f}(re^{i\theta})|\,\mathrm{d}\theta.$$

Since $N(r,\infty,\ell\circ\mathbf{f})=0$, Jensen's formula in 13.2.6 proves

$$T_{f,H}(r) = \frac{1}{2\pi}\int_0^{2\pi} \log\max\{|f_0(re^{i\theta})|,\ldots,|f_n(re^{i\theta})|\}\,\mathrm{d}\theta - \log|c(\ell\circ\mathbf{f},0)|.$$

Example 13.4.6. We specialize further in Example 13.4.5 by setting $n=1$. Let $\ell(\mathbf{x})=x_0$ and hence $H=[\infty]$ on $\mathbb{P}^1_{\mathrm{an}}$. Then we have

$$N_{f,H}(r) = N(r,\infty,f)$$

and

$$m_{f,H}(r) = m(r,\infty,f),$$

where on the right-hand side, we have the quantities from the definitions in 13.2.4 and 13.2.7. We conclude that $T_{f,H}(r)=T(r,f)$.

Example 13.4.7. Let us consider again the case $n=1$ taking now $\ell(\mathbf{x})=x_1-ax_0$ for some $a\in\mathbb{C}$, hence $H=[a]$ on $\mathbb{P}^1_{\mathrm{an}}$. Again, we have

$$N_{f,H}(r) = N(r,a,f).$$

To get the old proximity function from Definition 13.2.4, we use the new metric

$$\|x_1-ax_0\| = \frac{|x_1-ax_0|}{\max\{|x_0|,|x_1-ax_0|\}}$$

on $L=O_{\mathbb{P}^1}(1)$. We easily compute from Example 13.4.5 that

$$m_{f,H}(r) = m(r,a,f).$$

Example 13.4.8. Going back to $X=\mathbb{P}^n_{\mathrm{an}}$, let us consider the **Fubini–Study metric**

$$\|\ell(\mathbf{x})\| = \frac{|\ell(\mathbf{x})|}{(|x_0|^2+\cdots+|x_n|^2)^{1/2}}$$

on $L=O_{\mathbb{P}^n}(1)$. Using the standard differential operator

$$\mathrm{d}^c = \frac{1}{4\pi i}(\partial-\overline{\partial}),$$

the Chern form

$$\mathrm{dd}^c\log\|\ell(\mathbf{x})\|^{-2} = \omega$$

is the generalization to higher dimension of the Fubini–Study form from 13.3.1. First, we assume $f(0) \notin H = \operatorname{div}(\ell(\mathbf{x}))$. We compute the proximity function

$$\begin{aligned} m_{f,H}(r) &= -\frac{1}{2\pi}\int_0^{2\pi} \log \|\ell \circ \mathbf{f}(re^{i\theta})\| \, d\theta \\ &= -\int_0^r \frac{\partial}{\partial t}\left(\frac{1}{2\pi}\int_0^{2\pi} \log \|\ell \circ \mathbf{f}(te^{i\theta})\| \, d\theta\right) dt - \log \|\ell \circ \mathbf{f}(0)\|. \end{aligned}$$

The differential operator d^c is given in polar coordinates $z = re^{i\theta}$ by

$$d^c g = \frac{1}{4\pi}\left(r\frac{\partial g}{\partial r}\, d\theta - \frac{\partial g}{\partial \theta}\frac{dr}{r}\right)$$

for any differentiable function g. Therefore, by interchanging differentiation and integration, we get

$$m_{f,H}(r) = -\int_0^r \int_{|z|=t} d^c \log \|\ell \circ \mathbf{f}(z)\|^2 \, \frac{dt}{t} - \log \|\ell \circ \mathbf{f}(0)\|.$$

By Stokes's theorem, we have

$$\int_{|z|=t} d^c \log \|\ell \circ \mathbf{f}(z)\|^2 = \int_{|z|\le t} dd^c \log \|\ell \circ \mathbf{f}(z)\|^2 + n(t, 0, \ell \circ \mathbf{f}),$$

where the enumerating function $n(t, 0, g)$ is the number of zeros, counted with multiplicities, of the holomorphic function $g = \ell \circ \mathbf{f}$ in the open disk $\{|z| < t\}$. By

$$f^*\omega = dd^c \log \|\ell \circ \mathbf{f}\|^{-2}$$

and

$$\int_0^r n(g, 0, t)\, \frac{dt}{t} = N(r, 0, g),$$

we get

$$m_{f,H}(r) = \int_0^r \int_{|z|\le r} f^*\omega \, \frac{dt}{t} - N(r, 0, \ell \circ \mathbf{f}) - \log \|\ell \circ \mathbf{f}(0)\|.$$

This proves our final result

$$T_{f,H}(r) = \int_0^r \int_{|z|\le r} f^*\omega \, \frac{dt}{t} - \log \|\ell \circ \mathbf{f}(0)\| \tag{13.19}$$

provided $f(0) \notin H$. Note that

$$T_f^\omega(r) := \int_0^r \int_{|z|\le r} f^*\omega \, \frac{dt}{t}$$

does not depend on the choice of ℓ. It is called the **Ahlfors–Shimizu characteristic** of f generalizing the construction in 13.3.3. The equivalence with $T_{f,H}(r)$ is a consequence of (13.19).

If $f(0) \in H$, then $T_{f,H}$ is still equivalent to T_f^ω by using the second part of the following **first main theorem** in the higher-dimensional case.

Theorem 13.4.9. *Let L be a line bundle on the compact complex variety X with an invertible meromorphic section s and a continuous metric $\| \ \|$ giving rise to the characteristic function $T_{f,\mathrm{div}(s)}(r)$. For a holomorphic map $f : \mathbb{C} \to X$ with $f(\mathbb{C}) \not\subset |\mathrm{div}(s)|$, the following results hold:*

(a) *If we replace the metric $\| \ \|$ by a continuous metric $\| \ \|'$ getting the characteristic function $T'_{f,\mathrm{div}(s)}$, then*

$$T_{f,\mathrm{div}(s)}(r) - T'_{f,\mathrm{div}(s)}(r) = O(1)$$

with bound independent of f.

(b) *If we replace s by the invertible meromorphic section s' such that $f(\mathbb{C}) \not\subset |\mathrm{div}(s')|$, then*

$$T_{f,\mathrm{div}(s)}(r) - T_{f,\mathrm{div}(s')}(r) = \log |c((s'/s) \circ f, 0)|.$$

Proof: Changing the metric does not influence the counting function. For the corresponding proximity functions, we get

$$m_{f,\mathrm{div}(s)}(r) - m'_{f,\mathrm{div}(s)}(r) = \frac{1}{2\pi} \int_0^{2\pi} \log \frac{\|s\|'}{\|s\|} \circ f(re^{i\theta}) \, d\theta.$$

Now $\|s\|'/\|s\|$ is a continuous function on X without zeros. By compactness of X, we conclude that $\log \frac{\|s\|'}{\|s\|}$ is bounded proving (a).

Now we exchange s. Then we have

$$m_{f,\mathrm{div}(s)}(r) - m_{f,\mathrm{div}(s')}(r) = \frac{1}{2\pi} \int_0^{2\pi} \log \left| \frac{s'}{s} \circ f(re^{i\theta}) \right| \, d\theta$$

and

$$N_{f,\mathrm{div}(s)}(r) - N_{f,\mathrm{div}(s')}(r) = \\ - \mathrm{ord}_0 \left(\frac{s'}{s} \circ f \right) \log r - \sum_{0<|z|<r} \mathrm{ord}_0 \left(\frac{s'}{s} \circ f \right) \log \left| \frac{r}{z} \right|.$$

By Jensen's formula from 13.2.6, we get (b). □

Remark 13.4.10. Considering f as fixed, we see that the characteristic function is determined by the isomorphism class of L up to bounded functions. This makes the analogy with Theorem 2.3.8 in diophantine geometry perfect. In particular, we denote by $T_{f,L}$ any characteristic function, it is determined up to $O(1)$. Moreover, the proof shows that $m_{f,D}$ is determined by the Cartier divisor D up to a bounded quantity.

Proposition 13.4.11. *Let D_1, D_2 be Cartier divisors on the compact complex variety X and let $f : \mathbb{C} \to X \setminus (\mathrm{supp}(D_1) \cup \mathrm{supp}(D_2))$ be a holomorphic map. Then*

$$m_{f,D_1+D_2} = m_{f,D_1} + m_{f,D_2} + O(1), \quad N_{f,D_1+D_2} = N_{f,D_1} + N_{f,D_2}$$

and

$$T_{f,D_1+D_2} = T_{f,D_1} + T_{f,D_2} + O(1).$$

Proof: This follows as for local heights. The details are left to the reader. □

Proposition 13.4.12. *Let $\varphi : X \to X'$ be a holomorphic map of compact complex varieties and let L' be a line bundle on X'. Then*

$$T_{\varphi\circ f,L'} = T_{f,\varphi^* L'} + O(1).$$

Moreover, if s' is an invertible meromorphic section of L' with $D' := \mathrm{div}(s')$ such that neither $f(\mathbb{C})$ nor any component of X are mapped into $\mathrm{supp}(D')$, then

$$N_{\varphi\circ f,D'} = N_{f,\varphi^* D'}, \quad m_{\varphi\circ f,D'} = m_{f,\varphi^* D'} + O(1).$$

Proof: Using the pull-back metric from L', all claims are immediately clear from the definitions and hold even without the $O(1)$ term. The first main theorem implies the claim in general. □

Proposition 13.4.13. *If the line bundle L on the compact complex variety X is generated by global sections, then $T_{f,L}$ is bounded below.*

Proof: If $X = \mathbb{P}^n_{\mathrm{an}}$ and $L = O_{\mathbb{P}^n}(1)$, then, by choosing the Fubini–Study metric and $s = x_j$ with $f(0) \notin \mathrm{div}(x_j)$, it is clear that $m_{f,\mathrm{div}(s)} \geq 0$ and $N_{f,\mathrm{div}(s)} \geq 0$. In general, L is a pull-back of a line bundle $O_{\mathbb{P}^n}(1)$ (cf. A.6.8) and the claim follows from Proposition 13.4.12. □

Concerning the generalization of the second main theorem, not much is known apart from the linear case in $\mathbb{P}^n_{\mathrm{an}}$ which we study next.

Definition 13.4.14. *The set $\{H_1, \dots, H_q\}$ of hyperplanes in $\mathbb{P}^n_{\mathbb{C}}$ is said to be in **general position** if any sublist of less than $n+2$ hyperplanes corresponds to linearly independent linear forms. Equivalently, for any subset $I \subset \{1, \dots, q\}$ of cardinality $|I| \leq n+1$, we have*

$$\dim\Big(\bigcap_{i\in I} H_i\Big) = n - |I|.$$

13.4.15. Let $D := H_1 + \dots + H_q$ be the associated divisor. Let $f : \mathbb{C} \to \mathbb{P}^n_{\mathrm{an}}$ be a holomorphic map whose image is not contained in any hyperplane of $\mathbb{P}^n_{\mathrm{an}}$. By the Weierstrass factorization theorem ([8], Ch.5, Th.7), we have $f = (f_0 : \cdots : f_n)$ for entire functions $f_0, \dots, f_n$ without common zero. We define the **Wronskian**

of $(f_0 : \cdots : f_n)$ by

$$W(f_0, \ldots, f_n) = \det \begin{pmatrix} f_0 & f_1 & \cdots & f_n \\ f_0' & f_1' & \cdots & f_n' \\ \cdot & \cdot & \cdots & \cdot \\ f_0^{(n)} & f_1^{(n)} & \cdots & f_n^{(n)} \end{pmatrix}.$$

Clearly this is determined by f up to entire functions without zeros, hence the counting function

$$N_{f,\mathrm{ram}}(r) := N(r, 0, W(f_0, \ldots, f_n))$$

is well defined. In this situation, we have **Cartan's second main theorem**:

Theorem 13.4.16. *For any $\varepsilon > 0$ and any given hyperplane H, the inequality*

$$m_{f,D}(r) + N_{f,\mathrm{ram}}(r) \leq (n+1)T_{f,H}(r) + O(\log T_{f,H}(r)) + O(\log r)$$

holds for all r outside of a set E of finite Lebesgue measure.

Proof: If we add additional hyperplanes to $H_1, \ldots, H_q$, then the left-hand side increases up to bounded functions (see the proof of Theorem 13.4.9). So we may assume $q \geq n+2$ and we set $p := q - n - 1$. Let $\ell_j(\mathbf{x})$ be a linear form with $H_j = \mathrm{div}(\ell_j(\mathbf{x}))$. For $K = \{k_1, \ldots, k_p\} \subset \{1, \ldots, q\}$ of cardinality p, we define

$$\ell_K(\mathbf{x}) := \ell_{k_1}(\mathbf{x}) \cdots \ell_{k_p}(\mathbf{x}).$$

We consider the morphism

$$\varphi : \mathbb{P}^n_{\mathrm{an}} \longrightarrow \mathbb{P}^{\binom{p}{q}}_{\mathrm{an}}, \quad \mathbf{x} \mapsto (\ell_K(\mathbf{x}))_{|K|=p}.$$

Using $\varphi^* O_{\mathbb{P}^{\binom{p}{q}}}(1) \cong O_{\mathbb{P}^n}(p)$, Proposition 13.4.12 implies

$$T_{\varphi \circ f, O(1)}(r) = pT_{f,H}(r) + O(1). \tag{13.20}$$

Let $I = \{i_1, \ldots, i_{n+1}\} := \{1, \ldots, q\} \setminus K$ and $g_i := \ell_i(f_0, \ldots, f_n)$. Then we have

$$W(g_{i_1}, \ldots, g_{i_{n+1}}) = d_I \cdot W(f_0, \ldots, f_n), \tag{13.21}$$

where d_I is the determinant of the $(n+1) \times (n+1)$ matrix formed with the coefficients of $(\ell_i)_{i \in I}$. We define the logarithmic Wronskian by

$$\lambda(f_0, \ldots, f_n) := \det \begin{pmatrix} 1 & 1 & \cdots & 1 \\ f_0'/f_0 & f_1'/f_1 & \cdots & f_n'/f_n \\ \cdot & \cdot & \cdots & \cdot \\ f_0^{(n)}/f_0 & f_1^{(n)}/f_1 & \cdots & f_n^{(n)}/f_n \end{pmatrix} = \frac{W(f_0, \ldots, f_n)}{f_0 \cdots f_n}.$$

From (13.21), we deduce

$$\ell_K(\mathbf{f}) = \frac{g_1 \cdots g_q}{W(f_0, \ldots, f_n)} \cdot d_I^{-1} \cdot \lambda(g_{i_1}, \ldots, g_{i_{n+1}}). \tag{13.22}$$

To compute (13.20), we will use equation (13.22). The first factor is independent of K and will be handled by Jensen's formula. The second factor is independent of r and hence contributes only a bounded amount to the height. The third factor is small by the lemma on the logarithmic derivative in 13.2.23.

The details are as follows. For a moment, we fix $r > 0$. Given a meromorphic function g on $\mathbb{C}$ and $v \in \mathbb{C}$, $|v| \leq r$, it is notationally convenient to use

$$|g|_v = \begin{cases} |g(v)| & \text{if } |v| = r, \\ \left|\frac{v}{r}\right|^{\mathrm{ord}_v(g)} & \text{if } v \neq 0, \\ r^{-\mathrm{ord}_v(g)} & \text{if } v = 0. \end{cases}$$

The functions $|\ |_v$ behave almost like absolute values and Jensen's formula may be interpretated as a product formula (see 14.2.2). Later, this notation is helpful for translating the argument to the function field case.

Let $v \in \mathbb{C}$, $|v| = r$ with $v \notin f^{-1}(D)$. Then we have

$$\begin{aligned} \log|\lambda(g_{i_1}, \ldots, g_{i_{n+1}})|_v &\leq \log \max_\pi |g_{i_1}^{(\pi_0)}/g_{i_1} \cdots g_{i_{n+1}}^{(\pi_n)}/g_{i_{n+1}}|_v + \log|(n+1)!|_v \\ &\leq \sum_{j=0}^{n} \sum_{i=1}^{q} \log^+ |g_i^{(j)}/g_i|_v + \log|(n+1)!|_v, \end{aligned} \tag{13.23}$$

where π ranges over all bijective maps of $\{0, \ldots, n\}$. By Example 13.4.5 and Jensen's formula in 13.2.6, we have

$$T_{\varphi \circ f, O(1)}(r) = \sum_{|v|<r} \max_K \log|\ell_K(\mathbf{f})|_v + \frac{1}{2\pi} \int_{|v|=r} \max_K \log|\ell_K(\mathbf{f})|_v \, \mathrm{d}\theta + O(1). \tag{13.24}$$

For $|v| < r$, we use $|\ell_K(\mathbf{f})|_v \leq 1$. For $|v| = r$, equation (13.22) implies

$$\begin{aligned} &\frac{1}{2\pi} \int_{|v|=r} \max_K \log|\ell_K(\mathbf{f})|_v \, \mathrm{d}\theta \\ &\leq \frac{1}{2\pi} \int_{|v|=r} \left(\log\left| \frac{g_1 \cdots g_q}{W(f_0, \ldots, f_n)} \right|_v - \log|d_I|_v + |\lambda(g_{i_1}, \ldots, g_{i_{n+1}})|_v \right) \mathrm{d}\theta. \end{aligned} \tag{13.25}$$

Applying the lemma on the logarithmic derivative (see Lemma 13.2.23) to

$$g_i^{(j)}/g_i = (g_i^{(j)}/g_i^{(j-1)}) \cdots (g'/g)$$

in (13.23), we get

$$\frac{1}{2\pi} \int_{|v|=r} \log|\lambda(g_{i_1}, \ldots, g_{i_{n+1}})|_v \, \mathrm{d}\theta \leq O(\log T_{f,H}(r)) + O(\log r) \tag{13.26}$$

for all r outside of a set E of finite Lebesgue measure. Here, we have to note that $T_{g_i} = O(T_{f,H})$ and similarly for the derivatives of g_i. Using (13.25) and (13.26) in (13.24), we get

$$T_{\varphi\circ f, O(1)}(r) \leq \frac{1}{2\pi} \int_{|v|=r} \log \left| \frac{g_1 \cdots g_q}{W(f_0, \ldots, f_n)} \right|_v d\theta + O(\log T_{f,H}(r)) + O(\log r)$$

for $r \notin E$. By Jensen's formula in 13.2.6, we deduce

$$\begin{aligned} T_{\varphi\circ f, O(1)}(r) &\leq \sum_{|v|<r} \log \left| \frac{W(f_0, \ldots, f_n)}{g_1 \cdots g_q} \right|_v + O(\log T_{f,H}(r)) + O(\log r) \\ &= N_{f,D}(r) - N_{f,\mathrm{ram}}(r) + O(\log T_{f,H}(r)) + O(\log r) \end{aligned}$$

outside of E. If we insert this in (13.20) on page 471, we get the claim from $T_{f,D} = m_{f,D} + N_{f,D}$. □

13.4.17. Let X be a connected projective complex manifold with an ample line bundle H. The highest exterior power $K_X := \Lambda^{\dim} T_X^*$ of the cotangent bundle is called the **canonical line bundle** of X.

We consider a **divisor** D on X **with normal crossings**, i.e. for every $x \in X$, there is a holomorphic coordinate system $\mathbf{w}$ around x such that D is given by the equation $w_1 \cdots w_k = 0$. The generalization of the second main theorem is the following **conjecture of Griffiths**:

Conjecture 13.4.18. *There is a closed algebraic subset $Z \neq X$ such that for any holomorphic map $f : \mathbb{C} \to X$ with $f(\mathbb{C}) \not\subset Z$, the estimate*

$$m_{f,D}(r) + T_{f,K_X}(r) = O(\log T_{f,H}(r)) + O(\log r)$$

holds for all r outside of a set of finite Lebesgue measure.

Remark 13.4.19. In the case $X = \mathbb{P}^n_{\mathrm{an}}$ and $D = \sum_{j=1}^q H_j$, we compare the Griffiths conjecture with Cartan's second main theorem in 13.4.16. The assumption that D has normal crossings means that the hyperplanes are in general position. For H, we may choose a hyperplane and we have $K_X = O(-(n+1)H)$ (cf. [**148**], Example II.8.20.1). If we neglect the ramification term $N_{f,W}$, then the Griffiths conjecture matches with Cartan's second main theorem up to the exceptional set Z. P. Vojta [**316**] removed this discrepancy and proved that Griffiths's conjecture holds for $X = \mathbb{P}^n_{\mathrm{an}}$ and D a sum of hyperplanes in general position, with an exceptional set Z equal to a finite union of proper linear subspaces.

13.4.20. Let $f : C \to C'$ be an analytic map of Riemann surfaces and we assume that f is not constant on any connected component of C. Then the **ramification divisor** R_f is locally given as a Cartier divisor by f', where the derivative is with respect to any charts on C and C'. Then Example B.4.8 shows that the notion agrees in the algebraic framework with the definition in B.4.4. If $g : C' \to C''$ is a further analytic map of Riemann surfaces, the chain rule gives

$$R_{g\circ f} = R_f + f^* R_g. \tag{13.27}$$

13.4.21. For a non-constant analytic map $f : \mathbb{C} \to C$ to a Riemann surface C with ramification divisor $R_f = \sum m_z[z]$, we have the counting function

$$N_{R_f}(r) = m_0 \log r + \sum_{0<|z|<r} m_z \log\left|\frac{r}{z}\right|.$$

13.4.22. Next, we prove Griffiths's conjecture for an irreducible smooth projective curve C. Let D be a sum of different points $a_1, \dots, a_q \in C$. We choose also an ample line bundle H on C.

Theorem 13.4.23. *Let $f : \mathbb{C} \to C \setminus \{a_1, \dots, a_q\}$ be a non-constant holomorphic map. Then*

$$m_{f,D}(r) + T_{f,K_C}(r) + N_{R_f}(r) = O(\log T_{f,H}(r)) + O(\log r)$$

for r outside of a set of finite Lebesgue measure.

Proof: This follows formally from Nevanlinna's second main theorem using a non-constant rational function on C viewed as a morphism $g : C \to \mathbb{P}^1_{\mathrm{an}}$ and functorial properties.

The details are as follows: We choose an effective divisor D_0 on $\mathbb{P}^1_{\mathrm{an}}$ such that $D \subset g^{-1}(D_0)$ and such that g is unramified outside of $g^{-1}(D_0)$. Let $D' := g^*(D_0)$, $D_1 := (D')_{\mathrm{red}}$ (namely, the sum of the components of D') and $D_2 := D_1 - D$. Nevanlinna's second main theorem in 13.2.24 implies

$$\begin{aligned} &T_{g\circ f,D_0}(r) + T_{g\circ f,K_{\mathbb{P}^1}}(r) + N_{R_{g\circ f}}(r) \\ &\leq N_{g\circ f,D_0}(r) + O(\log T_{g\circ f,O_{\mathbb{P}^1}(1)}(r)) + O(\log r) \end{aligned} \tag{13.28}$$

outside of a set of finite Lebesgue measure. First, note that Proposition B.4.7 implies that $R_g = D' - D_1$ and hence Propositions 13.4.11 and 13.4.12 yield

$$\begin{aligned} N_{g\circ f,D_0}(r) &= N_{f,D'}(r) = N_{f,D_1}(r) + N_{f,R_g}(r) \\ &= N_{f,D}(r) + N_{f,D_2}(r) + N_{f,R_g}(r). \end{aligned} \tag{13.29}$$

Note that (13.27) implies

$$N_{R_{g\circ f}}(r) = N_{R_f}(r) + N_{f,R_g}(r).$$

Hurwitz's theorem in B.4.5 gives

$$K_C \cong g^*(K_{\mathbb{P}^1}) \otimes O(R_g).$$

By the first main theorem in 13.4.9 and $R_g = D' - D_1$, we get

$$\begin{aligned} T_{g\circ f,D_0}(r) + T_{g\circ f,K_{\mathbb{P}^1}}(r) &= T_{f,D'}(r) + T_{f,g^*K_{\mathbb{P}^1}}(r) + O(1) \\ &= T_{f,D}(r) + T_{f,D_2}(r) + T_{f,K_C}(r) + O(1). \end{aligned}$$

Finally, for n sufficiently large, we have that $H^{\otimes n} \otimes g^* O_{\mathbb{P}^1}(-1)$ is base-point-free (cf. A.6.10) and hence Proposition 13.4.13 shows

$$T_{g\circ f, O_{\mathbb{P}^1}(1)}(r) = T_{f, g^* O_{\mathbb{P}^1}(1)}(r) + O(1) = O(T_{f,H}(r)) \tag{13.30}$$

for $r \to \infty$. Here we have to note that $T_{f,H}$ is unbounded because f is not a constant (see Proposition 13.2.17). If we use (13.29) and (13.30) in (13.28), we get easily the claim because $N_{f,D_2} \leq T_{f,D_2} + O(1)$ and $m_{f,D} + N_{f,D} = T_{f,D} + O(1)$. □

Remark 13.4.24. The generalization of the second main theorem to equidimensional non-degenerate holomorphic maps $f : \mathbb{C}^n \to X$ is due to J. Carlson and P. Griffiths [**58**]. It motivates the Griffiths conjecture. Note that the ramification term only occurs in the equidimensional setting. The error term may be improved in the equidimensional case in a similar way as in Remark 13.2.25. For details, we refer to [**174**], Ch.2, or [**249**], Ch.5.

Back to $X = C$, the case of genus $g = 0$ is just Nevanlinna's second main theorem from 13.2.24. If $g = 1$, then K_C is trivial (cf. Proposition 8.2.26 or A.13.6) and we conclude that the deficient value satisfies

$$\delta_f(a) := \liminf_{r\to\infty} \frac{m_{f,[a]}(r)}{T_{f,[a]}(r)} \leq 0$$

for every $a \in C$. The first main theorem in 13.4.9 proves $N_{f,[a]}(r) > 0$ for r sufficiently large, hence every f is surjective if the genus is $g = 1$.

For genus $g \geq 2$, K_C is ample (see A.13.6, A.13.7). Theorem 13.4.23 implies that $T_{f,[a]} = O(\log r)$. It follows from Proposition 13.2.17 that f induces an algebraic morphism $\mathbb{P}^1_{\mathbb{C}} \to C$, which is impossible by Hurwitz's theorem in B.4.6.

Remark 13.4.25. The above results for curves may be proved without Nevanlinna techniques. To get non-trivial results in the case $g \geq 1$, we may prove a similar inequality as in Theorem 13.4.23 for holomorphic maps $f : \{|z| \leq R\} \to C$ and we get a upper bound for R (cf. S. Lang and W. Cherry [**174**], p.93 or [**64**], §5.8).

13.4.26. For an effective reduced divisor D on C, we define the **truncated counting function** $N^{(1)}_{f,D}(r)$ by

$$N^{(1)}_{f,D}(r) := \min\{1, \mathrm{ord}_0^+(f^*D)\} \log r + \sum_{0<|z|<r} \min\{1, \mathrm{ord}_z^+(f^*D)\} \log \left|\frac{r}{z}\right|.$$

Lemma 13.4.27. $N^{(1)}_{f,D}(r) \geq N_{f,D}(r) - N_{R_f}(r)$

Proof: This follows from $f^*D - (f^*D)_{\mathrm{red}} \leq R_f$ proved similarly as in Proposition B.4.7 (or just count zeros locally) and from Proposition 13.4.13. □

As an immediate consequence, we obtain the **truncated second main theorem for curves.**

Theorem 13.4.28. *For every* $\varepsilon > 0$, *we have*

$$T_{f,D}(r) + T_{f,K_C}(r) \le N^{(1)}_{f,D}(r) + O(\log T_{f,H}(r)) + O(\log r).$$

Proof: This follows from $T_{f,D} = m_{f,D} + N_{f,D}$, Theorem 13.4.23 and Lemma 13.4.27. □

13.4.29. Theorem 13.4.23 holds also if we replace $\mathbb{C}$ by a finite ramified covering. In diophantine geometry, this corresponds to passing from $\mathbb{Q}$ to a number field.

Let Y be a connected Riemann surface and let $p : Y \to \mathbb{C}$ be a ramified covering. As we have seen in 12.3.8, this extends to a finite algebraic morphism $\overline{p} : \overline{Y} \to \mathbb{P}^1_{\mathbb{C}}$ of the compactifications.

Let D be a divisor on a complex variety X and let $f : Y \to X$ be a non-constant holomorphic map with image not contained in $\mathrm{supp}(D)$. For $r > 0$, we define the **counting function**

$$N_{f,D}(r) :=$$
$$\frac{1}{\deg(p)}\left(\sum_{p(y)=0} \mathrm{ord}_y(f^*D)\log r + \sum_{0<|p(y)|<r} \mathrm{ord}_y(f^*D)\log\left|\frac{r}{p(y)}\right|\right)$$

and the **proximity function,** using a continuous metric on $O(D)$ as before, is

$$m_{f,D}(r) := -\frac{1}{2\pi\deg(p)}\int_{p^{-1}\{|z|=r\}} \log\|s_D(y)\|\, p^*(d\theta).$$

Note the analogy with the normalizations in 1.3.6. They lead to the fact that the **characteristic function**

$$T_{f,D}(r) := N_{f,D}(r) + m_{f,D}(r)$$

is invariant under base change to a finite ramified covering of Y. Then the generalization of Theorem 13.4.23 is

Theorem 13.4.30. *Let* $a_1, \dots, a_q$ *be different points of a complex projective curve* C *with an ample line bundle* H *and let* $f : Y \to C \setminus \{a_1, \dots, a_q\}$ *be a holomorphic map. Then*

$$\sum_{j=1}^{q} m_{f,[a_j]}(r) + T_{f,K_C}(r) + N_{R_f}(r) - N_{R_p}(r) \le O(\log(rT_{f,H}(r)))$$

holds for all $r > 0$ *outside of a set of finite Lebesgue measure.*

This is a special case of the second main theorem of P. Griffiths and J. King [**131**] and W. Stoll [**292**] for parabolic coverings in the equidimensional setting analogous to Remark 13.4.24. For a proof, we refer to [**292**], Th.18.13E.

Remark 13.4.31. This result indicates that N_{R_p} measures the dependence on the covering p. However, to make sense of this statement, we should look for a second main theorem uniform in f. This is done explicitly by W. Cherry (cf. [**174**] and [**63**] for details) and his result implies that N_{R_p} indeed gives the contribution of the covering p to the second main theorem.

The following is a reformulation of the abc-theorem in 13.2.39:

Theorem 13.4.32. *Let f, g, h be non-constant entire functions without common zeros such that $f + g = h$ and let*

$$T(r, f, g, h) := \frac{1}{2\pi} \int_0^{2\pi} \log \max\{|f(re^{i\theta})|, |g(re^{i\theta})|, |h(re^{i\theta})|\} \, d\theta.$$

Then

$$T(r, f, g, h) \leq N^{(1)}(r, 0, fgh) + O(\log T(r, f, g, h)) + O(\log r)$$

holds for all $r > 0$ outside of a set E of finite Lebesgue measure. If f, g, h have finite order, we may choose E bounded.

Proof: By Example 13.4.5, $T(r, f, g, h)$ and $T_{(h:f:g),H}$ differ only by a bounded function for any hyperplane H of $\mathbb{P}^2_{\mathrm{an}}$. Using Proposition 13.4.12 for the map $\varphi((x_0 : x_1)) = (x_0 : x_1 : (x_0 - x_1))$ and Example 13.4.6, we conclude that the left-hand side may be replaced by $T(r, f/h)$. On the other hand, we have

$$\mathrm{cond}\left(r, \frac{f}{h} \cdot \frac{g}{h}\right) = N^{(1)}(r, 0, fgh)$$

and the first claim follows from Theorem 13.2.39 applied to the relation $\frac{f}{h} + \frac{g}{h} = 1$. If f, h are of finite order, then we have seen in 13.2.20 that f/h has also finite order and the last claim follows as well. □

13.5. Bibliographical notes

For more details about Nevanlinna theory in one variable, we refer the reader to the classic books of R. Nevanlinna [**221**] and W. K. Hayman [**149**]. If the reader is interested in a finer analysis of the error term, he is referred to W. Cherry and Z. Ye [**64**].

The foundations of the theory are given in R. Nevanlinna's article [**219**]. He proved the lemma on the logarithmic derivative to deduce his second main theorem. A little later, his brother F. Nevanlinna gave a proof of the second main theorem based on differential geometry and differential equations. Ahlfors simplified this proof further and expanded its geometric interpretation leading to the presentation in Section 13.3. In a breakthrough work [**6**], L. Ahlfors interpreted and extended Nevanlinna theory as a geometric theory of covering surfaces.

Cartan's formula appears in H. Cartan [**59**]. In his thesis (see H. Cartan [**60**]), he proved his second main theorem for hyperplanes.

References for Section 13.4 are the books of S. Lang [**170**], S. Lang and W. Cherry [**174**] and M. Ru [**249**]. For a generalized abc-theorem in Nevanlinna theory similar to Theorem 12.4.4, we refer to [**249**], Theorem A.3.2.6. The reader may also consult, in connexion with Cartan's version of Nevanlinna's theory and its applications, the expository article by G.G. Gundersen and W. K. Hayman [**142**].

R. Nevanlinna asked for a second main theorem with moving targets, i.e. the constants a_i are replaced by meromorphic functions g_i with $\log T(r, g_i) = o(\log T(r, f))$. He treated the case of three targets by elementary means, using a fractional linear transformation to reduce it to the constant case. The general case remained open for a long time. The weaker form of the second fundamental theorem without the contribution due to ramification was then obtained independently by C.F. Osgood [**232**] and N. Steinmetz [**290**], see [**249**] for details and further extensions. This was Vojta's motivation for Roth's theorem with moving targets (see Section 6.5).

Finally, in a major paper K. Yamanoi [**333**] obtained a second fundamental theorem with moving targets in full generality, with the expected contribution coming from ramification. However, the error terms here are far weaker than those in earlier works, because of the use of Ahlfors's [**6**] geometric theory of covering surfaces as a main tool.

14 THE VOJTA CONJECTURES

14.1. Introduction

Ch. Osgood [**231**], [**232**], was the first to observe, in his researches on diophantine approximation in differential fields, that the corresponding Roth's theorem in that setting could be viewed as analogous to Nevanlinna's second main theorem, with the exponent 2 in Roth's theorem and the coefficient 2 in $2T(r,f)$ having the same significance. To P. Vojta, in his landmark Ph.D. thesis, goes the credit of finding a solid connexion between classical diophantine geometry over number fields and Nevanlinna theory, thereby leading to far-reaching conjectures, which unified and motivated much further research, see [**306**], [**307**].

This final chapter is dedicated to the Vojta conjectures. They may be considered as an arithmetic counterpart of the Nevanlinna theory discussed in Chapter 13 and of which the abc-conjecture, which was the subject of a detailed analysis in Chapter 12, turns out to be an important special case.

The first two sections of this chapter develop Vojta's dictionary establishing a parallel between diophantine approximation and Nevanlinna theory, leading to his conjectures over number fields in Section 14.3. Schmidt's subspace theorem and the theorems of Roth, Siegel, and Faltings now appear as special cases of Vojta's conjectures without the ramification term. This lends support to the validity of Vojta's conjectures and also shows that the crux of the matter in attacking the general case consists precisely in controlling the ramification. The next Section 14.4 contains a generalization of the strong abc-conjecture to curves over number fields and we show its equivalence to the Vojta conjecture with ramification for curves, due to Elkies, van Frankenhuysen, and Vojta. In particular, the abc-conjecture implies the Mordell conjecture. Section 14.5 deals with the analogue of the strong abc-conjecture over function fields of characteristic 0, concluding with the general abc-theorem over function fields due to Voloch and Brownawell-Masser.

As it is already clear from the above, this chapter uses many results from previous chapters. The reader is assumed to be familiar with the geometric theory of heights from Chapter 2, with the abc-conjecture for integers and polynomials from Chapter 12 and with the basic results from Nevanlinna theory in Chapter XIII. We will use frequently the results about ramification from Appendix B.

14.2. The Vojta dictionary

In this short section, we introduce Vojta's dictionary between basic objects of Nevanlinna theory and corresponding concepts in diophantine approximation.

14.2.1. In diophantine approximation, we deal with an infinite sequence $(\beta_j)_{j\in\mathbb{N}}$ of distinct elements of a number field K approximating given numbers $a_1, \ldots, a_q \in K$ and arranged by increasing height. The sequence corresponds, in Nevanlinna theory, to a non-constant meromorphic function $f : \mathbb{C} \to \mathbb{P}^1_{\mathrm{an}}$. The role of the elements β_j is played by the series of maps

$$f_r : \overline{D(r)} = \{z \in \mathbb{C} \mid |z| \leq r\} \longrightarrow \mathbb{P}^1_{\mathrm{an}}$$

for varying $r > 0$. The numbers a_i are now given constants in $\mathbb{C}$.

14.2.2. For $r > 0$, the role of places in Nevanlinna theory is played by the closed disk $\overline{D(r)}$. Let F_r be the field of meromorphic functions on $\overline{D(r)}$, i.e. every element extends to a meromorphic function in some neighbourhood of $\overline{D(r)}$. For $v \in D(r)$, we have the discrete valuation ord_v giving rise to the normalized absolute value

$$|g|_v := \begin{cases} \left|\frac{v}{r}\right|^{\mathrm{ord}_v(g)} & \text{if } v \neq 0, \\ r^{-\mathrm{ord}_v(g)} & \text{if } v = 0, \end{cases}$$

for any $g \in F_r$. For v on the boundary $\partial D(r)$, we set

$$|g|_v := |g(v)|_v.$$

It is only well defined outside the poles of g contained in $\partial D(r)$, although $|\ |_v$ remains almost an absolute value in the sense that the triangle inequality $|g_1 + g_2|_v \leq |g_1|_v + |g_2|_v$ and $|g_1 g_2|_v = |g_1|_v |g_2|_v$ hold outside a finite set. Moreover, $|g|_v = 0$ for infinitely many v if and only if $g = 0$. The important point is that $v \mapsto \log |g|_v$ is integrable on $\partial D(r)$ for any $g \in F_r \setminus \{0\}$. Thus the boundary has to be considered as the set of archimedean places.

The product formula in K corresponds to Jensen's formula in 13.2.6 written as

$$\sum_{v\in D(r)} \log |g|_v + \frac{1}{2\pi} \int_{\partial D(r)} \log |g|_v \, \mathrm{d}\theta = \log |c(g, 0)|.$$

14.2.3. Thus $|\beta_j|_v$ corresponds in Nevanlinna theory to

$$|f_r|_v = \begin{cases} |f(v)| & \text{for } |v| = r, \\ \left|\frac{v}{r}\right|^{\mathrm{ord}_v(f)} & \text{for } 0 < |v| < r, \\ r^{-\mathrm{ord}_v(f)} & \text{for } v = 0. \end{cases}$$

A reader puzzled by the fact that the set of places varies with $r > 0$ should identify $\overline{D(r)}$ with the closed unit disk using the dilatation $v \mapsto w := v/r$ and thus f_r with $f_{(r)}(w) := f(rw)$. Then we have

$$|f_{(r)}|_w = \begin{cases} |f(rw)| & \text{for } |w| = 1, \\ |w|^{\operatorname{ord}_w(f_{(r)})} = |w|^{\operatorname{ord}_{rw}(f)} & \text{for } 0 < |w| < 1, \\ r^{-\operatorname{ord}_w(f)} & \text{for } w = 0. \end{cases}$$

For $w = 0$, note that the normalization of the absolute value still depends on r and not only on $f_{(r)}$. This is the reason why it is better to consider the set of places as variable. Ignoring the place $w = 0$ would lead to an additional error term $O(\log r)$.

14.2.4. By the above analogy between normalized absolute values in number theory and Nevanlinna theory, it is clear that the height $h(\beta)$ corresponds to the characteristic function $T(r, f)$. For $a \in K$, the analogue of the proximity function in number theory is

$$m_S(a, \beta) := -\sum_{v \in S} \log \min\{1, |\beta - a|_v\} = \sum_{v \in S} \log^- |\beta - a|_v$$

which was used on the left-hand side of Roth's theorem in 6.2.3, where S is any finite set of places containing the archimedean ones. The counting function in this number theoretic setting would be

$$N_S(a, \beta) := \sum_{v \notin S} \log^- |\beta - a|_v.$$

Note that in Nevanlinna theory the proximity function is defined by integration over $\partial D(r)$ which, by comparing the product formula with Jensen's formula, only corresponds to the archimedean places.

14.2.5. The first main theorem

$$m_S(a, \beta) + N_S(a, \beta) = h(\beta) + O(1)$$

with a constant independent of β clearly holds also for a number field K. For a proof, note that the left-hand side is simply $h(\beta - a)$ and use the standard inequality $h(x + y) \leq h(x) + h(y) + \log 2$, see Proposition 1.5.15.

To deal with the analogue of the second main theorem, we give the following variant of Roth's theorem in 6.2.3.

Theorem 14.2.6. *Let K be a number field, let $S \subset M_K$ be a finite set of places containing all archimedean places, and let $a_1, \ldots, a_q$ be distinct elements of K. Then for any fixed $\varepsilon > 0$ there are only finitely many $\beta \in K$ such that*

$$\sum_{i=1}^{q} m_S(a_i, \beta) \geq (2 + \varepsilon) h(\beta). \tag{14.1}$$

Proof: For every $v \in S$, there is at most one $i \in \{1, \ldots, q\}$ such that $\beta \in K$ is a good approximation of a_i, in the sense that

$$|\beta - a_i|_v < c_v := \frac{1}{2} \min_{j \neq k} |a_j - a_k|_v.$$

Now let $\beta \in K$ be a solution of (14.1) which may be rewritten as

$$\prod_{i=1}^{q} \prod_{v \in S} \min\left(1, |\beta - a_i|_v\right) \leq H(\beta)^{-2-\varepsilon}. \tag{14.2}$$

For $v \in S$, we choose $a_v \in \{a_1, \ldots, a_q\}$ such that $|\beta - a_i|_v$ is minimal, then the above shows

$$\prod_{i=1}^{q} \min\left(1, |\beta - a_i|_v\right) \geq \min(1, c_v)^{q-1} |\beta - a_v|_v.$$

If we substitute this in (14.2), we get the inequality in Roth's theorem with a multiplicative constant. If there were infinitely many solutions β of (14.1), then Northcott's theorem in 1.6.8 would yield $H(\beta) \to \infty$ and the multiplicative constant could be eliminated by passing to a smaller $\varepsilon > 0$, and Roth's theorem in 6.2.3 would lead to a contradiction. □

Proposition 14.2.7. *Roth's theorem in 6.2.3 is equivalent to Theorem 14.2.6.*

Proof: We have seen above that Roth's theorem implies Theorem 14.2.6. In order to see the converse, it is enough to deduce Theorem 6.4.1 for a normal finite extension E of K. Given $\alpha_v \in E$ and extensions $|\ |_{v,K}$ of $|\ |_v$ to E for v in a finite subset S of M_K, we have to prove that

$$\prod_{v \in S} \min(1, |\beta - \alpha_v|_{v,K}) \leq H(\beta)^{-2-\varepsilon} \tag{14.3}$$

has only finitely many solutions $\beta \in K$. By enlarging S, we may assume that S contains all archimedean places. By Corollary 1.3.5, $\mathrm{Gal}(E/K)$ operates transitively on $\{w \in M_E \mid w|v\}$, hence the local degrees $[E_w : K_v]$ remain constant and Corollary 1.3.2 gives

$$\min(1, |\beta - \alpha_v|_{v,K}) = \prod_{w|v} \min(1, |\beta - \sigma_w(\alpha_v)|_w)$$

for suitable $\sigma_w \in \mathrm{Gal}(E/K)$. So we may assume $E = K$. Applying Theorem 14.2.6 with $\{a_1, \ldots, a_q\} = \{\alpha_v \mid v \in S\}$, we get finiteness of (14.3). □

14.2.8. Hence Roth's theorem is the analogue of the second main theorem without the ramification term (see 13.2.24). In order to obtain a perfect correspondence, one should have

$$\sum_{i=1}^{N} m_S(a_i, \beta) \leq 2h(\beta) + O(\log h(\beta))$$

for $h(\beta) \to \infty$ instead of $(2+\epsilon)h(\beta)$ in Theorem 14.2.6. This was conjectured by Lang. In the particular case $K = \mathbb{Q}$ and $S = \infty$, computations by Lang and Trotter [**175**] of continued fraction expansions of certain algebraic numbers give some limited support to this statement.

14.3. Vojta's conjectures

In this section we translate the notions from the geometric part 13.4 of Nevanlinna theory to projective varieties over a number field K. Griffiths's conjecture leads us to Vojta's conjecture which would imply Roth's theorem, Siegel's theorem on integral points, Faltings's theorem, as well as several other outstanding conjectures. If we allow the points to vary in number fields of bounded degree, then the ramification term analogous to Nevanlinna theory is given by the absolute logarithmic discriminant. The consequences of Vojta's conjecture with ramification will be discussed in the next Section 14.4.

14.3.1. Let X be a projective variety over K and let $s = s_D$ be an invertible meromorphic section of a line bundle $L = O(D)$ with corresponding Cartier divisor D. We choose a presentation of L (or more generally a bounded M-metric) to get a local height $\lambda_D(\cdot, v)$ for every $v \in M_K$. For a finite subset S of M_K (usually containing the archimedean primes), the **counting function** is

$$N_{S,D}(P) := \sum_{w|v \in M_K \setminus S} \lambda_D(P, w)$$

and the **proximity function** is

$$m_{S,D}(P) := \sum_{w|v \in S} \lambda_D(P, w),$$

where w ranges over $M_{K(P)}$. This is well defined for $P \in X \setminus \mathrm{supp}(D)$. As a change of presentations (or M-metrics) changes these quantities only up to bounded functions, which does not really matter here, we omit in what follows explicit references to the presentations.

Based on his analogies between Nevanlinna theory and diophantine geometry, discussed here in Section 14.2, Vojta translated Griffiths's conjectural second main theorem from 13.4.18 into the following

Conjecture 14.3.2. *Let X be an irreducible smooth projective variety over K. Let D be a divisor with normal crossings, let H be an ample line bundle and let K_X be the canonical line bundle on X. For any $\varepsilon > 0$ and any finite subset $S \subset M_K$, there is a closed subset $Z \neq X$ such that for all $P \in X(K) \setminus Z$, we have*

$$m_{S,D}(P) + h_{K_X}(P) \leq \varepsilon h_H(P) + O(1).$$

Here, D is said to have **normal crossings** if the base change to $\mathbb{C}$ has normal crossings in the analytic sense of 13.4.17. The above inequality is called **Vojta's height inequality**.

Remark 14.3.3. Since Theorem 14.2.6 easily extends to the case $a_i \in \overline{K}$ (see below for a generalization), Proposition 14.2.7 shows that the case $X = \mathbb{P}^1_K$ in Vojta's conjecture above is equivalent to Roth's theorem.

More generally, for $X = \mathbb{P}^n_K$ Schmidt's subspace theorem yields the following analogue of Vojta's version of Cartan's second main theorem (see Remark 13.4.19):

Theorem 14.3.4. *Vojta's conjecture in 14.3.2 holds for $X = \mathbb{P}^n_K$ and D a divisor equal to a finite union of hyperplanes in general position defined over $\overline{K}$. The exceptional subset Z may be chosen as a finite union of linear subspaces defined over K.*

Proof: Since all quantities in Conjecture 14.3.2 are invariant under base change to a larger number field, we may assume that all hyperplane components $\{L_i = 0\}$ of D are defined over K. For $\mathbf{x} \in \mathbb{P}^n_K(K)$, we have the local height

$$-\log\left(|L_i(\mathbf{x})|_v/|\mathbf{x}|_v\right), \quad |\mathbf{x}|_v := \max_j |x_j|_v,$$

with respect to $\{L_i = 0\}$ and hence

$$m_{S,D}(\mathbf{x}) = -\sum_{v\in S}\sum_i \log\left(\frac{|L_i(\mathbf{x})|_v}{|\mathbf{x}|_v}\right) + O(1).$$

By $K_{\mathbb{P}^n_K} \cong O_{\mathbb{P}^n_K}(-n-1)$ ([**148**], Example II.8.20.1), we easily deduce the claim from Theorem 7.2.9. □

Remark 14.3.5. As in Nevanlinna theory, Vojta's conjecture in 14.3.2 is known for curves $X = C$ of genus g. Note that in this case, the exceptional set is finite and hence may be omitted by enlarging the bound. We sketch the argument and we give additional explanations to the meaning of Vojta's conjecture:

If $g = 0$, then we know that Conjecture 14.3.2 is equivalent to Roth's theorem.

If $g = 1$ and $D \neq 0$, then Vojta's conjecture is a special case of the approximation theorem for abelian varieties (cf. [**277**], §7.3) which is an intermediate step in the standard proof of Siegel's theorem on finiteness of S-integral points (see Remark 7.3.10 and Serre [**277**], §7.5). The latter claims that for a geometrically irreducible smooth projective curve of genus $g \geq 1$ and for a reduced divisor $D \neq 0$ (or $g = 0$ and $|\mathrm{supp}(D)| \geq 3$), the S-integral points of $C \setminus \mathrm{supp}(D)$ are finite. Note that the complement of D is always affine ([**148**], Exercise IV.1.3).

In order to see that the Vojta height inequality directly implies Siegel's theorem, note that S-integral means that $N_{S,D}(P) = O(1)$ and hence $m_{S,D}(P) =$

$h_D(P)+O(1)$. If $g = 0$ and using $K_C = -2[\infty]$, $H = [\infty]$ and $|\mathrm{supp}(D)| \geq 3$, we get $h_H(P)$ bounded proving the claim by Northcott's theorem in 2.4.9. If $g = 1$, then K_C is trivial (cf. Proposition 8.2.26 or A.13.6) and $H := D \neq 0$ implies again $h_H(P)$ bounded.

For $g \geq 2$, Vojta's conjecture is equivalent to Faltings's theorem in 11.1.1 (use $K_X = H$ ample and that we may assume $\lambda_D(P, v) \geq 0$ for all $v \in M_K$ by Proposition 2.3.9).

In [**312**], Vojta has given a direct proof of his height inequality for curves in 14.3.2. This gives a unified proof of Faltings's, Roth's, and Siegel's theorems.

Remark 14.3.6. The assumption that D has normal crossings is really necessary. This will be shown in a series of examples where we always assume that S is a finite subset of M_K containing the archimedean places:

If $X = \mathbb{P}^1_K$ and $D = 2[0] + [\infty]$, then Vojta's height inequality would give

$$h(P) \leq N_{S,D}(P) + \varepsilon h(P) + O(1) \tag{14.4}$$

for all $P \in \mathbb{G}_m(K)$. The set of S-integral points in $\mathbb{G}_m(K)$ is equal to the set of S-units. If $|S| \geq 2$, then Dirichlet's unit theorem in 1.5.13 shows that there are infinitely many S-units, and they satisfy $N_{S,D}(P) = O(1)$. By Northcott's theorem in 2.4.9, this contradicts (14.4), hence no multiple components are allowed for D.

Next, we consider the example $X = \mathbb{P}^2_K$ and

$$D = [x_1 = 0] + [x_2 = 0] + [(x_1 - x_2)x_0 = (x_1 + x_2)^2].$$

Then $(1 : 0 : 0)$ is an ordinary triple point, hence D has not normal crossings. The coordinate ring of the affine variety $\mathbb{P}^2_K \setminus D$ is

$$K\left[z, x, \frac{1}{x}, \frac{1}{(x-1)z - (x+1)^2}\right], \quad z := \frac{x_0}{x_2}, \quad x := \frac{x_1}{x_2}.$$

The goal is to construct infinitely many S-integral points $P = (z : x : 1)$ in $\mathbb{P}^2_K \setminus D$ lying Zariski dense. Here a point P is S-integral if and only if z is an S-integer and $x, (x-1)z - (x+1)^2$ are S-units. Note that

$$(x-1)z - (x+1)^2 = -4x^n \tag{14.5}$$

has a unique solution $z_n(x) \in K[x, x^{-1}]$ for every $n \in \mathbb{Z}$. So we assume that S contains all places over 2. By Dirichlet's unit theorem, there is an S-unit a which is not a root of unity. We conclude that the set

$$Y := \{(z_n(a^m) : a^m : 1) \mid m, n \in \mathbb{Z}\}$$

is S-integral in $\mathbb{P}^2_K \setminus D$. For fixed $n \in \mathbb{Z}$, equation (14.5) describes a rational projective curve in $\mathbb{P}^2_K$ such that $\{(z_n(a^m) : a^m : 1) \mid m \in \mathbb{Z}\}$ is dense. For $n \geq 3$, the irreducible curves have distinct degrees and hence they are not contained in a proper closed subvariety of $\mathbb{P}^2_K$. We conclude that Y is a dense S-integral subset in $\mathbb{P}^2_K \setminus D$. As before, we conclude that Vojta's height inequality does not hold.

Finally, we show that no bad singularities of the components are allowed either. Let us consider the plane curve $D = [x_0 x_1^{d-1} = x_2^d]$ for $d \geq 4$. Then D is irreducible and has again not normal crossings in the singularity $(1:0:0)$. For an S-integer a and an S-unit u, the point $P = (a^d + u : 1 : a)$ is S-integral on $\mathbb{P}^2_K \setminus D$, hence

$$N_{S,D}(P) = \sum_{v \notin S} \log \frac{\max |x_i|_v}{|x_0 x_1^{d-1} - x_2^d|_v} = 0.$$

For $|S| \geq 2$, again Dirichlet's unit theorem implies easily that such points are dense in $\mathbb{P}^2_K$, in contradiction with Vojta's height inequality.

Remark 14.3.7. A smooth projective variety X over K is called of **general type** if there is a positive integer n, an ample line bundle L on X and an effective divisor E on X with $K_X^{\otimes n} \cong L \otimes O(E)$ (cf. [**307**], §1.2). If X is of general type, then Vojta's conjecture, applied with $D = 0$ and $L = H$, implies

$$(1 - \varepsilon) h_H(P) + h_E(P) \leq O(1).$$

By Proposition 2.3.9, we may assume $h_E(P) \geq 0$. Then Northcott's theorem in 2.4.9 shows that $X(K)$ is not Zariski dense in X. This is the **Bombieri–Lang conjecture.**

For any projective variety X over K, the **special set** Sp_X is defined as the Zariski closure of the union of the images of all non-constant rational maps from irreducible group varieties to X. Clearly, in the special set one may have infinitely many K-rational points by considering the images of $1, g, g^2, \cdots$ in X for a K-rational point g of the group variety. Then the **general Lang conjecture** claims that X is of general type if and only if $(X \setminus \mathrm{Sp}_X)(K')$ is finite for all finitely generated extensions K'/K (see [**171**], Ch.I, §3 for a further discussion).

Faltings proved this for a subvariety X of an abelian variety, which is **Faltings's big theorem** (see Theorem 11.10.1 for the number field case and [**115**] for the general case). In this case the special set is the union of all translates of abelian subvarieties of dimension ≥ 1 contained in $X_{\overline{K}}$ (see [**171**], Ch.I, §6).

14.3.8. In Conjecture 14.3.2, we have only considered points rational in a fixed number field K. But what happens if we allow P to vary over all $\overline{K}$-rational points? The analogous situation in Nevanlinna theory was considered in Theorem 13.4.30. The additional effect of finite field extension was measured by the counting function of the ramification divisor of the covering. In order to get an analogy with the number field case, we use the language of schemes. If the reader is not familiar with the latter, then he may pass directly to Definition 14.3.9.

Let F/K be a finite dimensional field extension (say $F \subset \overline{K}$). Then an F-rational point P of the projective variety X may be identified with a morphism $\mathrm{Spec}(F) \to X$ over K with image P. Instead of the finite coverings $p : Y \to \mathbb{C}$, we use here $\mathrm{Spec}(O_F) \to \mathrm{Spec}(O_K)$ as models for our fields which may be seen as arithmetic curves. The ramification divisor of p may be written in the form

$$R_p = \sum_{y \in Y} \ell(\Omega_{Y/\mathrm{Spec}(\mathbb{C}),y}) \cdot [y],$$

where Ω denotes the sheaf of relative differentials (see A.7.29) and ℓ is the length (see B.4.4). So it is natural to define the ramification divisor of $F/\mathbb{Q}$ by

$$R_{F/K} = \sum_{\mathcal{P} \in \mathrm{Spec}(O_F)} \ell(\Omega_{O_F/O_K,\mathcal{P}})\mathcal{P}.$$

By B.1.18, Ω_{O_F/O_K} is a principal O_F-module with annihilator equal to the different $\mathcal{D}_{F/K}$, hence

$$\ell(\Omega_{O_F/O_K,\mathcal{P}}) = v_{\mathcal{P}}(\mathcal{D}_{F/K}),$$

where $v_{\mathcal{P}}$ is the discrete valuation associated to $\mathcal{P}$ and where $v_{\mathcal{P}}(I) := \min\{v_{\mathcal{P}}(a) \mid a \in I\}$ for any ideal I of O_F. Note that after localization at $\mathcal{P}$, it is always the valuation of a principal generator. Thus the counting function of $R_{F/K}$ should be

$$N_{R_{F/K}} = -\sum_{\mathcal{P} \in \mathrm{Spec}(O_F)} \ell(\Omega_{O_F/O_K,\mathcal{P}}) \log|\mathcal{P}|_{v_{\mathcal{P}}} = -\sum_{\mathcal{P} \in \mathrm{Spec}(O_F)} \log|\mathcal{D}_{F/K}|_{v_{\mathcal{P}}}.$$

Since the norm $N_{F/\mathbb{Q}}$ of the different $\mathcal{D}_{F/K}$ is the discriminant $\mathfrak{d}_{F/K}$ (see B.1.17), Lemma 1.3.7 and the product formula lead to

$$N_{R_{F/K}} = -\frac{1}{[F:K]} \sum_{\wp \in \mathrm{Spec}(O_K)} \log|\mathfrak{d}_{F/K}|_{\wp} = \frac{1}{[F:\mathbb{Q}]} \log|N_{K/\mathbb{Q}}\mathfrak{d}_{F/K}|.$$

Definition 14.3.9. *For a number field K, we define the* **absolute logarithmic discriminant** *d_K by*

$$d_K := \frac{1}{[K:\mathbb{Q}]} \log|D_{K/\mathbb{Q}}|$$

where $D_{K/\mathbb{Q}} \in \mathbb{N}$ is the discriminant (see B.1.14). The **absolute logarithmic discriminant of a point** *P in a K-variety is defined by $d(P) := d_{K(P)}$.*

Proposition 14.3.10. *Let F/K be a finite extension of number fields. Then*

$$0 \le d_F - d_K = \frac{1}{[F:\mathbb{Q}]} \log|N_{K/\mathbb{Q}}\mathfrak{d}_{F/K}| = -\frac{1}{[F:K]} \sum_v \log|\mathfrak{d}_{F/K}|_v.$$

where v ranges over all non-archimedean places of M_K.

Proof: Recall that $|\mathfrak{d}_{F/K}|_v$ is the absolute value of a principal generator of the localization of $\mathfrak{d}_{F/K}$ in the prime ideal corresponding to v. It follows from the approximation theorem in 1.2.13 that we may choose the same principal generator for a finite set of places. Then the claim follows from Proposition B.1.19 and Lemma 1.3.7. □

From 14.3.8, we get $N_{R_{F/K}} = d_F - d_K$. The analogy with Theorem 13.4.30 leads to **Vojta's conjecture with ramification.**

Conjecture 14.3.11. *Let X, D, H, S, ε be as in Conjecture 14.3.2 and let $d \in \mathbb{N}$. Then there is a closed subset $Z \neq X$ such that for all $P \in X \setminus Z$ with $[K(P) : K] \leq d$ it holds*

$$m_{S,D}(P) + h_{K_X}(P) - d(P) \leq \varepsilon h_H(P) + O(1).$$

Remark 14.3.12. Even for $C = \mathbb{P}^1_{\mathbb{Q}}$, Vojta's conjecture with ramification is unknown. The case of curves and its relations to the abc-conjecture are studied in the next section.

14.4. A general abc-conjecture

A natural thing to ask is what form the abc-conjecture should take over a number field K. Also, the question arises whether there is any special feature in the structure of the equation $a + b = c$, and what may be an appropriate higher-dimensional generalization of it. To this end, we may view the abc-conjecture as a statement about points on the model $x + y = 1$, $x, y \neq 0, 1$ of the affine curve $\mathbb{P}^1 \setminus \{0, 1, \infty\}$. There is nothing special about this particular model, for example we could have worked instead with the affine curve $x(x-1)y = 1$. What matters here is that the abc-conjecture is a statement about ramification, both arithmetic and geometric.

In this section, C denotes an irreducible smooth projective curve of genus g defined over K. Let D be an effective reduced divisor on C with local height λ. Then we will formulate a conjecture on C for D and λ similar to the truncated second main theorem in Nevanlinna theory. The case $D = [0] + [1] + [\infty]$ will give us the desired generalization of the strong abc-conjecture to number fields. Based on the work of Elkies and Vojta, we will show that this strong abc-conjecture, the conjectural truncated second main theorem for arbitrary C, and Vojta's conjecture with ramification for curves, are all equivalent.

14.4.1. Clearly, the left-hand side $\max\{|a|, |b|, |c|\}$ of the abc-conjecture in 12.2.2 corresponds logarithmically to the height h_λ on C. So we have to deal with

$$\log \operatorname{rad}(abc) = \sum_{p|abc} \log |1/p|_p.$$

For every natural prime p, there is a contribution $\log p$ if the local height of the point $(a : b : c)$ of the curve $x + y = z$ is non-zero. This leads to the following generalization of the radical. For a precise explanation, we refer to Example 14.4.4.

Definition 14.4.2. *The* ***conductor*** *of $P \in C \setminus \mathrm{supp}(D)$ with respect to λ is defined by*

$$\mathrm{cond}_\lambda^K(P) := \sum_{v \in M_{K(P)}^{\mathrm{fin}}} \chi(\lambda(P,v)) \log |1/\pi_v|_v,$$

where $v \in M_{K(P)}^{\mathrm{fin}}$ denotes the discrete valuations of $K(P)$ with local parameters π_v and

$$\chi(t) = \begin{cases} 0 & \text{if } t \le 0, \\ 1 & \text{if } t > 0. \end{cases}$$

Remark 14.4.3. Note that the conductor is completely analogous to the truncated counting function in Nevanlinna theory from 13.4.26. It makes also sense for non-reduced effective divisors. The following example shows how to recover the special conductor for the simple minded formulation of the abc-conjecture we have considered in Chapter 12.

Example 14.4.4. Consider the case in which $C = \mathbb{P}_K^1$ and $D = [0] + [1] + [\infty]$. Let $(x_0 : x_1)$ be standard homogeneous coordinates on $\mathbb{P}_K^1$, which we view as global sections of $O_{\mathbb{P}^1}(1)$. Then D has a presentation

$$\mathcal{D} = (x_0 x_1 (x_0 - x_1);\ O_{\mathbb{P}^1}(3), x_0^3, x_0^2 x_1, x_0 x_1^2, x_1^3;\ O_{\mathbb{P}^1}, 1)$$

with associated height

$$h_{\mathcal{D}}(P) = 3h(P),$$

where $h(P)$ is the standard projective height in $\mathbb{P}_K^1$. Moreover, $\Omega_{\mathbb{P}^1}^1 \cong O_{\mathbb{P}^1}(-2)$ (use A.13.6), thus there is a presentation $\mathcal{K}$ of a canonical divisor such that

$$h_{\mathcal{K}}(P) = -2h(P).$$

It follows that

$$h_{\mathcal{D}}(P) + h_{\mathcal{K}}(P) = h(P).$$

The local height function $\lambda(P,v) := \lambda_{\mathcal{D}}(P,v)$ at $v \in M_{K(P)}$ is given by

$$\begin{aligned} \lambda(P,v) &= \max_{i=0,\dots,3} \log \left| \frac{x_0^i x_1^{3-i}}{x_0 x_1 (x_0 - x_1)} \right|_v \\ &= \max\left(\log \left| \frac{x_0^2}{x_1(x_0 - x_1)} \right|_v, \log \left| \frac{x_1^2}{x_0(x_0 - x_1)} \right|_v \right). \end{aligned}$$

If $x_0 = c$, $x_1 = a$, $x_0 - x_1 = b$, where $O_{K(P)}a$, $O_{K(P)}b$, $O_{K(P)}c$ are coprime ideals, which is always possible if $O_{K(P)}$ is a principal ideal domain, then $a + b = c$ and

$$\begin{aligned} \lambda(P,v) &= \max\left(\log \left| \frac{c^2}{ab} \right|_v, \log \left| \frac{a^2}{cb} \right|_v \right) \\ &= \max\left(\log \left| \frac{1}{a} \right|_v, \log \left| \frac{1}{b} \right|_v, \log \left| \frac{1}{c} \right|_v \right); \end{aligned}$$

therefore

$$\mathrm{cond}_\lambda^K(P) = \sum_{v(abc)>0} \log|1/\pi_v|_v.$$

Another way of writing this conductor, which works in any case, is as follows. For $x = x_1/x_0$, let

$$\mathrm{cond}_{[0]}^K(x) = \sum_{v(x)>0} \log|1/\pi_v|_v,$$

$$\mathrm{cond}_{[1]}^K(x) = \sum_{v(1-x)>0} \log|1/\pi_v|_v,$$

$$\mathrm{cond}_{[\infty]}^K(x) = \sum_{v(1/x)>0} \log|1/\pi_v|_v.$$

Clearly, they correspond to conductors of the divisors $[0]$, $[1]$ and $[\infty]$ on $\mathbb{P}_K^1$. We have

$$\mathrm{cond}_\lambda^K(P) = \mathrm{cond}_{[0]}^K(x(P)) + \mathrm{cond}_{[1]}^K(x(P)) + \mathrm{cond}_{[\infty]}^K(x(P)).$$

Proposition 14.4.5. *Let D, D' be effective Cartier divisors on C:*

(a) *If $D \leq D'$, then there are presentations $\mathcal{D}, \mathcal{D}'$ with associated local heights $0 \leq \lambda \leq \lambda'$ and hence $\mathrm{cond}_\lambda^K \leq \mathrm{cond}_{\lambda'}^K$ outside of $\mathrm{supp}(D')$.*

(b) *If $\mathrm{supp}(D) = \mathrm{supp}(D')$, then $\mathrm{cond}_\lambda^K - \mathrm{cond}_{\lambda'}^K$ is a bounded function outside of $\mathrm{supp}(D)$, for any local heights λ, λ' of D and D'.*

(c) *Let $\lambda, \lambda', \lambda''$ be local heights associated to D, D' and $D + D'$. Then $\mathrm{cond}_{\lambda''}^K \leq \mathrm{cond}_\lambda^K + \mathrm{cond}_{\lambda'}^K + O(1)$ outside of $\mathrm{supp}(D) \cup \mathrm{supp}(D')$.*

Proof: By Proposition 2.3.9, there are presentations $\mathcal{D}, \mathcal{D}''$ of D and $D' - D$ such that the corresponding local heights λ, λ'' are non-negative functions outside of the supports of D and $D' - D$, respectively. Setting $\mathcal{D}' = \mathcal{D} + \mathcal{D}''$ (cf. 2.2.4), we get (a).

It follows from Theorem 2.2.11 and Remark 2.2.13 (resp. Theorem 2.7.14 in the case of Néron divisors) that the conductor is independent of the choice of the local height up to bounded functions on $C \setminus \mathrm{supp}(D)$. If $\mathrm{supp}(D) = \mathrm{supp}(D')$, then there is $n \in \mathbb{N}$ such that $D \leq nD'$ and (b) follows from (a).

To prove (c), we may assume by (a) and (b) that $\lambda, \lambda' \geq 0$ and $\lambda'' = \lambda + \lambda'$. Then the claim is obvious from the definition of the conductor. $\square$

The conductor $\mathrm{cond}_\lambda^K(P)$, as well as the absolute logarithmic discriminant $d(P)$, does not have a good functorial behaviour with respect to base change. As we will show next, their sum $\mathrm{cond}_\lambda^K(P) + d(P)$ does not suffer from this defect and it is this quantity which will appear on the right-hand side of the conjectural second main theorem.

Proposition 14.4.6. *Let $\varphi : C' \to C$ be a finite morphism of irreducible smooth projective curves over K. Let also λ be a local height relative to the effective divisor D on C and $\lambda' := \lambda \circ \varphi$ be the local height relative to $\varphi^* D$. Then for $P' \in C'$ with $P := \varphi(P') \notin \operatorname{supp}(D)$ the following statements hold:*

(a) $\operatorname{cond}_{\lambda'}^K(P') \leq \operatorname{cond}_{\lambda}^K(P)$;

(b) $d(P') + \operatorname{cond}_{\lambda'}^K(P') \geq d(P) + \operatorname{cond}_{\lambda}^K(P)$;

(c) *if φ is unramified outside of $\varphi^{-1}(\operatorname{supp}(D))$, then*

$$d(P') + \operatorname{cond}_{\lambda'}^K(P') \leq d(P) + \operatorname{cond}_{\lambda}^K(P) + O(1).$$

Proof: By Lemma 1.3.7, we get

$$\operatorname{cond}_{\lambda}^K(P) - \operatorname{cond}_{\lambda'}^K(P') = \sum_{v \in M_{K(P)}^{\mathrm{fin}}} \chi(\lambda(P, v)) \sum_{w|v} \left(1 - \frac{1}{e_{w/v}}\right) \log |1/\pi_v|_w, \tag{14.6}$$

where w ranges over $M_{K(P')}^{\mathrm{fin}}$ and $e_{w/v}$ is the ramification index. This proves (a). On the other hand, Proposition 14.3.10 gives

$$d(P') - d(P) = -\frac{1}{[K(P') : K(P)]} \sum_{v \in M_{K(P)}^{\mathrm{fin}}} \log |\mathfrak{d}_{K(P')/K(P)}|_v. \tag{14.7}$$

By Dedekind's discriminant theorem in B.2.12, we have

$$\log |\mathfrak{d}_{K(P')/K(P)}|_v = \sum_{w|v} f_{w/v}(e_{w/v} - 1 + \delta'_w) \log |\pi_v|_v,$$

where $f_{w/v}$ is the residue degree and

$$\delta'_w \begin{cases} \in [1, e_{w/v} v(e_{w/v})] & \text{if } v(e_{w/v}) \geq 1, \\ = 0 & \text{else.} \end{cases}$$

As usual the discrete valuation v is normalized by $v(\pi_v) = 1$. The case in which $v(e_{w/v}) \geq 1$ is called the case of wild ramification.

By Proposition 1.2.11, $e_{w/v} f_{w/v}$ is the local degree of $w|v$ and hence

$$-\frac{1}{[K(P') : K(P)]} \log |\mathfrak{d}_{K(P')/K(P)}|_v = \sum_{w|v} \left(1 + \frac{\delta'_w - 1}{e_{w/v}}\right) \log |1/\pi_v|_w.$$

If we substitute this in (14.7), comparison with (14.6) leads to (b).

To prove (c), we note that (14.7) minus (14.6) and the above considerations yield

$$d(P') + \operatorname{cond}_{\lambda'}^K(P') \leq d(P) + \operatorname{cond}_{\lambda}^K(P) + C_1 + C_2,$$

where

$$C_1 = -\frac{1}{[K(P') : K(P)]} \sum_v \log |\mathfrak{d}_{K(P')/K(P)}|_v$$

with v ranging over all non-archimedean places of $K(P)$ with $\lambda(P, v) \leq 0$ and

$$C_2 = \sum_{v \in M_{K(P)}^{\mathrm{fin}}} \sum_{w|v} v(e_{w/v}) \log |1/\pi_v|_w.$$

Note that only places $w \in M_{K(P')}^{\mathrm{fin}}$ wildly ramified over v contribute to C_2. Let p be the natural prime with $v|p$ and let v_p be the corresponding discrete valuation normalized by $v_p(p) = 1$. Using

$$v(e_{w/v}) = e_{v/p} v_p(e_{w/v}) \leq e_{v/p} \log[K(P') : K(P)] / \log p,$$

we get

$$\begin{aligned} \sum_{v|p} \sum_{w|v} v(e_{w/v}) \log |1/\pi_v|_w &\leq \frac{\log[K(P') : K(P)]}{\log p} \sum_{w|p} \log |1/p|_w \\ &= \log[K(P') : K(P)], \end{aligned}$$

where the last step was done by appealing to Lemma 1.3.7. Let S_0' be the set of natural primes p such that $K(P')/K(P)$ is wildly ramified at a place over p. For every $p \in S_0'$, Example 1.4.12 shows

$$p \mid e_{w/v} \leq [K(P') : K(P)] \leq \deg(\varphi),$$

hence the cardinality $|S_0'|$ is bounded by a constant depending on $\deg(\varphi)$. We conclude

$$C_2 \leq |S_0'| \log \deg(\varphi) = O(1).$$

It remains to bound C_1. By Proposition 14.4.5, we may assume $\lambda \geq 0$. Let M be the set of places of $\overline{K}$ represented by the extensions of $|\ |_p, p \in M_{\mathbb{Q}}$. It is easy to show that $E^u := \{x \in C \setminus \mathrm{supp}(D) \mid \lambda(x, u) = 0\}$, $u \in M$, is an M-bounded family in $C \setminus \mathrm{supp}(D)$ (use Proposition 2.6.17). The Chevalley–Weil theorem in the form of Theorem 10.3.5 gives the existence of a non-zero $\alpha \in \mathbb{Z}$ such that

$$\alpha \in \mathfrak{d}_{K(P')_w / K(P)_v}$$

for all $P' \in C \setminus \mathrm{supp}(D)$ and $w \in M_{K(P')}^{\mathrm{fin}}$ with $\lambda'(P', w) = 0$. By Corollary 1.3.2, B.1.20 and B.1.21, we get

$$|\mathfrak{d}_{K(P')/K(P)}|_v = \prod_{w|v} |\mathfrak{d}_{K(P')_w / K(P)_v}|_v \geq |\alpha|_v^{[K(P'):K(P)]}.$$

Therefore

$$C_1 \leq -\sum_{v \in M_{K(P)}^{\mathrm{fin}}} \log |\alpha|_v = \log |\alpha|$$

by Lemma 1.3.7 and the product formula. This finishes the proof of (c). □

Corollary 14.4.7. *Let F/K be a finite subextension of $\overline{K}/K$. Then:*

(a) $\mathrm{cond}_\lambda^F(P) \leq \mathrm{cond}_\lambda^K(P)$ *with equality if F/K is unramified.*

(b) $d_{K(P)} + \mathrm{cond}_\lambda^K(P) \leq d_{F(P)} + \mathrm{cond}_\lambda^F(P) \leq d_{K(P)} + \mathrm{cond}_\lambda^K(P) + O(1)$.

Proof: Let us consider the base change C_F as a curve C' over K. Then it is clear that the natural morphism $\varphi : C' \to C$ is finite and unramified. We have $P' \in C'$ in the fibre over P such that $K(P') = F(P)$ and the claim follows from Proposition 14.4.6. □

Proposition 14.4.8. *Let S_0 be a finite set of natural primes. Then the contribution of the places v lying over the primes of S_0 to $\mathrm{cond}_\lambda^K(P)$ satisfies the bound*

$$\sum_{v|p\in S_0} \chi(\lambda(P,v)) \log|1/\pi_v|_v \leq \sum_{p\in S_0} \log p$$

on $C \setminus \mathrm{supp}(D)$.

Proof: This is obvious from $|\pi_v|_v \geq |p|_v$ and Lemma 1.3.7. □

Proposition 14.4.9. *Let λ be a local height relative to the effective divisor D on C. Then*

$$\mathrm{cond}_\lambda^K(P) \leq h_\lambda(P) + O(1)$$

for all $P \notin \mathrm{supp}(D)$.

Proof: By Proposition 14.4.5, we may assume that $\lambda \geq 0$ and that λ is given by a presentation. Then $\lambda(P,v) \geq \log|1/\pi_v|_v$ or $\lambda(P,v) = 0$ and hence

$$h_\lambda(P) = \sum_{v\in M_{K(P)}} \lambda(P,v) \geq \sum_{v\in M_{K(P)}^{\mathrm{fin}}} \chi(\lambda(P,v)) \log|1/\pi_v|_v = \mathrm{cond}_\lambda^K(P),$$

proving the claim. □

In Nevanlinna theory, we have seen that the second main theorem for curves in 13.4.23 easily implies the truncated second main theorem in 13.4.28. Similarly, we can show that the second main theorem for coverings in 13.4.30 implies a corresponding truncated second main theorem. The contribution of a finite ramified covering $p : Y \to \mathbb{C}$ is measured by N_{R_p} (cf. 13.4.31), which is the analogue of the absolute logarithmic discriminant $d(P)$ from diophantine geometry (cf. 14.3.8). Since the conductor corresponds to the truncated counting function, the truncated second main theorem for coverings leads to the following:

Conjecture 14.4.10. *Let D be a reduced effective divisor on C with local height λ, let H be an ample line bundle on C and let $\varepsilon > 0$. Then*

$$h_D(P) + h_{K_C}(P) \leq \mathrm{cond}_\lambda^K(P) + d(P) + \varepsilon h_H(P) + O_{[K(P):K]}(1)$$

for all $P \in C \setminus \mathrm{supp}(D)$.

Here, as in the other conjectures of this chapter, in the bound $O_{[K(P):K]}(1)$ *which may depend on* ε *and a whole complex of other data*, it is only the dependence on the point P which matters here and we give it in terms of the degree $[K(P):K]$.

14.4.11. If we specialize to the case $C = \mathbb{P}^1_K$ and $D = [0] + [1] + [\infty]$, then $K_C \cong O_{\mathbb{P}^1}(-2)$ (cf. A.13.6) and hence

$$h_D(P) + h_{K_C}(P) = h(P) + O(1).$$

By Example 14.4.4, we get the **strong abc-conjecture** of Elkies generalizing the strong abc-conjecture in 12.2.2 to number fields:

Conjecture 14.4.12. *For every* $\varepsilon > 0$, *it holds*

$$(1-\varepsilon)h(x) \leq \mathrm{cond}^{\mathbb{Q}}_{[0]}(x) + \mathrm{cond}^{\mathbb{Q}}_{[1]}(x) + \mathrm{cond}^{\mathbb{Q}}_{[\infty]}(x) + d_{\mathbb{Q}(x)} + O_{[\mathbb{Q}(x):\mathbb{Q}]}(1)$$

for all $x \in \overline{\mathbb{Q}} \setminus \{0,1\}$, *where* $O_{[\mathbb{Q}(x):\mathbb{Q}]}(1)$ *depends only on* ε *and* $[\mathbb{Q}(x):\mathbb{Q}]$.

We will show that both conjectures 14.4.10 and 14.4.12 are equivalent to Conjecture 14.3.11 restricted to curves. For the latter, the exceptional set Z is finite and may be omitted by enlarging the $O(1)$-term. Hence **Vojta's conjecture with ramification for curves** reads as:

Conjecture 14.4.13. *Let* S *be a finite subset of* M_K. *With the same hypothesis as in Conjecture 14.4.10, the estimate*

$$m_{S,D}(P) + h_{K_C}(P) \leq d(P) + \varepsilon h_H(P) + O_{[K(P):K]}(1)$$

holds for all $P \in C \setminus \mathrm{supp}(D)$.

14.4.14. Vojta's idea to deduce the strong abc-conjecture from his conjecture is that, by passing to a finite covering $\pi : C' \to C$, we may improve the height inequality in 14.4.13. In fact, Conjecture 14.4.13 applied to C' and $D' := \pi^*(D)_{\mathrm{red}}$ (namely, the sum of the irreducible components) gives

$$m_{S,D'}(P') + h_{K_{C'}}(P') - d(P') \leq \varepsilon' h_{\pi^* H}(P') + O_{[K(P'):K]}(1) \qquad (14.8)$$

for all $P' \notin \mathrm{supp}(D')$. Now $K_{C'} \cong \pi^* K_C + R_\pi$ from Theorem B.4.5 implies

$$h_{K_{C'}}(P') = h_{K_C}(P) + m_{S,R_\pi}(P') + N_{S,R_\pi}(P') + O(1) \qquad (14.9)$$

for $P := \pi(P')$. By Proposition B.4.7, we have $D' \geq \pi^*(D) - R_\pi$ and hence

$$m_{S,D'}(P') \geq m_{S,D}(P) - m_{S,R_\pi}(P') + O(1). \qquad (14.10)$$

By (14.9) and (14.10) in (14.8), we get

$$m_{S,D}(P) + h_{K_C}(P) + N_{S,R_\pi}(P') - d(P') \leq \varepsilon' h_H(P) + O_{[K(P'):K]}(1).$$

This leads to the improvement $N_{S,R_\pi}(P') + d(P) - d(P')$ on the left-hand side of the original Vojta height inequality on C. Moreover, by a Chevalley–Weil type argument, this improvement is always bounded from below ([**307**], Th.5.1.6), but we will not need this result here.

If the morphism π is unramified outside of $\pi^{-1}(\mathrm{supp}(D))$, we may apply Proposition 14.4.6, getting

$$\begin{aligned} & h_D(P) + h_{K_C}(P) - d(P) + N_{S,R_\pi}(P') \\ & \qquad \leq \mathrm{cond}_\lambda^K(P) + N_{S,D}(P) + \varepsilon' h_H(P) + O_{[K(P'):K]}(1). \quad (14.11) \end{aligned}$$

Example 14.4.15. A **Fermat curve** is a plane projective curve given by the projective equation

$$x_0^n + x_1^n = x_2^n$$

for some $n \geq 1$. By the Jacobi criterion in A.7.15, the Fermat curve is smooth. There is a finite covering $\pi : C_n \to C_1 \cong \mathbb{P}^1_{\mathbb{Q}}$ given by mapping $(x_0 : x_1 : x_2)$ to $(x_0^n : x_1^n : x_2^n)$. Local analytically over 0, 1, or ∞, the morphism is given by $z \mapsto z^n$, hence $R_\pi = (n-1)D'$, where $D' = \pi^{-1}\{0, 1, \infty\}$ (use Example B.4.8). As a consequence of the Hurwitz theorem in B.4.6, we note that C_n has genus $\frac{1}{2}(n-1)(n-2)$ (also clear from A.13.4).

Theorem 14.4.16. *The following conjectures are all equivalent:*

(a) *strong abc-conjecture in 14.4.12;*

(b) *Conjecture 14.4.10 for all curves C over any number field;*

(c) *Vojta's conjecture in 14.4.13 for all curves C over any number field;*

(d) *Vojta's conjecture in 14.4.13 for all Fermat curves over $\mathbb{Q}$.*

The implication (a) $\Rightarrow$ (b) follows from an argument of Elkies [**98**] which was elaborated by M. van Frankenhuysen [**305**]. The claims (b) $\Rightarrow$ (c) $\Rightarrow$ (d) are trivial and (d) $\Rightarrow$ (a) is due to P. Vojta (see [**307**], [**317**]).

Proof: (a) $\Rightarrow$ (b): Let C be an irreducible smooth projective curve over the number field K with local height λ relative to the reduced divisor D. The proof is based on Elkies's idea of using a Belyĭ function $f : C \to \mathbb{P}^1_K$ for D, in other words with $\mathrm{supp}(D) \subset f^{-1}\{0, 1, \infty\}$ and unramified outside of $f^{-1}\{0, 1, \infty\}$ (see Lemma 12.2.7). We use $D_0 = [0] + [1] + [\infty]$, $D' = f^*(D_0)$, $D_1 = D'_{\mathrm{red}}$ and $D_2 = D_1 - D$ with corresponding local heights $\lambda_0, \lambda', \lambda_1$ and λ_2. By Proposition B.4.7, the ramification divisor R_f satisfies

$$R_f = D' - D_1. \quad (14.12)$$

For $P \notin f^{-1}\{0, 1, \infty\}$, the strong abc-conjecture, Example 14.4.4 and Corollary 14.4.7 imply

$$(1 - \varepsilon')h(f(P)) \leq \mathrm{cond}_{\lambda_0}^K(f(P)) + d(f(P)) + O_{[K(f(P)):\mathbb{Q}]}(1).$$

By Proposition 14.4.6, we have

$$\mathrm{cond}_{\lambda_0}^K(f(P)) + d(f(P)) \leq \mathrm{cond}_{\lambda'}^K(P) + d(P).$$

By Proposition 14.4.5, we get

$$(1-\varepsilon')h(f(P)) \leq \mathrm{cond}_\lambda^K(P) + \mathrm{cond}_{\lambda_2}^K(P) + d(P) + O_{[K(P)):K]}(1).$$

Note that only the dependence on P is indicated in the bound, which may depend on K as well. By Proposition 14.4.9 and $h(f(P)) = h_{f^*O_{\mathbb{P}^1}(1)}(P) + O(1)$, we deduce

$$(1-\varepsilon')h_{f^*O_{\mathbb{P}^1}(1)}(P) \leq \mathrm{cond}_\lambda^K(P) + h_{D_2}(P) + d(P) + O_{[K(P):K]}(1). \tag{14.13}$$

We have proved this only for $P \notin \mathrm{supp}(D_1)$, but by increasing the constants we may assume that it holds for all $P \notin \mathrm{supp}(D)$. By Theorem B.4.5 and (14.12), we have

$$K_C \cong f^*K_{\mathbb{P}^1} \otimes O(R_f) \cong f^*O_{\mathbb{P}^1}(1) \otimes O(-D_1).$$

By this equation, inequality (14.13), and Theorem 2.3.8, we get

$$(1-\varepsilon')\left(h_{K_C}(P) + h_{D_1}(P)\right) \leq \mathrm{cond}_\lambda^K(P) + h_{D_2}(P) + d(P) + O_{[K(f(P)):K]}(1).$$

Finally, $D = D_1 - D_2$ leads to

$$h_{K_C}(P) + h_D(P) - \varepsilon'\left(h_{K_C}(P) + h_{D_1}(P)\right) \leq \mathrm{cond}_\lambda^K(P) + d(P) + O_{[K(P):K]}(1).$$

By A.6.10, there is $n \in \mathbb{N}$ such that $nH - D_1 - K_C$ is ample, hence

$$h_{D_1} + h_{K_C} \leq nh_H + O(1).$$

Choosing $\varepsilon' = \varepsilon/n$, we get (b).

(b) $\Rightarrow$ (c): This is obvious from $h_D(P) = m_{S,D}(P) + N_{S,D}(P)$ and from

$$\mathrm{cond}_\lambda^K(P) \leq N_{S,D}(P) + O(1)$$

easily deduced from Proposition 14.4.8.

(c) $\Rightarrow$ (d) is trivial.

(d) $\Rightarrow$ (a): Let $P \in \mathbb{P}^1_{\mathbb{Q}} \setminus \{0, 1, \infty\}$ with affine coordinate $x = x(P) \in \overline{\mathbb{Q}}$. We identify $\mathbb{P}^1_{\mathbb{Q}}$ with $C := C_1$ and we consider the covering $\pi : C' \to C$ of Fermat curves with $C' := C_n$ for suitable $n \geq 1$ (see Example 14.4.15). We choose $P' \in C'$ with $P = \pi(P')$. For $D := [0] + [1] + [\infty]$ and $D' := \pi^{-1}\{0, 1, \infty\}$, Example 14.4.15 yields easily $\pi^*D = nD'$ and hence $R_\pi = (n-1)D'$ proves

$$N_{S,R_\pi}(P') = \left(1 - \frac{1}{n}\right) N_{S,\pi^*D}(P') + O(1) = \left(1 - \frac{1}{n}\right) N_{S,D}(P) + O(1).$$

Applying (14.11) on page 495 to the covering π, we get

$$h_D(P) + h_{K_C}(P) - d(P) \leq \mathrm{cond}_\lambda^{\mathbb{Q}}(P) + \frac{1}{n} N_{S,D}(P) + \varepsilon' h_H(P) + O_{[\mathbb{Q}(P):\mathbb{Q}]}(1).$$

By Proposition 2.3.9, we have $N_{S,D}(P) \leq h_D(P) + O(1)$ and hence

$$\left(1 - \frac{1}{n}\right) h_D(P) + h_{K_C}(P) - \varepsilon' h_H(P) \leq \mathrm{cond}_\lambda^{\mathbb{Q}}(P) + d(P) + O_{[\mathbb{Q}(P):\mathbb{Q}]}(1).$$

Since $C = \mathbb{P}^1_{\mathbb{Q}}$, $K_C \cong O_{\mathbb{P}^1}(-2)$ (cf. A.13.6), $H = O_{\mathbb{P}^1}(1)$ with h_H the standard height and Example 14.4.4, we get

$$\left(1 - \frac{3}{n} - \varepsilon'\right) h(P) \leq \operatorname{cond}^{\mathbb{Q}}_{[0]}(x(P)) + \operatorname{cond}^{\mathbb{Q}}_{[1]}(x(P)) + \operatorname{cond}^{\mathbb{Q}}_{[\infty]}(x(P)) \\ + d(P) + O_{[\mathbb{Q}(P):\mathbb{Q}]}(1)$$

for all $P \in \mathbb{P}^1_{\mathbb{Q}} \setminus \{0, 1, \infty\}$ with affine coordinate $x(P)$. If we choose $\varepsilon' = \varepsilon/2$ and $n \geq 6/\varepsilon$, we get (a). □

Remark 14.4.17. If we are only interested in $x \in K \setminus \{0, 1\}$ for a *fixed* number field K, then dependence on K plays no role and the strong abc-conjecture in 14.4.12 and Corollary 14.4.7 imply the **K-rational abc-conjecture**

$$(1 - \varepsilon)h(x) \leq \operatorname{cond}^K_{[0]}(x) + \operatorname{cond}^K_{[1]}(x) + \operatorname{cond}^K_{[\infty]}(x) + O(1).$$

The proof of Theorem 14.4.16 can be adapted to show that the K-rational abc-conjecture implies the K-rational version of Conjecture 14.4.10, namely

$$h_D(P) + h_{K_C}(P) \leq \operatorname{cond}^K_\lambda(P) + \varepsilon h_H(P) + O(1)$$

for all $P \in C(K) \setminus \operatorname{supp}(D)$. In particular, choosing $K = \mathbb{Q}$, $C = \mathbb{P}^1_{\mathbb{Q}}$ and $D = \operatorname{div}(F)$, we get immediately Theorem 12.2.12. Indeed, we have

$$\log(\operatorname{rad}(F(m, n))) = \operatorname{cond}^{\mathbb{Q}}_\lambda((m : n))$$

for the local height

$$\lambda(\mathbf{x}, p) = \log\left(\frac{\max\{|x_0|^d_p, |x_1|^d_p\}}{|F(x_0, x_1)|_p}\right).$$

14.4.18. In Section 12.2 we have shown that the strong abc-conjecture over $\mathbb{Q}$ implies the classical Roth theorem over $\mathbb{Q}$. It is now clear, as pointed out by Elkies, that the K-rational abc-conjecture implies Roth's theorem over the number field K in 6.2.3.

Indeed, the proof of Theorem 14.4.16 shows that the K-rational abc-conjecture implies Vojta's height inequality in 14.3.2 for K-rational points of an irreducible smooth projective curve X over K. Hence Remark 14.3.3 proves the claim.

By Remark 14.3.5, the same argument proves the following result of Elkies [**98**]:

Theorem 14.4.19. *The K-rational abc-conjecture implies Faltings's theorem in 11.1.1 for the number field K.*

Remark 14.4.20. The proof we have given actually shows that an effective version of the K-rational abc-conjecture implies effective versions of Roth's and Faltings's theorems.

Remark 14.4.21. Equivalently, we may formulate the strong abc-conjecture as

$$\begin{aligned} h(x) \leq\ & C_1 \cdot \left(\mathrm{cond}_{[0]}^{\mathbb{Q}}(x) + \mathrm{cond}_{[1]}^{\mathbb{Q}}(x) + \mathrm{cond}_{[\infty]}^{\mathbb{Q}}(x)\right) \\ & + C_2 \cdot d_{\mathbb{Q}(x)} + O_{C_1, C_2, [\mathbb{Q}(x):\mathbb{Q}]}(1) \end{aligned} \tag{14.14}$$

for $x \in \overline{\mathbb{Q}} \setminus \{0, 1\}$ with $C_1 = C_2 > 1$.

Such a result would be quite useful for any constants C_1, C_2. Note also that D. Masser **[194]** proved that (14.14) is false for $C_2 = 1$ and any C_1, with a method similar as the one used by Stewart–Tijdeman in proving Theorem 12.4.6.

14.5. The abc-theorem for function fields

We have seen in Example 12.4.1 that the abc-conjecture holds for complex polynomials. In this section, we extend this to a function field K of characteristic 0 proving the theorem of Stothers and Mason. But first, we transfer the results of the last section to the case of function fields of characteristic 0. Then we prove the abc-conjecture and also the Vojta height inequality in the split function field case where no ε-term is necessary. In this situation, we can also transfer Cartan's second main theorem with similar arguments as in Nevanlinna theory. As a corollary, we obtain the result of Voloch and Brownawell-Masser bounding the non-degenerate solutions of the unit equation with several summands.

14.5.1. In this section, $K = k(B)$ denotes a function field of an irreducible projective variety B over a field k of characteristic zero. We assume that B is regular in codimension 1 and we fix an ample class $\mathbf{c}$ on B.

Let M_K be the set of prime divisors on B. We recall from Proposition 1.4.7 that the discrete absolute values

$$|f|_v := e^{-\deg_{\mathbf{c}}(v)\mathrm{ord}_v(f)} \quad (f \in K, v \in M_K)$$

satisfy the product formula.

14.5.2. By Lemma 1.4.10, every finite extension F/K is a function field $F = k(Y)$ for a variety Y regular in codimension 1 and a finite morphism $p : Y \to B$. We have seen in Remark 1.4.11 that the set of places M_F is independent of the choice of the model Y and the argument shows that two such models are isomorphic outside subsets of codimension at least 2. In fact, we may choose the normalization of B in F as a canonical model. By Example 1.4.13, the absolute values on M_F are normalized according to 1.3.6 by

$$|f|_w := e^{-\deg_{p^*\mathbf{c}}(w)\mathrm{ord}_w(f)/[F:K]} \quad (f \in F, w \in M_F).$$

14.5.3. In what follows, we assume that the reader is familiar with Appendix B.4. In analogy to 14.3.8, we define the counting function of the ramification divisor

R_p by

$$N_{R_p} = -\sum_{v \in M_F} \ell(\Omega_{Y/B,v}) \log |\pi_v|_v.$$

Note that π_v is a local parameter of v, hence

$$N_{R_p} = \frac{1}{[F:K]} \deg_{p^*\mathbf{c}}(R_p).$$

Let K_Y (resp. K_B) be the canonical line bundle on the smooth part Y_{reg} (resp. B_{reg}). Since the complement of the smooth part has codimension at least 2, the corresponding canonical divisors are well-defined in the Chow groups by passing to the Zariski closures of the components, hence their degrees are also well-defined. By Theorem B.4.5 and projection formula (A.13) on page 558, we get

$$N_{R_p} = \frac{1}{[F:K]} \left(\deg_{p^*\mathbf{c}}(K_Y) - \deg_{p^*\mathbf{c}}(p^* K_B)\right) = \frac{\deg_{p^*\mathbf{c}}(K_Y)}{[F:K]} - \deg_{\mathbf{c}}(K_B).$$

This suggests the following analogue of the absolute logarithmic discriminant

$$d_F := \frac{\deg_{\mathbf{c}}(K_Y)}{[F:K]}.$$

By the arguments in 14.5.2, all these quantities do not depend on the choice of Y.

Example 14.5.4. If B is a geometrically irreducible smooth projective curve and $\mathbf{c}$ is the equivalence class of a point, then A.13.6 shows that $d_F = (2g(Y) - 2)/[F:K]$, where $g(Y)$ is the genus of Y.

Definition 14.5.5. *For a complete variety X over K with any local height λ relative to a Cartier divisor D and for a finite subset $S \subset M_K$, the* ***counting function*** *$N_{S,D}$ and the* ***proximity function*** *$m_{S,D}$ are defined as in 14.3.1. By 14.5.3, we define $d(P) := d_{K(P)}$ for any $P \in X$.*

If D is effective, then we define the ***conductor*** cond_λ^K *as in Definition 14.4.2, where the sum now ranges over all $M_{K(P)}$.*

Example 14.5.6. A point $P = (f_0 : \cdots : f_n) \in \mathbb{P}_k^n(F)$ induces a rational map $f_P : Y \dashrightarrow \mathbb{P}_k^n$ defined over k. By Example 2.4.11, the standard height satisfies

$$h(P) = \frac{1}{[F:K]} \deg_{p^*\mathbf{c}} f_P^* H$$

for every hyperplane H of $\mathbb{P}_k^n$ with $P \notin H$.

Example 14.5.7. More generally, we consider a complete variety X with a Cartier divisor D defined over the constant field k. Then $P \in X(F) \setminus \mathrm{supp}(D)(F)$ induces a rational map $f_P : Y \dashrightarrow X$ (locally defined as in Example 14.5.6), but note that f_P is intrinsically defined because X is defined over k. We claim that

$$\lambda(P, v) = \frac{1}{[F:K]} \mathrm{ord}_v(f_P^* D) \deg_{p^*\mathbf{c}}(v) \quad (v \in M_F)$$

is a local height relative to D. If D is a hyperplane section, then this follows similarly as in Example 14.5.6. On a projective variety, D is the difference of two hyperplane sections in suitable embeddings (cf. A.6.10, A.6.11), thus leading to the claim. For complete varieties, we have to use the theory of local heights from Section 2.7. Indeed, $O(D)$ has a canonical metric $\| \ \|_v$ given by $\|s\|_v := 1$ for any local nowhere vanishing section s defined over k and it is immediate that $\lambda(P, v)$ is the corresponding local height. We conclude that

$$h_D(P) = \frac{1}{[F:K]} \deg_{p^*\mathbf{c}}(f_P^*(D)).$$

By the valuative criterion of properness, the complement of the domain of f_P has codimension at least 2 in Y (see A.11.10), hence $f_P^*(D)$ may be viewed as a Weil divisor on Y respecting rational equivalence. Let S be a finite subset of M_K, then $f_P^*(D) = f_P^*(D)_S + f_P^*(D)_T$, where the first summand is supported over S and the second over the complement T of S. We conclude

$$m_{S,D}(P) = \frac{1}{[F:K]} \deg_{p^*\mathbf{c}}(f_P^*(D)_S), \quad N_{S,D}(P) = \frac{1}{[F:K]} \deg_{p^*\mathbf{c}}(f_P^*(D)_T).$$

For $F = K(P)$, it is clear that

$$\mathrm{cond}_D^K(P) = \frac{1}{[F:K]} \deg_{p^*\mathbf{c}}(f_P^*(D)_{\mathrm{red}}),$$

where we recall that E_{red} denotes the sum of the components of the divisor E.

14.5.8. We introduce a technique to reduce the canonical quantities from Example 14.5.7 to the case of a function field of a curve. By intersection theory, all quantities are homogeneous in $\mathbf{c}$ of degree $\dim(B) - 1$. So we may assume that $\mathbf{c}$ is very ample giving rise to a closed embedding $B \to \mathbb{P}_K^n$. Let $F := K(P)$ and let Y be a model for F as in 14.5.2. By Bertini's theorem (cf. [**148**], Cor.III.10.9), we may choose (generic) hyperplanes $H_1, \ldots, H_{\dim(B)-1}$ of $\mathbb{P}_K^n$ such that the proper intersection product

$$Y_{\mathbf{c}} := p^*H_1 \ldots p^*H_{\dim(B)-1}.Y$$

is an irreducible smooth projective curve over k. Bertini's theorem is usually stated for k algebraically closed and X regular, but this implies easily the claim for any field of characteristic zero and the singularities of Y do not disturb the applications of Bertini's theorem because the dimension of the singular part decreases at least by 1 every time we intersect with a generic hyperplane. Similarly, we may assume that

$$B_{\mathbf{c}} := H_1 \ldots H_{\dim(B)-1}.B$$

is an irreducible smooth projective curve over k. Let $K_{\mathbf{c}}$, $F_{\mathbf{c}}$ be the function fields of $B_{\mathbf{c}}$ and $Y_{\mathbf{c}}$. Then the projection formula (A.13) on page 558 gives

$$[F:K] = \deg_{p^*\mathbf{c}} Y / \deg_{\mathbf{c}} B = \deg_{p^*\mathbf{c}} Y_{\mathbf{c}} / \deg_{\mathbf{c}} B_{\mathbf{c}} = [F_{\mathbf{c}} : K_{\mathbf{c}}]. \qquad (14.15)$$

Moreover, it is clear that P may be viewed as a point $P_{\mathbf{c}} \in X(F_{\mathbf{c}})$ (see below). We conclude that $K_{\mathbf{c}}(P_{\mathbf{c}}) = F_{\mathbf{c}}$. The goal is to show that the quantities from Example 14.5.7 for P are the same as the corresponding quantities for $P_{\mathbf{c}}$ relative to the function field $K_{\mathbf{c}}$ of the curve $B_{\mathbf{c}}$.

Let U be the largest open subset of Y_{reg} where f_P is defined. Because $Y \setminus U$ has codimension at least 2 in Y and choosing $H_1, \ldots, H_{\dim(B)-1}$ generic, we may assume that $p^*H_1, \ldots, p^*H_{\dim(B)-1}$ and the closure of f_P^*D in Y intersect properly (even generically transversely) and that the proper intersection product is supported in U. We may assume also that the restriction g_P of f_P to $Y_{\mathbf{c}}$ is a well-defined morphism. By commutativity of the proper intersection product (generalizing A.9.20, see [**125**], Th.2.4), we have

$$p^*H_1 \ldots p^*H_{\dim(B)-1}.f_P^*D = f_P^*D.Y_{\mathbf{c}} = g_P^*D$$

and hence

$$h_D(P) = \frac{1}{[F:K]} \deg_{p^*\mathbf{c}}(g_P^*D) = h_D(P_{\mathbf{c}}).$$

For $v_1, \ldots, v_m \in M_F$ and choosing $H_1, \ldots, H_{\dim(B)-1}$ generic, we may assume that

$$p^*H_1 \ldots p^*H_{\dim(B)-1}.v_i = \sum_{j=1}^{\deg_{p^*\mathbf{c}}(v_i)} [y_{ij}]$$

and that all the points y_{ij} are different (here, we use that the intersections are generically transverse and A.9.22). If we choose for $v_1, \ldots, v_m$ the components of f_P^*D, then we get

$$p^*H_1 \ldots p^*H_{\dim(B)-1}.(f_P^*D)_{\text{red}} = (g_P^*D)_{\text{red}},$$

proving easily

$$\operatorname{cond}_D^K(P) = \operatorname{cond}_D^{K_{\mathbf{c}}}(P_{\mathbf{c}}).$$

Applying the above to a finite subset $S = \{v_1, \ldots, v_m\}$ of M_K and to $p^{-1}(S)$, we get a finite subset $S_{\mathbf{c}}$ of $M_{K_{\mathbf{c}}}$ formed by the set of prime divisors, defined over K, containing a point y_{ij}. As above, we may assume that

$$N_{S,D}(P) = N_{S_{\mathbf{c}},D}(P_{\mathbf{c}}), \quad m_{S,D}(P) = m_{S_{\mathbf{c}},D}(P_{\mathbf{c}}).$$

As we may work on Y_{reg} (the complement has codimension at least 2), Proposition II.8.20 of [**148**] implies

$$\operatorname{div}(K_{Y_{\mathbf{c}}}) \sim (\operatorname{div}(K_Y) + p^*H_1 + \cdots + p^*H_{\dim(B)-1}).Y_{\mathbf{c}}$$

and hence

$$d(P_{\mathbf{c}}) = d(P) + (\dim(B) - 1)\deg_{\mathbf{c}}(B)$$

using (14.15) and

$$\deg\left((p^*H_1 + \cdots + p^*H_{\dim(B)-1}).Y_{\mathbf{c}}\right) = (\dim(B) - 1)\deg_{p^*\mathbf{c}}(Y).$$

Remark 14.5.9. Now we restrict our attention to the case $X = C$ an irreducible smooth projective curve over K. The same arguments prove that the results 14.4.5–14.4.9 hold also in the context of function fields. In order to prove Proposition 14.4.6, we have to use Proposition B.4.9 replacing Dedekind's discriminant theorem. In Proposition 14.4.8, we have to replace the right-hand side by

$$\sum_{v \in S_0} \log |1/\pi_v|_v = \deg(S_0).$$

We leave the details to the reader. By the way, we may easily extend these results to higher-dimensional complete varieties X, but we do not need that in the sequel.

14.5.10. We may pose the conjectures 14.4.10 and 14.4.13 also in the function field case. However, as no Belyĭ function exists in this case, we are unable to show that the strong abc-conjecture implies the other conjectures. As a substitute for the Fermat coverings in 14.4.15, we will use the following result:

Lemma 14.5.11. *Let C be an irreducible smooth projective curve over K and let $n \in \mathbb{N}\setminus\{0\}$. Suppose that we have disjoint non-zero rationally equivalent reduced divisors D_1, D_2 on C. Then there is an irreducible smooth projective curve C' over K and a finite morphism $\pi : C' \to C$ of degree n which has ramification divisor $R_\pi = (n-1)D'$ for $D' := (\pi^*(D_1 + D_2))_{\mathrm{red}}$.*

Proof: The argument works for any field K of characteristic zero. There is $f \in K(C) \setminus \{0\}$ with

$$\operatorname{div}(f) = D_1 - D_2.$$

Let C' be the irreducible smooth projective curve with function field $K(C)(\sqrt[n]{f})$ and let $\pi : C' \to C$ be the natural finite morphism induced by the extension of function fields (use Lemma 1.4.10). First note that

$$\deg(\pi) = [K(C') : K(C)] \leq n.$$

By B.1.9, we verify easily that the discriminant of $x^n - f$ is $\pm n^n f^{n-1}$. By Lemma B.2.2, we conclude that $K(C')/K(C)$ is unramified over any $v \in M_{K(C)}$ which is no component of D_1 or D_2. For $w \in M_{K(C')}$ with $v := \pi(w)$ equal to a component of D_1, the ramification index satisfies

$$e_{w/v} = \operatorname{ord}_w(f) = n \cdot \operatorname{ord}_w(\sqrt[n]{f}) \geq n.$$

Similarly, we obtain $e_{w/v} \geq n$ if v is a component of D_2. By Example 1.4.12, we have

$$[K(C') : K(C)] = \sum_w e_{w/v} f_{w/v} = \sum_w e_{w/v} [K(w) : K(v)],$$

where w ranges now over all components of $\pi^{-1}(v)$ for a given component v of $D_1 + D_2$. We conclude that the fibre consists only of one component w and

$$\deg(\pi) = [K(C') : K(C)] = e_{w/v} = n, \quad [K(w) : K(v)] = 1.$$

By Proposition B.4.9, we get the claim. □

Theorem 14.5.12. *The following conjectures are equivalent for the function field K:*

(a) *Conjecture 14.4.10 for $C = \mathbb{P}^1_K$;*

(b) *Conjecture 14.4.10 for all curves C over K;*

(c) *Vojta's conjecture in 14.4.13 for all curves C over K.*

Proof: (a) $\Rightarrow$ (b): We have still a finite covering $f : C \to \mathbb{P}^1_K$ using any non-constant rational function. We choose a reduced effective divisor D_0 on $\mathbb{P}^1_K$ such that $\mathrm{supp}(D) \subset f^{-1}(D_0)$ and such that f is unramified outside of $f^{-1}(D_0)$. Then (b) follows from (a) along the same lines as in Theorem 14.4.16.

(b) $\Rightarrow$ (c): This is analogous to the proof of Theorem 14.4.16.

(c) $\Rightarrow$ (a): The goal is to prove Conjecture 14.4.10 for an effective reduced divisor D on $\mathbb{P}^1_K$. By the analogues of Propositions 14.4.5 and 14.4.9, we may replace D by a larger effective reduced divisor. We choose any effective reduced divisor D_1 disjoint from D with $\deg(D_1) = \deg(D)$. Note that $D \sim D_1$. We may replace D by $D + D_1$. For $n \in \mathbb{N} \setminus \{0\}$, Lemma 14.5.11 gives a finite morphism $\pi : C' \to \mathbb{P}^1_K$ of degree n with ramification divisor

$$R_\pi = (n-1)\,(\pi^*(D))_{\mathrm{red}}\,.$$

Using this covering instead of Example 14.4.15, the proof of (a) is completely similar as the implication (d) $\Rightarrow$ (a) in the proof of Theorem 14.4.16. □

If everything is defined over the constant field k, then we have seen in Example 14.5.7 that we have canonical local heights induced by geometry. In this special case, we may prove the abc-conjecture, even without the ε-term. The corresponding result is the following **theorem of Stothers and Mason.**

Theorem 14.5.13. *Let C be an irreducible smooth projective curve over k and let D be an effective reduced divisor on C defined over the constant field k. If we use the canonical local heights from Example 14.5.7 relative to D and K_C, then*

$$h_{D+K_C}(P) \le \mathrm{cond}_D^K(P) + d(P) + (\dim(B) - 1)\deg_{\mathbf{c}}(B)$$

for all $P \in C \setminus \big(\mathrm{supp}(D) \cup C(\overline{k})\big)$.

Proof: The basic idea is to use Bertini's theorem to reduce the problem to the case of a function field of a curve and then apply Hurwitz's theorem.

By 14.5.8, we may assume that B is a curve. Let $F := K(P)$ with a model $p : Y \to B$ as in 14.5.2. Taking into account the equivalence

$$\mathrm{div}(f_P^* K_C) \sim \mathrm{div}(K_Y) - R_{f_P}\,,$$

from Theorem B.4.5, we conclude

$$h_{K_C}(P) = \frac{1}{[F:K]}\big(\deg_{p^*\mathbf{c}}(K_Y) - \deg_{p^*\mathbf{c}}(R_{f_P})\big)\,.$$

From Proposition B.4.7, we get

$$R_{f_P} \geq f_P^* D - (f_P^* D)_{\mathrm{red}} .$$

This proves

$$\begin{aligned} h_{D+K_C} &\leq \frac{1}{[F:K]} \left(\deg_{p^*\mathbf{c}} (f_P^* D)_{\mathrm{red}} + \deg_{p^*\mathbf{c}}(K_Y)\right) \\ &= \mathrm{cond}_D^K(P) + d(P) \end{aligned}$$

proving the claim. □

Example 14.5.14. In particular, if B is a geometrically irreducible smooth projective curve over k and $P \in C(k(B)) \setminus C(\overline{k})$, then

$$h_{D+K_C}(P) \leq \mathrm{cond}_D^{k(B)}(P) + 2g(B) - 2. \tag{14.16}$$

We consider the special case $C = \mathbb{P}_k^1$ and $D = [0] + [1] + [\infty]$. There are $a(x), c(x) \in k(B)$ with $P = (c(x) : a(x))$. We define $b(x)$ by $a(x) + b(x) = c(x)$. Since $K_{\mathbb{P}^1} \cong O_{\mathbb{P}^1}(-2)$, the left-hand side of (14.16) is the standard height

$$h_{D+K_C}(P) = h(P) = h((a:b:c)).$$

We set $S := f^{-1}(D)$, hence $|S| = \mathrm{cond}_D^{k(B)}(P)$. Dividing a, b, c by c, we may assume that a, b, c are S-units and (14.16) yields

$$h((a:b:c)) \leq |S| + 2g(B) - 2,$$

which was already mentioned in Remark 12.4.3 and which was proved by Mason [**192**]. Assuming also $B = \mathbb{P}_k^1$, we may assume that a, b, c are coprime polynomials in $k[x]$. Let n_0 be the number of different zeros of $a(x)b(x)c(x)$. We get

$$|S| = \mathrm{cond}_D^{k(B)}(P) = \begin{cases} n_0 & \text{if } \deg(a) = \deg(b) = \deg(c), \\ n_0 + 1 & \text{otherwise.} \end{cases}$$

Hence we recover Theorem 12.4.1, which was proved by Stothers [**293**], from (14.16).

The following simple lemma is the substitute of the lemma on the logarithmic derivative in the function field case.

Lemma 14.5.15. *Let Y be an irreducible smooth curve over k, let $f \in k(Y)$, let $v \in M_{k(Y)}$ and let $j \in \mathbb{N}$. Then*

$$\mathrm{ord}_v \left(f^{-1} \left(\frac{\mathrm{d}}{\mathrm{d}\pi_v} \right)^j f \right) \geq -j$$

for any local parameter π_v in $\mathcal{O}_{Y,v}$. Moreover, if f is a unit in $\mathcal{O}_{Y,v}$ it holds

$$\mathrm{ord}_v \left(f^{-1} \left(\frac{\mathrm{d}}{\mathrm{d}\pi_v} \right)^j f \right) \geq 0.$$

Proof: By the definition of $\mathrm{d}/\mathrm{d}\pi_v$ in A.7.25 and noting that $\mathrm{d}\pi_v$ is a basis of $\Omega^1_{C,v}$ (see proof of Proposition B.4.7), we conclude that $\mathrm{d}/\mathrm{d}\pi_v$ is a derivative with

$$\frac{\mathrm{d}}{\mathrm{d}\pi_v}\mathcal{O}_{Y,v} \subset \mathcal{O}_{Y,v}.$$

Then the claim follows easily from

$$f = u\pi_v^{\mathrm{ord}_v(f)}$$

for a unit u in $\mathcal{O}_{Y,v}$ and from Leibniz's rule. □

14.5.16. Our next goal is an analogue of **Cartan's second main theorem for function fields**. As we will see the arguments are very similar as for Theorem 13.4.16.

The basic assumptions are the following: Let S be a finite subset of M_K, let $H_1, \ldots, H_q$ be hyperplanes of $\mathbb{P}^n_K$ in general position, let $D := \sum_{j=1}^q H_j$ and let P be a point of $\mathbb{P}^n_K$ not lying in any hyperplane defined over k. As in 14.5.2, we may choose a model Y of $F := K(P)$.

14.5.17. We first assume that K is the function field of a curve. This case resembles most the field of meromorphic functions on $\mathbb{C}$ considered in Nevanlinna theory. It enables us to define the ramification divisor of P in the following way:

The point P is given by a rational map $f_P : Y \dashrightarrow \mathbb{P}^n_k$ defined over k. For $v \in M_F$, we may choose relatively prime elements $f_0^v, \ldots, f_n^v$ of the discrete valuation ring $\mathcal{O}_{Y,v}$ with $f_P = (f_0^v : \cdots : f_n^v)$. By assumption, $f_0^v, \ldots, f_n^v$ are linearly independent over k and so we may define $\mathrm{ord}_v(R_P)$ as the order of the Wronskian of $f_0^v, \ldots, f_n^v$ in v, where the derivatives are taken with respect to a local parameter π_v in v. By Leibniz's rule and the multilinearity of the determinant, this is independent of the choices of $f_0^v, \ldots, f_n^v$ and π_v. Then we define the ramification divisor to be

$$R_P := \sum_{v \in M_F} \mathrm{ord}_v(R_P) v,$$

with counting function

$$N_{S,\mathrm{ram}}(P) := \sum_{v \notin p^{-1}S} \mathrm{ord}_v(R_P) \log |1/\pi_v|_v.$$

Theorem 14.5.18. *Let K be the function field of a curve in characteristic 0 and let $\varepsilon > 0$. Under the assumptions in 14.5.16, it holds*

$$m_{S,D}(P) + N_{S,\mathrm{ram}}(P) \leq (n+1)h(P) + \frac{n(n+1)}{2}\left(d(P) + \sum_{v \in p^{-1}S} \log |1/\pi_v|_v\right),$$

where we use the canonical proximity function from Example 14.5.7.

Proof: We proceed as in the proof of Theorem 13.4.16. We choose fixed $f_0, \ldots, f_n \in F$ with $f_P = (f_0 : \cdots : f_n)$. Of course, we cannot assume that they are regular and relatively prime. Another difficulty is that no global coordinate z is available on Y. In order to overcome this problem, we choose $z \in F \setminus \overline{k}$ as a reference for differentiation. Then we use the Wronskian

$$W(f_0, \ldots, f_n) := \det\left(\left(\frac{\mathrm{d}}{\mathrm{d}z}\right)^j f_i\right)_{i,j=0,\ldots,n}.$$

Again, we may assume $q \geq n+2$ and we set $p' := q - n - 1$. Let $H_j := \operatorname{div}(\ell_j(\mathbf{x}))$ for linear forms $\ell_j(\mathbf{x})$ and let $g_j := \ell_j(f_0, \ldots, f_n)$. First, it is clear that

$$h\left((g_{k_1} \cdots g_{k_{p'}})_{k_1 < \cdots < k_{p'}}\right) = p' \cdot h(P). \tag{14.17}$$

No $O(1)$-term is necessary because the heights are canonical. Let $I = \{i_1 < \cdots < i_{n+1}\}$ be the complement of $\{k_1 < \cdots < k_{p'}\}$ in $\{1, \ldots, q\}$. For the logarithmic Wronskian

$$\lambda(g_{i_1}, \ldots, g_{i_{n+1}}) := \frac{W(g_{i_1}, \ldots, g_{i_{n+1}})}{g_{i_1} \cdots g_{i_{n+1}}},$$

we get again the fundamental identity

$$g_{k_1} \cdots g_{k_{p'}} = \frac{g_1 \cdots g_q}{W(f_0, \ldots, f_n)} \cdot d_I^{-1} \cdot \lambda(g_{i_1}, \ldots, g_{i_{n+1}}), \tag{14.18}$$

where d_I is the determinant formed with the coefficients of $(\ell_i)_{i \in I}$.

Let $v \in p^{-1}S$. We choose a local parameter π_v for v. Leibniz's rule and the multilinearity of the determinant give

$$\lambda(g_{i_1}, \ldots, g_{i_{n+1}}) = (\mathrm{d}\pi_v/\mathrm{d}z)^{\frac{n(n+1)}{2}} \lambda_v(g_{i_1}, \ldots, g_{i_{n+1}}), \tag{14.19}$$

where λ_v denotes the logarithmic Wronskian with respect to the differential operator $\mathrm{d}/\mathrm{d}\pi_v$ instead of $\mathrm{d}/\mathrm{d}z$. By Lemma 14.5.15, we get

$$|\lambda(g_{i_1}, \ldots, g_{i_{n+1}})|_v \leq |\mathrm{d}\pi_v/\mathrm{d}z|_v^{\frac{n(n+1)}{2}} \cdot |1/\pi_v|_v^{\frac{n(n+1)}{2}}.$$

Together with the product formula, (14.18) implies

$$\begin{aligned}
&\sum_{v \in p^{-1}S} \max_{k_1 < \ldots < k_{p'}} \log |g_{k_1} \cdots g_{k_{p'}}|_v \\
&\quad \leq \sum_{v \notin p^{-1}S} \log \left|\frac{W(f_0, \ldots, f_n)}{g_1 \cdots g_q}\right|_v \\
&\quad\quad + \frac{n(n+1)}{2} \sum_{v \in p^{-1}S} \left(\log |\mathrm{d}\pi_v/\mathrm{d}z|_v + \log |1/\pi_v|_v\right).
\end{aligned}$$

For $i = 0, \ldots, n$ the functions

$$f_i^v := \pi_v^{-\min_{j=0,\ldots,n} \operatorname{ord}_v(f_j)} f_i$$

are regular and relatively prime in v. As in (14.19), we get

$$W(f_0, \ldots, f_n) = (\mathrm{d}\pi_v/\mathrm{d}z)^{\frac{n(n+1)}{2}} \pi_v^{(n+1)\min_{i=0,\ldots,n} \operatorname{ord}_v(f_i)} W_v(f_0^v, \ldots, f_n^v),$$

where W_v denotes the Wronskian with respect to the differentiation $\mathrm{d}/\mathrm{d}\pi_v$. We conclude

$$\sum_{v \notin p^{-1}S} \log |W(f_0, \ldots, f_n)|_v = -N_{S,\mathrm{ram}}(P)$$
$$+ \sum_{v \notin p^{-1}S} \left(\frac{n(n+1)}{2} \log \left| \frac{\mathrm{d}\pi_v}{\mathrm{d}z} \right|_v + (n+1) \log \max_i |f_i|_v \right).$$

Moreover, we have

$$\begin{aligned} \lambda_{H_j}(P, v) &= -\log |\ell_j(f_0^v, \ldots, f_n^v)|_v \\ &= -\log |g_j|_v + \log \max_i |f_i|_v. \end{aligned}$$

We conclude that

$$\sum_{v \in p^{-1}S} \max_{k_1 < \cdots < k_{p'}} \log |g_{k_1} \cdots g_{k_{p'}}|_v \leq$$
$$\sum_{j=1}^{q} N_{S,H_j}(P) - N_{S,\mathrm{ram}}(P) - \sum_{v \notin p^{-1}S} p' \log \max_i |f_i|_v$$
$$+ \frac{n(n+1)}{2} \left(\sum_{v \in M_F} \log \left| \frac{\mathrm{d}\pi_v}{\mathrm{d}z} \right|_v + \sum_{v \in p^{-1}S} \log |1/\pi_v|_v \right).$$

Note that $\mathrm{d}z$ may be viewed as a meromorphic section of K_Y and hence

$$\sum_{v \in M_F} \log \left| \frac{\mathrm{d}\pi_v}{\mathrm{d}z} \right|_v = d(P).$$

From (14.17), on page 506 we have

$$p'h(P) = \sum_{v \in M_F} \log \max_{k_1 < \cdots < k_{p'}} |g_{k_1} \cdots g_{k_{p'}}|_v.$$

For $v \notin p^{-1}S$, we use

$$\log |g_{k_1} \cdots g_{k_{p'}}|_v \leq p' \cdot \log \max_i |f_i|_v.$$

This leads to

$$p' \cdot h(P) \leq \sum_{j=1}^{q} N_{S,H_j}(P) - N_{S,\mathrm{ram}}(P) + \frac{n(n+1)}{2}\Big(d(P) + \sum_{v \in p^{-1}S} \log|1/\pi_v|_v\Big)$$

proving immediately the claim. □

Remark 14.5.19. To generalize Cartan's second main theorem to a higher-dimensional function field $K = k(B)$, we introduce the nth **truncated counting function**

$$N_D^{(n)}(P) := \frac{1}{[F:K]} \sum_{v \in M_F} \min\left(\mathrm{ord}_v(f_P^* D), n\right) \deg_{p^*\mathbf{c}}(v)$$

for $n \in \mathbb{N}$ and where $F := K(P)$ is as in 14.5.2. By Example 14.5.7, we have $N_D^{(1)}(P) = \mathrm{cond}_D^K(P)$. Then the desired generalization is

$$\begin{aligned}(q-n-1)h(P) \leq & \sum_{j=1}^{q} N_{H_j}^{(n)}(P) + \\ & + \frac{n(n+1)}{2}\left(d(P) + (\dim(B)-1)\deg_{\mathbf{c}}(B) + \frac{\deg_{p^*\mathbf{c}}(p^{-1}S)}{[F:K]}\right).\end{aligned} \tag{14.20}$$

To give a sketch of proof, we first note that we may assume B to be a curve by the techniques of 14.5.8. Then the claim follows from Theorem 14.5.18 and from the identity

$$qh(P) - N_{\emptyset,\mathrm{ram}}(P) \leq \sum_{j=1}^{q} N_{H_j}^{(n)}(P),$$

which is easily deduced from the properties of the Wronskian (similarly as in [**249**], Lemma A.3.2.1).

In the following application to the generalized unit equation, we will not use (14.20), but we will directly reduce it to Theorem 14.5.18, which is sharper.

Let S be a finite subset of M_K. Recall that $u \in K$ is called an S-unit if $\mathrm{ord}_v(u) = 0$ for all $v \in M_K \setminus S$. As a consequence of the above theorem, we obtain:

Corollary 14.5.20. *Suppose that $u_0, \ldots, u_n$ are S-units in K satisfying*

$$u_0 + \cdots + u_n = 0 \tag{14.21}$$

and assume that the elements $(u_i)_{i\in I}$ are linearly independent over k for every proper subset I of $\{0,\dots,n\}$. Then

$$h((u_0:\cdots:u_n)) \le \frac{n(n-1)}{2}\left(d_K + \deg_{\mathbf{c}}(S) + (\dim(B)-1)\deg_{\mathbf{c}}(B)\right).$$

Proof: By the techniques introduced in 14.5.8, we may assume that B is a curve. We note also that

$$\deg_{\mathbf{c}}(S) = \sum_{v\in S} \log|1/\pi_v|_v.$$

We apply Theorem 14.5.18 to $P := (u_1:\cdots:u_n) \in \mathbb{P}^{n-1}(K)$ and to the hyperplanes $H_j := \{x_j = 0\}$ for $j = 0,\dots,n-1$ and $H_n := \{x_0+\cdots+x_{n-1} = 0\}$. Because the u_is are S-units, we have $N_{S,H_j}(P) = 0$ and hence $m_{S,H_j}(P) = h(P)$. Omitting the term $N_{S,\mathrm{ram}}(P) \ge 0$ in Theorem 14.5.18 and using

$$h(P) = h((u_0:\cdots:u_n))$$

as an easy consequence of (14.21), we get the claim. □

The special case in which K is the function field of a curve B of genus g is the **theorem of Voloch and Brownawell-Masser** ([**319**], [**52**] Th.A).

Theorem 14.5.21. *Let K be the function field of a geometrically irreducible smooth projective curve B of genus g over the field of constants k of characteristic 0. For $\mathbf{u} := (u_0,\dots,u_n) \in K^{n+1}\setminus\{\mathbf{0}\}$, define*

$$h(\mathbf{u}) := -\sum_{v\in M_K} \deg(v)\min(\mathrm{ord}_v(u_0),\dots,\mathrm{ord}_v(u_n)).$$

Suppose that $u_0+\ldots+u_n = 0$, that for any proper subset $I \subset \{0,\dots,n\}$ the elements u_i, $i \in I$, are linearly independent over k, and that $u_0,\dots,u_n$ are S-units for some finite set $S \subset M_K$. Then

$$h(\mathbf{u}) \le \frac{1}{2}n(n-1)\{\deg(S) + 2g - 2\}.$$

In fact, Brownawell and Masser ([**52**], Th.B) proved a bit more, substantially relaxing the linear independence condition of subsets of $u_0,\dots,u_n$. This is useful for specific applications.

Theorem 14.5.22. *Let $K = k(B)$ be a function field in characteristic 0. Suppose that $u_0+\ldots+u_n = 0$, that no non-empty proper subsum vanishes, and that $u_0,\dots,u_n$ are S-units for some finite set $S \subset M_K$. Then*

$$h(\mathbf{u}) \le \frac{(n-1)n}{2}\max\{d_K + \deg_{\mathbf{c}}(S) + (\dim(B)-1)\deg_{\mathbf{c}}(B), 0\}.$$

For the proof, we may not use Cartan's second theorem for function fields because the functions $u_1,\dots,u_n$ need not be linearly independent over k. The following lemma enables us to define an analogue of the Wronskians such that we can transfer the steps from the proof of Cartan's second main theorem. The proof does not use the linearly independent case and hence reproves Corollary 14.5.20.

A subset I of $\{0, \ldots, n\}$ is called **minimal** if the set $u_I := \{u_i \mid i \in I\}$ is linearly dependent over k but every proper subset of u_I is linearly independent over k.

Lemma 14.5.23. *There are disjoint non-empty subsets $I_1, \ldots, I_l$ of $\{0, \ldots, n\}$ and non-empty subsets $J_1, \ldots, J_{l-1}$ with*

$$\{0, \ldots, n\} = I_1 \cup \cdots \cup I_l, \quad J_\nu \subset I_1 \cup \cdots \cup I_\nu \ (\nu = 1, \ldots, l-1)$$

such that $I_1, J_1 \cup I_2, J_2 \cup I_3, \ldots, J_{l-1} \cup I_l$ are minimal.

For the elementary proof using simple linear algebra, we refer to [**52**], Lemma 6, or [**249**], Lemma A.3.2.7.

Proof of Theorem 14.5.22: By the usual arguments from 14.5.8, we may assume that B is a curve. By renumbering, we may assume that $I_j = \{N_{j-1}, \ldots, N_j - 1\}$ for a sequence $N_0 = 0 < N_1 < \cdots < N_l = n + 1$. For convenience, we set $J_0 = \emptyset$. For every $\nu \in \{1, \ldots, l\}$, we have a linear relation

$$c_{\nu,0} u_0 + \cdots + c_{\nu,n} u_n = 0$$

with $c_{\nu,i} \in k^\times$ for $i \in J_{\nu-1} \cup I_\nu$ and $c_{\nu,i} = 0$ else. We set $n_\nu := |I_\nu|$ and let $z \in k(B) \setminus \overline{k}$. We consider the $(n_1 - 1) \times (n+1)$ matrix

$$A_1 := \left(c_{1,j} \left(\frac{\mathrm{d}}{\mathrm{d}z} \right)^i u_j \right)_{i=0,\ldots,n_1-2; j=0,\ldots,n}$$

and the $n_\nu \times (n+1)$ matrices

$$A_\nu := \left(c_{\nu,j} \left(\frac{\mathrm{d}}{\mathrm{d}z} \right)^i u_j \right)_{i=0,\ldots,n_\nu-1; j=0,\ldots,n}$$

for $\nu = 2, \ldots, l$. Finally, let us consider the $n \times (n+1)$ matrix

$$A := \begin{pmatrix} A_1 \\ \vdots \\ A_l \end{pmatrix}.$$

For $j = 0, \ldots, n$, let $W_j(u_0, \ldots, u_n)$ be the determinant of the matrix obtained from A by deleting its jth column. They are analogues of the Wronskians in the proof of Theorem 14.5.18. Since the sum of columns of A is 0, we get

$$W_j(u_0, \ldots, u_n) = (-1)^j W_0(u_0, \ldots, u_n). \tag{14.22}$$

The block nature of the matrix A shows that $W_0(u_0, \ldots, u_n)$ is $k^\times$-proportional to the product

$$W(u_1, \ldots, u_{N_1-1}) W(u_{N_1}, \ldots, u_{N_2-1}) \cdots W(u_{N_{l-1}}, \ldots, u_n)$$

of usual Wronskians. By minimality of the I_ν in the decomposition of Lemma 14.5.23, these Wronskians and hence all $W_j(u_0,\ldots,u_n)$ are non-zero. For

$$\lambda_j(u_0,\ldots,u_n) := \frac{W_j(u_0,\ldots,u_n)}{\prod_{i\neq j} u_i},$$

the analogue of the fundamental identity (14.18) on page 506 is

$$u_j = (-1)^j \frac{u_0\cdots u_n}{W_0(u_0,\ldots,u_n)} \lambda_j(u_0,\ldots,u_n) \tag{14.23}$$

obtained immediately from (14.22). For

$$a := \sum_{i=0}^{n_1-2} i + \sum_{\nu=2}^{l} \sum_{i=0}^{n_\nu-1} i,$$

Lemma 14.5.15 yields as in the proof of Theorem 14.5.18 the bound

$$|\lambda_j(u_0,\ldots,u_n)|_v \leq |\mathrm{d}\pi_v/\mathrm{d}z|_v^a \cdot |1/\pi_v|_v^a .$$

If we combine this inequality with (14.23) and sum over $v \in S$, the product formula yields

$$\begin{aligned} \sum_{v\in S} \log \max_{j=0,\ldots,n} |u_j|_v \leq \sum_{v\notin S} \log \left| \frac{W_0(u_0,\ldots,u_n)}{u_0\cdots u_n} \right|_v + \\ + a \sum_{v\in S} \left(\log |\mathrm{d}\pi_v/\mathrm{d}z|_v + \log |1/\pi_v|_v\right). \end{aligned} \tag{14.24}$$

Replacing differentiation $\mathrm{d}/\mathrm{d}z$ by $\mathrm{d}/\mathrm{d}\pi_v$, we get

$$\log |W_0(u_0,\ldots,u_n)|_v \leq a \log |\mathrm{d}\pi_v/\mathrm{d}z|_v .$$

Since $u_0,\ldots u_n$ are S-units, we conclude from (14.24) that

$$h(\mathbf{u}) \leq a\left(d_K + \sum_{v\in S} \log |1/\pi_v|_v\right).$$

Obviously, we have $a \leq (n-1)n/2$ proving the claim. □

Remark 14.5.24. Note that Theorem 14.5.22 includes the general abc-theorem for polynomials mentioned in Theorem 12.4.4. Indeed, we just have to choose S as the union of ∞ and the zeros of the polynomials. Similarly, we get the **general *abc*-theorem for function fields** in characteristic zero:

Theorem 14.5.25. *Let $f_0,\ldots,f_n \in K = k(B)$ with $f_0+\cdots+f_n = 0$ such that no proper subsum vanishes identically. For $\mathbf{f} := (f_0 : \cdots : f_n) \in \mathbb{P}^n(K)$ and $D := \sum_{j=0}^n \{x_j = 0\}$, we get*

$$h(\mathbf{f}) \leq \frac{(n-1)n}{2} \max\left(\mathrm{cond}_D^K(\mathbf{f}) + d_K + (\dim(B)-1)\deg_{\mathbf{c}}(B), 0\right).$$

Proof: We know that $\mathbf{f}$ induces a rational map $f : B \dashrightarrow \mathbb{P}^n_k$. Let S be the set of prime divisors of B contained in the closure of $f^{-1}(D)$. Dividing by f_0, we may assume that $f_0, \ldots, f_n$ are S-units. We claim that $f^*(D)_{\text{red}} = S$. For $v \in S$, the local equation of $f^*(\{x_j = 0\})$ is $f_j^v := f_j \pi_v^{-\min_j \text{ord}_v(f_j)}$. Hence $v \in f^*(D)_{\text{red}}$ if and only if at least one f_j^v is in the maximal ideal of $\mathcal{O}_{B,v}$ proving immediately the claim. Hence we have

$$\text{cond}_D^K(\mathbf{f}) = \deg_{\mathbf{c}}(f^*(D)_{\text{red}}) = \deg_{\mathbf{c}}(S)$$

and Theorem 14.5.25 follows from Theorem 14.5.22. □

14.5.26. There are interesting unresolved questions connected with Theorem 14.5.25, best illustrated in the simplest case in which $K = \mathbb{C}(t)$ is the field of rational functions in one variable and $f_i \in \mathbb{C}[t]$ are polynomials, not all constant and without a common factor. As shown in 12.4.5, the coefficient $n(n-1)/2$ is sharp if $n = 2$ or 3, and is at least $2n - 3$ for $n \geq 4$. Browkin and Brzeziński [**51**] conjecture that $2n - 3$ is the correct constant in the theorem, in the slightly stronger form

$$\max \deg(f_i) \leq (2n-3) \max \big\{ \deg(\text{rad}(f_0 \cdots f_n)) - 1, 0 \big\}. \qquad (14.25)$$

Another intriguing question has been posed by Vojta, namely whether for any fixed positive $\varepsilon > 0$ and any $n \geq 2$ there is a closed subvariety $V \subsetneq \{x_0 + \cdots + x_n = 0\}$ in projective space $\mathbb{P}^n_{\mathbb{C}}$, depending only on n and ε, with the following property: If $f_0(t), \ldots, f_n(t)$ are polynomials in $\mathbb{C}[t]$ without common zeros and $f_0 + \cdots + f_n = 0$, then

$$\max \deg(f_i) \leq (1+\varepsilon) \max\{\deg(\text{rad}(f_0 \ldots f_n)) - 1, 0\}$$

unless the holomorphic curve $\{(f_0(t) : \cdots : f_n(t)) \mid t \in \mathbb{C}\}$ is contained in V. We illustrate Vojta's idea by showing how to exclude the example by Browkin and Brzeziński (see 12.4.5) for the case $n = 3$. Consider the identity

$$(a+b+c)^3 = a^3 + b^3 + c^3 - 3abc + 3(ab+ac+bc)(a+b+c),$$

which reduces to the four term identity

$$a^3 + b^3 + c^3 - 3abc = 0$$

if $a + b + c = 0$. The example of Browkin and Brzeziński is obtained by taking $a = 1$, $b = -t$, $c = t - 1$ and $f_0 = a^3$, $f_1 = b^3$, $f_2 = c^3$, $f_3 = -3abc$, so that equality holds in (14.25). Clearly, no subsum vanishes. The relations to avoid are simply

$$(f_i + f_j + f_k)^3 - 27 f_i f_j f_k = 0$$

for any choice of distinct indices i, j, k.

14.6. Bibliographical notes

Sections 14.2 and 14.3 are taken from [**307**] and [**318**], where the reader will find further informations. The degeneracy of K-rational points in a variety of general type was posed as an open problem in Bombieri's lecture at the university of Chicago in 1980. For function fields in characteristic 0, this was solved by J. Noguchi [**225**] under the stronger assumption that the cotangent bundle is ample. S. Lang gave more general conjectures relating the structure of K-rational points also to hyperbolicity and connecting the special sets from Nevanlinna theory and diophantine approximation (see [**172**]).

The zero-dimensional part of the exceptional set Z in the K-rational Vojta conjecture must depend on ε, but it may be that the higher-dimensional part is independent. At least, this holds in Schmidt's subspace theorem and in Vojta's version of Cartan's second main theorem, both proved by P. Vojta in [**309**], [**316**].

Vojta's conjecture with ramification does imply Conjecture 14.4.10 also in higher dimension. This is due to P. Vojta [**317**], true over number fields and function fields of characteristic zero.

Cartan's second main theorem holds also in the linearly degenerate case, with the factor $n+1$ replaced by $n+t+1$, where t is the codimension of the linear span of the image of f. This was done by E.I. Nochka (see [**223**], [**224**]). Ru and Wong have worked out the number theoretic analogue, see also P. Vojta [**316**] for an alternative proof. For function fields in characteristic 0 (with hyperplanes defined over the constant field), this is due independently to J. Noguchi [**226**] and J.T.-Y. Wang [**320**]. In [**320**], there is also a generalization of the linearly non-degenerate case to characteristic p.

In the non-split case (when varieties and divisors are not defined over the constant field), very few results are known. For a function field K of characteristic 0 and a curve C of genus $g \geq 2$ over K, P. Vojta [**311**] proved

$$h_{K_C}(P) \leq (2+\varepsilon)d(P) + O(1)$$

for all $P \in C(\overline{K})$ using the methods of Grauert [**129**] in the proof of the Mordell conjecture over function fields. M. Kim [**160**] generalized it to characteristic p under an assumption on the Kodaira spencer map. Note that Vojta's conjecture ($H = K_C$ ample, $D = 0$) would predict a factor $1+\varepsilon$ instead of $2+\varepsilon$, at least for points of bounded degree.

The proof of the general abc-theorem for polynomials is similar to the one of Brownawell and Masser [**52**], it is inspired by Cartan's proof of its second main theorem in Nevanlinna theory. The proof of Voloch [**319**] is more geometric and relies on the Brill–Segre formula, which is a generalization of Hurwitz's genus formula for non-degenerate maps from a curve into projective space.

APPENDIX A ALGEBRAIC GEOMETRY

A.1. Introduction

We collect here some definitions and results from algebraic geometry needed in the text. For most of our purposes, it is enough to work with varieties over a base field K and to consider points rational in a fixed algebraic closure $\overline{K}$. Thus we may neglect the modern language of schemes in order to keep the exposition elementary. In some side remarks or proofs, not essential for the basic understanding of the book, it will still be convenient to use the theory of schemes on the level of [**148**], Ch.II.

Arguments are only given if they are easy and instructive or if we have not found an appropriate reference, otherwise we freely quote from standard text books up to the volumes of Grothendieck. Most of the quoted results are true for more general classes of schemes, but we formulate them just for varieties using the dictionary mentioned in A.2.8.

No knowledge of algebraic geometry is required for reading Appendix A. However, the presentation is too brief for learning the subject and it would be useful, from an educational point of view, if the reader is familiar with the theory of varieties over an algebraically closed field as in the books of R. Hartshorne [**148**], Ch. I, D. Mumford [**213**], Ch. I, or I.R. Shafarevich [**279**].

We advise the reader to work through Sections A.2–A.4 to gather the most frequently used definitions, notations, terminology, and results. Also if you look up to the definition of a projective variety in Section A.6, then you will be ready to start the book, coming back to Appendix A only when required in the text.

A.2. Affine varieties

Let K be a field and $\overline{K}$ an algebraic closure of K.

A.2.1. The **affine n-space** $\mathbb{A}_K^n$ is equal to $\overline{K}^n$ endowed with the following topology: A subset Y is closed if and only if there is $T \subset K[x_1, \ldots, x_n]$ with

zero set $Y = Z(T)$, where

$$Z(T) := \{\alpha \in \overline{K}^n \mid f(\alpha) = 0 \, \forall f \in T\}.$$

This defines a topology on $\mathbb{A}_K^n$ depending on the ground field K. It is called the **Zariski topology.** Note that $x_1, \ldots, x_n$ are the coordinate functions on $\mathbb{A}_K^n$. On an affine n-space, we always fix a set of coordinate functions and set $\mathbf{x} := (x_1, \ldots, x_n)$.

A.2.2. Let Y be a closed subset of $\mathbb{A}_K^n$. The **ideal of vanishing** is given by

$$I(Y) := \{f \in K[\mathbf{x}] \mid f(Y) = \{0\}\}.$$

Then

$$K[Y] := K[\mathbf{x}]/I(Y)$$

is the **coordinate ring** of Y. It is a **reduced K-algebra** meaning that $K[Y]$ has no nilpotent elements. For an ideal J in $K[\mathbf{x}]$, let

$$\sqrt{J} := \{f \in K[\mathbf{x}] \mid \exists n \in \mathbb{N} \text{ with } f^n \in J\}$$

be the **radical** of J. Note that $I(Y)$ is a **radical ideal,** i.e. $I(Y) = \sqrt{I(Y)}$. By the trivial fact $Z(I(Y)) = Y$, every closed subset is the zero set of a radical ideal. **Hilbert's Nullstellensatz** says

$$I(Z(J)) = \sqrt{J}.$$

For a proof, we refer to **[157]**, Th.7.15, where it is assumed that K is algebraically closed but the same proof applies to our situation.

A.2.3. The Zariski topology on $\mathbb{A}_K^n$ induces a topology on Y. Let U be an open subset of Y. Then a function $f : U \longrightarrow \overline{K}$ is called regular in $\alpha \in U$ if there are $p, q \in K[\mathbf{x}]$ with $q(\alpha) \neq 0$ and $f = p/q$ in a neighbourhood of α. If f is regular in all points of U, then f is called a **regular function.** The ring of regular functions on U is denoted by $\mathcal{O}_Y(U)$. Clearly, any element of the coordinate ring may be viewed as a regular function on Y. In fact, we can prove $K[Y] = \mathcal{O}_Y(Y)$. The pair $(Y, \mathcal{O}_Y(Y))$ is called an **affine variety over K**. By abuse of notation, it is usually denoted by Y.

A.2.4. Let V, V' be open subsets of affine K-varieties X, X'. A **K-morphism** (or a **morphism over K**) is a continuous map $\varphi : V \longrightarrow V'$ such that for all open subsets U' of V', we have a well-defined map

$$\varphi^\sharp : \mathcal{O}_{X'}(U') \longrightarrow \mathcal{O}_X(\varphi^{-1}(U')), \quad f' \mapsto f' \circ \varphi.$$

This means simply that the function $f' \circ \varphi$ is regular on $\varphi^{-1}(U')$. If the ground field K is fixed, then we often simply speak about varieties and morphisms meaning always that they are defined over K.

Note that we may reconstruct φ from $\varphi^\sharp$: We may assume that X, X' are closed subsets of $\mathbb{A}_K^n, \mathbb{A}_K^m$ with coordinates $\mathbf{x}$ and $\mathbf{x}'$, respectively. Then let $\varphi_i :=$

$\varphi^\sharp(x_i')$. We get $\varphi(\boldsymbol{\alpha}) = (\varphi_1(\boldsymbol{\alpha}), \ldots, \varphi_n(\boldsymbol{\alpha}))$ for all $\boldsymbol{\alpha} \in V$. This way, we obtain a bijective correspondence between K-morphisms of varieties over K and K-algebra homomorphisms of the corresponding coordinate rings.

The morphism φ is called a **K-isomorphism** if and only if there is a K-morphism $\psi : X' \to X$ such that $\varphi \circ \psi$ and $\psi \circ \varphi$ are both the identity map.

A.2.5. For $x \in X$, we consider pairs (U, f), where U is an open neighbourhood of x and $f \in \mathcal{O}_X(U)$. Two pairs (U, f) and (U', f') are called equivalent if and only if $f = f'$ is on a neighbourhood of x. The set of equivalence classes is denoted by $\mathcal{O}_{X,x}$ and is called the **local ring of x in X**. In $K[X]$, we consider the maximal ideal $\overline{\mathfrak{m}}_x = \{f \in K[X] \mid f(x) = 0\}$. Then $\mathcal{O}_{X,x}$ is the localization of $K[X]$ in $\overline{\mathfrak{m}}_x$ with unique maximal ideal $\mathfrak{m}_x = \overline{\mathfrak{m}}_x \mathcal{O}_{X,x}$ (see A.2.10).

A.2.6. Let $K \subset L \subset \overline{K}$ be an intermediate field. Let X be an affine variety over K. Then the set of **L-rational points** of X is

$$\{x \in X \mid f(x) \in L \,\forall f \in \mathcal{O}_X(X)\}.$$

If we view X as a closed subset of $\mathbb{A}_K^n$, then this means that all coordinates of x are in L. For $x \in X$, the smallest L such that x is L-rational is denoted by $K(x)$. In fact, we have

$$\mathcal{O}_{X,x}/\mathfrak{m}_x \xrightarrow{\sim} K(x), \quad f + \mathfrak{m}_x \mapsto f(x).$$

A.2.7. Note that points in X have not to be closed. For $x \in X$, the closure of x is the set of **conjugates** of x. If X is a closed subset of $\mathbb{A}_K^n$, then the conjugates of $\mathbf{x} \in X$ are given by $\sigma\mathbf{x}$ applying $\sigma \in \mathrm{Gal}(\overline{K}/K)$ componentwise. Thus the local rings of x and of its conjugates are the same.

A.2.8. Now we relate affine varieties to affine schemes. This makes it possible to quote results from standard books about schemes. If the reader is not familiar with the language of schemes, he can skip the following remarks without any problems for the understanding of the book.

Let X be an affine variety over K with ring of regular functions $A = \mathcal{O}_X(X)$, Then we have a map

$$t : X \longrightarrow \mathrm{Spec}(A), \quad x \mapsto I(\{x\}).$$

Then $t(X)$ is the set of maximal ideals and this is dense in $\mathrm{Spec}(A)$. Moreover, the topology on X is the coarsest topology making t continuous, i.e. U is open on X if and only if it is the inverse image of an open subset V in $\mathrm{Spec}(A)$. Then we have

$$\mathcal{O}_{\mathrm{Spec}(A)}(V) \cong \mathcal{O}_X(t^{-1}(V))$$

or more formally $t_*\mathcal{O}_X \cong \mathcal{O}_{\mathrm{Spec}(A)}$. This follows immediately from the definitions and the density of $t(U)$ in V. Therefore any morphism of affine varieties over K extends to a morphism of affine schemes.

Conversely, for any reduced affine scheme $\mathrm{Spec}(A)$ of finite type over K, we can consider the $\overline{K}$-rational points of $\mathrm{Spec}(A)$ as an affine variety X over K with $\mathcal{O}_X(X) = A$. This may be used to translate results from affine varieties over K to reduced affine schemes of finite type over K and conversely.

A.2.9. Let X be an affine variety over K with coordinate ring $K[X]$. By Hilbert's basis theorem, $K[X]$ is a **noetherian ring**, i.e. any ideal is finitely generated or equivalently there exists no properly ascending chain of ideals in $K[X]$ (see for example [**157**], 7.9). Using A.2.2, there is a one-to-one correspondence between closed subsets of X and radical ideals in $K[X]$, given by considering the ideal of vanishing $I(Y)$ or passing to the zero set $Z(J)$. We conclude that there is no properly descending infinite chain of closed subsets in X. This implies that for any family $(U_\alpha)_{\alpha \in I}$ of open subsets of X, there is a finite $I_0 \subset I$ with

$$\bigcup_{\alpha \in I_0} U_\alpha = \bigcup_{\alpha \in I} U_\alpha.$$

In other words, every open subset of an affine variety over K is **quasicompact**.

A.2.10. Not every open subset of an affine K-variety X is an affine K-variety. But for any $f \in K[X]$, the open subset

$$X_f := \{x \in X \mid f(x) \neq 0\}$$

is an affine variety over K with coordinate ring isomorphic to the localization of $K[X]$ in the multiplicative monoid $\{f^n \mid n \in \mathbb{N}\}$ ([**148**], Prop.II.2.2). Note that this open subsets form a **basis** for the topology on X.

To prove this, let x be a point in an open subset U. There is $f \in I(X \setminus U)$ with $f(x) \neq 0$ and hence $x \in X_f \subset U$.

This is easily used to prove that the local ring $\mathcal{O}_{X,x}$ of X in x is isomorphic to the localization of $K[X]$ in the maximal ideal $\mathfrak{m}_x$ ([**148**], Prop.II.2.2).

A.2.11. Here, we explain how to pass from affine varieties over K to affine varieties over $\overline{K}$. Let X be an affine variety over K. It is a closed subset of $\mathbb{A}^n_K$ given by the ideal $I(X)$. As $\mathbb{A}^n_K$ and $\mathbb{A}^n_{\overline{K}}$ have the same underlying set $\overline{K}^n$, we may view X as a closed subset of $\mathbb{A}^n_{\overline{K}}$ given by the zeros of the set $I(X) \subset \overline{K}[\mathbf{x}]$. The corresponding affine variety over $\overline{K}$ is denoted by $X_{\overline{K}}$. Note that X and $X_{\overline{K}}$ have the same points but a different topology and a different ring of regular functions. It is easily seen that the variety $X_{\overline{K}}$ does not depend on the choice of the affine space. It is called the **base change** of X to $\overline{K}$.

For any extension field F of K, we define the base change X_F as the affine variety given by the closed subset $Z(I(X))$ in $\mathbb{A}^n_F$ and by the corresponding coordinate ring

$$F[X_F] = F[x_1, \ldots, x_n]/\sqrt{I(X)}.$$

A.3. Topology and sheaves

Let T be a topological space and K a field.

A.3.1. Then $Y \subset T$ is called **irreducible** if and only if for all closed subsets A and B different from Y we have $T \neq A \cup B$. The empty set is not irreducible.

Example A.3.2. If Y is a closed subset of $\mathbb{A}^n_K$, then Y is irreducible (with respect to the induced topology) if and only if $I(Y)$ is a prime ideal.

To prove it, assume first that Y is irreducible. Let $a \in K[\mathbf{x}] \setminus I(Y)$ and $b \in K[\mathbf{x}]$ with $ab \in I(Y)$. Then the union of $Z(a) \cap Y$ and $Z(b) \cap Y$ is Y, hence $Z(b) \supset Y$ proving $b \in I(Y)$. Thus $I(Y)$ is a prime ideal.

Conversely, assume that $I(Y)$ is a prime ideal. Let A and B be closed subsets of Y with $A \cup B = Y$. If $A \neq Y$, then A.2.2 shows that there is an $a \in I(A) \setminus I(Y)$. Since

$$I(A) \cap I(B) = I(A \cup B)$$

contains ab for every $b \in I(B)$, we conclude that $b \in I(Y)$. Therefore $I(B) = I(Y)$ proving $B = Y$ and the irreduciblity of Y.

For an affine variety X, we conclude that in the one-to-one correspondence between radical ideals of $K[X]$ and closed subsets of X mentioned in A.2.9, the prime ideals correspond to the irreducible closed subsets of X.

A.3.3. A maximal irreducible subset of T is called an **irreducible component** of T. As the closure of an irreducible subset is again irreducible, the irreducible components are closed. If T is irreducible, then any non-empty open subset is irreducible and dense. The proofs are immediate from the definitions.

Example A.3.4. If Y is a closed subset of $\mathbb{A}^n_K$, then the irreducible components of Y are the zero sets of the minimal prime ideals containing $I(Y)$. There are finitely many irreducible components and their union is equal to Y.

A.3.5. We define the **dimension** of T to be

$$\dim(T) := \sup\{n \mid A_0 \subsetneq A_1 \subsetneq \cdots \subsetneq A_n\},$$

where $A_0 \subsetneq A_1 \subsetneq \cdots \subsetneq A_n$ is ranging over all chains of irreducible closed subsets of T.

Example A.3.6. If X is an affine variety over K, then Example A.3.4 shows that the dimension of X is equal to the **Krull dimension** of the noetherian ring $K[X]$. By definition, the Krull dimension $\dim(A)$ of a commutative ring A is the supremum over the length of all prime ideal chains. So this follows from the one-to-one correspondence between prime ideals in $K[X]$ and irreducible closed subsets of X, which we deduce from Example A.3.2. If X is irreducible, then it is a consequence of commutative algebra that the dimension of X is equal to the transcendence degree of $K(X)$ over K, where $K(X)$ is the quotient field of $K[X]$ (H. Matsumura [**197**], Ch.5, §14). In particular, the dimension of $\mathbb{A}^n_K$ is n.

Example A.3.7. Let X be an affine variety over $\mathbb{C}$. We may view X as a closed subset of $\mathbb{A}^n_{\mathbb{C}}$. Instead of the Zariski topology on $\mathbb{C}^n$, we may use the complex topology where the balls form a basis. It induces the **complex topology** on X and we get a **complex analytic variety** X_{an}. Instead of regular functions, we use holomorphic functions. For details, we refer to Section A.14 or to [**280**], Chs VII, VIII. The complex topology is finer than the Zariski topology. In contrast to the Zariski topology, the complex topology is Hausdorff. We have mentioned in Example A.3.6 that $\dim(\mathbb{A}^n_K) = n$. However, if we use the complex topology on $\mathbb{C}^n$, the dimension introduced in A.3.5 is infinite for $n \geq 1$. So our notation is useful only for the Zariski topology, but this is the only case of interest for us.

A.3.8. We call an affine variety X over K **geometrically irreducible** if $X_{\overline{K}}$ is irreducible (cf. A.2.11). Note that a geometrically irreducible variety X is always irreducible since the topology on $X_{\overline{K}}$ is finer.

X is called **geometrically reduced** if $K[X] \otimes_K \overline{K}$ contains no nilpotent elements. If we consider X as a closed subset of $\mathbb{A}^n_K$, then this is equivalent to the assumption that the ideal in $\overline{K}[\mathbf{x}]$ generated by $I(X)$ is a radical ideal. By Hilbert's Nullstellensatz in A.2.2, we conclude that

$$\overline{K}[X_{\overline{K}}] = K[X] \otimes_K \overline{K}$$

for a geometrically reduced X.

Example A.3.9. Note that $Z(x^2+1) \subset \mathbb{A}^1_{\mathbb{R}}$ is an irreducible zero-dimensional $\mathbb{R}$-variety but not geometrically irreducible.

Example A.3.10. Let K be the quotient field of the polynomial ring $\mathbb{F}_2[t]$. Then $Z(x^2+t)$ is a zero-dimensional variety in $\mathbb{A}^1_K$, which is not geometrically reduced.

A.3.11. Let X and X' be affine varieties over K. On $X \times X'$, we define the following structure of an affine variety over K. We may view X, X' as closed subsets of $\mathbb{A}^n_K$ and $\mathbb{A}^m_K$, respectively. We use the coordinates $\mathbf{x}$ on $\mathbb{A}^n_K$ and $\mathbf{x}'$ on $\mathbb{A}^m_K$. We identify $\mathbb{A}^n_K \times \mathbb{A}^m_K$ with $\mathbb{A}^{n+m}_K$, hence the coordinates on the latter are given by $(\mathbf{x}, \mathbf{x}')$. We consider $I(X)$ and $I(X')$ as subsets of $K[\mathbf{x}, \mathbf{x}']$ by

$$I(X) \subset K[\mathbf{x}] \subset K[\mathbf{x}, \mathbf{x}'] \supset K[\mathbf{x}'] \supset I(X').$$

Then

$$X \times X' = Z(I(X) \cup I(X')) \subset \mathbb{A}^{n+m}_K$$

makes $X \times X'$ into an affine variety over K. We call $X \times X'$ the **product variety** of X and X'. It is easy to see that it does not depend on the choices of $\mathbb{A}^n_K$ and $\mathbb{A}^m_K$. Note that the topology on $X \times X'$ is finer than the product topology.

Example A.3.12. We consider the group $GL(n, \overline{K})$ of invertible $n \times n$ matrices with entries in $\overline{K}$. We identify the space of $n \times n$ matrices with entries in $\overline{K}$

with $\overline{K}^{n^2}$ by ordering the entries x_{ij} of the matrices lexicographically. Now we consider $GL(n, \overline{K})$ as a closed subset of $\mathbb{A}_K^{n^2+1}$ by using the map

$$GL(n, \overline{K}) \hookrightarrow \mathbb{A}_K^{n^2+1}, \quad (x_{ij}) \mapsto ((x_{ij}); \det(x_{ij})^{-1}).$$

Let $(y_{ij})_{1 \leq i,j \leq n}; y_{n^2+1}$ be the coordinates on $\mathbb{A}_K^{n^2+1}$. Then $GL(n, \overline{K})$ is identified with $Z(\det(y_{ij})y_{n^2+1} - 1)$. This makes $GL(n, \overline{K})$ into a n^2-dimensional affine variety over K, which we denote by $GL(n)_K$. Note that for any intermediate field L of K and $\overline{K}$, the L-rational points are equal to $GL(n, L)$. The base change of $GL(n)_K$ to $\overline{K}$ is equal to $GL(n)_{\overline{K}}$. The group operation gives a K-morphism

$$GL(n)_K \times GL(n)_K \longrightarrow GL(n)_K, \quad (g_1, g_2) \mapsto g_1 \cdot g_2.$$

Similarly, the inverse is a K-morphism. Hence $GL(n)_K$ is an example of a group variety handled in Section 8.2.

A.3.13. A **presheaf** of abelian groups on our topological space T is a map assigning to each open subset U of T an abelian group $\mathcal{F}(U)$ and to every open subset V of U a homomorphism $\rho_V^U : \mathcal{F}(U) \to \mathcal{F}(V)$ called the **restriction map** such that:

(a) $\mathcal{F}(\emptyset) = 0$;

(b) ρ_U^U is the identity;

(c) if $W \subset V \subset U$ are open subsets, then $\rho_W^U = \rho_W^V \circ \rho_V^U$.

Instead of abelian groups, we may consider rings, K-vector spaces or other algebraic structures.

Example A.3.14. For an open subset U of T, let $\mathcal{F}(U)$ be the set of continuous real functions on U. If V is an open subset of U, then we define ρ_V^U by restricting functions of U to V. Then $\mathcal{F}$ is a presheaf of $\mathbb{R}$-algebras on T. If T is a differentiable manifold, then the same construction works with C^∞-functions.

A.3.15. A presheaf $\mathcal{F}$ on T is called a **sheaf** if the following conditions are satisfied for every open subset U of T and every open covering $(U_i)_{i \in I}$ of U:

(a) if $s \in \mathcal{F}(U)$ with $\rho_{U_i}^U(s) = 0$ for all $i \in I$, then $s = 0$;

(b) if $s_i \in \mathcal{F}(U_i)$ for all $i \in I$ and $\rho_{U_i \cap U_j}^{U_i}(s_i) = \rho_{U_i \cap U_j}^{U_j}(s_j)$ for all $i, j \in I$, then there is $s \in \mathcal{F}(U)$ with $\rho_{U_i}^U(s) = s_i$ for each $i \in I$.

Note that s in (b) is unique by (a). The presheaves in Example A.3.14 are sheaves.

Example A.3.16. Let X be an affine variety over K. Then $\mathcal{O}_X$ is a presheaf of K-algebras. We define ρ_V^U again as the restriction map of functions. It is almost by definition of a regular function that $\mathcal{O}_X$ is a sheaf.

A.3.17. Let $\mathcal{F}$ and $\mathcal{G}$ be presheaves of abelian groups on T. A **homomorphism** $\varphi : \mathcal{F} \to \mathcal{G}$ **of presheaves** is a homomorphism $\varphi_U : \mathcal{F}(U) \to \mathcal{G}(U)$ for all open subsets U such that for all open subsets $V \subset U$, we have $\rho_V^U \circ \varphi_U = \varphi_V \circ \rho_V^U$.

A homomorphism of sheaves is the same as a homomorphism of presheaves. Obviously, φ is called an **isomorphism of presheaves** if there is a morphism $\psi : \mathcal{G} \to \mathcal{F}$ of presheaves such that $\varphi \circ \psi$ and $\psi \circ \varphi$ are both the identity on the corresponding presheaves.

For more details, we refer to **[148]**, II.1, or to **[134]**, 0.3.

A.4. Varieties

Let K be a field and $\overline{K}$ an algebraic closure of K.

A.4.1. A **prevariety** X over K is a topological space with a finite open covering $(U_\alpha)_{\alpha \in I}$ of X and homeomorphisms $\varphi_\alpha : U_\alpha \to X_\alpha$ for affine varieties X_α over K such that

$$\varphi_\alpha \circ \varphi_\beta^{-1} : \varphi_\beta(U_\beta \cap U_\alpha) \longrightarrow \varphi_\alpha(U_\beta \cap U_\alpha) \tag{A.1}$$

is a K-isomorphism for all $\alpha, \beta \in I$. Then $(U_\alpha, \varphi_\alpha)$ is called an **affine chart** of X.

Note that this concept is similar to the concept of manifold in differential geometry. Not to focus to much on the above covering, we should consider a maximal atlas compatible with the covering. We leave the details to the reader.

A.4.2. Let U be an open subset of the prevariety X. Then a function $f : U \to X$ is called regular in $x \in U$ if $f \circ \varphi_\alpha^{-1}$ is regular in $\varphi_\alpha(x)$ for all $\alpha \in I$ with $x \in U_\alpha$. By (A.1), we have to check that only for one α. Again, f is called **regular** if it is regular in all $x \in U$. The K-algebra of regular functions on U is denoted by $\mathcal{O}_X(U)$. Obviously, we get a sheaf $\mathcal{O}_X$ on X.

A.4.3. Let X, X' be prevarieties over K. A map $\varphi : X \to X'$ is called a **morphism over** K (or a K-morphism or simply a morphism), if φ is continuous and for each open subset U' of X' and $f' \in \mathcal{O}_{X'}(U')$, we have

$$\varphi^\sharp(f') := f' \circ \varphi \in \mathcal{O}_X(\varphi^{-1}U').$$

In analogy with the affine case, we define an **isomorphism** to be an invertible morphism.

A.4.4. Let X and X' be prevarieties over K, given by finite affine charts $(U_\alpha)_{\alpha \in I}$ and $(U'_\beta)_{\beta \in J}$, respectively. Let $\varphi_\alpha : U_\alpha \to X_\alpha$ and $\psi_\beta : U'_\beta \to X'_\beta$ be the homeomorphisms to affine varieties X_α, X'_β from A.4.1. On $U_\alpha \times U'_\beta$, we use the topology such that the map $\varphi_\alpha \times \psi_\beta$ to the affine product variety $X_\alpha \times X'_\beta$ is a homeomorphism. Clearly, the topologies coincide on overlappings, therefore the covering $(U_\alpha \times U'_\beta)_{\alpha \in I, \beta \in J}$ defines a unique topology on $X \times X'$ such that each $U_\alpha \times U'_\beta$ is an open subset of $X \times X'$. Then $X \times X'$ is a prevariety with affine charts $(U_\alpha \times U'_\beta, \varphi_\alpha \times \psi_\beta)$. The structure does not depend on the choice of the affine charts. The details are left to the reader.

A.4.5. A prevariety X over K is called a **variety** over K if the diagonal $\Delta := \{(x,x) \mid x \in X\}$ is closed in $X \times X$. Obviously, any affine variety over K is a variety. Regular functions and morphisms are the same as for prevarieties.

A.4.6. In the whole book, only the notion of a K-variety will occur. Let X be a variety over K. For an intermediate field L of K and $\overline{K}$, a point $x \in X$ is called **L-rational** if x is L-rational in one affine chart (and hence in all affine charts containing x). The set of L-rational points is denoted by $X(L)$. Similarly as in A.2.5, we define the **local ring** $\mathcal{O}_{X,x}$. It has a unique maximal ideal $\mathfrak{m}_x$ and the **residue field** $K(x) := \mathcal{O}_{X,x}/\mathfrak{m}_x$.

The **base change** X_F of X to an extension field F/K is the variety over F obtained by using base change of the affine charts from A.2.11. The set $X(F)$ of F-rational points of X is defined by $X(F) := X_F(F)$, i.e. an F-rational point of X is a point $x \in X_F$ with an affine neighbourhood, where x takes coordinates in F.

The **product** $X_1 \times X_2$ of varieties X_1, X_2 over K is again a variety over K. The easy proofs are left to the reader.

We call X **geometrically irreducible** if $X_{\overline{K}}$ is irreducible. A K-variety X is **geometrically reduced** if the affine open charts are geometrically reduced. Note that for X geometrically irreducible (resp. geometrically reducible), the base change X_F with respect to any field extension F/K is irreducible (resp. reduced). Moreover, if the characteristic of K is zero (or more generally if K is perfect), then all K-varieties are geometrically reduced ([**148**], Exercise II.3.15).

For the reader familiar with the language of schemes, A.2.8 shows that varieties over K are the "same" as reduced schemes of finite type and separated over K.

A.4.7. Let U be an open subset of the K-variety X. We choose an open covering $(U_\alpha, \varphi_\alpha)_{\alpha \in I}$ of X by affine charts. Then U is a K-variety with the induced topology and with the charts $U_\alpha \cap U$. There is a slight problem that $U_\alpha \cap U$ has not to be affine. By A.2.10, we may replace $U_\alpha \cap U$ by an open affine covering $(V_{\alpha\beta})_{\beta \in J_\alpha}$ meaning that the restriction of φ_α to $V_{\alpha\beta}$ gives a homeomorphism onto an affine open subset of X_α. Then the $(V_{\alpha\beta}, \varphi_\alpha|_{V_{\alpha\beta}})$ are the affine charts of the K-variety U and U is called an **open subvariety** of X. If U is isomorphic to an affine variety over K, then U is called an **affine open subset** of X. We have seen above that the affine open subsets of X form a **basis** for the topology of X.

A.4.8. Let Y be a closed subset of the K-variety X. Again, let $(U_\alpha, \varphi_\alpha)_{\alpha \in I}$ be affine charts covering X. Since any closed subset of an affine K-variety is again an affine K-variety, we conclude that Y is a K-variety with the induced topology and with the affine charts $(U_\alpha \cap Y, \varphi_\alpha|_{U_\alpha \cap Y})$. Clearly, the diagonal in $Y \times Y$ is closed. We always use this structure of K-variety on Y and we call Y a **closed subvariety** of X.

For each open subset U of X, let $\mathcal{J}_Y(U)$ be the ideal of regular functions on U vanishing on $Y \cap U$. Then $\mathcal{J}_Y$ is called the **ideal sheaf of** Y.

If $i : Y' \to X$ is a morphism of varieties over K such that i induces a K-isomorphism onto a closed subvariety of X, then i is called a **closed embedding** of Y' in X.

A.4.9. We apply the topological considerations from Section A.3 to the K-variety X. First, X has finitely many irreducible components which cover X. For affine X, this is Example A.3.4. In general, we apply this to the affine charts and then we take closure. As the closure of an irreducible subset is again irreducible, we get easily the claim.

On X, we use the dimension introduced in A.3.5. Let

$$A_0 \subsetneq A_1 \subsetneq \cdots \subsetneq A_n$$

be a chain of irreducible closed subsets. We choose an affine chart U_α with $A_0 \cap U_\alpha \neq \emptyset$. Using irreducibility, it is clear that $A_j \cap U_\alpha$ is dense in A_j. Therefore

$$\dim(X) = \max_{\alpha \in I} \dim(X_\alpha) < \infty.$$

For an irreducible closed subset A of X, the **codimension** of A in X is

$$\operatorname{codim}(A, X) := \sup\{n \mid A_0 \subsetneq A_1 \subsetneq \cdots \subsetneq A_n\},$$

where $A_0 \subsetneq A_1 \subsetneq \cdots \subsetneq A_n$ is ranging over all chains of irreducible closed subsets of X with $A = A_0$. For any closed subset B of X, the codimension of B in X is

$$\operatorname{codim}(B, X) := \inf\{\operatorname{codim}(A, X) \mid A \subset B\},$$

where A is ranging over all irreducible closed subsets of B. Clearly, it is enough to consider the irreducible components of B. If B and X are irreducible, then

$$\dim(B) + \operatorname{codim}(B, X) = \dim(X).$$

Let U_α be an affine chart with $U_\alpha \cap B \neq \emptyset$. Since U_α is dense, we have $\dim(U_\alpha) = \dim(X)$ and $\operatorname{codim}(U_\alpha \cap B, U_\alpha) = \operatorname{codim}(B, X)$. So we may assume that X is an irreducible affine variety over K. Then the claim follows from [**148**], Caution II.3.2.8.

A.4.10. A variety X over K is called **equidimensional** (or of **pure dimension** n) if all irreducible components of X have dimension n. If X and X' are K-varieties of pure dimension n and n', then $X_{\overline{K}}$ is a $\overline{K}$-variety of pure dimension n and $X \times X'$ is a K-variety of pure dimension $n + n'$ ([**137**], Cor.4.2.8, Prop.4.2.4). However, the intersection of two irreducible subsets need not to be pure dimensional.

A.4.11. Let X be an irreducible variety over K. We consider pairs (U, f) where U is a non-empty open subset and f is a regular function on U. Two pairs (U, f) and (U', f') are called equivalent if $f = f'$ on $U \cap U'$. An equivalence class is called a **rational function** on X. Clearly, for a rational function, there is a maximal representative (U, f) containing all other representatives. We usually identify a rational function with this representative. The rational functions form a field $K(X)$ called the **function field** of X. This follows from irreducibility of X. Any non-empty open subset U of X is irreducible and dense (see A.3.3), hence $K(U) = K(X)$. If U is affine with coordinate ring $K[U]$, then it follows that $K(X)$ is the quotient field of $K[U]$. By Example A.3.6, the dimension of X is equal to the transcendence degree of $K(X)$ over K.

We remark that X is geometrically irreducible (resp. geometrically reduced) if and only if K is separably algebraically closed in $K(X)$ (resp. $K(X)$ is separable over K). For a proof, we refer to [**213**], II.4, Prop.4.

If X is geometrically irreducible and Y is an irreducible variety over K, then the product variety $X \times Y$ is irreducible ([**137**], Cor.4.5.8). But the product of two irreducible varieties is not necessarily irreducible.

A.4.12. To define rational functions on any variety X over K, we consider pairs (U, f) with U open dense in X and f regular on U. The equivalence classes with respect to the above relation are called rational functions on X. They form a ring $K(X)$. If X is not irreducible, then $K(X)$ is no longer a field. We have

$$K(X) = \prod_j K(X_j),$$

where X_j is ranging over all irreducible components of X, i.e. $K(X)$ is a product of fields. To see this, note that any $(U, f) \in K(X)$ may be represented on $\bigcup_j (U \setminus \cup_{i \neq j} X_i)$ and this is a disjoint union of open subsets.

A.4.13. Let F be a subfield of $\overline{K}/K$ and let $\sigma : F \to \overline{K}$ be an embedding over K. For an affine variety Y over F given by polynomials $f_1(\mathbf{x}), \ldots, f_m(\mathbf{x})$ in $\mathbb{A}^n_F$, the conjugate Y^σ is the affine variety in $\mathbb{A}^n_{\sigma(F)}$ given by $\sigma(f_1), \ldots, \sigma(f_m)$. The **conjugate** Y^σ of any variety Y over F is defined by the conjugates of the local affine charts. Then Y^σ is a variety over $\sigma(F)$. If Y is projective subvariety of $\mathbb{P}^n_F$ (see Section A.6), then Y^σ is given by the conjugate homogeneous polynomials. Applying σ to the points of Y, we get a bijection onto Y^σ, which, however, is not a morphism over F.

Let R be the set of embeddings $\sigma : F \to \overline{K}$ over K. Let X be a variety over K and let Y be a closed subvariety of X_F. Then we claim that $Y_0 := \bigcup_{\sigma \in R} Y^\sigma$ is the smallest closed subvariety of X which is defined over K and contains Y. Moreover, if Y is irreducible, then Y_0 is also irreducible.

In a first step, we show that Y_0 is defined over K. To see this, we may assume that Y is affine given by $f_1(\mathbf{x}), \dots, f_m(\mathbf{x})$ in $\mathbb{A}^n_F$. Then Y_0 is given by all the polynomials of the form $\prod_{\sigma \in R} \sigma(f_{j_\sigma})$, where $1 \leq j_\sigma \leq m$. Since Y is defined over a finite subextension, we may assume that $[F:K] < \infty$. We consider first the case F/K separable. Passing to a finite extension, we may assume F/K Galois with Galois group R. It is enough to show that Y_0 is defined by the polynomials

$$\sum_{\rho \in R} \rho(\mu) \prod_{\sigma \in R} \rho \circ \sigma(f_{j_\sigma}) = \mathrm{Tr}_{F(\mathbf{x})/K(\mathbf{x})}\left(\mu \prod_{\sigma \in R} \sigma(f_{j_\sigma})\right) \quad (\mu \in F, 1 \leq j_\sigma \leq m)$$

defined over K. Clearly, Y_0 is contained in the zero set of these polynomials. If $x \notin Y_0$, there are $1 \leq j_\sigma \leq m (\sigma \in G)$ with $\sigma(f_{j_\sigma})(x) \neq 0$. By Artin's theorem on linear independence of characters ([**173**], Ch.VIII, Th.4.1), there is $\mu \in F$ with

$$\sum_{\rho \in R} \rho(\mu) \prod_{\sigma \in R} \rho \circ \sigma(f_{j_\sigma})(x) \neq 0,$$

proving the first step in the separable case.

In general, let E be the separable closure of K in F. Then F/E is purely inseparable. We may assume that $\mathrm{char}(K) = p > 0$, otherwise we are done by the separable case. By [**156**], Prop.8.13, there is a power p^e such that $\alpha^{p^e} \in E$ for all $\alpha \in F$. We conclude that $f_1^{p^e}, \dots, f_m^{p^e}$ have coefficients in E, hence Y is defined over E (as a closed subvariety of X_E). Now the separable case yields that Y_0 is defined over K.

Every closed subvariety defined over K containing Y has to contain also all conjugates, hence Y_0 is indeed minimal. If Y is irreducible and Y_0 is the union of two closed subsets, then Y is contained in one of them and irreducibility of Y_0 follows from minimality.

Example A.4.14. Let Z be a zero-dimensional irreducible closed subvariety of X. Then Z is contained in an affine open subset U with ideal of vanishing $I(Z)$ equal to a maximal ideal $\mathfrak{m}$ of $K[U]$. For $x \in Z$, A.2.5 and A.2.6 show that $\mathcal{O}_{X,x} \cong K[U]_\mathfrak{m}$ and that $K(x) \cong K(Z) \cong K[U]/\mathfrak{m}$. By A.2.7, Z is the set of conjugates of x.

Example A.4.15. Let X be an irreducible variety over K and let F be the separable algebraic closure of K in $K(X)$. By [**137**], Cor.4.5.10, the irreducible components of $X_{\overline{K}}$ are defined over the Galois closure E of F/K and their number is $[F:K]$. By A.4.13, they are conjugates with respect to $\mathrm{Gal}(E/K)$.

A.5. Vector bundles

Let X be a variety over a field K.

A.5.1. A **vector bundle** over X is a variety E over K with a morphism $\pi_E : E \to X$ over K and the following additional structure: There is an open covering $(U_\alpha)_{\alpha \in I}$ of X and isomorphisms $\varphi_\alpha : \pi_E^{-1}(U_\alpha) \to U_\alpha \times \mathbb{A}_K^{r_\alpha}$ of varieties over K such that $\pi_E = p_1 \circ \varphi_\alpha$ on $\pi_E^{-1}(U_\alpha)$, where p_1 is the first projection of

$U_\alpha \times \mathbb{A}_K^{r_\alpha}$. Moreover, we assume that for all $\alpha, \beta \in I$ and $x \in U_\alpha \cap U_\beta$, there is $g_{\alpha\beta}(x) \in GL(r_\alpha, \overline{K})$ with

$$\varphi_\alpha \circ \varphi_\beta^{-1}(x, \boldsymbol{\lambda}) = (x, g_{\alpha\beta}(x)\boldsymbol{\lambda}) \tag{A.2}$$

for all $\boldsymbol{\lambda} \in \mathbb{A}_K^{r_\alpha}$. Note that $r_\alpha = r_\beta$ on overlappings. If all r_α are equal to a fixed number r, then E is called a vector bundle of **rank r**.

A.5.2. For example, $p_1 : X \times \mathbb{A}_K^r \to X$ is a vector bundle of rank r. It is called the **trivial vector bundle** of rank r. In A.5.1, we have required that E is locally isomorphic to the trivial vector bundle. Hence $(U_\alpha, \varphi_\alpha)_{\alpha \in I}$ is called a **trivialization** of E. By abuse of notation, we often skip the morphism π_E.

The **fibres** of E are denoted by $E_x := \pi_E^{-1}(x)$. Note that $\pi_E = p_1 \circ \varphi_\alpha$ means that φ_α maps E_x isomorphically onto the fibre of $U_\alpha \times \mathbb{A}_K^r$ over x. We consider the restriction

$$E_x(K(x)) \xrightarrow{\varphi_\alpha} \{x\} \times K(x)^r$$

to $K(x)$-rational points. The right-hand side is a $K(x)$-vector space. We also endow $E_x(K(x))$ with the $K(x)$-vector space structure such that the above map is an isomorphism. We claim that this definition is independent of the choice of the trivialization. To see this we have just to note that $g_{\alpha\beta}(x) \in GL(r, K(x))$ in (A.2), because both φ_α and φ_β are isomorphisms of K-varieties.

Similarly, we see that E_x has a canonical $\overline{K}$-vector space structure such that φ_α induces an isomorphism of $\overline{K}$-vector spaces onto $\{x\} \times \overline{K}^r$. Clearly, we have $E_x = E_x(K(x)) \otimes_{K(x)} \overline{K}$.

Remark A.5.3. If we assume in A.5.1 that E is only a prevariety with all the other properties of a vector bundle, then local triviality easily implies that E is a variety.

A.5.4. We call $g_{\alpha\beta}(x)$ the **transition matrix**. Since the trivializations $\varphi_\alpha, \varphi_\beta$ are morphisms over K, it is easy to see that we have a morphism

$$U_\alpha \cap U_\beta \longrightarrow GL(r)_K, \quad x \mapsto g_{\alpha\beta}(x),$$

where $GL(r)_K$ is the group variety introduced in Example A.3.12. On $U_\alpha \cap U_\beta \cap U_\gamma$, we have the **cocycle rule**

$$g_{\alpha\beta} g_{\beta\gamma} = g_{\alpha\gamma}. \tag{A.3}$$

A.5.5. Let E and F be vector bundles over X. A map $\varphi : E \to F$ is called a **homomorphism of vector bundles** if it is a morphism of varieties over K such that for all $x \in X$, there is a linear map $\varphi_x : E_x \to F_x$ of $\overline{K}$-vector spaces with $\varphi(x) = (x, \varphi_x(x))$. Note that this implies $\pi_F \circ \varphi = \pi_E$.

An **isomorphism of vector bundles** is an invertible homomorphism of vector bundles. If a homomorphism $\varphi : E \to F$ of vector bundles is a closed embedding, then E is a **subbundle** of F and we may identify E with $\varphi(E)$.

Example A.5.6. Let E be a vector bundle over K and let $\mu \in K$. Then we have a homomorphism $[\mu] : E \to E$ given by using multiplication with μ on the fibre E_x. To check that $[\mu]$ is a K-morphism, we use that $[\mu]$ is given in a trivialization $U_\alpha \times \mathbb{A}_K^r$ by $(x, \boldsymbol{\lambda}) \mapsto (x, \mu\boldsymbol{\lambda})$.

A.5.7. In this remark, we show how to give a vector bundle by its transition matrices. Let $(U_\alpha)_{\alpha \in I}$ be an open covering of X. For all $\alpha, \beta \in I$, we consider K-morphisms $g_{\alpha\beta} : U_\alpha \cap U_\beta \to GL(r)_K$ satisfying the cocycle rule (A.3). Then we glue the trivial bundles $U_\beta \times \mathbb{A}_K^r$ and $U_\alpha \times \mathbb{A}_K^r$ along the isomorphisms

$$\varphi_{\alpha\beta} : (U_\alpha \cap U_\beta) \times \mathbb{A}_K^r \xrightarrow{\sim} (U_\alpha \cap U_\beta) \times \mathbb{A}_K^r$$

given by $\varphi_{\alpha\beta}(x, \boldsymbol{\lambda}) = (x, g_{\alpha\beta}(x)\boldsymbol{\lambda})$. For more on glueing, see [**148**], Exercise II.2.12. We obtain a vector bundle E over X with trivializations $\varphi_\alpha : \pi_E^{-1}(U_\alpha) \to U_\alpha \times \mathbb{A}_K^r$ such that the transition matrices are equal to $g_{\alpha\beta}$.

Conversely, if we start with a vector bundle and apply this process to its transition matrices, we get a new vector bundle isomorphic to the original one.

Example A.5.8. Let E and E' be vector bundles over X. As an abstract set, the **direct sum** $E \oplus E'$ is given as the disjoint union of $(E_x \oplus E'_x)_{x \in X}$. We get $\pi_{E \oplus E'}$ by mapping $E_x \oplus E'_x$ onto x. To define a vector bundle structure on it, we choose trivializations $(U_\alpha, \varphi_\alpha)_{\alpha \in I}$ and $(U_\alpha, \varphi'_\alpha)_{\alpha \in I}$ of E and E', respectively. Note that we may always assume that the open coverings are the same by passing to a common refinement. We claim that there is a unique vector bundle structure on $E \oplus E'$ such that

$$\varphi_\alpha \oplus \varphi'_\alpha : \pi_{E \oplus E'}^{-1}(U_\alpha) \longrightarrow U_\alpha \times \mathbb{A}_K^{r_\alpha + r'_\alpha} = U_\alpha \times \left(\overline{K}^{r_\alpha} \oplus \overline{K}^{r'_\alpha}\right),$$

defined *a priori* on fibres, is an isomorphism of vector bundles. Implicitly, this is used to define $E \oplus E'$ as a K-variety. To see the claim, we note that the transition matrices of $E \oplus E'$ are given by

$$g''_{\alpha\beta} := \begin{pmatrix} g_{\alpha\beta}(x) & 0 \\ 0 & g'_{\alpha\beta}(x) \end{pmatrix} \in GL(r_\alpha + r'_\alpha, K(x)),$$

where $g_{\alpha\beta}, g'_{\alpha\beta}$ are the transition matrices of E and E', respectively. Clearly, $g''_{\alpha\beta}$ gives a morphism $U_\alpha \cap U_\beta \to GL(r_\alpha + r'_\alpha)_K$ proving well-definedness of the vector bundle. Obviously, we have

$$(E \oplus E')_x(K(x)) = E_x(K(x)) \oplus E'_x(K(x)).$$

Note that we have a homomorphism $E \oplus E \to E$ of vector bundles given by addition on each fibre.

A.5.9. Let E be a vector bundle over X. A **section** of E over an open subset U of X is a K-morphism $s : U \to E$ such that $\pi_E \circ s$ is the identity map on U. If $U = X$, then s is called a **global section** of E. Using Examples A.5.6 and A.5.8, it is clear that the set $\Gamma(U, E)$ of sections of E over U is a K-vector space. Then we get a sheaf $\mathcal{E}$ of K-vector spaces on X by setting $\mathcal{E}(U) := \Gamma(U, E)$ for each open subset U of X and using restriction of morphisms. Then $\mathcal{E}$ is called the **sheaf of sections** of E.

Example A.5.10. Let $O_X := X \times \mathbb{A}^1_K$ be the trivial bundle of rank 1 over X, then we identify the sheaf of sections with the sheaf of regular functions $\mathcal{O}_X$ on X in the following way. Let f be a regular function, then $x \mapsto (x, f(x))$ gives a section of O_X and every section has this form.

Example A.5.11. Not every sheaf of K-vector spaces is the sheaf of sections of a vector bundle. For example, if $\mathcal{J}_Y$ is the ideal sheaf of a closed subvariety Y of codimension at least 2 in X, then it is not ot this form. Otherwise, $\mathcal{J}_Y$ would be associated to a vector bundle of rank 1, since $\mathcal{J}_Y = \mathcal{O}_X$ is outside of the subvariety. But then $\mathcal{J}_Y$ would be locally a principal ideal in $\mathcal{O}_X$, which is a contradiction to codimension 2.

For the precise relation between sheaves and vector bundles, see A.10.13.

A.5.12. Let E and F be vector bundles over K of rank r and r'. Then the **tensor product** $E \otimes E'$ of E and E' is constructed similarly as the direct sum in Example A.5.8. As an abstract set, it is the disjoint union of $E_x \otimes E'_x, x \in X$. Let $(U_\alpha, \varphi_\alpha)_{\alpha \in I}, (U_\alpha, \varphi'_\alpha)_{\alpha \in I}$ be trivializations of E and E', respectively. Let $\mathbf{x}$ (resp. $\mathbf{x}'$) be the coordinates of $\mathbb{A}^r_K$ (resp. $\mathbb{A}^{r'}_K$). We order $(x_i \otimes x'_j)_{1 \leq i \leq r, 1 \leq j \leq r'}$ lexicographically to get an identification of $\overline{K}^r \otimes_{\overline{K}} \overline{K}^{r'}$ with $\mathbb{A}^{rr'}_K$. There is a unique structure on $E \otimes E'$ as a vector bundle over X such that all

$$\varphi_\alpha \otimes \varphi'_\alpha : \pi^{-1}_{E \otimes E'}(U_\alpha) \longrightarrow U_\alpha \times \mathbb{A}^{rr'}_K,$$

defined *a priori* on fibres, are isomophisms of vector bundles. This is clear since the transition matrices of $E \otimes E'$ are given by $g_{\alpha\beta} \otimes g'_{\alpha\beta}$. Clearly, we have

$$(E \otimes E')_x(K(x)) = E_x(K(x)) \otimes_{K(x)} E'_x(K(x))$$

and

$$\Gamma(U_\alpha, E \otimes E') = \Gamma(U_\alpha, E) \otimes \Gamma(U_\alpha, E').$$

However, this has not to be true for all open subsets of X. Here, equal means up to canonical isomorphism.

A.5.13. Similarly, we construct the **dual vector bundle** E^* of E. It is the disjoint union of the dual vector spaces E^*_x of $E_x, x \in X$. The transition matrices of E^* are given by the transposes $h_{\alpha\beta} := g^t_{\beta\alpha}$.

We can also extend other constructions from linear algebra to vector bundles. It is always the same pattern. First, we define the underlying set using the construction fibrewise. Then we choose the evident trivializations. We use it to define the vector bundle structure on the abstract set. We have to show that it fits on overlappings, which becomes clear by considering transition matrices. Note that pointwise, they are the same as the transformation matrices in linear algebra.

For example, we can define **exterior products** $E \wedge E'$ of vector bundles, Hom(E, E') or the **quotient vector bundles** E/F for a subbundle F of E. The details are left to the reader.

A.5.14. Let $\varphi : X' \to X$ be a morphism of varieties over K and let E be a vector bundle over X. Suppose that E is given by the transition matrices $g_{\alpha\beta}$ with respect to the open covering $(U_\alpha)_{\alpha \in I}$. Then the **pull-back** $\varphi^*(E)$ of E is the vector bundle over X' given by the transition matrices $g_{\alpha\beta} \circ \varphi$ with respect to the open covering $(\varphi^{-1}U_\alpha)_{\alpha \in I}$.

If φ is a closed embedding or an inclusion of an open subvariety, then $\varphi^* E$ is simply denoted by $E|_{X'}$ called the **restriction** of E to X'.

A.5.15. Let F be an extension field of K and let E be a vector bundle on X. Then the **base change** E_F is the vector bundle on X_F given by the same transition functions as E. If $F \subset \overline{K}$, then the fibre of E_F over $x \in X$ is $E_x \otimes_K F$.

A.5.16. A **line bundle** L on X is a vector bundle of rank 1 over X. Note that the tensor product or the pull-back of line bundles are again line bundles. We use the following notation: The n-fold tensor product of L is denoted by $L^{\otimes n}$ and $L^{-1} = L^*$ for the dual. For negative n, we define $L^{\otimes n} := (L^{-1})^{\otimes |n|}$.

The set of isomorphism classes of line bundles on X form a group under $\otimes$. It is called the **Picard group** Pic(X). An element $\mathbf{c}$ of Pic(X) is always written boldly. The Picard group is abelian and so it is written additively. The isomorphism class of a line bundle L is denoted by cl(L). We have $\mathbf{0} = \mathrm{cl}(O_X)$ and $-\mathrm{cl}(L) = \mathrm{cl}(L^{-1})$. To check this, note that the **transition functions** of $L \otimes L^{-1}$ are given by $g_{\alpha\beta} g_{\beta\alpha}^t = 1$.

The following evident functoriality rule is important. Given morphisms $X'' \xrightarrow{\psi} X' \xrightarrow{\varphi} X$ of varieties over K, we have

$$\psi^* \varphi^*(\mathbf{c}) = (\varphi \circ \psi)^*(\mathbf{c}) \in \mathrm{Pic}(X'')$$

for every $\mathbf{c} \in \mathrm{Pic}(X)$.

A.5.17. The zero section may be the only global section of a line bundle L on X. To bypass this difficulty, we introduce meromorphic sections. We consider pairs (U, s_U), where U is an open dense subset of X and $s_U \in \Gamma(U, L)$. Two pairs (U, s_U) and (V, s_V) are called equivalent if $s_U = s_V$ on $U \cap V$. This gives an equivalence relation. The corresponding equivalence classes are called **meromorphic sections** of L. By abuse of notation, we simply denote them by s. Note that there is a maximal (W, s_W) and we identify the equivalence class with s to have a concrete section at hand. We call s an **invertible meromorphic section** of L if there is an open dense subset U of X such that s is a regular function on U without zeros.

Example A.5.18. Similarly as in Example A.5.10, we may identify a rational function f with a meromorphic section $x \mapsto (x, f(x))$ of O_X and conversely.

A.5.19. Let E be a vector bundle on X given by the transition matrices $g_{\alpha\beta}$ with respect to the open covering $(U_\alpha)_{\alpha \in I}$. For an open subset U of X, the restriction of $s \in \Gamma(U, E)$ may be identified with $s_\alpha \in \mathcal{O}_X(U_\alpha)$ satisfying

$$s_\alpha = g_{\alpha\beta} s_\beta$$

on $U_\alpha \cap U_\beta \cap U$. Conversely, any set $s_\alpha \in \mathcal{O}_X(U_\alpha), \alpha \in I$, with this transition rule gives rise to a section $s \in \Gamma(U, E)$ using the sheaf property.

If $\varphi : X' \to X$ is a morphism of K-varieties and $s \in \Gamma(U, E)$, then $s_\alpha \circ \varphi = (g_{\alpha\beta} \circ \varphi)(s_\beta \circ \varphi)$ and hence we get a section $\varphi^*(s)$ of $\varphi^* E$ over $\varphi^{-1} U$.

A.5.20. A vector bundle over X is said to be **generated by global sections** $(s_\lambda)_{\lambda \in I}$ if for all $x \in X$, the vectors $\{s_\lambda(x) \mid \lambda \in I\}$ generate the $\overline{K}$-vector space E_x. In particular, if E is a line bundle, this means that $s_\lambda(x) \neq 0$ for at least one $\lambda \in I$.

We say that E is **generated by global sections** if it is generated by the family of all global sections.

If E, E' are vector bundles over X generated by global sections $(s_\lambda)_{\lambda \in I}$ and $(s'_\mu)_{\mu \in J}$, then $E \otimes E'$ is generated by global sections $(s_\lambda \otimes s'_\mu)_{\lambda \in I, \mu \in J}$. Moreover, if $\varphi : X' \to X$ is a morphism of varieties over K, then $\varphi^*(E)$ is generated by the global sections $(\varphi^* s_\lambda)_{\lambda \in I}$.

A **base-point** of a line bundle L is a point where every global section of L vanishes. A line bundle generated by global sections is called **base-point-free.**

A.6. Projective varieties

Let K be a field and $\overline{K}$ an algebraic closure of K.

A.6.1. On $\overline{K}^{n+1}$, two non-zero vectors are called equivalent if they lie in the same one-dimensional linear subspace. The set of equivalence classes is denoted by $\mathbb{P}^n(\overline{K})$. It may be viewed as the set of one-dimensional linear subspaces in $\overline{K}^{n+1}$. We denote points in $\mathbb{P}^n(\overline{K})$ by $(\alpha_0 : \cdots : \alpha_n)$, where $\boldsymbol{\alpha} \in \overline{K}^{n+1} \setminus \{0\}$ is a representative. We should always be aware that the coordinate vector $\boldsymbol{\alpha}$ is only determined up to multiples. For $T \subset K[x_0, \ldots, x_n]$ consisting of homogeneous polynomials, let

$$Z(T) := \{\boldsymbol{\alpha} \in \mathbb{P}^n(\overline{K}) \mid f(\boldsymbol{\alpha}) = 0 \quad \forall f \in T\}$$

be the **zero set** of T. All subsets of this form are called closed in $\mathbb{P}^n(\overline{K})$. The **Zariski topology** on $\mathbb{P}^n(\overline{K})$ is the topology with exactly these closed subsets. It depends on K, so we denote the corresponding topological space by $\mathbb{P}^n_K$. It is called the **projective n-space** over K. In fact, it is canonically a variety, as we will show below. On $\mathbb{P}^n_K$, we always fix a set of coordinates $\mathbf{x} = (x_0 : \cdots : x_n)$ and we call them **homogeneous coordinates.**

A.6.2. The **standard affine open subsets** of $\mathbb{P}^n_K$ are

$$U_i := \{\boldsymbol{\alpha} \in \mathbb{P}^n_K \mid \alpha_i \neq 0\} \qquad (i = 0, \ldots, n).$$

Then we have homeomorphisms

$$\varphi_i : U_i \longrightarrow \mathbb{A}^n_K, \ \boldsymbol{\alpha} \mapsto \left(\frac{\alpha_0}{\alpha_i}, \ldots, \frac{\alpha_{i-1}}{\alpha_i}, \frac{\alpha_{i+1}}{\alpha_i}, \ldots, \frac{\alpha_n}{\alpha_i}\right).$$

So $\mathbb{P}^n_K$ is a K-variety with affine charts $(U_i, \varphi_i)_{i=0,\ldots,n}$. For the proof that the diagonal is closed, we refer to A.6.4. Clearly, the K-variety $\mathbb{P}^n_K$ does not depend on the choice of coordinates, i.e. if we change coordinates by $g \in GL(n+1, K)$, then we get the same K-variety $\mathbb{P}^n_K$.

A.6.3. Let Y be a closed subset of $\mathbb{P}^n_K$. By A.4.8, it is a closed subvariety of $\mathbb{P}^n_K$. We call it a **projective variety** over K. The **homogeneous ideal** $I(Y)$ of Y is the ideal in $K[x_0, \ldots, x_n]$ generated by the homogeneous polynomials vanishing on Y. The **homogeneous coordinate ring** $S(Y)$ of Y is defined by

$$S(Y) := \bigoplus_{d \in \mathbb{N}} S(Y)_d := K[\mathbf{x}]/I(Y),$$

where the graduation is induced by the degree of polynomials. If Y is irreducible, then the field of rational functions $K(Y)$ is the subfield of the quotient field of $S(Y)$ given by

$$\{f/g \mid \exists d \in \mathbb{N} \text{ with } f, g \in S(Y)_d\}.$$

To sketch the proof, we view Y as a closed subset of $\mathbb{P}^n_K$. Then there is a standard affine open subset U_i with $Y \cap U_i \neq \emptyset$. By A.4.11, the quotient field of $K[U_i \cap Y]$ is equal to $K(Y)$. Since U_i is isomorphic to affine n-space, any rational function may be written as the quotient of two polynomials in $x_0, \ldots, x_{i-1}, x_{i+1}, \ldots, x_n$. By passing to the homogenizations, we get the claim.

A.6.4. The product of projective varieties over K is again a projective variety over K. To prove it, let Y (resp. Y') be a closed subvariety of $\mathbb{P}^n_K$ (resp. $\mathbb{P}^m_K$). We consider the **Segre embedding**

$$\iota : \mathbb{P}^n_K \times \mathbb{P}^m_K \longrightarrow \mathbb{P}^N_K, \quad (\mathbf{x}, \mathbf{x}') \mapsto (x_i x'_j)_{0 \leq i \leq n, 0 \leq j \leq m},$$

where $N = (n+1)(m+1) - 1$. As $Y \times Y'$ is a closed subvariety of $\mathbb{P}^n_K \times \mathbb{P}^m_K$, it is enough to show that the Segre embedding is a closed embedding. Let (y_{ij}) be the coordinates on $\mathbb{P}^N_K$ (say ordered lexicographically) such that ι maps $(\mathbf{x}, \mathbf{x}')$ to $(y_{ij}) = (x_i x'_j)$. We have to prove that ι maps $\mathbb{P}^n_K \times \mathbb{P}^m_K$ isomorphically onto the closed subvariety

$$Z := Z(\{y_{ij} y_{kl} - y_{kj} y_{il} \mid i, k \in \{0, \ldots, n\}, j, l \in \{0, \ldots, m\}\}).$$

Clearly, the image of ι is contained in Z. Let $V_{kl} := \{y_{kl} \neq 0\}$ be a standard affine open subset of $\mathbb{P}_K^N$. Then $\iota^{-1}(V_{kl})$ is the product of the standard affine open subsets $U_k := \{x_k \neq 0\}$ and $U_l' := \{x_l' \neq 0\}$. Then we define a K-morphism

$$V_{kl} \cap Z \longrightarrow U_k \times U_l', \quad (y_{ij}) \mapsto ((y_{0l} : \cdots : y_{nl}), (y_{k0} : \cdots : y_{km})).$$

It is easily checked that it is inverse to $\iota|_{U_k \times U_l'}$. Therefore ι is an isomorphism onto Z.

If $n = m$, then the diagonal in $\mathbb{P}_K^n \times \mathbb{P}_K^n$ is the intersection of $\mathbb{P}_K^n \times \mathbb{P}_K^n$ with $Z(\{y_{ij} - y_{ji} \mid i, j = 0, \ldots, n\})$ in $\mathbb{P}_K^N$, hence the diagonal is closed. We conclude that every projective variety is indeed a variety.

A.6.5. Let L be the subbundle of the trivial vector bundle $E := \mathbb{P}_K^n \times \mathbb{A}_K^{n+1}$ given by

$$L := \{(\boldsymbol{\lambda}, \boldsymbol{\mu}) \mid \boldsymbol{\mu} \in \overline{K}\boldsymbol{\lambda}\}.$$

This means that the fibre over $\boldsymbol{\lambda}$ is just the line through $\boldsymbol{\lambda}$ in $\overline{K}^{n+1}$. We denote the standard coordinates of $\mathbb{P}_K^n$ and $\mathbb{A}_K^{n+1}$ by $x_0, \ldots, x_n$ and $y_0, \ldots, y_n$, respectively. Then L is given in the trivial bundle by the equations

$$x_i y_j - x_j y_i \quad (i, j = 0, \ldots, n).$$

Hence L is a closed subset of E. We use that to define L as a closed subvariety over K. Moreover, $\pi_L := \pi_E|_L$ is a K-morphism. To define trivializations, we use the standard affine open subsets $U_\alpha := \{x_\alpha \neq 0\}$ of $\mathbb{P}_K^n$. Let

$$\varphi_\alpha : \pi_L^{-1}(U_\alpha) \longrightarrow U_\alpha \times \mathbb{A}_K^1, \quad (\boldsymbol{\lambda}, \boldsymbol{\mu}) \mapsto (\boldsymbol{\lambda}, \mu_\alpha).$$

Then it is easy to see that φ_α is an isomorphism of K-varieties which is linear in the fibres. Therefore L is a line bundle on $\mathbb{P}_K^n$ called the **tautological line bundle**. It is denoted by $O_{\mathbb{P}_K^n}(-1)$. The above trivializations lead to the transition functions

$$g_{\alpha\beta}(\mathbf{x}) = \frac{x_\alpha}{x_\beta}$$

on $U_\alpha \cap U_\beta$. For $m \in \mathbb{Z}$, let

$$O_{\mathbb{P}_K^n}(m) := O_{\mathbb{P}_K^n}(-1)^{\otimes(-m)}.$$

Then its transition functions are $g_{\alpha\beta}(\mathbf{x}) = \left(\frac{x_\beta}{x_\alpha}\right)^m$ on $U_\alpha \cap U_\beta$.

A.6.6. We claim that the global sections of $O_{\mathbb{P}_K^n}(m)$ may be identified with the homogeneous polynomials in $K[x_0, \ldots, x_t]$ of degree d.

For $s \in \Gamma(\mathbb{P}_K^n, O_{\mathbb{P}_K^n}(m))$ and with respect to the above trivializations, there are regular functions s_α on U_α such that

$$(\varphi_\alpha \circ s)(x) = (x, s_\alpha(x)) \tag{A.4}$$

label for all $x \in U_\alpha$. They satisfy

$$s_\alpha = g_{\alpha\beta} s_\beta \tag{A.5}$$

on $U_\alpha \cap U_\beta$, where $g_{\alpha\beta}(\mathbf{x}) = (x_\beta/x_\alpha)^m$ are the transition functions of $O_{\mathbb{P}^n_K}(m)$. Conversely, any regular functions $s_\alpha \in O_{\mathbb{P}^n_K}(U_\alpha)$ for $\alpha = 0, \dots, n$ satisfying (A.5) determine a unique global section s with (A.4). Since U_α is a standard affine open subset of $\mathbb{P}^n_K$, we may identify the s_α with polynomials. The rule (A.5) means that there is a homogeneous polynomial of degree m such that s_α is obtained by inserting 1 for x_α. Hence the global sections of $O_{\mathbb{P}^n_K}(m)$ may be identified with the homogenous polynomials of degree m in the variables $x_0, \dots, x_m$ with coefficients in K.

A.6.7. Let E be a vector bundle over a projective K-variety X. Then $\Gamma(X, E)$ is a finite-dimensional K-vector space ([**148**], Th.II.5.19). In particular, if K is algebraically closed and X is irreducible, then $O_X(X) = \Gamma(X, O_X)$ has to be a finite-dimensional field extension of K and hence $K = O_X(X)$.

A.6.8. Let L be a line bundle on a K-variety X and assume that it is generated by a set of global sections $\{s_0, \dots, s_n\}$. Then we have a K-morphism

$$\varphi : X \longrightarrow \mathbb{P}^n_K, \quad x \mapsto (s_0(x) : \cdots : s_n(x)).$$

To be more precise, we choose a trivialization of L around x such that we can identify $s_0, \dots, s_n$ with regular functions as in Example A.5.10. Then the above definition makes sense because not all $s_j(x)$ vanish. Since we have homogeneous coordinates on $\mathbb{P}^n_K$, this definition does not depend on the choice of the trivialization. We claim that $L = \varphi^* O_{\mathbb{P}^n_K}(1)$ and $s_j = \varphi^*(x_j)$ hold for $j = 0, \dots, n$.

In order to prove this statement, let $V_j := \{x \in X \mid s_j(x) \neq 0\}$. Since L is generated by these global sections, $(V_j)_{j=0,\dots,n}$ is an open covering of X. Clearly, we have $V_j = \varphi^{-1}(\{x_j \neq 0\})$. Moreover, s_j defines a trivialization

$$\varphi_j : \pi_L^{-1}(V_j) \longrightarrow V_j \times \mathbb{A}^1_K, \quad \lambda s_j(x) \mapsto (x, \lambda).$$

The transition functions of L with respect to this trivialization are

$$g_{ij}(x) = \frac{s_j(x)}{s_i(x)} = \frac{x_j \circ \varphi}{x_i \circ \varphi}(x).$$

This proves $L = \varphi^* O_{\mathbb{P}^n_K}(1)$. To check $s_j = \varphi^*(x_j)$, note that x_j (resp. s_j) is given in the trivialization over $\{x_i \neq 0\}$ (resp. V_i) by the regular function $\frac{x_j}{x_i}$ (resp. $\frac{s_j}{s_i}$). Because of $\frac{x_j}{x_i} \circ \varphi = \frac{s_j}{s_i}$, we get the claim.

A.6.9. Conversely, let $\varphi : X \to \mathbb{P}^n_K$ be a K-morphism for a K-variety X. Then $L := \varphi^* O_{\mathbb{P}^n_K}(1)$ is generated by the global sections $s_0 := \varphi^*(x_0), \dots, s_n := \varphi^*(x_n)$. If we construct now the morphism determined by these global sections as in A.6.8, then we get back our original morphism φ.

A.6.10. Let X be a projective variety over K. Then a line bundle L on X is called **very ample** if there is a closed embedding $i : X \to \mathbb{P}^n_K$ such that $L \cong i^* O_{\mathbb{P}^n_K}(1)$. Obviously, very ample line bundles are generated by its global sections, which may be viewed as **hyperplane sections**. Note however that not every global section of L has to be a hyperplane section with respect to the embedding i. But we can construct an i such that this is true (see Remark A.6.11 below).

A line bundle L is called **ample** if $L^{\otimes n}$ is very ample for some $n \in \mathbb{N}$.

For line bundles L and M, the following holds ([**148**], Exercise II.7.5):

(a) If M is ample, then $L \otimes M^{\otimes n}$ is very ample for n sufficiently large.

(b) If L is generated by global sections and M is ample (resp. very ample), then $L \otimes M$ is ample (resp. very ample).

(c) If L and M are both ample, then $L \otimes M$ is ample.

Remark A.6.11. Let L be a very ample line bundle on X and let $s_0, \dots, s_n$ be a basis of $\Gamma(X, L)$. Then we claim that the morphism

$$\varphi : X \longrightarrow \mathbb{P}^n_K, \quad x \mapsto (s_0(x) : \cdots : s_n(x))$$

from A.6.8 is a closed embedding.

To prove this, we choose a closed embedding $i : X \to \mathbb{P}^m_K$ such that $i^* O_{\mathbb{P}^m_K}(1) = L$. We denote the coordinates on $\mathbb{P}^m_K$ by $y_0, \dots, y_m$ and those of $\mathbb{P}^n_K$ by $x_0, \dots, x_n$. The morphism i is determined by the global sections $t_0 = i^*(y_0), \dots, t_m = i^*(y_m)$ as in A.6.9. We adjoin some $t_{m+1}, \dots, t_M \in \Gamma(X, L)$ to get generators of the vector space and a morphism

$$j : X \longrightarrow \mathbb{P}^M_K, \quad x \mapsto (t_0(x) : \cdots : t_M(x)).$$

First note that $j(X)$ is closed in $\mathbb{P}^M_K$ (see A.6.15 below). Clearly, we have a morphism $\psi : j(X) \to i(X)$ with $i = \psi \circ j$. By definition of a closed embedding, i maps X isomophically onto $i(X)$ and so we check easily that ψ is an isomorphism. This proves that j is a closed embedding. In particular, the morphism

$$\varphi' : X \longrightarrow \mathbb{P}^{m+n+1}_K, \quad x \mapsto (t_0(x) : \cdots : t_m(x) : s_0(x) : \cdots : s_n(x))$$

is a closed embedding. Since $t_0, \dots, t_m$ are linear combinations of $s_0, \dots, s_n$, we have a morphism $\psi' : \mathbb{P}^n_K \to \mathbb{P}^{n+m+1}_K$ with $\varphi' = \psi' \circ \varphi$. Since φ' is a closed embedding, we conclude as above that φ is a closed embedding.

Remark A.6.12. A line bundle L on X is ample (resp. very ample) if and only if the base change $L_{\overline{K}}$ on $X_{\overline{K}}$ is ample (resp. very ample). To see the ample case, use the cohomological criterion of ampleness from [**148**], Prop.III.5.3, and the compatibility of cohomology and base change (see A.10.28). The very ample case follows from [**137**], Prop.2.7.1, implying of course also the ample case. In fact, it is proved that a morphism is a closed embedding if and only if its base change is a closed embedding.

Example A.6.13. A **multiprojective space** is a product

$$\mathbb{P}_K := \mathbb{P}^{n_1}_K \times \cdots \times \mathbb{P}^{n_r}_K$$

of projective spaces. The projection to the ith factor $\mathbb{P}_K^{n_i}$ is denoted by p_i. Now for $d_1, \ldots, d_r \in \mathbb{Z}$, let

$$O_{\mathbb{P}}(d_1, \ldots, d_r) := p_1^* O_{\mathbb{P}^{n_1}}(d_1) \otimes \cdots \otimes p_r^* O_{\mathbb{P}^{n_r}}(d_r).$$

Let $\mathbf{x}_i = (x_{i0} : \cdots : x_{in_i})$ be the coordinates on $\mathbb{P}_K^{n_i}$. By generalizing A.6.6, we see that the global sections of $O_{\mathbb{P}}(d_1, \ldots, d_r)$ may be identified with the multihomogeneous polynomials in $\mathbf{x}_1, \ldots, \mathbf{x}_r$, homogeneous of degree d_i in $\mathbf{x}_i$.

The Segre embedding may be extended to include several factors, thereby proving that $O_{\mathbb{P}}(1, \ldots, 1)$ is very ample. On the other hand, $p_i^* O_{\mathbb{P}^{n_i}}(1)$ is generated by global sections (but certainly not ample for two or more factors). By A.6.10, we conclude that $O(d_1, \ldots, d_r)$ is very ample if $d_1, \ldots, d_r \geq 1$.

A.6.14. A variety X over K is called **complete** if for all varieties Y over K, the second projection $p_2 : X \times Y \to Y$ is closed (i.e. maps closed sets to closed sets). In algebraic geometry, this is the analogue of compact complex manifolds.

There is a relative version of this notion called proper morphisms. Every closed embedding and every morphism from a complete variety will be proper. This notion is important in algebraic geometry, many finiteness results are related to it. For our book, it plays only a minor role and is used to state the results properly. The reader may always think of a morphism of complete varieties.

For $j = 1, 2$, let $\varphi_j : X_j \to S$ be a morphism of varieties over K. Then

$$X_1 \times_S X_2 := \{(x_1, x_2) \in X_1 \times X_2 \mid \varphi_1(x_1) = \varphi_2(x_2)\}$$

is called the **fibre product** of X_1 and X_2 over S. It is easily seen that the fibre product is a closed subset of $X_1 \times X_2$, so we may view $X_1 \times_S X_2$ as a closed subvariety of $X_1 \times X_2$. Let $\varphi : X \to X'$ be a morphism of varieties over K. For every morphism $\psi : Y' \to X'$, we define the base change $\varphi_{Y'} : X \times_{X'} Y' \to Y'$ by $\varphi_{Y'}(x, y') = y'$. The morphism φ is called **proper** if all base changes of φ are closed.

Clearly, a variety X is complete if and only if the constant map $X \to \mathbb{A}_K^0$ is a proper morphism. For details about proper morphisms, we refer to [**148**], II.4. (However, the reader has to translate the results from the category of schemes. Note that the fibre product of varieties is the variety associated to the fibre product of schemes. Since any morphism of varieties is of finite type and separated, our definition of a proper map agrees with the one in [**148**].)

A.6.15. We mention some properties of complete varieties over K. Most of them can be deduced from the corresponding properties of proper morphisms in [**148**], Cor.4.8.

(a) A closed subvariety of a complete variety over K is complete because any closed embedding of varieties is proper and the composition of proper morphisms is proper.

(b) The base change $X_{\overline{K}}$ of a complete K-variety X is a complete variety over $\overline{K}$. More generally, the base change of a proper morphism is again proper. Thus the fibre product of proper morphisms over S is proper over S. In particular, the product of complete varieties over K is a complete variety over K.

(c) If φ is a morphism from a complete variety X over K to an arbitrary K-variety X', then φ is closed and the image of X is a complete variety over K. To deduce it, note that if a composition $\psi_1 \circ \psi_2$ of morphisms is proper, then ψ_2 is also proper. This proves that φ is closed and we get the claims.

(d) A complete affine variety consists of finitely many points. For a proof, see A.10.18.

(e) Any projective variety over K is complete ([**148**], Th.II.4.9). However, the converse does not hold (cf. [**280**], Ch.VI, 2.3).

A.7. Smooth varieties

Let K be a field, $\overline{K}$ an algebraic closure of K, and let X be a variety over K with sheaf of regular functions $\mathcal{O}_X$.

A.7.1. Let A be a commutative K-algebra with identity and let M be an A-module. Then a **K-derivative** of A into M is a K-linear map $\partial : A \to M$ satisfying **Leibniz's rule**

$$\partial(ab) = a\partial(b) + b\partial(a)$$

for all $a, b \in A$. Note that $K \subset A$ and $\partial(K) = \{0\}$ by the Leibniz's rule. We denote the A-module of K-derivations of A into M by $\mathrm{Der}_K(A, M)$.

A.7.2. The goal is to define the tangent space. Recall from differential geometry that this is done by using derivations. A similar concept is used here.

For $x \in X$, let $K(x) = \mathcal{O}_{X,x}/\mathfrak{m}_x$ be the residue field of x. Then the **tangent space** $T_{X,x}$ of x is the $K(x)$-vector space

$$T_{X,x} := \mathrm{Der}_K(\mathcal{O}_{X,x}, K(x)).$$

Example A.7.3. Let $A = K[x_1, \ldots, x_n]$ and let M be an A-module. Then

$$\mathrm{Der}_K(A, M) \xrightarrow{\sim} M^n, \quad \partial \mapsto (\partial x_1, \ldots, \partial x_n).$$

This is easily deduced from Leibniz's rule. In particular, if $M = K[x_1, \ldots, x_n]$, then we have

$$\mathrm{Der}_K(K[x_1, \ldots, x_n], K[x_1, \ldots, x_n]) = \bigoplus_{i=1}^{n} K[x_1, \ldots, x_n]\frac{\partial}{\partial x_i}.$$

For $\alpha \in \mathbb{A}_K^n$, let $\frac{\partial}{\partial x_i}|_\alpha \in \mathrm{Der}_K(K[\mathbf{x}], K(\alpha))$ be the partial derivative evaluated at α. Using $M = K(\alpha)$ above, we get

$$\mathrm{Der}_K(K[\mathbf{x}], K(\alpha)) = \bigoplus_{i=1}^n K(\alpha) \left. \frac{\partial}{\partial x_i} \right|_\alpha .$$

Recall that $\mathcal{O}_{\mathbb{A}_K^n,\alpha}$ is the localization of $K[\mathbf{x}]$ in the maximal ideal $I(\alpha)$. Again by Leibniz's rule, we get

$$\mathrm{Der}_K(\mathcal{O}_{\mathbb{A}_K^n,\alpha}, K(\alpha)) = \mathrm{Der}_K(K[\mathbf{x}], K(\alpha))$$

and hence

$$T_{\mathbb{A}_K^n,\alpha} = \bigoplus_{i=1}^n K(\alpha) \left. \frac{\partial}{\partial x_i} \right|_\alpha .$$

A.7.4. Let $\pi : A \to A'$ be a homomorphism of commutative K-algebras with identity. Note that an A'-module M' is naturally an A-module using π to define multiplication by elements of A. Then we have an A-linear map

$$\mathrm{Der}_K(A', M') \longrightarrow \mathrm{Der}_K(A, M'), \quad \partial \mapsto \partial \circ \pi.$$

If π is surjective, then this map is one-to-one. Moreover, the image consists of the K-derivatives of A into M vanishing on $\ker(\pi)$.

Let $\varphi : X \to Y$ be a morphism. We consider $x \in X$ and we assume that $K(x) = K(\varphi(x))$. Then we have a homomorphism $\varphi^\sharp : \mathcal{O}_{Y,\varphi(x)} \to \mathcal{O}_{X,x}$ inducing the **differential** $(\mathrm{d}\varphi)_x : T_{X,x} \to T_{Y,\varphi(x)}, \partial \to \partial \circ \varphi^\sharp$.

A.7.5. For a point y of a closed subvariety Y of X, we use this trivial remark to identify $T_{Y,y}$ with a subspace of $T_{X,y}$. Note that the restriction of regular functions gives a surjective homomorphism $\mathcal{O}_{X,y} \to \mathcal{O}_{Y,y}$ and hence an injective $K(y)$-linear map $T_{Y,y} \to T_{X,y}$.

A.7.6. As a corollary of Example A.7.3, we see that the tangent space in $x \in X$ is a finite-dimensional $K(x)$-vector space. More precisely, we may assume X a closed affine subvariety of $\mathbb{A}_K^n$ given by polynomials $f_1, \ldots, f_r$. By A.7.4 and A.7.5, the tangent space of x in X is a $K(x)$-linear subspace of $T_{\mathbb{A}_K^n,x}$ given by

$$T_{X,x} = \{\partial \in T_{\mathbb{A}_K^n,x} \mid \partial(f_1) = \cdots = \partial(f_r) = 0\} .$$

Let F/K be a field extension. By A.7.4 again, we see that $\mathrm{Der}_F(K[X] \otimes_K F, F(x))$ is the subspace of $T_{\mathbb{A}_F^n,x} = \mathrm{Der}(F[\mathbf{x}], F(x))$ characterized by $\partial(f_1) = \cdots = \partial(f_r) = 0$. We conclude

$$T_{X,x} \otimes_{K(x)} F(x) = \mathrm{Der}_F(K[X] \otimes_K F, F(x)) . \tag{A.6}$$

The coordinate ring of the base change X_F is $K[X] \otimes_K F/\sqrt{0}$, where the radical ideal $\sqrt{0}$ consists of the nilpotent elements (see A.2.11). By A.7.4, we conclude that $T_{X_F,x}$ is an $F(x)$-linear subspace of $T_{X,x} \otimes_{K(x)} F(x)$. Equality occurs if X is geometrically reduced. This holds certainly in $\mathrm{char}(K) = 0$ (see A.4.6).

A.7.7. We use A.7.6 to extend the definition of the differential in A.7.4 assuming no longer $K(x) = K(y)$. Let $\varphi : X \to Y$ be a morphism mapping x to y. Since we deal with a local problem, we may assume X, Y affine. By base change, $\varphi^\sharp$ induces a homomorphism $K[Y] \otimes_K K(x) \to K[X] \otimes_K K(x)$ of $K(x)$-algebras and hence a $K(x)$-linear map

$$\mathrm{Der}\,(K[X] \otimes_K K(x), K(x)) \to \mathrm{Der}\,(K[Y] \otimes_K K(x), K(x))\,, \partial \mapsto \partial \circ (\varphi^\sharp \otimes 1)\,.$$

By A.7.6, this is a map $T_{X,x} \to T_{Y,y} \otimes_{K(y)} K(x)$, which we call the **differential** $(\mathrm{d}\varphi)_x$.

A.7.8. Let $x \in X$ and let $\mathfrak{m}_x$ be the maximal ideal of $\mathcal{O}_{X,x}$. Note that $\mathfrak{m}_x/\mathfrak{m}_x^2$ is a $K(x) = \mathcal{O}_{X,x}/\mathfrak{m}_x$-vector space whose dual is denoted by $(\mathfrak{m}_x/\mathfrak{m}_x^2)^*$. Then we have a $K(x)$-linear map

$$\rho : T_{X,x} \longrightarrow (\mathfrak{m}_x/\mathfrak{m}_x^2)^*,$$

where $\rho(\partial)$ is defined for $\partial \in T_{X,x}$ by

$$\rho(\partial)(f) := \partial(f) \in K(x)$$

for $f \in \mathfrak{m}_x$. By Leibniz's rule, $\rho(\partial)$ vanishes on $\mathfrak{m}_x^2$ and is $K(x)$-linear, hence we may view $\rho(\partial)$ as an element of $(\mathfrak{m}_x/\mathfrak{m}_x^2)^*$.

Now we assume that x is K-rational. Then ρ is an isomorphism. Its inverse maps $\ell \in (\mathfrak{m}_x/\mathfrak{m}_x^2)^*$ to $\partial \in T_{X,x}$ given by

$$\partial(f) := \ell(f - f(x)), \quad f \in \mathcal{O}_{X,x}.$$

The space $(\mathfrak{m}_x/\mathfrak{m}_x^2)^*$ is Zariski's definition for the tangent space at $x \in X(K)$.

A.7.9. For $x \in X$, let $\dim_x(X) := \max_Y \dim(Y)$, where Y ranges over all irreducible components containing x. Since the prime ideals of $\mathcal{O}_{X,x}$ are in one-to-one correspondence with the closed irreducible subsets containing x, we get

$$\dim_x(X) = \dim\,(\mathcal{O}_{X,x}).$$

From commutative algebra ([**197**], p.78), we know

$$\dim_{K(x)}(\mathfrak{m}_x/\mathfrak{m}_x^2) \geq \dim\,(\mathcal{O}_{X,x}).$$

We call $\mathcal{O}_{X,x}$ a **regular local ring** if equality occurs. More generally, this holds for any noetherian local ring.

If $\mathcal{O}_{X,x}$ is regular, then we say that x is a **regular point** of X. A **singular point** is a point which is not regular. If all points of X are regular, then X is called a **regular variety**.

A.7.10. For $x \in X$, we have

$$\dim_{K(x)}(T_{X,x}) \geq \dim_x(X).$$

This follows from A.7.8 and A.7.9 if x is K-rational. In general, we use base change to $\overline{K}$. The right-hand side does not change (see A.4.10). Then we use A.7.6 to get the claim.

A.7.11. A point $x \in X$ is called **smooth** if

$$\dim_{K(x)}(T_{X,x}) = \dim_x(X).$$

If all points are smooth, then X is a **smooth variety over K.**

A.7.12. Let $x \in X$ be a smooth point. We claim that x is a regular point of $X_{\overline{K}}$ and hence also of X ([**137**], Prop.0.17.3.3, for the descent).

Let $\overline{\mathfrak{m}}_x$ be the maximal ideal of x in $K[U] \otimes_K \overline{K}$. As in A.7.8, we have

$$\mathrm{Der}_{\overline{K}}\left(\left(K[U] \otimes_K \overline{K}\right)_{\overline{\mathfrak{m}}_x}, \overline{K}\right) \xrightarrow{\sim} \left(\overline{\mathfrak{m}}_x / \overline{\mathfrak{m}}_x^2\right)^*.$$

By A.7.6, the left-hand side may be identified with

$$\mathrm{Der}_{\overline{K}}\left(K[U] \otimes_K \overline{K}, \overline{K}\right) = T_{X,x} \otimes_{K(x)} \overline{K}.$$

By A.4.10, the Krull dimension of $\mathcal{O}_{X,x}$ is equal to the one of $\mathcal{O}_{X_{\overline{K}},x}$ and hence

$$\dim(\mathcal{O}_{X_{\overline{K}},x}) = \dim_{K(x)}(T_{X,x}) = \dim_{\overline{K}}(\overline{\mathfrak{m}}_x / \overline{\mathfrak{m}}_x^2).$$

But $\mathcal{O}_{X_{\overline{K}},x}$ is the quotient of $\left(K[U] \otimes_K \overline{K}\right)_{\overline{\mathfrak{m}}_x}$ by the nilpotent elements, hence they have the same Krull dimension. We conclude that

$$\dim\left(K[U] \otimes_K \overline{K}\right)_{\overline{\mathfrak{m}}_x} = \dim_{\overline{K}}(\overline{\mathfrak{m}}_x / \overline{\mathfrak{m}}_x^2)$$

and hence $\left(K[U] \otimes_K \overline{K}\right)_{\overline{\mathfrak{m}}_x}$ is a regular local ring. In commutative algebra, we prove that a regular local ring is a unique factorization domain ([**197**], Th.48, p.142). In particular, it is an integral domain proving

$$\mathcal{O}_{X_{\overline{K}},x} = \left(K[U] \otimes_K \overline{K}\right)_{\overline{\mathfrak{m}}_x}.$$

We conclude that x is a regular point of $X_{\overline{K}}$.

A.7.13. A regular point of X is not necessarily smooth. For K perfect however, smooth and regular points of X are the same (use [**137**], Prop.6.7.4).

A.7.14. If X is a smooth variety over K, then A.7.12 shows that X is geometrically reduced. Conversely, if X is a geometrically reduced variety over K such that $X_{\overline{K}}$ is a smooth variety over $\overline{K}$, then X is a smooth variety over K. This follows from

$$T_{X,x} \otimes_{K(x)} \overline{K} = T_{X_{\overline{K}},x}$$

as we have seen in A.7.6.

It follows also from A.7.12 that the irreducible components of a smooth variety over K are disjoint because the local ring $\mathcal{O}_{X,x}$ in an intersection point x of two irreducible components is not an integral domain contradicting regularity. Therefore, X is the disjoint union of irreducible open subvarieties. Hence the connected components are the same as the irreducible components. If X is an irreducible smooth variety over K with at least one K-rational point, then X is geometrically irreducible ([**137**], Cor.4.5.14).

A.7.15. Let X be a closed subset of $\mathbb{A}_K^n$. For $x \in X$, we have the **Jacobi criterion** of smoothness: Let $f_1, \ldots, f_r$ be generators of $I(X)$. Then x is a smooth point of X if and only if the Jacobi-matrix

$$\left(\frac{\partial}{\partial x_i} (f_j)(x) \right)_{1 \leq i \leq n, 1 \leq j \leq r}$$

has rank $n - \dim_x(X)$.

To prove this, we consider the $K(x)$-linear map

$$\varphi : T_{\mathbb{A}_K^n, x} \longrightarrow K(x)^r, \quad \partial \mapsto (\partial f_1, \ldots, \partial f_r).$$

By A.7.3, the dimension of the image of φ is equal to the rank of the Jacobi-matrix $J(x)$. Using A.7.6, the kernel may be identified with $T_{X,x}$. We conclude

$$\dim(T_{X,x}) + \operatorname{rank}(J(x)) = n.$$

A.7.16. By the Jacobi criterion, the smooth points of X form an open subset characterized by the non-vanishing of a suitable minor. If the variety is geometrically reduced, this subset is dense ([**137**], Cor.17.15.13).

A.7.17. If X, X' are smooth varieties over K, then it follows immediately from the Jacobi criterion that $X \times X'$ is a smooth variety over K. For $x \in X(K), x' \in X'(K)$, we have

$$T_{X \times X', (x,x')} = T_{X,x} \oplus T_{X',x'}$$

using Leibniz's rule.

A.7.18. Our next goal is to define the tangent bundle for X. First, we handle $\mathbb{A}_K^n$. Then the **tangent bundle** $T_{\mathbb{A}_K^n}$ is the disjoint union of $(T_{\mathbb{A}_{\overline{K}}^n, x})_{x \in \mathbb{A}_{\overline{K}}^n}$ as a set. We get a map $\pi : T_{\mathbb{A}_K^n} \to \mathbb{A}_K^n$, mapping the $\overline{K}$-vector space $T_{\mathbb{A}_K^n, x}$ to x. There is a bijective map φ from $T_{\mathbb{A}_K^n}$ to the trivial bundle $\mathbb{A}_K^n \times \mathbb{A}_K^n$ over $\mathbb{A}_K^n$ with inverse given by $\varphi^{-1}(\boldsymbol{\lambda}) := \sum_{i=1}^n \lambda_i \left. \frac{\partial}{\partial x_i} \right|_{\boldsymbol{\alpha}}$ in the fibre over $\boldsymbol{\alpha} \in \mathbb{A}_K^n$. Then $T_{\mathbb{A}_K^n}$ is a vector bundle by requiring that φ is a trivialization. By Example A.7.3, we may identify the global sections of $T_{\mathbb{A}_K^n}$ with $\operatorname{Der}(K[\mathbf{x}], K[\mathbf{x}])$.

A.7.19. Let X be a smooth affine variety over K. An element

$$\partial \in \operatorname{Der}(K[X], K[X])$$

is called a **vector field** over X. We have the evaluation $\partial|_x \in T_{X,x}$ for all $x \in X$ given by

$$\partial|_x(f) = \partial(f)(x).$$

(By Leibniz's rule, ∂ extends to a derivative on $\mathcal{O}_{X,x}$.) If we view X as a closed subset of $\mathbb{A}_K^n$, then $T_{X,x}$ is a subspace of $T_{\mathbb{A}_K^n, x}$. Let $f_1, \ldots, f_r$ be generators of $I(X)$. We assume that X is of pure dimension d. Then the Jacobi criterion shows that X is covered by the open subsets

$$U_{I,J} := \left\{ \det \left(\frac{\partial}{\partial x_i} f_j \right)_{i \in I, j \in J} \neq 0 \right\},$$

where $I \subset \{1, \ldots, n\}$, $J \subset \{1, \ldots, r\}$ are both of cardinality $n - d$.

A.7.20. Let X be a pure dimensional affine variety over K of dimension d given as a closed subset of $\mathbb{A}_K^n$ by $I(X) = (f_1, \ldots, f_r)$. We assume that there are subsets $I \subset \{1, \ldots, n\}$, $J \subset \{1, \ldots, r\}$ of cardinality $n - d$ with

$$\det\left(\frac{\partial}{\partial x_i} f_j(x)\right)_{i \in I, j \in J} \neq 0$$

for all $x \in X$. Then we claim that $\mathrm{Der}(K[X], K[X])$ is a free $K[X]$-module with a basis $\partial_1, \ldots, \partial_d$. Moreover, $\partial_1|_x, \ldots, \partial_d|_x$ is a $K(x)$-basis of $T_{X,x}$ for every $x \in X$.

We sketch the proof: For notational simplicity, we assume $I = J = \{1, \ldots, n - d\}$. For $x \in X$, we consider

$$\sum_{i=1}^{n} g_i(x) \left.\frac{\partial}{\partial x_i}\right|_x \in T_{\mathbb{A}_K^n, x}.$$

By A.7.5, it is an element of $T_{X,x}$ if and only if

$$\sum_{i=1}^{n} g_i(x) \frac{\partial}{\partial x_i}(f_j)(x) = 0 \quad (j = 1, \ldots, r). \tag{A.7}$$

We consider it as a system of homogeneous linear equations in the unknowns $g_i(x)$. Let us write the Jacobi matrix in block form

$$\left(\frac{\partial}{\partial x_i}(f_j)(x)\right)_{1 \leq i \leq n, 1 \leq j \leq r} = \begin{pmatrix} A(x) & B(x) \\ C(x) & D(x) \end{pmatrix},$$

where $A(x)$ is an invertible $(n - d) \times (n - d)$ matrix. Then (A.7) is equivalent to

$$(g_1(x), \ldots, g_n(x)) \begin{pmatrix} A(x) \\ C(x) \end{pmatrix} = \mathbf{0}.$$

By linear algebra, a basis of solutions is given by the rows of the $d \times n$ matrix

$$(h_{ij}(x)) := \left(-C(x)A^{-1}(x) \quad I_d\right).$$

Clearly, the entries of A and C are regular functions on X. By the formula for $A^{-1}(x)$ using determinants, this also holds for the entries of $A^{-1}(x)$. We conclude that $h_{ij} \in K[X]$. For $i = 1, \ldots, d$, let $\partial_i := \sum_{j=1}^{n} h_{ij} \frac{\partial}{\partial x_j}$. Since its evaluation at every point of X satisfies (A.7), this is a well-defined vector field on X, i.e. $\partial_i \in \mathrm{Der}(K[X], K[X])$. Then $\partial_1|_x, \ldots, \partial_d|_x$ is a basis of $T_{X,x}$ for all x, therefore $\partial_1, \ldots, \partial_d$ are $K[X]$-linearly independent. For any $\partial \in \mathrm{Der}(K[X], K[X])$, we have

$$\partial = \sum_{j=1}^{d} \partial(x_{n-d+j})\partial_j,$$

because $\partial_i(x_{n-d+j}) = \delta_{ij}$. Hence $\partial_1, \ldots, \partial_d$ is a $K[X]$-basis of $\mathrm{Der}(K[X], K[X])$.

A.7.21. Now we are ready to define the **tangent bundle** T_X of a smooth K-variety X. As a set, T_X is the disjoint union of the tangent spaces $T_{X_{\overline{K}},x}, x \in X$. We consider pure dimensional affine open subsets U of X such that the module $\mathrm{Der}(K[U], K[U])$ is a free $K[U]$-module of rank $\dim(U)$. Since X is the disjoint union of irreducible open subvarieties, we see that X has a basis of irreducible affine open subsets. By A.7.20, X is covered by open subsets U considered above. For such an U, we choose a $K[U]$-basis $\partial_1, \dots, \partial_d$ of $\mathrm{Der}(K[U], K[U])$. Then the coordinate map φ_U gives an isomorphism from the fibres of T_X over U to the fibres of the trivial bundle $U \times \mathbb{A}^d_K$. We claim that there is a unique vector bundle structure on T_X such that the φ_U's are trivializations.

Proof: Clearly, we may assume that X is irreducible. Let $d := \dim(X)$ and suppose that (U, φ_U) is given by $\partial_1, \dots, \partial_d$ as above. Let $(U', \varphi_{U'})$ be another trivialization given by another basis $\partial'_1, \dots, \partial'_d \in \mathrm{Der}(K[U'], K[U'])$. Using Leibniz's rule, any vector field on U extends to an element of $\mathrm{Der}(\mathcal{O}_{X,x}, \mathcal{O}_{X,x})$ for all $x \in U$. Moreover, $\partial_1, \dots, \partial_d$ (resp. $\partial'_1, \dots, \partial'_d$) form a basis of $\mathrm{Der}(\mathcal{O}_{X,x}, \mathcal{O}_{X,x})$ for all $x \in U \cap U'$. Since the evaluations at x form also a basis of $T_{X,x}$, there are $g_{ij}(x) \in K(x)$ with

$$\partial'_i\big|_x = \sum_{j=1}^{d} g_{ji}(x) \partial_j|_x.$$

We conclude that g_{ij} is a regular function on $U \cap U'$. Since $\varphi_U \circ \varphi_{U'}^{-1}$ is given by $(x, \boldsymbol{\lambda}) \mapsto (x, (g_{ij}) \cdot \boldsymbol{\lambda})$, it is a morphism of varieties. It has the inverse $\varphi_{U'} \circ \varphi_U^{-1}$, hence it is an isomorphism. This proves the claim. □

In fact, we have used φ_U to define the vector bundle structure over U by requiring that φ_U is an isomorphism of vector bundles. Then the above argument shows that the structure fits on overlappings.

Note it is the same construction as in A.5.7, where we have constructed a vector bundle from given transition matrices. However, here the vector bundle is given as an abstract set.

A.7.22. Let U be an affine open subset of X. Then we have

$$\mathrm{Der}(K[U], K[U]) = \Gamma(U, T_X).$$

By the construction of the tangent bundle, this is true locally. Using the behaviour of derivatives with respect to localizations, we get easily the claim. It is also immediate that the vector space of $K(x)$-rational points in the fibre T_X over x is equal to the tangent space in x.

A.7.23. Let X be a smooth variety over K. We apply the various constructions of Section A.5 to the tangent bundle T_X. The **cotangent space** $T^*_{X,x}$ at x is the dual $\mathrm{Hom}_{K(x)}(T_{X,x}, K(x))$ of $T_{X,x}$. The **cotangent bundle** T^*_X is the dual bundle of the tangent bundle. The sections of T_X (resp. $\wedge^k T^*_X$) are called **vector fields** (resp. k**-forms**). The **sheaf of** k**-forms** is denoted by Ω^k_X. If X is of pure dimension d, then $\wedge^k T^*_X$ is a vector bundle of rank $\binom{d}{k}$. In particular, $\wedge^d T^*_X$ is a line bundle called the **canonical line bundle** of X, which we denote by K_X.

A.7.24. Let U be an open subset of the smooth K-variety X. For $f \in \mathcal{O}_X(U)$, we define the **differential** $\mathrm{d}f \in \Omega^1_X(U)$ by the following procedure: For $x \in X$, $\mathrm{d}f(x) \in T^*_{X,x}$ is given by $\partial \in T_{X,x} \mapsto \partial(f)(x)$.

A.7.25. Let C be a smooth curve over K. For every $f \in K(C)$, we get a meromorphic section $\mathrm{d}f$ of T^*_C. If $g \in K(C)$ has differential not identically 0, then we get a rational function $\frac{\mathrm{d}f}{\mathrm{d}g} \in K(C)$ defined by $\mathrm{d}f = \frac{\mathrm{d}f}{\mathrm{d}g} \cdot \mathrm{d}g$.

If $\mathrm{char}(K) = 0$, then $\mathrm{d}g$ is identically zero if and only if g is locally constant on $C_{\overline{K}}$. In $\mathrm{char}(K) = p \neq 0$, this may fail but it follows from A.7.8 that $\mathrm{d}\pi$ does not vanish identically for any local parameter π in a point $x \in C$.

A.7.26. Let $\varphi : X \to X'$ be a morphism of smooth varieties over K. The dual of the differential $\mathrm{d}\varphi$ from A.7.6 induces a **pull-back** $\varphi^* : \Omega^k_{X'}(U') \to \Omega^k_X(\varphi^{-1}(U'))$ **of k-forms** for every open subset U' of X'. We leave the details to the reader.

A.7.27. Let Y be a closed subvariety of the smooth variety X over K. We assume that Y is also smooth. Then T_Y is a subbundle of T_X. The quotient vector bundle $T_X|_Y/T_Y$ is called the **normal bundle** of Y in X and it is denoted by $N_{Y/X}$. The **conormal bundle** $N^*_{Y/X}$ has fibre over $y \in Y$ equal to $\{\ell \in T^*_{X,y} \mid \ell(\partial) = 0 \ \forall \partial \in T_{Y,y}\}$.

A.7.28. Using the theory of coherent sheaves (see Section A.10 on cohomology), it is possible to define Ω^1_X also on singular varieties. Let $\Delta : X \to X \times X$ be the diagonal morphism and let $\mathcal{J}_\Delta$ be the ideal sheaf of the diagonal in $X \times X$. Then $\Omega^1_X := \Delta^*(\mathcal{J}_\Delta/\mathcal{J}^2_\Delta)$ is a coherent sheaf on X. If U is an affine open subset, then it follows from **[148]**, Rem.II.8.9.2, that

$$\mathrm{Der}_K(K[U], M) = \mathrm{Hom}(\Omega^1_X(U), M) \tag{A.8}$$

for any $K[U]$-module M. Applying this with $M = K(x)$ for $x \in U$ or with $M = K[U]$, we see easily that the old definition of Ω^1_X agrees with the above one for a smooth variety. Moreover, if X is of pure dimension n, then X is smooth over K if and only if Ω^1_X is locally free of rank n. This follows immediately from (A.8) for $M = K(x)$.

A.7.29. More generally, if $\varphi : X \to Y$ is a morphism of varieties (or schemes), then we define the **sheaf of relative differentials** $\Omega^1_{X/Y}$ by

$$\Omega^1_{X/Y} := \Delta^*(\mathcal{J}_\Delta/\mathcal{J}^2_\Delta),$$

where $\Delta : X \to X \times_Y X$ is the diagonal morphism and $\mathcal{J}_\Delta$ is the sheaf of ideals of $\Delta(X)$. If X and Y are affine, then $\Omega^1_{X/Y}(X)$ is the module of Kähler differentials defined in B.1.18. In general, $\Omega^1_{X/Y}$ is a coherent sheaf. If X, Y, φ are defined over K, then we have a natural exact sequence

$$\varphi^*\Omega^1_Y \to \Omega^1_X \to \Omega^1_{X/Y} \to 0.$$

For details, we refer to **[148]**, II.8.

A.8. Divisors

In this section, X denotes a variety over the field K.

A.8.1. A **Weil divisor** on X is a formal linear combination $\sum_{i=1}^{n} n_i Y_i$ of irreducible closed subvarieties Y_i of codimension 1. The **multiplicities** n_i are assumed to be in $\mathbb{Z}$. In more abstract terms, a Weil-divisor is an element of the free abelian group with basis the set of irreducible closed subvarieties of X of codimension 1. The Weil divisors form an abelian group with addition

$$\sum_Y n_Y Y + \sum_Y n'_Y Y = \sum_Y (n_Y + n'_Y) Y,$$

where Y ranges over all irreducible closed subvarieties of codimension 1. Note that only finitely many n_Y and n'_Y are different from 0.

A.8.2. An irreducible closed subvariety of codimension 1 is called a **prime divisor**. Non-negative linear combinations of prime divisors are called **effective Weil divisors.** We define a partial order on the space of Weil divisors by setting $D \geq D'$ if and only if $D - D'$ is effective.

The **support** $\operatorname{supp}(D)$ of a Weil divisor D is the union of all prime divisors with non-zero multiplicity. The latter are called the **components** of D. The components Y with multiplicity $n_Y > 0$ (resp. $n_Y < 0$) are called the **zeros** (resp. **poles**) of D and $\sum_{n_Y > 0} Y$ (resp. $\sum_{n_Y < 0} Y$) is the **zero-** (resp. **pole-**) **divisor** of D.

Example A.8.3. Let C be a curve over an algebraically closed field. Then the prime divisors of C are just the sets with one element and we denote the prime divisor associated to $x \in C$ by $[x]$. Then the Weil divisors on C are of the form $\sum_{x \in C} n_x [x]$.

A.8.4. Let F be a field. A surjective function $v : F \to \mathbb{Z} \cup \{\infty\}$ is called a **discrete valuation** if

(a) $v(\alpha) = \infty \Leftrightarrow \alpha = 0$;

(b) $v(\alpha\beta) = v(\alpha) + v(\beta)$;

(c) $v(\alpha + \beta) \geq \min(v(\alpha), v(\beta))$.

The ring $R_v := \{\alpha \in F \mid v(\alpha) \geq 0\}$ is called a **discrete valuation ring.** It is a local ring with maximal ideal $\mathfrak{m}_v := \{\alpha \in F \mid v(\alpha) > 0\}$. We can show easily that R_v is a principal ideal domain. A principal generator π of $\mathfrak{m}_v$ is called a **local parameter** and it is unique up to multiplication with units. It is characterized by $v(\pi) = 1$ or equivalently by π irreducible.

The valuation v may be reconstructed from the discrete valuation ring R_v by the unique factorization property: For $\alpha \in K$ and a local parameter π, there is a unique unit $u \in R_v$ such that $\alpha = u\pi^{v(\alpha)}$. By passing to the discrete absolute value $|\ |_v := e^{-v}$, we see that our definitions agree with 1.2.9.

Theorem A.8.5. *Let R be a commutative noetherian local ring with 1 and maximal ideal $\mathfrak{m}$. Then the following conditions are equivalent:*

(a) $\dim_{R/\mathfrak{m}}(\mathfrak{m}/\mathfrak{m}^2) = \dim(R) = 1$;

(b) *R is a unique factorization domain of Krull dimension 1;*

(c) *$\mathfrak{m}$ is a principal ideal and $\dim(R) = 1$;*

(d) *R is a principal ideal domain which is not a field;*

(e) *R is a discrete valuation ring;*

(f) *R is an integrally closed domain of Krull dimension 1.*

Proof: The first equality in (a) means that R is a regular local ring and hence a unique factorization domain ([**197**], Th.48, p.142) yielding (b). The implications (b) $\Rightarrow$ (a), (c), (f) are obvious.

To deduce (d) from (c), let π be a principal generator of $\mathfrak{m}$. By Krull's intersection theorem $\bigcap_j \mathfrak{m}^j = \{0\}$ ([**157**], Th.7.21), every $\alpha \in R\setminus\{0\}$ may be written in the form $\alpha = u\pi^{v(\alpha)}$ for a unit u in R and $v(\alpha) \in \mathbb{N}$. We deduce that R is an integral domain, otherwise π would be nilpotent contradicting $\dim(R) = 1$. Hence the above factorization of α is unique. If I is an ideal of R, then it is generated by any α with $v(\alpha)$ minimal. This proves (d).

For (d) $\Rightarrow$ (e), we deduce from the assumptions that R has an irreducible element π unique up to units. Then the above unique factorization holds for every non-zero α in the quotient field F of R if we allow $v(\alpha) \in \mathbb{Z}$. This gives the discrete valuation v on F with valuation ring R. Conversely, we get (e) $\Rightarrow$ (b) using the local parameter π. Finally, we quote (f) $\Rightarrow$ (b) from the theory of Dedekind domains ([**157**], Th.10.6). □

Example A.8.6. Let f be a non-zero rational function on the irreducible smooth curve C over an algebraically closed field. For every $x \in C$, the local ring $\mathcal{O}_{C,x}$ is regular (see A.7.12). From Example A.3.6, we easily deduce $\dim(\mathcal{O}_{C,x}) = 1$. By Theorem A.8.5, $\mathcal{O}_{C,x}$ is a discrete valuation ring with a canonical discrete valuation v on $K(C)$. We define the **order of f in x** by $\mathrm{ord}_x(f) := v(f)$. The **Weil divisor associated to f** is defined by

$$\mathrm{div}(f) = \sum_{x \in C} \mathrm{ord}_x(f)[x].$$

We have to prove that $\{x \in C \mid \mathrm{ord}_x(f) \neq 0\}$ is finite. We may assume that C is an affine variety. As f is the quotient of two regular functions, we may assume $f \in K[C]$. Since f is an invertible regular function outside the closed subset $Z(\{f\})$, we get the claim.

From A.8.4 (b), we get the property $\mathrm{div}(fg) = \mathrm{div}(f) + \mathrm{div}(g)$ for $f, g \in K(C)\setminus\{0\}$.

A.8.7. In order to extend this construction to a higher dimension, we need the local ring in a prime divisor. Let Y be an irreducible closed subset of the K-variety X. Then we consider pairs (U, f), where U is an open subset of X with $U \cap Y \neq \emptyset$ and $f \in \mathcal{O}_X(U)$. Two pairs (U, f) and (U', f') are called equivalent if $f = f'$ on an open subset $U'' \subset U \cap U'$ with $U'' \cap Y \neq \emptyset$. (Since Y is irreducible, $U \cap Y$ and $U' \cap Y$ are both open dense subsets of Y and hence $U \cap U' \cap Y$ is not empty.) The equivalence classes form a ring $\mathcal{O}_{X,Y}$ called the **local ring of X in Y**. It is a local ring with maximal ideal $\mathfrak{m}_Y$ formed by the classes of (U, f) with $f(Y \cap U) = \{0\}$.

If Y is of dimension 0, then we have seen in A.2.7 that Y is the set of conjugates of a point $x \in X$. Just by definition, we have $\mathcal{O}_{X,x} = \mathcal{O}_{X,Y}$.

A.8.8. To study the local ring in a prime divisor Y, we may restrict to any open subset U of X with $U \cap Y \neq \emptyset$. So we may assume that X is an affine variety over K. By Example A.3.2, the ideal $\{f \in K[X] \mid f(Y) = \{0\}\}$ of Y is a prime ideal $\wp$ in $K[X]$. We have a homomorphism

$$K[X] \longrightarrow \mathcal{O}_{X,Y}, \quad f \mapsto (X, f),$$

which induces a homomorphism $K[X]_\wp \to \mathcal{O}_{X,Y}$. The latter is an isomorphism by definition of localization and since f is locally the quotient of two regular functions on X. Note that the prime ideals of $K[X]_\wp$ are in one-to-one correspondence with prime ideals in $K[X]$ contained in $\wp$. A prime ideal $\tilde{\wp}$ of $K[X]_\wp$ corresponds to its inverse image in $K[X]$ (see [**157**], Prop.7.9). Using the one-to-one correspondence between prime ideals in $K[X]$ and irreducible closed subsets of X, we conclude that

$$\dim(\mathcal{O}_{X,Y}) = \operatorname{codim}(Y, X). \tag{A.9}$$

This holds also for non-affine varieties X. We leave the details to the reader.

A.8.9. A variety X over K is called **regular in codimension 1** if $\mathcal{O}_{X,Y}$ is a regular local ring for all prime divisors Y of X.

A.8.10. A variety X over K is said to be **normal** if $\mathcal{O}_{X,x}$ is an integrally closed domain for all $x \in X$. An easy exercise shows that any localization of an integrally closed domain remains integrally closed. For a prime divisor Y of X and $y \in Y$, $\mathcal{O}_{X,Y}$ is the localization of $\mathcal{O}_{X,x}$ in the prime ideal $\{f \in \mathcal{O}_{X,Y} \mid f|_Y = 0\}$. We conclude that $\mathcal{O}_{X,Y}$ is an integrally closed domain of Krull dimension 1.

By Theorem A.8.5, a normal variety is regular in codimension 1. Moreover, every regular variety is normal. This is an easy consequence of the fact that a regular local ring is a unique factorization domain ([**197**], Th.48, p.142).

A.8.11. Let X be a K-variety which is regular in codimension 1. For any prime divisor Y, the regular local ring $\mathcal{O}_{X,Y}$ has Krull dimension 1 by (A.9). By

Theorem A.8.5, $\mathcal{O}_{X,Y}$ is a discrete valuation ring for a canonical valuation ord_Y on its field of fractions Q. We call ord_Y the **order** of $f \in Q$ in Y.

If X is irreducible, then $Q = K(X)$ and we define the **Weil divisor of a non-zero rational function** f on X by

$$\mathrm{div}(f) := \sum_Y \mathrm{ord}_Y(f)Y,$$

where Y ranges over all prime divisors of X. We call them also **principal Weil divisors**. We have to prove that $\mathrm{ord}_Y(f) \neq 0$ only for finitely many prime divisors Y. As in Example A.8.6, we may assume that X is affine and that $f \in K[X]$. Then f is an invertible regular function outside $Z(\{f\})$, hence $\mathrm{ord}_Y(f) \neq 0$ only for irreducible components of $Z(\{f\})$ proving the claim.

Two Weil divisors D, D' are called **rationally equivalent**, denoted by $D \sim D'$, if $D - D'$ is a principal divisor.

A.8.12. We still assume that X is a K-variety regular in codimension 1. We generalize the above construction to an invertible meromorphic section s of a line bundle L on X. Let Y be any prime divisor of X. We choose a trivialization $(U_\alpha, \varphi_\alpha)$ of L with $U_\alpha \cap Y \neq \emptyset$. Then we define **the order of** s **in** Y by $\mathrm{ord}_Y(s) := \mathrm{ord}_{Y \cap U}(s_\alpha)$, where $s_\alpha = \varphi_\alpha \circ (s|_{U_\alpha})$ is considered as a rational function on U. If (U_β, φ_β) is another trivialization, then we have $s_\alpha = g_{\alpha\beta} s_\beta$ for the transition function $g_{\alpha\beta}$. Since $g_{\alpha\beta}$ is invertible on $U_\alpha \cap U_\beta$ and hence in $\mathcal{O}_{X,Y}$, we see that the definition of $\mathrm{ord}_Y(s)$ does not depend on the choice of the trivialization. The **Weil divisor associated to s** is defined by

$$\mathrm{div}(s) := \sum_Y \mathrm{ord}_Y(s)Y,$$

where Y ranges over all prime divisors of X. Since X is covered by finitely many trivializations and on a trivialization, the Weil-divisor of s is the Weil-divisor of the corresponding rational function, we see that $\mathrm{ord}_Y(s) \neq 0$ only for finitely many prime divisors Y. Moreover, if we view s as a section of L defined on an open subset U (say with U maximal), then $\mathrm{ord}_Y(s) \neq 0$ only for some irreducible components of $(X \setminus U) \cup \{x \in U \mid s(x) = 0\}$.

A.8.13. Now we consider two line bundles L, L' on X. Let s, s' be invertible meromorphic sections of L and L', respectively. Then $s \otimes s'$ is an invertible meromorphic section of $L \otimes L'$ and we have

$$\mathrm{div}(s \otimes s') = \mathrm{div}(s) + \mathrm{div}(s')$$

using property (b) of the order function.

A.8.14. Clearly, if $\varphi : L \to L'$ is an isomorphism of line bundles and $s' = \varphi \circ s$, then we have $\mathrm{div}(s') = \mathrm{div}(s)$. To connect line bundles with sections, we should identify sections up to such isomorphisms.

A.8.15. Let X be a variety over K. On X, we consider the following group $D(X)$. It is the set of equivalence classes of pairs (L, s), where s is an invertible meromorphic section of L and $(L, s) \sim (L', s')$ if there is an isomorphism $\varphi : L \to L'$ with $s' = \varphi \circ s$. The group operation is given by

$$(L, s) \cdot (L', s') = (L \otimes L', s \otimes s').$$

The identity element is represented by $(O_X, 1)$ and the inverse of (L, s) is represented by (L^*, s^*), where s^* is given on an open dense subset by $s^*(x)(s(x)) = 1$. This explains the notion of invertible meromorphic section. Up to now, we write ss' for $s \otimes s'$, s^{-1} for s^* and s/s' for $s \otimes (s')^{-1}$. Clearly, $D(X)$ is an abelian group.

A.8.16. We have a surjective homomorphism of $D(X)$ onto $\mathrm{Pic}(X)$, given by $(L, s) \mapsto \mathrm{cl}(L)$.

To see surjectivity, we have to prove that every line bundle L has an invertible meromorphic section. For any irreducible component X_j of X, we choose a non-empty trivialization (U_j, φ_j) of L such that U_j is disjoint from the other irreducible components. Then we have $s_j \in \Gamma(U_j, L)$ given by $s_j(x) = \varphi_j^{-1}(x, 1)$. The union U of all U_js is disjoint and there is $s \in \Gamma(U, L)$ defined by $s|_{U_j} = s_j$. Since U is dense, (U, s) is an invertible meromorphic section of L.

A.8.17. The elements of $D(X)$ may be described by Cartier divisors. The idea behind this concept is that divisors should locally given by single equations. To make it precise, a **Cartier divisor** on X is given by the data $(U_\alpha, f_\alpha)_{\alpha \in I}$, where $(U_\alpha)_{\alpha \in I}$ is an open covering, f_α is a unit in $K(U_\alpha)$, and f_α / f_β is a unit in $\mathcal{O}_X(U_\alpha \cap U_\beta)$ for all α, β. We identify two Cartier divisors $(U_\alpha, f_\alpha)_{\alpha \in I}$, $(U'_\beta, f'_\beta)_{\beta \in J}$ if f_α / f'_β is a unit in $\mathcal{O}_X(U_\alpha \cap U'_\beta)$ for all α, β. Given two Cartier divisors D and D', it is always possible to pass to a common refinement. So we add two Cartier divisors D and D' by choosing representatives $(U_\alpha, f_\alpha)_{\alpha \in I}$ and $(U_\alpha, f'_\alpha)_{\alpha \in I}$ and then $D + D'$ is given by $(U_\alpha, f_\alpha f'_\alpha)_{\alpha \in I}$. Clearly, they form an abelian group.

A.8.18. We will show below that $D(X)$ is isomorphic to the group of Cartier divisors. The line bundle associated to the Cartier divisor D will be denoted by $O(D)$ occuring with a distinguished invertible meromorphic section s_D. We will speak about certain properties of D as ample or base-point-free if $O(D)$ has the corresponding properties.

Let s be an invertible meromorphic section of a line bundle L and let us choose a trivialization $(U_\alpha, \varphi_\alpha)_{\alpha \in I}$ of L. Then $s_\alpha := \varphi_\alpha \circ s$ may be viewed as a rational function on U_α. If $g_{\alpha\beta}$ is the transition function, then we have $s_\alpha = g_{\alpha\beta} s_\beta$ on $U_\alpha \cap U_\beta$. Since $g_{\alpha\beta}$ is a unit, we see that $D(s) := (U_\alpha, s_\alpha)_{\alpha \in I}$ is a Cartier divisor on X. It does not depend on the choice of the trivialization. Then $(L, s) \mapsto D(s)$ induces a homomorphism from $D(X)$ to the group of Cartier divisors.

It remains to give an inverse of this map. So let $D = (U_\alpha, f_\alpha)_{\alpha \in I}$ be a Cartier divisor. Then $g_{\alpha\beta} := f_\alpha / f_\beta$ is a unit in $\mathcal{O}_X(U_\alpha \cap U_\beta)$. Let $O(D)$ be the line bundle on X given by the transition functions $g_{\alpha\beta}$ (see A.5.7). Recall its construction: We glue the trivial bundles $U_\alpha \times \mathbb{A}^1_K$ along the isomorphisms given by $g_{\alpha\beta}$. This gives a trivialization over U_α. If we consider f_α as a meromorphic section of $U_\alpha \times \mathbb{A}^1_K$, then $f_\alpha = g_{\alpha\beta} f_\beta$ shows that they fit on overlappings, i.e. we get an invertible meromorphic section s_D of $O(D)$. It is easy to check that $D \mapsto (O(D), s_D)$ gives the inverse.

A.8.19. Let f be a rational function on X, which is not identically 0 on any irreducible component of X. In other words, f is an invertible meromorphic section of O_X. Then the associated Cartier divisor $D(f)$ is called **principal**. The above considerations show that $(L, s) \mapsto D(s)$ induces an isomorphism from $\operatorname{Pic}(X)$ onto the group of Cartier divisors modulo principal Cartier divisors.

A.8.20. Let D be a Cartier divisor on a variety X over K assumed to be regular in codimension 1. Then the **Weil divisor associated to** D is given by $\operatorname{cyc}(D) = \operatorname{div}(s_D)$. If D is given by the local data $(U_\alpha, f_\alpha)_{\alpha \in I}$, then the restriction of $\operatorname{cyc}(D)$ to U is equal to $\operatorname{div}(f_\alpha)$. Here, restriction means that we replace all prime divisors Y of X by $Y \cap U$ letting the coefficients invariant.

A.8.21. Let X be a regular variety over K. Then $D \mapsto \operatorname{cyc}(D)$ is an isomorphism from the group of Cartier divisors onto the group of Weil divisors ([**148**], Prop.II.6.11, Rem.II.6.11.1A). So on regular varieties, we do not distinguish between Cartier divisors and Weil divisors and we simply speak about **divisors.**

We claim that a divisor D is effective if and only if s_D is a global section of $O(D)$. As a special case, we obtain that a non-zero rational function f on an irreducible regular variety is regular if and only if the pole-divisor of f is zero.

Clearly, if s_D is a global section, then $D = \operatorname{div}(s_D)$ is effective. On the other hand, let D be a prime divisor. For any $x \in X$, the local ring $\mathcal{O}_{X,x}$ is a unique factorization domain ([**197**], Th.48, p. 142). The ideal I_x of D in $\mathcal{O}_{X,x}$ is defined by $I_x := \{(U, f) \in \mathcal{O}_{X,x} \mid f = 0 \text{ on } U \cap D\}$. If $x \notin D$, then we have $I_x = \mathcal{O}_{X,x}$. If $x \in D$, then we choose an affine neighbourhood V of x. Then the ideal I of D in $K[V]$ is a prime ideal not containing smaller prime ideals (see A.3.2). So the same holds for I_x in $\mathcal{O}_{X,x}$. We conclude that I_x is a principal ideal generated by a prime $\pi_x \in \mathcal{O}_{X,x}$. Since I is finitely generated, we deduce that π_x is regular and generates I in a neighbourhood U_x of x. For $x \notin D$, we set $U_x := X \setminus D$ and $\pi_x = 1$. Then $(U_x, \pi_x)_{x \in X}$ is a Cartier divisor of X. The associated Weil divisor is clearly supported in the irreducible D. Moreover, for any $x \in D$, (U_x, π_x) generates the maximal ideal in $\mathcal{O}_{X,D}$. Hence the associated Weil divisor is D. This proves the claim. Note that this proves also surjectivity of the cycle map.

A.8.22. On an irreducible regular complete variety X over an algebraically closed field K, it is sometimes more convenient to work with divisors than with sections.

For a divisor D on X, we define the **complete linear system**

$$|D| := \{D' \mid D \sim D' \geq 0\}$$

in the space of divisors. Then we have a surjective map

$$\Gamma(X, O(D)) \setminus \{0\} \longrightarrow |D|, \quad s \mapsto \text{div}(s).$$

Using K algebraically closed, two non-trivial global sections s, s' have the same divisor if and only if $s/s' \in K^\times$. In fact, s/s' has to be a rational function without poles, hence regular (see A.8.21) and thus constant (use A.6.15(c) and (d)). We may identify the complete linear system $|D|$ with the projective linear space given by the one-dimensional linear subspaces in $\Gamma(X, O(D))$. Hence we have

$$\dim(|D|) = \dim(\Gamma(X, O(D))) - 1.$$

A **base-point** of $|D|$ is $x \in X$ with $x \in \text{supp}(D')$ for all $D' \in |D|$. It is the same as a base-point for $O(D)$ (see A.5.20). More generally, a subspace of the projective space $|D|$ is called a **linear system**, but in this book we consider only complete linear systems.

A.8.23. Let $D = (U_\alpha, f_\alpha)_{\alpha \in I}$ be a Cartier divisor on any K-variety X. The **support** of D is the closed subset

$$\text{supp}(D) := \bigcup_\alpha \{x \in U_\alpha \mid f_\alpha \notin O_{X,x}^\times\},$$

where $O_{X,x}^\times$ denotes the group of invertible elements in $O_{X,x}$. The Cartier divisor D is called **effective** if and only if $f_\alpha \in O_X(U_\alpha)$ for all $\alpha \in I$. If D is effective, then it follows from Krull's principal ideal theorem ([**157**], p.449) that $\text{supp}(D)$ is of codimension 1 in X. For non-effective Cartier divisors, this does not necessarily hold. However, if X is regular, then we may identify Cartier- and Weil-divisors (see A.8.21) and their supports agree, thus the support is of codimension 1.

A.8.24. Let X be a regular variety over K with an irreducible closed subset Y and a closed subset Z of codimension 1 with $Y \not\subset Z$. As a consequence of A.8.21 and A.8.23, we remark that $Y \cap Z$ has codimension 1 in Y and hence

$$\text{codim}(Y \cap Z, X) = \text{codim}(Y, X) + 1$$

following from additivity of codimensions (use A.4.9).

A.8.25. We mention the following generalization. If Y and Z are irreducible closed subsets of a smooth variety X over K, then

$$\text{codim}(Y \cap Z, X) \leq \text{codim}(Y, X) + \text{codim}(Z, X).$$

For a proof, we refer to [**125**], Sec. 8.2.

A.8.26. The pull-back of a Cartier divisor is not always well defined as a Cartier divisor. Let $\varphi : X' \to X$ be a morphism of varieties over K and let $D = (U_\alpha, f_\alpha)_{\alpha \in I}$ be a Cartier divisor on X. We have to assume that the image of no irreducible component of X' is contained in $\mathrm{supp}(D)$. Then the **pull-back** of D is the Cartier divisor

$$\varphi^*(D) := (\varphi^{-1}(U_\alpha), f_\alpha \circ \varphi)_{\alpha \in I}.$$

Our assumptions imply that $f_\alpha \circ \varphi$ are well-defined rational functions on X'. It is easy to check that $(O(\varphi^* D), s_{\varphi^*(D)}) = (\varphi^* O(D), \varphi^*(s_D))$.

A.9. Intersection theory of divisors

The basic reference for intersection theory is the book of W. Fulton [**125**]. We need only the first two chapters, namely properties of the intersection product of a divisor with a closed subvariety. We collect here the most important results, not going into full generality, assuming often that the ambient variety is smooth and neglecting the theory of refined intersections which takes care about supports.

In this section, X is a variety over a field K.

A.9.1. We extend the concept of divisors to higher codimension. A **cycle of dimension d** is a formal linear combination of irreducible closed subvarieties of X of dimension d with coefficients in $\mathbb{Z}$. They form an abelian group $Z_d(X)$. The elements of $Z(X) := \bigoplus_{d \in \mathbb{N}} Z_d(X)$ are called **cycles**. A basis of this abelian group is formed by the irreducible closed subvarieties called **prime cycles**. If $Z = \sum_Y n_Y Y$ is a cycle, then the prime cycles Y with **multiplicity** $n_Y \neq 0$ are called the **components** of Z. By definition, they are finite in number. Their union is called the **support** of Z. Note that a Weil divisor is a cycle of pure codimension 1.

A.9.2. We assume that X is irreducible and Y is a prime divisor on X. The goal is to define the order of a non-zero function f of X in Y. This will be a generalization of the construction in A.8.11. In A.8.7, we have introduced the local ring $\mathcal{O}_{X,Y}$. By A.8.8, it is an integral domain of Krull dimension 1 with quotient field $K(X)$.

For any non-zero f in the maximal ideal $\mathfrak{m}_Y$ of $\mathcal{O}_{X,Y}$, the ring $A = \mathcal{O}_{X,Y}/f\mathcal{O}_{X,Y}$ has Krull dimension 0. Since the localization of a noetherian ring remains noetherian ([**157**], Th.7.10), it follows from A.8.8 that $\mathcal{O}_{X,Y}$ is a noetherian ring. In commutative algebra, a theorem of Krull says that the intersection of all prime ideals in a commutative ring is the ideal of nilpotent elements ([**157**], Th.7.1). Since A is a noetherian local ring whose maximal ideal $\mathfrak{m} = \mathfrak{m}_Y/f\mathcal{O}_{X,Y}$ is the unique prime ideal, we conclude that $\mathfrak{m}$ is nilpotent, i.e. the ideal $\mathfrak{m}^n$ generated by the n-fold products in $\mathfrak{m}$ is 0 for some $n \in \mathbb{N}$. Now we have a chain of ideals

$$0 = \mathfrak{m}^n \subset \mathfrak{m}^{n-1} \subset \cdots \subset \mathfrak{m} \subset A. \tag{A.10}$$

Since A is noetherian, the $K(Y)$-vector space $\mathfrak{m}^j/\mathfrak{m}^{j+1}$ is finite dimensional, where $K(Y)$ is equal to the residue field $A/\mathfrak{m}$. Then we define the **order** of f in Y by

$$\operatorname{ord}_Y(f) := \sum_{j=0}^{n-1} \dim_{K(Y)}(\mathfrak{m}^j/\mathfrak{m}^{j+1}).$$

If $f \in \mathcal{O}_{X,Y} \setminus \mathfrak{m}_Y$, then f is a unit in $\mathcal{O}_{X,Y}$ and we define $\operatorname{ord}_Y(f) = 0$.

Example A.9.3. If X is regular in codimension 1, then $\mathcal{O}_{X,Y}$ is a principal ideal domain (see A.8.11). Then $\mathfrak{m}_Y$ is generated by the local parameter π_Y. If $f = u\pi_Y^n$ is the prime factorization of $f \in \mathcal{O}_{X,Y}$, u a unit, then $\mathfrak{m}^j$ is generated by the image of π_Y^j and $\mathfrak{m}^n = 0$. Moreover, multiplication by π_Y^j gives an isomorphism of $K(Y) = A/\mathfrak{m}$ onto $\mathfrak{m}^j/\mathfrak{m}^{j+1}$. This proves that our new definition of the order agrees with the one in A.8.11.

A.9.4. The concept of order is best understood in terms of composition series. Let Λ be a ring and M be a Λ-module. Then M is said to have **finite length** if there is a chain

$$0 = M_0 \subsetneq M_1 \subsetneq \cdots \subsetneq M_r = M$$

of submodules without possible refinement. Such a chain is called a **composition series** and r is called its length. Then the Jordan Hölder theorem ([**157**], p.108) implies that all composition series have the same length. This number is called the **length of** M and is denoted by $\ell_\Lambda(M)$. If N is a submodule of M, then it is easy to prove that M has finite length if and only if N and M/N have finite length. In this case, we have ([**157**], Exercise 2, p.109)

$$\ell_\Lambda(N) + \ell_\Lambda(M/N) = \ell_\Lambda(M). \tag{A.11}$$

If X is an irreducible K-variety with prime divisor Y, we apply this to the integral domain and noetherian local ring $\Lambda = \mathcal{O}_{X,Y}$. In the notation of A.9.2, we consider the Λ-module $\mathfrak{m}^j/\mathfrak{m}^{j+1}$ for any non-zero $f \in \mathcal{O}_{X,Y}$. It is clear that a Λ-submodule is the same as a $K(Y)$-subspace. Therefore

$$\ell_\Lambda(\mathfrak{m}^j/\mathfrak{m}^{j+1}) = \dim_{K(Y)}(\mathfrak{m}^j/\mathfrak{m}^{j+1}).$$

With an inductive application of (A.11) to the chain (A.10), we get

$$\operatorname{ord}_Y(f) = \ell_\Lambda(\Lambda/f\Lambda).$$

A.9.5. Let f, g be non-zero elements of $\mathcal{O}_{X,Y}$. We claim

$$\operatorname{ord}_Y(fg) = \operatorname{ord}_Y(f) + \operatorname{ord}_Y(g). \tag{A.12}$$

This follows from (A.11) and the exact sequence

$$0 \longrightarrow \Lambda/g\Lambda \xrightarrow{\cdot f} \Lambda/fg\Lambda \longrightarrow \Lambda/f\Lambda \longrightarrow 0$$

for $\Lambda = \mathcal{O}_{X,Y}$. Therefore we have a unique extension of ord_Y to a function $\operatorname{ord}_Y : K(X)^\times \to \mathbb{Z}$ satisfying (A.12).

A.9.6. Recall that, if $\mathcal{O}_{X,Y}$ is regular, then we have

$$\mathrm{ord}_Y(f+g) \geq \min\{\mathrm{ord}_Y(f), \mathrm{ord}_Y(g)\}$$

for all $f, g \in K(X)^\times$. Conversely, if this holds, then $\mathcal{O}_{X,Y}$ is regular (as a discrete valuation ring, see Theorem A.8.5).

A.9.7. Let X be an irreducible K-variety and f a non-zero rational function on X. There is a non-empty open subset U of X such that f is a unit in $\mathcal{O}_X(U)$. By definition of the local ring $\mathcal{O}_{X,Y}$, we have $\mathrm{ord}_Y(f) = 0$ for all prime divisors Y with $Y \cap U \neq \emptyset$. Therefore $\mathrm{ord}_Y(f) \neq 0$ is possible only for the irreducible components of $X \setminus U$ of codimension 1. As they are finite in number, we get a well-defined Weil divisor

$$\mathrm{div}(f) := \sum_Y \mathrm{ord}_Y(f) Y,$$

where Y ranges over all prime divisors of X. It is called the **Weil divisor of** $\boldsymbol{f}$ (or **principal Weil divisor**). By A.9.5, we have

$$\mathrm{div}(fg) = \mathrm{div}(f) + \mathrm{div}(g)$$

for non-zero $f, g \in K(X)$.

The assumption X irreducible was just made for simplicity. By the same construction, we can define the Weil divisor of a rational function, which is not identically zero on any irreducible component of X.

A.9.8. Let X be a K-variety. We consider the subgroup $R(X)$ of $Z(X)$ generated by $\mathrm{div}(f)$, where f ranges over all non-zero rational functions on prime cycles of X. Note that we view the Weil divisor $\mathrm{div}(f)$ of Y as a cycle on X. Two cycles are called **rationally equivalent** if their difference is in $R(X)$. This gives an equivalence relation on $Z(X)$ denoted by $\sim$. The quotient $CH(X) := Z(X)/R(X)$ is called the **Chow group.** We grade it by dimension.

A.9.9. Let $\varphi : X \to X'$ be a morphism of varieties with X complete (or more generally a proper morphism). Then the image of a closed subset of X is a closed subset of X' (see A.6.15). Let Y be a prime cycle of X. Then $\varphi(Y)$ is a prime cycle of X' and we may view $K(\varphi(Y))$ as a subfield of $K(Y)$ by the map $f' \to f' \circ \varphi$. Let

$$\varphi_*(Y) := \begin{cases} [K(Y) : K(\varphi(Y))]\varphi(Y) & \text{if } [K(Y) : K(\varphi(Y))] < \infty, \\ 0 & \text{else.} \end{cases}$$

For any cycle $Z = \sum n_Y Y$, we define the **push-forward** of Z by

$$\varphi_*(Z) := \sum n_Y \varphi_*(Y) \in Z(X'),$$

where Y ranges over all the prime divisors of X.

A.9.10. Note that $[K(Y) : K(\varphi(Y))] < \infty$ if and only if Y and $\varphi(Y)$ have the same dimension. This follows from the fact that the dimension is equal to the transcendence degree of the function field (see A.4.11). If $K(Y)$ is separable over $K(\varphi(Y))$, then $[K(Y) : K(\varphi(Y))]$ is the number of points in the fibre of a generic point of $\varphi(Y)$ (see A.12.9).

A.9.11. Let $\varphi : X \to X'$ be a surjective proper morphism of irreducible varieties over K and let f be a non-zero rational function on X. We have seen in A.9.9 that $K(X)$ is a field extension of $K(X')$. If this is a finite extension, then

$$\varphi_*(\mathrm{div}(f)) = \mathrm{div}(N(f)),$$

where $N : K(X) \to K(X')$ is the norm. If the extension is infinite, the push-forward is 0. For a proof, see [**125**], Prop.1.4.

A.9.12. Let $\varphi : X \to X'$ be a proper morphism of varieties over K. Then A.9.11 shows that $\varphi_* R(X) \subset R(X')$ and so we get a push-forward map

$$\varphi_* : CH(X) \longrightarrow CH(X'),$$

mapping the class of a cycle to the class of its push-forward.

A.9.13. Let X be a K-variety and let L be a line bundle on X. For an invertible meromorphic section s of L, the **Weil divisor $\mathrm{div}(s)$ associated to s** is defined similarly as in A.8.12 and A.8.13 still holds.

A.9.14. Let Y be a prime cycle on a K-variety X. Then we define $c_1(L).Y \in CH(X)$ to be the rational equivalence class of $\mathrm{div}(s_Y)$, where s_Y is any invertible meromorphic section of $L|_Y$. We have seen in A.8.16 that such an s_Y always exists. If s'_Y is another choice, then s_Y/s'_Y is a rational function and hence $\mathrm{div}(s_Y)$ and $\mathrm{div}(s'_Y)$ are rationally equivalent.

By additivity, we define $c_1(L).Z$ for all cycles Z on X. If Z is rationally equivalent to 0, then $c_1(L).Z = 0$ (see [**125**], Cor.2.4.1). Therefore $c_1(L).\alpha$ is well-defined for $\alpha \in CH(X)$ by using representatives. Then the homomorphism

$$CH(X) \longrightarrow CH(X), \quad \alpha \mapsto c_1(L).\alpha$$

is called the **first Chern class operation of L**. Clearly, it does not depend on the isomorphism class of the line bundle.

A.9.15. If L and L' are line bundles on X and $\alpha \in CH(X)$, then

$$c_1(L \otimes L').\alpha = c_1(L).\alpha + c_1(L').\alpha.$$

This follows immediately from A.8.13. Moreover, we have

$$c_1(L).\left(c_1(L').\alpha\right) = c_1(L').\left(c_1(L).\alpha\right).$$

For a proof of commutativity, we refer to [**125**], Cor.2.4.2.

A.9.16. If $\varphi : X \to X'$ is a proper morphism over K and L' is a line bundle on X', $\alpha \in CH(X)$, then

$$\varphi_* \left(c_1(\varphi^* L').\alpha \right) = c_1(L').\varphi_*(\alpha).$$

This is the **projection formula**. For a proof, see [**125**], Prop.2.5(c).

A.9.17. The following remark is for readers familiar with the basics of schemes. If X is any scheme of finite type over the field K with irreducible components $X_1, \dots, X_r$, then the obvious generalization of (A.9) on page 546 proves $\dim(\mathcal{O}_{X,X_j}) = 0$. Since $\mathcal{O}_{X,X_j}$ is noetherian (as in A.9.2), we conclude that $\mathcal{O}_{X,X_j}$ is of finite length ([**157**], Th.7.12) and we may define the multiplicity of X_j in X by $\ell(\mathcal{O}_{X,X_j})$. The **cycle of X** is

$$\mathrm{cyc}(X) := \sum_j \ell(\mathcal{O}_{X,X_j}) \cdot X_j.$$

If D is an effective Cartier divisor on X, then D may be viewed as a closed subscheme of X given by the local equations. By construction, the cycle of D is just the Weil divisor associated to D.

This is useful to handle base change to a field extension F over K. Let X_F be the base change of X to F as a scheme. Then there is a unique base change homomorphism $Z(X) \to Z(X_F), Z \mapsto Z_F$, such that for any closed subscheme Y of X, we have $\mathrm{cyc}(Y_F) = \mathrm{cyc}(Y)_F$. To see this, define the base change first for prime cycles in the obvious way and then extend by linearity. By [**125**], Lemma A.4.1, it has the required property (the argument is as in [**125**], Lemma 1.7.1).

It is easy to see that the base change of cycles descends to the Chow groups. For $\alpha \in CH(X)$ and a line bundle L on X, we claim that $c_1(L_F).\alpha_F$ is equal to the base change of $c_1(L).\alpha$. To see this, we may assume that α is prime and even equal to X. By A.8.18, it is enough to show that $\mathrm{cyc}(D_F) = \mathrm{cyc}(D)_F$ for a Cartier divisor D on X. This is a local question, so we may assume D effective and the claim follows from the above.

Note that for an irreducible closed subvariety Y of a variety X over K, the base change to F as a scheme may be non-reduced and hence may be different from the base change as a variety. Only the use of schemes leads to the above compatibilities.

A.9.18. Now let X be a smooth variety over K. Then Cartier divisors and Weil divisors are the same (cf. A.8.21). It follows immediately that we have an isomorphism

$$\mathrm{Pic}(X) \longrightarrow CH^1(X), \quad \mathrm{cl}(L) \mapsto c_1(L).X.$$

Hence we get an intersection theory with divisors: For a divisor D on X and a cycle Z on X, the **intersection product** is

$$D.Z := c_1(O(D)).Z \in CH(X).$$

Then the intersection product is bilinear, compatible with rational equivalence and satisfies commutativity for divisors. If $\varphi : X \to X'$ is a morphism of smooth varieties, then we have a **pull-back** $\varphi^*(D') := c_1(\varphi^* O(D')).X' \in CH(X)$ of a divisor D' on X'. If φ is a proper morphism, then we have the **projection formula**

$$\varphi_*(\varphi^*(D').Z) = D'.\varphi_*(Z)$$

for a cycle Z on X and a divisor D' on X'.

Proposition A.9.19. *Let X be a smooth variety over K and let D be a prime divisor which is smooth over K. Then the self-intersection $D.D \in CH(X)$ is represented by the divisor of any invertible meromorphic section of the normal bundle $N_{D/X}$.*

Proof: It is enough to prove $O(D)|_D \cong N_{D/X}$. Let D be equal to $(U_\alpha, f_\alpha)_{\alpha\in I}$ as a Cartier divisor. Since D is effective, we have $f_\alpha \in \mathcal{O}(U_\alpha)$ (see A.8.21). The conormal bundle $N^*_{D/X}$ is generated by $\mathrm{d}f_\alpha$ as a subbundle of T^*_X over $D \cap U_\alpha$ (use A.7.27). In particular, $\mathrm{d}f_\alpha$ is a non-vanishing section defining a trivialization of the line bundle $N^*_{D/X}$ on D by mapping $\lambda \cdot \mathrm{d}f_\alpha$ to λ. The transition function $g_{\alpha\beta}$ is computed by

$$g_{\alpha\beta} \cdot \mathrm{d}f_\alpha = \mathrm{d}f_\beta = \frac{f_\beta}{f_\alpha} \cdot \mathrm{d}f_\alpha$$

on $D \cap U_\alpha \cap U_\beta$. Hence $N^*_{D/X}$ has the same transition functions $g_{\alpha\beta} = f_\beta/f_\alpha$ as $O(-D)|_D$ proving the claim. □

A.9.20. Let X be a smooth variety over K. For some purposes, it is necessary to define the intersection product of a divisor D with a prime cycle Y of codimension p as a honest cycle (not only as a rational equivalence class as above). However, this is only possible if Y is not contained in the support $\mathrm{supp}(D)$ of D. Under this assumption, we see that the invertible meromorphic section s_D of $O(D)$ corresponding to D restricts to an invertible meromorphic section $s_D|_Y$ of $O(D)|_Y$. Then the **proper intersection product** of D and Y is the cycle

$$D.Y := \mathrm{div}(s_D|_Y) \in Z^{p+1}(X).$$

Clearly, the rational equivalence class of the proper intersection product induces the intersection product in the Chow group. By additivity, we define the proper intersection product of a Cartier divisor D and a cycle Z under the hypothesis that no component of Z is contained in $\mathrm{supp}(D)$. Then the proper intersection product is of the form $D.Z = \sum_W n_W W$, where W is ranging over all prime cycles. The number n_W is called the **intersection multiplicity** of W in $D.Z$.

If additionaly $Z = D'$ is also a divisor, we can prove $D.D' = D'.D$ as an identity of cycles ([**125**], Th.2.4).

A.9.21. Let X be a smooth variety over K. We consider a prime cycle Y not contained in the support of a divisor D. For a prime cycle W different from the irreducible components of $\mathrm{supp}(D) \cap Y$, the intersection multiplicity of W in the proper intersection product $D.Y$ is obviously zero. Now we assume additionaly that D is effective. Then we claim that the intersection multiplicity of an irreducible component W of $\mathrm{supp}(D) \cap Y$ is at least 1.

To prove this, note that D is given as a divisor by a local equation $\gamma \in \mathcal{O}_X(U)$ for some open subset U intersecting W. Then the intersection multiplicity of $D.Y$ in W is the order of γ in $\mathcal{O}_{Y,W}$. Since γ vanishes on W, it is contained in the maximal ideal $\mathfrak{m}_{Y,W}$ of $\mathcal{O}_{Y,W}$. Hence the length of $\mathcal{O}_{Y,W}/\gamma\mathcal{O}_{Y,W}$ is at least 1.

As a corollary of proof, we see that the intersection multiplicity of W in $D.Y$ is 1 if and only if the maximal ideal $\mathfrak{m}_{Y,W}$ of $\mathcal{O}_{Y,W}$ is generated by a local equation of D. In this case, $\mathcal{O}_{Y,W}$ is a regular local ring since it follows that $\mathfrak{m}_{Y,W}/\mathfrak{m}_{Y,W}^2$ is a one-dimensional $K(W) = \mathcal{O}_{Y,W}/\mathfrak{m}_{Y,W}$-vector space and the Krull dimension of $\mathcal{O}_{Y,W}$ is $\operatorname{codim}(W, Y) = 1$ (cf. A.8.8).

Example A.9.22. Assume that D and Y are both smooth. Then the intersection of D and Y is called **transversal** if $T_{Y,y} \not\subset T_{D,y}$ for all $y \in D \cap Y$. We claim that the intersection multiplicity of an irreducible component W of $D \cap Y$ in $D.Y$ is 1.

To prove it, we may assume that K is algebraically closed because the intersection product is compatible with base change (see A.9.17). We may assume that D is given by a single equation $\gamma \in \mathcal{O}_X(X)$. Let $\gamma = u\pi^n$ for a unit u in $\mathcal{O}_{Y,W}$ and a local parameter π in $\mathcal{O}_{Y,W}$ (which is a principal ideal domain since Y is smooth). We know that $n = \operatorname{ord}_W(\gamma|_Y) \geq 1$. There is a $y \in W$ such that π vanishes in y and u is regular in y. Using Zariski's definition of the tangent space (see A.7.8), $T_{Y,y} \not\subset T_{D,y}$ means that $\gamma \in \mathfrak{m}_{Y,y} \setminus \mathfrak{m}_{Y,y}^2$. It follows $n = 1$ proving the claim.

Example A.9.23. Let C be a smooth curve over K. Then $C \times C$ is a smooth surface (see A.7.17). We claim that for $x, y \in C(K)$, we have

$$(C \times \{y\}).(\{x\} \times C) = \{(x, y)\}$$

as a proper intersection product. By Example A.9.22, this follows from transversality of the intersection. Let Δ be the diagonal of $C \times C$. Similarly, we deduce that

$$\Delta.(\{x\} \times C) = \{(x, x)\} = (C \times \{x\}).\Delta$$

for $x \in C(K)$.

A.9.24. Let X be a complete K-variety. For $Z \in Z_0(X)$ of the form $Z = \sum_Y n_Y Y$ with Y ranging over the irreducible closed subvarieties of dimension 0, we define the **degree** of Z by

$$\deg(Z) := \sum_Y n_Y [K(Y) : K].$$

Note that $K(Y)$ is a finite-dimensional field extension of K (see A.9.10). The unique map $\pi : X \to \mathbb{A}_K^0$ is a proper morphism. The degree is characterized by

$$\pi_*(Z) = \deg(Z) \cdot \mathbb{A}_K^0.$$

It follows from A.9.11 that $\deg(Z) = 0$ for Z rationally equivalent to 0. Therefore the degree may be viewed as an additive homomorphism on $CH_0(X)$.

Let X be a smooth complete variety over K of pure dimension d and let $D_1, \ldots,$ D_d be divisors on X, not necessarily distinct. We define their **intersection number** by

$$D_1 \cdots D_d := \deg(D_1 \ldots D_d) \in \mathbb{Z}.$$

Here $D_1 \ldots D_d$ denotes their intersection product in $CH_0(X)$. Clearly, the intersection number depends only on the rational equivalence classes in the Chow group and it is invariant under permutation of the divisors. In other words, the intersection number induces a symmetric multilinear form on the Chow group with values in $\mathbb{Z}$.

If $\varphi : X' \to X$ is a surjective morphism of irreducible smooth complete varieties over K, both of dimension d, then we have

$$\varphi^* D_1 \cdots \varphi^* D_d = [K(X') : K(X)] D_1 \cdots D_d \tag{A.13}$$

for divisors $D_1, \ldots, D_d$ on X. This follows from

$$\varphi_*(X') = [K(X') : K(X)] X \in Z_d(X)$$

and the projection formula.

If $D_1, \ldots, D_d$ are effective divisors, it may happen that $D_1 \cdots D_d$ is negative. But, if $D_1, \ldots, D_{d-1}$ are base-point free and $D_d \geq 0$, then A.9.33 below shows that

$$D_1 \cdots D_d \geq 0.$$

A.9.25. Let F/K be a field extension and $D_1, \ldots, D_d$ divisors on the smooth complete K-variety X of pure dimension d. Using the same set of equations, we get a divisor $(D_j)_F$ on the base change X_F and we claim that

$$(D_1)_F \cdots (D_d)_F = D_1 \cdots D_d.$$

By A.9.17, the intersection product is compatible with base change of cycles. Now the claim follows from the general fact that the degree of a zero-dimensional cycle is invariant under base change. To see this, it is enough to consider $Y \in Z_0(X)$ prime and even $Y = X$. Then the base change Y_F is the cycle associated to $\mathrm{Spec}(K(Y) \otimes_K F)$. Since $B := K(Y) \otimes_K F$ is a finite-dimensional F-algebra, it is the product of the local rings in the prime ideals ([**157**], Th.7.13) and we conclude

$$\deg(Y) = [K(Y) : K] = [B : F] = \sum_{\mathfrak{p} \in \mathrm{Spec}(B)} [B_{\mathfrak{p}} : F] = \deg(Y_F),$$

where the last step follows from $[B_{\mathfrak{p}} : F] = \ell(B_{\mathfrak{p}})[B/\mathfrak{p} : F]$ (see [**125**], Lemma A.1.3).

A.9.26. Let X be a projective variety over K and let L be an ample line bundle on X. For a prime cycle $Z \in Z_d(X)$, the **degree of Z with respect to L** is

$$\deg_L(Z) := \deg(c_1(L) \ldots c_1(L).Z) \in \mathbb{Z},$$

where $c_1(L)$ occurs d times. By additivity, we extend the degree to all cycles. In particular, we define the **degree of X with respect to L** by

$$\deg_L(X) := \sum_j \deg_L(X_j),$$

where X_j is ranging over all irreducible components of X. Clearly, the degree of a cycle depends only on its rational equivalence class, hence we may view it as an additive homomorphism

$$\deg_L : CH(X) \longrightarrow \mathbb{Z}.$$

Example A.9.27. For affine space, we claim that $CH^1(\mathbb{A}^n_K) = \{0\}$.

Let X be a prime divisor of $\mathbb{A}^n_K$. First, we show that the ideal $I(X)$ of X in $K[\mathbf{x}]$ is generated by an irreducible polynomial. We choose any non-zero $f(\mathbf{x}) \in I(X)$. Clearly, $f(\mathbf{x})$ is not a constant. By considering the prime factorization of $f(\mathbf{x})$ and since $I(X)$ is a prime ideal, we see that $I(X)$ contains an irreducible polynomial $g(\mathbf{x})$. Since X has codimension 1, we have $X = Z(g)$ and Hilbert's Nullstellensatz shows that $g(\mathbf{x})$ generates $I(X)$. Next, we have to prove $X = \operatorname{div}(g)$. To see it, note that g is invertible outside of X. This proves $\operatorname{div}(g) = mY$ for some $m \in \mathbb{N}$. The maximal ideal of the local ring $\mathcal{O}_{\mathbb{A}^n_K, X}$ is generated by g, proving $X = \operatorname{div}(g)$. Since this holds for any prime divisor, we get $CH^1(\mathbb{A}^n_K) = \{0\}$.

As a corollary of A.9.18, we see that any line bundle on $\mathbb{A}^n_K$ is isomorphic to the trivial bundle, i.e. $\operatorname{Pic}(\mathbb{A}^n_K) = \{0\}$.

Example A.9.28. We deduce from the above that

$$\operatorname{Pic}(\mathbb{P}^n_K) \cong CH^1(\mathbb{P}^n_K) \cong \mathbb{Z}.$$

To see it, let Y be a prime divisor of $\mathbb{P}^n_K$. We choose a standard affine open subset $U_j = \{x_j \neq 0\}$ with $U_j \cap Y \neq \emptyset$. By the example above, $Y \cap U_j = \operatorname{div}(f)$ for some $f \in \mathcal{O}_{\mathbb{P}^n_K}(U_j)$. Using $U_j \cong \mathbb{A}^n_K$, f comes from an irreducible polynomial of degree d in n variables. Let $\tilde{f}$ be the homogenization of f, i.e.

$$\tilde{f}(x_0, \ldots, x_n) = x_j^d f\left(\frac{x_0}{x_j}, \ldots, \frac{x_{j-1}}{x_j}, \frac{x_{j+1}}{x_j}, \ldots, \frac{x_n}{x_j}\right).$$

We may view $\tilde{f}$ as a global section of $O_{\mathbb{P}^n_K}(d)$ (see A.6.6). We claim that $\operatorname{div}(\tilde{f}) = Y$. This follows immediately from $\operatorname{div}(f) = Y \cap U_j$ and the non-vanishing of $\tilde{f}$ outside of U_j. So we conclude that $d \mapsto c_1(O_{\mathbb{P}^n_K}(d)).\mathbb{P}^n_K$ is a surjective homomorphism of $\mathbb{Z}$ onto $CH^1(\mathbb{P}^n_K)$. Since the line bundles $O_{\mathbb{P}^n_K}(d)$ are pairwise not isomorphic (compare the dimensions of $\Gamma(\mathbb{P}^n_K, O_{\mathbb{P}^n_K}(\pm d))$) and since $\operatorname{Pic}(X) \cong CH^1(X)$ (see A.9.18), we get the claim.

For a multiprojective space $\mathbb{P}_K := \mathbb{P}^{n_1}_K \times \cdots \times \mathbb{P}^{n_r}_K$ (see A.6.13), we can similarly show that $\operatorname{Pic}(\mathbb{P}_K) \cong \mathbb{Z}^r$. We simply have to replace homogeneous polynomials in the above consideration by multihomogeneous polynomials.

Proposition A.9.29. Let X be a variety over K, then $CH^p(X) \xrightarrow{\sim} CH^p(X \times \mathbb{A}_K^n)$, where the isomorphism is given by mapping a prime cycle Y on X to the prime cycle $Y \times \mathbb{A}_K^n$ on X (see A.4.11).

Proof: For simplicity, we only prove surjectivity (see [**125**], Th.3.3, for a whole proof). By induction, we may assume that $n = 1$. Let Y' be a prime cycle on $X \times \mathbb{A}_K^1$. We have to prove that Y' is rationally equivalent to a linear combination of cycles of the form $Y \times \mathbb{A}_K^1$. Replacing X by the closure of $p_1(Y')$, where p_1 denotes the first projection, we may assume $X = \overline{p_1(Y')}$. If $\dim(Y') > \dim(X)$, then $Y' = X \times \mathbb{A}_K^1$. So we may assume $\dim(Y') = \dim(X)$. Then Y' is a divisor in $X \times \mathbb{A}_K^1$. Let U be a non-empty affine open subset of X. As a closure of an irreducible subset, X is irreducible and hence the same holds for U. We consider the ideal $I(Y' \cap U')$ in $K[U'] = K[U][x]$ for $U' = U \times \mathbb{A}_K^1$. The ideal in $K(U)[x]$ generated by $I(Y' \cap U')$ is generated by a polynomial f with coefficients in $K(U)$. By shrinking U, we may assume that $f \in K[U][x]$ and that f generates $I(Y' \cap U')$ in $K[U']$. This shows that $\mathrm{div}(f)$ and Y' agree on $Y' \cap U'$, hence $Y' = \mathrm{div}(f) + \sum_j n_j(Y_j \times \mathbb{A}_K^1)$, where Y_j is ranging over the irreducible components of $X \setminus U$ (of codimension 1). This proves surjectivity. □

Remark A.9.30. In particular, we have $CH(\mathbb{A}_K^n) = \{0\}$. Note that we need only surjectivity in the above statement.

Example A.9.31. We claim that we have an isomorphism $CH(\mathbb{P}_K^n) \cong \mathbb{Z}[x]/[x^{n+1}]$ of abelian groups. This can be seen as follows. A $d+1$-dimensional subspace of K^{n+1} is the same as the intersection of the kernels of $p = n - d$ linearly independent linear forms $\ell_1(\mathbf{x}), \ldots, \ell_p(\mathbf{x})$. Then $L_d = Z(\{\ell_1(\mathbf{x}), \ldots, \ell_p(\mathbf{x})\})$ is called a d-dimensional **projective linear subspace** of $\mathbb{P}_K^n$. If $p = 1$, then we call it a **hyperplane**. Since the hyperplanes intersect transversally, we easily get $L_d = \mathrm{div}(\ell_1) \ldots \mathrm{div}(\ell_p)$ as a proper intersection product (use Example A.9.22). Since all hyperplanes are rationally equivalent, the same holds for all projective linear subspaces of dimension $0, \ldots, n$. We prove that their classes form a basis of $CH(\mathbb{P}_K^n)$. Let Y be a prime cycle on $\mathbb{P}_K^n$ of dimension d. Let $U_j = \{x_j \neq 0\}$ be a standard affine open subset of $\mathbb{P}_K^n$ intersecting Y. By Remark A.9.30, $Y \cap U_j$ is rationally equivalent to 0 on U_j. This proves that Y is rationally equivalent to a cycle contained in $\mathbb{P}_K^n \setminus U_j \cong \mathbb{P}_K^{n-1}$. By induction on n, we see that Y is rationally equivalent to a linear combination of $L_0, \ldots, L_{n-1}$. It remains to prove that the classes of $L_0, \ldots, L_n$ are linearly independent in $CH(\mathbb{P}_K^n)$. By dimensionality reasons, it is enough to show that no class is zero. This follows from $\deg_{O_{\mathbb{P}_K^n}(1)} L_j = 1$, which is obvious from transversal intersection.

We use the above isomorphism to define a ring structure on $CH(\mathbb{P}_K^n)$. The multiplication is called the **intersection product of cycles** on $\mathbb{P}_K^n$ (see [**125**] for an intersection product on any smooth variety). Note that this extends the intersection product of divisors. The intersection product is determined by

$$L_j.L_k = \begin{cases} L_{j+k-n} & \text{if } j+k \geq n, \\ 0 & \text{else.} \end{cases} \tag{A.14}$$

On $\mathbb{P}_K^n$, the **degree of a cycle** Z is always with respect to $O_{\mathbb{P}_K^n}(1)$ and is denoted by $\deg(Z)$. If $Z \in Z_d(\mathbb{P}_K^n)$, then Z is rationally equivalent to $\deg(Z)\, L_d$. This follows immediately from $\deg(L_d) = 1$. If $Z' \in Z_{d'}(X)$ and if $d + d' \geq n$, then we get **Bézout's theorem**

$$\deg(Z.Z') = \deg(Z)\deg(Z').$$

Example A.9.32. If Z is an effective divisor on $\mathbb{P}_K^n$, then Example A.9.28 shows that $Z = \operatorname{div}(\tilde{f})$ for some homogeneous polynomial $\tilde{f}(x_0, \ldots, x_n)$ of degree d. By rational equivalence, Z has the same degree as $d \cdot \operatorname{div}(x_0)$ and hence $\deg(Z) = d$.

A.9.33. We mention some positivity properties of the intersection product. Let X be a variety over K with a line bundle L and let Z be a cycle on X with non-negative multiplicities. Then Z is called an **effective cycle.** If L is generated by global sections, then $c_1(L).Z$ may be represented by an effective cycle. To see it, we may assume that Z is a prime cycle. By assumption, there is a global section s whose restriction to Z is not identically zero. Then $\operatorname{div}(s|_Z)$ is an effective representative of $c_1(L).Z$.

We conclude that $\deg_L(Z) \geq 0$. In particular, this holds for very ample line bundles. But for L ample and Z a non-zero effective cycle, we claim even that $\deg_L(Z) > 0$.

For a proof, we may assume that L is very ample (replace L by a suitable power). But then, we may assume that X is a closed subvariety of $\mathbb{P}_K^n$ and $L = O_{\mathbb{P}_K^n}(1)|_X$. Then $\deg_L(Z)$ is the degree of Z in $\mathbb{P}_K^n$. We may assume that Z is a prime cycle of dimension d. Let H be a hyperplane in $\mathbb{P}_K^n$ not containing the prime cycle Z. If $d > 0$, then Z is not contained in the affine space $\mathbb{P}_K^n \setminus H$ (see A.6.15). Therefore $H \cap Z$ is not empty and, since the multiplicities of $H.Z$ are at least 1 in every irreducible component of $H \cap Z$ (see A.9.21), we get the claim by induction on d.

A.9.34. Next, we introduce algebraic equivalence. It is a coarser equivalence than rational equivalence and it is similar to homotopy in homology. We use it only for divisors.

Let X be a variety over K and let L_1, L_2 be line bundles on X. We say that L_1 and L_2 are **algebraically equivalent** if there is an irreducible smooth variety T called the parameter space and a line bundle L on $X \times T$ such that

$$L_1 \cong L|_{X_{t_1}} \quad \text{and} \quad L_2 \cong L|_{X_{t_2}}$$

for some $t_1, t_2 \in T(K)$. Here, we identify the fibres $X_{t_j} := X \times \{t_j\}$ with X. This is possible because of the K-rationality of t_j.

It is easy to show that it is indeed an equivalence relation. To prove transitivity, we pass to the product of parameter spaces. Clearly, isomorphic line bundles are algebraically equivalent, hence algebraic equivalence makes sense on $\mathrm{Pic}(X)$.

A.9.35. Let L_1, L_2, M be line bundles on X such that L_1 is algebraically equivalent to L_2. Then it is easy to see that $L_1 \otimes M$ is algebraically equivalent to $L_2 \otimes M$. In particular, the elements in $\mathrm{Pic}(X)$ algebraically equivalent to $\mathbf{0}$ form a subgroup of $\mathrm{Pic}(X)$ denoted by $\mathrm{Pic}^0(X)$.

A.9.36. Let $\varphi : X' \to X$ be a morphism of varieties over K and let L_1, L_2 be algebraically equivalent line bundles on X. Then $\varphi^* L_1$ is algebraically equivalent to $\varphi^* L_2$. In the notation of the definition in A.9.34, we use the same parameter space and the pull-back of L to $X' \times T$.

A.9.37. Let L_1, L_2 be algebraically equivalent ample line bundles on a projective variety over K. Then we claim $\deg_{L_1}(X) = \deg_{L_2}(X)$.

By passing to a suitable tensor power, we may assume that L_1 and L_2 are both very ample (see A.6.10). Then $\deg_{L_j}(X)$ is determined by the leading coefficient of the Hilbert polynomial of L_j (see A.10.33). By definition, we have a line bundle L on $X \times T$ with T an irreducible smooth variety over K such that $L_{t_1} \cong L_1, L_{t_2} \cong L_2$ for some $t_1, t_2 \in T(K)$. Since the projection of $X \times T$ onto X is flat (see A.12.11) and X is complete, the Hilbert polynomial of L_t does not depend on the choice of $t \in T$ (see A.10.35). This proves the claim.

Chow's lemma ([137], Th.5.6.1) says that every complete K-variety X is image of a birational surjective morphism $\varphi : X' \to X$ from a projective K-variety X'. By projection formula, the invariance of the degree under algebraic equivalence holds more generally for complete varieties over K.

A.9.38. Let X be a complete K-variety with line bundles $L_1, \ldots, L_r$ and let $Z \in Z_r(X)$. Then $\deg(c_1(L_1) \ldots c_1(L_r).Z)$ depends only on the algebraic equivalence classes of $L_1, \ldots, L_r$.

For L_1' algebraically equivalent to L_1, it is enough to show that

$$\deg(c_1(L_1) \ldots c_1(L_r).Z) = \deg(c_1(L_1').c_1(L_2) \ldots c_1(L_r).Z).$$

By multilinearity of the intersection product and by A.6.10, we may assume L_1, L_1' both ample. Let us choose a representative Y of $c_1(L_2) \ldots c_1(L_r).Z$. Clearly, we may assume that Y is a prime cycle, i.e. an irreducible curve. By A.9.36, we know that $L_1|_Y$ is algebraically equivalent to $L_1'|_Y$. Using A.9.37, we get

$$\deg(c_1(L_1).Y) = \deg(c_1(L_1|_Y)) = \deg(c_1(L_1'|_Y)) = \deg(c_1(L_1').Y).$$

A.9.39. Let X be a smooth K-variety. Since $\mathrm{Pic}(X)$ is isomorphic to $CH^1(X)$ (see A.9.18), we can translate the above theory to divisors. Two divisors D_1 and D_2 on X are **algebraically equivalent** if $O(D_1)$ and $O(D_2)$ are algebraically equivalent. We have the following properties:

(a) If D_1 and D_2 are rationally equivalent, then they are algebraically equivalent.

(b) If D_1 and D_2 are algebraically equivalent and D is a further divisor on X, then $D_1 + D$ is algebraically equivalent to $D_2 + D$.

(c) Algebraic equivalence is preserved by pull-back of divisors with respect to a morphism of smooth varieties over K.

(d) Algebraically equivalent divisors on a smooth projective curve have the same degree.

A.9.40. Let X be an irreducible smooth projective curve over an algebraically closed field. Then the converse of (d) also holds, i.e. if $\deg(D_1) = \deg(D_2)$, then D_1 and D_2 are algebraically equivalent.

To prove it, note that D_1 and D_2 are sums of $\pm[x]$ for some $x \in X$. Since they have the same degree and algebraic equivalence is compatible with sum of divisors, we may assume $D_1 = [x_1], D_2 = [x_2]$. Note that $X \times X$ is smooth (see A.7.17) and the diagonal Δ is a divisor on $X \times X$. Moreover, for any $x \in X$, we have $O(\Delta)|_{X \times \{x\}} \cong O([x])$. This follows from transversality of $X \times x$ and Δ in $X \times X$. So we choose $T := X$ as parameter space, $t_1 := x_1, t_2 := x_2$ and $L = O(\Delta)$ to get the claim.

A.9.41. Two line bundles L_1, L_2 on a smooth complete variety X over K are called **numerically equivalent** if $c_1(L) \cdot \alpha = c_1(L_2) \cdot \alpha$ for every $\alpha \in CH_1(X)$. By A.9.38, algebraically equivalent line bundles are numerically equivalent. As above, we define divisors to be numerically equivalent if and only if the associated line bundles are numerically equivalent.

A.10. Cohomology of sheaves

This section gives a brief introduction to sheaf cohomology on varieties. For more details, we refer to **[148]**, Ch.III. First, we need some additional constructions of sheaves. We consider a topological space T. On T, all (pre-)sheaves will be (pre-)sheaves of abelian groups. Later, we pass to a K-variety X.

A.10.1. For a presheaf $\mathcal{F}$ on T, there is a canonical way to associate a sheaf $\mathcal{F}^+$ on T and a homomorphism $\iota : \mathcal{F} \to \mathcal{F}^+$ such that for any sheaf $\mathcal{G}$ on T and any homomorphism $\varphi : \mathcal{F} \to \mathcal{G}$, there is a unique homomorphism $\varphi^+ : \mathcal{F}^+ \to \mathcal{G}$ with $\varphi = \varphi^+ \circ \iota$. Then $\mathcal{F}^+$ is called the **sheaf associated to $\mathcal{F}$**. For the easy construction, we refer to **[148]**, Prop.-Def.II.1.2.

A.10.2. Let $\mathcal{F}$ be a sheaf on T. A **subsheaf** of $\mathcal{F}$ is a sheaf $\mathcal{G}$ such that $\mathcal{G}(U)$ is a subgroup of $\mathcal{F}(U)$ for all open subsets U of T and such that the restriction maps of $\mathcal{G}$ are induced by the restriction maps of $\mathcal{F}$.

Example A.10.3. If $\varphi : \mathcal{F} \to \mathcal{F}'$ is a homomorphism of sheaves, then the **kernel** of φ is the subsheaf of $\mathcal{F}$ given by $\ker(\varphi)(U) := \ker(\varphi_U)$ for all open subsets U. However, the abelian groups $\varphi_U(U)$ form only a subpresheaf $\mathcal{G}$ of $\mathcal{F}'$. We define the **image** $\mathrm{im}(\varphi) = \varphi(\mathcal{F})$ to be the sheaf associated to the presheaf $\varphi(\mathcal{F}(U))$ with U ranging over the open subsets of T. We call φ **surjective** if $\varphi(\mathcal{F}) = \mathcal{F}'$. In general, this does not mean that $\varphi_U(\mathcal{F}(U)) = \mathcal{F}'(U)$ for all open subsets U.

A.10.4. A sequence

$$\cdots \longrightarrow \mathcal{F}^{p-1} \xrightarrow{\varphi^{p-1}} \mathcal{F}^p \xrightarrow{\varphi^p} \mathcal{F}^{p+1} \longrightarrow \cdots$$

of sheaves is said to be **exact** if all maps φ^p are homomorphisms of sheaves such that $\mathrm{im}(\varphi^{p-1}) = \ker(\varphi^p)$ for all p. An exact sequence of the form

$$0 \longrightarrow \mathcal{F}' \xrightarrow{\varphi'} \mathcal{F} \xrightarrow{\varphi''} \mathcal{F}'' \longrightarrow 0$$

is called a **short exact sequence** of sheaves.

Example A.10.5. Again, we consider a sheaf $\mathcal{F}$ on T and a subsheaf $\mathcal{F}'$. For an open subset U of T, let $\mathcal{G}(U) := \mathcal{F}(U)/\mathcal{F}'(U)$. Then $\mathcal{G}$ with the restriction maps induced by $\mathcal{F}$ is a presheaf of abelian groups on T. The sheaf associated to $\mathcal{G}$ is called the **quotient sheaf** $\mathcal{F}/\mathcal{F}'$ and we get a short exact sequence

$$0 \longrightarrow \mathcal{F}' \xrightarrow{\varphi'} \mathcal{F} \xrightarrow{\varphi''} \mathcal{F}/\mathcal{F}' \longrightarrow 0.$$

Conversely, for any short exact sequence as in A.10.4, we have a canonical isomorphism $\mathcal{F}/\varphi'(\mathcal{F}') \cong \mathcal{F}''$. For any homomorphism $\varphi : \mathcal{F}_1 \to \mathcal{F}_2$ of sheaves, the sheaf $\mathrm{coker}(\varphi) := \mathcal{F}_2/\varphi(F_1)$ is called the **cokernel** of φ.

A.10.6. Let $\mathcal{F}$ be a sheaf on T. We fix an open covering $\mathcal{U} = (U_\alpha)_{\alpha \in I}$ of X. We fix a well-ordering on I. This will not cause a problem later, because all our coverings will be finite. For $p \in \mathbb{N}$, we define an abelian group

$$C^p(\mathcal{U}, \mathcal{F}) := \prod_{i_0 < \cdots < i_p} \mathcal{F}(U_{i_0} \cap \cdots \cap U_{i_p}).$$

The coboundary map is a homomorphism

$$d_p : C^p(\mathcal{U}, \mathcal{F}) \longrightarrow C^{p+1}(\mathcal{U}, \mathcal{F}).$$

For $\sigma \in C^p(\mathcal{U}, \mathcal{F})$, the components of $d_p(\sigma)$ are given by

$$(d_p(\sigma))_{i_0,\ldots,i_{p+1}} = \sum_{k=0}^{p+1} (-1)^k \rho^{U_{i_0} \cap \cdots \cap U_{i_{k-1}} \cap U_{i_{k+1}} \cap \cdots \cap U_{i_p}}_{U_{i_0} \cap \cdots \cap U_{i_p}} (\sigma_{i_0,\ldots,i_{k-1},i_{k+1},\ldots,i_{p+1}}),$$

where ρ is the restriction homomorphism of the sheaf $\mathcal{F}$. Then we get the **Čech complex**

$$0 \xrightarrow{d^{-1}} C^0(\mathcal{U}, \mathcal{F}) \xrightarrow{d^0} C^1(\mathcal{U}, \mathcal{F}) \xrightarrow{d^1} \cdots$$

of abelian groups. Complex means $d^p \circ d^{p-1} = 0$ for all p. The elements in the kernel of d^p are called **Čech cocycles** with respect to $\mathcal{U}$ of degree p and the elements in the image of d^{p-1} are **Čech boundaries** with respect to $\mathcal{U}$ of degree p. The pth **Čech cohomology group** with respect to $\mathcal{U}$ is

$$\check{H}^p(\mathcal{U}, \mathcal{F}) := \ker(d^p)/\mathrm{im}(d^{p-1}).$$

By the sheaf axioms, we get

$$\check{H}^0(\mathcal{U}, \mathcal{F}) = \mathcal{F}(X).$$

This is clear since a Čech cocycle with respect to $\mathcal{U}$ is a collection of local sections, which agree on overlappings.

A.10.7. Let $\varphi : \mathcal{F} \to \mathcal{F}'$ be a homomorphism of sheaves. Then we get homomorphisms

$$\varphi^p : C^p(\mathcal{U}, \mathcal{F}) \longrightarrow C^p(\mathcal{U}, \mathcal{F}'), \quad (\alpha_{i_0,\ldots,i_p}) \to (\varphi^p(\alpha_{i_0,\ldots,i_p}))$$

such that $\varphi^{p+1} \circ d^p = d^p \circ \varphi^p$, i.e. φ is a **homomorphism of complexes**. Hence, it induces homomorphisms $\check{H}^p(\mathcal{U}, \mathcal{F}) \to \check{H}^p(\mathcal{U}, \mathcal{F}')$ of Čech cohomology groups, which we denote also by φ^p or simply by φ.

A.10.8. From a formal point of view, it is nicer to define Čech cocycles without choosing a well-ordering. We define $C^p(\mathcal{U}, \mathcal{F})$ as the set of elements

$$\sigma_{i_0,\ldots,i_p} \in \prod_{(i_0,\ldots,i_p) \in I^{p+1}} \mathcal{F}(U_{i_0} \cap \cdots \cap U_{i_p})$$

with $\sigma_{i_0,\ldots,i_p} = 0$ if two components of $(i_0, \ldots, i_p)$ are equal and

$$\sigma_{i_{\pi(0)},\ldots,i_{\pi(p)}} = (-1)^{\mathrm{sign}(\pi)} \sigma_{i_0,\ldots,i_p}$$

if π is a permutation of $\{0, \ldots, p\}$. The same definitions as in A.10.6 may be used to define the Čech complex. Clearly, the Čech cocycles and boundaries may be identified with the old ones.

Now let $\mathcal{U}' = (U'_\beta)_{\beta \in J}$ be a **refinement** of $\mathcal{U}$, i.e. $\mathcal{U}'$ is also an open covering of X and for every $\beta \in J$ there is $\alpha \in I$ with $U'_\beta \subset U_\alpha$. We choose a map $\alpha : J \to I$ with $U'_\beta \subset U_{\alpha(\beta)}$. Then we define a homomorphism $\theta : C^p(\mathcal{U}, \mathcal{F}) \longrightarrow C^p(\mathcal{U}', \mathcal{F})$ by

$$\theta^p(\sigma)_{j_0,\ldots,j_p} := \rho^{U_{\alpha(j_0)} \cap \cdots \cap U_{\alpha(j_p)}}_{U'_{j_0} \cap \cdots \cap U'_{j_p}}(\sigma_{\alpha(j_0),\ldots,\alpha(j_p)}).$$

Then θ is a homomorphism of Čech complexes inducing a homomorphism

$$\check{H}^p(\mathcal{U}, \mathcal{F}) \to \check{H}^p(\mathcal{U}', \mathcal{F})$$

of Čech cohomology groups. It does not depend on the choice of the refinement map ([**148**] Exercise III.4.4 or F. Warner [**321**], 5.33).

Example A.10.9. Let X be a variety over K. We consider the sheaf $\mathcal{K}$ of rational functions on X, i.e. $\mathcal{K}(U) = K(U)$ for any open subset U of X. The sheaf $\mathcal{K}$ is easy to understand. If we denote the irreducible components of X by X_j, then $K(U)$ is the product of the fields $K(X_j)$, where j is ranging over all components with $X_j \cap U \neq \emptyset$ (see A.4.12). Note that $\mathcal{O}_X$ and $\mathcal{K}$ are sheaves of K-algebras. We denote by $\mathcal{O}_X^\times(U)$ and $\mathcal{K}^\times(U)$ the group of multiplicative units for any open subset U of X. Clearly, $\mathcal{O}_X^\times$ is a subsheaf of the sheaf $\mathcal{K}^\times$ of abelian groups.

Then any open covering $\mathcal{U} = (U_\alpha)_{\alpha \in I}$ and $f_\alpha \in K(U_\alpha)$ give a Cartier divisor on X if and only if

$$\left(f_\alpha \cdot \mathcal{O}_X^\times(U_\alpha)\right)_{\alpha \in I} \in \check{H}^0(\mathcal{U}, \mathcal{K}^\times/\mathcal{O}_X^\times).$$

Moreover, we see that the group of Cartier divisors is isomorphic to $(\mathcal{K}^\times/\mathcal{O}_X^\times)(X)$.

Example A.10.10. Recall that we can give a line bundle on X by a covering $\mathcal{U} = (U_\alpha)_{\alpha \in I}$ and transition functions $g_{\alpha\beta} \in \mathcal{O}_X^\times(U_\alpha \cap U_\beta)$ satisfying $g_{\alpha\beta} g_{\beta\gamma} = g_{\alpha\gamma}$ on $U_\alpha \cap U_\beta \cap U_\gamma$. We see that $(g_{\alpha\beta})$ is a Čech cocycle of $\mathcal{O}_X^\times$ with respect to the covering $\mathcal{U}$.

Two line bundles L, L' are isomorphic if and only if there is a common trivialization $\mathcal{U} = (U_\alpha)_{\alpha \in I}$ and $h_\alpha \in \mathcal{O}(U_\alpha)^\times$ for $\alpha \in I$ such that $g'_{\alpha\beta}/g_{\alpha\beta} = h_\alpha/h_\beta$, i.e. a Čech coboundary with respect to the covering $\mathcal{U}$. The function h_α gives the matrix of the isomorphism from the trivialization of L to the one of L' on U_α. We conclude that

$$\mathrm{Pic}(X) \cong \varinjlim_{\mathcal{U}} \check{H}^1(\mathcal{U}, \mathcal{O}_X^\times),$$

where the direct limit is over all open coverings directed with respect to refinements.

A.10.11. An **$\mathcal{O}_X$-module** $\mathcal{F}$ is a sheaf of abelian groups on X such that $\mathcal{F}(U)$ is an $\mathcal{O}_X(U)$-module with

$$\rho_V^U(\xi) \cdot \rho_V^U(f) = \rho_V^U(\xi \cdot f)$$

for all open subsets $V \subset U$ in X and $\xi \in \mathcal{O}_X(U), f \in \mathcal{F}(U)$. A homomorphism of $\mathcal{O}_X$-modules is a homomorphism $\varphi : \mathcal{F} \to \mathcal{F}'$ of sheaves such that φ_U is a homomorphism of $\mathcal{O}_X(U)$-modules for each open subset U.

A.10.12. An $\mathcal{O}_X$-module $\mathcal{F}$ on X is called **free** of rank $r \in \mathbb{N}$ if there is an isomorphism $\mathcal{F} \to \mathcal{O}_X^r$ of $\mathcal{O}_X$-modules. We call an $\mathcal{O}_X$-module **locally free** if, for all $x \in X$, there is an open neighbourhood U of x such that $\mathcal{F}|_U$ is a free $\mathcal{O}_X$-module of rank $r_U \in \mathbb{N}$.

A.10.13. Let E be a vector bundle on X. Then the sheaf of sections $\mathcal{E}$ is a locally free $\mathcal{O}_X$-module. Just the trivializations give us the isomorphisms to $\mathcal{O}_U^{r_U}$. We claim that every locally free $\mathcal{O}_X$-module $\mathcal{F}$ arises this way and we will carry over the terminology of vector bundles (especially of line bundles) to locally free $\mathcal{O}_X$-modules (of rank 1).

We have an open covering $(U_\alpha)_{\alpha\in I}$ of X such that $\mathcal{F}|_{U_\alpha}$ is free of rank r_α. We have an $\mathcal{O}_X(U)$-basis $s_{\alpha 1},\dots,s_{\alpha r_\alpha}$ in $\mathcal{F}(U_\alpha)$. On non-empty $U_\alpha\cap U_\beta$,

$$\rho^{U_\alpha}_{U_\alpha\cap U_\beta}(s_{\alpha 1}),\dots,\rho^{U_\alpha}_{U_\alpha\cap U_\beta}(s_{\alpha r_\alpha})$$

and

$$\rho^{U_\beta}_{U_\alpha\cap U_\beta}(s_{\beta 1}),\dots,\rho^{U_\beta}_{U_\alpha\cap U_\beta}(s_{\beta r_\beta})$$

are both an $\mathcal{O}_X(U_\alpha\cap U_\beta)$-basis of $\mathcal{F}(U_\alpha\cap U_\beta)$. Therefore, we have $r_\alpha=r_\beta$ and $g_{\alpha\beta}\in GL(r_\alpha,\mathcal{O}_X(U_\alpha\cap U_\beta))$ with

$$\rho^{U_\beta}_{U_\alpha\cap U_\beta}(s_{\beta i})=\sum_{j=1}^{r_\alpha}(g_{\alpha\beta})_{ji}\rho^{U_\alpha}_{U_\alpha\cap U_\beta}(s_{\alpha j}) \tag{A.15}$$

for $i=1,\dots,r_\alpha$.

An element of $GL(r_\alpha,\mathcal{O}_X(U_\alpha\cap U_\beta))$ is an invertible $r_\alpha\times r_\alpha$-matrix with entries in $\mathcal{O}_X(U_\alpha\cap U_\beta)$. Therefore $g_{\alpha\beta}$ may be viewed as a morphism $U_\alpha\cap U_\beta\to GL(r_\alpha)_K$. Clearly, we have $g_{\alpha\beta}g_{\beta\gamma}=g_{\alpha\gamma}$ on $U_\alpha\cap U_\beta\cap U_\gamma$. By the construction in A.5.7, we get a vector bundle E on X with transition matrices $g_{\alpha\beta}$. Let $\varphi_\alpha:\pi_E^{-1}(U_\alpha)\to U_\alpha\times\mathbb{A}^r_K$ be the corresponding trivialization. We consider $e_1,\dots,e_{r_\alpha}\in\Gamma(U_\alpha,U_\alpha\times\mathbb{A}^r_K)$, pointwise equal to the standard basis. Then $\varphi_\alpha^{-1}\circ e_1,\dots,\varphi_\alpha^{-1}\circ e_{r_\alpha}$ form an $\mathcal{O}_X(U_\alpha)$-basis of $\Gamma(U_\alpha,E)$. Since they satisfy the same transition rule on $U_\alpha\cap U_\beta$ as in (A.15), we may identify them with $s_{\alpha 1},\dots,s_{\alpha r_\alpha}$. Then the sheaf of sections $\mathcal{E}$ of E coincides with $\mathcal{F}$.

Example A.10.14. Let X be a K-variety. In Example A.10.9, we have introduced the sheaf $\mathcal{K}$ of rational functions on X. For any Cartier divisor D on X, we realize the sheaf of sections $\mathcal{O}_X(D)$ of the line bundle $O_X(D)$ as a subsheaf of $\mathcal{K}$: Let $D=(U_\alpha,f_\alpha)_{\alpha\in I}$. For an open subset U of X, we define $\mathcal{F}(U):=\{f\in K(U)\mid f\cdot f_\alpha\in\mathcal{O}_X(U\cap U_\alpha)\ \forall\alpha\in I\}$. Clearly, this is a subsheaf of $\mathcal{K}$. The claim follows from the isomorphism

$$\Gamma(U,\mathcal{O}_X(D))\longrightarrow\mathcal{F}(U),\quad s\mapsto s/s_D.$$

By A.9.17, an effective Cartier divisor D may be viewed as a closed subscheme of X and the subsheaf of $\mathcal{K}$ corresponding to $\mathcal{O}_X(-D)$ is equal to the ideal sheaf $\mathcal{J}_D$.

A.10.15. We have seen in Example A.5.11 that not every sheaf is locally free. We now introduce the notion of coherent sheaves, which includes almost all sheaves of importance for our book. A **coherent sheaf** is an $\mathcal{O}_X$-module $\mathcal{F}$ on X, which is locally isomorphic to the cokernel of free sheaves, i.e. for all $x\in X$ there is an open neighbourhood U of x and an $\mathcal{O}_U$-module homomorphism $\varphi:\mathcal{O}_U^{r_U}\to\mathcal{O}_U^{s_U}$ for some $r_U,s_U\in\mathbb{N}$ such that $\mathcal{O}_U^{s_U}/\mathrm{im}(\varphi)$ is isomorphic to $\mathcal{F}$. Obviously, every locally free $\mathcal{O}_X$-module is coherent.

A.10.16. If $\varphi : \mathcal{E} \to \mathcal{F}$ is a homomorphism of $\mathcal{O}_X$-modules, then $\ker(\varphi), \mathrm{im}(\varphi)$ and $\mathrm{coker}(\varphi)$ are $\mathcal{O}_X$-modules. They are all coherent when $\mathcal{E}$ and $\mathcal{F}$ are coherent ([**134**], (0.5.3.4)).

A.10.17. We introduce now some basic operations on $\mathcal{O}_X$-modules. Let $\mathcal{E}$ and $\mathcal{F}$ be $\mathcal{O}_X$-modules. Then $\mathcal{H}om_{\mathcal{O}_X}(\mathcal{E}, \mathcal{F})$ is the sheaf which is given on an open subset U by the homomorphisms $\mathcal{E}|_U \to \mathcal{F}|_U$ of $\mathcal{O}_U$-modules and whose restriction maps are the restrictions of homomorphisms.

We define $\mathcal{E} \otimes_{\mathcal{O}_X} \mathcal{F}$ to be the sheaf associated to the presheaf $\mathcal{E}(U) \otimes_{\mathcal{O}_X(U)} \mathcal{F}(U)$, U open in X, with the obvious restriction maps.

If $\mathcal{E}$ and $\mathcal{F}$ are coherent, then $\mathcal{H}om_{\mathcal{O}_X}(\mathcal{E}, \mathcal{F})$ and $\mathcal{E} \otimes_{\mathcal{O}_X} \mathcal{F}$ are both coherent ([**134**], (0.5.3.5). Suppose that $\mathcal{E}, \mathcal{F}$ are the sheaves of sections of vector bundles E and F. Then it is easy to see that $\mathcal{H}om_{\mathcal{O}_X}(\mathcal{E}, \mathcal{F})$ (resp. $\mathcal{E} \otimes_{\mathcal{O}_X} \mathcal{F}$) is the sheaf of sections of $\mathrm{Hom}(E, F)$ (resp. $E \otimes F$).

A.10.18. Let $\varphi : X \to X'$ be a morphism of K-varieties and let $\mathcal{F}$ be an $\mathcal{O}_X$-module on X. For an open subset U' of X', we define

$$\varphi_*(\mathcal{F})(U') := \mathcal{F}(\varphi^{-1}U').$$

Together with the restriction maps induced from $\mathcal{F}$, we get an $\mathcal{O}_X$-module $\varphi_*(\mathcal{F})$ on X' called the **direct image** of $\mathcal{F}$. If X is a complete variety, then the direct image of a coherent sheaf is coherent. In fact, this holds more generally for proper morphisms ([**136**], Cor.3.2.2).

We deduce that a complete affine variety X is finite. Let $\varphi : X \to \mathbb{A}^0_K$ be the constant map. Then $\varphi_*(\mathcal{O}_X)(\mathbb{A}^0_K) = K[X]$ is a finite-dimensional K-vector space. We conclude that φ is a finite morphism proving the claim (see A.12.4).

A.10.19. Again, we consider a morphism $\varphi : X \to X'$ of K-varieties. For an $\mathcal{O}_{X'}$-module $\mathcal{F}'$ on X', we define the presheaf $\mathcal{F}$ on X in the following way. We fix an open subset U of X. Then we consider pairs (U', s'), where U' is an open neighbourhood of $\varphi(U)$ and $s' \in \mathcal{F}'(U')$. Two pairs (U', s') and (V', t') are called equivalent if the restrictions of s' and t' agree on an open neighbourhood of $\varphi(U)$. Then the set of equivalence classes is denoted by $\mathcal{F}(U)$. We use the restriction maps induced from $\mathcal{F}'$ to get a presheaf $\mathcal{F}$.

The same thing can be done with $\mathcal{O}_{X'}$ instead of $\mathcal{F}'$. Then we get a presheaf $\mathcal{G}$ of K-algebras on X, where the elements of $\mathcal{G}(U)$ are equivalence classes of pairs (U', f') with U' as above and f' a regular function on U'. Clearly, $\mathcal{F}(U)$ is a $\mathcal{G}(U)$-module. Using composition with φ, we see that $\mathcal{O}_X(U)$ is also a $\mathcal{G}(U)$-module. Then we define the **inverse image** (or **pull-back**) $\varphi^*\mathcal{F}'$ to be the sheaf associated to the presheaf on X given on an open subset U by $\mathcal{F}(U) \otimes_{\mathcal{G}(U)} \mathcal{O}_X(U)$. Note that this construction is necessary to get an $\mathcal{O}_X$-module. If $\mathcal{F}'$ is coherent, then $\varphi^*(\mathcal{F}')$ is also coherent ([**148**], Prop.II.5.8).

Obviously, we have $\varphi^*\mathcal{O}_{X'} = \mathcal{O}_X$. We deduce that the inverse image sheaf of a (locally) free $\mathcal{O}_{X'}$-module is (locally) free. If $\mathcal{E}'$ is the sheaf of sections of a vector bundle E', this shows easily that $\varphi^*(\mathcal{E}')$ is the sheaf of sections of $\varphi^*(E')$. If X is an open or closed subvariety of X', then we simply write $\mathcal{F}'|_X$ for the pull-back.

Example A.10.20. Let Y be a closed subvariety of X with ideal sheaf $\mathcal{J}_Y$. If i denotes the inclusion map $Y \subset X$, then we have a short exact sequence

$$0 \longrightarrow \mathcal{J}_Y \longrightarrow \mathcal{O}_X \longrightarrow i_*\mathcal{O}_Y \longrightarrow 0. \tag{A.16}$$

Locally, this is obvious and, by the sheaf property, we get the claim. Since $\mathcal{O}_X$ and $i_*\mathcal{O}_Y$ are coherent, it follows that $\mathcal{J}_Y$ is coherent. Note that, if we tensor the short exact sequence with a locally free sheaf $\mathcal{E}$ on X, then we obtain a short exact sequence

$$0 \longrightarrow \mathcal{J}_Y \otimes_{\mathcal{O}_X} \mathcal{E} \longrightarrow \mathcal{E} \longrightarrow i_*\mathcal{O}_Y \otimes_{\mathcal{O}_X} \mathcal{E} \longrightarrow 0.$$

By the projection formula ([**148**], Exercise II.5.1(d), we have

$$i_*\mathcal{O}_Y \otimes_{\mathcal{O}_X} \mathcal{E} \cong i_*\left(\mathcal{O}_Y \otimes_{\mathcal{O}_Y} i^*\mathcal{E}\right) \cong i_*i^*\mathcal{E}.$$

If Y is just a K-rational point P of X, then $i_*\mathcal{O}_Y$ is called the **skyscraper sheaf** $\mathcal{K}_P$ of P. Note that $\mathcal{K}_P$ is the sheaf given by

$$\mathcal{K}_P(U) = \begin{cases} K & \text{if } P \in U, \\ \{0\} & \text{if } P \notin U. \end{cases}$$

A.10.21. Now we are ready to define the cohomology groups of a coherent $\mathcal{O}_X$-module $\mathcal{F}$ on a K-variety X. We choose a covering $\mathcal{U}$ of X by affine open subsets. For $p \in \mathbb{N}$, the pth **cohomology group** is $H^p(X,\mathcal{F}) := \check{H}^p(\mathcal{U},\mathcal{F})$. It does not depend on the choice of $\mathcal{U}$; i.e., if $\mathcal{U}'$ is also an affine open covering of X, then there exists a common refinement $\mathcal{U}''$ also by affine open subsets (use A.2.10) and the canonical homomorphisms $\check{H}^p(\mathcal{U},\mathcal{F}) \to \check{H}^p(\mathcal{U}'',\mathcal{F})$ and $\check{H}^p(\mathcal{U}',\mathcal{F}) \to \check{H}^p(\mathcal{U}'',\mathcal{F})$ from A.10.8 are isomorphisms ([**148**], Th.III.4.5). Note that the cohomology groups are K-vector spaces. By A.10.6, we have $H^0(X,\mathcal{F}) = \mathcal{F}(X)$.

A.10.22. Let $0 \longrightarrow \mathcal{F}' \xrightarrow{\varphi'} \mathcal{F} \xrightarrow{\varphi''} \mathcal{F}'' \longrightarrow 0$ be a short exact sequence of coherent $\mathcal{O}_X$-modules. Then we get a **long exact sequence** of cohomology groups

$$0 \longrightarrow H^0(X,\mathcal{F}') \xrightarrow{\varphi'_X} H^0(X,\mathcal{F}) \xrightarrow{\varphi''_X} H^0(X,\mathcal{F}'') \xrightarrow{\delta} H^1(X,\mathcal{F}') \longrightarrow \cdots$$

$$\cdots \xrightarrow{\delta} H^p(X,\mathcal{F}') \longrightarrow H^p(X,\mathcal{F}) \longrightarrow H^p(X,\mathcal{F}'') \xrightarrow{\delta} H^{p+1}(X,\mathcal{F}') \longrightarrow \cdots.$$

We briefly sketch the argument assuming some familiarity with homological algebra (see for example [**157**], Ch.6). By [**148**], Prop.II.5.6, the sequence

$$0 \longrightarrow \mathcal{F}'(U) \xrightarrow{\varphi'} \mathcal{F}(U) \xrightarrow{\varphi''} \mathcal{F}''(U) \longrightarrow 0$$

is exact for any *affine* open subset U. Since the intersection of finitely many affine open subsets remains affine ([**148**], Exercise II.4.3), we see that the sequence

$$0 \longrightarrow C^*(\mathcal{U},\mathcal{F}') \xrightarrow{\varphi'} C^*(\mathcal{U},\mathcal{F}) \xrightarrow{\varphi''} C^*(\mathcal{U},\mathcal{F}'') \longrightarrow 0$$

is also exact. By homological algebra, we get the long exact sequence of cohomology groups.

A.10.23. If X is an affine K-variety and $\mathcal{F}$ is a coherent sheaf of $\mathcal{O}_X$-modules, then $H^p(X,\mathcal{F}) = 0$ for all $p \geq 1$([**148**], Th.III.3.7). This explains the use of affine open coverings in A.10.21.

A.10.24. For any K-variety X and $p > \dim(X)$, we have $H^p(X,\mathcal{F}) = 0$ ([**148**], Th.III.2.7).

A.10.25. If Y is a closed subvariety of X and $\mathcal{F}$ is a coherent $\mathcal{O}_Y$-module on Y, then we have

$$H^p(Y,\mathcal{F}) = H^p(X, i_*\mathcal{F}) \quad (p \in \mathbb{N}),$$

where $i : Y \to X$ is the inclusion. This follows immediately from

$$C^p(\mathcal{U}, i_*\mathcal{F}) = C^p(\mathcal{U} \cap Y, \mathcal{F}),$$

where $\mathcal{U}$ is an affine open covering of X.

A.10.26. For a projective variety X over K, we have the following generalization of A.6.7. If $\mathcal{F}$ is a coherent $\mathcal{O}_X$-module, then $H^p(X,\mathcal{F})$ is a finite-dimensional K-vector space for all $p \in \mathbb{N}$ ([**148**], Th.III.5.2). In fact, this is more generally true for complete varieties ([**136**], 3.2.1).

A.10.27. The sheaf associated to a line bundle is called **invertible**. Let $\mathcal{L}$ be an invertible sheaf on a projective variety X over K. If $\mathcal{L}$ is ample, then for each coherent $\mathcal{O}_X$-module $\mathcal{F}$, there is $n_0 \in \mathbb{N}$ such that

$$H^p(X, \mathcal{F} \otimes \mathcal{L}^{\otimes n}) = 0$$

for all $p \geq 1$ and all $n \geq n_0$ ([**148**], Prop.III.5.3).

A.10.28. Now we handle **base change**. Let L be a field extension of K. We consider a coherent sheaf $\mathcal{F}$ on a geometrically reduced K-variety X. There is a unique coherent sheaf $\mathcal{F}_L$ on X_L such that $\mathcal{F}_L(U_L) = \mathcal{F}(U) \otimes_K L$ and $\rho_{V_L}^{U_L} = \rho_V^U \otimes 1$ for all affine open subsets $V \subset U$ of X. This follows easily from the fact that on an affine variety U, there is a one-to-one correspondence between coherent sheaves and finitely generated K-modules, given by $\mathcal{G} \mapsto \mathcal{G}(U)$ (use [**148**], Cor.II.5.5, Prop.II.5.2).

Anyway, we need the base change only for locally free sheaves $\mathcal{E}$. Then $\mathcal{E}$ is the sheaf of sections of a vector bundle E. We have seen that E_L is a vector bundle over X_L (see A.5.15). It is given by the same transition functions as E. We claim that $\mathcal{E}_L$ is the sheaf of sections of E_L. Since X is geometrically reduced, this is clear for the restriction to trivializations and shows immediately the claim.

Cohomology is compatible with base change, i.e.

$$H^p(X,\mathcal{F}) \otimes_K L = H^p(X_L, \mathcal{F}_L)$$

for all p. To prove that we choose a finite affine open covering $\mathcal{U} = (U_\alpha)_{\alpha \in I}$ of X. Then the base change $\mathcal{U}_L$ of the covering is affine. Clearly, the Čech complex $C^*(\mathcal{U}_L, \mathcal{F}_L)$ is obtained from $C^*(\mathcal{U}, \mathcal{F})$ by tensoring with L. Thereby, kernels and images go to kernels and images proving the claim.

A.10.29. Let X be an irreducible smooth projective variety over $\overline{K}$ of dimension d. We have the canonical line bundle $K_X = \wedge^d T_X^*$ on X (see A.7.23). We denote its sheaf of sections (i.e. the d-forms) by ω_X. For a locally free sheaf $\mathcal{E}$ of X, we have the **Serre duality**

$$H^i(X, \mathcal{E}) \cong H^{d-i}(X, \check{\mathcal{E}} \otimes \omega_X)^*,$$

where $\check{\mathcal{E}} = \mathcal{H}om_{\mathcal{O}_X}(\mathcal{E}, \mathcal{O}_X)$ is the dual of $\mathcal{E}$ ([**148**], Cor.III.7.7, Cor.III.7.12).

A.10.30. Let $\mathcal{F}$ be a coherent sheaf on a projective variety X over K. The **Euler characteristic** of $\mathcal{F}$ is

$$\chi(\mathcal{F}) := \sum_{j=0}^{\dim(X)} (-1)^j \dim H^j(X, \mathcal{F}).$$

Let us consider an exact sequence

$$0 \longrightarrow \mathcal{F}_0 \xrightarrow{\varphi_0} \mathcal{F}_1 \xrightarrow{\varphi_1} \cdots \xrightarrow{\varphi_{n-1}} \mathcal{F}_n \longrightarrow 0$$

of coherent sheaves on X. Then we have

$$\sum_{j=0}^{n} (-1)^j \chi(\mathcal{F}_j) = 0. \tag{A.17}$$

For a short exact sequence, this follows from the long exact sequence of cohomology groups and the fact that the alternate sum of dimensions is zero for a finite exact sequence of finite-dimensional vector spaces. In general, we have

$$\begin{aligned}\sum_{j=0}^{n} (-1)^j \chi(\mathcal{F}_j) &= \sum_{j=0}^{n} (-1)^j \left(\chi(\ker(\varphi_j)) + \chi(\operatorname{im}(\varphi_j))\right) \\ &= \sum_{j=0}^{n} (-1)^j \chi(\operatorname{im}(\varphi_{j-1})) + \sum_{j=0}^{n} (-1)^j \chi(\operatorname{im}(\varphi_j)) = 0.\end{aligned}$$

Let $\mathcal{L}$ be a very ample invertible sheaf on X. We define $\mathcal{F}(n) := \mathcal{F} \otimes \mathcal{L}^{\otimes n}$ for all $n \in \mathbb{Z}$. Then there is $P \in \mathbb{Q}[x]$ called the **Hilbert polynomial of $\mathcal{F}$ with respect to $\mathcal{L}$** given by $P(n) = \chi(\mathcal{F}(n))$ for all $n \in \mathbb{Z}$.

To prove it, we may assume that X is a closed subvariety of $\mathbb{P}_K^n$ with $\mathcal{L} = \mathcal{O}_{\mathbb{P}_K^n}(1)|_X$. Let U be the set of points $x \in X$ such that $\mathcal{F}(V) = 0$ for every sufficiently small neighbourhood V of x. The **support** of $\mathcal{F}$ is $\operatorname{supp}(\mathcal{F}) := X \setminus U$. Since $\mathcal{F}$ is coherent, we can show that the support is closed ([**148**], Exercise II.5.6). Let $d(\mathcal{F})$ be the dimension of $\mathrm{supp}(\mathcal{F})$ and let $n(\mathcal{F})$ be the number of $d(\mathcal{F})$-dimensional irreducible components of $\mathrm{supp}(\mathcal{F})$. We order the pairs $(d(\mathcal{F}), n(\mathcal{F}))$ lexicographically and we use induction with respect to this order. If $\mathcal{F} \neq 0$, then the support of $\mathcal{F}$ is not empty and we find a standard open subset $U_j = \{x_j \neq 0\}$ intersecting $\mathrm{supp}(\mathcal{F})$. Consider the exact sequence

$$0 \longrightarrow \mathcal{K} \longrightarrow \mathcal{F}(-1) \xrightarrow{\varphi} \mathcal{F} \longrightarrow \mathcal{C} \longrightarrow 0, \tag{A.18}$$

where the homomorphism φ is tensoring with x_j. Note that the restriction of φ to U_j is an isomorphism. We conclude that the supports of the kernel $\mathcal{K}$ and the cokernel $\mathcal{C}$ are contained in $\mathrm{supp}(\mathcal{F}) \cap \{x_j = 0\}$. By induction, we may assume that $\chi(\mathcal{K}(n))$ and $\chi(\mathcal{C}(n))$ have the form $\sum_{j=0}^{d} a_j \binom{n}{j}$ for some $a_j \in \mathbb{Z}$. Note that (A.18) remains exact after tensoring with $\mathcal{L}^{\otimes n}$ and the supports do not change. By (A.17), we see that $\chi(\mathcal{F}(n)) - \chi(\mathcal{F}(n-1)) = \sum_{j=0}^{d(\mathcal{F})-1} b_j \binom{n}{j}$ for some $b_j \in \mathbb{Z}$. Then $\chi(\mathcal{F}(n)) = \chi(\mathcal{F}(-1)) + \sum_{j=0}^{d(\mathcal{F})-1} b_j \binom{n+1}{j+1}$.

As a corollary of the proof, we see that the Hilbert polynomial has the form

$$P(x) = \sum_{j=0}^{d(\mathcal{F})} a_j \binom{x}{j}$$

with integer coefficients a_j. By [**135**], Prop.5.3.1, the degree of the Hilbert polynomial is equal to the dimension of $\mathrm{supp}(\mathcal{F})$.

A.10.31. Let $i : X \to \mathbb{P}_K^m$ be a closed embedding with $\mathcal{O}_{\mathbb{P}_K^m}(1)|_X = \mathcal{L}$. From A.10.25 and the projection formula ([**148**], Exercise II.5.1(d)

$$i_*(\mathcal{F} \otimes \mathcal{L}^{\otimes n}) = i_*(\mathcal{F} \otimes i^* \mathcal{O}_{\mathbb{P}_K^m}(n)) \cong i_*(\mathcal{F}) \otimes \mathcal{O}_{\mathbb{P}_K^m}(n),$$

we deduce that the Hilbert polynomial of $\mathcal{F}$ with respect to $\mathcal{L}$ is the same as the Hilbert polynomial of $i_*(\mathcal{F})$ with respect to $\mathcal{O}_{\mathbb{P}_K^m}(1)$. We conclude that without loss of generality, we can always assume $X = \mathbb{P}_K^m, \mathcal{L} = \mathcal{O}_{\mathbb{P}_K^m}(1)$.

So let $\mathcal{F}$ be a coherent sheaf on $\mathbb{P}_K^m$. Then we use $\mathcal{F}(n) := \mathcal{F} \otimes \mathcal{O}_{\mathbb{P}_K^m}(n)$. Note that

$$\Gamma_*(\mathcal{F}) := \bigoplus_{n=0}^{\infty} H^0(\mathbb{P}_K^m, \mathcal{F}(n))$$

is a graded $K[x_0, \ldots, x_n]$-module. It follows from [**148**], Exercise II.5.9 and from A.10.26 that $\Gamma_*(\mathcal{F})$ is a finitely generated graded $K[\mathbf{x}]$-module. Let P be the Hilbert polynomial of $\mathcal{F}$ (always with respect to $\mathcal{O}_{\mathbb{P}_K^m}(1)$). By A.10.27, we get

$$P(n) = \dim H^0(\mathbb{P}_K^m, \mathcal{F}(n))$$

for all $n \gg 0$. Hilbert has shown that for any finitely generated $K[\mathbf{x}]$-module M, there is a polynomial $P_M \in \mathbb{Q}[x]$ with $P_M(n) = \dim M_n$ for $n \gg 0$ (see [**157**], Th.7.23). We conclude that the Hilbert polynomial of $\mathcal{F}$ is equal to $P_{\Gamma_*(\mathcal{F})}$.

Example A.10.32. The Hilbert polynomial of $\mathcal{O}_{\mathbb{P}_K^m}$ is $P(x) = \binom{x+m}{m}$ by A.6.6.

A.10.33. Let X be a closed subvariety of $\mathbb{P}_K^m$ with homogenization $I(X)$ and let P be the Hilbert polynomial of $\mathcal{O}_X$ with respect to $\mathcal{L} = \mathcal{O}_{\mathbb{P}_K^m}(1)|_X$. We call it the **Hilbert polynomial** of X. Then for n sufficiently large, $P(n)$ is the dimension of the space of homogeneous polynomials of degree n restricted to X, i.e.

$$P(n) = \binom{m+n}{n} - \dim I(X)_n.$$

This follows from the long exact cohomology sequence applied to (A.16) on page 569, with arguments as those in A.10.31.

The Hilbert polynomial P of X has degree $d = \dim(X)$ (see A.10.30). By [**125**], Example 2.5.2, the highest coefficient of P is $a_d/d!$ with a_d equal to the degree of X. This is illustrated in the following example.

A.10.34. Assume that the homogeneous ideal $I(X)$ of X is generated by a homogenous non-zero polynomial f of degree d. By A.10.14, we have $\mathcal{O}_{\mathbb{P}^m}(-X) = \mathcal{J}_X$ giving rise to

$$0 \longrightarrow \mathcal{O}_{\mathbb{P}_K^m} \xrightarrow{f\cdot} \mathcal{O}_{\mathbb{P}_K^m}(d) \longrightarrow \left(\mathcal{O}_{\mathbb{P}_K^m}/\mathcal{J}_X\right)(d) \cong i_*\mathcal{O}_X(d) \longrightarrow 0.$$

The associated long exact sequence, A.10.25 and Example A.10.32 show that the Hilbert polynomial of X is

$$P(x) = \binom{x+m}{m} - \binom{x+m-d}{m}.$$

Then P is a polynomial of degree $m-1$ with leading coefficient $d/(m-1)!$.

A.10.35. An important property of the Hilbert polynomial is that it is invariant under flat perturbation. This is useful for modular problems (see [**280**], Ch.VI, 4). To formulate it properly, we have to use the language of schemes. The reason is that for a morphism $\varphi : X \to X'$ and $y \in X'$, the fibre X_y may be different in the sense of schemes. However, in our applications, there is no difference, i.e. the fibres will be reduced.

Let $\varphi : X \to X'$ be a proper morphism of noetherian schemes. Consider a coherent sheaf $\mathcal{F}$ on X such that $\mathcal{F}$ is flat over X', i.e. $\mathcal{F}(\varphi^{-1}U')$ is a flat $\mathcal{O}_{X'}(U')$-module (see A.12.10) for every affine open subset U' of X'. Then the Euler characteristic of $\mathcal{F}_y := \mathcal{F}|_{X_y}$ is locally constant in $y \in X'$ ([**212**], II.5). If $\mathcal{L}$ is a very ample invertible sheaf on X, we conclude that the Hilbert polynomial of $\mathcal{F}_y$ with respect to $\mathcal{L}_y$ is locally constant on X'. If X' is connected, it is independent of the choice of the fibre X_y.

Example A.10.36. Let $K = \mathbb{F}_2$, let X be the closed subvariety of $\mathbb{P}_K^1 \times \mathbb{A}_K^1$ given by $0 = x^2 + yx + 1$, where x is an affine coordinate on $\mathbb{P}_K^1$ and y is the coordinate on $\mathbb{A}_K^1$. Let $\varphi : X \to \mathbb{A}_K^1$ be the second projection. Then the fibre of φ over y as a scheme is given by the equation $x^2 + yx + 1 = 0$, hence the Hilbert polynomial of X_y is $\binom{x+1}{1} - \binom{x-1}{1} = 2$. On the other hand, the fibre over 0 is as a variety given by $x = 1$, i.e. the Hilbert polynomial is equal to 1. This shows that it is necessary to consider the fibres in the sense of schemes.

A.10.37. It is possible to relate the cohomology of the product to the cohomologies of the factors. This is done in the **Künneth formula**: Let X_1 and X_2 be varieties over K. Let $\mathcal{E}$ be a locally free sheaf on X_1 and let $\mathcal{F}$ be a coherent sheaf on X_2. We denote the projection of $X_1 \times X_2$ onto X_i by p_i. Then

$$H^n(X_1 \times X_2, p_1^*\mathcal{E} \otimes p_2^*\mathcal{F}) \cong \bigoplus_{p+q=n} H^p(X_1, \mathcal{E}) \otimes_K H^q(X_2, \mathcal{F}).$$

For the proof (of a much more general result), we refer to [**136**], Th.6.7.8.

A.10.38. Let $\mathcal{F}$ be a coherent sheaf on a multiprojective space $\mathbb{P}_K : \mathbb{P}_K^{n_1} \times \cdots \times \mathbb{P}_K^{n_r}$ (see A.6.13). Then there is $k \in \mathbb{N}$ such that

$$H^i(\mathbb{P}_K, \mathcal{F} \otimes \mathcal{O}_{\mathbb{P}}(d_1, \ldots, d_r)) = 0$$

for all $d_1, \ldots, d_r \geq k$ and $i > 0$.

Note first that the claim holds for $\mathcal{F} = \mathcal{O}_X$ with $k = 1$. This is clear from the case $r = 1$ ([**148**], Th.III.5.1) and the Künneth formula. For general $\mathcal{F}$, there is a short exact sequence

$$0 \longrightarrow \mathcal{F}' \longrightarrow \mathcal{E} \longrightarrow \mathcal{F} \longrightarrow 0,$$

where $\mathcal{E}$ is a finite direct sum of sheaves $\mathcal{O}_{\mathbb{P}}(q_i, \ldots, q_i)$ for various integers q_i (because of [**148**], Cor.II.5.18 and $O_{\mathbb{P}}(1, \ldots, 1)$ very ample). Applying descending induction on i in the associated long exact cohomology sequence, we deduce the claim from the case above because cohomology is compatible with direct sums ([**148**], Rem.III.2.9.1).

A.11. Rational maps

A.11.1. Let X, X' be varieties over K. We consider pairs (U, φ), where U is an open dense subset in X and $\varphi : U \to X'$ is a morphism. Two pairs are called equivalent if the morphisms agree on the intersection. To show that it is an equivalence relation, we use that a morphism is determined by its restriction to an open dense subset. This follows from the fact that the diagonal is closed. An equivalence class is called a **rational map**. A rational map $\varphi : X \dashrightarrow X'$ may be represented by a pair $(U_{\max}, \varphi_{\max})$, where $U_{\max}$ is maximal, in fact the union of all possible open subsets where φ is defined. We call $U_{\max}$ the **domain** of φ. If $X' = \mathbb{A}_K^0$, then rational maps are the same as rational functions.

A.11.2. Let $\varphi : X \dashrightarrow X'$ and $\psi : X' \dashrightarrow X''$ be rational maps. The **image** $\mathrm{im}(\varphi)$ of a rational map is defined as the image of its domain. Note that the composition $\psi \circ \varphi$ does not always make sense, since $\mathrm{im}(\varphi)$ may be contained in the complement of the domain of ψ. However, if X is irreducible and $\mathrm{im}(\varphi)$ is dense in X', then the composition $\psi \circ \varphi$ makes sense as a rational map from X to X''. A rational map of varieties is called **dominant** if the image is dense.

A.11.3. A rational map $\varphi : X \dashrightarrow X'$ of irreducible varieties over K is said to be a **birational map** if it is dominant and if there is a dominant rational map $\psi : X' \dashrightarrow X$ which is inverse to φ, i.e. $\varphi \circ \psi$ and $\psi \circ \varphi$ are both (equivalent to) the identity on X and X', respectively. In this case, X is said to be **birational** to X'. Clearly, the composition of birational maps is again birational.

A.11.4. Let $\varphi : X \dashrightarrow X'$ be a dominant rational map of irreducible varieties. Then we get a homomorphism $\varphi^\sharp : K(X') \to K(X), f \mapsto f \circ \varphi$, of function fields. Conversely, any homomorphism of function fields arises uniquely this way because on suitable affine open dense subsets, it induces a homomorphism of coordinate rings and then we may use A.2.4. In particular, φ is birational if and only if $\varphi^\sharp$ is an isomorphism of function fields.

Up to birationality, we have seen that X is determined by its function field. Note that any finitely generated field extension F of K is a function field of an irreducible (affine) variety over K.

A.11.5. Let X be an irreducible K-variety of dimension r which is geometrically reduced. We claim that X is birational to a hypersurface in $\mathbb{A}_K^{r+1}$.

We sketch the proof: By A.4.11, $K(X)$ is separable over K, i.e. there are algebraically independent $f_1, \dots, f_r \in K(X)$ such that $K(X)$ is a finite-dimensional separable field extension of $F := K(f_1, \dots, f_r)$. By the primitive element theorem ([**156**], Sec.4.14), $K(X)$ is generated over F by a rational function f_{r+1}. It follows that the vector $\varphi(x) := (f_1(x), \dots, f_{r+1}(x))$ gives a rational map to $\mathbb{A}_K^{r+1}$. Let p be the minimal polynomial of f_{r+1} over F. By clearing denominators, we may assume that $p = q(f_1, \dots, f_r, \cdot)$ for some $q \in K[x_1, \dots, x_{r+1}]$. The hypersurface $Z(\{q\})$ in $\mathbb{A}_K^{r+1}$ has a function field isomorphic to $K(X)$. It follows easily that φ gives a birational map from X to $Z(\{q\})$.

A.11.6. Now we assume that K is infinite and that X is an irreducible geometrically reduced projective variety in $\mathbb{P}_K^n$ of dimension $r < n$. Then we realize the birational map into a hypersurface of $\mathbb{P}_K^{r+1}$ by a suitable projection with centre outside of X:

By a linear change of coordinates, we may assume that all $r+1$-codimensional projective linear subspaces of the form $\{x_{j_0} = \dots = x_{j_r} = 0\}$ are disjoint from X. In particular, $X \not\subset \{x_0 = 0\}$. Then $K(X)$ is generated by $\frac{x_1}{x_0}, \dots, \frac{x_n}{x_0}$ over K. Since this extension is separable, we may assume after a renumbering of the coordinates that $K(X)$ is separable over $F := K(\frac{x_1}{x_0}, \dots, \frac{x_r}{x_0})$ (see [**157**], proof of Th.8.37). By separability, $F \subset K(X)$ has only finitely many intermediate fields ([**156**], Th.4.28). Since K is infinite, we conclude that there are linearly independent linear forms $\ell_{r+1}, \dots, \ell_n \in K[x_{r+1}, \dots, x_n]$ such that the intermediate fields $E_j := F(\ell_j(\frac{x_{r+1}}{x_0}, \dots, \frac{x_n}{x_0}))$ are the same for all $j \in \{r+1, \dots, n\}$. Let A be the square matrix whose rows are the coefficients of the ℓ_j. Then the primitive elements of the E_j are the components of $A \cdot (\frac{x_{r+1}}{x_0}, \dots, \frac{x_n}{x_0})^t$. Multiplying the latter with A^{-1}, we see that $E_j = K(X)$ for all $j \in \{0, \dots, n\}$. Now consider the projection

$$\pi : \mathbb{P}_K^n \dashrightarrow \mathbb{P}_K^{r+1}, \quad \mathbf{x} \mapsto (x_0 : \dots : x_r : \ell_{r+1}(\mathbf{x})).$$

As in A.11.5, we conclude that π is a birational map to a hypersurface in $\mathbb{P}_K^{r+1}$. By assumption, we have $\{x_0 = \dots = x_r = 0\} \cap X = \emptyset$. Therefore the centre of π lies outside of X.

A.11.7. Let π be a projection as above with K still infinite. It maps X birationally onto $Z(\{f\})$ for some homogeneous polynomial $f \in K[x_0, \dots, x_{r+1}]$. The degree d of f is equal to the degree of X. To see this, note that pull-back π^* maps a hyperplane to a hyperplane and hence the claim follows from projection formula and Example A.9.32. After a linear change of coordinates, we may assume that f has the form

$$f(\mathbf{x}) = f_0(x_0, \dots, x_r) + \cdots + f_{d-1}(x_0, \dots, x_r)x_{r+1}^{d-1} + x_{r+1}^d.$$

A.11.8. Let $\varphi : X \dashrightarrow X'$ be a rational map of varieties, defined as a morphism on the open dense subset U of X. If φ is defined as a morphism in an open neighourhood of $x \in X$ with $x' = \varphi(x)$, then x' is contained in the closure of $\varphi(U)$ and we have $\varphi^\sharp(\mathcal{O}_{X',x'}) \subset \mathcal{O}_{X,x}$.

We claim that the converse of this evident fact also holds.

Let $x \in X$ and $x' \in X'$. We may assume that all the irreducible components of X pass through x. We suppose that x' is contained in the closure of $\varphi(U)$. Then we get a well-defined homomorphism of $\mathcal{O}_{X',x'}$ to the ring of rational functions $K(X)$, given by $f' \mapsto f' \circ \varphi$. Finally, we assume that the range of this homomorphism is contained in $\mathcal{O}_{X,x}$. Then the claim is that φ is defined in x with image x'.

Proof: We choose an affine neighbourhood V' of x' with coordinates $x_1', \dots, x_n'$. Then there is an affine open neighbourhood V of x such that all $x_j := x_j' \circ \varphi$ are regular functions on V. We define a morphism $V \to V'$ by $v \mapsto (x_1(v), \dots, x_n(v))$. By considering $\varphi^\sharp$, it is clear that φ agrees with this morphism on $U \cap V$ proving the claim. □

A.11.9. Let $\varphi : X \dashrightarrow X'$ be a rational map of K-varieties with X smooth. If the base change $\varphi_{\bar{K}}$ extends to a morphism $X_{\bar{K}} \to X'_{\bar{K}}$, then φ extends to a morphism $X \to X'$.

Proof: Assume that $\varphi_{\bar{K}}$ extends to a morphism $\bar{\varphi} : X_{\bar{K}} \to X'_{\bar{K}}$ and let $x \in X$. We need to prove that φ extends to a morphism in a neighbourhood of x with image $x' := \bar{\varphi}(x)$.

We may assume that all irreducible components of X pass through x. Let U be an open dense subset of X, where φ is defined as a morphism.

Clearly, x' is in the closure of $\varphi(U)$. By A.11.8, we have to prove that, for any regular function f' in an open neighbourhood of x', the rational function $f' \circ \varphi$ is regular in a neighbourhood of x.

Since $f' \circ \bar{\varphi}$ is regular in x over $\bar{K}$, we know that no poles of $\mathrm{div}(f' \circ \bar{\varphi})$ pass through x. Since $\mathrm{div}(f' \circ \bar{\varphi})$ is the base change of $\mathrm{div}(f' \circ \varphi)$ to $\bar{K}$ (use that X is smooth and work with Cartier divisors), we conclude that $f' \circ \varphi$ has no poles through x.

Now we know that a rational function on a smooth variety without poles is regular (cf. A.8.21). Therefore, the function $f' \circ \varphi$ is regular in an open neighbourhood of x. □

A.11.10. Let $\varphi : X \dashrightarrow X'$ be a rational map of a K-variety X regular in codimension 1 to a complete K-variety X' with domain U. By the valuative criterion of properness ([**148**], Th.II.4.7) and the scheme-theoretic analogue of A.11.8, we have

$$\operatorname{codim}(X \setminus U, X) \geq 2.$$

In particular, if X is a regular curve, then φ extends to a morphism.

A.12. Properties of morphisms

A.12.1. Let $\varphi : X \to X'$ be a dominant morphism of irreducible varieties over K. Then for any $y \in \varphi(X)$, every irreducible component of $X_y = \varphi^{-1}(y)$ has dimension $\geq \dim(X) - \dim(X')$ and there is an open dense subset U of X' such that

$$\dim(X_y) = \dim(X) - \dim(X')$$

for all $y \in U$. This is the **dimension theorem**.

The inequality follows from [**137**], Th.5.5.8, and the existence of U is a consequence of generic flatness (see A.12.13 below). The whole statement is part of [**148**], Exercise II.3.22.

A.12.2. Let $\varphi : X \to X'$ be a morphism of K-varieties. Then φ is called a **finite morphism** if there is an open affine covering $(U'_\alpha)_{\alpha \in I'}$ of X' such that $U_\alpha := \varphi^{-1}(U'_\alpha)$ is also affine and such that $K[U_\alpha]$ is a finitely generated $K[U'_\alpha]$-module using A.2.4.

Lemma A.12.3. *A finite morphism has finite fibres.*

Proof: We may assume X and X' both affine.

For $x' \in X'$, the fibre over x' is an affine variety over $K(x')$ with coordinate ring $K(x') \otimes_{K[X']} K[X]$ divided by the radical ideal $\sqrt{0}$. We may assume that $K[X]$ is a finite $K[X']$-module. Therefore, the coordinate ring of $X_{x'}$ is a finite-dimensional $K(x')$-vector space.

Suppose now that Y is an irreducible component of $X_{x'}$. Then $K[Y]$ is a quotient of $K[X_{x'}]$. We conclude that $K[Y]$ is also a finite-dimensional $K(x')$-vector space and hence a field. Its transcendence degree over $K(x')$ is 0 proving $\dim(Y) = 0$ (see Example A.3.6). □

A.12.4. There is the following converse: A morphism $\varphi : X \to X'$ of varieties over K is finite if and only if φ is a proper morphism with finite fibres ([**136**], Prop.4.4.2).

A.12.5. Let $\varphi : X \to X'$ be a finite morphism and let U' be any affine open subset of X'. Then $U := \varphi^{-1}(U')$ is affine and $K[U]$ is a finitely generated $K[U']$-module ([**148**], Exercise II.3.4). Therefore, the definition does not depend on the choice of the affine open covering.

Moreover, we conclude that the composition of finite morphisms remains finite.

Example A.12.6. For every irreducible variety X over K, there is a canonical irreducible normal variety X' called the **normalization of** X.

To sketch the construction, we first assume that X is affine.

The integral closure A of $K[X]$ in $K(X)$ is a finite $K[X]$-module ([**338**], Th.9, p.267). We conclude that there is an irreducible affine variety X' with $K[X'] = A$. It follows easily that the localization of an integrally closed domain remains integrally closed and hence X' is normal.

The inclusion $K[X] \subset K[X']$ leads to a canonical birational finite morphism $\pi : X' \to X$ (using A.2.4, A.11.4). For general X, the irreducible normal variety X' and the canonical finite birational morphism $\pi : X' \to X$ is obtained by a gluing process (for details, see [**135**], Sec.6.4).

A.12.7. Let $\varphi : X \to X'$ be a finite surjective morphism of varieties over K and let L' be a line bundle on X'. Then L' is ample if and only if $\varphi^*(L')$ is ample on X ([**148**], Exercise III.5.7(d)).

A.12.8. Let $\varphi : X \to X'$ be a proper morphism of varieties over K. Then there is a variety X'', a finite morphism $\psi : X'' \to X'$ and a surjective morphism $\rho : X \to X''$ with connected fibres such that $\varphi = \psi \circ \rho$. These properties do not determine X'', ρ and ψ uniquely. However, the construction below is canonical, called the **Stein factorization**:

For an affine open subset U' of X', the sheaf $\mathcal{O}_X(\varphi^{-1}(U'))$ is a finitely generated $K[U]$-module (see A.10.18). Although $\varphi^{-1}U'$ may not be affine, there is an affine variety U'', unique up to isomorphism, with $K[U''] = \mathcal{O}_X(\varphi^{-1}(U'))$. This follows from the fact that the K-algebra $\mathcal{O}_X(\varphi^{-1}(U'))$ is finitely generated and without nilpotents, so we may use the generators to define coordinates. Using the homomorphism $\varphi^\sharp : K[U'] \to K[U'']$, we get a finite morphism $\psi : U'' \to U'$ with $\psi^\sharp = \phi^\sharp$ (see A.2.4). Using the affine open subsets $U'_g, g \in K[U']$, we prove easily that the morphisms $\psi : U'' \to U'$ agree on overlappings, i.e. with varying U' we can paste the U'' and the morphisms $U'' \to U'$ to get a K-variety X'' and a morphism $\psi : X'' \to X'$. By construction, ψ is finite. To define $\rho : X \to X''$ on $\varphi^{-1}(U')$, let $x_1, \ldots, x_n$ be generators of $K[U''] = \mathcal{O}_X(\varphi^{-1}(U'))$ as a K-algebra. We may view them as coordinates on U'' and we define $\rho(x) := (x_1(x), \ldots, x_n(x))$. It is easy to see that we get a well-defined morphism $\rho : X \to X''$ with $\rho^{-1}(U'') = \varphi^{-1}(U')$ for all U'. By construction, $\rho^\sharp : K[U''] \to \mathcal{O}_X(\varphi^{-1}(U'))$ is the identity. We conclude that $\varphi^\sharp = \rho^\sharp \circ \psi^\sharp$ on $K[U']$ proving $\varphi = \psi \circ \rho$ on X. By the factorization property and A.6.15(c), we conclude that ρ is proper. Since $\rho^\sharp$ is one-to-one, it is easy to see that $\rho(X)$ is dense in X'' (use [**148**], Exercise II.2.18). Since a proper map is closed, we conclude $\rho(X) = X''$. For the proof of connected fibres and more details, we refer to [**136**], 4.3.

A.12.9. Suppose that $\varphi : X \to X'$ is a morphism of irreducible varieties over K. We assume that $\dim(X) = \dim(X')$ and that φ is dominant. The **degree** of φ is defined by $\deg(\varphi) := [K(X) : K(X')]$ (finite by A.9.10). Then there is an open dense subset U' of X' such that the restriction of φ gives a finite morphism $\varphi^{-1}(U') \to U'$ whose fibres have cardinality equal to the separable degree of $K(X)/K(X')$ also called the **separable degree of** φ.

We sketch the proof. We may assume that X and X' are affine. Using the correspondence between finitely generated fields and varieties up to birationality, we may assume that $K(X)$ is a primitive extension of $K(X')$, which is either separable or purely inseparable. Let $q(t)$ be the minimal polynomial of a primitive element over $K(X')$. Again by shrinking X and X', we may assume that $q(t) \in K[X'][t]$ and that

$$K[X] \cong K[X'][t]/(q(t)).$$

This proves already finiteness. If q is separable, then q and $\frac{d}{dt}q$ are coprime in $K(X')[t]$, hence there are $a, b \in K(X')[t]$ with $aq + b\frac{d}{dt}q = 1$. Shrinking X, X' again, we may assume that $a, b \in K[X'][t]$. It follows that the identity survives after specializing in any $x' \in X'$, hence the specialization of q in x' is separable. We conclude that the fibre over x' has exactly $\deg(q) = [K(X) : K(X')]$ points.

If $K(X)/K(X')$ is purely inseparable, then K has characteristic $p \neq 0$ and $q(t) = t^{p^e} - h$ for some $h \in K(X')$ ([**157**], Prop.8.13). Again, we may assume $h \in K[X']$ and all the fibres have exactly one element.

A.12.10. In algebra, there is an important generalization of free modules. Let A be a commutative ring with 1 and let M be an A-module. Then M is called **flat** if for every injective A-module homomorphism $N' \to N$, the tensor homomorphism $M \otimes_A N' \to M \otimes_A N$ is also injective. Flatness is preserved under base change and localization. Moreover, M is flat if and only if the localization $M_\wp$ is a flat $A_\wp$-module for all prime ideals $\wp$ of A. Clearly, the same holds if we consider only maximal ideals. For proofs and more details, we refer to [**197**], Ch.2, §3.

A.12.11. We transfer this concept to algebraic geometry by calling a morphim $\varphi : X \to X'$ of varieties over K **flat in $x \in X$** if $\varphi^\sharp : \mathcal{O}_{X',\varphi(x)} \to \mathcal{O}_{X,x}$ makes $\mathcal{O}_{X,x}$ into a flat $\mathcal{O}_{X',\varphi(x)}$-module. If φ is flat in all points of X, then φ is called **flat**. If X and X' are affine, then φ is flat if and only if $K[X]$ is a flat $K[X']$-module. This follows from A.12.10 and the one-to-one correspondence between maximal ideals and points modulo conjugates (see A.2.7).

A.12.12. Flatness is important since many properties of one fibre hold for all fibres. For example, if $\varphi : X \to X'$ is a flat morphism of pure dimensional varieties over K, then for $x' \in \varphi(X)$, we have

$$\dim \varphi^{-1}(x') = \dim(X) - \dim(X').$$

[**148**], Cor.III.9.6. Without flatness, the dimension theorem only ensures $\geq$. For example, if X is the blow up of a two-dimensional variety in a point, then the inequality is strict. If φ is a flat morphism of projective varieties, then equality follows from the more general fact that the Hilbert polynomial is stable with respect to the scheme-theoretic fibres (use A.10.33 and A.10.35).

A.12.13. The theorem on generic flatness states that for any morphism $\varphi : X \to X'$ of varieties over K, there is an open dense subset U' of X' such that the restriction of φ to a map $\varphi^{-1}(U') \to U'$ is flat ([**137**], Th.6.9.1).

Note that if φ is not dominant, this is not interesting since we may choose $U' := X' \setminus \overline{\varphi(X)}$. On the other hand, every flat morphism of K-varieties is open ([**137**], Th.2.4.6) and hence dominant.

A.12.14. A morphism $\varphi : X \to X'$ of varieties over K is called **unramified in** $x \in X$ if the following conditions are satisfied:

(a) The maximal ideal $\mathfrak{m}_x$ of $\mathcal{O}_{X,x}$ is generated by the image of $\mathfrak{m}_{\varphi(x)}$ under the map $\varphi^\sharp : \mathcal{O}_{X',\varphi(x)} \to \mathcal{O}_{X,x}, f' \mapsto f' \circ \varphi$.

(b) $\varphi^\sharp$ induces a separable extension $K(\varphi(x)) \subset K(x)$ of residue fields.

If φ is unramified in all points of X, then it is called an **unramified morphism**. If φ is unramified and flat, then it is called **étale**.

Example A.12.15. We assume that X and X' are both affine varieties over K and $K[X] = K[X'][t]/(f(x', t))$ where $f(x', t)$ is a monic polynomial in the variable t with coefficients in $K[X']$. There is a unique morphism $\varphi : X \to X'$ with $\varphi^\sharp$ equal to the canonical homomorphism $K[X'] \to K[X]$ (see A.2.4). Note that $\varphi^\sharp$ makes $K[X]$ into a free $K[X']$-module of rank $\deg(f)$. Hence φ is a finite flat morphism. We claim that the set of unramified (even étale) points in X is equal to $X \setminus Z(\{\frac{\partial}{\partial t} f\})$.

We compute first the fibre over $x'_0 \in X'$. Let

$$f(x'_0, t) = \prod_{j=1}^{r} f_j(x'_0, t)^{n_j}$$

be the decomposition of $f(x'_0, t)$ into different irreducible polynomials $f_j(x'_0, t)$ in the polynomial ring $K(x'_0)[t]$. Then the fibre $\varphi^{-1}(x'_0)$ has coordinate ring

$$K[\varphi^{-1}(x'_0)] \cong \prod_{j=1}^{r} K(x'_0)[t]/(f_j(x'_0, t))$$

by the Chinese remainder theorem. The fibre points are in one-to-one correspondence with the zeros of $f_1(x'_0, \cdot), \ldots, f_r(x'_0, \cdot)$. Let $x_0 = (x'_0, t_0)$ be a fibre point over x'_0. Without loss of generality, we may assume that t_0 is a zero of $f_1(x'_0, t)$ (and hence of no other $f_j(x'_0, t)$). Let $I(x'_0)$ be the ideal of vanishing of x'_0 in $K[X']$, hence $K(x'_0) = K[X']/I(x_0)$. By the Chinese remainder theorem again, we have

$$K[X]/K[X]I(x'_0) \cong \prod_{j=1}^{r} K(x'_0)[t]/(f_j(x'_0, t)^{n_j}).$$

Using localization in $I(x_0)$, we conclude that

$$\mathcal{O}_{X,x_0}/\mathcal{O}_{X,x_0}\mathfrak{m}_{x'_0} \cong K(x'_0)[t]/(f_1(x'_0, t)^{n_1}).$$

On the other hand, we have

$$\mathcal{O}_{X,x_0}/\mathfrak{m}_{x_0} = K(x_0) \cong K(x_0')[t]/(f_1(x_0',t)).$$

We conclude that $\mathfrak{m}_{x_0'}$ generates $\mathfrak{m}_{x_0}$ if and only if $n_1 = 1$. Moreover, the residue field $K(x_0)$ is separable over $K(x_0')$ if and only if $f_1(x_0',t)$ has no multiple roots. We conclude that φ is unramified in x_0 if and only if t_0 is a simple zero of $f(x_0',t)$ which proves the claim.

A.12.16. In the situation of Example A.12.15, let V be an open subset of $X \setminus Z(\{\frac{\partial}{\partial t} f\})$. Then the restriction $\psi : V \to X'$ of φ is called **a standard étale morphism**. Next, we will see that étale morphisms are locally build up by standard étale morphisms.

A.12.17. Let $\varphi : X \to X'$ be an étale morphism. For every $x \in X$, there are open affine neighbourhoods U, V' of x and $\varphi(x)$, respectively, such that φ restricts to a standard étale morphism $\psi : U \to V'$. For a unramified morphism, there is a closed embedding $i : U \to V$ and a standard étale morphism $\psi : V \to V'$ such that $\varphi = \psi \circ i$. This follows from a theorem of Chevalley (see [**137**], Th.18.4.6 and the proof of Cor.18.4.7).

In the theory of complex manifolds, an étale morphism is just a local isomorphism. However, this does not hold in algebraic geometry, but we have the following result which we deduce easily from [**44**], Prop.2.2.8:

Proposition A.12.18. *Let $\varphi : X \to X'$ be a morphism of smooth varieties over K. Then φ is étale in $x \in X$ if and only if the differential $\mathrm{d}\varphi$ induces an isomorphism*

$$T_{X,x} \xrightarrow{\sim} T_{X',\varphi(x)} \otimes_{k(\varphi(x))} k(x).$$

A.12.19. A dominant morphism $\varphi : X \to X'$ of irreducible varieties over K is called **separable** if $K(X)$ is a separable extension of $K(X')$. If φ is separable and $\dim(X) = \dim(X')$, the proof of A.12.9 shows that φ is generically unramified over X', i.e. there is an open dense subset U' of X' such that the restriction of φ to $\varphi^{-1}(U') \to U'$ is étale.

A.13. Curves and surfaces

A.13.1. A **curve** (resp. a **surface**) over the field K is a pure dimensional K-variety of dimension 1 (resp. 2).

A.13.2. Any curve over K is birational to a regular projective curve over K.

To see this, we note first that the disjoint union of projective curves is projective. Hence we may assume C irreducible. By passing to an open affine part and then to the projective closure, we may assume C projective. The normalization $\pi : C' \to C$ is a birational finite morphism (see A.12.6). By A.12.7, C' is projective. Since a normal curve is regular (Theorem A.8.5), we get the claim.

A.13.3. If K is perfect, then a regular curve is smooth (see A.7.13). For non-perfect K, this does not necessarily hold (see [**148**], Exercise III.10.1). For any curve C over K, the above implies that $C_{\overline{K}}$ is birational to a smooth projective curve over $\overline{K}$. Clearly, this holds also over a finite-dimensional subextension of $\overline{K}/K$.

A.13.4. Let C be a geometrically irreducible smooth projective curve over K. The dual of the tangent bundle is a line bundle which we call the **canonical line bundle** of C. It is denoted by K_C. The **genus** of C is

$$g(C) := \dim \Gamma(C, K_C).$$

In other words, it is the dimension of the space $\Omega^1_C(C)$ of globally defined 1-forms on C. For example, if C is a smooth **plane curve** of degree d in $\mathbb{P}^2_K$, i.e. C is the zero set of a geometrically irreducible homogeneous polynomial $f(x_0, x_1, x_2)$ of degree d, then the genus formula

$$g(C) = \frac{1}{2}(d-1)(d-2)$$

holds ([**148**], Exercise II.8.4(f)).

We have seen in A.9.24 that the degree of a zero-dimensional cycle does not depend on its rational equivalence class. Using $CH_0(C) \cong \mathrm{Pic}(C)$ from A.9.18, the **degree** $\deg(L)$ of a line bundle L on C is defined as the degree of the corresponding rational equivalence class. Most important is the **Riemann–Roch theorem for curves:**

Theorem A.13.5. *Let L be a line bundle on the geometrically irreducible smooth projective curve C over K. Then we have*

$$\dim \Gamma(C, L) - \dim \Gamma(C, K_C \otimes L^{-1}) = \deg(L) + 1 - g(C).$$

Proof: For K algebraically closed, this is proved in [**148**], Th.IV.1.3. By base change and the following remarks, we reduce to this case. First, note that the base change of the tangent bundle is the tangent bundle of the base change (see A.7.6). So the same holds for the canonical line bundle. The cohomology on a geometrically reduced variety is compatible with base change (see A.10.28). We have seen in the proof of A.9.25 that degree of a divisor (and hence of a line bundle) is invariant under base change. This proves the claim in general. □

A.13.6. As an application (setting $L = K_C$), we conclude $\deg(K_C) = 2g(C) - 2$. Using cohomology, we can reformulate the Riemann–Roch theorem by

$$\chi(L) = \deg(L) + 1 - g(C),$$

where $\chi(L) = H^0(C, \mathcal{L}) - H^1(C, \mathcal{L})$ is the Euler characteristic of the sheaf of sections $\mathcal{L}$ of L.

To see this, we may assume K algebraically closed as above. Then the claim follows from Serre duality (see A.10.29), i.e. $H^1(C,\mathcal{L})$ is the dual space of $H^0(C,\Omega^1_C\otimes\mathcal{L}^{-1})$.

A.13.7. Let C be a smooth geometrically irreducible projective curve over K and let L be a line bundle on C. Then L is ample if and only if $\deg(L) > 0$. If $\deg(L)\geq 2g(C)+1$, then L is very ample.

For K algebraically closed, this is proved in [**148**], Cor.IV.3.2 and the general case follows by base change using A.6.12.

A.13.8. Let X be a smooth geometrically irreducible projective variety over K. Then the tangent bundle of X is a vector bundle of rank $d := \dim(X)$ and $K_X := \wedge^d T^*_X$ is called the **canonical line bundle** of X. Its sections are d-forms on X. A divisor of a meromorphic section of K_X is denoted by $\operatorname{div}(K_X)$ and is called a **canonical divisor**. The number $p_a(X) := (-1)^d(\chi(\mathcal{O}_X)-1)$ is **the arithmetic genus** of X, where χ is the Euler characteristic from A.10.30. If X is a smooth irreducible projective curve, then the arithmetic genus agrees with the genus defined in A.13.4 by Serre duality A.10.29.

The **Riemann–Roch theorem for surfaces** says:

Theorem A.13.9. *Let D be a divisor on the irreducible smooth projective surface X. Then*

$$\chi(\mathcal{O}_X(D)) = \frac{1}{2}D\cdot(D-\operatorname{div}(K_X))+1+p_a(X).$$

Proof: For K algebraically closed, this is [**148**], Th.V.1.6. The reduction to this case is by base change similarly as for curves by noting that intersection numbers are invariant under base change (see A.9.25). □

A.13.10. By Serre duality again (see A.10.29), we may rephrase this as

$$\dim\Gamma(X,O(D)) - \dim H^1(X,O(D)) + \dim\Gamma(X,O(K_X-D)) =$$
$$= \frac{1}{2}D\cdot(D-\operatorname{div}(K_X))+1+p_a(X).$$

The proof is the same as for curves.

A.14. Connexion to complex manifolds

This section is for a reader which is more familiar with the theory of complex manifolds (as in [**130**]) than with the algebraic side. We explain the meaning of the other sections of the appendix in terms of complex analysis, which could be helpful for the understanding. For more details and proofs, we refer to [**148**], App. B, [**280**], Ch.s VII, VIII, and J.-P. Serre [**275**].

In this section, K denotes a field contained in $\mathbb{C}$ and $\overline{K}$ is the algebraic closure of K in $\mathbb{C}$. As usual, $\mathbf{x} = (x_1, \ldots, x_n)$.

A.14.1. Let X be a Zariski-closed subset of $\mathbb{A}^n_K$, i.e. there are $f_1(\mathbf{x}), \ldots, f_r(\mathbf{x}) \in K[\mathbf{x}]$ such that $X \subset \overline{K}^n$ is the zero set of these polynomials. Then X was called an affine variety over K and regular functions on X were restrictions of polynomials from $K[\mathbf{x}]$. If we consider the zero set of $f_1, \ldots, f_r$ in $\mathbb{C}^n$, then we get an affine variety $X_{\mathbb{C}}$ over $\mathbb{C}$. It is closed in $\mathbb{C}^n$ with respect to the Zariski topology.

A.14.2. On $\mathbb{C}^n$, we have the **complex topology** given by the basis formed by the open balls. Clearly, the complex topology is finer than the Zariski topology. We define X_{an} to be the complex space with the same underlying set as $X_{\mathbb{C}}$ and with the topology induced by the complex topology of $\mathbb{C}^n$.

A.14.3. More generally, if X is a variety over K, then we can use an affine atlas $(U_\alpha, \varphi_\alpha)_{\alpha \in I}$ of X to define an atlas on $X_{\mathbb{C}}$. Passing to the corresponding complex spaces, we get an atlas of a **complex space** X_{an}. The original variety X is often called an **algebraic variety** and X_{an} is called the **associated complex analytic variety**. The dimension of the algebraic variety is the same as the dimension of the associated complex space. It follows from the Jacobi criterion that X is smooth if and only if X_{an} is a complex manifold.

For the rest of the section, we assume for simplicity that X is a smooth variety over K.

A.14.4. First, note that every regular function on an open subset U of X is the restriction of a unique analytic function on U_{an}. A morphism of smooth varieties over K extends uniquely to an analytic map of the associated complex manifolds. The smooth variety X is geometrically irreducible if and only if X_{an} is connected. The dimension of X is equal to the dimension of the complex manifold X_{an}. If X_1 and X_2 are smooth varieties over K, then $(X_1 \times X_2)_{\mathrm{an}} = (X_1)_{\mathrm{an}} \times (X_2)_{\mathrm{an}}$. A rational function on X gives rise to a meromorphic function on X_{an}.

A.14.5. If E is a vector bundle over X, then E_{an} is a holomorphic vector bundle on X_{an}. They have the same transition matrices. Clearly, every section of E extends uniquely to a holomorphic section of E_{an}. The basic operations of vector bundles are compatible with passing to the associated holomorphic vector bundles. Moreover, $(T_X)_{\mathrm{an}}$ is the tangent bundle of the complex manifold X_{an}.

A.14.6. The variety X is complete if and only if X_{an} is compact. More generally, a morphism φ of algebraic varieties is proper if and only if inverse images of compact subsets with respect to φ_{an} remain compact.

A.14.7. We have seen above that to every algebraic object, we have canonically an analytic one. On complete varieties, we can often reverse the procedure due to the **GAGA principle** of Serre (see [**275**]; [**140**], XII).

Suppose that X is a smooth complete variety over $\mathbb{C}$. Then meromorphic functions on X_{an} are the same as rational functions on X. More generally, a holomorphic map $\varphi : X_{\mathrm{an}} \to X'_{\mathrm{an}}$ for smooth complete varieties X, X' over $\mathbb{C}$ gives a morphism $X \to X'$. The procedure $Y \mapsto Y_{\mathrm{an}}$ maps the set of closed subsets of $\mathbb{P}^n_{\mathbb{C}}$ bijectively onto the set of closed subsets of $(\mathbb{P}^n_{\mathbb{C}})_{\mathrm{an}}$. If E, E' are vector bundle on X, then $E \cong E'$ if and only if $E_{\mathrm{an}} \cong E'_{\mathrm{an}}$ as holomorphic vector bundles. Moreover, the cohomology groups $H^p(X, \mathcal{E})$ agree with the cohomology groups $H^p(X_{\mathrm{an}}, E_{\mathrm{an}})$.

A.14.8. Let X be a smooth variety over K. Any divisor D on X gives rise to a divisor D_{an} on X_{an}. Principal divisors pass to principal divisors on X_{an}. We have a canonical homomorphism $CH_r(X) \to H_{2r}(X_{\mathrm{an}}, \mathbb{Z})$ mapping the prime cycle Y to Y_{an} in the homology group. If D is a divisor on X, then $D.Y$ maps to the cup product of D_{an} and Y_{an}. Hence it is the same to compute intersection numbers on X or in homology on X_{an}. Two divisors on X are algebraically equivalent if and only if they are homologically equivalent on X_{an}, i.e. if they have the same image in $H_{\dim(X)-2}(X_{\mathrm{an}}, \mathbb{Z})$. For details, we refer to [**125**], Ch.19.

A.14.9. If X is an irreducible smooth projective curve over $\mathbb{C}$, then X_{an} is a connected projective manifold of dimension 1. The genus of X is equal to the genus of X_{an}. A complex manifold of dimension 1 is called a **Riemann surface**. We conclude that X_{an} is a connected compact Riemann surface. Conversely, **Riemann's existence theorem** says that every connected compact Riemann surface has this form ([**148**], Th.B.3.1).

APPENDIX B RAMIFICATION

B.1. Discriminants

The discriminant is a measure for ramification. In Section B.1, we begin with the study of discriminants from a purely algebraic point of view. The arithmetic aspect enters by applying it to the local rings of Dedekind domains. This will be continued in Section B.2 to study unramified field extensions in some given places and we will mention the classical discriminant theorems from algebraic number theory. In Section B.3, we compare these notions with the concept of unramified morphisms from algebraic geometry. The final Section B.4 introduces the ramification divisor of a morphism of algebraic varieties in characteristic 0 and we will prove Hurwitz's theorem which is important at various places in this book.

B.1.1. Let R be a commutative ring with 1 and let A be a commutative R-algebra free as an R-module of finite rank n. The **discriminant of A with respect to the basis $e_1, \dots, e_n$** is given by

$$D_{A/R}(e_1, \dots, e_n) := \det(\mathrm{Tr}_{A/R}(e_i e_j)) \in R.$$

Recall that the **trace** and the **norm** of $a \in A$ are given by

$$\mathrm{Tr}_{A/R}(a) = \sum_{i=1}^{n} \alpha_{ii} \in R, \quad N_{A/R} = \det(\alpha_{ij}) \in R,$$

where $a \cdot e_j = \sum_{i=1}^{n} \alpha_{ij} e_i$ for coordinates $\alpha_{ij} \in R$.

B.1.2. If $e_1', \dots, e_n'$ is another basis of A, then there is a $n \times n$ matrix λ with entries in R such that $e_i' = \sum_{j=1}^{n} \lambda_{ji} e_j$. The transformation matrix λ is invertible and we have

$$D_{A/R}(e_1', \dots, e_n') = \det(\lambda)^2 D_{A/R}(e_1, \dots, e_n).$$

Since $\det(\lambda)$ is a unit in R, this proves that the principal ideal $\mathfrak{d}_{A/R}$ in R generated by $D_{A/R}(e_1, \dots, e_n)$ does not depend on the choice of the basis. Then $\mathfrak{d}_{A/R}$ is called the **discriminant ideal of A/R** or simply the **discriminant**.

B.1.3. Let $f \in R[t]$ be a monic polynomial of degree n. Then

$$A := R[t]/[f(t)]$$

is an R-algebra with R-module basis $1, t, \ldots, t^{n-1}$. We call

$$D_f := D_{A/R}(1, t, \ldots, t^{n-1})$$

the **discriminant of** f.

Example B.1.4. Let $R := \mathbb{Z}[x_1, \ldots, x_n]$. We consider the polynomial

$$f(t) := t^n + x_1 t^{n-1} + \cdots + x_n \in R[t].$$

Let $y_1, \ldots, y_n$ be the zeros of $f(t)$. It is known from Galois theory that $y_1, \ldots, y_n$ generate a Galois extension $K = \mathbb{Q}(y_1, \ldots, y_n)$ of degree $n!$ over the fraction field $\mathbb{Q}(x_1, \ldots, x_n)$ of R. We have $f(t) = (t - y_1) \cdots (t - y_n)$. For the ith elementary symmetric polynomial s_i, we get

$$x_i = s_i(-y_1, \ldots, -y_n). \tag{B.1}$$

Clearly, the discriminants of f considered as a polynomial in $R[t]$ and $K[t]$, respectively, are the same. For $B := K[t]/[f(t)]$, we conclude that

$$D_f = D_{B/K}(1, t, \ldots, t^{n-1}).$$

By the Chinese remainder theorem, we have a canonical isomorphism

$$B \xrightarrow{\sim} \prod_{j=1}^{n} K[t]/(t - y_j), \quad t \mapsto (y_1, \ldots, y_n).$$

Let e_j be the jth idempotent, i.e.

$$e_j = \prod_{k \neq j} (y_j - y_k)^{-1} \prod_{k \neq j} (t - y_k).$$

Then $e_1, \ldots, e_n$ is a K-basis of B and we have

$$t^i = \sum_{j=1}^{n} y_j^i e_j.$$

Hence the transformation matrix $\mathbf{a}$ is a Vandermonde matrix. By B.1.2, this leads to

$$D_{B/K}(1, t, \ldots, t^{n-1}) = \det(\mathbf{a})^2 D_{B/K}(e_1, \ldots, e_n).$$

Since $D_{B/K}(e_1, \ldots, e_n) = 1$, we get

$$D_f = \prod_{i>j} (y_i - y_j)^2 = (-1)^{\frac{n(n-1)}{2}} \prod_{i \neq j} (y_i - y_j). \tag{B.2}$$

Note that the right-hand side is a symmetric polynomial in $y_1, \ldots, y_n$ and hence it is a polynomial in $x_1, \ldots, x_n$ with coefficients in $\mathbb{Z}$ ([**156**], Th.2.19).

Remark B.1.5. Note that Example B.1.4 may be used to compute the discriminant of a monic polynomial f whenever the zeros of f are known. More precisely, let R be again an arbitrary commutative ring with 1 and let $f \in R[t]$ be monic and of degree n. Suppose that R is a subring of a commutative ring R' and that $f(t) = (t - \xi_1) \ldots (t - \xi_n)$ for some $\xi_1, \ldots, \xi_n \in R'$. If we specialize $y_1 = \xi_1, \ldots, y_n = \xi_n$ in Example B.1.4, we get easily from (B.2)

$$D_f = \prod_{i>j} (\xi_i - \xi_j)^2 = (-1)^{\frac{n(n-1)}{2}} \prod_{i \neq j} (\xi_i - \xi_j).$$

Remark B.1.6. Even if we do not know the zeros, we get information from Example B.1.4. Let $f(t) := t^n + a_1 t^{n-1} + \cdots + a_n \in R[t]$. By the specialization $x_i = a_i$ in Example B.1.4, we see that the discriminant D_f is a polynomial in $a_1, \ldots, a_n$ with coefficients in $\mathbb{Z}$. The right-hand side of (B.2) is homogeneous of degree $n(n-1)$ in $y_1, \ldots, y_n$. Hence if we consider a_i of degree i, then (B.1) shows that the discriminant is a homogeneous polynomial in the weighted variables $a_1, \ldots, a_n$ of degree $n(n-1)$.

B.1.7. Let $f(t) = a_m + a_{m-1}t + \cdots + a_1 t^{m-1} + a_0 t^m$, $g(t) = b_n + b_{n-1}t + \cdots + b_1 t^{n-1} + b_0 t^n$ be polynomials with coefficients in R. For $k \in \mathbb{Z} \setminus \{0, \ldots, m\}$, we set $a_k = 0$ and, similarly, we proceed with the coefficients of g. Then we form the $(m+n) \times (m+n)$ matrix M by the rules

$$M_{ij} = \begin{cases} a_{i-j} & \text{if } 1 \leq j \leq n, \\ b_{i-j+n} & \text{if } n+1 \leq j \leq m+n. \end{cases}$$

For $m = 2, n = 3$, the matrix is

$$\begin{pmatrix} a_0 & 0 & 0 & b_0 & 0 \\ a_1 & a_0 & 0 & b_1 & b_0 \\ a_2 & a_1 & a_0 & b_2 & b_1 \\ 0 & a_2 & a_1 & b_3 & b_2 \\ 0 & 0 & a_2 & 0 & b_3 \end{pmatrix}.$$

Then the **resultant of f, g with respect to m, n** is $\mathrm{res}_{m,n}(f,g) := \det(M)$. If we make the natural choice $m = \deg(f)$ and $n = \deg(g)$, then we skip the reference to m and n.

B.1.8. Suppose that f is monic and g is an arbitrary polynomial in $R[t]$, then $\mathrm{res}(f,g)$ is invertible in R if and only if the ideal generated by f and g is $R[t]$ (N. Bourbaki [**49**], Ch.IV, §6, No.6, Prop.7, Cor.1). In particular for a field R, we have $\mathrm{res}(f,g) \neq 0$ if and only if f and g are coprime (i.e. if and only if they have no common root in an algebraic closure).

B.1.9. The resultant may be used to compute the discriminant in terms of the coefficients. Let $f'(t)$ be the formal derivative of the monic polynomial $f(t) = t^n + a_1 t^{n-1} + \cdots + a_n \in R[t]$ of degree n. Then

$$D_f = (-1)^{\frac{n(n-1)}{2}} \mathrm{res}_{n,n-1}(f, f').$$

([**49**], Ch.IV, §6, No.7, Prop.11). This shows that D_f is a polynomial of degree $\leq 2n-2$ in $a_1, \ldots, a_n$ with coefficients in R and we have certainly equality in "most" cases. Note that in contrast to Remark B.1.6, the $a_1, \ldots, a_n$ are not weighted, i.e. we consider $a_1, \ldots, a_n$ as variables of degree 1.

B.1.10. The considerations in B.1.9 make it natural to define the **discriminant of any polynomial** $f(t) := a_0 t^n + a_1 t^{n-1} + \cdots + a_n$ by

$$D_f := a_0^{2n-2} D_g,$$

where $g(t) := t^n + \frac{a_1}{a_0}t^{n-1} + \cdots + \frac{a_n}{a_0}$. In this way, D_f is a homogeneous polynomial of degree $2n-2$ in $a_0, \ldots, a_n$ with coefficients in $\mathbb{Z}$. Obviously, this definition extends the one in B.1.3 for monic polynomials. Note that the definition even makes sense in the case $a_0 = 0$ if we consider first the right-hand side as a homogeneous polynomial in the variables $a_0, \ldots, a_n$ and then we specialize. To avoid notational problems in this case, we have to emphasise that the discriminant of f is with respect to n and we denote it by $D_{f,n}$. From B.1.9, we get

$$D_{f,n} = (-1)^{\frac{n(n-1)}{2}} \mathrm{res}_{n,n-1}(f, f')/a_0$$

for an arbitrary polynomial f.

B.1.11. Let $f, g \in R[t]$ be of degree bounded by m and n, respectively. Then

$$D_{fg,m+n} = D_{f,m} D_{g,n} \mathrm{res}_{m,n}(f, g)^2$$

([**49**], Ch.IV, §6, No.7, Prop.11, Cor.1).

B.1.12. Let $\varphi : R \to R'$ be a homomorphism of commutative rings with 1. For $f(t) := a_0 t^n + a_1 t^{n-1} + \cdots + a_n$, let $\varphi(f) := \varphi(a_0)t^n + \varphi(a_1)t^{n-1} + \cdots + \varphi(a_n)$. Since forming the resultant is compatible with this operation, we have $D_{\varphi(f),n} = \varphi(D_{f,n})$. This property was already used in the specialization of Remark B.1.5.

B.1.13. Let $f, g \in R[t]$ with f monic and let $A := R[t]/[f]$ be as in B.1.3. Let $m := \deg(f)$ and $\deg(g) \leq n$. Then the resultant may be expressed in terms of the norm by

$$\mathrm{res}_{m,n}(f, g) = N_{A/R}(g).$$

([**49**], Ch.IV, §6, No.6, Prop.7). In particular, we get

$$D_f = (-1)^{\frac{m(m-1)}{2}} N_{A/R}(f').$$

B.1.14. Now we study discriminants in the case of Dedekind domains. The main reference is the book of J.-P. Serre [**276**]. A **Dedekind domain** is an integrally closed domain of Krull dimension ≤ 1. Let R be a Dedekind domain with field of fractions K. We consider a finite-dimensional separable field extension L/K. In most of our applications, R will be the ring of algebraic integers in a number field. The integral closure of R in L is denoted by $\bar{R}^L$. Note that $\bar{R}^L$ is not necessarily a free R-module. The integral closure is only an R-lattice, i.e. a finitely generated R-module which generates L as a K-vector space (see [**276**], Ch.I, §4, Prop.8). Let $n := [L : K]$. Then the **discriminant of $\bar{R}^L$ over** R is the ideal $\mathfrak{d}_{\bar{R}^L/R}$ in R generated by

$$\{\det(\mathrm{Tr}_{L/K}(a_i b_j)) \mid a_1, \ldots, a_n, b_1, \ldots, b_n \in \bar{R}^L\}.$$

Since the K-bilinear form $\langle \alpha, \beta \rangle := \mathrm{Tr}_{L/K}(\alpha\beta)$ on L is not degenerated (as L/K is separable, see [**49**], Ch.V, §10, no.6, Prop.10), the discriminant is not zero.

If no confusion is possible, we use the notion $\mathfrak{d}_{L/K} := \mathfrak{d}_{\bar{R}^L/R}$. Moreover, if the discriminant ideal is principal, $D_{L/K}$ denotes a principal generator of $\mathfrak{d}_{L/K}$. For $K = \mathbb{Q}$, the choice can be made unique by using the positive principal generator.

Remark B.1.15. If we suppose additionally that $\bar{R}^L$ is a free R-module with basis $e_1, \ldots, e_n$, then for every $a_1, \ldots, a_n, b_1, \ldots, b_n \in R_L$, we have

$$\det(\mathrm{Tr}_{L/K}(a_i b_j)) = \det(A)\det(B)\det(\mathrm{Tr}_{L/K}(e_i e_j)),$$

where $a_i = \sum_k A_{ki} e_k$ and $b_j = \sum_k B_{kj} e_k$. Hence the definitions in B.1.2 and B.1.14 agree. Since L/K is separable, we have exactly n different embeddings $\sigma_1, \ldots, \sigma_n$ of L into the algebraic closure $\bar{K}$ (see [**156**], Th.4.26). Since the trace is the sum of the conjugates, we get

$$\mathrm{Tr}_{L/K}(e_i e_j) = \sum_{k=1}^{n} \sigma_k(e_i e_j)$$

and hence $\mathfrak{d}_{L/K}$ is the principal ideal of R generated by $\det(\sigma_i(e_j))^2$.

Example B.1.16. Let d be a square free integer, $|d| \geq 2$. We consider the quadratic number field $L := \mathbb{Q}(\sqrt{d})$. We first determine the algebraic integers in L. Let $\alpha := r + s\sqrt{d} \in L$ with $r, s \in \mathbb{Q}$. Then

$$\alpha^2 - 2r\alpha + r^2 - s^2 d = 0.$$

Hence α is an algebraic integer if and only if $2r, r^2 - s^2 d \in \mathbb{Z}$. For α to be an algebraic integer, it is necessary that $r = m/2, s = n/2$ for $m, n \in \mathbb{Z}$. If m is even, then $r^2 - s^2 d = (m^2 - n^2 d)/4$ is an integer if and only if n is even. If m is odd, then $r^2 - s^2 d \in \mathbb{Z}$ if and only if n is odd and $d \equiv 1 \pmod 4$. This proves that O_L has $\mathbb{Z}$-basis

$$\begin{cases} 1, \frac{1+\sqrt{d}}{2} & \text{if } d \equiv 1 \pmod 4, \\ 1, \sqrt{d} & \text{if } d \equiv 2, 3 \pmod 4. \end{cases}$$

In the first case, Remark B.1.15 shows

$$\mathfrak{d}_{L/\mathbb{Q}} = \left[\det \begin{pmatrix} 1 & \frac{1+\sqrt{d}}{2} \\ 1 & \frac{1-\sqrt{d}}{2} \end{pmatrix}^2 \right] = [d].$$

Similarly, for $d \equiv 2, 3 \pmod 4$, we prove

$$\mathfrak{d}_{L/\mathbb{Q}} = \left[\det \begin{pmatrix} 1 & \sqrt{d} \\ 1 & -\sqrt{d} \end{pmatrix}^2 \right] = [4d].$$

B.1.17. Let L/K be still a finite-dimensional separable extension of the quotient field K of a Dedekind domain R. The **codifferent** $\{a \in L \mid \mathrm{Tr}_{L/K}(a\bar{R}^L) \subset R\}$ is a fractional ideal of L containing $\bar{R}^L$. Its inverse is an ideal in $\bar{R}^L$ called the **different** of $\bar{R}^L$ over R. It is denoted by $\mathfrak{D}_{\bar{R}^L/R}$ or simply by $\mathfrak{D}_{L/K}$. We have

$$\mathfrak{d}_{L/K} = N_{L/K}(\mathfrak{D}_{L/K}).$$

For proofs, we refer to [**276**], Ch.III, §3.

B.1.18. The **Kähler differentials** build an $\bar{R}^L$-module $\Omega^1_{\bar{R}^L/R} := I/I^2$, where I is the kernel of tensor multiplication $\bar{R}^L \otimes_R \bar{R}^L \to \bar{R}^L$ (compare with A.7.29). If all residue extensions of L/K are separable, then $\Omega^1_{\bar{R}^L/R}$ is generated by one element as an $\bar{R}^L$-module and $\mathfrak{D}_{L/K}$ is its annihilator (see [**276**], Ch.III, §7, Prop.14).

Proposition B.1.19. *Let K be the fraction field of the Dedekind domain R and let L/K and M/L be finite-dimensional separable field extensions. Then*

$$\mathfrak{D}_{M/K} = \mathfrak{D}_{M/L} \cdot \mathfrak{D}_{L/K}, \quad \mathfrak{d}_{M/K} = N_{L/K}(\mathfrak{d}_{M/L}) \cdot \mathfrak{d}_{L/K}^{[M:L]}.$$

Proof: See [**276**], Ch.III, §4, Prop.8. □

B.1.20. By localizing in maximal ideals, we can always reduce to the free case. More generally, let S be a multiplicative submonoid of $R \setminus \{0\}$. Then the localization $S^{-1}R$ is still a Dedekind domain with integral closure $S^{-1}\bar{R}^L$ in L and we have

$$\mathfrak{D}_{S^{-1}\bar{R}^L/S^{-1}R} = S^{-1}\mathfrak{D}_{\bar{R}^L/R}, \quad \mathfrak{d}_{S^{-1}\bar{R}^L/S^{-1}R} = S^{-1}\mathfrak{d}_{\bar{R}^L/R}.$$

In particular, if $\mathfrak{p}$ is a non-zero maximal ideal of R, we get a discrete valuation ring $R_\mathfrak{p}$ and the integral closure of $R_\mathfrak{p}$ in L is $(\bar{R}^L)_\mathfrak{p}$. Since $(\bar{R}^L)_\mathfrak{p}$ is a finitely generated torsion free module over the principal ideal domain $R_\mathfrak{p}$, we conclude that $(\bar{R}^L)_\mathfrak{p}$ is a free $R_\mathfrak{p}$-module of rank $[L:K]$ ([**156**], Th.3.10) and

$$\mathfrak{d}_{(\bar{R}^L)_\mathfrak{p}/R_\mathfrak{p}} = (\mathfrak{d}_{\bar{R}^L/R})_\mathfrak{p}.$$

Note that $(\mathfrak{d}_{\bar{R}^L/R})_\mathfrak{p}$ is a power of $\mathfrak{p}R_\mathfrak{p}$, hence

$$\mathfrak{d}_{\bar{R}^L/R} = \prod_\mathfrak{p} \left(\mathfrak{d}_{(\bar{R}^L)_\mathfrak{p}/R_\mathfrak{p}} \cap R \right)$$

is the prime factorization of the discriminant. Similarly, we may proceed for the different using prime ideals of $\bar{R}^L$.

B.1.21. The formation of the discriminant and the different is also compatible with completions. Let $\mathfrak{p}$ be a non-zero maximal ideal of R and let $\mathfrak{P}$ be a maximal ideal of $S = \bar{R}^L$ with $\mathfrak{P}|\mathfrak{p}$ (i.e. $\mathfrak{P} \cap R = \mathfrak{p}$). The completions of the discrete valuation rings $R_\mathfrak{p}$ and $S_\mathfrak{P}$ are denoted by $\hat{R}_\mathfrak{p}$ and $\hat{S}_\mathfrak{P}$, respectively. They are still discrete valuation rings and $\hat{S}_\mathfrak{P}$ is a free $\hat{R}_\mathfrak{p}$-module of rank equal to the local degree. By [**276**], Ch.III, §4, (iii), we have

$$\mathfrak{D}_{\hat{S}_\mathfrak{P}/\hat{R}_\mathfrak{p}} = \hat{R}_\mathfrak{P}\mathfrak{D}_{\bar{R}^L/R}, \quad \prod_{\mathfrak{P}|\mathfrak{p}} \mathfrak{d}_{\hat{S}_\mathfrak{P}/\hat{R}_\mathfrak{p}} = \hat{R}_\mathfrak{p}\mathfrak{d}_{\bar{R}^L/R}.$$

B.2. Unramified field extensions

In this section, L/K is a finite-dimensional field extension and v, w are non-archimedean places of K and L with $w|v$.

B.2.1. We call L/K **unramified in** w if L/K and the extension of residue fields $k(w)/k(v)$ are separable and if the local degree $[L_w : K_v]$ is equal to the residue degree $[k(w) : k(v)]$. Otherwise, L/K is called **ramified in** w. Note that by passing to completions, the residue degree $f_{w/v}$ and the ramification index $e_{w/v}$ do not change, hence

$$e_{w/v} f_{w/v} \leq [L_w : K_v] \tag{B.3}$$

by Proposition 1.2.11. Hence $e_{w/v} = 1$ if L/K is unramified in w. If the valuation is discrete, then we have equality in (B.3) and hence $[L_w : K_v] = [k(w) : k(v)]$ is equivalent to $e_{w/v} = 1$. We say that L/K is **unramified over** v if L/K is unramified in every place of L lying over v.

Lemma B.2.2. *Assume that L is generated over K by a root of a monic $f \in R_v[t]$, where R_v is the valuation ring of v. If the discriminant D_f is a unit in R_v, then L/K is unramified over v.*

Proof: We call the root α. Using B.1.12, we deduce that α and hence L/K are separable. A standard trick using the ultrametric inequality or Gauss's lemma (see Lemma 1.6.3) proves $\alpha \in R_v$. By B.1.12, we have $D_{\bar{f}} = \overline{D_f} \neq \bar{0}$, where the bar denotes reduction to the residue field. By Remark B.1.5, the polynomial $\bar{f}$ is separable. By Gauss's lemma and B.1.11, we may assume that f is also irreducible. Let $f = f_1 \cdots f_r$ be the factorization into irreducible polynomials $f_j \in K_v[t]$. By Gauss's lemma again, we may assume that the coefficients of these polynomials are contained in the discrete valuation ring of the completion. By separabiltiy of $\bar{f}$, the factors f_j are pairwise coprime. The polynomial $f_1, \ldots, f_r$ are in one-to-one correspondence with the places of L over v (see Proposition 1.3.1), we denote the latter by $w_1, \ldots, w_r$. By Hensel's lemma (see Lemma 1.2.10), $\bar{f_j}$ is a power of an irreducible polynomial in $k(v)[t]$. Using separability of $\bar{f}$, we conclude that $\overline{f_j}$ is indeed irreducible over $k(v)$. Since $K_v[t]/[f_j(t)] = L_{w_j}$ (see Proposition 1.3.1), we get

$$[L_{w_j} : K_v] \geq f_{w_j/v} \geq \deg(\overline{f_j}) = \deg(f_j) = [L_{w_j} : K_v]$$

and hence equality occurs everywhere. We conclude that $\bar{\alpha}$ is a primitive separable element of $k(w_j)/k(v)$. Hence L/K is unramified in every w_j proving the claim. □

Proposition B.2.3. *Let E/L be a finite-dimensional field extension and let u be a place of E with $u|w$. Then E/K is unramified in u if and only if E/L is unramified in u and L/K is unramified in w.*

Proof: By [**49**], Ch.V, §7, no.5, Prop.9, the residue field extension $k(u)/k(v)$ is separable if and only if $k(u)/k(w)$ and $k(w)/k(v)$ are separable. The same transitivity holds for $K \subset L \subset E$. By $f_{u/v} = f_{u/w} f_{w/v}$ and $[E_u : K_v] = [E_u : L_w][L_w : K_v]$, we get the claim. □

Proposition B.2.4. *Let L'/K be any field extension. We suppose that L, L' are contained in a field E to define the composite LL' inside E. Let u be a place of LL' with $u|w$. If L/K is unramified in w, then LL'/L' is unramified in u.*

Proof: Since LL'/L is generated by separable elements (from L), the extension is separable. Note that the completion $(LL')_u$ is the composite of L_w and L'. This follows since L_w is the closure of L in $(LL')_u$ and L_wL' is a finite-dimensional field extension of the complete field L_w, hence L_wL' is also complete (Proposition 1.2.7). For the restriction w' of u to L', we get

$$(LL')_u = L_wL' = L_wL'_{w'}.$$

Because of the invariance of the residue fields under completions, we may assume that L and L' are complete with respect to w and w', respectively. Since $k(w)/k(v)$ is separable, the theorem of the primitive element (see [**156**] §4.14) gives $k(w) = k(v)(\bar{\alpha})$ for a suitable α in the valuation ring R_w of w. As a consequence of unramifiedness and completeness, we have $[L : K] = [k(w) : k(v)]$, hence α is a primitive element of L/K. Let f be the minimal polynomial of α over K, then our above considerations also show that we may assume f to be a monic polynomial with coefficients in R_v and that the reduction $\bar{f}$ is the minimal polynomial of $\bar{\alpha}$ over $k(v)$. Obviously, α is a primitive element of LL'/L'. By Remark B.1.5, the discriminant $D_{\bar{f}}$ of the separable polynomial $\bar{f}$ is not zero. Using $D_{\bar{f}} = \overline{D_f}$ (see B.1.12), we conclude that D_f is a unit in R_v. Finally, Lemma B.2.2 proves that LL'/L' is unramified in u. □

From the propositions in B.2.3 and B.2.4, we deduce:

Corollary B.2.5. *Let L'/K be a finite-dimensional field extension such that L and L' are contained in a field E. Let w' be a place of L' with $w'|v$ and let u be a place of the composite LL' inside E lying over w and w'. Then LL'/K is unramified in u if and only if L/K is unramified in w and L'/K is unramified in w'.*

In the next lemma, we consider v as a **valuation** of K, i.e. we fix an absolute value $|\ |_v$ in the non-archimedean place v and we set $v := -\log |\ |_v$ (by abuse of notation). The **value group** of v is defined by $v(K^\times)$, it is an additive subgroup of $\mathbb{R}$.

Lemma B.2.6. *Let $m \in \mathbb{N} \setminus \{0\}$ and let $\alpha^{1/m}$ be an mth root of $\alpha \in K^\times$. If $K(\alpha^{1/m})/K$ is unramified in a valuation w of L with $w|_K = v$, then m divides $v(\alpha)$ in the value group of v. Conversely, if the characteristic of the residue field $k(v)$ is prime to m and if $m|v(\alpha)$, then $K(\alpha^{1/m})/K$ is unramified over v.*

Proof: Suppose that $K(\alpha^{1/m})/K$ is unramified in w. Then $e_{w/v} = 1$ (see B.2.1) and hence

$$v(\alpha) = mw(\alpha^{1/m}) \in mv(K^\times).$$

Conversely, we assume that m and $\mathrm{char}(k(v))$ are coprime and that $m|v(\alpha)$. The latter implies that we may assume $v(\alpha) = 0$. Obviously, $\alpha^{1/m}$ is a root of $f(t) := t^m - \alpha$. Using B.1.9, we easily see that

$$D_f = (-1)^{\frac{m(m-1)}{2}} m^m \alpha^{m-1}.$$

By assumption, this is a unit in the valuation ring R_v. Then Lemma B.2.2 proves the claim. □

Example B.2.7. Let $K := \mathbb{Q}_2(t)$ be the quotient field of the polynomial ring over the field $\mathbb{Q}_2$ of 2-adic numbers. On K, we have the discrete valuation v induced by

$$\left|\sum_i a_i t^i\right|_v := \max_i |a_i|_2 ,$$

which we considered in Definition 1.6.1. Clearly, the same construction gives a discrete valuation w on $K(\sqrt{t})$ extending v. Obviously, we have $e_{w/v} = 1$ and the residue field $k(w) = \mathbb{F}_2(\sqrt{t})$ is a non-separable extension of $k(v) = \mathbb{F}_2(t)$ of degree 2. Since $f_{w/v} = 2 = [K(\sqrt{t}) : K]$, we conclude that w is the unique valuation over v. Although $2|v(t) = 0$, the extension $K(\sqrt{t})/K$ is ramified over v.

Definition B.2.8. Let E/K be an algebraic field extension which is not necessarily finite-dimensional and let v be a non-archimedean place of K. By Corollary B.2.5, the union of all intermediate fields L of E/K with $[L : K] < \infty$ and with L/K unramified over v is still a subfield of E denoted by K_v^{nr}. It follows from Corollary B.2.5 that a finite-dimensional subextension L/K of E/K is unramified over v if and only if L is contained in K_v^{nr}.

B.2.9. Let R be a Dedekind domain with field of fractions $K \neq R$ and let L/K be a finite-dimensional separable field extension. Then the integral closure $\bar{R}^L$ of R in L is also a Dedekind domain (see B.1.14). By Theorem A.8.5, the localization $R_\mathfrak{p}$ in a maximal ideal $\mathfrak{p}$ of R is a discrete valuation ring, inducing a place $v_\mathfrak{p}$ of K. We say that L/K is unramified in a maximal ideal $\mathfrak{P}$ of $\bar{R}^L$ if the extension is unramified in the corresponding place $v_\mathfrak{P}$ of L. A fundamental property of a Dedekind domain is that every non-zero ideal has a unique factorization into maximal ideals (see [**157**], Th.10.1). Let $\mathfrak{p} := \mathfrak{P} \cap R$ with factorization

$$\mathfrak{p}\bar{R}^L = \mathfrak{P}_1^{e_1} \dots \mathfrak{P}_r^{e_r}$$

into different maximal ideals $\mathfrak{P}_1, \dots, \mathfrak{P}_r$ of $\bar{R}^L$. Note that e_j is the ramification index of $v_{\mathfrak{P}_j}$ over $v_\mathfrak{p}$ and that every place of L over $v_\mathfrak{p}$ has this form. We conclude that L/K is unramified in $\mathfrak{P}$ if and only if $\mathfrak{P}$ occurs in the prime factorization with exact power 1 and if $\bar{R}^L/\mathfrak{P}$ is a separable field extension of $R/\mathfrak{p}$.

B.2.10. Under the assumptions of B.2.9, let $\mathfrak{P}$ be a non-zero prime ideal of $\bar{R}^L$ and let $\mathfrak{p} := \mathfrak{P} \cap R$. We assume that $\bar{R}^L/\mathfrak{P}$ is a separable field extension of $R/\mathfrak{p}$. Then $v_\mathfrak{P}$ is said to be **wildly ramified** over $v_\mathfrak{p}$ if and only if $\mathfrak{p}|e_{v_\mathfrak{P}/v_\mathfrak{p}}$. Otherwise, $v_\mathfrak{P}$ is called **tamely ramified** over $v_\mathfrak{p}$. Note that unramified is considered as a special case of tame ramification.

By abuse of notation, we identify $v_\mathfrak{p}$ with the discrete valuation on K normalized by $v_\mathfrak{p}(K^\times) = \mathbb{Z}$ (see A.8.4). Similarly $v_\mathfrak{P}(L^\times) = \mathbb{Z}$ leading to $v_\mathfrak{P} = e_{v_\mathfrak{P}/v_\mathfrak{p}} v_\mathfrak{p}$ on K. Important in this context is **Dedekind's different theorem**:

Theorem B.2.11. *Under the assumptions of B.2.10, we have*

$$e_{v_\mathfrak{P}/v_\mathfrak{p}} - 1 \leq v_\mathfrak{P}(\mathfrak{D}_{L/K}) < e_{v_\mathfrak{P}/v_\mathfrak{p}} + v_\mathfrak{P}\left(e_{v_\mathfrak{P}/v_\mathfrak{p}}\right)$$

with equality on the left if and only if $v_{\mathfrak{P}}$ is tamely ramified over $v_{\mathfrak{p}}$.

For a proof, we refer to [**162**], Th.3.12.9 and Th.4.6.7. As an immediate consequence of B.1.17 and of $N_{L/K}(\mathfrak{P}) = \mathfrak{p}^{f_{v_{\mathfrak{P}}/v_{\mathfrak{p}}}}$ (see [**162**] §3.10), we obtain **Dedekind's discriminant theorem**:

Theorem B.2.12. *Under the assumptions of B.2.9, let $\mathfrak{p}$ be a non-zero prime ideal of R such that the residue extension of every prime ideal of $\bar{R}^L$ over $\mathfrak{p}$ is separable. Then we have*

$$v_{\mathfrak{p}}\left(\mathfrak{d}_{L/K}\right) = \sum_{\mathfrak{P}|\mathfrak{p}} \left(e_{v_{\mathfrak{P}}/v_{\mathfrak{p}}}(1+\delta_{\mathfrak{P}}) - 1\right) f_{v_{\mathfrak{P}}/v_{\mathfrak{p}}},$$

where $\delta_{\mathfrak{P}} \in \{1, \ldots, v_{\mathfrak{p}}(e_{v_{\mathfrak{P}}/v_{\mathfrak{p}}})\}$ if $v_{\mathfrak{P}}$ is wildly ramified over $v_{\mathfrak{p}}$ and $\delta_{\mathfrak{P}} = 0$ if $v_{\mathfrak{P}}$ is tamely ramified over $v_{\mathfrak{p}}$.

B.2.13. As a corollary of B.2.11 (resp. B.2.12), we obtain that L/K is unramified in $v_{\mathfrak{P}}$ (resp. over $v_{\mathfrak{p}}$) if and only if $\mathfrak{P}$ (resp. $\mathfrak{p}$) is not a divisor of the different $\mathfrak{D}_{L/K}$ (resp. of the discriminant $\mathfrak{d}_{L/K}$). This holds even without assuming *a priori* that the residue field $\bar{R}^L/\mathfrak{P}$ is separable over $R/\mathfrak{p}$ (see [**276**], Ch.III, §5, Th.1).

For a lot of results in diophantine geometry, **Hermite's discriminant theorem** is crucial:

Theorem B.2.14. *For any $D > 0$, there are only finitely many number fields $K \subset \overline{\mathbb{Q}}$ with discriminant $D_{K/\mathbb{Q}}$ bounded in absolute value by D.*

Proof: We give the proof under the assumption that the degree of the number field is also bounded. In all our applications of Hermite's discriminant theorem, this is *a priori* clear. In fact, it follows from Minkowski's discriminant theorem (see Theorem B.2.16) that the degree is bounded in terms of the discriminant.

By Corollary C.2.7, O_K induces a lattice in $K_\infty := \prod_{v|\infty} K_v$. We assume first that K has a place w with $K_w = \mathbb{R}$. For a given constant $C > 0$, we consider the open convex subset

$$S_\infty := \{\alpha \in K_\infty \mid |\alpha_w|_w < |C|_w,\ |\alpha_v|_v < |\tfrac{1}{2}|_v\ \forall v|\infty, v \neq w\}.$$

By the proof of Proposition C.1.10, the volume of a fundamental domain of the lattice with respect to the normalized Haar measure on K_∞ introduced in C.1.9 is equal to $|D_{K/\mathbb{Q}}|^{1/2}$. Choosing C sufficiently large depending on $D_{K/\mathbb{Q}}$, Minkowski's first theorem from C.2.19 gives a non-zero $\alpha \in O_K$ with corresponding lattice point in S_∞. By the product formula, we have $|\alpha|_w \geq 1$ and hence one (real) conjugate of α is different from all other conjugates (with respect to the embeddings of K into $\mathbb{C}$). A given embedding $\mathbb{Q}(\alpha) \to \mathbb{C}$ has exactly $[K : \mathbb{Q}(\alpha)]$ extensions to K, hence we get $K = \mathbb{Q}(\alpha)$. Since $h(\alpha)$ is bounded in terms of C, Northcott's theorem (see Theorem 1.6.8) shows that only finitely many α may occur proving the claim.

If K has only complex archimedean places, we fix a place w and we consider the open convex subset

$$S_\infty := \{\alpha \in K_\infty \mid |\Re(\alpha_w)|_w < |\tfrac{1}{2}|_w, \, |\Im(\alpha_w)|_w < |C|_w, \, |\alpha_v|_v < |\tfrac{1}{2}|_v \; \forall v|\infty, v \neq w\}.$$

Then the same argument as above yields the claim. □

Corollary B.2.15. *Let K be a number field and let S be a finite set of places of K containing all archimedean ones. Then there are only finitely many number fields L in $\overline{K}$ of bounded degree which are unramified outside of S.*

Proof: The transitivity rule of discriminants (see B.1.19) implies

$$\mathfrak{d}_{L/\mathbb{Q}} = N_{K/\mathbb{Q}}(\mathfrak{d}_{L/K}) \cdot \mathfrak{d}_{K/\mathbb{Q}}^{[L:K]}.$$

By Dedekind's discriminant theorem (see Theorem B.2.12), the norm is bounded. We conclude that $D_{L/\mathbb{Q}}$ is also bounded and hence the above theorem proves the claim. □

For completeness, we state **Minkowski's discriminant theorem:**

Theorem B.2.16. *Let K be a number field of degree d with exactly s complex places. Then*

$$|D_{K/\mathbb{Q}}| \geq \left(\frac{\pi}{4}\right)^{2s} \cdot \frac{d^{2d}}{(d!)^2}.$$

The proof is another application of Minkowski's first theorem. For details, we refer to [**162**], Th.2.13.5.

Remark B.2.17. Note that the function $f(d) := \left(\frac{\pi}{4}\right)^d \frac{d^d}{d!}$ satisfies

$$\frac{f(d+1)}{f(d)} = \frac{\pi}{4}\left(1 + \frac{1}{d}\right)^d \geq \frac{\pi}{2} > 1,$$

hence Minkowski's discriminant theorem shows that $|D_{K/\mathbb{Q}}| > 1$ for $K \neq \mathbb{Q}$ and that d is bounded in terms of the discriminant.

B.2.18. Let L/K be a finite-dimensional Galois extension and let w be a place of L lying over the complete discrete valuation v of K. We assume also that the residue field extension $k(w)/k(v)$ is separable. We will find a maximal unramified subextension L^I/K such that L/L^I is **totally ramified**, meaning that the ramification index is equal to the degree.

By Proposition 1.2.7 and Corollary 1.3.5, the Galois group $D = \mathrm{Gal}(L/K)$ operates isometrically on L with respect to $| \;|_w$, hence reduction modulo w gives a homomorphism $\varepsilon : D \to \mathrm{Gal}(k(w)/k(v))$. The kernel of ε is called the **inertia group** and is denoted by I. From Hensel's lemma in 1.2.10, Gauss's lemma in 1.6.3 and the primitive element theorem ([**156**], §4.14), we deduce:

(a) The extension $k(w)/k(v)$ is Galois and ε is surjective.

For the fixed field L^I of I, Galois theory and (a) yield

$$\mathrm{Gal}(L^I/K) \xrightarrow{\sim} \mathrm{Gal}(k(w)/k(v))$$

and hence the theory of finite fields (see [**156**], §4.13) proves:

(b) $\mathrm{Gal}(L^I/K)$ is a cyclic group of order equal to the residue degree $f_{w/v}$.

Let w_I be the restriction of w to L^I. Applying (a) with L^I instead of K and using $I = \mathrm{Gal}(L/L^I)$, we get:

(c) $k(w_I) = k(w)$ and hence $f_{w_I/v} = f_{w/v}, f_{w/w_I} = 1$.

By Galois theory and (b), we have $[D : I] = [L^I : K] = f_{w/v}$. By Proposition 1.2.11 and (c), we deduce:

(d) $e_{w/v} = [L : L^I] = |I| = e_{w/w_I}$ and $e_{w_I/v} = 1$.

Since the compositum of unramified extensions remains unramified (Corollary B.2.5) and since the ramification index is bounded by the degree (see Proposition 1.2.11), it is clear from (d) that L^I/K is the maximal unramified subextension of L/K. If the residue characteristic is not a divisor of $e_{w/v}$, then w is called tamely ramified over v (see B.2.10). The following result may be deduced from (d) and [**276**], Ch.IV, §2, Cor.1-3.

(e) If w is tamely ramified over v, then I is a cyclic group.

B.2.19. To apply B.2.18 in the situation of a Galois extension L/K of number fields with non-archimedean places $w|v$, we make the following observations: By Corollary 1.3.5 the Galois group $G := \mathrm{Gal}(L/K)$ operates transitively on the set of places lying over v. We call $D_w := \{\sigma \in G \mid \sigma w = w\}$ the **decomposition group of w over v**. It is immediate that decomposition groups of places over v are conjugate subgroups in G.

Since the elements of D_w are isometric with respect to $|\ |_w$ and since L/K is generated by the zeros of a polynomial with coefficients in K, we deduce easily that D_w is isomorphic to the Galois group of the completions L_w/K_v, which is clearly a Galois extension. Using this natural isomorphism for identification, the inertia subgroup of L_w/K_v from B.2.18 may be seen as a subgroup of G, which we call the **inertia group of w over v**.

Let H be any subgroup of D_w with fixed field L^H. Let w_H be the restriction of w to L^H. To apply the results of B.2.18, it is useful to note that forming the completion is compatible with forming fixed fields, i.e. $(L^H)_{w_H} = (L_w)^H$.

It is clear that $(L^H)_{w_H} \subset (L_w)^H$. On the other hand, $\alpha \in (L_w)^H$ may be approximated by a sequence $\alpha_n \in L$. By continuity of the elements in D_w, we may replace α_n by

$$\frac{1}{|H|} \sum_{\sigma \in H} \sigma(\alpha_n) \in L^H,$$

proving the claim.

B.3. Unramified morphisms

In this section, we assume that the reader is familiar with the language of schemes. It will be of minor importance in the book and serves mainly to connect the results about unramified morphisms from A.12 with Appendix B. We consider a morphism $\varphi : X \to X'$ of finite type between noetherian schemes.

B.3.1. The morphism φ is called **unramified** in $x \in X$ if the following conditions are satisfied:

(a) The maximal ideal $\mathfrak{m}_x$ of $\mathcal{O}_{X,x}$ is generated by $\varphi^\sharp(\mathfrak{m}_{\varphi(x)})$.

(b) $\varphi^\sharp$ induces a separable extension $k(x)/k(\varphi(x))$ of residue fields.

If φ is unramified in all points of X, then φ is called an **unramified morphism.**

B.3.2. In the same way, we translate the definition of a **flat morphism** (see A.12.11) to the case of schemes. Moreover, φ is called **étale** if φ is flat and unramified. The set of unramified (resp. flat, resp. étale) points of X' with respect to φ is an open subset of Y (see **[137]**, Th.17.4.1, Th.11.3.1, Th.17.6.1). This makes it clear that a morphism of varieties over a field is unramified (resp. flat, resp. étale) if and only if the corresponding property holds for the associated morphism of schemes.

Remark B.3.3. In the situation of B.2.9, L/K is unramified in the maximal ideal $\mathfrak{P}$ (i.e. in $v_{\mathfrak{P}}$) if and only if the morphism $\mathrm{Spec}(\bar{R}^L) \to \mathrm{Spec}(R)$ is unramified in $\mathfrak{P}$. In fact, the assumption L/K separable was only used to ensure that $\bar{R}^L$ is a finitely generated R-module. This is also satisfied if R is a complete discrete valuation ring (see **[276]**, Ch.II, §2, Prop.3) or the coordinate ring of an affine irreducible regular curve (see A.12.6).

Example B.3.4. Let $X' = \mathrm{Spec}(A)$ be an affine (noetherian) scheme and let $X = \mathrm{Spec}(B)$, where $B = A[t]/[f(t)]$ for a monic polynomial $f(t) \in A[t]$. There is a unique morphism $\varphi : X \to X'$ with $\varphi^\sharp$ equal to the canonical homomorphism $A \to B$. Then Example A.12.15 still holds in this context leading to the same definition of **standard étale morphisms** as in A.12.16.

Every étale morphism is again locally of standard étale type and every unramified morphism is locally the composition of a closed embedding with a standard étale morphism.

B.3.5. Let X, X' be schemes of finite type over the discrete valuation ring R_v and let $\varphi : X \to X'$ be a morphism over R_v. Let $Q \in X'(K)$ and let $P \in X(\overline{K})$ with $\varphi(P) = Q$. We consider a place w of $K(P)$ and we assume that P extends to an R_w-integral point $\overline{P}$ of X. This means an R_v-morphism $\overline{P} : \mathrm{Spec}(R_w) \to X$ with $P = \overline{P}(\{0\})$. Using $R_w \cap K = R_v$, we conclude that Q extends to an R_v-valued point $\overline{Q}$ of X'. We denote the image of the maximal ideal of R_w (resp. R_v) by $P(w)$ (resp. $P(v)$).

Proposition B.3.6. *If φ is unramified in $P(w)$, then $K(P)/K$ is unramified in w.*

Proof: Since the set of unramified points of X with respect to φ is open (see B.3.2), we conclude that P is also an unramified point and hence $K(P)/K$ is separable. Using the local nature of unramified morphisms, we may assume that φ is a standard étale morphism (see Example B.3.4).

Now we proceed as in the proof of Proposition 10.3.3 using the same notation. Since $\overline{Q}$ is an R_v-integral point of X', we have $a_i(Q) \in R_v$ and hence we may choose $a = 1$. Similarly, $f'(P)$ is a unit in R_w. So we may assume $\alpha = 1$. This proves that the discriminant of the completions $K(P)_w/K_v$ is equal to the valuation ring $\hat{R}_v$ of K_v. Using base change to $\hat{R}_v$, the argument at the beginning of the proof shows that the extension $K(P)_w = K_v(P)/K_v$ is separable. Now B.2.13 implies that $K(P)_w/K_v$ is unramified over v. Since residue fields do not change by passing to completions (see Proposition 1.2.11), it follows that $K(P)/K$ is also unramified. □

B.4. The ramification divisor

In this section, k denotes a field of characteristic 0. Let B, Y be irreducible k-varieties regular of codimension 1 and let $p : Y \to B$ be a finite morphism over k. The goal is to study the divisor measuring the ramification of p. The reader is assumed to have some familiarity with function fields (see Section 1.4).

B.4.1. We have seen in A.8.11 that $\mathcal{O}_{Y,w}$ is a discrete valuation ring for every prime divisor w of Y. In particular, we may consider w as a place of $k(Y)$ and $v = p(w)$ as a place of $k(B)$ with $w|v$ (see Example 1.4.13).

B.4.2. The canonical line bundle K_Y is well defined on the smooth part Y_{reg} of Y by $K_Y := \Lambda^{\dim(Y)} T_Y^*$. By passing to closures of associated Weil divisors and using that $Y \setminus Y_{\mathrm{reg}}$ has codimension at least 2, we get a **canonical divisor** $\mathrm{div}(K_Y)$ well defined in $CH^1(Y)$.

Proposition B.4.3. On $V := Y_{\mathrm{reg}} \cap p^{-1}(B_{\mathrm{reg}})$, the pull-back induces an injective homomorphism $\nu : p^*\Omega^1_B|_V \to \Omega^1_Y|_V$ of locally free sheaves of rank $\dim(B)$.

Proof: By A.7.23, $p^*\Omega^1_B|_V$ and $\Omega^1_Y|_V$ are locally free of rank $\dim(Y)$. Anyway, we have an exact sequence

$$p^*\Omega^1_B|_V \xrightarrow{\nu} \Omega^1_Y|_V \longrightarrow \Omega^1_{Y/B}|_V \longrightarrow 0 \tag{B.4}$$

of coherent sheaves (see A.7.29). Let ξ be the generic point of Y. Then $\Omega^1_{Y/B,\xi} = \Omega^1_{k(Y)/k(B)} = 0$ by separability of $k(Y)/k(B)$ ([148], Th.8.6A). Hence $\Omega^1_{Y/B}$ is a **torsion sheaf**, i.e. it is supported in a closed subset $R \neq Y$. The image of ν is a locally free subsheaf of $\Omega^1_Y|_V$ of rank r'. The sheaves are equal outside of R, hence $r = r'$ proving immediately the claim. □

B.4.4. We have seen in the proof above that $\Omega^1_{Y/B}$ is a torsion sheaf, hence the stalk $\Omega^1_{Y/B,v}$ is of finite length for a prime divisor v of Y (following also from the proof below). So we may define the **ramification divisor** of p by

$$R_p := \sum_v \ell(\Omega^1_{Y/B,v})v,$$

where ℓ is the length. It is supported in the set R considered above.

Theorem B.4.5. *We have*

$$\mathrm{div}(K_Y) \sim p^*\mathrm{div}(K_B) + R_p.$$

Proof: Let v be a prime divisor on Y. Since B, Y are regular in codimension 1 and p is finite, the dimension theorem (see A.12.1) shows that the complement of the set V from Proposition B.4.3 has codimension at least 2 in Y. Considering the stalks of (B.4) in v, we get an exact sequence

$$0 \longrightarrow \left(p^*\Omega^1_B\right)_v \xrightarrow{\nu_v} \Omega^1_{Y,v} \longrightarrow \Omega^1_{Y/B,v} \longrightarrow 0$$

of modules over the discrete valuation ring $\mathcal{O}_{Y,v}$. By the theorem of elementary divisors ([**49**], Ch.VII, §4, no.3, Th.1), we deduce easily

$$\ell(\Omega^1_{Y/B,v}) = \mathrm{ord}_v(\det(\nu_v)). \tag{B.5}$$

Let $W \subset V$ be a trivalization of $p^*\Omega^1_B$ and Ω^1_Y. With respect to the trivialization, $\gamma_W := \det(\nu)$ is a regular function on W. Clearly, (γ_W) is a well-defined Cartier divisor on V. By (B.5), its associated Weil divisor is the ramification divisor R_p. On the other hand, we may consider $\det(\nu)$ as an injective homomorphism

$$\det(\nu) : p^*K_B|_V \longrightarrow K_Y|_V$$

induced by pull-back. For an invertible meromorphic section s of $K_{B_{\mathrm{reg}}}$, we get an invertible meromorphic section $s' := \det(\nu) \circ p^*(s)$ of $K_Y|_V$. Working locally with respect to the trivializations, it is easy to check that

$$\mathrm{div}(s') = p^*(\mathrm{div}(s)) + R_p|_V$$

on V. This proves the claim. □

As a corollary, we immediately obtain **Hurwitz's theorem.**

Theorem B.4.6. *Let $\varphi : C' \to C$ be a surjective morphism of irreducible smooth projective curves over k of genus $g(C')$ and $g(C)$. Then we have*

$$2g(C') - 2 = \deg(\varphi)(2g(C) - 2) + \deg(R_\varphi).$$

Proof: By A.6.15 and A.12.4, the morphism is finite. For any divisor D on C, we have seen in Example 1.4.12 that $\deg(\varphi^*D) = \deg(\varphi)\deg(D)$. We apply Theorem B.4.5 and A.13.6 to get the claim. □

For an effective divisor D, the sum of its components is denoted by D_{red}. The divisor is called **reduced** if $D = D_{\mathrm{red}}$.

Proposition B.4.7. *Let $\phi : C' \to C$ be a surjective morphism of irreducible smooth projective curves over k and let D be an effective reduced divisor on C. Then we have the inequality*

$$\phi^*(D) - \phi^*(D)_{\mathrm{red}} \leq R_\phi$$

of divisors on C' with equality if and only if ϕ is unramified outside of $\phi^{-1}(D)$.

Proof: We check the identity over an irreducible component v of $\mathrm{supp}(D)$. Since D is reduced, it is given by a local parameter π_v in v. Let $\pi_{v'}$ be a local parameter in $v' \in \phi^{-1}(v)$. Then we have

$$\pi_v = \pi_{v'}^e u' \in \mathcal{O}_{C',v'}, \quad u' \in \mathcal{O}_{C',v'}^\times,$$

where e is the multiplicity of $\phi^*(D)$ in v'. We claim that $\Omega^1_{C,v}$ is generated by $\mathrm{d}\pi_v$ and similarly $\Omega^1_{C',v'}$ is generated by $\mathrm{d}\pi_{v'}$. If v is k-rational, we argue as follows. By A.7.8, we may identify the fibre of Ω^1_C over v with $\mathfrak{m}_v/\mathfrak{m}_v^2$. As π_v generates the latter and corresponds to $\mathrm{d}\pi_v$ in the former, we conclude that $\mathrm{d}\pi_v$ generates also the stalk $\Omega^1_{C,v}$. In general, we may use base change to $\overline{k}$. Using that v is geometrically reduced (see A.4.6), π_v is a local parameter in all points of $X_{\overline{k}}$ lying in v. The special case considered above and the compatibility of Ω^1_C with base change (use A.7.6) show that $\mathrm{d}\pi_v$ is a basis of $\Omega^1_{C_{\overline{k}},x}$ for all $x \in v$. The stalk $\Omega^1_{C,v}$ is free of rank 1 over the discrete valuation ring $\mathcal{O}_{C,v}$ proving that $\mathrm{d}\pi_v$ is a basis of $\Omega^1_{C,v}$.

Returning to the proof of our original claim, Leibniz's rule gives

$$\mathrm{d}\pi_v = e\pi_{v'}^{e-1}u'\mathrm{d}\pi_{v'} + \pi_{v'}^e \mathrm{d}u'.$$

By the short exact sequence (B.4) on page 599 and using $\mathrm{char}(k) = 0$, we conclude that $\Omega^1_{C'/C,v}$ has length $e-1$ proving immediately the claim. □

Example B.4.8. If $k = \mathbb{C}$ in Proposition B.4.7, then we may consider the corresponding analytic map $\phi_{\mathrm{an}} : C'_{\mathrm{an}} \to C_{\mathrm{an}}$ of compact Riemann surfaces. Note that the local parameters $\pi_{v'}$ in $v' \in C'$ and π_v in $v := \phi(v') \in C$ correspond to analytic coordinates. Hence the above proof shows that R_ϕ has multiplicity $e-1$ in v' where e is the order of zeros of the holomorphic function $\pi_v \circ \phi \circ \pi_{v'}^{-1}(z)$ in $z = 0$.

Proposition B.4.9. *Let w be a prime divisor of Y with ramification index $e_{w/v}$ over $v := p(w)$ (see Example 1.4.13). Then the multiplicity of the ramification divisor R_p in w is equal to $e_{w/v} - 1$.*

Proof: We have $\Omega^1_{Y/B,w} = \Omega^1_{\mathcal{O}_{Y,w}/\mathcal{O}_{B,v}}$. By B.1.18, this module over the discrete valuation ring $\mathcal{O}_{Y,w}$ is generated by one element and its annihilator is the different of $\mathcal{O}_{Y,w}/\mathcal{O}_{B,v}$. We conclude that $\ell(\Omega^1_{Y/B,w})$ is the order of the different in the place w. By Dedekind's different theorem (see B.2.11), we get the claim. □

APPENDIX C GEOMETRY OF NUMBERS

C.1. Adeles

We first recall the existence and uniqueness of a Haar measure on a locally compact group. In the special case of a completion of a number field K of degree d, we prove this by an explicit construction. Afterwards we introduce adeles, which are an important tool in number theory, following A. Weil [**328**]. Section C.2, with McFeat's version [**198**] of Minkowski's second theorem in an adelic setting, is modeled after the exposition of Bombieri and Vaaler [**35**]. Section C.3 presents J.D. Vaaler's cube slicing inequality [**302**] using also techniques from A. Prékopa [**236**].

Theorem C.1.1. *Let G be a locally compact group. There is a non-zero positive left-invariant Borel measure μ_G on G, i.e.*

$$\int_G f(yx)\,\mathrm{d}\mu_G(x) = \int_G f(x)\,\mathrm{d}\mu_G(x)$$

for all $y \in G$ and all continuous complex functions f with compact support. The measure μ_G is uniquely determined up to positive multiples and is called a ***Haar measure*** *on G.*

We refer to N. Bourbaki [**46**], Ch.7, §1, no.2, Th.1 for a proof. Now let H be a normal closed subgroup and let $\pi : G \longrightarrow G/H$ be the quotient morphism. By definition, a subset U of G/H is open if and only if $\pi^{-1}(U)$ is open in G. Note that π is an open map and hence G/H is a locally compact group.

Corollary C.1.2. *Given Haar measures μ_G, μ_H on G and H, there is a unique Haar measure $\mu_{G/H}$ on G/H such that*

$$\int_G f(x)\,\mathrm{d}\mu_G(x) = \int_{G/H} \mathrm{d}\mu_{G/H}(\pi(x)) \int_H f(xy)\,\mathrm{d}\mu_H(y)$$

for all continuous complex functions f with compact support. Moreover, this formula continues to hold for any $f \in L^1(G, \mu_G)$.

Proof: We prove easily that

$$G/H \longrightarrow \mathbb{C}, \quad Hx \mapsto \int_H f(xy)\mathrm{d}\mu_H(y)$$

is a continuous function with compact support. Thus

$$\int_{G/H} \mathrm{d}\mu_{G/H}(\pi(x)) \int_H f(xy)\,\mathrm{d}\mu_H(y)$$

is a positive linear functional on the space of continuous complex functions with compact support. By the Riesz representation theorem ([**249**], Th.2.14), there is a unique Borel measure μ on G such that

$$\int f(x)\,\mathrm{d}\mu(x) = \int_{G/H} \mathrm{d}\mu_{G/H}(\pi(x)) \int_H f(xy)\,\mathrm{d}\mu_H(x)$$

for every continuous complex function f on G with compact support. Clearly, μ is left invariant and hence it is equal to μ_G up to a positive multiple (Theorem C.1.1). By normalization, we get the first claim. In order not to go into too many details of measure theory, we only refer for the last claim to [**46**], Ch.7, §2, no.3, Prop.5. The argument proceeds by showing that, for $\mu_{G/H}$-almost every $\pi(x) \in G/H$, the function $y \mapsto f(xy)$ is μ_H-integrable on H, then that the function $x \mapsto \int_H f(xy)\,\mathrm{d}\mu_H(x)$ is $\mu_{G/H}$-integrable on G/H, and finally that the desired formula holds. □

Example C.1.3. If $v \in M_K$, the completion K_v is locally compact (see 1.2.12). We will construct a Haar measure μ_v on K_v with the property

$$\mu_v(\alpha\Omega) = \|\alpha\|_v\,\mu_v(\Omega) \tag{C.1}$$

for any $\alpha \in K_v$ and any Borel measurable subset Ω of K. For the normalization of the absolute value, we refer to 1.3.6.

If v is archimedean, then $K_v = \mathbb{R}$ or $\mathbb{C}$, the Lebesgue measure is a Haar measure satisfying (C.1), and there is nothing else to prove. Hence let v be a non-archimedean place with valuation ring R_v in K_v, residue field $k(v)$, and local parameter π (i.e. π is a generator of the maximal ideal in R_v). We denote by e_v, f_v the ramification index and the residue degree of v over $p := \mathrm{char}(k(v))$. We consider the closed balls

$$B_\varepsilon(x) := \{y \in K_v \mid \|y - x\|_v \leq \varepsilon\},$$

where the "centres" x range over K_v. Note that we need only consider balls of radius δ^n $(n \in \mathbb{Z})$, where $\delta := \|\pi\|_v$. By the ultrametric triangle inequality, two balls are either disjoint or one is contained in the other. Every open subset of K_v is a countable disjoint union of such closed balls. We claim that $B_1(0)$ is the disjoint union of p^{f_v} balls $B_\delta(x)$. This follows from $B_1(0) = R_v$ by noting that the balls $B_\delta(x)$ are the fibres with respect to reduction. Thus

$$\mu_v(B_{\delta^n}(x)) := p^{-nf_v}$$

is a σ-additive and translation invariant set function on these balls. Obviously, μ_v extends uniquely to a function on the compact open subsets of K_v with the same properties.

By standard arguments of measure theory, μ_v extends uniquely to a translation invariant Borel measure. By Proposition 1.2.11, we have

$$\delta = \|\pi\|_v = \|p\|_v^{1/e_v} = p^{-f_v}.$$

This proves (C.1) first for $\Omega = B_{\delta^n}(0)$. By translation invariance, we get (C.1) for any closed ball and by uniqueness of the extension we get it for all Borel measurable subsets Ω of K_v.

Definition C.1.4. *The* ***adele ring*** *of K is the subring*

$$K_{\mathbb{A}} := \left\{ \mathbf{x} \in \prod_{v \in M_K} K_v \mid x_v \in R_v \text{ up to finitely many } v \right\}$$

of the additive group $\prod_{v \in M_K} K_v$.

Remark C.1.5. We never use the topology on $K_{\mathbb{A}}$, which is induced by the product topology because it is not locally compact. However, for every finite $S \subset M_K$ containing all archimedean places, the product topology makes

$$H_S := \prod_{v \in S} K_v \times \prod_{v \notin S} R_v$$

into a locally compact topological group. Then there is a unique structure on $K_{\mathbb{A}}$ as a topological group such that the groups H_S are open topological subgroups of $K_{\mathbb{A}}$. In fact, $K_{\mathbb{A}}$ is a locally compact topological ring.

We identify K with a subgroup of $K_{\mathbb{A}}$ by means of the diagonal map

$$K \longrightarrow K_{\mathbb{A}}, \quad x \mapsto (x)_{v \in M_K}.$$

Theorem C.1.6. *The subgroup K is a discrete closed subgroup of $K_{\mathbb{A}}$ and $K_{\mathbb{A}}/K$ is compact.*

Proof: We first show that K is discrete in $K_{\mathbb{A}}$. It is enough to prove that 0 is an isolated point. We choose $w \in M_K$ and we consider the neighbourhood

$$U := \{x \in K_w \mid |x|_w < 1\} \times \prod_{v \neq w} \{x \in K_v \mid |x|_v \leq 1\}$$

of 0 in $K_{\mathbb{A}}$. By the product formula, 0 is the only point of $K \cap U$.

Every discrete subgroup is closed. To see it, we choose a neighbourhood V of 0 in $K_{\mathbb{A}}$ such that $V - V \subset U$. Then it is clear that, for every $\mathbf{x} \in K_{\mathbb{A}}$, there is at most one point in $K \cap (V + \mathbf{x})$, proving closedness.

The compactness of $K_{\mathbb{A}}/K$ is an immediate consequence of the next lemma and Tychonov's theorem. □

Remark C.1.7. Let $\omega_1, \ldots, \omega_d$ be a basis of O_K over $\mathbb{Z}$. By equation (1.1) on page 8, there is a canonical isomorphism

$$K \otimes_{\mathbb{Q}} \mathbb{R} \cong \prod_{v|\infty} K_v. \tag{C.2}$$

For simplicity, we may identify both spaces. Let

$$\Omega_\infty := \left\{ \mathbf{x} \in \prod_{v|\infty} K_v \mid \exists a_j \in [0,1),\ \mathbf{x} = \sum_{j=1}^{d} \omega_j \otimes a_j \right\}.$$

If K is a number field, it is customary to call the non-archimedean places also **finite places**, while the archimedean places are called **infinite places.**

Lemma C.1.8. *The subset $\Omega := \Omega_\infty \times \prod_{v \text{ finite}} R_v$ of $K_{\mathbb{A}}$ is a fundamental domain of $K_{\mathbb{A}}/K$, i.e. every class in $K_{\mathbb{A}}/K$ has exactly one representative in Ω.*

Proof: First, we show uniqueness. Let $\mathbf{x}, \mathbf{x}' \in \Omega_\infty$ with $\alpha = \mathbf{x} - \mathbf{x}' \in K$ and with projections $\sum_{j=1}^d \omega_j \otimes a_j, \sum_{j=1}^d \omega_j \otimes a_j'$ to Ω_∞. For every finite place v, we conclude that $\alpha \in R_v$ and hence $\alpha \in O_K$. Now looking at the infinite places we see that

$$\alpha = \sum_{j=1}^{d} \omega_j \otimes (a_j - a_j')$$

with $a_j - a_j' \in (-1,1)$. Then $\alpha \in O_K$ yields $a_j = a_j'$ and hence $\alpha = 0$.

Now we show that any class $K + \mathbf{x} \in K_{\mathbb{A}}/K$ has a representative in Ω. Let S be the set of non-archimedean places with $|x_v|_v > 1$. By the strong approximation theorem (see Theorem 1.4.5) there is $\alpha \in K$ such that $|x_v - \alpha|_v < 1$ for all $v \in S$ and $|\alpha|_v \leq 1$ for all non-archimedean places $v \notin S$. Therefore, replacing $\mathbf{x}$ by $\mathbf{x} - \alpha$, we may assume that $\mathbf{x}_v \in R_v$ for all finite v. We have

$$(x_v)_{v|\infty} = \sum_{j=1}^{n} \omega_j \otimes a_j$$

for some $a_j \in \mathbb{R}$. There are $b_j \in \mathbb{Z}$ such that $0 \leq a_j - b_j < 1$ for $j = 1, \ldots, d$. Then $\mathbf{x} - \sum_{j=1}^d \omega_j \otimes b_j$ is the desired representative in Ω. □

C.1.9. Now we fix the Haar measures β_v on K_v and on the adeles in the following way:

(a) If v is not archimedean, then we normalize the Haar measure by

$$\beta_v(R_v) = \left| D_{K_v/\mathbb{Q}_p} \right|_p^{1/2},$$

where $D_{K_v/\mathbb{Q}_p}$ is the discriminant of $v|p \in M_{\mathbb{Q}}$.

(b) If $K_v = \mathbb{R}$, then β_v is the ordinary Lebesgue measure.

(c) If $K_v = \mathbb{C}$, then β_v is twice the ordinary Lebesgue measure.

For a finite subset S of M_K containing all the archimedean primes, the product measure

$$\beta_S := \prod_{v \in S} \beta_v \times \prod_{v \notin S} \beta_v|_{R_v}$$

is a Haar measure on the open topological subgroup H_S of $K_{\mathbb{A}}$ introduced in Remark C.1.5. The measures β_S fit together to give a Haar measure β on $K_{\mathbb{A}}$. Clearly, the counting measure is a Haar measure on the discrete subgroup K. By Corollary C.1.2, we get a uniquely determined Haar measure $\beta_{K_{\mathbb{A}}/K}$ on $K_{\mathbb{A}}/K$.

Proposition C.1.10. *The volume of* $K_{\mathbb{A}}/K$ *with respect to the Haar measure* $\beta_{K_{\mathbb{A}}/K}$ *is* 1.

Proof: The fundamental domain Ω is measurable. Since it is contained in a compact subset, Ω has finite measure. By Corollary C.1.2, we have

$$\beta(\Omega) = \beta_{K_{\mathbb{A}}/K}(K_{\mathbb{A}}/K).$$

By definition, we have

$$\beta(\Omega) = \left(\prod_{v|\infty} \beta_v \right) (\Omega_\infty) \prod_{v \text{ finite}} \left| D_{K_v/\mathbb{Q}_p} \right|_p^{1/2}. \tag{C.3}$$

From B.1.20 and B.1.21, we deduce

$$\left| D_{K/\mathbb{Q}} \right|_p = \prod_{v|p} \left| D_{K_v/\mathbb{Q}_p} \right|_p$$

and hence the product formula gives

$$|D_{K/\mathbb{Q}}|^{-1} = \prod_{v \text{ finite}} \left| D_{K_v/\mathbb{Q}_p} \right|_p. \tag{C.4}$$

Let $d = [K : \mathbb{Q}]$ be the degree of the number field K. We consider

$$K \otimes_{\mathbb{Q}} \mathbb{R} = \prod_{v|\infty} K_v = \mathbb{R}^d$$

as a d-dimensional real vector space. By the identification of $K_v = \mathbb{R}$ or $K = \mathbb{C}$, we get an embedding σ_v of K into $\mathbb{R}$ or $\mathbb{C}$. Let $\sigma_1, \dots, \sigma_r$ be the real embeddings and $\sigma_{r+1}, \dots, \sigma_{r+s}$ be the chosen complex embeddings. Then $\alpha \in K$ gives rise to the vector

$$\mathbf{v}(\alpha) := (\sigma_1(\alpha), \dots, \sigma_r(\alpha), \Re\sigma_{r+1}(\alpha), \Im\sigma_{r+1}(\alpha), \dots, \Re\sigma_{r+s}(\alpha), \Im\sigma_{r+s}(\alpha)).$$

The volume of Ω_∞ with respect to the Lebesgue measure on $\mathbb{R}^d$ is

$$|\det(\mathbf{v}(\omega_1), \dots, \mathbf{v}(\omega_n))| = 2^{-s} |\det(\sigma_i(\omega_j))|,$$

where on the right-hand side we have the determinant of a $d \times d$ matrix by using all complex embeddings, namely $\sigma_{r+s+1} := \overline{\sigma}_{r+1}, \dots, \sigma_n := \overline{\sigma}_{r+s}$. By Remark B.1.15, we conclude that

$$\left(\prod_{v|\infty} \beta_v \right) (\Omega_\infty) = |D_{K/\mathbb{Q}}|^{1/2}.$$

Now (C.3) and (C.4) show that $\beta(\Omega) = 1$, proving the claim. □

C.1.11. To motivate our normalizations of the Haar measures in C.1.9, we have to use duality and Fourier theory. The following considerations are not used in the sequel. For proofs, we refer to N. Bourbaki **[48]**, E. Hewitt and K.A. Ross **[150]**, **[151]**, W. Rudin **[253]**, and A. Weil **[326]**.

Let G be a locally compact abelian group. A **character** of G is a continuous homomorphism of G into $\mathbb{T} := \{z \in \mathbb{C} \mid |z| = 1\}$. Together with the compact-open topology (also called the topology of uniform convergence on compact sets), the set of characters $\widehat{G}$ is a locally compact abelian group. For $x \in G$ and $\gamma \in \widehat{G}$, let $\langle x, \gamma \rangle := \gamma(x)$. Then we get a perfect duality between G and $\widehat{G}$, i.e. the canonical homomorphism of G into the characters of $\widehat{G}$ is an isomorphism. For a continuous complex function f on G with compact support, the **Fourier transform** is the continuous complex function $\widehat{f}$ on $\widehat{G}$ given by

$$\widehat{f}(\gamma) := \int_G f(x)\overline{\langle x, \gamma \rangle}\, \mathrm{d}\mu(x),$$

where μ is a fixed Haar measure on G. By Plancherel's theorem, there is a unique Haar measure $\widehat{\mu}$ on $\widehat{G}$ such that the Fourier transform extends to an isometry of $L^2(G, \mu)$ onto $L^2(\widehat{G}, \widehat{\mu})$. It is called the Haar measure on $\widehat{G}$ associated to μ. Moreover, it is the unique Haar measure such that the Fourier inversion formula is true.

If G is self-dual and if we identify G with $\widehat{G}$ by a given isomorphism, then there is a unique Haar measure μ on G which is equal to its associated Haar measure $\widehat{\mu}$. This measure is called self-dual. If G is the additive group of K_w for a number field K and a place $w \in M_K$, then G is self-dual. Let χ_w be any non-trivial character, then we may identify $y \in K_w$ with the character $\langle x, y \rangle = \chi_w(xy)$ on K_w. Thus the choice of a non-trivial character leads to a well-determined Haar measure on K_w.

For $\mathbb{Q}_\infty = \mathbb{R}$, any character has the form $\chi(x) = \exp(2\pi i a x)$ for some non-zero real number a. We normalize χ_∞ by using $a = 1$.

If p is a prime number, we choose χ_p as follows. Any $x \in \mathbb{Q}_p$ has a representation of the form $\sum_{n=n_0}^{\infty} a_n p^n$ for some rational integers a_n and for some $n_0 \in \mathbb{Z}$. Let x_p be the sum restricted to indices $n < 0$; then any character on $\mathbb{Q}_p$ has the form $\chi(x) := \exp(-2\pi i (ax)_p)$ for some non-zero $a \in \mathbb{Q}_p$. Again, we choose $a = 1$ for χ_p.

Our normalizations are justified by the fact that the annihilator of $\mathbb{Z}$ (resp. $\mathbb{Z}_p$) with respect to the pairing $\langle \cdot, \cdot \rangle$ is again $\mathbb{Z}$ (resp. $\mathbb{Z}_p$) in the infinite (resp. finite) case.

For the place w of our number field K, let v be its restriction to $\mathbb{Q}$. We choose $\chi_w := \chi_v \circ \mathrm{Tr}_{K_w/\mathbb{Q}_v}$ leading to the formula $\prod_w \chi_w(x) = 1$ for any $x \in K$. Then our normalized Haar measure β_w is the unique self-dual measure on K_w with respect to the duality induced by χ_w.

C.2. Minkowski's second theorem

In this section, we prove Minkowski's second theorem over the adeles. We begin by introducing various types of lattices. This material is mainly borrowed from [**328**].

We fix notation in the following way. Let K be a number field of degree d, $N \in \mathbb{N}$, $v \in M_K$. By E, E_v, E_∞, $E_{\mathbb{A}}$, we denote the euclidean spaces K^N, K_v^N, $\prod_{v|\infty} K_v^N$, and $K_{\mathbb{A}}^N$. We fix the Haar measure on $E_{\mathbb{A}}$ by using the N-fold product of the Haar measure β on $K_{\mathbb{A}}$ introduced in C.1.9.

Definition C.2.1. *For a finite place v of K, a **K_v-lattice** in E_v is an open and compact R_v-submodule of E_v.*

Proposition C.2.2. *Let Λ_v be an R_v-submodule of E_v. Then Λ_v is a K_v-lattice in E_v if and only if Λ_v is a finitely generated R_v-module which generates E_v as a K_v-vector space.*

Proof: Let Λ_v be a K_v-lattice. Since Λ_v is open, it is clear that E_v is generated by Λ_v as a K_v-vector space. The R_v-span of N linearly independent vectors over K is an open and compact R_v-submodule of E_v. Note that Λ_v is covered by such submodules contained in Λ_v. Since Λ_v is compact, we can select a finite subcovering. This leads to a finite set of generators for Λ_v.

Conversely, let Λ_v be a finitely generated R_v-submodule which generates E_v as a K_v-vector space. As a continuous image of some R_v^M, the space Λ_v is compact. There is a K_v-basis of E_v contained in Λ_v. Let U be its R_v-span. Then U is an open neighbourhood of 0 and $U \subset \Lambda_v$. By a translation argument, we conclude that Λ_v is open in E_v and hence Λ_v is a K_v-lattice in E_v. □

Definition C.2.3. *An **$\mathbb{R}$-lattice** (or simply a **lattice**) in a finite-dimensional real (or complex) vector space V is a discrete subgroup Λ of V such that V/Λ is compact.*

The analogue of Proposition C.2.2 is the following well-known fact.

Proposition C.2.4. *Let Λ be a subgroup of the finite-dimensional real vector space V. Then the following conditions are equivalent:*

(a) *Λ is an $\mathbb{R}$-lattice in V;*

(b) *Λ is discrete in V and contains an $\mathbb{R}$-basis of V;*

(c) *Λ has a $\mathbb{Z}$-basis which is an $\mathbb{R}$-basis for V.*

Proof: (a) $\Rightarrow$ (b). We claim that Λ generates V as an $\mathbb{R}$-vector space. Let W be a complementary subspace of $\mathbb{R}\Lambda$. Then W is contained in the compact V/Λ, hence $W = 0$. We conclude that Λ spans V over $\mathbb{R}$, hence it contains an $\mathbb{R}$-basis of V.

(b) $\Rightarrow$ (c). We may assume that $V = \mathbb{R}^N$ and that Λ contains the standard basis $e_1, \ldots,$ e_N. Obviously, Λ is generated by $S := \{\boldsymbol{\lambda} \in \Lambda \mid \max_i |\lambda_i| \leq 1\}$. Since Λ is discrete in V, it is a closed subgroup (as in the proof of Theorem C.1.6). As an intersection of a compact cube and a discrete closed subset, our S has to be finite. Thus Λ is a finitely generated abelian group without torsion, hence free of finite rank $r \geq N$. Since every element of $\Lambda / \bigoplus_j \mathbb{Z}e_j$ has a representative in the finite set S, we conclude that $r = N$, proving (c).

Finally, (c) $\Rightarrow$ (a) is obvious. □

Definition C.2.5. *A* ***K*-lattice** *in E is a finitely generated O_K-submodule Λ of E which generates E as a K-vector space.*

Proposition C.2.6. *The K-lattices have the following characterization:*

(a) *If Λ is a K-lattice in E, then the closure Λ_v of Λ in E_v is a K_v-lattice in E_v for any non-archimedean $v \in M_K$. Moreover, we have $\Lambda_v = R_v^N$ up to finitely many $v \in M_K$.*

(b) *Conversely, if for any non-archimedean $v \in M_K$ we have a K_v-lattice Λ_v of E_v and if $\Lambda_v = R_v^N$ up to finitely many v, then there is a unique K-lattice Λ in E such that Λ_v is the closure of Λ in E_v. Moreover, we have*

$$\Lambda = \bigcap_{v \text{ finite}} (E \cap \Lambda_v).$$

Proof: Let Λ be a K-lattice generated by $x_1, \ldots, x_m$ as an O_K-module. Then $x_1, \ldots, x_m$ generate the closure Λ_v as an R_v-module (use the compactness of R_v) and they generate E_v as a K_v-vector space. By Proposition C.2.2, we know that Λ_v is a K_v-lattice in E_v. If we express our generators in terms of the standard basis of $E = K^n$ and *vice-versa*, then it is clear that the coordinates are in the valuation ring R_v up to finitely many v. This proves (a).

Now let Λ_v be a K_v-lattice in E_v, and assume $\Lambda_v = R_v^N$ up to finitely many v. We define

$$\Lambda := \bigcap_{v \text{ finite}} (E \cap \Lambda_v) \quad \text{and} \quad \Lambda_{\text{fin}} := \prod_{v \text{ finite}} \Lambda_v.$$

Clearly, Λ_{fin} is an open subgroup of $\prod_{v \text{ finite}} K_v$. We have to show that Λ is a K-lattice in E. By means of the canonical embedding of E into E_∞, this subgroup Λ is mapped onto the projection Λ_∞ of

$$\Gamma := E \cap (E_\infty \times \Lambda_{\text{fin}})$$

to E_∞. We claim that Λ_∞ is an $\mathbb{R}$-lattice in E_∞. Since E is a discrete closed subgroup in $E_{\mathbb{A}}$ (see Theorem C.1.6), we conclude that Γ is a discrete closed subgroup of $E_\infty \times \Lambda_{\text{fin}}$. The first projection of the latter has the property that the inverse image of a compact subset of E_∞ is compact. This proves easily that the image of a discrete closed subset is discrete. In particular, Λ_∞ is a discrete subgroup of E_∞. In order to prove compactness of $E_\infty / \Lambda_\infty$, it is enough to show that there is a compact subset C of $E_\infty \times \Lambda_{\text{fin}}$ such that

$$E_\infty \times \Lambda_{\text{fin}} = C + \Gamma. \tag{C.5}$$

Note that the first projection maps C onto E_∞/Λ_∞, hence this proves compactness of the latter.

We have an isomorphism

$$(E_\infty \times \Lambda_{\text{fin}})/\Gamma \cong (E + (E_\infty \times \Lambda_{\text{fin}}))/E \tag{C.6}$$

of locally compact groups. Since $E_\infty \times \Lambda_{\text{fin}}$ is an open subgroup of $E_{\mathbb{A}}$, we conclude that $E + (E_\infty \times \Lambda_{\text{fin}})$ and also its complement are open subgroups of $E_{\mathbb{A}}$. This is easily seen by writing them as a union of cosets of $E_\infty \times \Lambda_{\text{fin}}$. Hence the right-hand side of (C.6) is a closed subset of the compact quotient $E_{\mathbb{A}}/E$ (Theorem C.1.6). This proves that the left-hand side of (C.6) is also compact. Since the quotient homomorphism is an open map, we can cover $(E_\infty \times \Lambda_{\text{fin}})/\Gamma$ by the images of finitely many open subsets U_i which are relatively compact in $E_\infty \times \Lambda_{\text{fin}}$. Choosing $C = \cup_i \overline{U_i}$, we get (C.5). This finishes the proof that Λ_∞ is an $\mathbb{R}$-lattice in E_∞.

Now Proposition C.2.4 shows that every $\mathbb{Z}$-basis of Λ_∞ is an $\mathbb{R}$-basis of E_∞. Hence Λ is a free abelian group of finite rank which generates E as a $\mathbb{Q}$-vector space. Obviously, Λ is a K-lattice in E.

Next, we prove that the closure of Λ in E_v equals Λ_v for every finite v. Let $x_v \in \Lambda_v$ and let $\varepsilon > 0$, then the strong approximation theorem (see Theorem 1.4.5) applied to the coordinates of x_v shows that it exists $x \in E$ such that $|x - x_v|_v < \varepsilon$ and $x \in \Lambda_w$ for all finite $w \neq v$. If ε is sufficiently small, then the openness of Λ_v implies also $x \in \Lambda_v$. Thus $x \in \Lambda$ proving the density of Λ in Λ_v.

It remains to prove uniqueness. Let Λ' be another K-lattice with closure Λ_v in E_v for every finite v. By construction, we have $\Lambda' \subset \Lambda$. Since both are free abelian groups of rank Nd, the index is finite. Hence there is a non-zero $m \in \mathbb{Z}$ such that $m\Lambda \subset \Lambda'$. The strong approximation theorem shows that Λ' is dense in Λ_{fin} with respect to the diagonal embedding (choose O_K-generators of Λ' and then apply it to the coordinates). Let $\lambda \in \Lambda$. Since $m\Lambda_{\text{fin}}$ is an open subgroup of Λ_{fin}, there is $\lambda' \in \Lambda'$ such that $\lambda - \lambda' \in m\Lambda_{\text{fin}}$. We deduce

$$\lambda - \lambda' \in m\Lambda_{\text{fin}} \cap E = m\Lambda \subset \Lambda'$$

proving $\lambda \in \Lambda'$. We conclude $\Lambda = \Lambda'$. □

Corollary C.2.7. *The image Λ_∞ of a K-lattice Λ under the diagonal embedding $E \to E_\infty$ is an $\mathbb{R}$-lattice.*

C.2.8. For $\lambda \in \mathbb{R}$ and $\mathbf{x} = (x_v) \in E$, let $\lambda\mathbf{x} \in E$ be given by

$$(\lambda\mathbf{x})_v = \begin{cases} \lambda x_v & \text{if } v \text{ is archimedean,} \\ x_v & \text{if } v \text{ is not archimedean.} \end{cases}$$

For $v|\infty$, we consider a non-empty open convex symmetric bounded subset S_v of E_v. Here, symmetric means $-S_v = S_v$. Let Λ be a K-lattice in E and let us consider the subset

$$S := \prod_{v|\infty} S_v \times \prod_{v \text{ finite}} \Lambda_v$$

of $E_{\mathbb{A}}$.

Definition C.2.9. *For $n = 1, \dots, N$, the nth successive minimum is*

$$\lambda_n = \inf \{ t > 0 \mid tS \text{ contains } n \text{ linearly independent vectors of } \Lambda \text{ over } K \}.$$

Remark C.2.10. Note that $\lambda S \cap E$ is a discrete relatively compact subset of $E_{\mathbb{A}}$, hence finite. The closure of $\lambda_n S$ is the intersection of all λS with $\lambda > \lambda_n$. We conclude that this closure contains n linearly independent vectors. We easily verify

$$0 < \lambda_1 \leq \lambda_2 \leq \dots \leq \lambda_N < \infty.$$

Now we are ready to state **Minkowski's second theorem**.

Theorem C.2.11. *Let Λ be a K-lattice of E with closure Λ_v in E_v for finite places v. For $v|\infty$, let S_v be a non-empty open convex symmetric bounded subset S_v of E_v and let $S := \prod_{v|\infty} S_v \times \prod_{v \text{ finite}} \Lambda_v \subset E_{\mathbb{A}}$. Then*

$$(\lambda_1 \lambda_2 \cdots \lambda_N)^d \mathrm{vol(S)} \leq 2^{\mathrm{dN}},$$

where the volume is computed with respect to the Haar measure on $E_{\mathbb{A}}$ given by the product of the normalized Haar measures on $K_{\mathbb{A}}$ from C.1.9.

Moreover, suppose that for every complex place v the bounded set S_v is symmetric in the stronger sense that $\alpha S_v = S_v$ for $|\alpha| = 1$. Let r and s be the number of real and complex places of K, respectively. Then

$$\frac{2^{dN} \pi^{sN}}{(N!)^r ((2N)!)^s} |D_{K/\mathbb{Q}}|^{-N/2} \leq (\lambda_1 \cdots \lambda_N)^d \mathrm{vol(S)},$$

where $D_{K/\mathbb{Q}}$ is the discriminant of K.

For the proof we follow [35], obtaining it as a consequence of the Davenport–Estermann theorem stated below. For $v|\infty$, let T_v be a non-empty bounded convex open subset of E_v and let $T := \prod_{v|\infty} T_v \times \prod_{v \text{ finite}} \Lambda_v$.

Definition C.2.12. *For $n = 1, \dots, N$, let μ_n be the supremum of all $\mu \geq 0$ satisfying the following condition: If $\mathbf{x}, \mathbf{y} \in \mu T$ with $\mathbf{x} - \mathbf{y} \in E = K^N$, then the last $N - n + 1$ coordinates of $\mathbf{x}$ and $\mathbf{y}$ coincide, i.e. $\mathbf{x}_j = \mathbf{y}_j$ for $j = n, \dots, N$.*

Remark C.2.13. In the same way as in Remark C.2.10, we have

$$0 < \mu_1 \leq \mu_2 \leq \dots \leq \mu_N < \infty.$$

We leave the details to the reader (hint: consider $S := T - T$).

Theorem C.2.14. *With the hypothesis above, we have*

$$(\mu_1 \mu_2 \cdots \mu_N)^d \mathrm{vol}(T) \leq 1.$$

This is the **Davenport–Estermann theorem**. We prove it first and then we the deduce Minkowski's second theorem as a consequence.

Lemma C.2.15. *Let us consider the homomorphism*

$$\Phi_n : E_{\mathbb{A}} = K_{\mathbb{A}}^N \longrightarrow (K_{\mathbb{A}}/K)^n \times K_{\mathbb{A}}^{N-n}, \quad \mathbf{x} \mapsto (\overline{\mathbf{x}}_1, \ldots, \overline{\mathbf{x}}_n, \mathbf{x}_{n+1}, \ldots, \mathbf{x}_N),$$

where $\overline{\mathbf{x}}_j$ denotes the class of $\mathbf{x}_j$. For $\mu \geq 1$, we have

$$\mathrm{vol}\,(\Phi_n(\mu T)) \geq \mu^{d(N-n)} \mathrm{vol}\,(\Phi_n(T)),$$

where the volumes are with respect to a Haar measure.

Proof: Suppose first $n = N$ and pick $\mathbf{y} \in T$. The origin is contained in $T - \mathbf{y}$ (algebraic minus), hence the convexity shows $\mu(T - \mathbf{y}) \supset T - \mathbf{y}$ and so we have

$$\mathrm{vol}\,(\Phi_n\,(\mu(T - \mathbf{y}))) \geq \mathrm{vol}\,(\Phi_n\,(T - \mathbf{y})).$$

By invariance of the volume under translations, we get the claim for $n = N$.

Thus we may assume $n < N$. We write $K_{\mathbb{A}}^N = K_{\mathbb{A}}^n \times K_{\mathbb{A}}^{N-n}$ and for $\mathbf{y} \in K_{\mathbb{A}}^{N-n}$, let

$$T(\mathbf{y}) = \{\mathbf{w} \in K_{\mathbb{A}}^n \mid (\mathbf{w}, \mathbf{y}) \in T\}.$$

We denote the variables of $(K_{\mathbb{A}}/K)^n$ and $K_{\mathbb{A}}^{N-n}$ by $\overline{\mathbf{w}}$ and $\mathbf{y}$, respectively. The corresponding Haar measures will be called $\mathrm{d}\overline{\mathbf{w}}$ and $\mathrm{d}\mathbf{y}$. By Fubini's theorem, we have

$$\mathrm{vol}\,(\Phi_n(\mu T)) = \int_{K_{\mathbb{A}}^{N-n}} \mathrm{d}\mathbf{y} \int_{\Phi_n((\mu T)(\mathbf{y}))} \mathrm{d}\overline{\mathbf{w}}.$$

Using the substitution $\mathbf{y} = \mu \mathbf{y}'$ and noting that $(K_{\mathbb{A}}^{N-n})_\infty$ has real dimension $d(N-n)$, we get

$$\mathrm{vol}\,(\Phi_n(\mu T)) = \mu^{d(N-n)} \int_{K_{\mathbb{A}}^{N-n}} \mathrm{vol}\,(\Phi_n\,((\mu T)(\mu \mathbf{y}')))\,\mathrm{d}\mathbf{y}'. \tag{C.7}$$

By $(\mu T)(\mu \mathbf{y}') = \mu T(\mathbf{y}')$ and the case $N = n$, we conclude

$$\mathrm{vol}\,(\Phi_n\,((\mu T)(\mu \mathbf{y}'))) \geq \mathrm{vol}\,(\Phi_n\,(T(\mathbf{y}'))). \tag{C.8}$$

From (C.7) and (C.8), we obtain

$$\mathrm{vol}\,(\Phi_n(\mu T)) \geq \mu^{d(N-n)} \int_{K_{\mathbb{A}}^{N-n}} \mathrm{vol}\,(\Phi_n\,(T(\mathbf{y}')))\,\mathrm{d}\mathbf{y}' = \mu^{d(N-n)} \mathrm{vol}\,(\Phi_n(T))$$

proving the lemma. □

C.2.16. *Proof of the Davenport–Estermann theorem:* Let us apply Lemma C.2.15 with $\mu_n T$ in place of T and with $\mu = \mu_{n+1}/\mu_n$. For $n = 1, \ldots, N-1$, we get

$$\mathrm{vol}\,(\Phi_n(\mu_{n+1} T)) \geq \left(\frac{\mu_{n+1}}{\mu_n}\right)^{d(N-n)} \mathrm{vol}\,(\Phi_n(\mu_n T)). \tag{C.9}$$

We claim that the map

$$(K_{\mathbb{A}}/K)^n \times K_{\mathbb{A}}^{N-n} \longrightarrow (K_{\mathbb{A}}/K)^{n+1} \times K_{\mathbb{A}}^{N-n-1}$$

given by

$$(\overline{\mathbf{x}}_1, \ldots, \overline{\mathbf{x}}_n, \mathbf{x}_{n+1}, \ldots, \mathbf{x}_N) \mapsto (\overline{\mathbf{x}}_1, \ldots, \overline{\mathbf{x}}_{n+1}, \mathbf{x}_{n+2}, \ldots, \mathbf{x}_N)$$

is one-to-one on $\Phi_n(\mu_{n+1}T)$. If Φ_0 denotes the identity on $K_{\mathbb{A}}^N$, then this also holds for $n = 0$. To prove it, let $\mathbf{x}, \mathbf{y} \in \mu_{n+1}T$ with $\Phi_{n+1}(\mathbf{x}) = \Phi_{n+1}(\mathbf{y})$. This means that $\mathbf{x}_m = \mathbf{y}_m$ for $m > n+1$ and $\mathbf{x}_m - \mathbf{y}_m \in K$ for $m \leq n+1$. Especially, we have $\mathbf{x} - \mathbf{y} \in E$. Since T_v is open for $v|\infty$, there is $\mu < \mu_{n+1}$ with $\mathbf{x}, \mathbf{y} \in \mu T$. By definition of μ_{n+1}, we have $\mathbf{x}_{n+1} = \mathbf{y}_{n+1}$. This proves $\Phi_n(\mathbf{x}) = \Phi_n(\mathbf{y})$ which means injectivity on $\Phi_n(\mu_{n+1}T)$.

Now we use the normalized Haar measures on $K_{\mathbb{A}}$ and on $K_{\mathbb{A}}/K$ introduced in C.1.9. The volumes in (C.9) should be understood with respect to the product measures. Since $\Phi_n(\mu_{n+1}T)$ is mapped bijectively onto $\Phi_{n+1}(\mu_{n+1}T)$, Corollary C.1.2 shows

$$\operatorname{vol}(\Phi_n(\mu_{n+1}T)) = \operatorname{vol}(\Phi_{n+1}(\mu_{n+1}T)). \tag{C.10}$$

For $n = 1, \ldots, N$, we deduce from (C.9) and (C.10) that

$$\operatorname{vol}(\Phi_{n+1}(\mu_{n+1}T)) \geq \left(\frac{\mu_{n+1}}{\mu_n}\right)^{d(N-n)} \operatorname{vol}(\Phi_n(\mu_n T)).$$

This leads to

$$\begin{aligned} \operatorname{vol}(\Phi_N(\mu_N T)) &\geq \prod_{n=1}^{N-1} \left(\frac{\mu_{n+1}}{\mu_n}\right)^{d(N-n)} \operatorname{vol}(\Phi_1(\mu_1 T)) \\ &= (\mu_1 \cdots \mu_N)^d \operatorname{vol}(T), \end{aligned} \tag{C.11}$$

where in the last step we have used the identity

$$\operatorname{vol}(\Phi_1(\mu_1 T)) = \mu_1^{dN} \operatorname{vol}(T),$$

which follows from (C.10) for $n = 0$ and the transformation formula. By Proposition C.1.10, the left-hand side of (C.11) is bounded by 1, thereby proving the Davenport–Estermann theorem. □

C.2.17. *Proof of Minkowski's second theorem:* Note that a change of coordinates in $E_{\mathbb{A}}$ by an invertible $N \times N$ matrix γ does not affect the statement, because Example C.1.3 shows that the volume changes by

$$\prod_{v \in M_K} \|\det(\gamma)\|_v = 1.$$

Thus we may assume that for $n = 1, \ldots, N$ the closure of $\lambda_n S$ contains the first n elements of the standard basis $\mathbf{e}_1, \ldots, \mathbf{e}_N$ of $E_{\mathbb{A}} = K_{\mathbb{A}}^N$.

Let us apply the Davenport–Estermann theorem with $T = S$. It is enough to prove $\mu_n \geq \frac{1}{2}\lambda_n$. We proceed by induction on n. Let $\mathbf{x}, \mathbf{y} \in \frac{1}{2}\lambda_n S$ with $\mathbf{x} - \mathbf{y} \in E$. Since S is convex and symmetric, we get

$$\mathbf{x} - \mathbf{y} = \frac{1}{2}(2\mathbf{x}) + \frac{1}{2}(-2\mathbf{y}) \in \lambda_n S.$$

If $n = 1$, then Remark C.2.10 shows $\mathbf{x} = \mathbf{y}$ and hence $\mu_1 \geq \frac{1}{2}\lambda_1$. So we may assume $n \geq 2$ and, by induction hypothesis and C.2.13, we may assume $\lambda_n > \lambda_{n-1}$. Therefore the closure of $\lambda_{n-1}S$ is contained in $\lambda_n S$. Our assumptions show that $\mathbf{e}_1, \ldots, \mathbf{e}_{n-1}$ and $\mathbf{x} - \mathbf{y}$ belong to $\lambda_n S$. Again by Remark C.2.10, $\mathbf{x} - \mathbf{y}$ must be a linear combination of $\mathbf{e}_1, \ldots, \mathbf{e}_{n-1}$. This proves $\mathbf{x}_m = \mathbf{y}_m$ for $m \geq n$ and therefore $\mu_n \geq \frac{1}{2}\lambda_n$, as wanted.

For the lower bound, for each infinite place we define

$$S'_v := \left\{ \mathbf{t} \in E_v \mid \sum_{i=1}^{N} \lambda_i |t_i| < 1 \right\},$$

where $|\ |$ is the usual absolute value on $\mathbb{R}$ or $\mathbb{C}$. Then by convexity and symmetry we verify that $S'_v \subset S_v$. (If v is complex, the condition of symmetry needed is exactly $\alpha S_v = S_v$ for $|\alpha| = 1$.)

Now let S' be defined by S'_v as above if $v|\infty$ and $S'_v = R_v^N$ otherwise. Then $S' \subset S$. By computing volumes (see [**162**], Th.2.13.3) we find

$$\beta_v^N(S') = \begin{cases} \frac{2^N}{N!}(\lambda_1 \cdots \lambda_N)^{-1} & \text{if } v \text{ is real,} \\ \frac{(4\pi)^N}{(2N)!}(\lambda_1 \cdots \lambda_N)^{-2} & \text{if } v \text{ is complex,} \\ \left|D_{K_v/\mathbb{Q}_p}\right|_p^{N/2} & \text{if } v|p \text{ for } p \text{ prime.} \end{cases}$$

Since $\mathrm{vol}(S') \leq \mathrm{vol}(S)$, the lower bound follows by multiplying the local volumes and by (C.4) on page 606.

□

C.2.18. We compare our adelic approach with the classical geometry of numbers. Let Λ_∞ be an $\mathbb{R}$-lattice in $\mathbb{R}^N$ and let S_∞ be a non-empty open convex symmetric bounded subset of $\mathbb{R}^N$. Let $\lambda'_1, \ldots, \lambda'_N$ be the classical successive minima of S_∞ with respect to Λ_∞ defined by

$$\lambda'_n := \inf\{t > 0 \mid tS_\infty \text{ contains } n \text{ linearly independent vectors of } \Lambda_\infty\}.$$

The **classical second theorem of Minkowski** (see [**181**], Ch.2, §9, Th.1) states that

$$\lambda'_1 \cdots \lambda'_N \mathrm{vol}(S_\infty) \leq 2^N \mathrm{vol}(\Lambda_\infty),$$

where the volumes are taken with respect to the Lebesgue measure on $\mathbb{R}^N$, with $\mathrm{vol}(\Lambda_\infty)$ the volume of a fundamental domain of $\mathbb{R}^N/\Lambda_\infty$.

We show that this is equivalent to Theorem C.2.11. In order to see that the adelic version implies the classical theorem we may assume, by a linear transformation, that $\Lambda_\infty = \mathbb{Z}^N$ and then we may apply Theorem C.2.11 in the case $K = \mathbb{Q}$.

Conversely, let Λ be a K-lattice in $E = K^N$ and let S_v be a non-empty open symmetric convex bounded subset of E_v, for every non-archimedean place v. We choose a $\mathbb{Z}$-basis of O_K. Then we may identify $E = K^N$ with $\mathbb{Q}^{Nd}$, and similarly $\prod_{v|p} K_v^N$ with $\mathbb{Q}_p^{Nd}$, for every $p \in M_{\mathbb{Q}}$ by (C.2) on page 605. By Proposition C.1.10, the normalized Haar measures on $K_{\mathbb{A}}^N$ and on $\mathbb{Q}_{\mathbb{A}}^{Nd}$ agree. Our K-lattice Λ may be viewed as an $\mathbb{R}$-lattice in $\mathbb{R}^{Nd}$. For the non-empty open convex symmetric bounded subset $S_\infty = \prod_{v|\infty} S_v$ of $\mathbb{R}^{Nd}$ and for $S = S_\infty \times \prod_{v \text{ finite}} \Lambda_v \subset E_{\mathbb{A}}$ as in Theorem C.2.11, we claim that

$$\mathrm{vol}(S) = \mathrm{vol}(S_\infty)/\mathrm{vol}(\Lambda_\infty), \tag{C.12}$$

where on the left-hand side of the equation the volume is taken with respect to the normalized Haar measure from C.1.9, while on the right the volume of S_∞ is taken with respect to the usual real Lebesgue measure on $E_\infty = \mathbb{R}^{Nd}$ and $\mathrm{vol}(\Lambda_\infty)$ is the volume of a fundamental domain of E_∞/Λ_∞.

We have already seen that in proving this we may assume $K = \mathbb{Q}$. By the argument in C.2.17, both sides are invariant under a transformation by $A \in GL(Nd, \mathbb{Q})$, so we can reduce everything to the case $\Lambda = \mathbb{Z}^{Nd}$, where the claim is obvious.

Let $\lambda_1', \ldots, \lambda_{Nd}'$ be the classical successive minima of S_∞ with respect to Λ_∞. Then it is clear that $\lambda_1 = \lambda_1'$ and $\lambda_j \leq \lambda_{d(j-1)+1}'$. Hence (C.12) shows that the classical second theorem of Minkowski implies Theorem C.2.11.

C.2.19. Let Λ_∞ and S_∞ be as in C.2.18. If

$$\mathrm{vol}(S_\infty) > 2^N \mathrm{vol}(\Lambda_\infty),$$

then **Minkowski's first theorem** states that there is at least one non-zero lattice point contained in S_∞.

This is a special case of the classical second theorem of Minkowski using $\lambda_1' \leq \lambda_n'$. It is also an immediate consequence of **Blichfeldt's principle [26]**, a special case which states that, if Σ is a measurable set, $k \in \mathbb{N}$, and $\mathrm{vol}(\Sigma) > k\,\mathrm{vol}(\Lambda_\infty)$, then there is a translate of Σ containing $k+1$ distinct points of Λ_∞. Birkhoff's elementary proof in H.F. Blichfeldt **[26]** (see **[181]**, pp.35, 40–43 for extensions and an alternative proof) may be presented as follows. By intersecting Σ with a sufficiently large ball, we may assume that Σ is bounded. Let R be a parallelepiped which is a fundamental domain for Λ_∞ and consider all lattice translates of R by $\mathbf{x} \in \Lambda_\infty$ which intersect Σ; they cover Σ. The sum of the volumes of the sets $\Sigma \cap (R + \mathbf{x}) - \mathbf{x} \subset R$ is $\mathrm{vol}(\Sigma) > k\,\mathrm{vol}(R)$; hence there is a point $\mathbf{z} \in R$ which belongs to at least $k+1$ such sets (the sum of the characteristic functions cannot be bounded by k). The translate $\Sigma - \mathbf{z}$ contains at least $k+1$ points of Λ_∞.

Minkowski's first theorem is immediate by applying Blichfeldt's principle to $\Sigma = (1/2)S_\infty$ with $k = 1$, because if Σ is a convex symmetric set about the origin and $\mathbf{y}_1, \mathbf{y}_2 \in \Sigma$ then $\mathbf{y}_1 - \mathbf{y}_2 \in 2\Sigma$.

C.3. Cube slicing

In this section, we prove Vaaler's theorem **[302]** that the slice of a linear subspace with the symmetric unit cube of volume 1 has volume at least 1. The proof uses some basic facts about log-concave functions, which we handle first.

Definition C.3.1. *A non-negative real function f on $\mathbb{R}^n$ is called* **log-concave** *if* $\log f$ *is concave, i.e. for any $\mathbf{x}, \mathbf{y} \in \mathbb{R}^n$ and real numbers $\lambda, \mu > 0$ with $\lambda + \mu = 1$, we have*

$$f(\lambda \mathbf{x} + \mu \mathbf{y}) \geq f(\mathbf{x})^{\lambda} f(\mathbf{y})^{\mu}.$$

Remark C.3.2. For a log-concave function f, the set $\{\mathbf{x} \in \mathbb{R}^n \mid f(\mathbf{x}) > 0\}$ is convex and it is easy to see that f is continuous in the interior of this set.

Lemma C.3.3. *Let f, g be log-concave functions on $\mathbb{R}^n$ and let $\lambda, \mu > 0$ with $\lambda + \mu = 1$. For $\mathbf{t} \in \mathbb{R}^n$, let*

$$r(\mathbf{t}) := \sup_{\lambda \mathbf{x} + \mu \mathbf{y} = \mathbf{t}} f(\mathbf{x}) g(\mathbf{y}),$$

where $\mathbf{x}, \mathbf{y}$ range over $\mathbb{R}^n$. Then

$$\int r(\mathbf{t}) \, d\mathbf{t} \geq \left(\int f(\mathbf{x})^{\frac{1}{\lambda}} \, d\mathbf{x} \right)^{\lambda} \left(\int g(\mathbf{y})^{\frac{1}{\mu}} \, d\mathbf{y} \right)^{\mu}.$$

This is a kind of converse of Hölder's inequality. In fact, it holds for all non-negative Borel measurable functions. The proof for log-concave functions is easier and needs just basic techniques from analysis. For details, we refer to Prékopa **[236]**, Th.3.

Lemma C.3.4. *Let $f : \mathbb{R}^n \times \mathbb{R}^m \longrightarrow \mathbb{R}_+$ be a log-concave function and let A be a convex set of $\mathbb{R}^m$. Then*

$$\mathbf{x} \mapsto \int_A f(\mathbf{x}, \mathbf{y}) \, d\mathbf{y}$$

is a log-concave function on $\mathbb{R}^n$ if the integral is always finite.

Proof: We fix $\mathbf{x}_1, \mathbf{x}_2 \in \mathbb{R}^n$ and let $\lambda_1, \lambda_2 > 0$ with $\lambda_1 + \lambda_2 = 1$. Let $\mathbf{x}_3 := \lambda_1 \mathbf{x}_1 + \lambda_2 \mathbf{x}_2$. For $i = 1, 2, 3$, we define a function f_i on $\mathbb{R}^m$ by

$$f_i(\mathbf{y}) := \chi_A(\mathbf{y}) f(\mathbf{x}_i, \mathbf{y}),$$

where χ_A is the characteristic function of A. Clearly, f_i is log-concave and

$$f_3(\mathbf{y}) \geq \sup_{\lambda_1 \mathbf{y}' + \lambda_2 \mathbf{y}'' = \mathbf{y}} f_1(\mathbf{y}')^{\lambda_1} f_2(\mathbf{y}'')^{\lambda_2}.$$

By Lemma C.3.3, we get

$$\int f_3(\mathbf{y}) \, d\mathbf{y} \geq \left(\int f_1(\mathbf{y}') \, d\mathbf{y}' \right)^{\lambda_1} \left(\int f_2(\mathbf{y}'') \, d\mathbf{y}'' \right)^{\lambda_2},$$

proving the claim. □

Remark C.3.5. Let μ be the probability measure on $\mathbb{R}^n$ corresponding to the Gauss normal density. For a Borel subset Ω, it is given by

$$\mu(\Omega) = \int_{\Omega} \exp(-\pi |\mathbf{x}|^2) \, d\mathbf{x}.$$

Note that the density $\exp(-\pi |\mathbf{x}|^2)$ is a symmetric continuous log-concave function on $\mathbb{R}^n$. For $0 < s < 1$, let us consider the compact convex symmetric subset of $\mathbb{R}^n$, with non-empty interior, defined by

$$K_s := \{\mathbf{x} \in \mathbb{R}^n \mid \exp(-\pi |\mathbf{x}|^2) \geq s\}.$$

An easy application of Fubini's theorem gives

$$\mu(\Omega) = \int_0^1 \int_{K_s \cap \Omega} \mathrm{d}\mathbf{x}\, \mathrm{d}s\,.$$

Remark C.3.6. Let $B_{\rho(n)}$ be the closed ball of volume 1 in $\mathbb{R}^n$ with centre 0 and radius $\rho(n)$. It is well known that

$$\rho(n) = \pi^{-\frac{1}{2}} \Gamma\left(\frac{n}{2}+1\right)^{\frac{1}{n}}\,.$$

Lemma C.3.7. *Let $N = n_1 + \cdots + n_r$ be a partition of N and define*

$$Q_N := B_{\rho(n_1)} \times \cdots \times B_{\rho(n_r)}\,.$$

Let A be a closed symmetric convex subset of $\mathbb{R}^N$. Then we have

$$\mu(A) \leq \mathrm{vol}(A \cap Q_N),$$

where μ is the Gauss measure and vol *is the Lebesgue measure on $\mathbb{R}^N$.*

Proof: We prove the lemma by induction on r. Suppose that $r = 1$ and let $N = n$. On the sphere $\mathrm{S}^{n-1} = \{\mathbf{x} \in \mathbb{R}^n \mid |\mathbf{x}| = 1\}$, we consider the Lebesgue measure λ_{n-1}. Then the polar decomposition $\mathbf{x} = r\mathbf{x}'$ with $r > 0$ and $\mathbf{x}' \in \mathrm{S}^{n-1}$ gives

$$\mu(A) = \int_{\mathrm{S}^{n-1}} \int_0^\infty \chi_A(r\mathbf{x}') \exp(-\pi r^2) r^{n-1}\, \mathrm{d}r\, \mathrm{d}\lambda_{n-1}(\mathbf{x}')$$

for every closed convex symmetric subset A of $\mathbb{R}^n$. We fix $\mathbf{x}' \in \mathrm{S}^{n-1}$. By convexity, either

$$\mathbb{R}\mathbf{x}' \cap B_{\rho(n)} \subset \mathbb{R}\mathbf{x}' \cap A$$

or

$$\mathbb{R}\mathbf{x}' \cap A \subset \mathbb{R}\mathbf{x}' \cap B_{\rho(n)}\,.$$

In the first case

$$\begin{aligned}
\int_0^\infty \chi_A(r\mathbf{x}') \exp(-\pi r^2) r^{n-1}\, \mathrm{d}r &\leq \int_0^\infty \exp(-\pi r^2) r^{n-1}\, \mathrm{d}r \\
&= \frac{1}{n} \pi^{-\frac{n}{2}} \Gamma\left(\frac{n}{2}+1\right) \\
&= \int_0^{\rho(n)} r^{n-1}\, \mathrm{d}r \\
&= \int_0^\infty \chi_{B_{\rho(n)}}(r\mathbf{x}') r^{n-1}\, \mathrm{d}r\,.
\end{aligned}$$

We conclude that

$$\int_0^\infty \chi_A(r\mathbf{x}') \exp(-\pi r^2) r^{n-1}\, \mathrm{d}r \leq \int_0^\infty \chi_{B_{\rho(n)} \cap A}(r\mathbf{x}') r^{n-1}\, \mathrm{d}r$$

holds in both cases. Thus

$$\begin{aligned}
\mu(A) &\leq \int_{\mathrm{S}^{n-1}} \int_0^\infty \chi_{B_{\rho(n)} \cap A}(r\mathbf{x}') r^{n-1}\, \mathrm{d}r\, \mathrm{d}\lambda_{n-1}(\mathbf{x}') \\
&= \mathrm{vol}(B_{\rho(n)} \cap A)
\end{aligned}$$

proving the claim for $r = 1$.

For notational simplicity, we do the induction step only in the case $r = 2$. Points on $\mathbb{R}^N = \mathbb{R}^{n_1} \times \mathbb{R}^{n_2}$ are denoted by $\mathbf{x} = (\mathbf{y}, \mathbf{z})$. For $\mathbf{y} \in \mathbb{R}^{n_1}$, let

$$A_{\mathbf{y}} := \{\mathbf{z} \in \mathbb{R}^{n_2} \mid (\mathbf{y}, \mathbf{z}) \in A\}.$$

Then Lemma C.3.4 shows that the symmetric function

$$f(\mathbf{y}) := \int_{B_{\rho(n_2)} \cap A_{\mathbf{y}}} \mathrm{d}\mathbf{z}$$

is log-concave on $\mathbb{R}^{n_1}$. Note that the functions

$$f_n := \sum_{k=0}^{\infty} \frac{1}{n} \cdot \chi_{\{\mathbf{x} \in \mathbb{R}^{n_1} \mid f(\mathbf{x}) \geq \frac{k}{n}\}}$$

decrease as $n \to \infty$ to f. Since the sets involved in the characteristic functions are closed, convex, and symmetric, the induction hypothesis and monotone convergence give

$$\int f(\mathbf{y})\, \mathrm{d}\mu(\mathbf{y}) \leq \int_{B_{\rho(n_1)}} f(\mathbf{y})\, \mathrm{d}\mathbf{y}.$$

Using Remark C.3.5, we get

$$\int_0^1 \int_{B_{\rho(n_2)}} \int_{K_s \cap A_{\mathbf{z}}} \mathrm{d}\mathbf{y}\, \mathrm{d}\mathbf{z}\, \mathrm{d}s \leq \mathrm{vol}(A \cap Q_N). \tag{C.13}$$

Now the same argument as above applied to the function

$$f(\mathbf{z}) := \int_{K_s \cap A_{\mathbf{z}}} \mathrm{d}\mathbf{y}$$

shows that

$$\int f(\mathbf{z}) \mathrm{d}\mu(\mathbf{z}) \leq \int_{B_{\rho(n_2)}} f(\mathbf{z})\, \mathrm{d}\mathbf{z}.$$

Using this in (C.13), the induction step is completed by Fubini's theorem and Remark C.3.5. □

Finally, we are ready to prove **Vaaler's cube-slicing theorem**. In the simplest case of a cube in $\mathbb{R}^N$, this simply states that the volume of a linear slice through the centre of a cube of volume 1 is bounded below by 1. In general, it states that the volume of a slice through the centre of a product of balls of volume 1 is bounded below by 1.

Theorem C.3.8. *Let $N = n_1 + \cdots + n_r$ be a partition and let $Q_N := B_{\rho(n_1)} \times \cdots \times B_{\rho(n_r)}$ as above. For a real $N \times M$ matrix B of rank M, we have*

$$\det(B^t B)^{-\frac{1}{2}} \leq \mathrm{vol}\left(\{\mathbf{y} \in \mathbb{R}^M \mid B\mathbf{y} \in Q_N\}\right).$$

Proof: Let L be the M-dimensional linear subspace of $\mathbb{R}^N$ given as the image of $\mathbb{R}^M$. By the transformation formula, it is clear that we have to show

$$1 \leq \mathrm{vol}(Q_N \cap L),$$

where the volume is computed on L. This is the **cube-slicing inequality**. For $\varepsilon > 0$, let

$$L_\varepsilon := \{\mathbf{x} \in \mathbb{R}^N \mid \inf_{\mathbf{y} \in L} |\mathbf{x} - \mathbf{y}| \leq \varepsilon\}.$$

With respect to the orthogonal decomposition $\mathbb{R}^N = L \times L^\perp$, we have $L_\varepsilon = L \times B_\varepsilon$. Note that L_ε is a closed convex symmetric subset of $\mathbb{R}^N$. By Lemma C.3.7, we get

$$\mathrm{vol}(B_\varepsilon)^{-1}\mu(L_\varepsilon) \le \mathrm{vol}(B_\varepsilon)^{-1}\int_{Q_N \cap L_\varepsilon} d\mathbf{y} \tag{C.14}$$

As $\varepsilon \to 0$, the left-hand side of this inequality tends to 1. Here, we have used that

$$\mu(L_\varepsilon) = \mu(L)\mu(B_\varepsilon) = \mu(B_\varepsilon)$$

with respect to the decomposition above and

$$\lim_{\varepsilon\to 0} \mu(B_\varepsilon)/\mathrm{vol}(B_\varepsilon) = 1\,.$$

Let $\mathbf{y} \in L$, not a boundary point of $L \cap Q_N$. Then

$$\lim_{\varepsilon\to 0}\left(\mathrm{vol}(B_\varepsilon)^{-1}\int_{L^\perp} \chi_{Q_N\cap L_\varepsilon}(\mathbf{y},\mathbf{z})\,d\mathbf{z}\right) = \chi_{L\cap Q_N}(\mathbf{y})$$

and, by the Lebesgue dominated convergence theorem, we conclude that the right-hand side of (C.14) tends to $\mathrm{vol}(Q_N \cap L)$. This proves the cube-slicing inequality. □

References

[1] A. Abbes, Hauteurs et discrétude (d'après L. Szpiro, E. Ullmo et S. Zhang), *Séminaire Bourbaki*, Exposé 825, Vol. 1996/97, Astérisque **245** (1997), 141–166.

[2] D. Abramovich, Uniformité des points rationnels des courbes algébriques sur les extensions quadratiques et cubiques, *C. R. Acad. Sci. Paris Sér. I Math.* **321** (1995), 755–758.

[3] L.V. Ahlfors, Beiträge zur Theorie der meromorphen Funktionen, *Skand. Mathematikerkongress* **7** (1930), 84–88.

[4] L.V. Ahlfors, Über eine in der neueren Wertverteilungstheorie betrachtete Klasse transzendenter Funktionen, *Acta Math.* **58** (1932), 375–406. Also *Collected Papers*. Vol. 1, 112–143. Birkhäuser, Boston-Basel-Stuttgart 1982. xx+520 pp.

[5] L.V. Ahlfors, Über eine Methode in der Theorie der meromorphen Funktionen, *Soc. Sci. Fenn. Comm. Phys.-Math.* **8**, No. 10 (1935), pp.1–14. Also *Collected Papers*. Vol. 1, 190–203. Birkhäuser, Boston-Basel-Stuttgart 1982. xx+520 pp.

[6] L.V. Ahlfors, Zur Theorie der Überlagerungsflächen, *Acta Math.* **65** (1935), 157–194.

[7] L.V. Ahlfors, The theory of meromorphic curves, *Acta Soc. Sci. Fennicae Nova Ser. A.* **3**, No. 4 (1941), 31 pp.

[8] L.V. Ahlfors, *Complex Analysis: An Introduction to the Theory of Analytic Functions of One Complex Variable*. Third edition. International Series in Pure and Applied Mathematics. McGraw-Hill Book Co., New York 1978. xi+331 pp.

[9] F. Amoroso and S. David, Le problème de Lehmer en dimension supérieure, *J. reine angew. Math.* **513** (1999), 145–179.

[10] F. Amoroso and S. David, Distribution des points de petite hauteur dans les groupes multiplicatifs, *Ann. Scuola Norm. Sup. Pisa Cl. Sci.* (5) **3** (2004), 325–348.

[11] F. Amoroso and S. Dvornicich, A lower bound for the height in abelian extensions, *J. Number Th.* **80** (2000), 260–272.

[12] F. Amoroso and U. Zannier, A relative Dobrowolski lower bound over abelian extensions, *Ann. Scuola Norm. Sup. Pisa Cl. Sci.* (4) **29** (2000), 711–727.

[13] E. Arbarello, M. Cornalba, P.A. Griffiths, and J. Harris, *Geometry of Algebraic Curves*. Vol. I. Grundlehren der mathematischen Wissenschaften **267**. Springer-Verlag, New York 1985. xvi+386 pp.

[14] A. Baker, *Transcendental Number Theory*. Cambridge University Press 1975. ix+147 pp.

[15] A. Baker, Logarithmic forms and the abc-conjecture, in *Number Theory: Diophantine, Computational and Algebraic Aspects*, 37–44. Györy, Kálmán *et al.* (eds), Proceedings of the international conference (Eger, Hungary, 1996). De Gruyter, Berlin 1998.

[16] A. Baker and G. Wüstholz, Logarithmic forms and group varieties, *J. reine angew. Math.* **442** (1993), 19–62.

[17] V.V. Batyrev and Yu. Tschinkel, Rational points on some Fano cubic bundles, *C. R. Acad. Sci. Paris Sér. I Math.* **323**, No. 1 (1996), 41–46.

[18] B. Beauzamy, E. Bombieri, P. Enflo, and H.L. Montgomery, Products of polynomials in many variables, *J. Number Th.* **36** (1990), 219–245.

[19] S. Beckmann, On extensions of number fields obtained by specializing branched coverings, *J. reine angew. Math.* **419** (1991), 27–53.

[20] A. Beilinson, Height pairing between algebraic cycles, in *K-theory, Arithmetic and Geometry*, 1–26. Yu. Manin (ed.), Semin. Moscow Univ. 1984–86. Lecture Notes in Mathematics **1289**. Springer-Verlag, Berlin 1987.

[21] G.V. Belyĭ, On Galois extensions of a maximal cyclotomic field, *Izv. Akad. Nauk SSSR, Ser. Mat.* **43** (1979), 267–276; English transl. in *Math. USSR Izv.* **14** (1980), 247–256.

[22] A. Bertram, L. Ein, and R. Lazarsfeld, Vanishing theorems, a theorem of Severi, and the equations defining projective varieties, *J. Amer. Math. Soc.* **4** (1991), 587–602.

[23] F. Beukers and H.P. Schlickewei, The equation $x + y = 1$ in finitely generated groups, *Acta Arith.* **78** (1996), 189–199.

[24] Yu.F. Bilu, Limit distribution of small points on algebraic tori, *Duke Math. J.* **89** (1997), 465–476.

[25] Yu.F. Bilu, Catalan's conjecture (after Mihăilescu), *Séminaire Bourbaki*, Exposé 909, Vol. 2002/03, Astérisque **294** (2004), 1–26.

[26] H.F. Blichfeldt, A new principle in the geometry of numbers, with some applications, *Trans. Amer. Math. Soc.* **15** (1914), 227–235.

[27] S. Bloch, Height pairings for algebraic cycles, *J. Pure Appl. Algebra* **34** (1984), 119–145.

[28] E. Bombieri, On Weil's "Théorème de décomposition," *Amer. J. Math.* **105** (1983), 295–308.

[29] E. Bombieri, The Mordell conjecture revisited, *Ann. Scuola Norm. Sup. Pisa. Cl. Sci.* (4) **17** (1990), 615–640. Errata-corrige: "The Mordell conjecture revisited," ibid. **18** (1991), 473.

[30] E. Bombieri, On the Thue–Mahler equation II, *Acta Arith.* **67** (1994), 69–96.

[31] E. Bombieri, Effective Diophantine approximation on $\mathbb{G}_m$, *Ann. Scuola Norm. Sup. Pisa Cl. Sci.* (4) **20** (1993), 61–89.

[32] E. Bombieri and P.B. Cohen, An elementary approach to effective Diophantine approximation on $\mathbb{G}_m$, in *Number Theory and Algebraic Geometry*, 41–62. M. Reid and A. Skorobogatov (eds), London Math. Soc. Lecture Note Ser. **303**. Cambridge University Press, Cambridge 2003.

[33] E. Bombieri, A. Granville and J. Pintz, Squares in arithmetic progressions, *Duke Math. J.* **66** (1992), 369–385.

[34] E. Bombieri and H.P.F. Swinnerton-Dyer, On the local zeta function of a cubic threefold, *Ann. Scuola Norm. Super. Pisa Sci. Fis. Mat.* (3) **21** (1967), 1–29.

[35] E. Bombieri and J.D. Vaaler, On Siegel's lemma, *Invent. Math.* **73** (1983), 11–32. Ibid., Addendum to "On Siegel's lemma," *Invent. Math.* **75** (1984), 377.

[36] E. Bombieri and A.J. van der Poorten, Some quantitative results related to Roth's theorem, *J. Austral. Math. Soc.* **45** (1988) 233–248. Corrigenda, *J. Austral. Math. Soc.* **48** (1990), 154–155.

[37] E. Bombieri, A.J. van der Poorten, and J.D. Vaaler, Effective measures of irrationality for cubic extensions of number fields, *Annali Sc. Norm. Sup. Pisa Cl. Sc.* (4) **23** (1996), 211–248.

[38] E. Bombieri and U. Zannier, Algebraic points on subvarieties of $\mathbb{G}_m^n$, *Internat. Math. Res. Notices* (1995), 333–347.

[39] E. Bombieri and U. Zannier, Heights of algebraic points on subvarieties of abelian varieties, *Ann. Scuola Norm. Sup. Pisa Cl. Sci.* (4) **23** (1996), 779–792 (1997).

[40] E. Bombieri and U. Zannier, A note on heights in certain infinite extensions of $\mathbb{Q}$, *Atti Accad. Naz. Lincei Cl. Sci. Fis. Mat. Natur. Rend. Lincei* (9) *Mat. Appl.* **12** (2001), 5–14.

[41] A. Borel, *Linear Algebraic Groups*. Second edition. Graduate Texts in Mathematics **126**. Springer-Verlag, New York 1991. xii+288 pp.

[42] A.I. Borevich and I.R. Shafarevich, *Number Theory*. Translated from the Russian by Newcomb Greenleaf. Pure and Applied Mathematics, Vol. 20. Academic Press, New York–London 1966. x +435 pp.

[43] S. Bosch, U. Güntzer, and R. Remmert, *Non-Archimedean Analysis. A systematic approach to rigid analytic geometry*. Grundlehren der mathematischen Wissenschaften **261**. Springer-Verlag, Berlin 1984. xii+436 pp.

[44] S. Bosch, W. Lütkebohmert, and M. Raynaud, *Néron Models*. Ergebnisse der Mathematik und ihrer Grenzgebiete (3) **21**. Springer-Verlag, Berlin 1990. x+325 pp.

[45] J.-B. Bost, H. Gillet, and C. Soulé, Heights of projective varieties and positive Green forms, *J. Amer. Math. Soc.* **7**, No. 4 (1994), 903–1027.

[46] N. Bourbaki, *Éléments de Mathématique. Fasc. XXIX. Livre VI: Intégration.* Chapitre 7: Mesure de Haar. Chapitre 8: Convolution et représentations. Actualités Scientifiques et Industrielles, No. 1306. Hermann, Paris 1963. 222 pp.

[47] N. Bourbaki, *Eléments de Mathématique. Fasc. XXX. Algèbre commutative.* Chapitre 5: Entiers. Chapitre 6: Valuations. Actualités Scientifiques et Industrielles, No. 1308. Hermann, Paris 1964. 207 pp.

[48] N. Bourbaki, *Éléments de Mathématique. Fasc. XXXII. Théories Spectrales.* Chapitre I: Algèbres normées. Chapitre II: Groupes localement compact commutatifs. Actualités Scientifiques et Industrielles, No. 1332. Hermann, Paris 1967. iv+166 pp.

[49] N. Bourbaki, *Éléments de Mathématique. Algèbre.* Chapitres 4 à 7. Masson, Paris 1981. vii+422 pp.

[50] B. Brindza, K. Györy, and R. Tijdeman, On the Catalan equation over algebraic number fields, *J. reine angew. Math.* **367** (1986), 90–102.

[51] J. Browkin and J. Brzeziński, Some remarks on the abc-conjecture, *Math. Comp.* **62** (1994), 931–939.

[52] W.D. Brownawell and D.W. Masser, Vanishing sums in function fields, *Math. Proc. Cambridge Philos. Soc.* **100** (1986), 427–434.

[53] Y. Bugeaud, Bornes effectives pour les solutions des équations en S-unités et des équations de Thue–Mahler, *J. Number Th.* **71** (1998), 227–244.

[54] Y. Bugeaud and M. Laurent, Minoration effective de la distance p-adique entre puissances de nombres algébriques, *J. Number Th.* **61** (1996), 311–342.

[55] G.S. Call and J.H. Silverman, Canonical heights on varieties with morphisms, *Compos. Math.* **89** (1993), 163–205.

[56] L. Caporaso, J. Harris, and B. Mazur, Uniformity of rational points, *J. Amer. Math. Soc.* **10** (1997), 1–35.

[57] C. Carathéodory, *Theory of Functions of a Complex Variable*. Vol. II. Translated by F. Steinhardt. Chelsea Publ. Company, New York 1954, 220 pp.

[58] J. Carlson and P. Griffiths, A defect relation for equidimensional holomorphic mappings between algebraic varieties, *Ann. of Math.* (2) **95** (1972), 557–584.

[59] H. Cartan, Sur la fonction de croissance attachée à una fonction méromorphe de deux variables et ses applications aux fonctions méromorphes d'une variable, *C. R. Acad. Sci. Paris Sér. I Math.* **189** (1929), 521–523.

[60] H. Cartan, Sur les zéros des combinaisons linéaires de p fonctions holomorphes données, *Mathematica* **7** (1933), 5–31.

[61] J.W.S. Cassels, Diophantine equations with special reference to elliptic curves, *J. London Math. Soc.* **41** (1966), 193–291.

[62] K. Chandrakekharan, *Elliptic Functions*. Grundlehren der mathematischen Wissenschaften **281**. Springer-Verlag, Berlin 1985. xi+189 pp.

[63] W. Cherry, The Nevanlinna error term for coverings, generically surjective case, in *Proceedings of the Symposium on Value Distribution Theory in Several Complex Variables*, 37–53. W. Stoll (ed.), Notre Dame Math. Lectures **12**. University of Notre Dame Press, Notre Dame IN 1992.

[64] W. Cherry and Z. Ye, *Nevanlinna's Theory of Value Distribution: The Second Main Theorem and its Error Terms*. Springer Monographs in Mathematics. Springer-Verlag, Berlin 2001. xii+201 pp.

[65] C. Chevalley and A. Weil, Un théorème d'arithmétique sur les courbes algèbriques, *C. R. Acad. Sci. Paris Sér. I Math.* **195** (1932), 570–572.

[66] K. K. Choi and J. D. Vaaler, Diophantine approximation in projective space, in *Number Theory* (Ottawa ON 1996), 55–65. CRM Proc. Lecture Notes **19**. Amer. Math. Soc., Providence RI 1999.

[67] W. L. Chow, The Jacobian variety of an algebraic curve, *Amer. J. Math.* **76** (1954), 453–476.

[68] C.H. Clemens and P.A. Griffiths, The intermediate Jacobian of the cubic threefold, *Ann. of Math.* (2) **95** (1972), 281–356.

[69] R. Coleman, Manin's proof of the Mordell conjecture over function fields, *Enseign. Math.* (2) **36**, No. 3–4 (1990), 393–427.

[70] J.B. Conway, *Functions of One Complex Variable. II.* Graduate Texts in Mathematics **159**. Springer-Verlag, New York 1995. xvi+394 pp.

[71] J.H. Conway and A.J. Jones, Trigonometric diophantine equations (On vanishing sums of roots of unity), *Acta Arith.* **30** (1976), 229–240.

[72] P. Corvaja and U. Zannier, A subspace theorem approach to integral points on curves, *C. R. Math. Acad. Sci. Paris Sér. I Math.* **334**, No. 4 (2002), 267–271.

[73] P. Corvaja and U. Zannier, On the greatest prime factor of $(ab+1)(ac+1)$, *Proc. Amer. Math. Soc.* **131**, No. 6 (2003), 1705–1709.

[74] M. Cugiani, Sull'approssimabilità di un numero algebrico mediante numeri algebrici di un corpo assegnato, *Boll. Un. Mat. Ital.* (3) **14** (1959), 151–162.

[75] L.V. Danilov, The Diophantine equation $x^3 - y^2 = k$ and a conjecture of M. Hall. (Russian) *Mat. Zametki* **32** (1982), 273–275, 425. English translation in *Math. Notes* **32** (1983), 617–618.

[76] H. Darmon, Faltings plus epsilon, Wiles plus epsilon, and the generalized Fermat equation, *C. R. Math. Rep. Acad. Sci. Canada* **19**, No. 1 (1997), 3–14. Corrigenda, *C. R. Math. Rep. Acad. Sci. Canada* **19**, No. 2 (1997), 64.

[77] H. Darmon, F. Diamond, and R. Taylor, *Fermat's Last Theorem.* R. Bott *et al.* (eds) Current Developments in Mathematics. International Press, Cambridge MA 1995. 1–154.

[78] H. Darmon and A. Granville, On the equation $z^m = F(x,y)$ and $Ax^p + By^q = Cz^r$, *Bull. London Math. Soc.* **27** (1995), 513–543.

[79] H. Davenport and K.F. Roth, Rational approximations to algebraic numbers, *Mathematika* **2** (1955), 160–167.

[80] H. Davenport and W.M. Schmidt, Approximation to real numbers by quadratic irrationals. *Acta Arith.* **13** (1967/1968), 169–176.

[81] H. Davenport and W.M. Schmidt, Approximation to real numbers by algebraic integers. *Acta Arith.* **15** (1968/1969), 393–416.

[82] S. David and P. Philippon, Minorations des hauteurs normalisés des sous-variétés de variétés abéliennes, in *Number Theory,* 333–364. V. Kumar Murty (ed.) *et al.*, Proceedings of the Int. Conference of the Ramanujan Mathematical Society, Providence, *Contemp. Math.* **210** (1998).

[83] S. David and P. Philippon, Minorations des hauteurs normalisées des sous-variétés des tores, *Ann. Scuola Norm. Sup. Pisa Cl. Sci.* (4) **28** (1999), 489–543. Errata: ibid. **29** (2000), 729–731.

[84] S. David and P. Philippon, Minorations des hauteurs normalisées des sous-variétés de variétés abéliennes II, *Comment. Math. Helv.* **77** (2002), 639–700.

[85] M. Demazure and P. Gabriel, *Groupes algébriques. Tome I: Géométrie Algébrique, Généralités, Groupes Commutatifs: Avec une appendice Corps de Classes Local par Michel Hazewinkel.* Masson & Cie, Éditeur, Paris; North-Holland Publishing Co., Amsterdam 1970. xxvi+700 pp.

[86] P. Dèbes, Quelques remarques sur un article de Bombieri concernant le Theorème de Décomposition de Weil, *Amer. J. Math.* **107** (1985), 39–44.

[87] P. Dèbes, G-fonctions et théorème d'irreducibilité de Hilbert, *Acta Arith.* **47** (1986), 371–402.

[88] P. Dèbes and U. Zannier, Hilbert's irreducibility theorem and G-functions, *Math. Ann.* **309** (1997), 491–503.

[89] L.E. Dickson, *History of the Theory of Numbers. Vol. II: Diophantine Analysis*. Reprinted Chelsea Publishing Co., New York 1966. xxv+803 pp.

[90] E. Dobrowolski, On a question of Lehmer and the number of irreducible factors of a polynomial, *Acta Arith.* **39** (1979), 391–401.

[91] C. Doche, On the spectrum of the Zhang–Zagier height, *Math. Comp.* **70** (2001), 419–430.

[92] D. Drasin, The inverse problem of the Nevanlinna theory, *Acta Math.* **138** (1977), 83–151.

[93] D. Drasin, Proof of a conjecture of F. Nevanlinna concerning functions which have deficiency sum two, *Acta Math.* **158** (1987), 1–94.

[94] R. Dvornicich and U. Zannier, On sums of roots of unity, *Monatsh. Math.* **129** (2000), 97–108.

[95] B.M. Dwork and A.J. van der Poorten, The Eisenstein constant, *Duke Math. J.* **65** (1992), 23–43. Corrigenda, *Duke Math. J.* **76** (1994), 669–672.

[96] B. Edixhoven and J.-H. Evertse (eds), *Diophantine Approximation and Abelian Varieties: Introductory Lectures*. Papers from the Conference held in Soesterberg, April 12–16, 1992. Lecture Notes in Mathematics **1566**. Springer-Verlag, Berlin 1993. xiv+127 pp.

[97] H. Edwards, *Fermat's Last Theorem: A Genetic Introduction to Algebraic Number Theory*. Graduate Texts in Mathematics **50**. Springer-Verlag, New York 1996. xvi+410 pp.

[98] N.D. Elkies, ABC implies Mordell, *Internat. Math. Res. Notices* (1991), 99–109.

[99] W.J. Ellison, Waring's problem, *Amer. Math. Monthly* **78** (1971), 10–36.

[100] A. Erdélyi, W. Magnus, F. Oberhettinger, and F.G. Tricomi, *Higher Transcendental Functions*. Vol I. With a preface by Mina Rees. With a foreword by E.C. Watson. Reprint of the 1953 original. Robert E. Krieger Publishing Co., Inc., Melbourne, Fla. 1981. xiii+302 pp. Also: H. van Haeringen, L.P. Kok, Table errata: Higher Transcendental Functions, Vol. I by A. Erdélyi, W. Magnus, F. Oberhettinger, and F.G. Tricomi, *Math. Comp.* **41** (1983), 778.

[101] P. Erdős, C.L. Stewart, and R. Tijdeman, Some diophantine equations with many solutions, *Compos. Math.* **66** (1988), 37–56.

[102] H. Esnault and E. Viehweg, Dyson's lemma for polynomials in several variables (and the theorem of Roth), *Invent. Math.* **78** (1984), 445–490.

[103] J.-H. Evertse, On equations in S-units and the Thue–Mahler equation, *Invent. Math.* **75** (1984), 561–584.

[104] J.-H. Evertse, On sums of S-units and linear recurrences, *Compos. Math.* **53** (1984), 225–244.

[105] J.-H. Evertse, The subspace theorem of W. M. Schmidt, in *Diophantine Approximation and Abelian Varieties. Introductory Lectures*, 31–50. B. Edixhoven and J.-H. Evertse (eds), papers from the Conference held in Soesterberg, April 12–16, 1992. Lecture Notes in Mathematics **1566**. Springer-Verlag, Berlin 1993.

[106] J.-H. Evertse, An explicit version of Faltings' product theorem and an improvement of Roth's lemma, *Acta Arith.* **73** (1995), 215–248.

[107] J.-H. Evertse, *Math. Reviews* **95g:11068**.

[108] J.-H. Evertse, An improvement of the quantitative subspace theorem, *Compos. Math.* **101** (1996), 225–311.

[109] J.-H. Evertse, The number of solutions of the Thue–Mahler equation, *J. reine angew. Math.* **482** (1997), 121–149.

[110] J.-H. Evertse and R.G. Ferretti, Diophantine inequalities on projective varieties, *Internat. Math. Res. Notices* (2002), 1295–1330.

[111] J.-H. Evertse and H.P. Schlickewei, A quantitative version of the absolute subspace theorem, *J. reine angew. Math.* **548** (2002), 21–127.

[112] J.-H. Evertse, H.P. Schlickewei, and W.M. Schmidt, Linear equations in variables which lie in a multiplicative group, *Ann. of Math.* (2) **155** (2002), 807–836.

[113] G. Faltings, Endlichkeitssätze für abelsche Varietäten über Zahlkörpern, *Invent. Math.* **73** (1983), 349–366. Erratum: ibid. **75** (1984), 381.

[114] G. Faltings, Diophantine approximation on abelian varieties, *Ann. of Math.* (2) **133** (1991), 549–576.

[115] G. Faltings, The general case of S. Lang's conjecture, in *Barsotti Symposium in Algebraic Geometry*, 175–182. V. Cristante and W. Messing (eds), papers from the symposium held in Abano Terme, 1991. Perspectives in Mathematics **15**, Academic Press, Inc., San Diego CA 1994.

[116] G. Faltings and G. Wüstholz (eds), *Rational Points*. Seminar Bonn/Wuppertal 1983/84. Third enlarged edition. Aspects of Mathematics **E6**. Vieweg, Braunschweig 1992. x+311 pp.

[117] G. Faltings and G. Wüstholz, Diophantine approximations on projective spaces, *Invent. Math.* **116** (1994), 109–138.

[118] G. Fano, Sul sistema ∞^2 di rette contenuto in una varietà cubica generale dello spazio a quattro dimensioni, *Atti R. Acc. Sc. Torino* **39** (1904), 778–792.

[119] R.G. Ferretti, An effective version of Faltings' product theorem, *Forum Math.* **8** (1996), 401–427.

[120] R.G. Ferretti, Mumford's degree of contact and Diophantine approximations, *Compos. Math.* **21** (2000), 247–262.

[121] R.H. Fox, On Fenchel's conjecture about F-groups, *Mat. Tidsskr. B.* **1952** (1952), 61–65.

[122] J. Franke, Yu.I. Manin, and Y. Tschinkel, Rational points of bounded height on Fano varieties, *Invent. Math.* **95** (1989), 421–435. Erratum: "Rational points of bounded height on Fano varieties," *ibid.* **102** (1990), 463.

[123] G. Frey, Links between stable elliptic curves and certain diophantine equations, *Ann. Univ. Sarav. Ser. Math.* **1** (1986), 1–40.

[124] M. Fried, On the Sprindžuk–Weissauer approach to universal Hilbert subsets, *Israel J. of Math.* **51** (1985), 347–363.

[125] W. Fulton, *Intersection Theory.* Second edition. Ergebnisse der Mathematik und ihrer Grenzgebiete, 3. Folge. Springer-Verlag, Berlin 1998. xiv+470 pp.

[126] H. Gillet and C. Soulé, Arithmetic intersection theory, *Publ. Math. IHES* **72** (1990), 93–174.

[127] H. Gillet and C. Soulé, An arithmetic Riemann–Roch theorem, *Invent. Math.* **110** (1992), 473–543.

[128] A. Granville, ABC allows us to count squarefrees. *Internat. Math. Res. Notices* (1998), 991–1009.

[129] H. Grauert, Mordells Vermutung über rationale Punkte auf algebraischen Kurven und Funktionenkörper. *Publ. Math. IHES* **25** (1965), 131–149.

[130] P. Griffiths and J. Harris, *Principles of Algebraic Geometry.* Pure and Applied Mathematics. Wiley-Interscience, New York 1978. xii+813 pp.

[131] P. Griffiths and J. King, Nevanlinna theory and holomorphic mappings between algebraic varieties, *Acta Math.* **130** (1973), 145–220.

[132] R. Gross, A note on Roth's theorem, *J. Number Th.* **36** (1990), 127–132.

[133] A. Grothendieck, *Fondements de la géométrie algébrique*, Séminaire Bourbaki, Exposé 236, Vol. 1961/62, Secrétariat Math., Paris 1962.

[134] A. Grothendieck, Eléments de géométrie algébrique I. Le Langage des Schémas. Rédigés avec la collaboration de J. Dieudonné. *Publ. Math. IHES* **4** (1960), 228 pp.

[135] A. Grothendieck, Eléments de géométrie algébrique II. Étude globale élémentaire de quelques classes de morphismes. Rédigés avec la collaboration de J. Dieudonné. *Publ. Math. IHES* **8** (1961), 222 pp.

[136] A. Grothendieck, Eléments de géométrie algébrique III. Étude cohomologique des faisceaux cohérents. Rédigés avec la collaboration de J. Dieudonné. I, *Publ. Math. IHES* **11** (1961), 167 pp.; II, *ibidem* **17** (1963), 91 pp.

[137] A. Grothendieck, Eléments de géométrie algébrique IV. Étude locale des schémas et des morphismes de schémas. Rédigés avec la collaboration de J. Dieudonné. I, *Publ. Math. IHES* **20** (1964), 259 pp.; II, ibid. **24** (1965), 231 pp.; III, ibid. **28** (1966), 255 pp.; IV, ibid. **32** (1967), 361 pp.

[138] A. Grothendieck, Esquisse d'un programme, in *Geometric Galois Actions. 1. Around Grothendieck's "Esquisse d'un programme,"* 5–48. Leila Schneps and Pierre Lochak (eds), London Math. Soc. Lecture Note Ser. **242**. Cambridge University Press, Cambridge 1997. English translation ibid., 243–283.

[139] A. Grothendieck and J.A. Dieudonné, *Eléments de Géométrie Algébrique I.* Grundlehren der mathematischen Wissenschaften **166**, Springer-Verlag, Berlin Heidelberg New York 1971, ix+466 pp.

[140] A. Grothendieck *et al.*, *Revêtements Étales et Groupe Fondemental. Séminaire de Géométrie Algébrique du Bois Marie 1960–1961 (SGA 1).* Dirigé par Alexandre Grothendieck. Augmenté de deux exposés de M. Raynaud. Lecture Notes in Mathematics **224**. Springer-Verlag, Berlin–New York 1971. xxii+447 pp.

[141] W. Gubler, Local and canonical heights of subvarieties, *Ann. Scuola Norm. Sup. Pisa Cl. Sci.* (5) **II** (2003), 711–760.

[142] G.G. Gundersen and W.K. Hayman, The strength of Cartan's version of Nevanlinna theory, *Bull. London Math. Soc.* **36** (2004), 433–454.

[143] J. Hadamard, Résolution d'une question relative aux déterminants, *Bull. Sci. Math.* (2) **17** (1893), 240–246.

[144] M. Hall, The diophantine equation $x^3 - y^2 = k$, in *Computers in Number Theory*, 173–198. Proc. Sci. Res. Council Atlas Sympos. No. 2, Oxford 1969. Academic Press, London 1971.

[145] G.H. Hardy and E.M. Wright, *An Introduction to the Theory of Numbers.* Fifth edition. Oxford at the Clarendon Press 1979. xvi+426 pp.

[146] J. Harris, *Algebraic Geometry: A First Course.* Graduate Texts in Mathematics **133**. Springer-Verlag, Berlin 1992. xix+328 pp.

[147] J. Harris and J.H. Silverman, Bielliptic curves and symmetric products, *Proc. Amer. Math. Soc.* **112** (1991), 347–356.

[148] R. Hartshorne, *Algebraic Geometry.* Graduate Texts in Mathematics **52**. Springer-Verlag, New York–Heidelberg 1977. xvi+496 pp.

[149] W.K. Hayman, *Meromorphic Functions.* Oxford Mathematical Monographs. Oxford at the Clarendon Press 1964. xiv+191 pp. Reprinted 1975 (with Appendix). xiv+195 pp.

[150] E. Hewitt and K.A. Ross, *Abstract Harmonic Analysis. Volume I: Structure of Topological Groups, Integration Theory, Group Representations.* Second edition. Grundlehren der mathematischen Wissenschaften **115**. Springer-Verlag, Berlin 1994. viii+519 pp.

[151] E. Hewitt and K.A. Ross, *Abstract Harmonic Analysis. Volume II: Structure and Analysis for Compact Groups: Analysis on Local Compact Abelian Groups.* Second printing. Grundlehren der mathematischen Wissenschaften **152**. Springer-Verlag, Berlin 1994. viii+771 pp.

[152] E. Hille, *Analytic Function Theory.* Vol. II. Introductions to Higher Mathematics. Ginn & Co., Boston MA–New York–Toronto, ON. 1962. xii+496 pp.

[153] M. Hindry and J.H. Silverman, *Diophantine Geometry: An Introduction.* Graduate Texts in Mathematics **201**. Springer-Verlag, New York 2000. xiv+558 pp.

[154] L.-K. Hua, On Waring's problem, *Quart. J. Math. Oxford Ser.* (2) **9** (1938), 199–202.

[155] D. Husemöller, *Elliptic Curves.* With an appendix by Ruth Lawrence. Graduate Texts in Mathematics **111**. Springer-Verlag, New York 1987. xvi+350 pp.

[156] N. Jacobson, *Basic Algebra I.* First edition. W. H. Freeman & Co., San Francisco 1974. xvi+472 pp. Second edition. W. H. Freeman & Company, New York 1985. xviii+499 pp.

[157] N. Jacobson, *Basic Algebra II.* First edition. W. H. Freeman & Co., San Francisco 1980. xix+666 pp. Second edition. W. H. Freeman & Company, New York 1989. xviii+686 pp.

[158] G.J.O. Jameson, *The Prime Number Theorem.* London Math. Soc. Student Texts **53**. Cambridge University Press 2003. x+252 pp.

[159] S. Katok, *Fuchsian Groups.* Chicago Lectures in Mathematics. The University of Chicago Press, Chicago IL 1992. x+175 pp.

[160] M. Kim, Geometric height inequalities and the Kodaira–Spencer map, *Compos. Math.* **105**, No. 1 (1997), 43–54. erratum: ibid **121**, No. 2 (2000), 219.

[161] M. Kim, D.S. Thakur, and J.F. Voloch, Diophantine approximation and deformation, *Bull. Soc. Math. France* **128** (2000), 585–598.

[162] H. Koch, *Number Theory: Algebraic Numbers and Functions*. Translated from the German by David Kramer. Graduate Studies in Mathematics **24**. AMS, Providence RI 2000. xviii+368 pp.

[163] M. Krasner, Nombre des extensions d'un degré donné d'un corps p-adique, in *Les Tendances Géom. en Algèbre et Théorie des Nombres*, 143–169, Editions du Centre National de la Recherche Scientifique, Paris 1966.

[164] E. Landau, *Vorlesungen über Zahlentheorie. Bd. II: Aus der analytischen und geometrischen Zahlentheorie*. Hirzel, Leipzig 1927. Reprinted, Chelsea Publ. Co. 1947. viii+308 pp.

[165] S. Lang, *Abelian Varieties*. Interscience Publishers, New York 1959. x+169 pp. Reprinted, Springer-Verlag, New York–Berlin 1983. xii+256 pp.

[166] S. Lang, Integral points on curves, *Publ. Math. IHES* **6** (1960), 319–335.

[167] S. Lang, Division points on curves, *Annali Mat. Pura Appl.* (4) **70** (1965), 229–234.

[168] S. Lang, *Introduction to Algebraic and Abelian Functions*. Second edition. Graduate Texts in Mathematics **89**. Springer-Verlag, New York–Berlin 1982. ix+169 pp.

[169] S. Lang, *Fundamentals of Diophantine Geometry*. Springer-Verlag, New York 1983. xviii+370 pp.

[170] S. Lang, *Introduction to Complex Hyperbolic Spaces*. Springer-Verlag, New York 1987. viii+271 pp.

[171] S. Lang, *Number Theory III: Diophantine Geometry*. Encyclopaedia of Mathematical Sciences, Vol. **60**. Springer-Verlag, Berlin 1991, xiv+296 pp.

[172] S. Lang, *Algebraic Number Theory*. Second edition. Graduate Texts in Mathematics **110**. Springer-Verlag, New York 1994. xiv+357 pp.

[173] S. Lang, *Algebra*. Revised third edition. Graduate Texts in Mathematics **211**. Springer-Verlag, New York 2002. xvi+914 pp.

[174] S. Lang and W. Cherry, *Topics in Nevanlinna Theory*. With an Appendix by Zhuan Ye. Lecture Notes in Mathematics **1433**. Springer-Verlag, Berlin 1990. 174 pp.

[175] S. Lang and H. Trotter, Continued fractions for some algebraic numbers, *J. reine angew. Math.* **255** (1972), 112–134; addendum, ibid. **267** (1974), 219–220.

[176] M. Langevin, Cas d'égalité pour le théorème de Mason et applications de la conjecture (abc), *C. R. Acad. Sci. Paris Sér. I Math.* **317** (1993), 441–444.

[177] M. Langevin, Liens entre le théorème de Mason et la conjecture (abc), in *Number Theory* (Ottawa ON 1996), 187–213. CRM Proc. Lecture Notes **19**, AMS, Providence RI 1999.

[178] M. Laurent, Équations diophantiennes exponentielles, *Invent. Math.* **78** (1984), 299–327.

[179] M. Laurent, M. Mignotte, and Yu. Nesterenko, Formes linéaires en deux logarithmes et déterminants d'interpolation, *J. Number Th.* **55** (1995), 285–321.

[180] D.H. Lehmer, Factorization of certain cyclotomic functions, *Ann. of Math.* **34** (1933), 461–479.

[181] C.G. Lekkerkerker, *Geometry of Numbers*. Bibliotheca Mathematica, Vol. VIII. Wolters-Noordhoff Publishing, Groningen; North-Holland Publishing Co., Amsterdam–London 1969. ix+510 pp.

[182] P. Liardet, Sur une conjecture de Serge Lang. (French) *Journées Arithmétiques de Bordeaux* (Conf., Univ. Bordeaux, Bordeaux 1974), 187–210. Astérisque **24–25**, Soc. Math. France, Paris 1975.

[183] R. Louboutin, Sur la mesure de Mahler d'un nombre algébrique, *C. R. Acad. Sci. Paris Sér. I Math.* **296** (1983), 707–708.

[184] H. Luckhardt, Herbrand-Analysen zweier Beweise des Satzes von Roth: Polynomiale Anzahlschranken, *J. Symbolic Logic* **54** (1989), 234–263.

[185] K. Mahler, On the fractional parts of the powers of a rational number (II), *Mathematika* **4** (1957), 122–124.

[186] K. Mahler, *Lectures on Diophantine Approximations. Part I: g-adic Numbers and Roth's Theorem*. Prepared from the notes by R. P. Bambah of my lectures given at the University of Notre Dame in the Fall of 1957. University of Notre Dame Press, Notre Dame, Ind. 1961. xi+188 pp.

[187] K. Mahler, On some inequalities for polynomials in several variables, *J. London Math. Soc.* **37** (1962), 341–344.

[188] K. Mahler, An inequality for the discriminant of a polynomial, *Michigan Math. J.* **11** (1964), 257–262.

[189] Yu.I. Manin, Rational points of algebraic curves over function fields, *Izv. Akad. Nauk SSSR Ser. Mat* **27** (1963), 1395–1440; *Amer. Math. Soc., Transl., II. Ser.* **50** (1966), 189–234.

[190] Yu.I. Manin, *Cubic Forms. Algebra, Geometry, Arithmetic*. Translated from the Russian by M. Hazewinkel. Second edition. North-Holland Mathematical Library, Vol. 4. North-Holland Publishing Co., Amsterdam 1986. x+326 pp.

[191] H.B. Mann, On linear relations between roots of unity, *Mathematika* **12** (1965), 107–117.

[192] R.C. Mason, The hyperelliptic equation over function fields, *Math. Proc. Cambridge Philos. Soc.* **93** (1983), 219–230.

[193] R.C. Mason, *Diophantine Equations over Function Fields*. London Math. Soc. Lecture Note Ser. **96**. Cambridge University Press 1984. x+125 pp.

[194] D. Masser, On *abc* and discriminants, *Proc. Amer. Math. Soc.* **130**, No. 11 (2002), 3141–3150.

[195] D. Masser and G. Wüstholz, Fields of large transcendence degree generated by values of elliptic functions, *Invent. Math.* **72** (1983), 407–464.

[196] W.S. Massey, *Algebraic Topology: An Introduction*. Reprint of the 1967 edition. Graduate Texts in Mathematics **56**. Springer-Verlag, New York–Heidelberg 1977. xxi+261 pp.

[197] H. Matsumura, *Commutative Algebra*. Second edition. Mathematics Lecture Note Series **56**. W.A. Benjamin, Inc., New York 1970. xii+262 pp. Benjamin/Cummings Publishing Co., Inc., Reading MA 1980. xv+313 pp.

[198] R.B. McFeat, Geometry of numbers in adele spaces, *Dissertationes Math. Rozprawy Mat.* **88** (1971), 49 pp.

[199] M. McQuillan, Division points on semiabelian varieties, *Invent. Math.* **120** (1995), 143–159.

[200] M. Mignotte, Sur l'Équation de Catalan, *C. R. Acad. Sci. Paris Sér. I Math.* **314** (1992), 165–168.

[201] M. Mignotte, Y. Roy, Minorations pour l'équation de Catalan. *C. R. Acad. Sci. Paris Sér. I Math.* **324** (1997), 377–380.

[202] P. Mihăilescu, A class number free criterion for Catalan's conjecture, *J. Number Th.* **99** (2003), 225–231.

[203] P. Mihăilescu, Primary cyclotomic units and a proof of Catalan's conjecture, *J. reine angew. Math.* **572** (2004), 167–195.

[204] J.S. Milne, Abelian varieties, in *Arithmetic Geometry*, 103–150. Cornell and Silverman (eds), papers from the conference at University of Connecticut, Storrs Conn. 1984. Springer, New York 1986.

[205] J.S. Milne, Jacobian varieties, in *Arithmetic Geometry*, 167–212. Cornell and Silverman (eds), papers from the conference at University of Connecticut, Storrs Conn. 1984. Springer, New York 1986.

[206] L. Mirsky, *An Introduction to Linear Algebra*. Oxford at the Clarendon Press 1955. xi+433 pp.

[207] L.J. Mordell, On the rational solutions of the indeterminate equations of the third and fourth degrees. *Proc. Cambridge Philos. Soc.* **21** (1922), 179–192.

[208] A. Moriwaki, Arithmetic height functions over finitely generated fields, *Invent. Math.* **140** (2000), 101–142.

[209] T. Muir, *The Theory of Determinants in the Historical Order of Development. Vol 4: The period 1880 to 1900*. Macmillan & Co. Limited, St. Martin Street, London 1923. xxxi+508 pp.

[210] D. Mumford, The topology of normal singularities of an algebraic surface and a criterion for simplicity, *Publ. Math. IHES* **9** (1961), 5–22.

[211] D. Mumford, A remark on Mordell's conjecture, *Amer. J. Math.* **87** (1965), 1007–1016.

[212] D. Mumford, *Abelian Varieties*. Published for the Tata Institute of Fundamental Research Studies in Mathematics, No. 5. Oxford University Press, London 1970. viii+242 pp.

[213] D. Mumford, *The Red Book of Varieties and Schemes*. Second, expanded edition. Includes the Michigan lectures (1974) on curves and their Jacobians. With contributions by Enrico Arbarello. Lecture Notes in Mathematics **1358**. Springer-Verlag, Berlin 1999. x+306 pp.

[214] D. Mumford, *Tata Lectures on Theta. I: Introduction and Motivation: Theta Functions in One Variable. Basic Results on Theta Functions in Several Variables. II: Jacobian Theta Functions and Differential Equations. III (with M. Nori, P. Norman)*. Progr. Math. **28**, **43**, **97**. Birkhäuser 1983, 1984, 1991. xiii+235 pp., xiv+272 pp., viii+202 pp.

[215] W. Narkiewicz, *Elementary and Analytic Theory of Algebraic Numbers*. Second Edition. PWN–Polish Scientific Publishers and Springer-Verlag, Warszawa 1990. xiv+746 pp.

[216] A. Néron, Problèmes arithmétiques et géométriques rattachés à la notion de rang d'une courbe algébrique dans un corps, *Bull. Soc. Math. France* **80** (1952), 101–166.

[217] A. Néron, Modèles minimaux des variétés abéliennes sur les corps locaux et globaux, *Publ. Math. IHES* **21** (1964), 361–482.

[218] A. Néron, Quasi-fonctions et hauteurs sur les variétés abéliennes, *Ann. of Math.* (2) **82** (1965), 249–331.

[219] R. Nevanlinna, Zur Theorie der meromorphen Funktionen, *Acta Math* **46** (1925), 1–99.

[220] R. Nevanlinna, Über Riemannsche Flächen mit endlich viele Windungspunkten, *Acta Math.* **58** (1932), 295–373.

[221] R. Nevanlinna, *Eindeutige Analytische Funktionen*. Zweite Auflage, Reprint. Grundlehren der mathematischen Wissenschaften **46**. Springer-Verlag, Berlin-New York 1974. x+379 pp.

[222] A. Nitaj, La conjecture *abc*, *Enseign. Math.* (2) **42**, No. 1–2 (1996), 3–24.

[223] E.I. Nochka, Defect relations for meromorphic curves (Russian) *Izv. Akad. Nauk Moldav. SSR Ser. Fiz.-Tekhn. Mat. Nauk* No. 1 **(1982)**, 41–47, 79.

[224] E.I. Nochka, On the theory of meromorphic functions, *Sov. Math., Dokl* **27** (1983), 377–381; transl. from *Dokl. Akad. Nauk SSSR* **269** (1983), 547–552.

[225] J. Noguchi, A higher dimensional analogue of Mordell's conjecture over function fields, *Math. Ann.* **258** (1981), 207–212.

[226] J. Noguchi, Nevanlinna–Cartan theory over function fields and a diophantine equation, *J. reine angew. Math.* **487** (1997), 61–83. Correction: *ibid.* **497** (1998), 235.

[227] D.G. Northcott, An inequality in the theory of arithmetic varieties, *Proc. Cambridge Philos. Soc.* **45** (1949), 502–509.

[228] D.G. Northcott, A further inequality in the theory of arithmetic varieties, *Proc. Cambridge Philos. Soc.* **45** (1949), 510–518.

[229] J. Oesterlé, Nouvelles approches du "théorème" de Fermat, *Séminaire Bourbaki*, Exposé 694, Vol. 1987/88, Astérisque **161/162** (1988), 165–186.

[230] A. Ogg, Elliptic curves and wild ramification, *Amer. J. Math.* **89** (1967), 1–21.

[231] Ch.F. Osgood, A number theoretic-differential equations approach to generalizing Nevanlinna theory, *Indian J. Math.* **23** (1981), 1–15.

[232] Ch.F. Osgood, Sometimes effective Thue–Siegel–Roth–Schmidt–Nevanlinna bounds, or better, *J. Number Th.* **21** (1985), 347–389.

[233] O. Perron, *Die Lehre von den Kettenbrüchen. Bd. I. Elementare Kettenbrüche*. 3. Aufl., Teubner Verlagsgesellschaft, Stuttgart 1954, vi+194 pp.

[234] E. Peyre, Points de hauteur bornée, topologie adélique et mesures de Tamagawa, 22nd Journées Arithmétiques (Lille 2001), *J. Théor. Nombres Bordeaux* **15**, No. 1 (2003), 319–349.

[235] B. Poonen, Mordell–Lang plus Bogomolov, *Invent. Math.* **137** (1999), 413–425.

[236] A. Prékopa, On logarithmic concave measures and functions, *Acta Sci. Math. (Szeged)* **34** (1973), 335–343.

[237] M. Raynaud, Courbes sur une variété abélienne et points de torsion, *Invent. Math.* **71** (1983), 207–233.

[238] M. Raynaud, Sous-variétés d'une variété abélienne et points de torsion, in *Arithmetic and Geometry I*, 327–352. J. Coates and S. Helgason (eds), Progr. Math. **35**, Birkhäuser Boston Inc. 1983.

[239] L. Rédei, *Algebra. Vol. 1*. International Series of Monographs in Pure and Applied Mathematics **91**. Oxford: Pergamon Press 1967. xviii+823 pp.

[240] G. Rémond, Décompte dans une conjecture de Lang, *Invent. Math.* **142** (2000), 513–545.

[241] G. Rémond, Sur le théorème du produit, 21st Journées Arithmétiques (Rome, 2001), *J. Théor. Nombres Bordeaux* **13**, No. 1 (2001), 287–302.

[242] G. Rémond, Approximation diophantienne sur les variétés semi-abéliennes, *Ann. Sci. Éc. Norm. Sup.* (4) **36**, No. 2 (2003), 191–212.

[243] P. Ribenboim, *Catalan's Conjecture. Are* 8 *and* 9 *the Only Consecutive Powers?* Academic Press, Inc., Boston MA 1994. xvi+364 pp.

[244] P. Roquette, *Analytic Theory of Elliptic Functions over Local Fields*. Hamburger mathematische Einzelschriften (N.F.), Heft 1. Vandenhoeck & Ruprecht, Göttingen 1970. 90 pp.

[245] J.B. Rosser and L. Schoenfeld, Approximate formulas for some functions of prime numbers, *Illinois J. Math.* **6** (1962), 64–94.

[246] D. Roy, Approximation to real numbers by cubic algebraic integers. II, *Ann. of Math.* (2) **158** (2003), 1081–1087.

[247] D. Roy and J. L. Thunder, An absolute Siegel's lemma, *J. reine angew. Math.* **476** (1996), 1–26.

[248] D. Roy and J. L. Thunder, Addendum and erratum to: "an absolute Siegel's lemma", *J. reine angew. Math.* **508** (1999), 47–51.

[249] M. Ru, *Nevanlinna Theory and its Relation to Diophantine Approximation*. World Scientific Publishing Co., Inc., River Edge NJ 2001. xiv+323 pp.

[250] M. Ru and P. Vojta, Schmidt's subspace theorem with moving targets, *Invent. Math.* **127** (1997), 51–65.

[251] W. Rudin, *Functional Analysis*. McGraw-Hill Series in Higher Mathematics. McGraw-Hill Book Co., New York–Düsseldorf–Johannesburg 1973. xiii+397 pp. Second edition. International Series in Pure and Applied Mathematics. McGraw-Hill, Inc., New York 1991. xviii+424 pp.

[252] W. Rudin, *Real and Complex Analysis*. Third edition. McGraw-Hill Book Co., New York 1987. xiv+416 pp.

[253] W. Rudin, *Fourier Analysis on Groups*. Reprint of the 1962 original. Wiley-Interscience, New York 1990. ix+285 pp.

[254] R. Rumely, On Bilu's equidistribution theorem, in *Spectral Problems in Geometry and Arithmetic. Iowa City IA 1997*, 159–166. Contemp. Math. **237**, AMS, Providence RI 1999.

[255] C. Runge, Ueber ganzzahlige Lösungen von Gleichungen zwischen zwei Veränderlichen, *J. reine angew. Math.* **100** (1887), 425–435.

[256] T. Saito, Conductor, discriminant, and the Noether formula of arithmetic surfaces, *Duke Math. J.* **57** (1988), 151–173.

[257] S.H. Schanuel, Heights in number fields, *Bull. Soc. Math. France* **107** (1979), 433–449.

[258] A. Schinzel, On the product of the conjugates outside the unit circle of an algebraic number, *Acta Arith.* **24** (1973), 385–399. Addendum ibid. **26** (1974/75), 329–331.

[259] A. Schinzel, *Polynomials with Special Regard to Reducibility*. With an Appendix by Umberto Zannier. Encyclopedia of Mathematics and its Applications **77**. Cambridge University Press, Cambridge 2000. x+558 pp.

[260] H.P. Schlickewei, S-unit equations over number fields, *Invent. Math.* **102** (1990), 95–107.

[261] H.P. Schlickewei, The quantitative subspace theorem for number fields, *Compos. Math.* **82** (1992), 245–273.

[262] H.P. Schlickewei, Equations in roots of unity, *Acta Arith.* **76** (1996), 99–108.

[263] W.M. Schmidt, On heights of algebraic subspaces and diophantine approximations, *Ann. of Math.* (2) **85** (1967), 430–472.

[264] W.M. Schmidt, On simultaneous approximations of two algebraic numbers by rationals, *Acta Math.* **119** (1967), 27–50.

[265] W.M. Schmidt, Asymptotic formulae for point lattices of bounded determinant and subspaces of bounded height, *Duke Math. J.* **35** (1968), 327–339.

[266] W.M. Schmidt, Simultaneous approximation to algebraic numbers by rationals, *Acta Math.* **125** (1970), 189–201.

[267] W.M. Schmidt, Norm form equations, *Ann. of Math.* (2) **96** (1972), 526–551.

[268] W.M. Schmidt, *Diophantine Approximation*. Lecture Notes in Mathematics **785**. Springer-Verlag, Berlin 1980. x+299 pp.

[269] W.M. Schmidt, The subspace theorem in Diophantine approximations, *Compos. Math.* **69** (1989), 121–173.

[270] W.M. Schmidt, Eisenstein's theorem on power series expansions of algebraic functions, *Acta Arith.* **56** (1990), 161–179.

[271] W.M. Schmidt, *Diophantine Approximations and Diophantine Equations*. Lecture Notes in Mathematics **1467**. Springer-Verlag, Berlin 1991. viii+217 pp.

[272] W.M. Schmidt, Heights of algebraic points lying on curves or hypersurfaces, *Proc. Amer. Math. Soc.* **124**, No. 10 (1996), 3003–3013.

[273] W.M. Schmidt, Heights of points on subvarieties of $\mathbb{G}_m^n$, in *Number Theory (Paris, 1993–1994)*, 157–187. London Math. Soc. Lecture Note Ser. **235**. Cambridge University Press, Cambridge 1996.

[274] H. Seifert and W. Threlfall, *Lehrbuch der Topologie*. B.G. Teubner VII, Leipzig und Berlin 1934. Also Herbert Seifert and William Threlfall, *Seifert and Threlfall: A Textbook of Topology*. Translated from the German edition of 1934 by M.A. Goldman. With a preface by J.S. Birman. With *Topology of 3-dimensional Fibered Spaces* by Seifert. Translated from the German by Wolfgang Heil. Pure and Applied Mathematics **89**. Academic Press, Inc.. New York–London 1980. xvi+437 pp.

[275] J.-P. Serre, Géométrie algébrique et géométrie analytique, *Ann. Inst. Fourier* **6** (1956), 1–42.

[276] J.-P. Serre, *Corps Locaux*. Deuxième édition. Publications de l'Université de Nancago, No. VIII. Hermann, Paris 1968. 245 pp.

[277] J.-P. Serre, *Lectures on the Mordell–Weil Theorem*. Translated from the French and edited by Martin Brown from notes by Michel Waldschmidt. Aspects of Mathematics, E15. Friedr. Vieweg & Sohn, Braunschweig 1989. x+218 pp.

[278] J.-P. Serre, *Galois Cohomology*. Translated from French by P. Ion. Springer-Verlag, Berlin 1997. x+210 pp.

[279] I.R. Shafarevich, *Basic Algebraic Geometry. 1. Varieties in Projective Space*. Second edition. Translated from the 1988 Russian edition and with notes by Miles Reid. Springer-Verlag, Berlin 1994. xx+303 pp.

[280] I.R. Shafarevich, *Basic Algebraic Geometry. 2. Schemes and Complex Manifolds*. Second edition. Translated from the 1988 Russian edition by Miles Reid. Springer-Verlag, Berlin 1994. xiv+269 pp.

[281] T. Shimizu, On the theory of meromorphic functions, *Japanese Journ. of Math.* **6** (1929), 119–171.

[282] X, The integer solutions of the equation $y^2 = ax^n + bx^{n-1} + \ldots + k$, *J. London Math. Soc.* **1** (1926), 66–68. Also Gesammelte Abhandlungen, Bd. I, Springer-Verlag, Berlin-Heidelberg-New York 1966, 207–208.

[283] C.L. Siegel, Über einige Anwendungen diophantischer Approximationen, Abh. Preuß. Akad. Wissen. Phys.-math. Klasse 1929, Nr. 1. Also Gesammelte Abhandlungen, Bd. I, Springer-Verlag, Berlin–Heidelberg–New York 1966, 209–274.

[284] J.H. Silverman, *The Arithmetic of Elliptic Curves*. Graduate Texts in Mathematics **106**. Springer-Verlag, New York 1986. xii+400 pp.

[285] J.H. Silverman, Rational points on K3 surfaces: A new canonical height, *Invent. Math.* **105** (1991), 347–373.

[286] J.H. Silverman, *Advanced Topics in the Arithmetic of Elliptic Curves*. Graduate Texts in Mathematics **151**. Springer-Verlag, New York 1994. xiv+525 pp.

[287] C.J. Smyth, On the product of the conjugates outside the unit circle of an algebraic integer, *Bull. London Math. Soc.* **3** (1971), 169–175.

[288] C.J. Smyth, On the measure of totally real algebraic numbers, I, *J. Austral. Math. Soc. Ser. A* **30** (1980/81), 137–149; II, *Math. Comp.* **37** (1981), 205–208.

[289] V.G. Sprindžuk, Arithmetic specializations in polynomials, *J. reine angew. Math.* **340** (1983), 26–52.

[290] N. Steinmetz, Eine Verallgemeinerung des zweiten Nevanlinnaschen Hauptsatzes, *J. reine angew. Math.* **368** (1986), 134–141.

[291] C.L. Stewart and R. Tijdeman, On the Oesterlé–Masser conjecture, *Monatsh. Math.* **102** (1986), 251–257.

[292] W. Stoll, *Value Distribution on Parabolic Spaces*. Lecture Notes in Mathematics **600**. Springer-Verlag, Berlin-New York 1977. viii+216 pp.

[293] W. W. Stothers, Polynomial identities and Hauptmoduln, *Quart. J. Math. Oxford Ser.* (2) **32** (1981), 349–370.

[294] T. Struppeck and J.D. Vaaler, Inequalities for heights of algebraic subspaces and the Thue–Siegel principle, in *Analytic Number Theory* (Allerton Park IL 1989), 493–528. *Progr. Math.* **85**, Birkhäuser Boston, Boston MA 1990.

[295] L. Szpiro (ed.), *Séminaire sur les pinceaux arithmétiques: la conjecture de Mordell*. Papers form the seminar held at the École Normale Supériore, Paris 1983–84. Astérisque **127**. Soc. Math. France, Paris 1985. x+287 pp.

[296] L. Szpiro, E. Ullmo, and S. Zhang, Equirépartition des petits points, *Invent. Math.* **127** (1997), 337–347.

[297] R. Taylor and A. Wiles, Ring-theoretic properties of certain Hecke algebras, *Ann. of Math.* (2) **141** (1995), 553–572.

[298] A. Thue, Über Annäherungswerte algebraischer Zahlen, *J. reine angew. Math.* **135** (1909), 284–305. Also *Selected Mathematical Papers of Axel Thue*. With an introduction by Carl Ludwig Siegel. Universitetsforlaget, Oslo-Bergen-Tromsø 1977, 232–253.

[299] J.L. Thunder, Asymptotic estimates for rational points of bounded height on flag varieties, *Compos. Math.* **88**, No. 2 (1993), 155–186.

[300] R. Tijdeman, On the equation of Catalan, *Acta Arith.* **29** (1976), 197–209.

[301] E. Ullmo, Positivité et discrétion des points algébriques des courbes, *Ann. of Math.* (2) **147** (1998), 167–179.

[302] J.D. Vaaler, A geometric inequality with applications to linear forms, *Pacific J. Math.* **83** (1979), 543–553.

[303] M. van der Put, The product theorem, in *Diophantine Approximation and Abelian Varieties. Introductory Lectures*, 77–82. Papers from the Conference held in Soesterberg, April 12–16, 1992. B. Edixhoven and J.-H. Evertse (eds), Lecture Notes in Mathematics **1566**. Springer-Verlag, Berlin 1993.

[304] M. van Frankenhuysen, A lower bound in the *abc* conjecture, *J. Number Th.* **82** (2000), 91–95.

[305] M. van Frankenhuysen, The ABC conjecture implies Vojta's height inequality for curves, *J. Number Th.* **95** (2002), 289–302.

[306] P. Vojta, A Diophantine conjecture over $\overline{\mathbb{Q}}$, in *Séminaire de théorie des nombres, Paris 1984–85*, 241–250. *Progr. Math.* **63**, Birkhäuser Boston, Boston MA 1986.

[307] P. Vojta, *Diophantine Approximations and Value Distribution Theory*. Lecture Notes in Mathematics **1239**. Springer-Verlag, Berlin 1987. x+132 pp.

[308] P. Vojta, Mordell's conjecture over function fields, *Invent. Math.* **98** (1989), 115–138.

[309] P. Vojta, A refinement of Schmidt's subspace theorem, *Amer. J. Math* **111**, No. 3 (1989), 489–518.

[310] P. Vojta, Siegel's theorem in the compact case, *Ann. of Math.* (2) **133** (1991), 509–548.

[311] P. Vojta, On algebraic points on curves, *Compos. Math.* **78**, No. 1 (1991), 29–36.

[312] P. Vojta, A generalization of theorems of Faltings and Thue–Siegel–Roth–Wirsing, *J. Amer. Math. Soc.* **5** (1992), 763–804.

[313] P. Vojta, Applications of arithmetic algebraic geometry to Diophantine approximations, in *Arithmetic Algebraic Geometry*, 164–208. E. Ballico (ed.), lectures from the Second C.I.M.E. Session held in Trento, 1991. Lecture Notes in Mathematics **1553**. Springer-Verlag, Berlin 1993.

[314] P. Vojta, Integral points on subvarieties of semiabelian varieties, I, *Invent. Math.* **126**, No. 1 (1996), 133–181.

[315] P. Vojta, Roth's theorem with moving targets, *Internat. Math. Res. Notices* (1996), 109–114.

[316] P. Vojta, On Cartan's theorem and Cartan's conjecture, *Amer. J. Math.* **119** (1997), 1–17.

[317] P. Vojta, A more general *abc* conjecture, *Int. Math. Res. Notices* (1998), 1103–1116.

[318] P. Vojta, Nevanlinna theory and diophantine approximation, in *Several Complex Variables*, 535–564. M. Schneider (ed.) *et al.*, Berkeley CA 1995–96, MSRI Publ. **37**, Cambridge University Press, Cambridge 1999.

[319] J.F. Voloch, Diagonal equations over function fields, *Bol. Soc. Brasil. Mat.* **16** (1985), 29–39.

[320] J.T.-Y. Wang, The truncated second main theorem of function fields, *J. Number Th.* **58** (1996), 139–157.

[321] F. Warner, *Foundations of Differentiable Manifolds and Lie Groups*. Corrected reprint of the 1971 edition. Graduate Texts in Mathematics **94**, Springer-Verlag, New York 1983. ix+272 pp.

[322] L.C. Washington, *Introduction to Cyclotomic Fields*. Second edition. Graduate Texts in Mathematics **83**. Springer-Verlag, New York 1997. xiv+487 pp.

[323] W.C. Waterhouse, *Introduction to Affine Group Schemes*. Graduate Texts in Mathematics **66**. Springer-Verlag, New York–Berlin 1979. xi+164 pp.

[324] A. Weil, L'arithmétique sur les courbes algébriques, *Acta Math.* **52** (1929), 281–315. Also *Œuvres Scientifiques – Collected Papers*. Vol. I, Corrected Second Printing, Springer-Verlag, New York–Heidelberg–Berlin 1980, 11–45.

[325] A. Weil, Sur un théorème de Mordell, *Bull. Sc. Math.* (2) **54** (1929), 182–191. Also *Œuvres Scientifiques–Collected Papers*. Vol. I, Corrected Second Printing, Springer-Verlag, New York–Heidelberg–Berlin 1980, 47–56.

[326] A. Weil, *Arithmétique et Géométrie sur les Variétés Algébriques*. Actualités Scientifiques et Industrielles, No. 206. Hermann, Paris 1935. 3–16. Also *Œuvres Scientifiques–Collected Papers*. Vol. I, Corrected Second Printing, Springer-Verlag, New York–Heidelberg–Berlin 1980, 87–100.

[327] A. Weil, *Variétes Abéliennes et Courbes Algébriques*. Hermann, Paris 1948. 163 pp.

[328] A. Weil, Arithmetic on algebraic varieties, *Ann. of Math.* (2) **53** (1951), 412–444. Also *Œuvres Scientifiques–Collected Papers*. Vol. I, Corrected Second Printing, Springer-Verlag, New York–Heidelberg–Berlin 1980, 454–486.

[329] R. Weissauer, Der Hilbertsche Irreduzibilitätssatz, *J. reine angew. Math.* **333** (1982), 203–220.

[330] A. Weitsman, A theorem on Nevanlinna deficiencies, *Acta Math.* **128** (1972), 41–52.

[331] A. Wiles, Modular elliptic curves and Fermat's Last Theorem, *Ann. of Math.* (2) **141** (1995), 443–551.

[332] P.M. Wong, On the second main theorem of Nevanlinna theory, *Amer. J. Math.* **111** (1989), 549–583.
[333] K. Yamanoi, The second main theorem for small functions and related problems, *Acta Math.* **192** (2004), 225–294.
[334] Z. Ye, A sharp form of Nevanlinna's second main theorem of several complex variables, *Math. Z.* **222** (1996), 81–95.
[335] K. Yu, p-adic logarithmic forms and group varieties. I, *J. reine angew. Math.* **502** (1998), 29–92; II, *Acta Arithmetica* **89** (1999), 337–378.
[336] D. Zagier, Algebraic numbers close to both 0 and 1, *Math. Comp.* **61** (1993), 485–491.
[337] U. Zannier, *Some Applications of Diophantine Approximation to Diophantine Equations, with Special Emphasis on the Schmidt Subspace Theorem.* Forum, Editrice Universitaria Udinese, Udine 2003. 69 pp.
[338] O. Zariski, P. Samuel, *Commutative Algebra.* Vol. II. Reprint of the 1960 edition. Graduate Texts in Mathematics **29**. Springer-Verlag, New York–Heidelberg–Berlin 1975. x+414 pp.
[339] S. Zhang, Positive line bundles on arithmetic surfaces, *Ann. of Math.* (2) **136** (1992), 569–587.
[340] S. Zhang, Positive line bundles on arithmetic varieties, *J. Amer. Math. Soc.* **8** (1995), 187–221.
[341] S. Zhang, Small points and adelic metrics, *J. Alg. Geom.* **4** (1995), 281–300.
[342] S. Zhang, Equidistribution of small points on abelian varieties, *Ann. of Math.* **147** (1998), 159–165.
[343] A. Zygmund, *Trigonometric Series.* Vols I, II. Third edition. With a foreword by Robert A. Fefferman. Cambridge Mathematical Library. Cambridge University Press, Cambridge 2002. xii; Vol. I: xiv+383 pp.; Vol. II: viii+364 pp.

Glossary of Notation

$A \subset B$	A is a subset of B (possibly equal), xv
$B \setminus A$	complement of A in B, xv
$\|A\|$	number of elements of A, xv
id	identity map, xv
$\mathbb{N}, \mathbb{Z}$	natural numbers (0 included), rational integers, xv
$\mathbb{Q}, \mathbb{R}, \mathbb{C}$	rational, real and complex numbers, xv
$\mathbb{R}_+$	non-negative real numbers, xv
δ_{ij}	Kronecker symbol, xv
$\mathbb{R}^X$	real functions on X, xv
$f = O(g), f = o(g)$	Landau symbols, xv
$f \ll g, g \gg f$	Vinogradov symbols, xv
$\lfloor x \rfloor$	$\max\{m \in \mathbb{Z} \mid m \leq x\}$, xv
$\lceil x \rceil$	$\min\{m \in \mathbb{Z} \mid m \geq x\}$, xv
$\mathrm{GCD}(a, b)$	greatest common divisor of $a, b \in \mathbb{Z}$. xvi
$a\|b$	a divides b in $\mathbb{Z}$, xvi
$\pi(x)$	number of primes up to $x \in \mathbb{R}_+$, xvi
$R^\times$	multiplicative units of a commutative ring R with 1, xvi
V^*	dual of a vector space V, xvi
$[g_1, \ldots, g_m]$	ideal generated by $g_1, \ldots, g_m$, 83
$\mathrm{char}(K)$	characteristic of a field K, xvi
$\mathbb{F}_q$	field with q elements, xvi
$K[x]$	polynomials in variable x and coefficients in K, xvi
$\deg(\alpha)$	degree of algebraic number α, xvi
$\mathbf{x}$	vector with entries x_i, xvi
$\overline{K}$	algebraic closure of the field K, xvi
$\|\ \|, \|\ \|_v$	absolute value (with respect to place v), 1, 2
$w\|v$	place w is an extension of place v, 2
K_v	completion of K at place v, 2
$\mathbb{Q}_p, \mathbb{Z}_p, \|\ \|_p$	p-adic numbers, integers and absolute value, 3
$N_{L/K}$	norm from L to K, 3
$R_v, k(v)$	valuation ring and residue field for place v, 3
$\overline{x}$	reduction of x to the residue field, 3
$f_{w/v}, e_{w/v}$	residue degree and ramification index, 4
O_K	algebraic integers of the number field K, 4
$\mathrm{Gal}(L/K)$	Galois group of field extension L/K, 6
$\\|x\\|_w, \|x\|_w$	normalizations for $w\|v$ on finite extension L/K, 6, 9
M_K	normalized absolute values of a number field K, 11

$\lvert\ \rvert_Z$	absolute value on function field in prime divisor Z, 12
M_X	standard absolute values on function field of X, 12
$h(P)$	absolute logarithmic height of $P \in \mathbb{P}^n_{\overline{K}}$, 16
$H(P)$	$e^{h(P)}$ multiplicative height of $P \in \mathbb{P}^n_{\overline{K}}$, 16
$\log^+ t$	$\max(0, \log t)$, 16
$h(\alpha)$	height of algebraic number α, 16
$O_{S,K}, U_{S,K}$	S-integers and S-units of number field, 17
$\lVert x \rVert_{w,K}$	$\lvert N_{L_w/K_v}(x)\rvert_v$, 20
$h(f)$	height of a polynomial f, 21
$\lvert f\rvert_v$	Gauss norm of f with respect to place v, 22
$M(f), M(\alpha)$	Mahler measure of f and algebraic number α, 22, 28
$\mathbb{T}$	the unit circle in $\mathbb{C}$, 22
$\ell_p(f)$	ℓ_p-norm of the coefficients of the polynomial f, 24
$[f]_p$	ℓ_p-norm of the hypercube representation, 31
$\mathrm{sh}(d,e)$	shuffle of type (d,e), 31
$\mathcal{D} = (s_D;\ L, \mathbf{s};\ M, \mathbf{t})$	presentation of Cartier divisor D, 35
$\lambda_{\mathcal{D}}(P)$	local height of P relative to $\mathcal{D}$, 35
$\lambda_f(P)$	local height relative to the rational function f, 36
$\lambda_{\mathcal{D}}(P,v), \lambda_{\mathcal{D}}(P,u)$	local heights relative to $v \in M_F$ and $u \in M_{\overline{K}}$, 39
$h_\lambda(P)$	global height of P relative to local height λ, 40
$\mathbf{h_c}$	class of height functions relative to $\mathbf{c} \in \mathrm{Pic}(X)$, 41
h_φ	Weil height relative to morphism φ, 43
$\varphi \# \psi$	join of morphisms φ, ψ to projective spaces, 43
$d(\mathbf{p}), h(\mathbf{p})$	degree and height of presentation $\mathbf{p}$, 47
$\lvert\boldsymbol{\alpha}\rvert$	$\alpha_0 + \cdots + \alpha_N$ for $\boldsymbol{\alpha} \in \mathbb{N}^{N+1}$, 49
$\mathbf{x}^{\boldsymbol{\alpha}}$	$x_0^{\alpha_0} \cdots x_N^{\alpha_N}$ for $\mathbf{x} \in \mathbb{R}^{N+1}$, 49
$\overline{L} = (L, \lVert\ \rVert)$	metrized line bundle, 58
$\widehat{D}, O(\widehat{D})$	Néron divisor and its metrized line bundle, 61, 62
$\lambda_{\widehat{D}}(P)$	local height of P relative to $\widehat{D}$, 61
$\lambda_{\widehat{D}}(P,u), \lambda_{\widehat{D}}(P,v)$	local heights relative to $v \in M_F$ and $u \in M_{\overline{K}}$, 62, 64
$h_{\overline{L}}(P)$	global height of P relative to $\overline{L}$, 64
$H_u(\mathbf{x}), H_u(A)$	Arakelov norm for vector $\mathbf{x}$ and matrix A, 66, 67
$h_{\mathrm{Ar}}, H_{\mathrm{Ar}}$	Arakelov height, multiplicative version, 66, 67, 69
$\wedge^m W$	mth exterior power of vector space W, 67
A^t	transpose of the matrix A, 69
$H_u^*(A)$	dual local height, 69
$\eta_u(\psi)$	distorsion factor of $\psi \in PGL(n+1, \overline{\mathbb{Q}}_u)$, 69
$\delta_u(\mathbf{x},\mathbf{y})$	projective distance, 69
$H_{\mathrm{Ar}}^{\mathrm{row}}(A)$	multiplicative Arakelov height with respect to rows, 75
$H(A)$	multiplicative height of matrix A, 75
I_M	$M \times M$ unit matrix, 78
$G := (\mathbb{G}_m^n, \cdot, \mathbf{1}_n)$	multiplicative algebraic group $\mathbb{A}^n_{\overline{K}} \setminus \{\mathbf{0}\}$, 82
$\mathbf{e}_j$	standard basis of $\mathbb{Z}^n$, 83
$\varphi_A(\mathbf{x})$	$(\mathbf{x}^{A\mathbf{e}_1}, \ldots, \mathbf{x}^{A\mathbf{e}_n})$ for integer matrix A, 83
$GL(n,\mathbb{Z}), SL(n,\mathbb{Z})$	general and special linear group over $\mathbb{Z}$, 83
$\widetilde{\Lambda}$	division group of subgroup Λ in $\mathbb{Z}^n$, 83

$\rho(\Lambda)$	$[\widetilde{\Lambda} : \Lambda]$, 83
H_Λ	$\{\mathbf{x} \in G \mid \mathbf{x}^{\boldsymbol{\lambda}} = 1\ \forall \lambda \in \Lambda\}$, 83
M_Λ	connected component of H_Λ, 83
$\delta(X)$	essential degree of subvariety X of G, 88
X^*	complement of all torsion cosets in X, 90
X°	complement of all torus cosets in X, 90
$X(H)$	union of $gH \subset X$ for given linear torus H, 91
$\widehat{h}(\mathbf{x})$	standard height of $\mathbf{x} \in \mathbb{G}_m^n$, 94
ε_v	$[K_v : \mathbb{R}]/[K : \mathbb{Q}]$ for $v \mid \infty$ and 0 else, 96
$\boldsymbol{\delta}_\xi$	average of Dirac measures over conjugates of $\xi \in \overline{\mathbb{Q}}$, 101
$C_c(X)$	continuous compactly supported functions on X, 101
$C_c(X)^*$	weak-* dual of $C_c(X)$, 102
(N)	Northcott property, 117
$\mathcal{A}(T)$	algebraic numbers in $\mathcal{A}$ with height $\leq T$, 117
$K^{(d)}$	compositum of degree d extensions of K, 117
$K_{ab}^{(d)}$	maximal abelian subfield of $K^{(d)}$, 117
(B)	Bogomolov property, 120
$V_p(\alpha; K)$	normalized variance of α in number field K, 122
$\mathrm{rank}_{\mathbb{Q}}(\Gamma)$	rank of abelian group Γ, 126
$\binom{\mathbf{m}}{\boldsymbol{\mu}}$	$\prod_{j=1}^m \binom{m_j}{\mu_j}$, 156
$\partial_{\boldsymbol{\mu}}$	$\frac{1}{\mu_1! \cdots \mu_m!} \left(\frac{\partial}{\partial x_1}\right)^{\mu_1} \cdots \left(\frac{\partial}{\partial x_m}\right)^{\mu_m}$, 156
$\mathrm{ind}(P; \mathbf{d}; \boldsymbol{\alpha})$	index of polynomial P at $\boldsymbol{\alpha}$, 156, 226
$V_m(t)$	volume of standard simplex $\sum x_i \leq t$ in $\mathbb{R}^m$, 157
$V_m(\mathbf{t})$	$\sum_{i=1}^n V_m(t_i)$, 157
$\Lambda(\beta)$	$\prod_{v \in S} \min(1, \lvert\beta - \alpha_v\rvert_{v,K})$, 164
$\mathcal{C}(\boldsymbol{\lambda}; N)$	approximation class of size $1/N$ and corner $\boldsymbol{\lambda}$, 164
$\lvert\mathbf{x}\rvert_v$	$\max_i \lvert x_i\rvert_v$, 177
$\mathbf{i}!$	$i_0! \cdots i_n!$, 197
ω^*	star-operator on exterior algebra $\wedge^k V$, 198
$h_{\mathrm{aff}}(\mathbf{x})$	affine height of $\mathbf{x} \in \mathbb{A}_K^n$, 16
$\Pi_v(Q), \Pi(Q)$	(v adic) approximation domain, 201
$V(Q), \mathrm{rank}(\Pi(Q))$	vector space spanned by $K^{n+1} \cap \Pi(Q)$, dimension, 203
$\mathbf{P}(\mathbf{d})$	multihomogeneous polynomials of multidegree $\mathbf{d}$, 204
$\partial_{\mathbf{I}}$	$\partial_{\mathbf{i}_1} \cdots \partial_{\mathbf{i}_m}$ for $\mathbf{I} = (\mathbf{i}_1, \ldots, \mathbf{i}_m)$, 204
$\mathbf{V}_i, (\mathbf{V}/\mathbf{d})$	$(v_{1i}, \ldots, v_{mi})$, $\frac{\lvert\mathbf{v}_1\rvert}{d_1} + \cdots + \frac{\lvert\mathbf{v}_m\rvert}{d_m}$, 204
ε	identity element of a group variety G, 232
m, ι	multiplication and inverse of G as morphisms, 232
$SL(n)_K$	special linear group as group variety, 233
τ_a	translation by a on abelian group A, 235
$[n], A[n]$	multiplication map by n on A and its kernel, 235
$\mathbf{c}_y$	restriction of $\mathbf{c} \in \mathrm{Pic}(X)$ to fibre X_y, 246
id_X	identity map of X, 248
$Pic^0(X)$	Picard variety of X, 249
$\widehat{\varphi}, \widehat{A}$	dual map and dual abelian variety, 251, 255
$\varphi_{\mathbf{c}}(a)$	$\tau_a^*(\mathbf{c}) - \mathbf{c} \in \widehat{A}$ for $a \in A$ and $\mathbf{c} \in \mathrm{Pic}(A)$, 252

$\mathrm{Mor}(X, A)$	morphisms from X to A, 260
$\theta(\mathbf{v}, Z)$	Riemann theta function for $\Lambda = \mathbb{Z}^g + Z \cdot \mathbb{Z}^g$, 262
$\mathbf{c}_+, \mathbf{c}_-$	even and odd part of $\mathbf{c} \in \mathrm{Pic}(A)$, 265
$\mathrm{Hom}(A_1, A_2)$	homomorphisms of abelian varieties, 268
$S_r, C^{(r)}$	symmetric group with r elements, C^r/S_r, 269
j	canonical embedding of C into Jacobian J, 271
Θ	theta divisor, 272
$L(D), \mathcal{L}(D)$	$L \otimes O(D), \mathcal{L} \otimes \mathcal{O}(D)$, 272
j_r	canonical maps from C^r to J, 274
$\Theta^-, \boldsymbol{\theta}, \boldsymbol{\theta}^-$	$[-1]^*\Theta$, classes of Θ, Θ^-, 275
$\widehat{h}_{\mathbf{c}}$	Néron–Tate height associated to $\mathbf{c}$, 287
$\langle a, a' \rangle, \lvert a \rvert$	canonical form of J and associated norm, 294
h_D	height function associated to $\mathrm{cl}(O(D))$, 294
$n_\Gamma(H)$	$\lvert \{Q \in C(\overline{K}) \cap j^{-1}(\Gamma) \mid Q \text{ large}, \lvert j(Q) \rvert \leq H\} \rvert$, 298
$d(\Vert\ \Vert, \Vert\ \Vert')$	distance of locally bounded metrics, 301
$(L, \nu), (\Vert\ \Vert_{\nu,u})_{u \in M}$	rigidified line bundle with canonical metrics, 303, 308
$[P]$	cycle in $Z_0(X_{\overline{K}})$ associated to $P \in X$, 304
$f(Z)$	$\prod_j f(P_j)^{n_j}$ for $Z = \sum_j n_j[P_j]$, 304
$B_0(X)$	kernel of the degree map on $Z_0(X_{\overline{K}})$, 304
$(D, Z)_u$	Néron symbol, 304, 308, 310
$B^1(X)$	divisors algebraically equivalent to zero, 307
$E(P), {}^tE(P')$	$(p_2)_*(E.(\{P\} \times X')), (p_1)_*(E.(X \times \{P'\}))$, 309
ε'_w	$\frac{[F_w:\mathbb{Q}_p]}{[F:\mathbb{Q}]}$ for $w \in M_F$, 315
$\widehat{\mathfrak{d}}_{P/Q}, \widehat{\mathfrak{d}}^u_{P/Q}$	local discriminant (in the place u), 336, 338
$\mathfrak{d}_{P/Q}$	global discriminant in the number field case, 339
$\frac{1}{m}A(K)$	m-division group of $A(K)$ in $A(\overline{K})$, 341
$K(S)$	smallest extension L/K with $S \subset A(L)$, 341
x^g	Galois action of $g \in \mathrm{Gal}(L/K)$ on $x \in X(L)$, 341
K^s	separable algebraic closure of K, 342
$\langle a, g \rangle$	Kummer pairing, 342, 344
$\mu_m(K)$	mth roots of unity in K, 344
$K^{\times m}, H^{\times m}$	mth powers, for subgroup H of $K^\times$, 344
$H^i(G, M)$	Galois cohomology group, 345
K^{ab}	maximal abelian extension of exponent dividing m, 346
K_v^{nr}, K_M^{nr}	max. subextension of K^{ab} unramified over v, M, 346, 346
H_M	$(K_M^{nr})^{\times m} \cap K^\times$ for a set of valuations M, 346
$\sum n_v[v] \in \mathrm{Div}_M(K)$	M-divisors, 346
$\mathrm{div}_M(f)$	M-divisor associated to $f \in K^\times$, 346
$\mathrm{Cl}_M(K), U_M$	M-class group, M-units, 346
Δ'	$\Delta - \{P_0\} \times C - C \times \{P_0\}$, 356
V	Vojta divisor for parameters d_1, d_2, d, 357
ψ	closed embedding of $C \times C$ into $\mathbb{P}^n_K \times \mathbb{P}^n_K$, 357
$O(d), O(\delta_1, \delta_2)$	$O_{\mathbb{P}^m}(d), O_{\mathbb{P}^n \times \mathbb{P}^n}(\delta_1, \delta_2)$, 357
φ_B	closed embedding of $C \times C$ into $\mathbb{P}^m_K$, 358
(V1), (V2), (V3)	conditions for the Vojta divisor, 358, 378

$K[[\mathbf{x}]]$ ring of formal power series, 362

$\|\sum a_{\boldsymbol{\alpha}}\mathbf{x}^{\boldsymbol{\alpha}}\|_{\mathbf{r}}$ $\sum |a_{\boldsymbol{\alpha}}|\mathbf{r}^{\boldsymbol{\alpha}}$, 362

$K\langle \mathbf{r}^{-1}\mathbf{x}\rangle$ $\{f \in K[[\mathbf{x}]] \mid \|f\|_{\mathbf{r}} < \infty\}$, 362

$\rho_{\mathbf{r}}(f)$, $\rho(f)$ spectral norm, for $\mathbf{r} = (1, \ldots, 1)$, 362, 364

$K\{\mathbf{r}^{-1}\mathbf{x}\}$ strictly convergent power series, 364

$\partial_{\boldsymbol{\alpha}} f(\mathbf{x}_0)$ Taylor coefficients at $\mathbf{x}_0$, 365

$\dot{f}$ $\frac{\partial}{\partial t} f$, 368

∂_i $\frac{1}{i!}\partial^i$, 371

(i_1^*, i_2^*) admissible pair in $(P, Q) \in C \times C$, 371

$N_L(X, T)$ number of points in $X(K)$ with $H_L(P) \le T$, 392

R_K, w_K, h_K regulator, number of roots of unity, class number, 392

ζ_K zeta function of number field K, 392

$N_{\mathrm{Ar}}(X, T)$ $N_{O(1)}(X, T)$ for the Arakelov height, 393

$N_{\mathrm{lines}}(X, T)$ contributions of points on rational lines, 395

$\mathrm{rad}(N)$ $\prod_{p|N} p$, radical of $N \in \mathbb{N}$, 402

$\pi_1(X, x_0)$ fundamental group of X, 411

$\Gamma\backslash Y$ quotient of Y by left action of Γ, 411

$\mathbb{H}$ upper half plane in $\mathbb{C}$, 413

$\mathrm{rad}(f)$ $\prod_{f(\alpha)=0}(x - \alpha)$, radical of $f \in K[x]$, 416

$\vartheta_{\mathcal{P}}$ $\sum_{p \in \mathcal{P}} \log p$, 420

$\Psi(N, \mathcal{P})$, $\Psi_{\mathrm{odd}}(N, \mathcal{P})$ number of (odd) $n \le N$ with all prime factors in $\mathcal{P}$, 420

$\overline{E}$ reduction of elliptic curve E, 426

$\mathrm{cond}(E)$ conductor of elliptic curve E, 429

$e_{x/\pi(x)}$ ramification index of x over $\pi(x)$, 436

$\mathbb{D}$ open unit disc in $\mathbb{C}$, 436

F^+, F^- $\max\{F, 0\}, -\min\{F, 0\}$, 445

$n(r, a, f)$ enumarating function for $r > 0, a \in \mathbb{P}^1_{\mathrm{an}}$, 446

$N(r, a, f), m(r, a, f)$ counting and proximity function, 446, 447

$c(f, 0)$ leading coefficient of the Laurent series at 0, 446

$T(r, f)$ Nevanlinna's characteristic function of f, 448

$\rho(f)$ order of a meromorphic function f, 450

$W(f_0, \ldots, f_n)$ Wronskian of entire functions $f_0, \ldots, f_n$, 450, 470

$N_{\mathrm{ram}}(r, f)$ counting function of ramification of f, 450

$N_{\mathrm{ram}}(r, a, f)$ counting function of ramification over a, 451

$\delta(a, f), \theta(a, f)$ defect and ramification defect of f, 452

$N^{(1)}(r, a, f), \mathrm{cond}(r, f)$ truncated counting function and conductor, 455

$k(z_1, z_2)$ chordal distance on the Riemann sphere, 457

ω Fubini–Study form on projective space, 457, 467

$\mathring{m}(r, a, f)$ spherical proximity function, 458

$\mathring{T}(r, f)$ Ahlfors–Shimizu characteristic, 458

$\mathrm{ord}_Y(D)$ multiplicity of divisor D in component Y, 466

$N_{f,D}(r), m_{f,D}$ counting function and proximity function, 466, 476

$T_{f,D}(r), T_{f,L}(r)$ characteristic function for holomorphic curve f, 466, 469, 476

$N_{f,\mathrm{ram}}$ counting function for ramification of f, 470

R_f ramification divisor of holomorphic map f, 473

$N^{(1)}_{f,D}(r)$	truncated counting function, 475
$m_S(a,\beta)$, $N_S(a,\beta)$	proximity and counting function for $a \in K$, 481
$N_{S,D}$, $m_{S,D}$	counting and proximity function for divisor D, 483, 499
d_K, $d(P)$	absolute logarithmic discriminant, 487, 499
$\mathrm{cond}^K_\lambda(P)$	conductor of P, 489, 499
M_K^{fin}	discrete valuations of number field K, 489
χ	characteristic function for $(0,\infty)$, 489
$N_{S,\mathrm{ram}}$	counting of ramification in function field case, 505
$N^{(n)}_D(P)$	higher truncated counting function, 508
$\mathbb{A}^n_K$, $x_1,\dots,x_n$	affine n-space over K, fixed coordinates, 514
$Z(T)$	zero set of $T \subset K[\mathbf{x}]$ in $\mathbb{A}^n_K$, 514
$I(Y)$	ideal of vanishing of $Y \subset \mathbb{A}^n_K$, 515
$K[X]$	coordinate ring of affine variety X, 515
$\sqrt{J}$	radical of ideal J, 515
$\varphi^\sharp$	composition with morphism φ, 515, 521, 574
$\mathcal{O}_{X,x}$, $\mathfrak{m}_x$	local ring of $x \in X$ and maximal ideal, 516, 522
$\overline{\mathfrak{m}}_x$	$\{f \in K[X] \mid f(x)=0\}$, 516
$X(L)$	L-rational points of X, 516, 522
$K(x)$	residue field of $\mathcal{O}_{X,x}$, 516, 522
$\mathrm{Spec}(A)$	Spectrum of ring A, 516
X_F	base change of X to extension F/K, 517, 522
$\dim(T)$	dimension of topological space T, 518
$\dim(A)$	Krull dimension of commutative ring A, 518
$GL(n)_K$, $GL(n,L)$	invertible matrices as K-variety, with entries in L, 519
ρ^U_V	restriction map from U to V for presheaf, 520
$\mathcal{O}_X$	sheaf of regular function on variety X, 521
$\mathcal{J}_Y$	ideal sheaf, 522
$\mathrm{codim}(B,X)$	codimension of B in X, 523
$K(X)$	function field of X, 524
$\pi_E : E \to X$	vector bundle E over X, 525
$g_{\alpha\beta}$	transition function of E, 525
E_x	fibre of E over x, 526
$E \oplus E'$, $E \otimes E'$	direct sum and tensor product of vector bundles, 527, 528
$\Gamma(U,E)$	space of sections of E over U, 527
$\mathcal{E}$	sheaf of sections of vector bundle E, 527
O_X	trivial bundle of rank 1 over X, 528
E^*, $E \wedge E'$, $\mathrm{Hom}(E,E')$	dual, wedge product and homomorphisms, 528
E/F	quotient vector bundle by subbundle F, 529
$\varphi^*(E)$, $E\|_{X'}$	pull-back and restriction of E, 529
$L^{\otimes n}$, L^{-1}	tensor power and dual of line bundle L, 529
$\mathbf{c} = \mathrm{cl}(L) \in \mathrm{Pic}(X)$	Picard group and class of L, 529
$\varphi^*(s)$	pull-back of section, 530
$\mathbb{P}^n_K$, $(x_0 : \cdots : x_n)$	projective space, fixed homogeneous coordinates, 530
$I(Y)$	homogeneous ideal of projective variety Y, 531
$S(Y) = K[\mathbf{x}]/I(Y)$	homogeneous coordinate ring of Y, 531
$O_{\mathbb{P}^n_K}(m)$	tensor powers of the tautological line bundle, 532

$\mathbb{P}_K, O_{\mathbb{P}}(d_1, \ldots, d_r)$	multiprojective space with line bundle, 534
$X_1 \times_S X_2$	fibre product of varieties X_1, X_2 over S, 535
$\mathrm{Der}_K(A, M)$	K-derivations of A into M, 536
$T_{X,x}$	tangent space of variety X in x, 536
$d\varphi$	differential of morphism or function, 537, 543
$\dim_x(X)$	dimension in $x \in X$, 538
$\partial\|_x$	evaluation of vector field ∂ at x, 540
T_X, T_X^*	tangent and cotangent bundle of X, 542, 542
Ω_X^k, φ^*	sheaf of k-forms with pull-back, 542, 543
$K_X, N_{Y/X}$	canonical line bundle, normal bundle, 542, 543
$\Omega_{X/Y}^1$	sheaf of relative differentials, 543
$D \geq D'$	partial order on Weil divisors, 544
$\mathrm{supp}(D)$	support of divisor, 544, 550
$[x]$	divisor of x in a curve, 544
$\mathrm{ord}_Y(f)$	order of $f \in K(X)$ in Y, 545, 546
$\mathrm{div}(f)$	Weil divisor associated to f, 545, 547, 553
$\mathcal{O}_{X,Y}, \mathfrak{m}_Y$	local ring of X in prime divisor Y, 546
$D \sim D'$	rational equivalence of Weil divisors, 547
$\mathrm{div}(s)$	Weil divisor of meromorphic section s, 547, 554
$ss', s^{-1}, s/s'$	multiplication, division of meromorphic sections, 548
$O(D), s_D$	line bundle and section for Cartier divisor D, 548
$D(s)$	Cartier divisor of meromorphic section s, 548
$\mathrm{cyc}(D)$	Weil divisor associated to D, 549
$\|D\|$	complete linear system, 549
$\varphi^*(D)$	pull-back of divisor D, 550, 555
$Z_*(X)$	group of cycles on X, graded by dimension, 551
$\ell_\Lambda(M)$	length of Λ-module M, 552
$Z \sim Z', CH_*(X)$	rational equivalence of cycles, Chow group, 553
$\varphi_*(Z)$	push-forward of cycle Z, 553
$c_1(L).Z$	first Chern class operation of line bundle L, 554
$\mathrm{cyc}(X)$	cycle of scheme X, 555
Z_F	base change of cycle Z to extension F/K, 555
$D.Z$	intersection product of divisor D with Z, 555, 556
$\deg(Z), \deg_L(Z)$	degree of Z (with respect to L), 557, 558, 561
$D_1 \cdots D_d$	intersection number of divisors, 557
$\mathrm{Pic}^0(X)$	classes of line bundles algebraically equivalent to $\mathbf{0}$, 562
$\ker(\varphi), \mathrm{im}(\varphi) = \varphi(\mathcal{F})$	kernel and image of sheaf homomorphism, 564
$\mathcal{F}/\varphi'(\mathcal{F}'), \mathrm{coker}(\varphi)$	quotient sheaf, cokernel of φ, 564
$C^*(\mathcal{U}, \mathcal{F}), \check{H}^*(\mathcal{U}, \mathcal{F})$	Čech complex and cohomology, 564, 565
$\mathcal{O}_X(D)$	sheaf of sections of $O_X(D)$, 567
$\mathcal{H}om_{\mathcal{O}_X}(\mathcal{E}, \mathcal{F})$	sheaf of $\mathcal{O}_X$-module homomorphisms, 568
$\mathcal{E} \otimes_{\mathcal{O}_X} \mathcal{F}$	tensor product of $\mathcal{O}_X$-modules, 568
$\varphi_*(\mathcal{F})$	direct image of $\mathcal{O}_X$-module $\mathcal{F}$, 568
$\varphi^*\mathcal{F}', \mathcal{F}'\|_X$	pull-back and restriction of $\mathcal{O}_X$-module, 568
$H^p(X, \mathcal{F})$	cohomology group of coherent $\mathcal{O}_X$-module, 569
ω_X	sheaf of sections of canonical line bundle, 571

$\chi(\mathcal{F})$	Euler characteristic of $\mathcal{F}$, 571
$\mathcal{F}(n)$	$\mathcal{F} \otimes \mathcal{L}^{\otimes n}$ for $\mathcal{L}$ very ample, 571
$\operatorname{supp}(\mathcal{F})$	support of $\mathcal{O}_X$-module $\mathcal{F}$, 571
$\varphi : X \dashrightarrow X'$	rational map, 574
X_y	fibre of $\varphi : X \to Y$ over $y \in Y$, 577
$\deg(\varphi)$	degree of equidimensional morphism, 578
$g(C)$	genus of curve C, 582
$\deg(L)$	degree of a line bundle on C, 582
$\operatorname{div}(K_X), p_a(X)$	canonical divisor, arithmetic genus, 583
X_{an}	complex analytic variety associated to X, 584
$D_{A/R}(e_1, \ldots, e_n), \mathfrak{d}_{A/R}$	discriminant of free R-algebra A, 586
$\operatorname{Tr}_{A/R}, N_{A/R}$	trace and norm of A over R, 586
D_f, $\operatorname{res}(f)$	discriminant and resultant of polynomial f, 586, 588
$\bar{R}^L$	integral closure of Dedekind domain R in L, 589
$\mathfrak{d}_{L/K} = \mathfrak{d}_{\bar{R}^L/R}$	discriminant of $\bar{R}^L$ for $K = \operatorname{Quot}(R)$, 589
$D_{L/K}$	principal generator of $\mathfrak{d}_{L/K}$, 589
$\mathfrak{D}_{\bar{R}^L/R} = \mathfrak{D}_{L/K}$	different of $\bar{R}^L$ for $K = \operatorname{Quot}(R)$, 590
K_v^{nr}	maximal unramified subextension, 594
$v_\mathfrak{p}$	place induced by maximal ideal $\mathfrak{p}$ of R, 594
R_p	ramification divisor of morphism p, 599
D_{red}	reduced divisor, 600
$K_\mathbb{A}$	adele ring of number field K, 604
β, β_v	normalized Haar measures on $K_\mathbb{A}$ and K_v, 605
$E, E_v, E_\infty, E_\mathbb{A}$	euclidean spaces K^N, K_v^N, $\prod_{v\mid\infty} K_v^N$ and $K_\mathbb{A}^N$, 608
λ_n	nth successive minimum, 610

Index

Printed in the United States
By Bookmasters